8th International Conference on Mechanical and Manufacturing Engineering 2017 (ICME'17)

MATEC Web of Conferences Volume 135 (2017)

Langkawi, Malaysia
22 - 23 July 2017

Editors:

M.N. Nik Hisyamudin
K. Amir
I. Al Emran
I. Mohd Rasidi
M.B. Mohd Faizal
Md J. Mohd Azlis Sani
T.A. Ahmad Mubarak
Z. Izzuddin
M. Sofian

ISBN: 978-1-5108-5206-8

Printed from e-media with permission by:

Curran Associates, Inc.
57 Morehouse Lane
Red Hook, NY 12571

Some format issues inherent in the e-media version may also appear in this print version.

Printed by Curran Associates, Inc. (2017)

For additional information, please contact EDP Sciences – Web of Conferences at the address below.

EDP Sciences – Web of Conferences
17, Avenue du Hoggar
Parc d'Activité de Courtabœuf
BP 112
F-91944 Les Ulis Cedex A
France

Phone: +33 (0) 1 69 18 75 75
Fax: +33 (0) 1 69 28 84 91

contact-edps@webofconferences.org

Additional copies of this publication are available from:

Curran Associates, Inc.
57 Morehouse Lane
Red Hook, NY 12571 USA
Phone: 845-758-0400
Fax: 845-758-2633
Email: curran@proceedings.com
Web: www.proceedings.com

TABLE OF CONTENTS

INTERNATIONAL CONFERENCE ON MECHANICAL AND MANUFACTURING ENGINEERING 2017

(ICME2017)

LANGKAWI, MALAYSIA

22 & 23 July 2017

ICME'17

8th International Conference on Mechanical and Manufacturing Engineering

SCIENTIFIC EDITORS

Nik Hisyamudin Muhd Nor

Al Emran Ismail

Amir Khalid

Mohd Rasidi Ibrahim

Mohd Faizal Mohideen Batcha

Ahmad Mubarak Tajul Arifin

Izzuddin Zaman @ Bujang

Sofian Mohd

Mohd Azlis Sani Md. Jalil

Waluyo Adi Siswanto

Azlan Ahmad

TECHNICAL REVIEWERS

Ahmad Mubarak Tajul Arifin

Ainun Rahmahwati Ainuddin

Akmal Nizam Mohammed

Anika Zafiah Mohd Rus

Aslam Abdullah

Azmahani Sadikin

Azriszul Md Amin

Azwan Sapit

Azzura Ismail

Badrul Aisham Md Zin

Chee Kiong Sia

Ching Teng Koh

Erween Abd Rahim

Hamidon Salleh
Hamimah Abd.Rahman
Ishkrizat Taib
Kamarul Ariffin Zakaria
Kamarul Azhar Kamarudin
Khir Harun
Mohamed Nasrul Bin Mohamed Hatta
Mohammad Kamil Abdullah
Mohammad Sukri Mustapa
Mohd Azahari Razali
Mohd Fadhli Zulkafli
Mohd Fahrul Hassan
Mohd Nasrull Abdol Rahman
Mohd Shahir Bin Yahya
Mohd Shukor Salleh
Mohd Yussni Hashim
Muhamad Zaini Yunos
Muhammad Akmal Johar
Muhammad Mujtaba Asad
Muhammed Nafis Osman Zahid
Musli Mohammad
Ng Chuan Huat
Nor Azali Bin Azmir
Noradila Abdul Latif
Nur Azam Badarulzaman
Omar Mohd Faizan Marwah
Othman Mohd Hilmi
Raja Zuraidah Raja Mohd Rasi
Reazul Haq Bin Abdul Haq
Rosniza Hussin
Ruzila Ismail
Saifulnizan Bin Jamian
Salihatun Bt Md Salleh
Sh Salleh Sh Ahmad
Shahiron Shahidan
Sharifah Adzila Syed Abu Bakar
Shazarel Shamsudin

Siti Aida Ibrahim
Sulaiman Hassan
Sulastri Sabudin
Suzairin Md Seri
Suzi Salwah Jikan
Syariful Syafiq Shamsudin
Wan Nur Azrina Binti Wan Muhammad
Yusri Yusof
Zakiah Kamdi
Zaleha Mohamad
Zamani Ngali
Zamri Noranai

Preface

Nik Hisyamudin Muhd Nor, Al Emran Ismail, Amir Khalid, Mohd Azlis Md Jalil, Mohd Faizal Mohideen Batcha, Izzuddin Zaman, Sofian Mohd, Mohd Rashidi Ibrahim, Ahmad Mubarak Tajul Arifin, Aida Mustapha.

The 8th International Conference on Mechanical and Manufacturing Engineering 2017 (ICME2017) was held successfully in Langkawi, Malaysia. It was aimed to provide a better platform for communication and discussion among academics, researchers, industrial players as well as post-graduate students from various research fields. It was also aimed to provide a place for sharing knowledge and ideas, networking, presenting research findings and exploring other potential benefits that might be contributed for a better future.

Faculty of Mechanical and Manufacturing Engineering (FKMP), Universiti Tun Hussein Onn Malaysia (UTHM) would like to express fullest gratitude and appreciation all participants, respective authors and technical reviewers for sharing their knowledge and experience for this conference.

Opening Address

Nik Hisyamudin Muhd Nor, Al Emran Ismail, Amir Khalid, Mohd Azlis Md Jalil, Mohd Faizal Mohideen Batcha, Izzuddin Zaman, Sofian Mohd, Mohd Rashidi Ibrahim, Ahmad Mubarak Tajul Arifin, Aida Mustapha.

First and foremost, on behalf of organizing committee, I would like to express our fullest and warmest gratitude to all participants, presenters, technical reviewers, and all collaborative partners for the supports and cooperation in ensuring the successful of the conference.

International Conference on Mechanical and Manufacturing Engineering 2017 (ICME2017) is organized by the Faculty of Mechanical and Manufacturing Engineering, UTHM. This conference is an annual conference started in year 2008 and have attracted thousands of participation from all around the world.

With the conference's theme of "Transformation Towards Translational Research", ICME2017 aims to bring all prominent scientists, researchers, inventors and innovators as well as industrial players together in one place so that we can share and exchange ideas, knowledge and expertise, widen networking among each other, present research findings, promote new technology and explore any potential benefits that might contribute to the new brighter and better future.

The introduction of Industrial Revolution 4.0 few years ago has expand the concept of world without border and the distance is no more a challenges or barriers. Thus it is a compulsory for the current researchers and academics to increase their collaboration and cooperation even through the online media or any media that does not require direct contact in ensuring researches can be done more effectively and efficiently.

Finally, we would like to thanks once again to all participants and all parties that involved in ICME2017 and we are looking forward to have you again in the following ICME.

Thank you.

Assc. Prof. Dr. Nik Hisyamudin Muhd Nor,
Chairman
International Conference on Mechanical and Manufacturing Engineering 2017 (ICME2017)
Faculty of Mechanical and Manufacturing Engineering,
Universiti Tun Hussein Onn Malaysia

Statement of Peer review

In submitting conference proceedings to MATEC Web of Conferences, I certify to the Publisher that I adhere to the **Policy on Publishing Integrity** of the journal in order to safeguard good scientific practice in publishing.

1. All articles have been subjected to peer review administered by the proceedings editors.
2. Reviews have been conducted by expert referees, who have been requested to provide unbiased and constructive comments aimed, whenever possible, at improving the work.
3. Proceedings editors have taken all reasonable steps to ensure the quality of the materials they publish and their decision to accept or reject a paper for publication has been based only on the merits of the work and the relevance to the journal.

Title, date and place of the conference

International Conference on Mechanical and Manufacturing Engineering 2017 (ICME2017)
22-23 July 2017
Langkawi, Malaysia

Proceedings editor(s):

Nik Hisyamudin Muhd Nor, Amir Khalid, Al Emran Ismail, Mohd Rasidi Ibrahim, Mohd Faizal Mohideen Bacha, Mohd Azlis Sani Md Jalil, Ahmad Mubarak Tajul Arifin, Izzuddin Zaman, Sofian Mohd

Date and editor's signature

23.10.2017

Combined-hole film cooling with the application of double flow control devices

Haswira Hassan and *Kamil* Abdullah*

Faculty of Mechanical and Manufacturing Engineering, Universiti Tun Hussein Onn Malaysia (UTHM), 86400 Parit Raja, Johor, Malaysia

Abstract. In the modern gas turbine, film cooling system was applied to the turbine blades to provide thermal protection from extreme turbine inlet temperatures. Recent literature discovers that adding the double flow control device (DFCD) and combining two cylindrical holes (combined-hole) are the ways to further enhance the film cooling performances. In the present simulation work, the influence of DFCD on the combined-hole film cooling was investigated at different blowing ratios. The comparison between two different computational domains of combined-hole film cooling with and without DFCD was further discussed in the present work. The results show that increase the blowing ratio, M rise the overall performance of combined-hole film cooling with and without DFCD. Based on the different computational domain, combined-hole with DFCD has better lateral spread at M = 0.5 and M = 0.75 in comparison with the combined-hole without DFCD. However, due to large counter clockwise vortex produced by the combined-hole with DFCD, unsymmetrical film cooling coverage was formed along the streamwise direction at M = 1.0 and M =1.5. As the conclusion, combined-hole film cooling with DFCD has better performance at M = 0.5 and M = 0.75 while combined-hole film cooling without DFCD has better performance at high blowing ratio of M = 1.0 and M = 1.5.

1 Introduction

The performance of the gas turbines can be improved with the increased of the turbine inlet temperature. Due to high turbine inlet temperature, very effective cooling techniques are required to protect the turbine components which exposed to extreme thermal loads to which also jeopardising its durability. Film cooling system was introduced especially on the turbine blades to provide enhance thermal protection and ensure its reliability. Film cooling provides external protection of the blade surfaces with formation of thin layer of cooled air. The cooled air which provided by the compressor bleed through the cooling holes and prevents the direct contact between the hot mainstream gases with the blade surfaces thus ensure lower temperature and heat transfer on the turbine blades. Goldstein [1], among the pioneers in the field has provided the fundamental understanding of film cooling.

Cylindrical film cooling hole has been introduced as basic geometry of the cooling hole and still being applied in modern gas turbine due to its simplicity, low maintenance and manufacturability. Several studies [2-4] were carried out and described the film cooling mechanism of the cylindrical film cooling hole. All the early studies agreed on the existence of kidney vortex in cylindrical film cooling hole system. The vortices stated to contribute to narrow coverage and degrading of the film cooling effectiveness. One of the effectiveness way to control the formation of the kidney vortex and improve the film cooling effect is by adding devices upstream or downstream of the cooling holes and changing the geometry of film cooling hole. Takisawa et al. [5] and Kawabata et al. [6] added the flow control devices upstream of the cylindrical film cooling hole which been reported to altered the formation of the kidney vortex while improving the film cooling performance. Meanwhile, other the research [7-9] have also reported improvement on the film cooling performance through combined-hole configuration which is achieved by combining two film cooling holes together as a unit. The combined-hole film cooling produced an anti-kidney vortex and keeps the cooling gas attached to the wall. The present work is aim to observe the influence of the flow controlling devices on the performance of the combined-hole film cooling at different blowing ratio.

2 Methodology

2.1 Computational domain and validation

The computational domain in the present study combined two main sections of mainstream and cooling gas duct embedded with the combined-hole film cooling. Basically, two computational domains have been considered; a) without Double Flow Control Devices (DFCD), and b) with Double Flow Control Devices (DFCD). Fig. 1 shows the dimensions of the computational domain with the value of D was based on the hole diameter which is equal to 3mm. Table 1 and Fig. 2 indicate the details on the combined-hole and DFCD applied in the present study and the origin was located at the middle between two centres of the combined-hole film cooling. The dimensions of computational domain and combined-hole film cooling were based on the research of Hassan and Abdullah [10] and the dimensions of DFCD was based on the research of Takisawa [5].

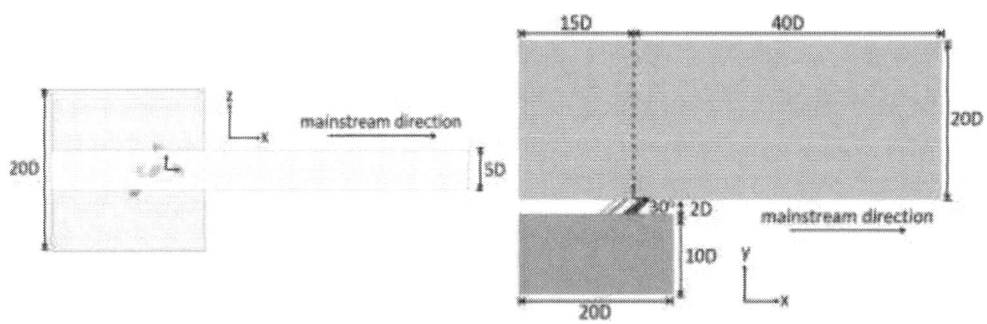

Fig. 1. Dimensions of the computational domain [10]

Table 1. Dimensions of the DFCD and combined-hole film cooling. [5, 10]

A	B	C	D	E	F	G	H	J

 DOI: 10.1051/matecconf/201713500003

1.0D	0.35D	0.3D	0.6D	20°	0.5D	1.5D	3.0D	45°

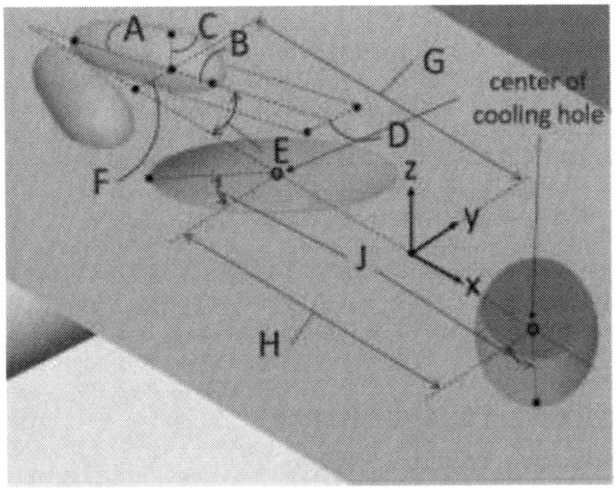

Fig. 2. Details on the DFCD and combined-hole film cooling [5, 10]

2.2 Numerical setup

Fig. 3 shows the boundary conditions on the computational domain of the present study. ANSYS CFX software involving steady state Reynolds Average Navier Stokes (RANS) analysis with the employment of Shear Stress Transport (SST) turbulence model was applied on the present study. The boundary conditions applied in the present study are similar to the work of Hassan and Abdullah [10]. Table 2 shows the flow conditions applied in the present study. The mass flow rate of the cooling gas for the combined-hole has been determined with the assumption that both cooling holes are operating at the same blowing ratio and the sum of the required mass flow rate of each hole has been applied at the cooling gas inlet of the computational domain.

Table 2. Flow conditions.

Reynold's Number, Re_D	4200
Mainstream Velocity, U_∞	22.065 m/s
Mainstream Temp., T_∞	300 K
Mainstream Turbulent. Intensity, Tu	6%
Blowing Ratio, M	0.5, 1.0, 1.5
Cooling Gas Inlet Temp., T_c	310 K

 DOI: 10.1051/matecconf/201713500003

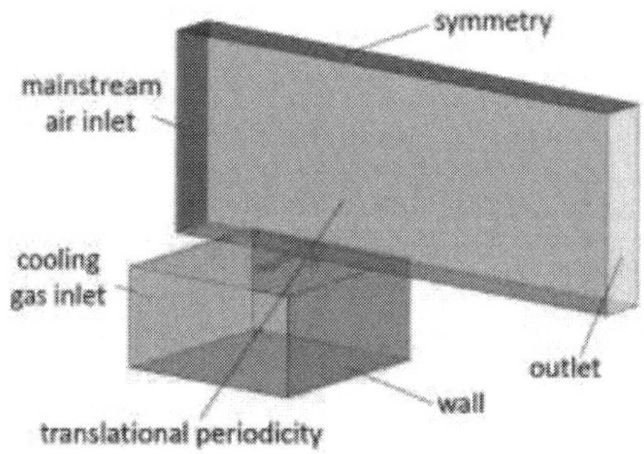

Fig. 3. Boundary conditions

2.3 Mesh dependency test and validation

Fig. 4 shows the lateral average film cooling effectiveness results of the mesh dependency test conducted at M = 0.5 with the use of hybrid meshing. The medium mesh with approximately 8.2 million nodes was applied to each case to minimize discretization error and computational time. Besides that, Fig. 4 also shows the validation between the numerical results of the present study with the experimental results of previous research reported by Han et al. [11]. The lateral results show good agreement along the streamwise direction. However, such divergent of the result might be caused by the incapability of the turbulent model to accurately predict the interaction of the mainstream air and the cooling air.

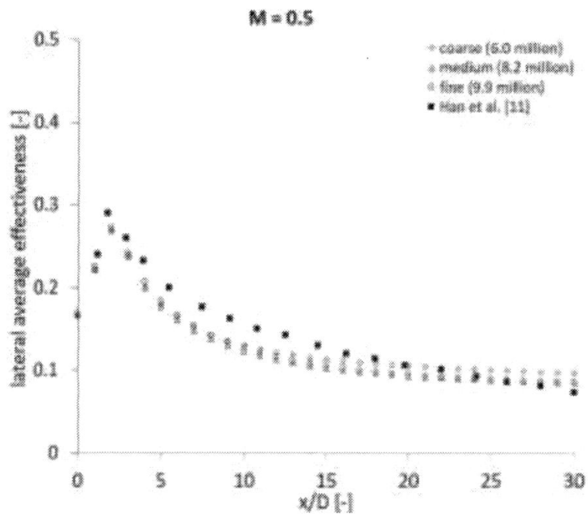

Fig. 4. Mesh dependency test and validation with the previous research

3 Results and discussion

Fig. 5 shows the film cooling effectiveness distributions for all considered cases. Based on figure, the film cooling effectiveness was completely altered with the application of DFCD at different blowing ratios. At M = 0.5, combined-hole with DFCD shows significant change with slightly wideer coverage of film cooling along the streamwise direction. The

DOI: 10.1051/matecconf/201713500003

lateral coverage also enhanced as the blowing ratio increases in comparison with the M = 0.5 case.

Fig. 6 and 7 show the vorticity contours on yz planes at x/D = 5, 10 and 15. In the case of combined-hole film cooling without DFCD, the kidney vortex was observed at the early downstream and reduced along the streamwise direction. In addition, increase the blowing ratio on the combined-hole film cooling without DFCD increase the size of kidney vortex. In the case of combined-hole film cooling with DFCD, an additional counter clockwise vortex was produced in early downstream at M = 0.5 and M = 0.75 which was preserved until x/D = 15 as shown in Fig. 6. Differed with the cases of M = 1.0 and M = 1.5, there are no sign of additional vortex is observed. However, in comparison with the case of M = 0.5 and 0.7, the size of the kidney vortex was further enhanced in the case of M = 1.0 and M = 1.5.

Based on Fig. 5 – 7, the application of DFCD on the combined-hole film cooling totally change the flow structure along the streamwise direction thus significantly influence the film cooling effectiveness. An additional counter clockwise vortex produces in the case of M = 0.5 and M = 0.75 with DFCD enhanced the film cooling coverage on the positive side of the z-axis. Besides that, improvement can also be observed at near cooling hole region in the case of M = 0.75 and M = 1.0 with DFCD. This is due to altered flow structure at early downstream. However, the cooling gas from both cooling holes tend to flow in their own path and produced a small area of low film cooling effectiveness from x/D = 4 to x/D = 8 in the case of M = 0.75 and M = 1.0 with DFCD. In addition, the large size of counter clockwise vortex observed at the early downstream region formed the unsymmetrical film cooling coverage which is preserved further downstream as shown in Fig. 5.

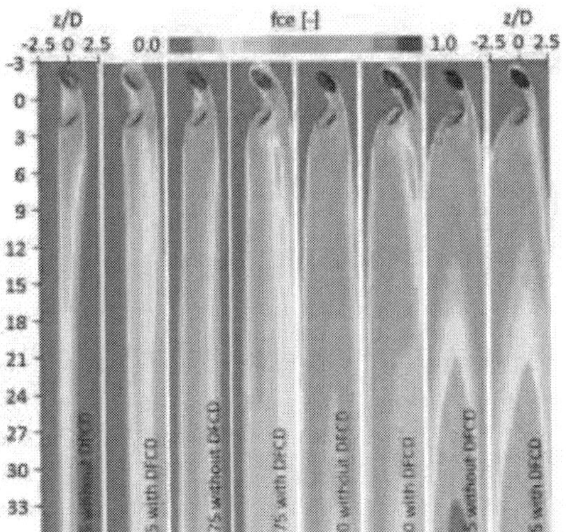

Fig. 5. Film cooling effectiveness distributions

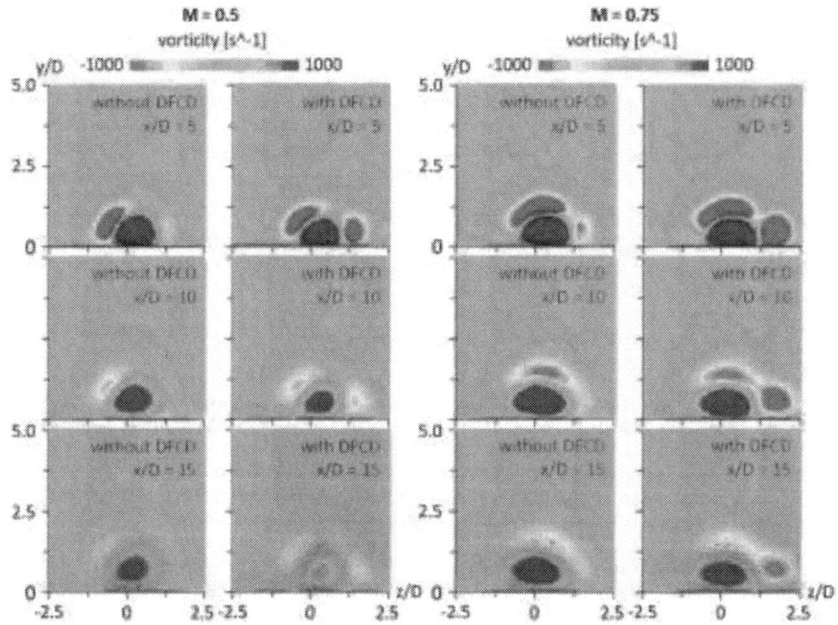

Fig. 6. Vorticity at M = 0.5 and M = 0.75

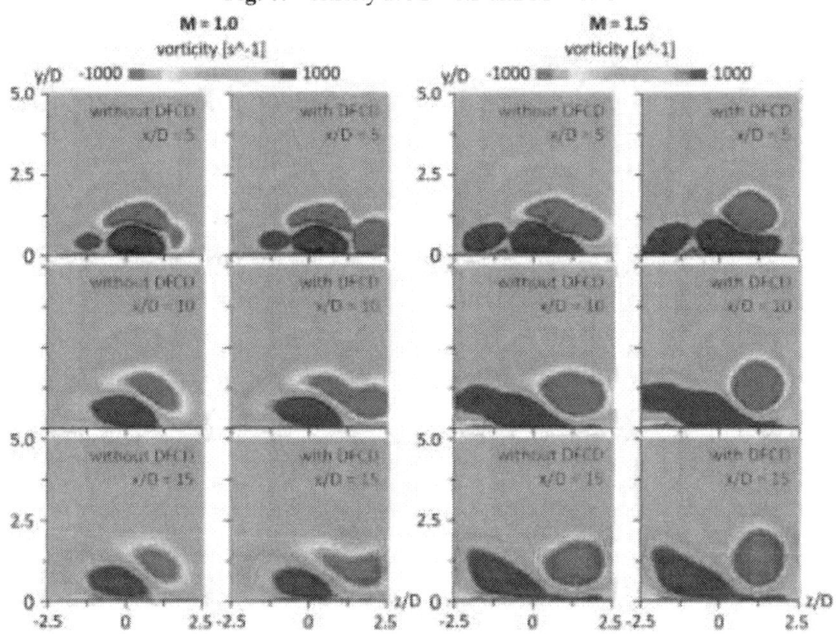

Fig. 7. Vorticity at M = 1.0 and M = 1.5

Fig. 8 shows the lateral average film cooling effectiveness graph for all considered cases. The performance of combined-hole film cooling with and without DFCD was increased as the blowing ratio increase until M = 1.0. However, the performance of combined-hole in the case of M = 1.5 is better at early downstream and reduced further downstream. At M = 0.5 and M = 0.75 cases, the combined-hole film cooling with DFCD has better lateral average film cooling effectiveness due to wide film cooling coverage as shown in Fig. 5. Meanwhile, the combined-hole film cooling without DFCD has better lateral average film cooling effectiveness in comparison with the combined-hole with

DFCD in the case of M = 1.0 and M = 1.5 due to narrow and unsymmetrical film cooling coverage along the streamwise direction.

Fig. 9 shows the area average film cooling effectiveness that generally evaluates the film cooling effect at different cases. The width of the area is the pitch of the combined-hole, 5D and the length is from x/D = 0 to x/D = 40D. At M = 0.5 and M = 0.75, combined-hole film cooling with the application of DFCD have better area average film cooling effectiveness compared to the combined-hole film cooling without DFCD. Meanwhile, combined-hole without DFCD configurations have better performance than combined-hole film cooling with DFCD.

In the case of M = 1.5 with DFCD, narrow coverage of film cooling effectiveness was observed in the early downstream due to the formation of counter clockwise vortex. The formation of the vortices in the case of M = 1.5 without DFCD covering the lateral direction along the streamwise direction which formed the wide coverage of film cooling effectiveness as shown in Fig. 5. Meanwhile, in the case of M = 1.5 with DFCD, the formation of vortices only covering the lateral direction until z/D = 2.0 as shown in Fig. 7 and consequently produce narrow coverage of film cooling effectiveness from x/D = 2 to x/D = 12 and unsymmetrical film cooling coverage further downstream as shown in Fig. 5.

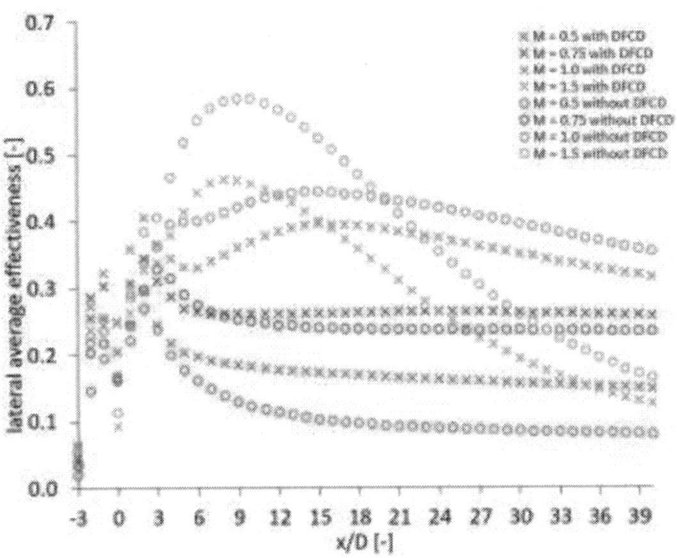

Fig. 8. Lateral average film cooling effectiveness graph

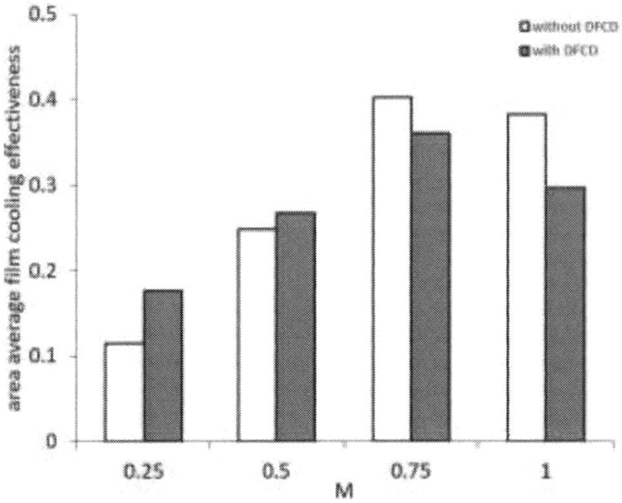

Fig. 9. Area average film cooling effectiveness graph

4 Conclusions

A batch of simulation focused on the application of DFCD on the combined-hole film cooling system had been carried out using steady state Reynolds Averaged Navier Stokes (RANS) method of ANSYS CFX, with Reynolds number, Re = 4200 at blowing ratios, M = 0.5, 0.75, 1.0 and 1.5. The conclusions for the present study are as follow:

• Increase the blowing ratio until M = 1.0 increase the overall film cooling performance. Meanwhile, increase the blowing ratio to M = 1.5 reduce the overall performance in comparison with the results of M = 1.0 due to faster dissipation occurred.

• In the case of M = 0.5 and M = 0.75, the combined-hole film cooling with DFCD show better overall performance compared to the combined-hole film cooling without DFCD. The existence of DFCD altered the formation of vortices in these cases and produced better separation of cooling gas along streamwise direction.

• In the case of M = 1.0 and M = 1.5, the combined-hole film cooling without DFCD show better overall performance compared to the combined-hole film cooling with DFCD. The existence of DFCD produced larger counter clockwise vortex and drag the cooling gas to the positive side of z-axis and formed unsymmetrical film cooling coverage alond streamwise direction.

Acknowledgment

The author would like to thank both Universiti Tun Hussien Onn Malaysia (UTHM) and Ministry of Higher Education of Malaysia for the financial support through Fundamentals Research Grant Scheme (Vot 1544).

References

1. R.J. Goldstein, *Film cooling. Advances in heat transfer*. 7, 321-379. (1971)
2. R.J. Goldstein, E.R.G Eckert, F. Burggraf. Effects of hole geometry and density on three-dimensional film cooling. *International Journal of Heat and Mass Transfer*. 17, 595-607. (1974)

3. M. Gritsch, A. Schulz, S. Wittig. Film-cooling holes with expanded exits: nearhole heat transfer coefficients. *International Journal of Heat and Fluid Flow.* 21, 146-155. (2000)
4. C. Han, J. Ren. Multi-parameter influence on combined-hole film cooling system. *International Journal of Heat and Mass Transfer.* 55, 4232-4240. (2012)
5. H. Takisawa, K.I. Funazaki, H. Sasaki, T. Kawamura, H. Tagawa, S. Nakano. Improvement of flat-plate film cooling performance by double flow control devices under high density ratio.
6. H. Kawabata, K.I. Funazaki, R. Nakata, D. Takahashi. Experimental and numerical investigations of effects of flow control devices upon flat-plate film cooling performance. *Journal of Turbomachinery.* 136. (2014)
7. K. Kusterer, D. Bohn, T. Sugimoto, R. Tanaka. Influence of blowing ratio on the double-jet ejection of cooling air. *ASME Turbo Expo 2007: Power for Land, Sea, and Air. American Society of Mechanical Engineers.* 305-315. (January, 2007)
8. K.D. Lee, D.W. Choi, K.Y. Kim. Optimization of ejection angles of double-jet film-cooling holes using RBNN model. *International Journal of Thermal Sciences.* 73,69-78. (2013)
9. D.W. Choi, K.D. Lee, K.Y. Kim. Analysis and optimization of double-jet film-cooling holes. *Journal of Thermophysics and Heat Transfer.* 27, 246-254. (2013)
10. H. Hassan, K. Abdullah. Effect of symmetrical compound-angle in combined-hole film cooling. (2006)
11. C. Han, Z. Chi, J. Ren, H. Jiang. Optimal arrangement of combined-hole for improving film cooling effectiveness. *Journal of Thermal Science and Engineering Applications.* (2015)

An Investigation of Viscosities, Calorific Values and Densities of Binary Biofuel Blends

Sharzali Che Mat[1,2,1], *M. Y.* Idroas[1], *Y. H.* Teoh[1] and *M. F.* Hamid[1]

[1]School of Mechanical Engineering, Engineering Campus, Universiti Sains Malaysia, 14300, Nibong Tebal, Pulau Pinang, Malaysia
[2]Faculty of Mechanical Engineering, Universiti Teknologi MARA Pulau Pinang, 13500, Permatang Pauh, Pulau Pinang, Malaysia

Abstract. Straight vegetable oil (SVO) biofuel is a promising alternative to petroleum diesel fuel primarily due to its comparable physical properties to that of petroleum diesel fuel. However, the relatively higher viscosity of SVO limits its direct application in diesel engine. To resolve this issue, binary biofuel blends was introduced in this study to reduce the viscosity of SVO. In this work, a novel biofuel namely Melaleuca Cajuputi oil (MCO) was used and blended with refined palm oil (RPO). A total of four blends with the mixing ratios of 20%, 40%, 50% and 60% of MCO were prepared. Various key properties of dynamic viscosity, calorific value and density of the blends were measured and benchmarked against the biodiesel standards based on ASTM D6751. It was found that viscosity and density of the blends decreased with the increase of MCO fraction. Meanwhile, the calorific value of the blends increased linearly as the MCO fraction increased. The blend of 40RPO60MCO was found to have comparable key properties of viscosity, calorific value and density to those of petroleum diesel fuel and ASTM D6751 standard.

1 Introduction

Recently, straight vegetable oils (SVOs) are gaining popularity as an alternative fuel for diesel engine due to its similar physical properties to that of petroleum diesel fuel. In addition, SVO is renewable, non-toxic, biodegradable and contain no sulphur that make it as an attractive candidate to replace diesel fuel [1]. However, the direct use of SVO in diesel engine will lead to severe carbon deposit, fuel injector clog and rapid wears of fuel pump components. These issues are primarily associated with a high viscosity of SVO that dramatically alters fuel spray characteristics, atomization quality and volatility [2].

There are several methods used to reduce the viscosity of SVO, such as micro-emulsion, preheat, blending, and transesterification. Among these different methods, the

*Corresponding author: sharzalm@yahoo.com

transesterification process is the most widely used to reduce the viscosity of SVO and convert triglycerides into fatty acid methyl ester (commonly known as biodiesel) and glycerol. Generally, the produced biodiesel has viscosity and other key properties close to petroleum diesel fuel [2]. However, this transesterification process involved with a complex yet expensive process where a specific equipment and instrumentation are required. Besides, the process also leads to a by-product formation of glycerol. This crude glycerol requires further expensive purification process to produce pure glycerol which has better value added product instead of crude glycerol [3].

Blending of SVO with other lower viscosity fuel was also found to be an effective and economical method to reduce its viscosity. This strategy has been studied by many researchers [4–6]. Generally, it was reported that engine fuelled with this blended fuel has comparable performance characteristics to those of diesel fuel, but with some penalties in brake specific fuel consumption (BSFC). Most of the authors have concluded that this is mainly associated with the low calorific value (CV) of the blend, which resulted in the reduction of engine performance and increase in BSFC. Meanwhile, on the bright side, NOx emission was reported to be slightly lower compared to diesel fuel.

Alcohol is another ingredient that is often used by researchers for blending with vegetable oil. Short-chain alcohols like ethanol and methanol are often used in the earlier studies [7–10]. These blends were found to reduce the viscosity and improve volatility of the vegetable oils, but the blends has limited miscibility [11]. Surfactant is required to be added into the blends to increase the miscibility and enhance the stability. Additionally, these alcohols have low cetane number and CV [12]. Thus, increase in the percentage of alcohol will lower the cetane number and CV of blended fuel. Recently, researchers have studied the blends of longer-chain alcohols like butanol and pentanol with SVO primarily due to their close properties to those of diesel fuel. Longer-chain alcohols has better CV, flash point, boiling point and cetane number as compared to short-chain alcohols. Moreover, Longer-chain alcohol is miscible with SVO thus, no additional surfactant is required to produce a stable blend [13]. Laza and Bereczky [14] analysed the properties of rapeseed oil (RSO) blended with higher alcohol (butanol and pentanol) at 10% and 20% by volume. They found that the viscosity, density, flash point, CV and cetane number of the blends were decreased as the percentage of alcohol increased. Viscosity of the blend also was found to be temperature dependent. Increase in temperature significantly decrease the viscosity. Another similar work is focussing with the blend of pongamia oil, ethanol and butanol and tested on diesel engine to analyse its performance and emissions [15]. It was reported that there was reduction in NO_x but increased in CO and HC was observed. The blends also exhibit higher BSFC compared to diesel fuel.

Recently, a less viscous biofuel such as pine oil, eucalyptus and camphor oil has also gained interest among researchers. This is mainly due to their relatively low viscosity and high calorific value as compared to those of SVO. As shown in Table 1 is the key properties of these less viscous biofuels and benchmarked against the biodiesel standards based on ASTM D6751. It was reported that the blend of these biofuels either with diesel or biodiesel were able to reduce viscosity and enhance the CV of the blends [16–18]. Apparently, the viscosity, CV and density of the blended fuel are the key properties to be analysed carefully to ensure their compatibility as an engine fuel. Therefore, a novel less viscous biofuel namely Melaleuca Cajuputi oil (MCO) was first introduced in this study for blending with refined palm oil (RPO). In the present study, MCO was chosen largely due to its commercially available as well as its comparable properties to those of eucalyptus and camphor oil. In this research works, the main objective is to obtain the optimum blend of RPO and MCO through binary biofuel blend strategy. This strategy proposed of blending of high viscosity RPO with low viscosity MCO as a method to reduce the viscosity and enhance the volatility of the blend.

DOI: 10.1051/matecconf/201713500004

Table 1. Various properties of less viscous biofuel [18–20].

Property	Eucalyptus oil	Pine oil	Camphor oil	Diesel	ASTM D6751
Kinematic viscosity at 40°C (mm^2/s)	2	1.3	1.9	3 - 4	1.9 - 6.0
Gross Calorific value (MJ/kg)	43.27	42.80	34.65	42.7 - 44.8	-
Density at 15°C (kg/m^3)	890 - 895.5	875	894.2	822 - 880	880
Flash point (°C)	54	52	50	50 - 74	93 min.
Boiling point (°C)	175	150 - 180	-	150 - 180	-

2 Material and method

2.1 Material

The test fuel of RPO used for the binary biofuel formulation was purchased from the local market meanwhile MCO was purchased from a private supplier. RPO is in liquid fraction at room temperature and have better cloud point as compared to crude palm oil. Besides, MCO has clear greenish liquid, low viscosity and highly volatile.

2.2 Blending preparation

Binary biofuels were prepared using RPO and MCO in volumetric percentage. Four blends at mixing ratios of 20% (80RPO20MC0), 40% (60RPO40MCO), 50% (50RPOMCO50) and 60% (40RPO60MCO) of MCO with RPO were prepared. All blends were prepared without emulsifier or surfactant. To ensure homogeneity, each blend of RPO-MCO was mixed thoroughly with a magnetic stirrer. The prepared samples were stored in glass bottle and kept at ambient temperature. The samples were monitored for phase separation for a month at a regular time interval.

2.3 Experimental

Dynamic viscosity of RPO, MCO and their blends were measured using Brookfield DV-III rheometer. Viscosity properties will determine the quality of fuel atomization, spray characteristic and volatility of the fuel. The measurement was performed at 40°C and the temperature was maintained by means of water bath. The viscosity was measured three times and the average value was obtained. The kinematic viscosity can be calculated by dividing the dynamic viscosity with the density of the blends.

Calorific values of the samples were determined using an adiabatic oxygen bomb calorimeter, Nenken 1013-B, Japan. This experiment was conducted according to ASTM D240 standard. CV is a measured of energy content in the fuel. Essentially, fuel with low CV will lead to the increase in BSFC of the engine.

The density of the fuel samples was determined by dividing the weight of the biofuel with the volume determined using 25 ml Gay-Lussac pycnometer. The measurement was taken at ambient temperature (~27 °C). The pycnometer used is certified according to DIN ISO 3507.

3 Results and discussion

3.1 The Effect of MCO blend on viscosity

Fuel viscosity is the major parameter that determine the ability of fluid to flow, atomization, spray characteristics and volatility of the fuel. The dynamic viscosity of neat RPO, MCO and their blends were present in Table 2. A total of four different biofuel blends, 80RPO20MCO (80%RPO, 20% MCO), 60RPO40MCO (60%RPO, 40% CMO), 50RPO50MCO (50% RPO, 50% MCO) and 40RPO60MCO (40% RPO, 60% MCO) were prepared and tested. Obviously, neat RPO has the highest viscosity as compared to neat MCO. Method of blending high viscosity oil with low viscosity oil significantly reduce the viscosity of RPO-MCO blend. Specifically, biofuel blend with highest MCO percentage (i.e. 40RPO60MCO) has remarkably reduced the viscosity down to 5.5 mPas. Besides, the experimental results also indicate that the higher the percentage of MCO in the blend, the lower the dynamic viscosity was observed. Thus, the proposed strategy of blending of less viscous biofuel with high viscosity RPO can effectively reduce the viscosity of the blend. The reduction in viscosity would enhance the volatility, thus enhance the spray characteristic of the fuel. Figure 1 present the correlation between dynamic viscosity with the MCO fraction. This correlation can be modelled by the mathematical relationship obtained from regression analysis of measured values, as given in Eq. (1). This relationship is in the form of exponential regression that best approximates all the measured data. Note that the obtained R^2 is 0.9986, which indicate a good fit between measured data and the regression model. Thus, Eq. (1) can be used to predict the viscosity of the blends.

$$v = 34.535e^{-0.03x} \quad (1)$$

Where v is dynamic viscosity and x is MCO fraction.

Table 2. Viscosity, calorific value and density of various RPO-MCO blends.

Blends ratio	Dynamic viscosity at 40°C (mPas)	Kinematic viscosity[a] (mm²/s)	Calorific value (MJ/kg)	Density (kg/m³)
100RPO	36	39.8	40.25	904.3
100MCO	1.8	2.1	43.16	867.1
80RPO20MCO	18.2	20.3	40.64	897.5
60RPO40MCO	10.5	11.8	41.27	890.6
50RPO50MCO	8	9.0	41.53	887.2
40RPO60MCO	5.5	6.2	41.84	883.7
Diesel	-	3 - 4	43 - 45	822 – 880
ASTM D6751	-	1.9 – 6.0	-	880

[a]calculated

 DOI: 10.1051/matecconf/201713500004

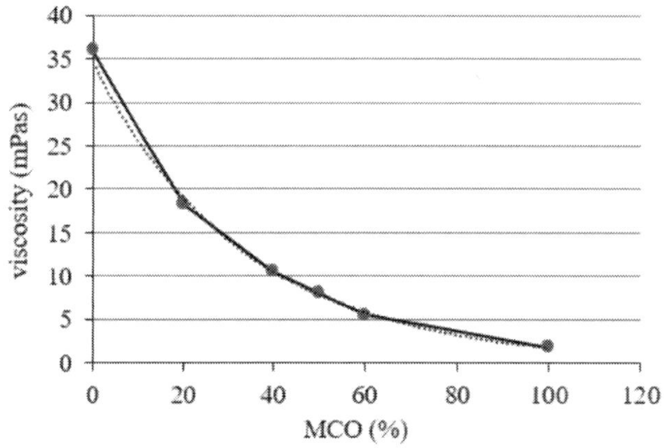

Fig. 1. The effect of RPO-MCO blends on dynamic viscosity.

3.2 The Effect of MCO blend on calorific value

Calorific values (CV) is an important parameter in the selection of a fuel. CV is the amount of heat released during the complete combustion of one gram of fuel. This value determines the chemical energy of the fuel which specify the combustion efficiency. Table 2 contains a description of the key properties of neat RPO, neat MCO and their blends. It can be observed that the CV of RPO was slightly lower compared to those of diesel fuel, but MCO interestingly has comparable CV to diesel fuel. As can be observed, the CV is directly increased as the percentage of MCO increased in the blends. Increase in CV is expected to improve the engine combustion efficiency, thus reduce the BSFC of the engine. Figure 2 present the correlation between CV with the MCO percentages. Mathematical model of the relationship is obtained in Eq. (2). Linear trendline with R^2 value 0. 9928 is the best fit to describe the measured data pattern. Eq. (2) can be used to predict the CV of the blends.

$$CV = 0.0295x + 40.121 \tag{2}$$

Where CV is calorific value and x is MCO fraction.

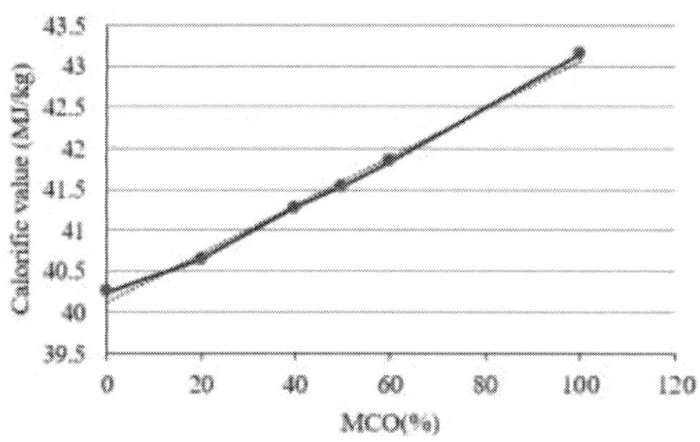

Fig. 2. The effect of RPO-MCO blends on calorific value.

3.3 The Effect of MCO blend on density

Generally, fuel density are closely related to viscosity, CV and cetane number of the fuel [21]. Accordingly, any change in density will affect the performance of the engine. Data tabulated in Table 2 shows that the neat RPO has greater density compared to MCO and diesel fuel. Clearly, binary blend method of RPO-MCO decreases the density close to diesel fuel. Figure 3 present the correlation between density and the MCO fraction in the binary blends. Notably, the density of the blends linearly decreases with the increase of MCO in the blends. This relationship can be modelled by mathematical Eq. (3). The linear regression having R^2 value of 0.9966, which reflects a good fit of the line to the experimental data.

$$\rho = 905.05 - 0.3701x \qquad (3)$$

Where ρ is density and x is MCO fraction.

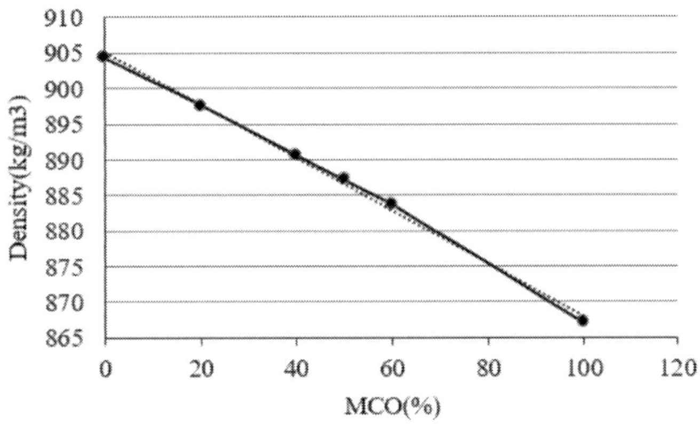

Fig. 3. The effect of RPO-MCO blends on density.

4 Conclusions

The key properties of dynamic viscosity, CV and density of binary biofuel blends (RPO-MCO) were successfully analysed and benchmarked against the biodiesel standards based on ASTM D6751. In the present study, a total of three mathematical model were developed from the regression analysis of measured data to predict the dynamic viscosities, CVs and densities of RPO-MCO blends. Experimental data shows that viscosity of the blend was reduced with the increased of MCO percentage. Meanwhile, the CV of the blends was linearly increased as the percentage of MCO increased in the blends. Notably, the density of the blends linearly decreases with the increase of MCO in the blends. Analysis of these 3 key properties of the fuel suggested that the optimum blend of 40RPO60MCO was found comparable to diesel fuel. Therefore, binary biofuel blend of RPO-MCO can be considered as a potential biofuel to replace fossil diesel fuel. While this research project represents an initial step in the long-term goal of exploring a new biofuel source, more comprehensive analysis of other key physiochemical properties of the fuel blends will be required in the future. In this regard, more intense key properties of fuels will be measured to ensure the fuel blends can comply with the biodiesel standards based on ASTM D6751.

Acknowledgement

We would like to acknowledge the financial support from Ministry of Higher Education Malaysia and Universiti Sains Malaysia under Fundamental Research Grant Scheme (FRGS/1/2016/TK07/USM/02/2; 203.PMEKANIK.6071356) and Research University Grant Scheme (project code of Mekanik/6071153).

References

[1] S.S. Sidibé, J. Blin, G. Vaitilingom, Y. Azoumah, Use of crude filtered vegetable oil as a fuel in diesel engines state of the art: Literature review, Renew. Sustain. Energy Rev. **14**, 2748–2759 (2010)

[2] L.C. Meher, D. Vidya Sagar, S.N. Naik, Technical aspects of biodiesel production by transesterification - A review, Renew. Sustain. Energy Rev. **10**, 248–268 (2006)

[3] M. Zhang, H. Wu, Effect of major impurities in crude glycerol on solubility and properties of glycerol/methanol/bio-oil blends, Fuel. **159**, 118–127 (2015)

[4] C.D. Rakopoulos, K.A. Antonopoulos, D.C. Rakopoulos, D.T. Hountalas, E.G. Giakoumis, Comparative performance and emissions study of a direct injection Diesel engine using blends of Diesel fuel with vegetable oils or bio-diesels of various origins, Energy Convers. Manag. **47**, 3272–3287 (2006)

[5] H. Hazar, H. Aydin, Performance and emission evaluation of a CI engine fueled with preheated raw rapeseed oil (RRO)–diesel blends, Appl. Energy. **87**, 786–790 (2010)

[6] T.F. Yusaf, B.F. Yousif, M.M. Elawad, Crude palm oil fuel for diesel-engines: Experimental and ANN simulation approaches, Energy. **36**, 4871–4878 (2011)

[7] M. Senthil Kumar, A. Ramesh, B. Nagalingam, An experimental comparison of methods to use methanol and Jatropha oil in a compression ignition engine, Biomass and Bioenergy. **25**, 309–318 (2003)

[8] P.J. Singh, J. Khurma, A. Singh, Preparation, characterisation, engine performance and emission characteristics of coconut oil based hybrid fuels, Renew. Energy. **35**, 2065–2070 (2010)

[9] S. Prabhakar, K. Annamalai, I.J. Lalvani, Experimental study of using hybrid vegetable oil blends in diesel engine, J. Sci. Ind. Res. **71**, 612–615 (2012)

[10] S. Bhimani, J.L. Alvarado, K. Annamalai, C. Marsh, Emission characteristics of methanol-in-canola oil emulsions in a combustion chamber, Fuel. **113**, 97–106 (2013)

[11] N. Arpornpong, C. Attaphong, A. Charoensaeng, D.A. Sabatini, S. Khaodhiar, Ethanol-in-palm oil/diesel microemulsion-based biofuel: Phase behavior, viscosity, and droplet size, Fuel. **132,** 101–106 (2014)

[12] S. Kumar, J.H. Cho, J. Park, I. Moon, Advances in diesel–alcohol blends and their effects on the performance and emissions of diesel engines, Renew. Sustain. Energy Rev. **22**, 46–72 (2013)

[13] B. Rajesh Kumar, S. Saravanan, Use of higher alcohol biofuels in diesel engines: A review, Renew. Sustain. Energy Rev. **60,** 84–115 (2016)

[14] T. Laza, Á. Bereczky, Basic fuel properties of rapeseed oil-higher alcohols blends, Fuel. **90**, 803–810 (2011)

[15] R. Senthil, E. Sivakumar, R. Silambarasan, Effect of butanol addition on performance and emission characteristics of a DI diesel engine fueled with pongamiaethanol blend., Int. J. ChemTech Res. **8,** 459–467 (2015)

[16] R. Vallinayagam, S. Vedharaj, W.M. Yang, P.S. Lee, K.J.E. Chua, S.K. Chou, Pine oil-biodiesel blends: A double biofuel strategy to completely eliminate the use of diesel in a diesel engine, Appl. Energy. **130**, 466–473 (2014)

[17] S. Kommana, B. Naik Banoth, K. Radha Kadavakollu, Eucalyptus-Palm Kernel Oil Blends: A Complete Elimination of Diesel in a 4-Stroke VCR Diesel Engine, J. Combust. **2015,** 1–7 (2015)

[18] G. Kasiraman, B. Nagalingam, M. Balakrishnan, Performance, emission and combustion improvements in a direct injection diesel engine using cashew nut shell oil as fuel with camphor oil blending, Energy. **47**, 116–124 (2012)

[19] P.K. Devan, N.V. Mahalakshmi, A study of the performance, emission and combustion characteristics of a compression ignition engine using methyl ester of paradise oil–eucalyptus oil blends, Appl. Energy. **86**, 675–680 (2009)

[20] R. Vallinayagam, S. Vedharaj, W.M. Yang, W.L. Roberts, R.W. Dibble, Feasibility of using less viscous and lower cetane (LVLC) fuels in a diesel engine: A review, Renew. Sustain. Energy Rev. **51**, 1166–1190 (2015)

[21] E. Alptekin, M. Canakci, Determination of the density and the viscosities of biodiesel–diesel fuel blends, Renew. Energy. **33**, 2623–2630 (2008)

Physical Characterization of Palm Fatty Acid Distillate (PFAD) Blends as Biofuel

Mantari M.H.A.R[11], *Hassim* H.M[1], Rahman R.A[1], *Zin* A.F.M [1], *Mohamad* M.A.H[1], *Asmuin*. N[2]

[1]Department of Mechanical Engineering, Centre for Diploma Studies, Universiti Tun Hussein Onn Malaysia

[2]Department of Energy and Thermofluid, Faculty of Mechanical and Manufacturing, Universiti Tun Hussein Onn Malaysia

Abstract. Palm oil has been accepted to be used as fuel with the introduction of petroleum diesel and Palm Methyl Ester (PME) in 2014. The high cost is the major drawback of PME. Not only PME is derived from an expensive low Free Fatty Acid (FFA) feedstock, the cost to convert FFA to PME through the process of trans esterification and purification of palm oil fatty acid is also expensive. Alternative feedstock of FFA is palm fatty acid distillate (PFAD). PFAD is a by-product of crude palm oil (CPO) refining. It is inedible and a low price palm oil product making it attractive as a feedstock for biofuel. The high viscosity of PFAD can be reduced by blending it with diesel fuel. Five blends of PFAD and diesel blends with 2%, 4%, 6%, 8% and 10% volumetric ratio were produced in this study. The physical properties of the blends such as density, viscosity, surface tension and flash point were determined and the results were compared to the Malaysian Standard for Diesel Fuel (MS123:2014). The results show that all properties of the blends are within the acceptable value for diesel fuel.

1 Introduction

Oil palm is one of the cash crops that is widely planted in Malaysia and Indonesia. Palm oil industry is a significant contributor to Malaysian economy providing both employment and income from exports. The history of palm oil as fuel started in the 1980s by Malaysia Palm Oil Board (MPOB) with their in house trial using a diesel engine and fully commercialized by government directive in 2014 as B5 that constitute a blend of 95% petroleum diesel and 5% palm methyl ester (Hamim, 2014). The palm methyl ester is produced from transesterification and purification of palm oil fatty acid and reacted with alcohol to produce palm methyl ester and glycerol (Krisnangkura, 1992).

* Corresponding author: mhanafi@uthm.edu.my

The main drawback of palm oil biodiesel is the high production cost. The high cost of producing the biodiesel product is due to its feedstock which come from high quality of virgin oil with low free fatty acid (FFA) contents (Chongkhong, 2008). In order to overcome this limitation, the biodiesel product needs to be produced by using cheaper feedstock either recycled or waste oils, or by product of refining vegetable oil that contains high FFA.

Source of feedstock that can be an alternative in producing the fuel from oil palm is by using palm fatty acid distillate (PFAD). The PFAD is a by-product of crude palm oil (CPO) refining. PFAD is inedible and low price palm oil product making it attractive as a feedstock for biofuel. PFAD contain high free fatty acid (FFA) that account up to 70%-80% and the remaining is triglyceride (Ping, 1992). The high content FFA requires two step operations that are esterification and trans-esterification reactions in order to convert it to become methyl ester (Chongkhong, 2007). Another method to utilize PFAD as a biofuel source is to direct blend it with petroleum diesel. This process eliminates the needs of the costly esterification and trans-esterification operation.

The use of a direct blend of palm oil product had already been done by several researchers. The blends of Refine, Bleach and Deodorized Palm Olein (RBDPO) with petroleum diesel gave a comparable performance to petroleum diesel in terms of spray and combustion characteristics especially at lower blend (10% blend) (Mantari, 2009).

The availability of PFAD as a low cost, non-food source feedstock is significant thus provide opportunity for biodiesel producers with excellent access to a low-cost, non-food source of feedstock (Cheah, 2010). The PFAD in comparison with other products of palm oil such as Refine Bleached Deodorized (RBD) Palm Oil, RBD Palm Olein and RBD Palm Stearin is always being traded at the lowest price as shown in Figure 1. Before the year 2012, the PFADs were traded at the discounted price that exceeds RM 500/MT and increases in the year 2011 with the discount exceeds RM700/MT. However, the gap between the prices was narrow from year of 2013, 2014 and 2015. The range of palm oil product's price for seven years are in between of -27 % to 18 %.

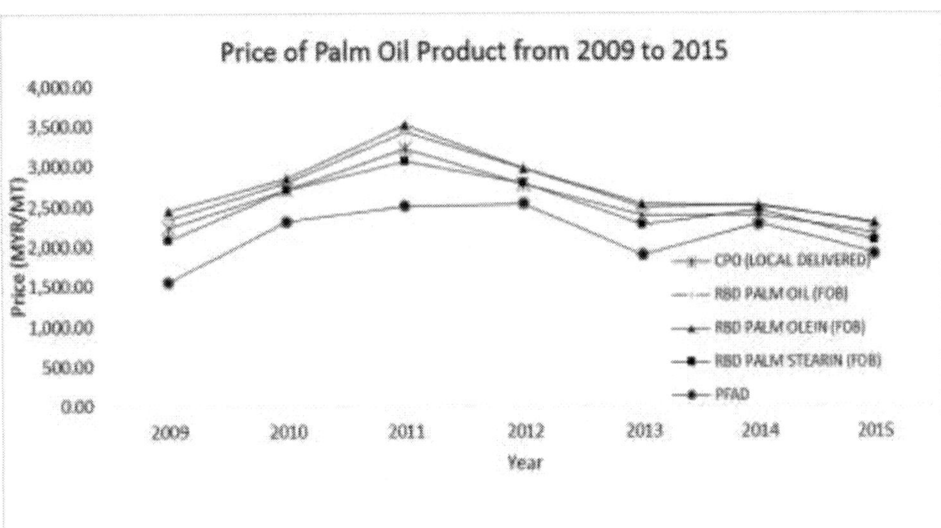

Fig. 1. Price comparison of PFAD with Others Palm Oil RPB product (MPOB, 2015)

DOI: 10.1051/matecconf/201713500005

2 Methodology

The quantity of each element in PFAD-diesel blend depends on its mixing ratio for instance if B2, then it contains 98% of petroleum diesel and 2% PFAD. The rest of the blends and their volumetric ratio are shown in Table 2.1. A total 5 litres of PFAD-diesel blend was produced according to the designated blending ratio. The blending was done using magnetic stirrer at temperature 50°C and was stirred for 15 minutes to allow the mixture to be homogenous.

Table 2.1. Composition of PFAD and Diesel for Each Blend

Blends	**Percentage of PFAD contain in the blend**	**PFAD** (litre)	**Diesel** (litre)	**Total** (litre)
B0	0%	0	5	5
B2	2%	0.1	4.9	5
B4	4%	0.2	4.8	5
B6	6%	0.3	4.7	5
B8	8%	0.4	4.6	5
B10	10%	0.5	4.5	5

The measurement of the physical properties was conducted according to the guideline prepared by world renowned standard bodies such as American Society and Testing Material (ASTM) and International Organization for Standardization (ISO). The density measurement was done according to ASTM D 4052 while for kinematics viscosity, the test was done according to ISO 3104. The surface tensions of these samples were measured according to ASTM D971. The flash points for these samples were tested according to Pensky-Martens closed cup (PMCC) as detailed in ISO 2719.

3 Results and Discussion

The physical characteristics of Palm Fatty Acid Distillate (PFAD) blend with diesel were investigated. The physical characteristics that have been tested include the dynamic viscosity and kinematic viscosity at 40 °C, surface tension at 27 °C and flash point of PFAD. The Malaysia MS-123-1:2014 have been chosen as the benchmark for comparison of the physical characteristics of diesel and PFAD blend.

 DOI: 10.1051/matecconf/201713500005

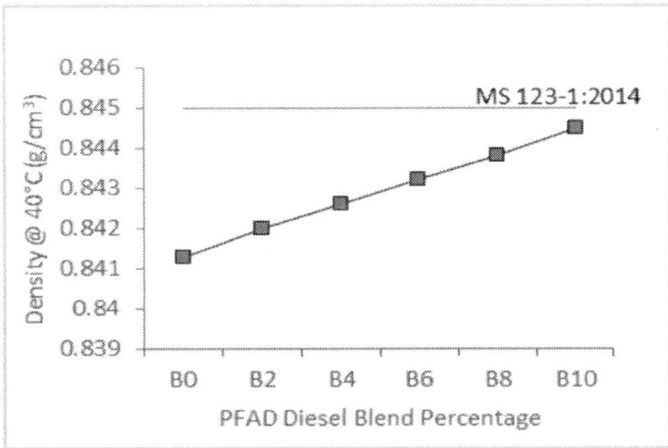

Fig. 2. Graph - Density of PFAD: Diesel Ratio

Figure 2 shows the density of diesel and PFAD blends obtained at 40 °C. It has been found that 100 % of diesel shows the lowest value of density. The value of density increases with the increment of the percentage of PFAD blend into the diesel mixture. The value of density still within the MS123:2014 Malaysian Standard for petroleum diesel even though the PFAD in the blend contain up to the 10% (B10).

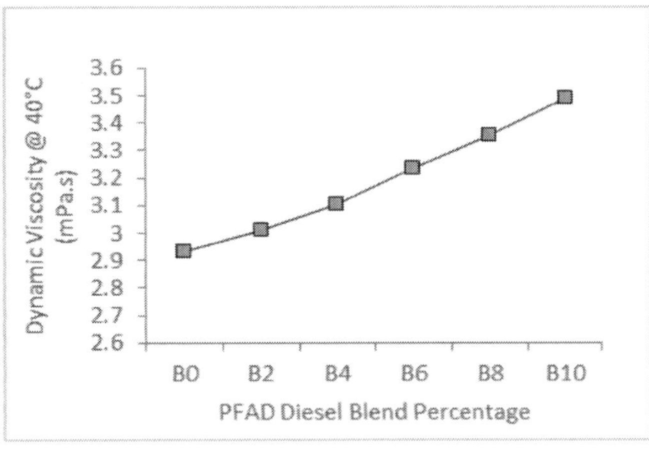

Fig. 3. Graph Dynamic viscosity of PFAD: Diesel Ratio

Figure 3 shows the dynamic viscosity of diesel and PFAD blends at 40 °C. The results indicate that at a lower percentage of PFAD blend will result in lower dynamic viscosity and vice versa. At diesel and PFAD mixture up to 10 % (B10), the dynamic viscosity increase 19 % from the dynamic viscosity of diesel (B0).

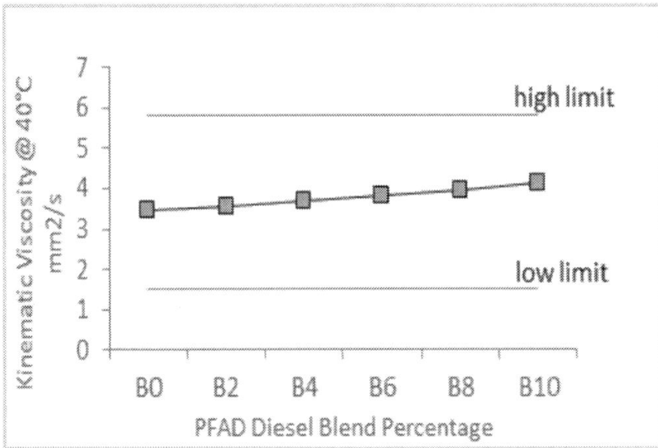

Fig. 4. Graph - Kinematic viscosity of PFAD: Diesel Ratio

Figure 4 shows the kinematic viscosity of the mixture tested at 40 °C. The result shows that the increment of kinematic viscosity of the mixture is up to 18.5 % with the increment of the percentage of diesel and PFAD blend in the mixture. However, the increments of kinematic viscosity are within the high and low limits (acceptable range limit) as described in MS123:2014.

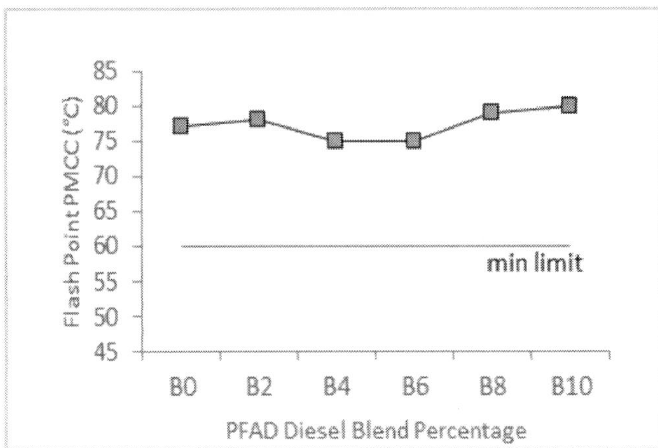

Fig. 5. Flash Point PMCC of PFAD: Diesel Ratio

The results of the flash point for diesel and PFAD blend are shown in Figure 5. In comparison with a standard value, the minimum flash point is 60°C. It is showed that the values of the flash point for diesel and PFAD blend are within 75°C to 80°C which is 25 % to 33 % higher than the standard value. Thus, the flash point for all the diesel and PFAD blends within the acceptable value prescribed in the Malaysian Standard for Diesel, MS123:2014.

DOI: 10.1051/matecconf/201713500005

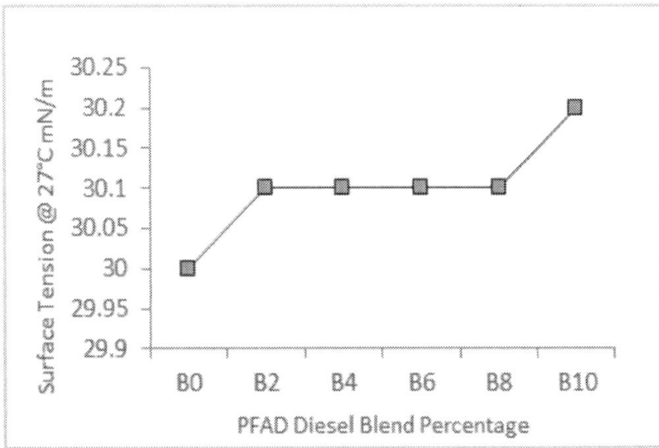

Fig. 6. Graph – Surface Tension of PFAD: Diesel Ratio

Figure 6 shows the surface tension of all diesel and PFAD blends. The surface tension of the blend was tested at 27°C which is around the ambient temperature. The values of all the blends differ by the range of 0.30 % to 0.67 % with respect to diesel (B0). Thus, it can be said that the surface tension of the blends did not have a much different with the diesel fuel.

Conclusion

Instead of palm methyl ester and envodiesel, Palm Fatty Acid Distillate is an alternative option for palm oil-based biofuel. By comparing PFAD with RBD palm oil, olein, and stearin, PFAD is the cheapest cost, making it suitable as a feedstock for fuel. The five diesel and PFAD blends (B2, B4, B6, B8 and B10) physical properties have been successfully measured. It is found that all the value for density, kinematics, viscosity and flash point were within the acceptable value of MS123:2014. The surface tension of the blends differ slightly (less than 1%) compared to the diesel fuel.

Acknowledgment

The authors would like to extend their deepest gratitude to Universiti Tun Hussein Onn Malaysia for awarding research grant to undertake this research (STG U145, STG 1307 and MDR 1321).

References

1) A.V. Metre, K. Nath Super Phosphoric Acid Catalyzed Esterification of Palm Fatty Acid Distillate for Biodiesel Production: Physicochemical Parameters and Kinetics. *Polish Journal of Chemical Technology*, 17(1): p. 88-96 (2015).

2) K.Y. Cheah, T.S. Toh, P.M. Koh, Palm Fatty Acid Distillate Biodiesel : Next Generation Palm Biodiesel. *Paper Presented at Palm and Lauric Oil Conference & Exhibition Price Outlook (POC)*, Kuala Lumpur, Malaysia, March 8-10 (2010).

3) S. Chongkhong, C. Tongurai, P. Chetpattananondh, Biodiesel Production by Esterification of Palm Fatty Acid Distillate.*Biomass and Bioenergy*, 31(8): p. 563-568 (2007).

4) S. Chongkhong, C. Tongurai, P. Chetpattananondh, Continuos Esterification for Biodiesel Production from Palm Fatty Acid Distillate Using Economical Process. *Renewable Energy*, 34, p.1059-1063 (2009).

5) A. Hamim, *Palm oil biodiesel programme to cover all of Malaysia by July*, in *The Star*, Star Media Group Berhad: Kuala Lumpur (2014).

6) K. Krisnangkura, R. Simamaharnnop, *Continuous transmethylation of palm oil in an organic solvent.* Journal of the American Oil Chemists' Society, 69(2): p. 166-169 (1992).

7) MPOB. *Overview of the Malaysian Oil Palm Industry 1999-2015.*

8) M.H.A.R. Mantari, M.N.M. Jaafar, Performance of Oil Burner System Utilizing Various Palm Biodiesel Blends, *International Journal of Mechanical and Materials Engineering (IJMME).* Vol. 4 (3), 273-278 (2009).

9) B.T.Y Ping, M. Yusof, *Characteristics and properties of fatty acid distillates from palm oil.* Oil Palm Bulletin, 59: p. 5-11 (2009).

DOI: 10.1051/matecconf/201713500006

Numerical Study on Self-Cleaning Canister Filter With Add-On Filter Cap

Akmal Nizam Mohammed[1,1], *Mohd Sahrizan* Sahrudin[1], *Mohd Hafiz* Zolkhaely[1], *Azwan* Sapit[1], *Mohd Kamil* Abdullah[1], *Suzairin* Md Seri[1], *Ishkrizat* Taib[1] and *Normayati* Nordin[2]

[1]Faculty of Mechanical and Manufacturing Engineering, Universiti Tun Hussein Onn Malaysia
[2]Flow Analysis, Simulation, and Turbulence Research Group, Universiti Tun Hussein Onn Malaysia

Abstract. Filtration in a turbo machinery system such as a gas turbine will ensure that the air entering the inlet is free from contaminants that could bring damage to the main system. Self-cleaning filter systems for gas turbines are designed for continuously efficient flow filtration. A good filter would be able to maintain its effectiveness over a longer time period, prolonging the duration between filter replacements and providing lower pressure drop over its operating lifetime. With this goal in mind, the current study is focused on the difference in pressure loss of the benchmark Salutary Avenue Self-cleaning filter in comparison to a new design with an add-on filter cap. Geometry for the add-on filter cap will be based from Salutary Avenue Manufacturing Sdn.Bhd. SOLIDWORKS software was used to model the geometry of the filter, while simulation analysis on the flow through the filter was done using Computational Fluid Dynamic (CFD) software. The simulations are based on a low velocity condition, in which the parameter for the inlet velocity are set at 0.032 m/s, 0.063 m/s, 0.094 m/s and 0.126 m/s respectively. From the simulation data obtained for the inlet velocities considered, the pressure drop reduction of the modified filter compared to the benchmark was found to be between 7.59% and 30.18%. All in all, the modification of the filter cap produced a lower pressure drop in comparison with the benchmark filter; an improvement of 27.02% for the total pressure drop was obtained.

1 Introduction

Filtration is one of the essential parts in a gas turbine. Filters are used to prevent foreign contaminants in air from entering the system in a gas turbine. Air is drawn into the turbine by using compressors until it reaches the combustion chamber, in which fuel is sprayed into the air and ignited to generate a high-temperature flow. In a gas turbine, its efficiency is its most important characteristic, and the factors that can affect its efficiency are many and various. Among the main factors that will affect the efficiency of a gas turbine is thru its

[1] Corresponding author: akmaln@uthm.edu.my.

inlet. As the gas turbine cycle operates, both its inlet and exhaust are open to the atmosphere, and hence the flow parameters would vary with atmospheric conditions. Contaminants in the air that consist of different types and from a variety of environments can cause several kinds of problems which can negatively impact the reliability, availability, and time between overhauls of gas turbine internal components thus affecting its efficiency.

The efficiency of the gas turbine can also be affected by the air inlet temperature of the gas turbine. Based from study of [1] among the method to increase the performance and efficiency of a gas turbine is by air cooling and filtration system in the inlet of a gas turbine. It is well known that ambient temperature, humidity and pressure are important factors in gas turbine performance. As mentioned in [2], thermodynamic analyses exposed that thermal efficiency and specific output decrease with an increase of humidity and ambient temperature.

The selection of the filtration system must be taken into consideration as there are many factors need to be considered in choosing filtration system in a gas turbine. As mentioned in [3] the system of filtration for a gas turbine are selected based on the operational philosophy and goals for the turbine, the contaminants that are present in the ambient air, and also based on expected changes in the contaminants in the future due to temporary emission sources or seasonal changes.

Increase in the pressure drop will causes an increase in energy consumption. Based from study of [4] Pressure loss indicates the resistance of airflow through a filter. High filter pressure drop reduces turbine output and a low pressure drops improves turbine output. To overcome the inlet system losses, the gas turbine will consume more fuel, and it also has to reduce its power output. The pressure loss will increases, while the power will decreases and the heat rate will increases linearly.

Self-cleaning filters are used in low approach velocity systems thus when normal flow is re-established the low air velocity assists in preventing dust that is being removed from the filter with air pulsing from re-entering again in the airflow stream. As the surface of the filter is loaded with particles, it decreases the available flow area through the filter which increases their efficiency. This type of filtration system is very useful in dry environments with high levels of dust. While in high moisture environments the self-cleaning filter is prone to high pressure losses and other problems.

Computational Fluid Dynamic (CFD) techniques have the potential to allow the effect of a proposed design change to be evaluated relatively quickly while the cost can be keep at a minimum rather than doing experimental investigation. As mentioned in [5], detailed information such as velocity, pressure, temperature, concentration and, turbulence intensity can also be obtain by using this method. The study aims to give further understanding about suitable design of a gas turbine canister filter so that an increase in gas turbine efficiency can be achieved. The analysis of the pressure drop in the gas turbine canister filter are made using CFD software so that the cost of producing prototype for the canister filter can be reduced.

2 Methodology

The simulation and numerical analysis method of the whole self-cleaning canister filter are made using Computational Fluid Dynamics software that is the Workbench16.1. In CFD, the process for the numerical analysis is divided into three stages which are pre-processing, solvers and post-processing. The type of analysis conducted using this software is Fluid Flow (CFX).

The parameter for the simulations are based from Salutary Avenue (SA) brochure, the data that are available from Salutary Avenue (SA) are the air flow rate, surface area of

DOI: 10.1051/matecconf/201713500006

the filter and also the pressure drop for the experiment of Salutary Avenue (SA) filter. The data that are obtained from Salutary Avenue (SA) are for the F9 type filter. From the data that available, Table 1 are build based from the data.

Table 1. Parameter for simulations of benchmark and modified filter.

Flowrate (m^3/s)	Pressure Drop (Pa)	Area (m^2)	Velocity (m/s)
0.236	25	7.5	0.032
0.472	47.23	7.5	0.063
0.708	77.1	7.5	0.094
0.944	121.81	7.5	0.126

Pre-processing is where the geometry of the item is determine than is drawn either using ANSYS Workbench or SOLIDWORKS software. For this study the geometry of the self-cleaning canister filter with add on filter cap is drawn by using SOLIDWORKS it is then generated into the Design Modeller (DM) of ANSYS workbench as showed in Figure-1 (a). Modelled of the geometry is then need to be computational mesh throughout the flow volume as in Figure-1 (b).

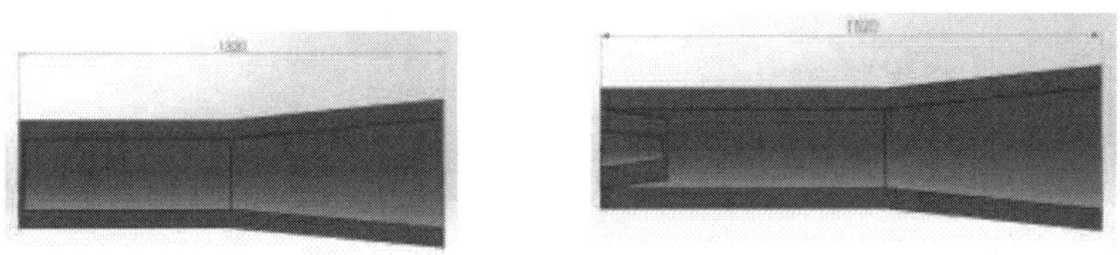

(b)

Fig. 1. Benchmark filter of original size is shown in (a), whilst the modified model is given in (b).

Mesh is the well organize arrangement of flow field at discrete point in a flow domain. Long solver runs may due to too many cells, while inaccurate results may cause by too few of cells. The last step in pre-processing is setup. In this step, parameters for the model need to be declared. Parameters such as domain, boundary condition and the medium of the material need to be specifically declared so that the setup process and solution solver of the model can be obtain. The model are split into two domain that is air domain and filter domain.

Solver is an unstructured grid finite volume solver for the compressible flow where the previous inputted data from the pre-processing step are calculated. The solver setting will calculate based on the number of iteration for the model geometry. The number of iteration can be set on the solver setting; once the number of target residuals are achieved, the solver will stop even though the number of maximum iterations are not achieved, the solver is converge at 0.0001 residual target with the number of iterations is generated on the output file.

The last step in ANSYS Fluent software is the post-processing where the simulation of the model are generated, calculated data in the solver is transferred into figures, meaningful graphics, animations and reports that make it easy to convey the fluid dynamic results. For the post-processing result, a line will generated based from the length dimensions for the filter, the data will be extracted and export that are based from the line that is generated.

Grid independency test is done to identify the best meshing size by saving the computation time without neglecting the accuracy of the results, meshing is a break-down of a single unit geometry into a small discrete control volume where the set of equations are applied on each cell and computed.. The grid independency test is done over a range of grid size to compare and obtain the best and minimum computational time. Table-2 show the different grid sizes with its number elements and nodes. The medium nodes of 337697 were selected as the results is not far apart to the higher nodes of 445034 and minimum computational time will occurred

Table 2. Grid size setting for coarse, medium, and fine meshes respectively.

Grid Size Setting		No of Nodes	No of Element
Min	Max		
8	13	207677	1118435
8	11	337697	1845530
7	10	445034	2451446

3 Results and Discussion

The pressure drop is calculated by subtracting the outlet pressure from the inlet pressure, as shown in Eqn. (1). The value of the pressure drop are calculated by getting the pressure value by assuming the inlet pressure is at the narrow end of the filter minus the outlet pressure at wider end.

$$\Delta P = P_{inlet} - P_{outlet} \quad (1)$$

Three dimensional velocity streamline for the benchmark and modified filter simulation can give a clear view of the air flow that enter the filter as showed in Figure 2, from the figure, it showed that the air that enter the filter is from the front and surrounding of the filter, by modifying the filter, a better flow of air is seen entering the filter, air can enter from the front for the modified filter rather than the benchmark filter, where wall is placed in front of the filter.

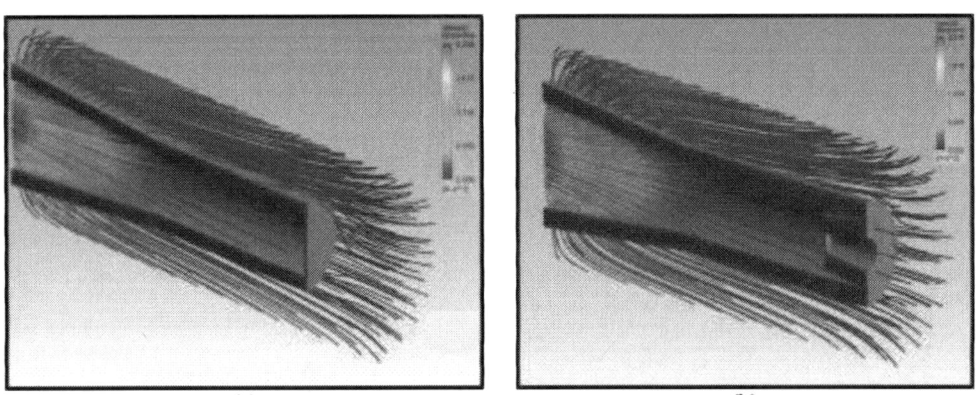

(a) (b)

Fig. 2. Benchmark filter of original size is shown in (a), whilst the modified model is given in (b).

 DOI: 10.1051/matecconf/201713500006

The simulation of benchmark filter, showed a lower pressure drop compared to the Salutary Avenue (SA) at lower inlet velocity of 0.032 m/s and 0.063 m/s, while at the last two velocity inlet parameter of 0.094 m/s and 0.126 m/s it showed that the Salutary Avenue (SA) has a lower pressure drop than simulation of benchmark filter as show in Figure 3a. The value of the result vary because the experimental data for Salutary Avenue (SA) is not fully revealed, thus leaving assumption need to be made at the boundary condition for simulation of the benchmark filter.

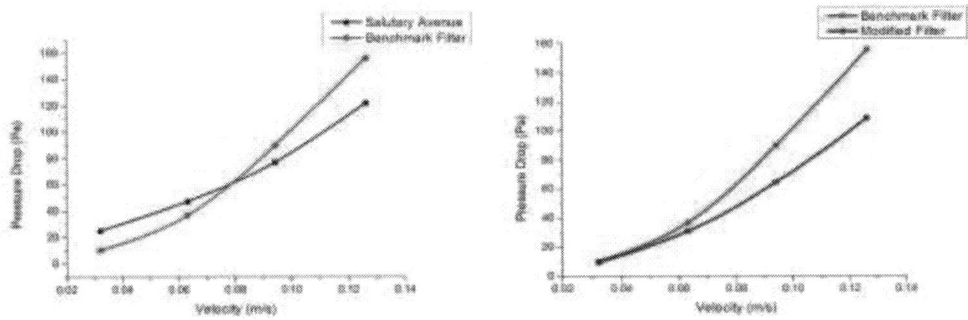

(a) (b)

Fig. 3. Benchmark filter of original size is shown in (a), whilst the modified model is given in (b).

For comparison of simulation for the benchmark and modified filter, it is more likely because all of the simulation setting are the same for the simulation except for the geometry of the filter, it showed that the pressure drop for the modified filter is better than the modified filter, the pressure drop for inlet velocity of 0.032 m/s is not far apart as benchmark filter recorded 10.191 Pascal of pressure drop while 9.418 Pascal for the modified filter, as the inlet velocity parameter became larger, the pressure drop increased with an increase of velocity.

Lastly, a comparison of all the data had been made for Salutary Avenue (SA) based from the graph on Figure 3b and Table 3 for benchmark and modified filter. Based from the plotted graph, it showed that the modified filter had a lower pressure drop compared with Salutary Avenue (SA) and the benchmark filter, the modified filter had a better pressure drop due to the additional area for the modified filter. The modified cap of the filter, enable air to flow from the front area of the filter, compared than benchmark filter where wall is place in front of the filter, restricted the airflow entering the filter. The additional surface area for the modified filter made the filter had a lower pressure drop than others.

Table 3. Tabulated values of pressure drop for the simulated filter compared with benchmarks.

Velocity (m/s)	**Pressure Drop (Pa)**		
	Salutary Avenue	Benchmark Filter	Modified Filter
0.032	25.00	10.19	9.42
0.063	47.23	36.63	30.85
0.094	77.10	89.88	64.44
0.126	121.81	155.77	108.75

4 Conclusion

It can be concluded that the objective of the research had been achieved. The cases are based on a low velocity system for velocity inlet parameter of 0.032, 0.063, 0.094 and 0.126 m/s. The parameter for obtaining a low pressure drop in a Salutary Avenue filter had been recognised, with an increase of its overall filter length of 20 cm, a total pressure drop of 213.456 Pascal can be achieved for simulation of modified filter, the result is 27.02% better than the simulation result from total pressure drop of benchmark filter that are 292.469 Pascal for total set parameter of inlet velocity. The range for pressure drop is between 7.59% and 30.18% for inlet velocity parameter range of 0.032 until 0.126 m/s for simulations comparison of benchmark and modified filter. The pressure drop can be lower if the area of the filter are increased, in this case is by increasing the overall filter length. The pressure drop for the filter is not only determined on the area of the filter, it also are due to the design of the inlet system ducting also, due to lack of experimental results and parameters from Salutary Avenue, the data that are obtain from simulation may not be highly precise, this is because the data that are not fully reveal for this study, and limited information that are able to be obtained.

Acknowledgement

This research was made possible through funding from UTHM IGSP Grant U413 and UTHM STG Grant U126. The authors would also like to acknowledge the contributions made by members of Flow Analysis, Simulation and Turbulence Research Group (FAST), The Centre for Energy and Industrial Environment Studies (CEIES) of UTHM, and Salutary Avenue Manufacturing Services Sdn. Bhd.

References

1. A. Ku, Comparative analysis of inlet air cooling techniques integrated to cooled gas turbine plant. Journal of the Energy Institute, 1–15 (2014).
2. M. Farzaneh-Gord, M. Deymi-Dashtebayaz, Effect of various inlet air cooling methods on gas turbine performance. Energy, 36(2), 1196–1205 (2011).
3. M. Wilcox, R. Kurz, K. Brun, Technology review of modern gas turbine inlet filtration systems. International Journal of Rotating Machinery (2012).
4. Z. Feng, Z. Long, Q. Chen, Assessment of various CFD models for predicting airflow and pressure drop through pleated filter system. Building and Environment, 75, 132–141 (2014).
5. C. Tsang, Analysis of pleated air filters using computational fluid dynamics. University of Toronto (1997).

Hybrid Micro-Hydro Power Generation Development in Endau Rompin National Park Johor, Malaysia

Azli *Yusop*[1,*], Izat *Yahaya*[1], Ghazali *Kadis*[1], Maizul *Ishak*[2], Mohd Sufyan *Abdullah*[2], Mohd Fadly *Razak*[2], Azmi *Sidek*[3], and Ismail *Abdul Rahman*[4]

[1]CeDS, Universiti Tun Hussein Onn Malaysia 86400, Batu Pahat, Johor, Malaysia
[2]FTK, Universiti Tun Hussein Onn Malaysia 86400, Batu Pahat, Johor, Malaysia
[3]FKEE, Universiti Tun Hussein Onn Malaysia 86400, Batu Pahat, Johor, Malaysia
[4]FKAAS, Universiti Tun Hussein Onn Malaysia 86400, Batu Pahat, Johor, Malaysia

Abstract. Micro-Hydro electrical power systems are very useful for remote area electrification which does not had supply from the national grid. On the contrary, this area has river streams with high potential for micro-hydro power generation. As such, the UTHM ECO-Hydro Team embarked on a project for erecting a micro-hydro power plant with collaboration with National Education Research Center (NERC), Johor National Park Corporation in Endau Rompin. The existing power generation in this area at present is by using diesel generator gives negative impact on finance and environment in the long run. It supplies power to several including library, offices, open laboratory, chalets and dorms.. At the moment, the micro-hydro system complements the diesel generator, thus becoming a hybrid power generation system.

1 Introduction

1.1 Location – NERC, Endau Rompin National Park Johor, Malaysia.

Endau Rompin lies at the headwaters of Endau River astride the Johor-Pahang border. It covers an area of 870 square km, about one and a half times the size of Singapore Island. It includes parts of the Labis and Endau-Mas forest reserves in Johor and the Lesong Forest reserve in Pahang.

The area contains a unique assemblage of lowland and hill rain forest with pristine rivers and varied wildlife. Besides possessing its own plants found nowhere else in the world, it is also home to the largest population of those highly endangered animals (e.g. Sumatran rhinoceros). This makes the area an important part of Malaysia's natural heritage that is badly in need of conservation.

The Nature Education and Research Centre (NERC), Endau Rompin became a reality with the collaboration and support of the Johor government, the Danish Corporation Environment and Development (DANCED) and the Malaysian Nature Society (MNS). Funding for the project was provided by DANCED while MNS is the consultant for the implementation of the NERC on behalf of Johor National Park Corporation (PTNJ) [1]. This centre is being developed as a focal point for nature education and research centre for visitors from all over the world.

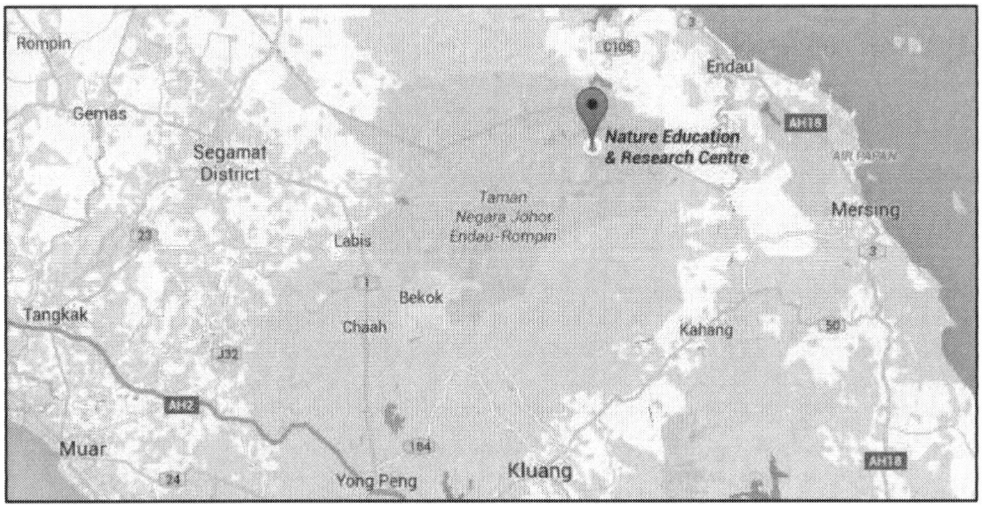

Fig. 1. Location of NERC, National Park Endau Rompin [2].

1.2 Micro-Hydro

Flowing and falling water have potential energy. Hydropower comes from converting energy in flowing water by means of a water wheel or through a turbine into useful mechanical power. This power is converted into electricity using an electric generator or is used directly to run milling machines. Small-scale hydropower systems, however, are receiving a great deal of public interest as a promising, renewable source of electrical power for homes, parks and remote communities [4].

A hydro power is generally categorized by six (6) types. The systems are classified as large, medium, small, mini, micro and pico according to their installed power generation capacity. The electrical power is measured in watts (W). General classification of hydro power pants were given in Table 1.

Table 1. A general classification of hydro power plants [5].

Type	Capacity
Large-Hydro	More than 100MW
Medium-Hydro	15 up to 100MW
Small-Hydro	1 up to 15 MW
Mini-Hydro	Above 100 kW, but below 1 MW
Micro-Hydro	From 5 kW up to 100 kW
Pico-Hydro	From few hundred watts up to 5 kW.

In this project, the maximum electrical power output was calculated to be 15 kW, hence it was categorized as a Micro-Hydro system with reference to Table 1. This system is ideal for power generation for small community in remote areas and away from the grid.

2 Planning

This multidiscipline project was funded by UTHM Contract Research Grants (C036), which combines researchers from three areas of expertise namely civil, mechanical and electrical engineering. On the other hand, this group was also supported by ecologists to study the water quality of Semawak River. Being located in a remote area with the hilly terrain, this project poses a special kind of challenge to the researcher's to develop and install the micro-hydro system with minimal environmental impact.

The development of micro hydro system in NERC began in June 2012 and fully operational in 18 October 2014. Early planning between the researchers and the Johor National Park Corporation was implemented carefully and periodical meetings were conducted to ensure the success of this project. Table 2 presents the timeline of activities that has been implemented.

Table 2. Timeline activities to completing the micro hydro system [6].

No.	Time	Activity
1.	June – Dis. 2012	Micro hydro fieldwork data acquisition for civil, mechanical, electrical and ecology (phase 1).
2.	Oct. – Nov. 2012	Preparation of model and prototype for the MOA signing ceremony on the 19th Convocation UTHM.
3.	Nov. 2012	MOA signing ceremony between UTHM and Johor National Park Corporation (PTNJ).
4.	Nov. – Dis. 2012	Technical specifications document preparation and workshop.
5.	Dis. 2012 – Mac 2013	Micro hydro fieldwork data acquisition for civil, mechanical, electrical and ecology (phase 2).
6.	Jan. – Feb. 2013	Purification and final correction of technical specifications.
7.	Mac – Mei 2013	Preliminary works started (site clearance , surveying contour and levelling)
8.	Jun – Sept. 2013	Turbine house construction work (soil work, concrete, frame, wall, ceiling, plastering & painting).
9.	Sept. – Oct. 2013	Clearing pipeline routes from forebay to turbine house.
10	Oct. – Dis. 2013	Forebay construction.
11.	Dis. 2013 – April 2014	Pipeline system installation (logistic, arrangement, pipe connection, valve installation, testing)
12.	April – Jun 2014	Turbine generator and electrical systems installation.
13.	Jun – July 2014	Electrical loads installation for UTHM dorm, UTHM open laboratory & toilets (underground cable lay, wiring, loads

		installation).
14.	July – Oct. 2014	Fully test for Micro Hydro system.
15.	18 Oct. 2014	Micro hydro turbine house opening ceremony by Yang Berhormat Datuk Hj. Ayub Rahmat (Johor state Health and Environment Executive Councilor).

Fig. 2. Turbine house frame construction [3].

Fig. 3. Turbine house [3].

3 Design

This system design takes into account its hilly landscape and plants. Any tree felling should be minimized and must get permission from the authorities. Clearing pipelines routes was done manually without using any heavy machinery that can damage flora or disturb wildlife in the area. The selected forebay has a gross area of 65 meter head with 720 meter downstream distance to the turbine house.

The average flow rate presumably was calculated at 45l/s. Catchment area is 6.8 meter x 5.5 meter x 1 meter. According to the head and flow rate measurement, this micro hydro plant can produce a maximum of 15 kW using Pelton turbine. In areas with adequate water resources, micro hydro power station is the best option compared with other renewable sources, especially in terms of costing [7].

3.1 Civil Design

The system design as shown in Figure 6. The piping size is 10" and the material is UPVC which can handle pressure up to 10 bar. There are three (3) valves installed, each in forebay area, cross road area and turbine house area. Carbon steel elbow is used as a support for pipes installed every 5 to 10 meters to further strengthen the position of the piping.

Discharge water from the turbine house are channeled into two storage tanks (holding tanks), so it can be used for daily needs. Next the water flows into the existing ecological pond in NERC where there were several species of small fish inside. Finally, after the pond was filled, the water directed into a river that is located downstream near the NERC jetty. The challenge during pipe laying is the bend and steep area as it is in the dangerous condition that can cause a person to slip and fall down.

Fig. 4. Piping logistics [3].

Fig. 5. Pipeline installed [3].

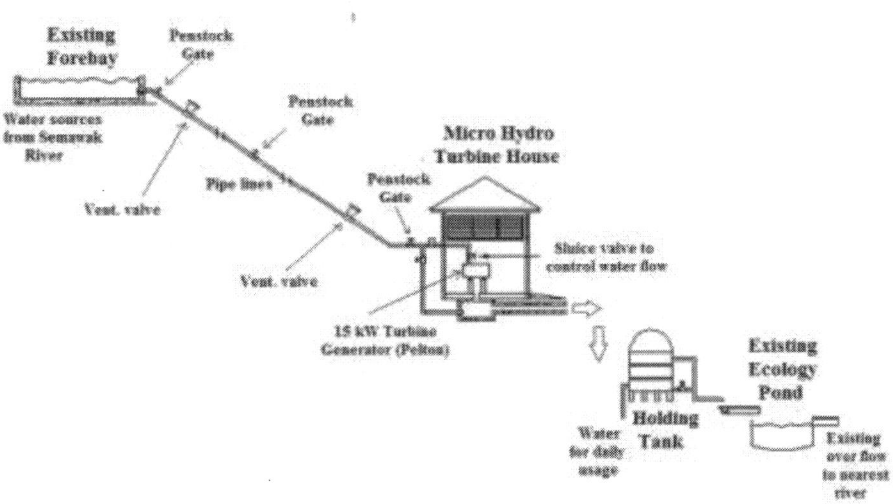

Fig. 6. NERC Micro Hydro system design [8].

Fig. 7. Forebay area [3].

Fig. 8. Holding tanks for daily usage [3].

3.2 Electrical Design

The electrical system diagram as shown Figure 9. The panelboard had displays such Voltmeter with a selector switch to read the voltage between phases and the line voltage, Ammeter on each phase to measure current, digital frequency meter and digital power meter. The system also being installed with short circuit, over current and earth leakage protective devices.

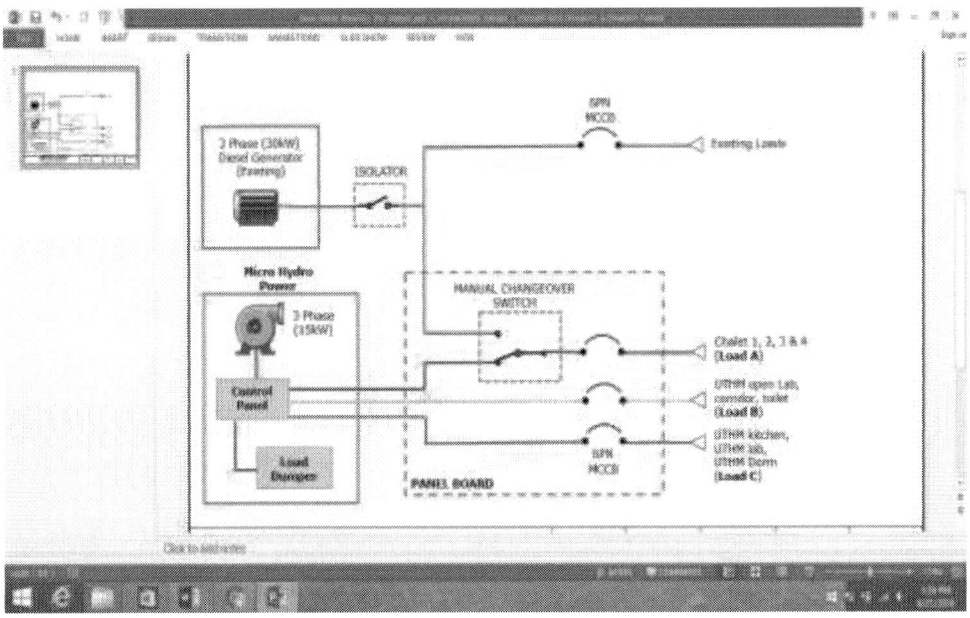

Fig. 9. Electrical system design [8].

The system composed of micro hydro generator, manual changeover switch, storage battery and inverter. The outline of the basic design of the principal equipment and operating system is detailed as following:

3.2.1 Micro Hydro Generator

The maximum output power is 15 kW (3 phase), with a voltage of 400/230 volt, 50 Hz frequency and turbine speed is 1500 rpm. The speed of the rotating turbine can be controlled using the intake gate manually depends on the loads current required.

3.2.2 Manual Changeover Switch

A manual changeover switch prevents the diesel generator power and the micro hydro power from powering the loads circuit at the same time. It is permanently installed in front of the panel board and operated by onsite personnel. In this project, the chalet loads (chalet 1, 2, 3 & 4) is shared by both supply source. At the time when micro hydro does not work, then the diesel generator become a backup supply, and vice versa.

DOI: 10.1051/matecconf/201713500007

3.2.3 Inverter

The inverter output power is 8 kW, with incoming voltage 240 volt and charging the storage battery at 48 volt. At the time when the battery is fully charged, the inverter will automatically stop.

3.2.4 Storage Battery

There is four (4) units of battery (12 volt) installed in series to create a total 48 volt with 120 aH each. The storage capacity was determined to be 5.76 kW. The battery life percentage are shown on the inverter display.

Fig. 10. Micro Hydro turbine installed [3].

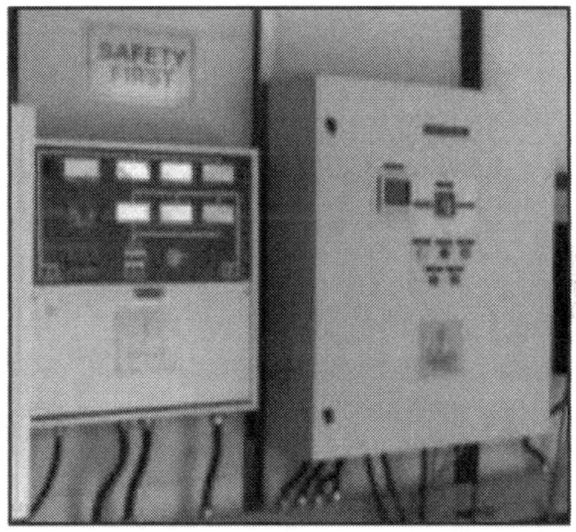

Fig. 11. Micro Hydro panel boards [3].

4.0 Loads Installation

Electrical energy from the micro hydro power plant was distributed to three loads (Load A, Load B and Load C). 'Load A' using red phase, is the existing loads which sharing power from diesel generator by using manual changeover switch. It supplies power to Chalet 1, Chalet 2, Chalet 3 and Chalet 4.

For 'Load B' (yellow phase), were used to supply power to UTHM open lab, corridor and toilets. Cable sizing is 2 x 6mm PVC with armored protection and being laid using underground method. The distance between the turbine house panel board to 'Load B' distribution board (DB) is about 55 meters long.

For 'Load C' (blue phase), was supplying power to UTHM kitchen, UTHM laboratory and UTHM dorm. Cable sizing is 2 x 10mm PVC with armored protection, also being laid underground. The distance between the turbine house panel board to 'Load C' distribution board (DB) is about 250 meters long. Total energy distribution for this system is provided in Table 3.

Table 3. Total electrical power distribution [6].

NO	LOCATION	LOAD TYPE	QUANTITY	CURRENT DEMAND WITH DIVERSITY FACTOR (A)	
1	Load A	Switch Socket	12 socket outlets	13.75 A	
		Lights + Fan	40 points	7.13 A	
2	Load B	Lights + Fan	34 points	3.03 A	
3	Load C	Lights + Fan	24 points	2.14 A	
				TOTAL	**26.05 A**
				TOTAL	**6,252 watt**

5.0 Conclusion

With the success of planning, designing, development and installation of the micro-hydro system the ECO Hydro Team plans to distribute the power supply to other blocks such library, NERC office and another two (2) dorm, as well as a few additional points on the existing loads. The team aims to fully utilize the energy consumption of micro-hydro. The main objective of the development project was achieved as following

i. Improvement in terms of operating time from 12 hours to 24 hours of electricity for domestic use in the NERC.
ii. Improvement of the quality for the ecosystems, environment and surrounding wildlife, including air quality in Endau Rompin National Park (air, water quality and noise level being studied and from time to time the study will continue to be pursued. - This result will be discuss in related papers).
iii. Savings in terms of operating cost, time and maintenance costs compared to diesel operating generator.

Therefore it can be concluded that the use of micro hydroelectric system was timely in the environmental conservation in which it is also upgrading the system in the Endau Rompin National Park.

References

1. Nature Education and Research Center (NERC) Endau Rompin Natinal Park Malaysia. Available at http://www.tourmalaysia.com.my/endau/nerc.htm.
2. Google maps Malaysia, available at https://www.google.com.my/maps.
3. ECO Hydro Team (authors) picture collections for micro hydro system project (2012 – 2013).
4. *Natural Resources Canada Renewable and Electrical: "Micro-Hydropower Systems: A Buyer's Guide" - ISBN 0-662-35880-5* (2004).
5. D. Singh: *"Micro Hydro Power", Resource Assessment Handbook* (2009).
6. ECO-Hydro Team (authors) record based on reports, surveys and meeting minutes (2012 – 2014).
7. K. Kusakana, J.L. Munda, A.A. Jimoh, *Economic and Environment Analysis of Micro Hydropower System for Rural Power Supply*. In: Proceeding of the 2nd IEEE Power and Energy Conference (2008).

DOI: 10.1051/matecconf/201713500007

8. ECO-Hydro Team initial drawing design based on technical documentation and reports (2013).

Detached-Eddy Simulation of Trailing-Edge (TE) Cutback Turbine Blade Cooling

Marwan Effendy[1], *Yufeng* Yao[2] , and *Denis* R. Marchant[3]

[1]Department of Mechanical Engineering, Universitas Muhammadiyah Surakarta, Jl. A.Yani Surakarta, Indonesia, 57102

[2]Department of Engineering Design and Mathematics, University of the West of England, Bristol BS16 1QY, UK

[3]School of Aerospace & Aircraft Engineering, Kingston University, Friars Avenue, London SW15 3DW, UK

Abstract. This research evaluates the cooling performance of trailing-edge cutback for gas turbine blade. By using DES based on SST k-ω turbulence model, numerical investigations were performed at two steps: first, to validate simulation results from an existing TE cutback cooling with staggered pin-fin arrays inside the cooling passage against experimental measurement. Three types structured mesh from coarse ($\Delta y^+ = 0.74$) to fine ($\Delta y^+ = 1.22$) were evaluated during this step; second, to investigate the TE cutback cooling performance on various blowing ratios. Simulations were performed by keeping the same initials and boundary conditions as the experiment. The result indicates that validation can be considered acceptable by controlling grid quality resolution near wall regions. Both computational data of the adiabatic film-cooling effectiveness and the discharge coefficient are in good agreement with available experimental measurements. The averaged film-cooling effectiveness along the cutback region is highly influenced by the blowing ratios, which is to be related to the turbulent flow structures formed at the mixing region as the impact of coolant flow ejection. The increase of coolant jet velocity triggers the heat transfer process up to the downstream region of TE cutback cooling.

1 Introduction

Gas turbine engines require a very high inlet temperature in order to maximise the power output to achieve a high thermal efficiency. Modern gas turbines typically operate at temperatures of around 1,200 – 1,500°C [1]. The temperature can be up

[1] Corresponding author: Marwan.Effendy@ums.ac.id

to 2,000°C for a double engine power in aircraft gas turbines [2]. It is significantly beyond the melting temperatures for turbine vane and blade materials. High turbine inlet temperature causes structural strength degradation that eventually results in vane and blade break down. It causes a serious implication on durability and safety. Therefore, turbine blades/vanes must be cooled down to the levels that are significantly below the melting point of the material using internal and external cooling techniques. The cooling can be done by combining convection and film cooling. A comprehensive review paper by Han *et al.* [2, 3] summarised recent advancements in this field including both internal-cooling passage and external film cooling jets.

The trailing edge is the most critical part in a vane/blade component due to its geometrical constraint, in combination with aerodynamic, thermal and structural requirements. The trailing edge is particularly vulnerable to high turbine inlet temperatures because of its thin structure, which is susceptible to heavy mechanical and thermal stress. An effective blade TE cooling system should keep vane/blade temperature within its limit without affecting both mechanical strength and aerodynamic losses. Trailing-edge blades become one of the most interesting topics due to the conflict of interest between the need of cooling and aerodynamic requirements. Moreover, the heat-transfer enhancement around the downstream of trailing-edge and thermal-shield by film-cooling behind the lip creates more complicated issues.

Numerical investigations of TE cutback cooling can be found in open literature, but only little information explains the dynamic flow interaction between the internal cooling and the external cooling in a comprehensive manner. So far, most of the numerical studies employ the steady RANS model that does not capture well in terms of the turbulent flow mixing. Therefore, the predicted CFD data does not match with the measurement data. For instance, the use of steady RANS predicts ideal results, which ignores a decay of effectiveness as reported by Holloway *et al.* [4], Martini *et al.* [5], Effendy *et al.* [6], and Egorov *et al.* [7].

It is known that the use of unsteady RANS calculation is capable on capturing the turbulent flow structures at the mixing region as discovered by Medic *et al.* [8] and Joo *et al.* [9]. Unfortunately, it does not optimally simulate the flow interactions between the mainstream flow and the coolant. Consequently, CFD prediction of film-cooling effectiveness does not match with the experimental data as found by Holloway *et al.* [4], Egorov *et al.* [7], Medic *et al.* [8], and Joo *et al.* [9].

Other unsteady methods have been applied by eddy simulation in order to gain a better prediction on capturing turbulent flow structures of mixing. For example, Scale-Adaptive Simulation (SST-SAS) was used by Egorov *et al.* [7], Detached-Eddy Simulation based on Spalart-Allmaras turbulence model was applied by Martini *et al.* [10] and Krueckel *et al.* [11], and Large-Eddy Simulation was utilized by Viswanathan *et al.* [12]] and Schneider *et al.* [13][14]. Moreover, Viswanathan *et al.* [15] employed eddy simulation for modelling forbody [16] and internal cooling [17]. A pressure-side bleed on the trailing edge with realistic turbine condition was numerically studied by Holloway *et al.* [18] for both steady and unsteady RANS.

Similar to TE cutback as experimentally investigated by Holloway *et al.* [4, 18] and Chen *et al.* [19], further numerical study has been undertaken by Medic *et al.* [8] using ANSYS-CFX commercial code. They applied the unsteady RANS with the shear-stress transport (SST) turbulence model computation using 1.25 million cells. In continuation of works from Medic *et al.* [8], Joo *et al.* [9] undertook their work using the Stanford University SUMB CFD code. They applied both the SST-RANS modelling and the SST-SAS (scale-adaptive simulation) for eddy-simulation on their numerical study. Based on their studies, both Medic *et al.* [8] and Joo *et al.* [9] noted that prediction by eddy simulation is appropriate with the available data of experiments, whereas steady/unsteady RANS predicts ideally the level of film-cooling effectiveness nearly unity along the trailing-edge surfaces, mainly for higher blowing ratios. The film-cooling effectiveness predicted by un-steady RANS was not consistent with tested data due to coolant air near the trailing-edge surface was protected from mixing process with main hot gas. A small chunk of main hot gas did not hit the bottom wall along the trailing-edge cutback. The over-prediction of film-cooling effectiveness by unsteady RANS was due to the deficiency of unsteady coherent energy at near wall of the TE cutback surfaces. Along this surface, both the total turbulence kinetic energy and the mean temperature predicted by unsteady RANS was significantly less than the eddy simulation. This finding indicated that there was a strong correlation between the level of turbulence kinetic energy and the profiles of the mean temperature, which caused the degree of mixing process. This research also noted that the three-dimensional (3D) vortex-shedding caused by the upper lip is a dominant aspect that influences a mixing flow over the TE cooling surface. Unsteadiness at the TE breakout was composed by two components namely coherent and broadband components. The coherent component was due to unsteadiness and the broadband component was denoted by a closure model [9].

In parallel with their experiments, Martini *et al.* [5, 20] evaluated numerically the performance of gas turbine blade trailing edge cutback cooling. Through studies of different TE configurations including "two-rows of long ribs" and "circular pin-fin arrays" inside the cooling passage, Martini *et al.* [5] developed a multi-block structured mesh up to 0.89 million cells with average of $\Delta y^+ < 2$, and simulated on three different blowing ratio (i.e. $M = 0.5$, 0.8, and 1.1). This prediction found that a highly complex turbulent mixing process was not captured correctly using 1 model with CFX Tasc-Flow. There was a significant deviation of film-cooling effectiveness between CFD data and measurements, particularly in the middle of the trailing-edge cutback region. Further numerical study showed that the prediction could be improved by applying the Spalart-Allmaras turbulence model for DES simulation. This approach gave a detailed insight into the unsteady film mixing on the blade TE breakout region [10]. The results revealed that the adiabatic film-cooling effectiveness and the heat-transfer coefficient were in a good agreement between numerical prediction and experiments at all blowing ratios.

Schneider *et al.* [13, 14] investigated computationally the interrelationship between turbulent heat-flux and large coherent structures in a film-cooled TE cutback of gas turbine blades. Their large-eddy simulation was found that the blowing ratio and the flow regime of the cool gas flow influence both type and the strength of large-scale structures. The change in large coherent structures had a significant consequence to the dynamic mixing process at the breakout region. Dominant clockwise-rotating flow structures existed in the region of blowing ratio in which the counter-intuitive behaviour occurred. A large coherent structure increased the upstream-and-wall-directed turbulent heat-flux. This turbulent heat-flux caused the enhanced thermal mixing process in the near wall region.

 DOI: 10.1051/matecconf/201713500008

Recently, Effendy et al. [21], demonstrated the capability of DES on the prediction of the cooling performance of a trailing edge cutback turbine blade cooling with various lip-thicknesses to slot height ratios (*t*/*H*) referring to the work of Martini *et al.* [10, 20]. The DES study was based on the shear-stress transport (SST) $k-\omega$ turbulence model in order to simulate the dynamic flow interaction between internal coolant and mainstream flow. The discharge coefficient, the adiabatic film-cooling effectiveness and the vortex shedding frequency were in a good agreement between numerical prediction and measurements. With respect to the previous works, the present study focuses on the numerical study of blade trailing-edge cutback cooling, which is particularly important and challenging for the susceptibility of the blade damage due to higher local thermal loadings and a relatively thinner structure. The study concentrates on the interaction mixing between the internal cooling and the TE cutback cooling as the effect of blowing ratio (M).

2 Numerical Treatment

2.1 Physical problem and flow conditions.

Numerical study considers an experimental configuration investigated by Martini *et al.* [10, 20]. The computational domain includes main hot gas domain of stream wise, as published by Effendy *et al.* [21]. The coolant passage contains five rows of staggered-array cylindrical pins located at the coolant flow region. The arrays are fitted in a 10° wedge-shaped duct, representing the shape of trailing edge that follow the angle of the coolant ejection slot. Spanwise pitch (*S*/*D*) and streamwise pitch (S_x/D) are kept the same ratio as the experiment. The blade trailing-edge model has a lip-thickness (*t*) of 4.8 mm, same as the coolant passage height (*H*), thus the ratio of lip thickness to coolant passage height (*t*/*H*) is 1 for this computational domain. The same *t*/*H* ratio of a trailing-edge model has been tested and simulated in previous studies using eddy simulation based on Spalart-Allmaras model [10].

Initial flow and boundary conditions use the same values as applied for the Martini's experiment as seen in table 1.

Table 1. The experimental test conditions [20] .

	Mainstream flow (hot gas)		**Coolant flow (cold gas)**
Re_{hg}	= 250,000	$T_{c'}$, $L_{2\,exit}$	= 295 – 330 K (cold gas temperature at the exit slot)
Ma_{hg}	= 0.125		
u_{hg}	= 56 m/s (inlet velocity)		
Tu_{hg}	= 7%	T_c, $L_{1\,inlet}$	= 293 K (cold gas temperature)
T_{hg}	= 500 K (hot gas temperature)	Tu_c	= 5%
P_{hg}	= 105 kPa (pressure)	M	= 0.2 – 1.25 (blowing ratio)

 DOI: 10.1051/matecconf/201713500008

2.2 Mesh generation

The mesh for this computation was constructed by high quality grids with $\Delta y^+ < 1$ on all surfaces, as shown in Fig. 1. In addition, the interesting area (i.e. adiabatic/protected wall surface) is constructed in higher quality meshes with average $\Delta y^+ < 0.5$, as addressed by Nishino, *et al.* [22] in an effort to assure a sufficiently fine spatial resolution from unsteady phenomenon at the mixing region. As suggested by Spalart *et al.* [23][24] and Joo *et al.* [9], grid must be refined in each of the *x*, *y* and *z* directions in order to satisfy the blending factor as addressed by Menter *et al.* [25]. It is to be related the consistency of local grid spacing in the 3-D domain. The growth of grid spacing must be controlled linking to the requirement of blending factor (*F*) for the DES method.

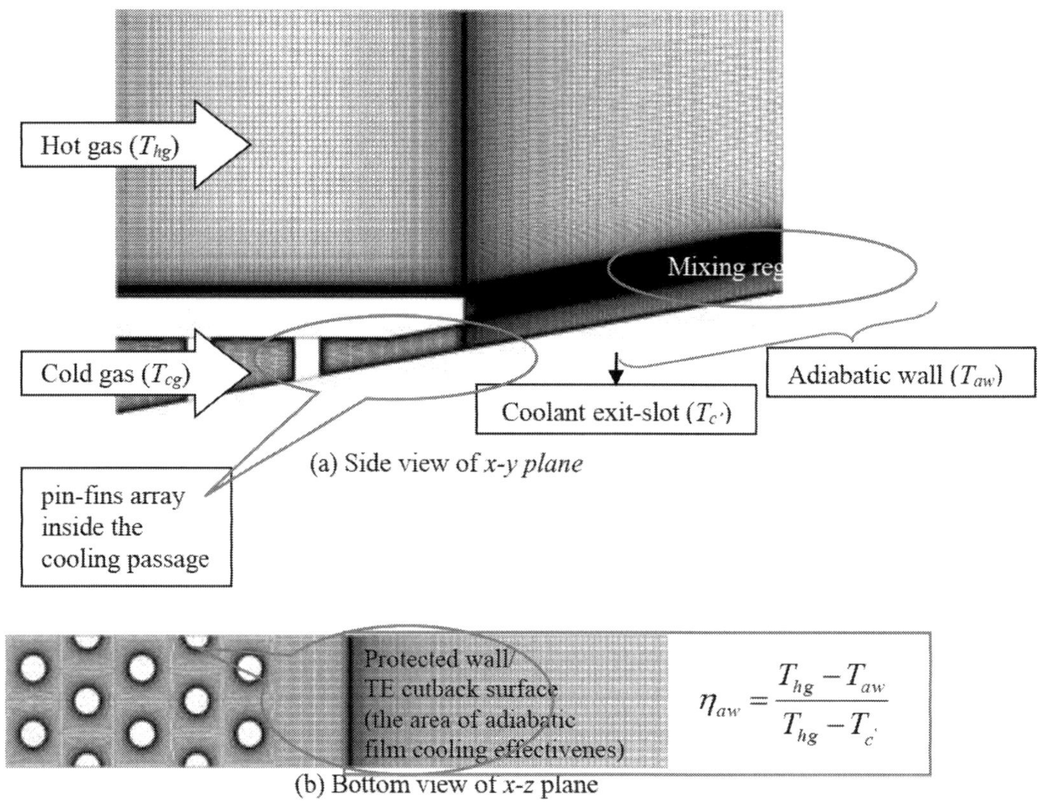

Fig. 1. Fine mesh for the baseline model.

2.3 Algorithm

A finite-volume method is utilised to solve the governing equations for incompressible flow. The equations are spatially computed using second-order accuracy on multi-block structured grids, whilst it is temporally calculated with a scheme of the second-order fully implicit. The Semi-Implicit Method for Pressure-

Linked Equations Consistent (SIMPLEC) algorithm is chosen with second-order numerical scheme applied for all flow equations (i.e. pressure, momentum, energy) of the unsteady RANS and DES calculations. Not only the grid, but also the time-step size must be small enough in an attempt to guarantee a sufficiently fine temporal resolution of the unsteady flow effects. With respect to these requirements, small time-step sizes from 5×10^{-5} to 1.25×10^{-5} seconds are applied in these computations with considering the previous work carried out by Martini *et al.* [10].

2.4 Adiabatic film-cooling effectiveness ($\overline{\eta}_{aw}$), coefficient of discharge (C_D), and blowing ratio (M)

Same formulas as seen in Martini et al. [10] are used to evaluate effectiveness, discharge coefficient and blowing ratios. These equations also have been presented in ref. [21] and [26].

3 Results and Discussion

3.1 Validation against test data

The grid refinement studies consider three successive meshes of the baseline domain from coarse to fine at a fixed-wall wall temperature (T_w) of 325 K, as published by Effendy *et al.* [21]. The aim was to assess the adiabatic film-cooling effectiveness for three types of meshes, with considering DES simulation at low-and-high blowing ratios (M = 0.5 and 1.1), in comparison with the experimental data reported by Martini *et al.* [10] and Horbach *et al.* [26]. It was realised to test the grid dependence for two different blowing ratios. In the paper highlighted that the DES modelling requires a fine resolution, which is fine enough to resolve the flow and heat transfer in the near end-wall region. Whilst results generated from the coarse mesh cause largely over-predict the film-cooling effectiveness near the downstream region. The use of the fine mesh produces results that are seen to be in a very good agreement with the test data along the protected/adiabatic wall surface from the slot-exit at x/H = 0 to the downstream region of x/H = 12, with the importance cooling effectiveness decay being captured successfully. The effect of meshes resolution is clear, indicated by low-blowing ratio case having largely deviation, as shown in ref. [21]. The deviation is reduced for the case with a higher blowing ratio.

3.2 Discharge Coefficients (C_D) predictions

The discharge coefficient represents the discharge behaviour of the blade trailing-edge cooling slots. This formula quantifies the global pressure loss within the cooling passage. It reveals the actual coolant mass flux to the ideal mass flux as the effect of an isentropic expansion from upstream region against free-stream at the ejection slot. Fig. 2 gives the predicted data of discharge coefficients for three different blowing ratios and various methods, in comparison with the experimental measurements. The predicted data are plotted versus the blowing ratios (M). It has been found that simulation results using both steady/unsteady RANS and DES are in good agreement with the experimental data. This implies that the same properties of coolant inside the cooling passage could be predicted well for the use of both methods. The proximity data predicted for both manners justify the studies conducted by Martini *et al.* [10], which emphasise that there is no need for a costly DES analysis. This is due to the use of both steady/unsteady RANS and DES being in a close agreement in terms of predicting the discharge coefficients; however, the prediction will be more reliable if DES simulation is done for C_D prediction. As the mainstream remains constant for all various simulations during this study, the increase of blowing ratio provided by raising the coolant flow velocity at the inlet causes the increase of Reynolds number at the slot-exit. The change of blowing ratio is proportional to the change of the real coolant mass flow. The discharge coefficient is increased slightly by increasing the blowing ratios. This finding is consistent with the works of Martini *et al.* [5], which show that the coefficient of discharge is increased by raising the Reynolds number at the coolant flow. This invention confirms the right finding of adiabatic film-cooling effectiveness, which also has a sound suitability result, as discussed above. Hence, three cases studies of validation, using the baseline, are reasonable for development for further investigation.

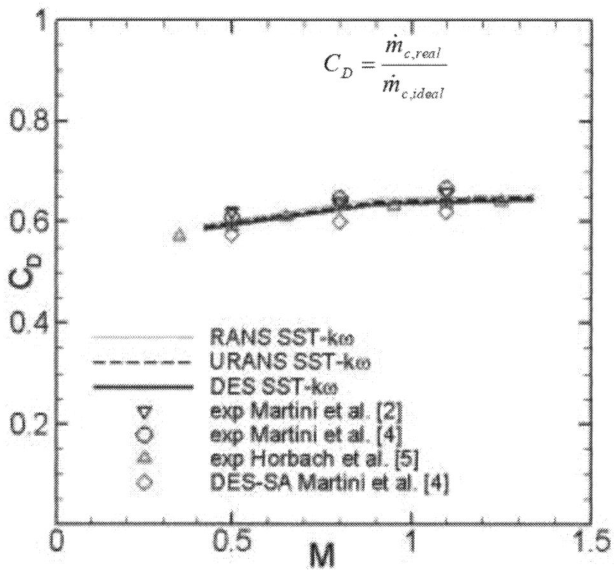

Fig. 2. Discharge coefficient.

3.3 Adiabatic film-cooling effectiveness ($\overline{\eta}_{aw}$) predictions

Fig. 3 shows the quantitative prediction data of the adiabatic film-cooling effectiveness along the adiabatic/protected wall surface at various blowing ratios, in comparison with experimental measurements and previous numerical results using DES-SA [10]. It has been found that the SST $k-\omega$ turbulence is capable of being applied for prediction on eddy simulation. CFD data follows a strong decay of the adiabatic film-cooling effectiveness, much the same as with the experimental measurements. The decay is reduced by increasing the blowing ratio. Anomalous phenomenon occurred when the blade TE cutback cooling is simulated around medium blowing ratio (M = 0.8). There is an opposite order of film-cooling effectiveness trend. The adiabatic film-cooling effectiveness is decreased at the downstream region by increasing a blowing ratio. This finding is similar to the previous simulation carried out by Martini *et al.* [20] and in the experiment by Horbach *et al.* [26] Both previous data indicated that the adiabatic film-cooling effectiveness, at the highest blowing ratio, is less than at the medium blowing ratio—mainly at the middle region $4 < x/H < 8$. Horbach *et al.* [26]] noted that the adiabatic film-cooling effectiveness dropped by increasing the blowing ratios from $M = 0.8$ to 1.25. This decrease is due to the intensified vortex shedding from the ejection lip. It occurs within a certain operating range with slot ejection, leading to an intensified mixing process between coolant and mainstream.

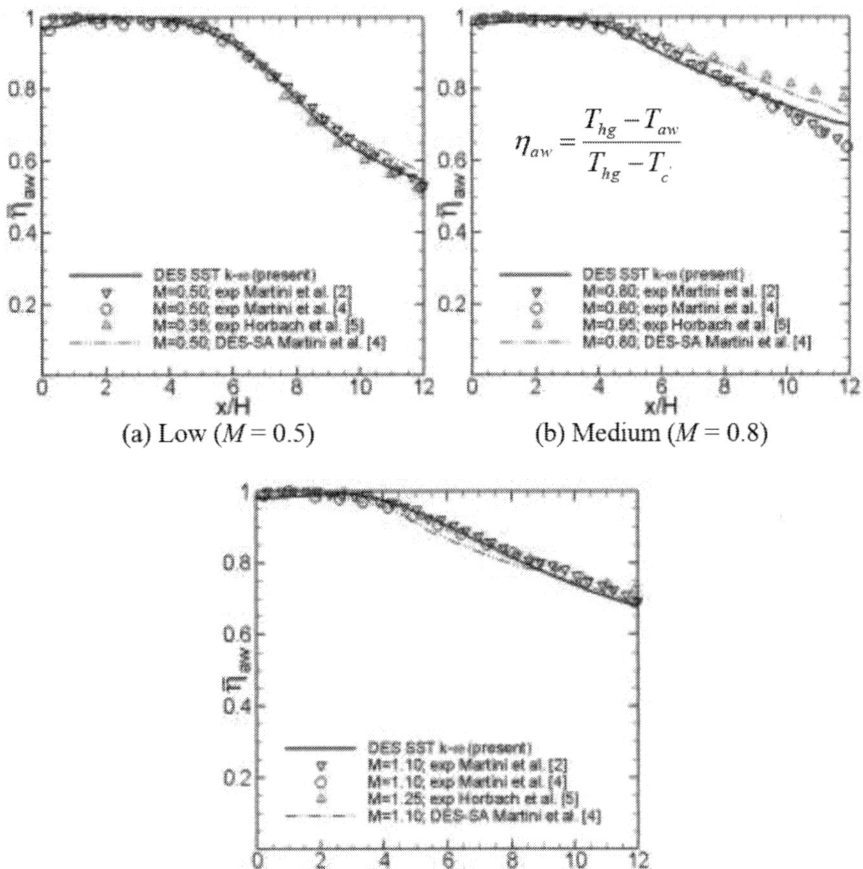

(a) Low ($M = 0.5$) (b) Medium ($M = 0.8$)

(c) High ($M = 1.1$)

Fig. 3. Cooling-film effectiveness at various blowing ratios.

Fig. 4 presents a quantitative comparison of the averaged film-cooling effectiveness for both TE cutbacks. It has been found that each case study has a typical characteristic of film-cooling effectiveness, which is depending on their blowing ratios. A rapid decay of film-cooling effectiveness occurs for the simulation at low blowing ratios (e.g. $M = 0.42$), whilst it tends to at a higher level for the simulation at high blowing ratios (e.g. $M = 1.83$). This finding is consistent with the previous eddy-simulation carried out by Martini *et al.* [26]. It is instructive to briefly recall that the results of verification and validation using a TE cutback with single-pitch domain, which have been described in Fig. 4. The CFD prediction matches with the experimental data for three different blowing ratios tested (i.e. $M = 0.5$, 0.8 and 1.1). The CFD prediction of the current case also agrees well compared to both the work of Martini *et al.* [10] and Horbach *et al.* [26].

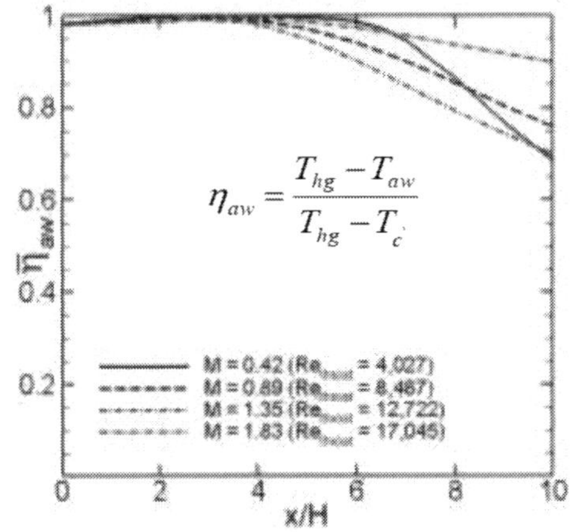

Fig. 4. Cooling-film effectiveness at various blowing ratios.

3.4 Velocity profiles

Fig. 5 shows the normalised u velocity profiles of the stream wise velocity at different blowing ratios within selected downstream distance of $x/H = 2$, 4, 6, 8, 10 and 12. These are based on the averaged u velocity at span wise direction (along the width of domain at a specific z-axis direction depending on the x/H position). It has been found that the blowing ratio is a key role parameter to play the dynamic mixing process along the breakout region. It is evidenced by a stronger flow velocity when blowing ratio is increased from 0.42 to 1.83. The left-curve and the right-curve discrepancy of the averaged u velocity near the protected wall are noticeable as indicated by Fig. 5(a). The change of u velocity profile is more pronounced around the slot exit region, for example at $x/H = 2$. The distance of the left-curve and the right-curve reduces near the downstream region, for instance at $x/H = 12$ (see Fig. 5(a)).

DOI: 10.1051/matecconf/201713500008

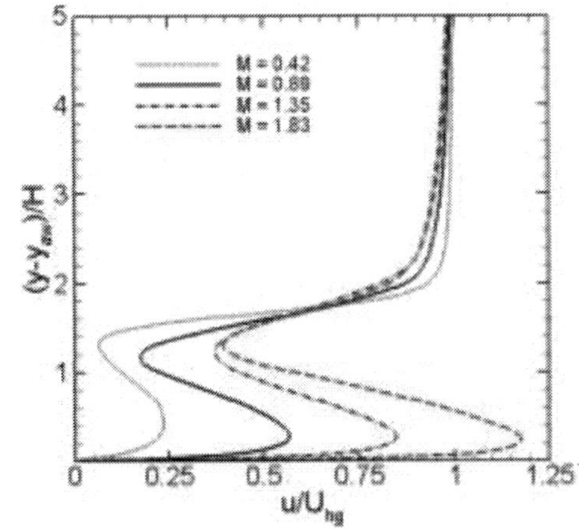

(a) $x/H = 2$

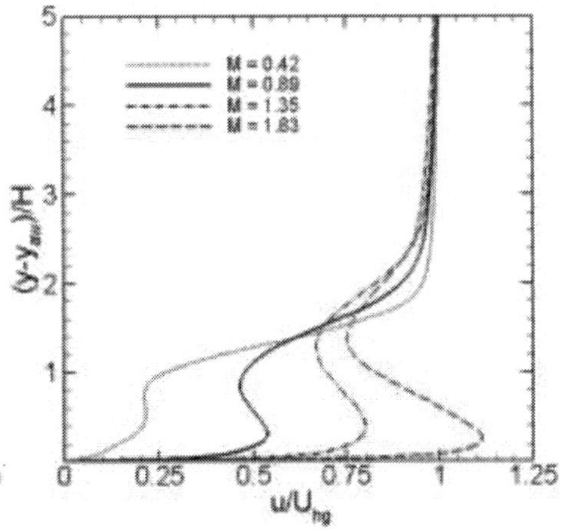

(b) $x/H = 4$

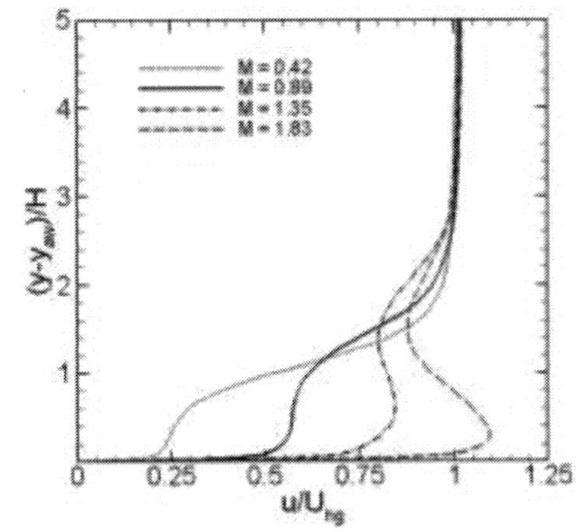

(c) $x/H = 6$

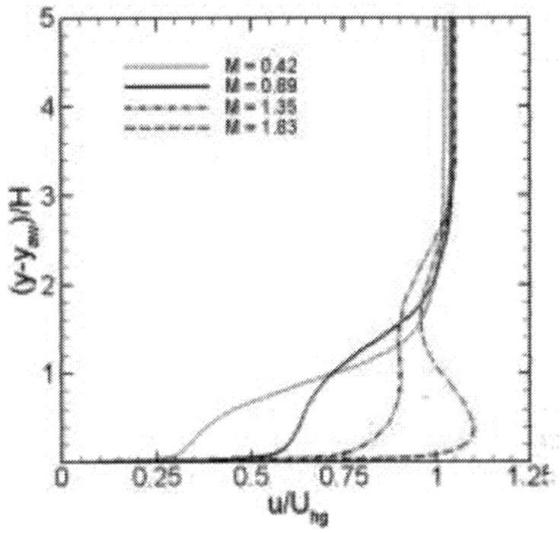

(d) $x/H = 8$

 DOI: 10.1051/matecconf/201713500008

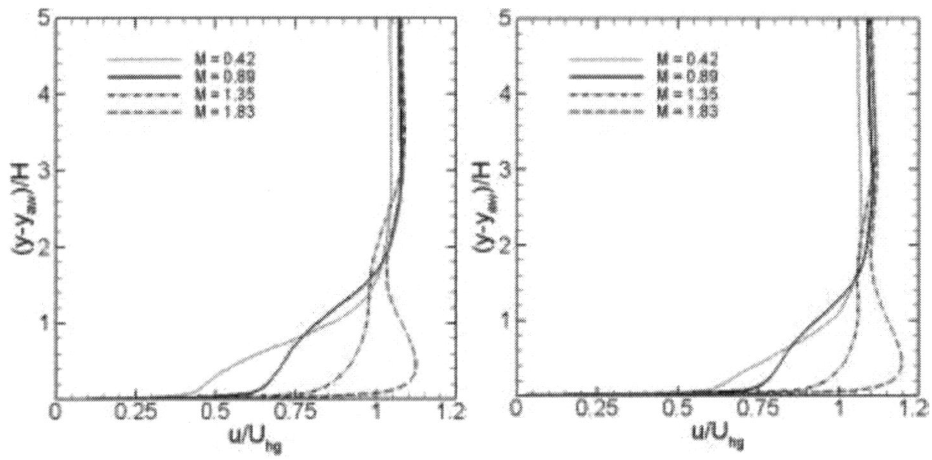

(e) $x/H = 10$ (f) $x/H = 12$

Fig. 5. Normalised u velocity profiles at different downstream locations.

3.5 Strouhal number

In terms of the Strouhal number, it has been recognised that a Strouhal number (S_t) of vortex shedding is approximately 0.22, which is corresponding to the shedding frequency of 2.21 kHz. It is suitable to previous computation study with St = 0.24. Sieverding *et al.* [25] report that vortex shedding of a turbulent boundary layer flow over a flat plate with a squared trailing edge is around $S_t = 0.21$. Boldman *et al.* [26] found that a Strouhal number of turbulent flow behind a blunt trailing edge is 0.2.

Conclusion

Overall, the predicted data of the adiabatic film-cooling effectiveness and the discharge coefficient are in good agreement for all various blowing ratios. It is proven by capturing well flow structures with the characteristic of shedding frequency and a Strouhal number approximately at f = 2.21 kHz and S_t = 0.22, respectively. DES based on the SST $k-\omega$ turbulence model, is strongly recommended for prediction simulations compared to both steady and/or unsteady RANS. It is due to the fact that the flow structures of gas-mixing between the coolant and the mainstream flow could be captured more realistically.

References

[1] J. Choi, S. Mhetras, J. -C. Han, S. Lau et R. Rudolph, «Film Cooling and Heat Transfer on Two Cutback Trailing Edge Model with Internal Performance Blockages,» *J. Heat Transfer,* vol. 130, no.11, p. 012201, 2008.

[2] J. -C. Han, «Fundamental Gas Turbine Heat Transfer,» *J. Thermal Science and Engineering Applications,* vol. 5, no.12, pp. 021007-1, 2013.

[3] J. -C. Han et A. P. Rallabandi, «Turbine Blade Film Cooling Using PSP

Technique,» *Frontier in Heat and Mass Transfer,* vol. 1, p. 1–21, 2010.

[4] S. D. Holloway, J. H. Leylek et F. A. Buck, «Pressure-Side Bleed Film Cooling: Part I - Steady Framework For Experimental And Computational Results,» chez *ASME paper No GT-2002-30471*, 2002.

[5] P. Martini, A. Schulz, C. F. Whitney et E. Lutum, «Experimental and Numerical Investigation of Trailing Edge Film Cooling Downstream of a Slot with Internal Rib Arrays,» chez *Proc. Inst. Mech. Eng., Part A, 217*, 2003.

[6] M. Effendy, Y. Yao et J. Yao, «Comparison Study of Turbine Blade with Trailing-Edge Cutback Coolant Ejection Designs,» chez *51st AIAA Aerospace Sciences Meeting, Texas, 2013-0548*, 2013.

[7] Y. Egorov, F. R. Menter, R. Lechner et D. Cokljat, «The Scale-Adaptive Simulation Method for Unsteady Turbulent Flow Predictions. Part 2: Application to Complex Flows,» *J. Flow, Turbulence and Combustion,* vol. 85, no.11, p. 139–165, 2010.

[8] G. Medic et P. A. Durbin, «Unsteady Effects on Trailing Edge Cooling,» *J. Heat Transfer,* vol. 127, no.14, p. 388–392, 2005.

[9] J. Joo et P. Durbin, «Simulation of Turbine Blade Trailing Edge Cooling,» *J. Fluids Engineering,* vol. 131, no.12, pp. 021102-1 - 021102-14, 2009.

[10] P. Martini, A. Schulz, H. -J. Bauer et C. F. Whitney, «Detached Eddy Simulation of Film Cooling Performance on The Trailing Edge Cutback of Gas Turbine Airfoils,» *J. Turbomachinery,* vol. 128, no.12, p. 292–300, 2006.

[11] J. Krueckels, M. Gritsch et M. Schnieder, «Design Consideration and Validation of Trailing Edge Pressure Side Bleed Cooling,» chez *ASME Paper No. GT-2009-59161*, 2009.

[12] K. Viswanathan, M. Schur, P. R. Spalart et M. Strelets, «Flow and Noise Predictions for Single and Dual Stream Beveled Nozzles,» *J. AIAA,* vol. 46, no.13, p. 601–626, 2008.

[13] H. Schneider, D. von Terzi et H. -J. Baurer, «Large-Eddy Simulations of Trailing-edge Cutback Film Cooling at Low Blowing Ratio,» *J. Heat and Fluid Flow,* vol. 31, no.15, p. 767–775, 2010.

[14] H. Schneider, D. von Terzi et H. -J. Baurer, «Turbulent Heat Transfer and Large Coherent Structures in Trailing-Edge Cutback Film Cooling,» *J. Flow, Turbulence and Combustion,* vol. 88, no.11-2, p. 101–120, 2012.

[15] A. Viswanathan et D. K. Tafti, «Detached-Eddy Simulation of Flow and Heat Transfer in Fully Developed Rotating Internal Cooling Channel with Normal Ribs,» *J. Heat and Fluid Flow,* vol. 27, no.11, p. 351–370, 2006.

[16] A. Viswanathan, K. Klismith, J. Forsythe et R. Squires, «Detached-Eddy Simulation around a Forbody at High Angle of Attack,» chez *AIAA Paper No. 2003-263*, 2003.

[17] A. Viswanathan et D. K. Tafti, «Detached-Eddy Simulation of Turbulent Flow and Heat Transfer in a Two-Pass Internal Cooling Duct,» *J. Heat and Fluid Flow,* vol. 27, no.11, p. 1–20, 2006.

[18] S. D. Holloway, J. H. Leylek et F. A. Buck, «Pressure-Side Bleed Film Cooling: Part II - UnSteady Framework For Experimental and Computational

Results,» chez *ASME paper No GT-2002-30472*, 2002.

[19] Y. Chen, C. G. Matalanis et J. K. Eaton, «High Resolution PIV Measurements around a Model Turbine Blade Trailing Edge Film Cooling Breakout,» *J. Experiments in Fluids,* vol. 44, no.12, p. 199–209, 2007.

[20] P. Martini, A. Schulz et H. -J. Bauer, «Film Cooling Effectiveness and Heat Transfer on The Trailing Edge Cutback of Gas Turbine Airfoils with Various Internal Cooling Designs,» *J. Turbomachinery,* vol. 128, no.11, p. 196–206, 2006.

[21] M. Effendy, Y. Yao, J. Yao et D. R. Marchant, «DES study of blade trailing edge cutback cooling performance with various lip-thicknesses,» *J. Applied Thermal Engineering,* vol. 99, pp. 434-445, 2016.

[22] T. Nishino, G. T. Roberts et X. Zhang, «Unsteady RANS and Detached-eddy Simulations of Flow around a Circular Cylinder in Ground Effect,» *J. Fluids and Structures,* vol. 24, no.11, p. 18–33, 2008.

[23] P. R. Spalart, «Young-Person's Guide to Detached-Eddy Simulations Grids,» Boeing Commercial Airplane Group, Seattle, WA USA, 2001.

[24] P. R. Spalart, S. Deck, M. L. Shur, K. D. Squires, M. K. Strelets et A. Travin, «A New Version of Detached Eddy Simulation, Resistant to Ambiguous Grid Densities,» *J. Theoretical and Computational Fluid Dynamics,* vol. 20, no.13, p. 181–195, 2006.

[25] F. R. Menter, M. Kuntz et R. Langtry, «Ten Years of Industrial Experience with the SST Turbulence Model,» *Turbulence, Heat and Mass Transfer 4 , edited by K. Hanjalic, Y. Nagano, and M. Tummers, Begell House, Inc,* p. 625–632, 2003.

[26] T. Horbach, A. Schulz et H. -J. Bauer, «Trailing Edge Film Cooling of Gas Turbine Airfoils – External Cooling Performance of Various Internal Pin Fin Configurations,» *J. Turbomachinery,* vol. 133, no.14, pp. 041006-1 – 041006-9, 2011.

[27] C. H. Sieverding et H. Heinemann, «The Influence of Boundary Layer State on Vortex Shedding from Flat Plates and Turbines Cascades,» J. Turbomachinery, vol. 112, no.12, p. 181–187, 1990.

[28] D. R. Boldman, P. F. Brinich et M. E. Goldstein, «Vortex Shedding from a Blunt Trailing Edge with Equal and Unequal External Mean Velocities,» J. Fluid Mechanics, vol. 75, no.14, p. 721–735, 1976.

Characterization of Bottom and Fly Ashes Generated Co-incineration of Biomass with Automotive Shredder Residue

Muhamad Fazli Othaman[*], *Sulastri Sabudin*, and *Mohd Faizal Mohideen*

Centre for Energy and Industrial Environment Studies (CEIES), Department of Energy and Thermofluid Engineering, Faculty Of Mechanical and Manufacturing Engineering Universiti Tun Hussien Onn Malaysia

Abstract. One of the viable techniques to reduce land filling of automotive shredder residue is by co-incinerating them with biomass. This study focuses on characterization of bottom and fly ashes produced from the co-incineration of the automotive shredded residue with oil palm biomass. The co-incineration was carried out in a pilot-scale fluidized bed incinerator. The oil palm biomass used was oil palm shell while the automotive shredded residue was obtained from a local recycling company. The characterization was done based on particle size distribution, morphology (SEM analysis) and chemical composition (EDS analysis). In term of chemical composition the ashes contain C (Carbon), O (Oxygen), Si (Silicon), K (Potassium), Ca (Calcium) and Fe (Ferum.

1 Introduction

Automotive shredder residue are basically waste obtained from end-of-life vehicles (ELV) which was unrecyclable and most from non-metallic parts[1]. These waste from ELV kept increasing every year with the growth of automotive industry. Annually the total amount of vehicles registration increases about 12% each year over the period of 5 years in Malaysia. It being estimated that in year 2020 around 31 million will be on Malaysian road [2]. These non–metal component from vehicles originates from various parts of the vehicle, mainly the dashboard, car seat, car carpet and car bumper. This various part mostly made up from plastic and PVC and most of these elements cannot be recycled nor reused, it will usually be disposes in landfills [3]. However, the increasing concern regarding the environmental issues regarding landfills has resulted more stringent regulations imposed by the local authorities and governments as landfilling affects directly the ecosystem, environment and population [4].

One of the viable techniques in handling automotive shredder residue (ASR) is by incineration and co-incineration with other wastes [3, 4]. Both of this processes may reduce significantly the total amount of ASR that is landfilled and thus ease the environmental impact. However, both incineration and co-incineration process produces fly ash and bottom ash which may contain concentrated amount of hazardous substance such as dioxin.

[*] Corresponding author: fazli.othaman@gmail.com

DOI: 10.1051/matecconf/201713500009

According to [5,6]biomass is one of the energy resource that was high potential to mitigate emission of greenhouse gases [5]. Furthermore from composition of biomass can reduces the carbon monoxide emission [6]. Significantly for incineration, fluid bed combustion technology have proven for high combustion efficiency and these technology have advantage of burning different type of fuel such as solid fuel in a combustor [7]. These technology increase air-fuel contact and result in the complete combustion can occur. Many researcher reported obtaining combustion efficiency up to 99% [8-10]. Moreover, with the higher combustion efficiency these process can also reduce the emission of NO_x [11]. Thus, in this study, the ASR and Oil Palm Shell (OPS) were blended together in the co-incineration process. This paper reports the characteristics of bottom and fly ashes generated from co-incineration process of oil palm biomass and automotive waste. The characteristic of bottom ash taken from combustion chamber and fly ash came out from the cyclone were reported in the following section.

Table 1. Composition of End Life Vehicle [3].

Material	% of total mass
Ferrous metal	65.4-71.0
Non- Ferrous metal	7.0-10.0
Plastics	7.0-9.3
Rubber	4.0-5.6
Glass	2.9-3.0
Fluid	0.9-6.0
Battery	1.0-1.1
Process polymers	1.0-1.1
Electrical/electronic	0.4-1.0
Other	1.0-5.9

2 Methodology

2.1 Fuel preparation

The fuel that used in this experiment is oil palm shell biomass from palm oil tree and the automotive shredded residue that came from end-life-vehicle (ELV) which cannot be recycle. For resident particles in the fluidized bed system, sand particle were used. The sand sieving range is between 650-850μm, and the average sand diameter was 750μm. The sieving was done using mechanical sieving

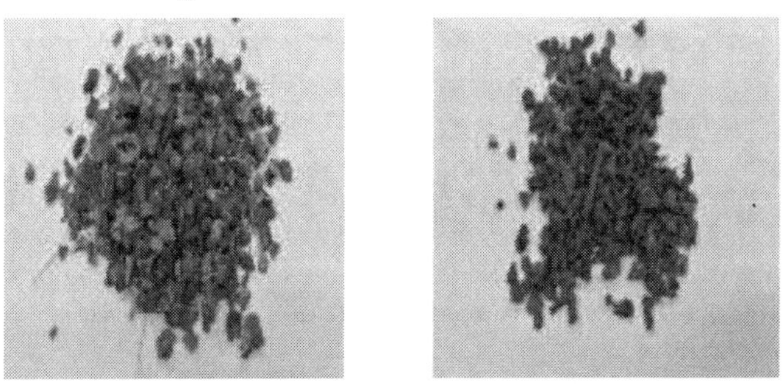

Fig. 1. (a) Oil palm shell and (b) automotive shredded residue

Fig. 2. Sieving mechanical shaking

2.2 Experimental setup

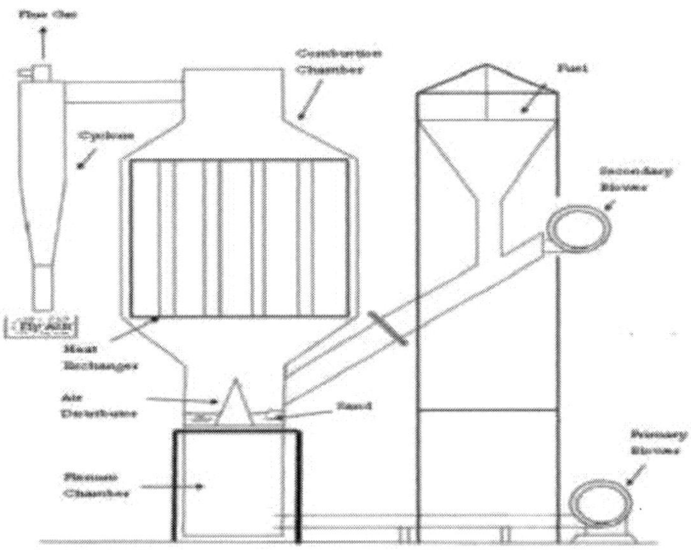

Fig. 3. Schematic diagram for experiment

Fig. 3 shows the schematic diagram for the experiment setup. Initially the resident material (sand) was preheated of at 600°C for about 20 minutes. Upon reaching this temperature, the fuel mixture (OPS and ASR) will be fed to the combustion chamber until incineration takes place steadily. Secondary air was also introduced into the bed upon attaining stable incineration process, temperature and emission gaseous produced were recorded using data logger as well as gas analyzer. Bottom ash was also collected from the cyclone and cooled before labelled. At the end of the co-incineration process, the system was allowed to cool for few hours before the bottom ash were recorded. Both of this ashes were characterized by particle size distribution, morphology and chemical composition. The particle size distribution was obtained by sieving through 6 sizes, 355µm, 250µm, 212 µm, 180µm, 125µm, and 90µm. Scanning Electron Microscope (SEM) was used to observe external morphology and crystalize structures while Energy Dispersive X-ray (EDS) was used to obtain the chemical composition.

DOI: 10.1051/matecconf/201713500009

3. Results and discussion

3.1 Particle size distribution

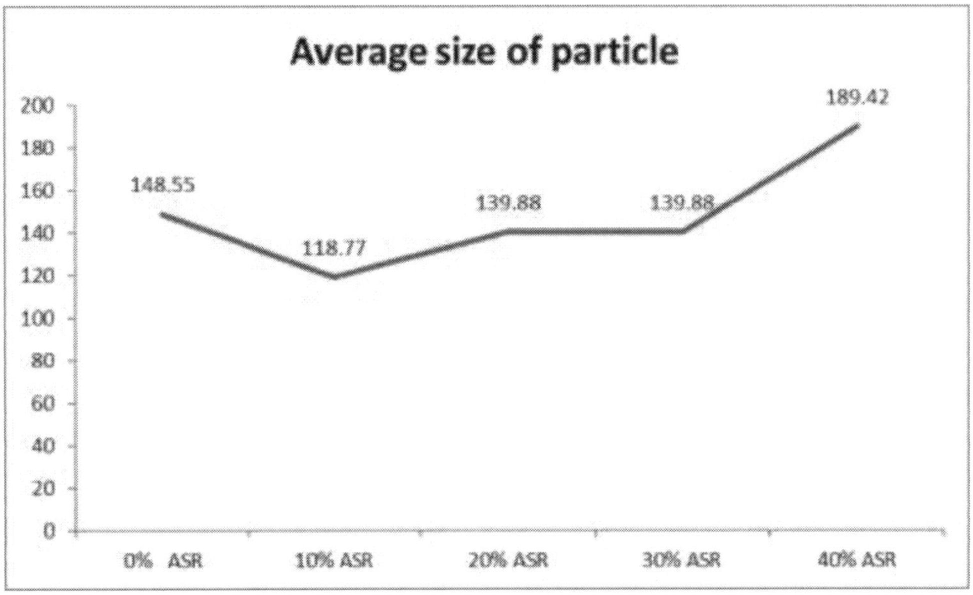

Fig.4. Average size particle size of fly ash

Referring to Fig.4, the average size of fly ash particles were between 118 μm to 190 μm. The sizes were found to slightly increase with the amount of ASR in the fuel mixture. The average fly ash particle sizes for 0%, 10%, 20%, 30% and 40% ASR were: 148.55 μm, 118.77mm, 139.88mm, 139.88mm and 189.42mm respectively. For bottom ash, the shapes were irregular and the particle sizes vary from 1 mm to 10 mm. The amount of ASR has little influence on the bottom ash. The morphology of the fly ash particles were shown via SEM as in the following section:

3.2 Morphology

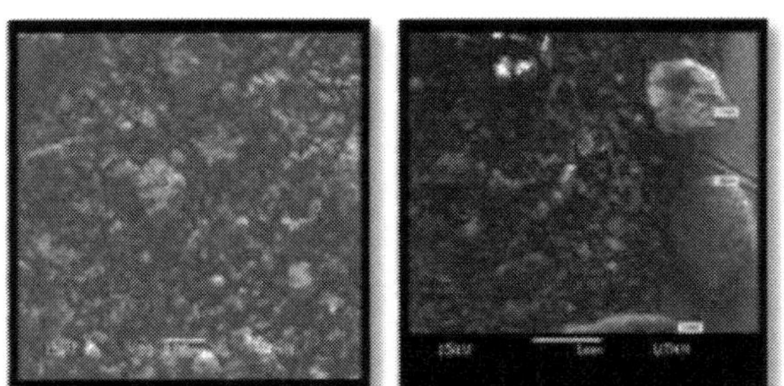

(a) 10% ASR (b) 20% ASR

 DOI: 10.1051/matecconf/201713500009

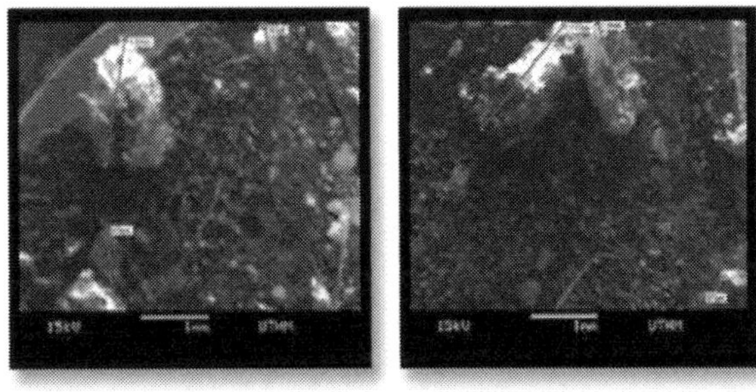

(c) 30% ASR (d) 40% ASR

Fig.5. SEM images for fly ash with different ASR content

In general, the fly ash particle size increased with amount of ASR. It was also found that the fly ash particles do not have a common feature and the shape was irregular and diverse. On the other hand, the fly ash from 0% ASR particle 100% OPS is given in Fig. 6 while the bottom ash for OPS and fuel mixture with ASR is shown in Fig. 7 and Fig. 8.

Fig.6. SEM image capture for OPS fly ash sample (0% ASR)

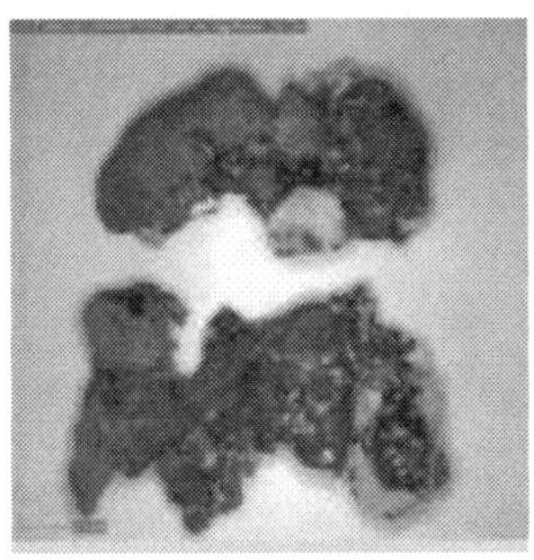

Fig.7. Bottom ash sample of OPS

According to both Fig. 7 and 8 above, sample 2 of bottom ash OPS also has irregular shape and doesn't have a fixed shape of bottom ash. Both of this samples shows that, the bottom ash of OPS doesn't have a fixed shape and has an irregular shape.

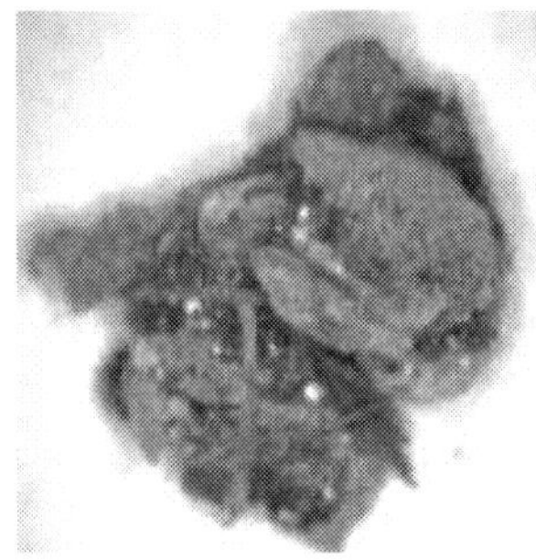

Fig.8. Bottom ash sample of OPS

3.3 Chemical composition

Table 2. Chemical composition in fly ash with certain ASR content

Point	Elements that exist (%)							
	10		20		30		40	
1	(C)	11388.14	(O)	3561.73	(C)	974.10	(O)	11391.27
	(S_i)	2296.05	(Fe)	13201.09	(O)	26435.98	(Fe)	16888.12
	(K)	6254.20	(Al)	14262.04	(Mg)	18409.72	(Zn)	486.70
	(Ca)	1495.70	(Mg)	1236.34	(Si)	27171.50	(Mg)	1381.64
	(O)	416.29	(Si)	12921.51	(S)	8712.88	(Si)	1955.39
					(Ca)	14311.07	(S)	1045.01
							(K)	1568.20
							(Ca)	1064.96
2	(C)	12930.24	(O)	11442.04	(C)	1625.09	(O)	17138.21
	(S_i)	3730.70	(C)	515.47	(O)	35193.30	(Fe)	13298.34
	(M_g)	420.36	(Fe)	17177.83	(Fe)	1607.53	(Mg)	6613.23
	(K)	1289.80	(Si)	506.00	(Mg)	29111.15	(Si)	10270.09
	(C_a)	888.26			(Si)	40044.45	(Ca)	957.25
	(O)	1761.96			(S)	2099.40	(Fe)	13298.34
					(K)	2083.84		
					(Ca)	3813.42		
3	(C)	86.76	(C)	1511.95	(C)	4607.29	(O)	17910.92
	(S_i)	186.79	(O)	20559.12	(O)	436.47	(Fe)	12053.11
	(M_g)	105.23	(Fe)	10301.90	(Mg)	446.37	(Mg)	6160.22
	(Fe)	2130.39	(Mg)	3935.38	(Si)	775.58	(Si)	8892.45
	(K)	169.88	(Si)	6550.09	(K)	201.21	(K)	819.80
	(C_a)	133.64	(S)	7899.74	(Ca)	379.45	(Ca)	755.35
	(O)	613.89	(K)	6440.19				
			(Ca)	3740.85				

The table indicates the chemical composition in the fly ash with particular amount of ASR. These chemical composition was obtained by scrutinizing 3 points which were randomly selected to indicate the percentage of elements exist all the samples. For fly ash sample containing 10% ASR, higher percentage of silicon (Si) was observed while for 20% ASR,

DOI: 10.1051/matecconf/201713500009

oxygen (O) was found to be dominant. As for fly ash sample with 30% ASR, higher percentages of Si was again found and finally for fly ash with 40% ASR, O content was higher. Silicon can be classified as gothic metal which is not harmful for the environment while oxygen can be classified as death metal. Both of these elements have higher percentages among blending ratio with the presence of ASR. From the table above, it also indicated that value of percentage ferum elements at 40% ASR is highest among other blending ratio. Since the ferum is in classic heavy metal classes, the sample in 40% of ASR is hazardous among others by comparing the value of percentages of counts mass. Thus, blending ratio of 10% ASR, 20% ASR and 30% ASR is safely landed for disposal while for 40% ASR blending ratio were not safe for disposal due to the present of ferum elements.

Table 3. Elements that exist in fly ash OPS

POINT	**ELEMENT PRESENT**	
	OPS	
1	(C)	5811.08
	(O)	328.96
	(Si)	147.96
	(K)	656.96
	(Ca)	309.42
2	(C)	104.73
	(O)	662.37
	(Fe)	99.98
	(Ca)	2325.33
3	(C)	602.36
	(O)	586.90
	(Si)	2057.47
	(K)	552.95
	(Ca)	418.39

According to the table of elements that exists in fly ash of OPS above, the existing elements were Carbon (C), Oxygen (O), Silicon (Si), Potassium (K), Calcium (Ca) and Ferum (Fe). There are 3 randomly point was selected to examine the composition of ashes. In each of point consist different of percentage in each elements exists. From the table above indicated that carbon elements have a higher percentage of count mass with 5811.08%. Element that have least percentage of mass count was ferum with 99.98%.

Table 4. Element that exist in bottom ash data

ELEMENT PRESENT (%)			
OPS + ASR		**OPS**	
(O)	1987.54	(O)	31158.14
(Mg)	160.64	(Fe)	39543.15
(Si)	7211.63	(Al)	14839.54
		(Si)	9707.14

In the table above, list of elements present in both of the bottom ash samples were presented. For sample mixture of OPS and ASR indicated only three elements which were Oxygen (O), Magnesium (Mg) and Silicon (Si) with Silicon in higher percentage of counts mass. For other bottom ash sample of OPS, the findings indicated four elements existed namely Oxygen (O), Ferum (Fe), Aluminium (Al) and Silicon (Si) with Ferum in higher percentages of mass count.

4 Conclusion

From the experimental results, it can be concluded that the objective of the scope was achieved. The experiments were carried out by several characterizations. There are four characterizations were analyzed in order to determine whether both of ashes were safe for disposal or not. First characterization was particle size distribution to determine average size of particle.as the result reviewed that increasing of percentage of ASR in blending ratio will increase the average size of particle. Second characterization was morphology purposely to compare the shape among of blending ratio and it was indicated that both of ashes was in irregular shape, spherical shape and doesn't have fixed shape. Third characterization was chemical composition is to determine the percentage of elements exist in each of bottom ash and fly ash according to their blending ratio. Lastly, a heavy metal test characterization is to identify whether the sample ash contain heavy metal elements and it's concentration.

First characterization was particle size distribution to determine average size of particle.as the result reviewed that increasing of percentage of ASR in blending ratio will increase the average size of particle. Second characterization was morphology purposely to compare the shape among of blending ratio and it was indicated that both of ashes was in irregular shape, spherical shape and doesn't have fixed shape. Third characterization is chemical composition is to determine the percentage of elements exist in each of bottom ash and fly ash according to their blending ratio. Results indicated that the percentage of ASR in blending ratio will increase the concentration of heavy metal elements.

5 Acknowledgement

The authors would like to express our deepest gratitude to Centre Graduate Studies (CGS) Universiti Tun Hussein Onn Malaysia (UTHM) for funding this research via Contract Research Fund, vot no. U 525 as well as providing the related equipment for our study.

6 References

1. L. Morselli, A. Santini, F. Passarini, I. Vassura, Automotive shredder residue (ASR) characterization for a valuable management, Waste Manag., **vol. 30, no. 11**, pp. 2228–34, (2010)
2. M. Azmi, M. Zameri, M. Saman, S. Sharif, 5 . 6 Proposed framework for End-Of-Life vehicle recycling system implementation in Malaysia, pp. 187–193, (2013).
3. I.Vermeulen, J.Van Caneghem, C.Block, J.Baeyens, C.Vandecasteele, Automotive shredder residue (ASR): reviewing its production from end-of-life vehicles (ELVs) and its recycling energy or chemicals' valorisation, Journal of Hazardous Materials, ***190*(1)**, pp.8-27, (2011)
4. A. Nzihou, N. J. Themelis, M. Kemiha, Y. Benhamou, Dioxin emissions from municipal solid waste incinerators (MSWIs) in France, Waste Manag, **vol. 32, no. 12**, pp. 2273–2277, (2012)
5. F. Okasha, G. Zaater, M. Awad, E. Zeidan, Co-combustion of biomass and gaseous fuel in a novel configuration of fluidized bed : Combustion characteristics FUEL, **vol. 133**, pp. 143–152, 2014.
6. M. Fazli othaman, S. Sabudin, M. Faizal Batcha, A Short Review on Biomass Thermo-Chemical Conversion : Recent Advances, **vol. 8, no. 6**, pp. 2494–2499, (2017).
7. M. Varol, A. T. Atimtay, Effect of biomass-sulfur interaction on ash composition and agglomeration for the co-combustion of high-sulfur lignite coals and olive cake in a

circulating fluidized bed combustor, Bioresour. Technol , **vol. 198**, pp. 325–331, (2015).

8. T. Madhiyanon, P. Sathitruangsak, S. Soponronnarit, Co-combustion of rice husk with coal in a cyclonic fluidized-bed combustor (ψ-FBC), Fuel, **vol. 88, no. 1**, pp. 132–138, Jan. (2009).
9. P. Ninduangdee, V. I. Kuprianov, Combustion of oil palm shells in a fluidized-bed combustor using dolomite as the bed material to prevent bed agglomeration, Energy Procedia, **vol. 52**, pp. 399–409, 2014.
10. T. Madhiyanon, P. Sathitruangsak, S. Soponronnarit, Co-firing characteristics of rice husk and coal in a cyclonic fluidized-bed combustor (Ψ-FBC) under controlled bed temperatures, Fuel, **vol. 90, no. 6**, pp. 2103–2112, Jun. (2011).
11. J. Van Caneghem, J, A. Brems, P. Lievens, C. Block, P. Billen, I. Vermeulen, R. Dewil, J. Baeyens, C. Vandecasteele, Fluidized bed waste incinerators: Design, operational and environmental issues, Prog. Energy Combust. Sci., **vol. 38, no. 4**, pp. 551–582, Aug. (2012).

Spray and Combustion Characteristics of a Novel Multi-circular Jet Plate in Air-assisted Atomizer

Shahrin Hisham Amirnordin[1,]*, *Amir* Khalid [1, 2], *David* Igram Patrick[1], *Mohd* Faizal Mohideen Batcha[1], and *Mas* Fawzi Mohd Ali[1]

[1]Centre for Energy and Industrial Environment Studies (CEIES), Faculty of Mechanical and Manufacturing Engineering, Universiti Tun Hussein Onn Malaysia, 86400 Parit Raja, Batu Pahat, Johor, Malaysia.

[2]Advanced Technology Centre (ATC), Faculty of Engineering Technology, Universiti Tun Hussein Onn Malaysia, 86400 Parit Raja, Batu Pahat, Johor, Malaysia.

Abstract. Atomization of liquid fuel in air-assisted atomizer is highly dependent on air mixing, which can be enhanced using turbulent generators, such as multi-circular jet (MCJ) plates and swirler. This study aims to determine the effects of novel MCJ plates on the spray and combustion characteristics of an air-assisted atomizer by evaluating spray and flame parameters, such as penetration length, cone angle, and cone area. MCJ 30 and MCJ 45, with inclined jets at 30° and 45°, respectively, were used in the experiment. A swirler was also used for comparison. The spray and flame images were recorded at different equivalence ratios through direct photography and analyzed using image J software. Flame temperature was determined using a thermal infrared camera, and burning chamber and flue gas temperatures were measured using thermocouples. The spray and flame characteristics of MCJ 30 exhibited performance comparable with those of the MCJ 45 and swirler. The integration of turbulence and swirling motion concept into the novel MCJ plates can enhance the mixing formation and thus improve the performance of burner combustion.

1 Introduction

Atomization is a process in which liquid fuel is converted from its bulk form into spray droplets and other physical dispersion of small particles in a gaseous environment through an atomizer. Atomization is important in combustion because most combustion systems, such as internal combustion engines and industrial furnace, work on fuel that should be

*Corresponding author : shahrin@uthm.edu.my

atomized prior to use. Spray quality and liquid distribution influence combustion efficiency, stability and pollution emissions, and increase the rate of heat and mass transfer [1, 2]. Atomization and spray formation can be characterized considering the improvements in measurement and instrumentation techniques. Spray characteristics can be categorized into macroscopic characteristics, including penetration, cone angle, radial dispersion, and axial velocity, and microscopic characteristics, including droplet diameter and distribution [3]. In addition, spray characteristics are affected by factors, such as pressure, atomizer geometry, gas to liquid ratio (GLR), fluid density, and viscosity [4]. Several studies [3, 5, 6] reported that spray penetration length is considerably affected by pressure supplied to the spray; moreover, the thermo-physical properties of the liquid do not control the spray characteristics. Khalid *et al.* [7] evaluated spray characteristics by using water-emulsified diesel fuel with various water percentages and indicated that increasing the quantity of water decreases the fuel viscosity, leading to high spray penetration. Khalid *et al.* [8] found that jatropha oil exhibits higher levels of spray penetration than diesel fuel [8]. Another study revealed that the spray cone angle increases due to high viscosity level and surface tension of the fluid [9, 10]. Moreover, the use of a swirler in an atomizer can significantly increase the spray cone angle and decrease the GLR [1, 4, 9, 11].

Atomizer geometry determines the spray characteristics and mixing effect on combustion [2]. Atomizer plates, such as fractal and swirl plates, possess different geometries and generate different motions of fluid inside the mixing chamber in an atomizer unit. Swirl plate is an atomizer commonly used in internal combustion and aero-engines. A simplex nozzle possesses a swirl chamber, which forces the liquid to leave the nozzle as an annular sheet. The sheet spreads out radially through a circular orifice and forms a hollow cone spray of finite film thickness [9]. Particularly, swirl flow exhibits potential to extend the residence time and enhances flow mixing by rotational movement of the flow field [12]. Swirling the mixture significantly increases the spray cone angle. However, secondary air beyond the exit orifice does not improve the spray characteristics but induces an undesirable contact between the liquid and the exit port [11–13].

Turbulent premixed flames are of great importance for technical combustion system because they exhibit high power density and low pollutant emission. Grids, such as perforated plates, are used for non-swirl stabilized flames to produce a certain level of standard turbulence [14, 15]. To generate turbulence in axisymmetric tubes, two approaches were available, namely, passive and active techniques. Passive approaches involve high-velocity streams and jet issuing into a mainstream, while active approaches involve oscillating grids, such as fractal grids. Fractal grids are a new type of low-blockage turbulence generator. Previous research found that fractal grids produce intense turbulence over a long range of streamwise distances. The frame brush thickness, flame angle, and turbulence burning increase with increasing turbulence level of the flow [16–19]. Fractal-shaped orifice also significantly affects pressure drop; that is, the pressure drop measured across the fractal-shape orifice is lower than that of the regular circular orifices from the same flow area [20, 21].

In this study, the effects of plate geometry on the spray and combustion characteristics of an air-assisted atomizer were investigated. An experiment was conducted in a burning chamber system through direct photography and thermal imaging.

DOI: 10.1051/matecconf/201713500010

2 Experimental Setup

An atomizer was equipped with an air compressor to supply primary air at the required pressure. A fuel pump was used to supply diesel to the atomizer and was controlled by the Ono Sokki mass flow meter. The schematic of the experiment is shown in Figure 1. The properties of diesel are presented in Table 1, and the experimental conditions are shown in Table 2.

Spray and flame videos were recorded using digital single-lens reflex camera Nikon D90. GOM player was used to obtain still images, and Image J was used for image processing. Flame temperature was recorded using thermal infrared (IR) camera FLIR T640. Flue gas and burning chamber temperatures were measured by K-type thermocouples, which were placed inside the chimney and burning chamber, respectively. Secondary air (from blower Hitachi RB40VA) was supplied into the atomizer to complete the combustion.

Diesel was sprayed via eight nozzles of the atomizer with a diameter of 1 mm by using three different plate geometries as turbulence generator in the premix chamber. Both plates were drilled with eight holes of 2 mm diameter at different angles, namely, 30° (MCJ 30) and 45° (MCJ 45). Both MCJ 30 and MCJ 45 (Figure 2) were introduced with a plate design that integrates the turbulence and swirl concept of fuel air mixing inside the premix chamber. The characteristics of both MCJ plates were compared with those of a swirler, a commercial turbulence generator. Figure 3 shows the actual images of the MCJ plates and the swirler used in the study. The parameters of spray, flame, and temperature are shown in Figure 4.

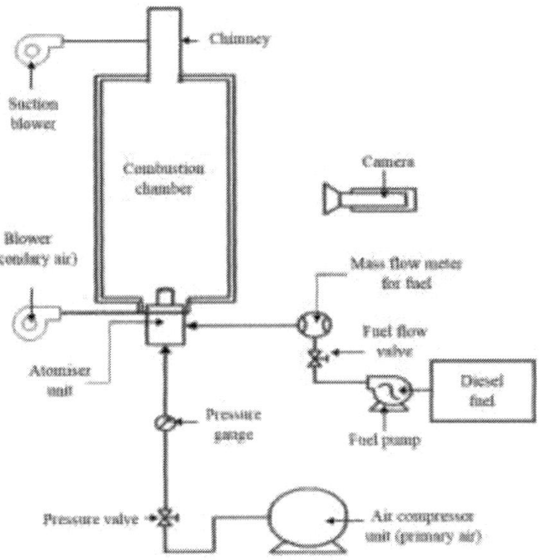

Fig. 1. Schematic diagram of experimental setup.

Table 1. Properties of diesel fuel (No. 2).

Property	Unit	Value
Density at 30° C	kg/m^3	0.8420
Flash point, minimum	°C	52
Cetane number, minimum		40
Kinematic viscosity 40°C, maximum	mm^2/sec	4.1

Table 2. Operating conditions.

Air pressure, MPa	0.1
Air density, kg/m^3	1.1770
Ambient temperature, K	300
Equivalence ratio	0.4 - 1.6

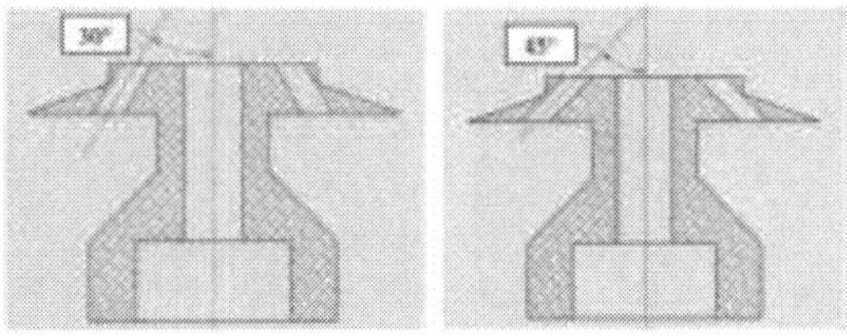

Fig. 2. Cross sectional area of MCJ 30 and MCJ 45.

Fig. 3. Turbulence generators.

(a) (b) (c)

Fig. 4. Parameters of (a) spray (b) flame and (c) temperature profile.

3 Results and discussion

This section describes the effects of different plates on the spray and flame characteristics of the air-assisted atomizer. Figure 5 shows the spray, flame, and thermal images acquired from three different types of plates at 1.0 equivalence ratio.

Subjects	Equivalence ratio, (ϕ)	Types of plate		
		MCJ 30	MCJ 45	Swirler
(a) Spray	1.0	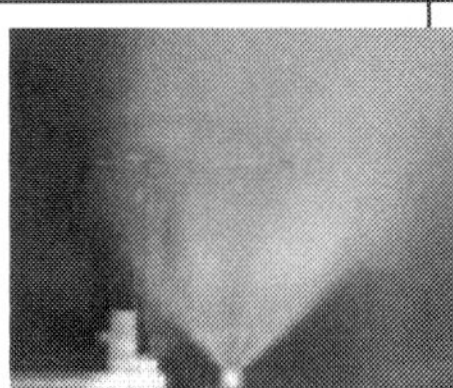	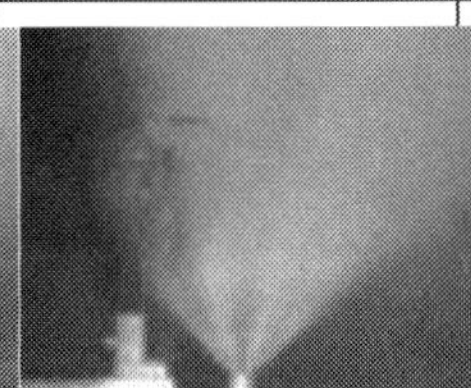	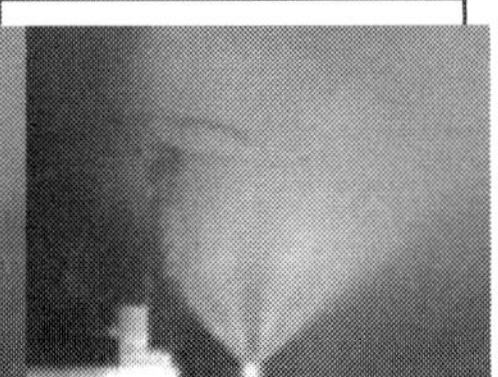
(b) Flame	1.0		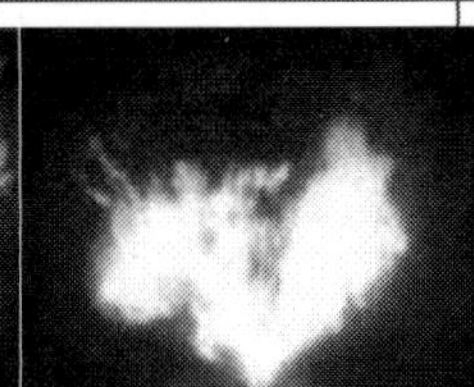	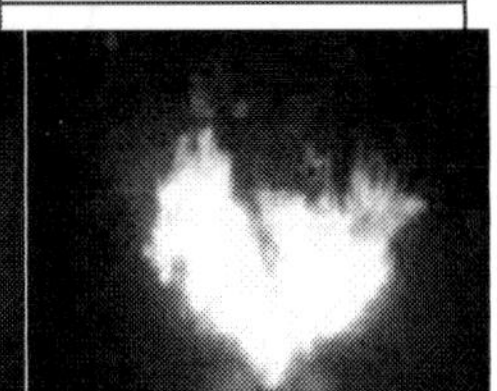
(c) Thermal image	1.0	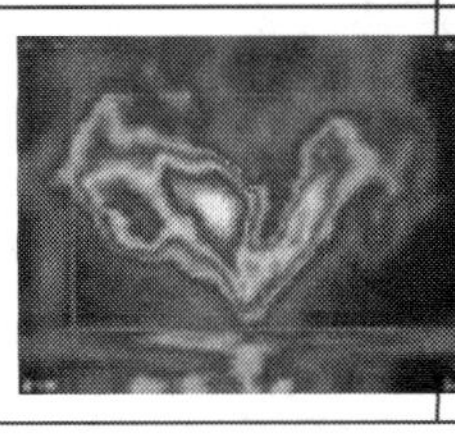	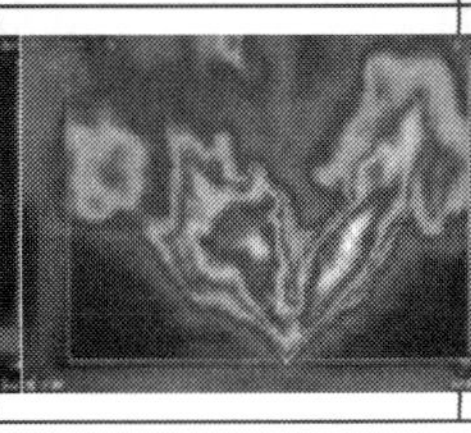	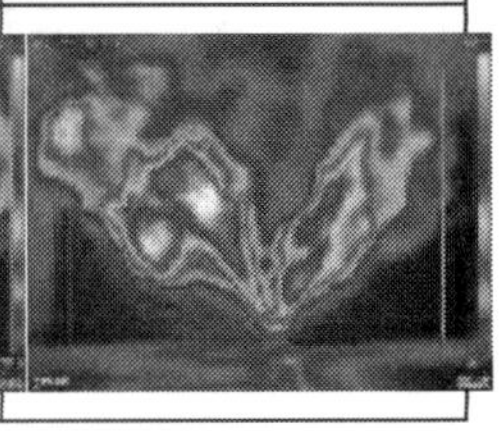

Fig. 5. Images of (a) spray (b) flame and (c) thermal at 1.0 equivalence ratio.

Figure 6(a) illustrates the spray penetration, spray cone angle, and spray area at different equivalence ratios. MCJ 30 produces the longest spray penetration (from 194 mm to 238 mm) compared with MCJ 45 (from 179 mm to 220 mm) and the swirler (from 173 mm to 211). MCJ 45 also exhibits the largest spray cone angle, followed by the swirler and

MCJ 30. Meanwhile, the swirler generates the smallest spray area (46,296 mm^2) compared with MCJ 30 and MCJ 45 (50,268 and 47,090 mm^2, respectively) at 0.4 equivalence ratio. As the equivalence ratio increases from 0.4 to 1.4, the spray penetration, spray cone angle, and spray area of all the plates also increase. Compared with MCJ 45 and the swirler, MCJ 30 exhibits better performance due to its 30° jet angle, which promotes higher mixing rate in the premix chamber and lower pressure losses [4].

Figure 6(b) shows the flame length, flame cone angle, and flame area against the equivalence ratios. Flame length and area are the highest in MCJ 30, followed by MCJ 45 and swirler. However, MCJ 45 generates the smallest flame cone angle, which increases gradually from 89° to 110°, whereas MCJ 45 and swirler produce wider spare cone areas at 100° to 115° and 102° to 113°, respectively. A misfire is detected in the swirler at $Ø = 0.4$. This result is in line with the spray characteristics shown in Figure 6 and is related to the geometrical advantages of the 30° jet angle in fuel air mixing [12].

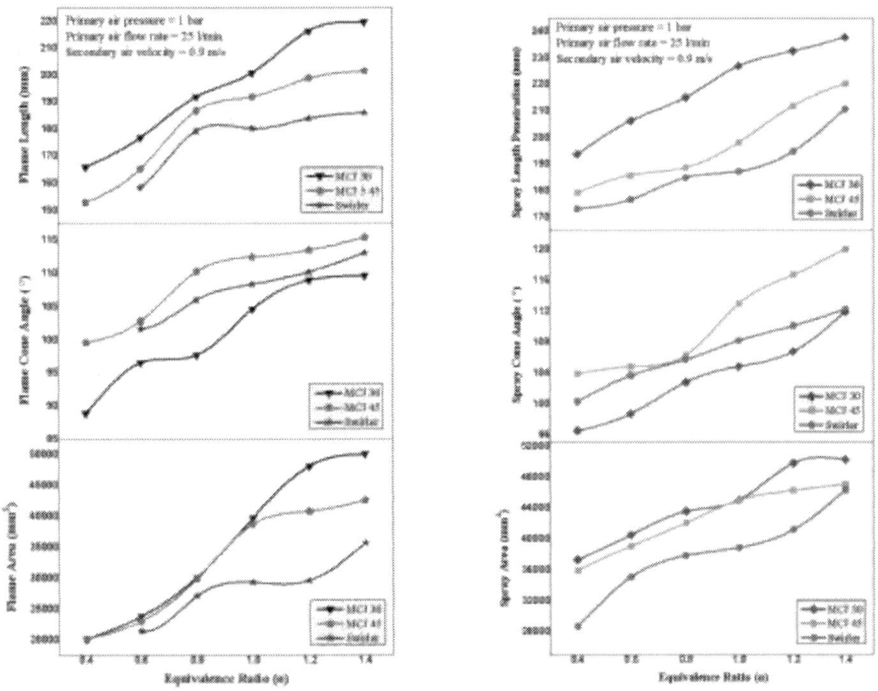

(a) (b)

Fig. 6. (a) Spray and (b) flame characteristics of three plates at different equivalence ratio.

In this study, temperature measurements were conducted using the thermal images of the flame. Data were collected from the K-type thermocouple and recorded using data logger. Figure 7 displays the maximum temperature of the flame and the temperatures of the burning chamber and flue gas. All the temperatures increase significantly with increasing equivalence ratio. High quantities of fuel and air mixtures were introduced into the atomizer at high equivalence ratios. Large spray and flame were created by the atomizer, resulting in increased temperatures of flame, burning chamber, and flue gas. These findings are consistent with those presented in Figure 6. Among all the plates, MCJ 30 exhibits the highest flame and flue gas temperatures, and MCJ 45 displays the highest temperature in the burning chamber. The high burning chamber temperature could be due to the high flame angle, which is closer to the thermocouple located at the center of the chamber [12]. Overall results show that MCJ 30 exhibits the highest performance for spray and combustion compared with MCJ 45 and swirler. The novel geometry of MCJ 30 not only produces high combustion temperature but also contributes to flame stability.

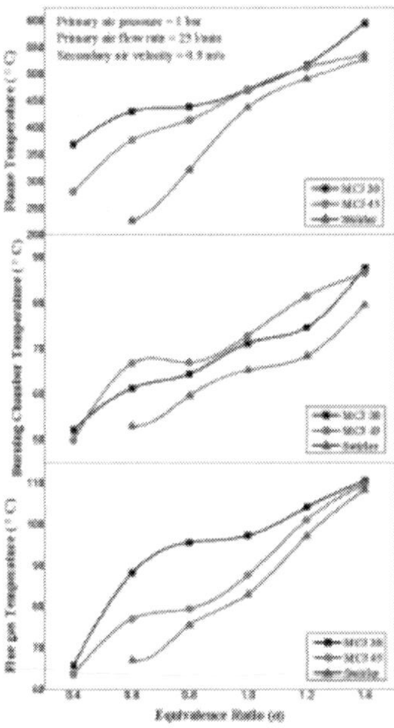

Fig. 7. Measured temperature of three plates at different equivalence ratio.

4 Conclusion

At the end of these investigations, the effects of geometrical conditions on the spray and combustion characteristics were determined. The following conclusions were derived.

1. Inclination of the jet plate at 30° allows high mixture formation in the mixing chamber, leading to improved performance of the air-assisted atomizer compared with those of the plate inclined at 45° and the swirler.
2. The inclined jet plate also contributes to flame stability during combustion.
3. The geometry of the turbulence generator in the atomizer should be considered in improving the efficiency of the burner system.

Acknowledgement

The authors would like to thank the Ministry of Higher Education Malaysia for supporting this research under Fundamental Research Grant Scheme (FRGS) Vot 1492 and Centre for Graduate Studies UTHM.

References

1. M. Rashad, H. Yong, and Z. Zekun, Int. J. Hydrogen Energy, **41**, 35, 15790–15799 (2016).
2. J. Jedelský and M. Jícha, Appl. Therm. Eng., **96**, 286–296 (2016).
3. A. M. E. Faik and Y. Zhang, *EPJ Web of Conferences*, **92**, 30–37 (2015).
4. M. Ochowiak, S. Rozanska, M. Matuszak, and S. Wlodarczak, Chem. Eng. Process. Process Intensi., **98**, 52–59 (2015).
5. T. N. C. Anand, A. M. Mohan, and R. V. Ravikrishna, Fuel, **102**, 613–623 (2012).
6. J.-K. Yeom and W.-S. Jung, Int. J. Precis. Eng. Manuf., **16**, 7, 1279–1291 (2015).
7. A. Khalid, S. H. Amirnordin, L. Lambosi, B. Manshoor, M. F. Sies, and H. Salleh, Advanced Materials Research, **408**, 1900–1911 (2013).
8. A. Khalid, L. Lambosi, M. Suardi, S. H. Amirnordin, I. Zaman, N. Mansor, and B. Manshoor, **774**, 496–500 (2015).
9. A. Davanlou, J. D. Lee, S. Basu, and R. Kumar, Chem. Eng. Sci., **131**, 243–255 (2015).
10. D. Sivakumar, S. K. Vankeswaram, R. Sakthikumar, B. N. Raghunandan, J. T. C. Hu, and A. K. Sinha, Fuel, **179**, 36–44 (2016).
11. J. Jedelsky and M. Jicha, *23rd Annu. Conf. Liq. At. Spray Syst.*, 1–5 (2010).
12. S. S. Hou, C. Y. Lee, and T. H. Lin, Energy Convers. Manag., **48**, 5, 1401–1410 (2007).
13. C. T. Chong and S. Hochgreb, Int. Biodeterior. Biodegrad., **102**, 1, 353–360 (2015).
14. T. Sponfeldner, S. Henkel, N. Soulopoulos, and F. Beyrau, In *European Combustion Meeting,* 1–19 (2011).
15. M. F. Othman, B. Manshoor, and A. Khalid , IOP Conf. Ser. Mater. Sci. Eng., **50**,1, 012050 (2013).
16. G. Coppola and A. Gomez, Exp. Therm. Fluid Sci., (33), 7, 1037–1048 (2009).
17. N. Soulopoulos, J. Kerl, F. Beyrau, Y. Hardalupas, and J. C. Vassilicos, Mech. Eng., 1–18 (2010).
18. A. A. Verbeek, T. W. F. M. Bouten, G. G. M. Stoffels, B. J. Geurts, and T. H. van der Meer, Combust. Flame, **162**, 1, 129–143 (2015).
19. R. Y. S. Chin, A. Khalid, A. M. A. Ghani, M. M. Issak, and S. H. Amirnordin, ARPN J. Eng. Appl. Sci., **11**, 12, 7491–7497 (2016).
20. A. Abou El-Azm Aly, A. Chong, F. Nicolleau, and S. Beck, Exp. Therm. Fluid Sci., **34**, 1, 104–111 (2010).
21. F. C. G. A. Nicolleau, S. M. M. Salim, and A. F. Nowakowski, J. Turb., **12,** N44.

Feasibility study on production of Metal Matrix Composite (MMC) material for Electrical Discharge Machining (EDM) tools using Rapid Prototyping (RP) technique

S. Shamsudin[1], and *M. A. Lajis*

Sustainable Manufacturing and Recycling Technology, Advanced Manufacturing and Materials Centre (SMART-AMMC), Universiti Tun Hussein Onn Malaysia, 86400 Parit Raja, Batu Pahat, Johor, Malaysia.

Abstract. In common practice, tools for EDM have traditionally been made by machining copper or graphite to the required profile using CNC machines. Increasing the degree of complexity of any tooling design for any operations results in a corresponding increase in time and cost required. With the advent of rapid prototyping techniques, the problem of making tools with complex shapes becomes much simpler and easy. The main aim of this research was to develop new EDM electrode material through a novel approach by rapid prototyping (RP) technique. In this study, the potential application of copper (Cu) reinforced alumina (Al_2O_3) fabricated with various compositions as an EDM electrode was investigated. The electrodes were fabricated by Canon PIXMA IP 1800 printer and underwent sintering temperature at 85 % and 95 % melting point of copper. The EDMed workpiece was aluminium and the electrodes surface was analyzed through scanning electron microscope (SEM). Findings showed that the electrode with Cu - 0 vol. %Al_2O_3 composite and sintered at temperature 977 °C resulted in highest metal removal rate (MRR) and lowest electrode wear rate (EWR) while Cu – 10 vol. %Al_2O_3 composite and sintered at temperature 977 °C revealed a better surface finish than other electrodes. An increase in Al_2O_3 content in general will increase the hardness of tool, as a trade-off, the conductivity was reduced.

1 Introduction

Electrical discharge machining (EDM) is a non-conventional machining process where material is removed electro thermally by a series of successive discrete discharges or sparks between two electrically conductive objects, which are electrode and workpiece. The

[1] Corresponding author: shazarel@uthm.edu.my

hardness of the workpiece has no influence on the process capability. Material for EDM electrode must possess basic properties such as electrical and thermal conductivity, high melting temperature, low wear rate, and good resistance to deformation during machining [1].

The recent development in the field of EDM have progressed due to the growing application of EDM process and the challenges being faced by the modern manufacturing industries, from the development of new materials that are hard and difficult to machine such as tool steels, composites, ceramics, super alloys, hastalloy, nitralloy, waspalloy, nemonics, carbides, stainless steel, heat resistant steel, and etc. that being widely used in die and mould making industries, aerospace, aeronautics and nuclear industries. EDM is used extensively to produce cavities in hardened tool steel for molding plastics. Complex geometries or profiles can often be more easily machined on an electrode than in a cavity, making EDM viable in terms of lead time and quality of cut [2, 3].

In EDM, copper and graphite tools are commonly used as electrodes. Nevertheless, high tool wear is the major drawback of these electrodes for prolonged machining. The low wear resistance electrodes like Cu, Cu alloys and graphite contribute problem in EDM operation. Increasing in the complexity of tooling for any EDM related operation results in a corresponding increase in time and costs required to develop such tooling.

Rapid prototyping (RP) is a process capable of fabricating prototype parts directly from three-dimensional CAD models [4]. One way to reduce time and to design tool with higher complexity in EDM is through rapid prototyping techniques. The techniques allow direct produced RP model used as an electrode for EDM. The tool should achieve greater metal removal rate with low tool wear characteristic. Multiple RP techniques such as StereoLithography (SL), fused deposition modeling, selective laser sintering, selective laser melting and 3D printing are possible to be adopted in the production environment of EDM tools fabrication.

Quite a lot investigations were devoted to study research related to metal powder added with ceramic materials to form a variety of composites. Study on sintering behavior of metal powder mixed with ceramic inclusions was also reported. The experiments were performed with mixtures of copper and alumina particles approximately with the same size and volume fractions of alumina between 5% and 30% are added together. Both experimental and simulation results on the effect of inert inclusions on the densification of metal-ceramic powder mixtures during free sintering was discussed. The densification was significantly delayed even for a volume fraction of inclusions as low as 5% and strongly hindered when this content reached up to 20% [5]. Other than that, study on copper composites containing O-3 wt% Al that were sintered in hydrogen at temperatures of 800°C and 900°C was investigated by [6]. The powders were prepared through the blending and mechanical alloying routes. The green compaction pressures were 500 and 560 MPa, respectively. Increasing in compaction pressure led to both sintered density and hardness increased. It was found that densification kinetics were faster for mechanically alloyed powder compacts than the blended ones. An increase in Al_2O_3 content in general increased hardness, with an associated loss in electrical conductivity.

Compounds of ZrB2 and TiSi with Cu at various compositions was also investigated for EDM electrodes by either using solid-state sintering or liquid phase sintering [4]. The performance of the newly formed material was compared with that of conventional electrode materials such as Cu, Graphite and the rapid tool produced from the selective laser sintering (SLS). While, the manufacture of EDM electrode from rapid prototype pattern made by the stereolithography technique was studied by [7]. The model was a square cavity with inner dimensions of 80 mm by 80 mm. A variety of features was added on the bottom face of this cavity. An SL pattern of the model was then successfully fabricated on the Stereolithography apparatus (SLA).

In this study, composites of copper mixed with alumina at various compositions for EDM electrodes were investigated. The tools were fabricated using rapid prototyping (RP) in particular 3D printing technique with performed post processing. The study was aimed to develop EDM tool from composite material that should possess higher material remover rate (MRR), lower wear ratio (WR) and superior surface finish. The alumina particles are known hard and thermally stable at high temperature. It's mostly used in the production of aluminium, though it is also used as an abrasive material due to its hardness and as a refractory material due to its high melting point. Its hardness and ability to retain its elevated temperature properties are expected to make it suitable for application in EDM operation. In order to enhance bonding between copper and alumina, maltodextrin powder with melting point 240 °C was also mixed. The maltodextrin powder will be dissolved and act to bind the alumina and copper during the fabrication process. While water based binder was used to dissolve the maltodextrin during the 3D printing.

2 Methodology

2.1 Material preparation

Powder metallurgy method was used to fabricate the composite electrodes. Mixtures of copper, alumina and maltodextrin (act as binder) were used to fabricate the electrodes. Distinct compositions of composite mixture were considered as given in Table 1. The mean particle size and shape of the mixtures were also estimated. It was observed that the copper powder has a spherical shape and maltodextrin powder possessed a round shape while the alumina powder revealed an angular shape in the composite. The size of copper powder was approximately 30 μm while both of alumina and maltodextrin powder were around 60 μm. The mixture of powders were initially mixed in ball mill for 1 hour to achieve homogenization.

Table 1. Percentage mixture of Cu powder and Al_2O_3

Sample No.	Composition		
	Copper, Cu (%)	Alumina, Al_2O_3 (%)	Maltodextrin, (%)
1	100	0	10
2	90	10	10
3	80	20	10

2.2 Electrode preparation

EDM electrodes were produced via Canon PIXMA IP 1800 printer which have been modified and has function similar to original 3D printer. The main reason of using this printer is to reduce the cost of material and printing. Figure 1 shows the microstructure of electrode before sinter. Sintering of the electrode was carried out in a tube furnace supplied with an argon gas. The temperature of sintering cycle is given in Figure 2. Sintering temperatures were set at two different levels. The temperature was first ramped to 400 ˚C to wipe out the binder (maltodextrin) and then raised to 874 ˚C and 977 ˚C respectively and stabilized for 4 hours.

(c)

Fig. 1. The microstructure of electrode before sinter; (a) 100% Cu / 10%Maltodextrin; (b) 90%Cu / 10%Al_2O_3 / 10%Maltodextrin; (c) 80%Cu / 20%Al_2O_3 / 10%Maltodextrin.

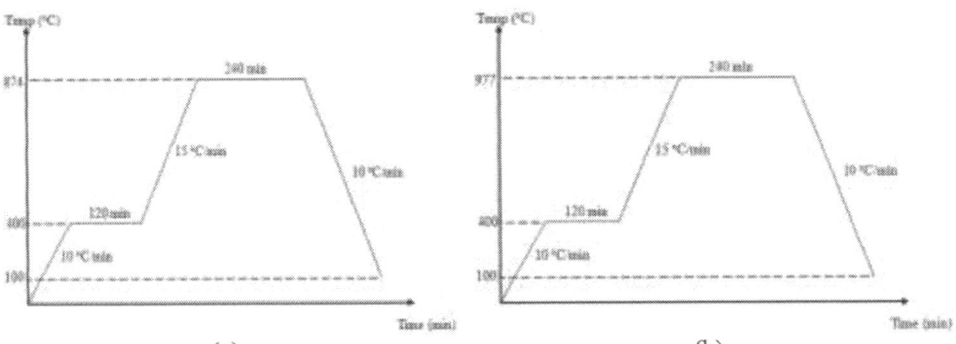

(a) (b)

Fig. 2. a) Sintering cycle for 85 %, b) sintering cycle for 95 %

DOI: 10.1051/matecconf/201713500011

3 Result and discussion

3.1 Microstructure

Figure 3 and 4 show microstructure of sintered electrodes. During sintering, maltodextrin was gradually vaporized because of its low melting point characteristic compared to copper and alumina. The melting point of maltodextrin was 240 °C while both copper and alumina powder were around 1028 °C and 2000 °C. The bonding was only took place among the copper powder and not visible within the alumina powder due to a huge different in melting temperature between both materials. The necking phenomena were only visible on copper particles and the boundary regions were created adjacent to copper inter-particles. It is also shows that more necking was visible at temperature 977 °C. Nevertheless, the bonds strength of copper powder was insufficient to hold the alumina particles firmly. The high melting point of alumina causes an extreme temperature was needed to overcome its surface energy to allow necking to form. Due to this reason, the electrode becomes fragile and difficult to handle during the EDM process.

Fig. 3. The microstructure of electrode after sintered at 874°C; (a) 100% Cu; (b) 90%Cu / 10%Al_2O_3; (c) 80%Cu / 20%Al_2O_3.

Fig.4. The microstructure of electrode after sintered at 977°C; (a) 100% Cu; (b) 90%Cu / 10%Al_2O_3; (c) 80%Cu / 20%Al_2O_3.

In all samples, the maltodextrin acts as a binder and used to hold temporarily the metallic compound during the 3D printing process. After sintering process, the binder vaporizes and this circumstance increases porosity in the fabricated electrode because the empty spaces left by the maltodextrin are not replaced immediately by the alumina and copper powders. As consequences, the porous samples that were already obtained after the electrodes fabrication via 3D printing become deteriorated.

3.2 EDM tests

The electrodes were tested for EDM and all parameters used in EDM was a default setting. Table 2 shows the parameters adopted for EDM. The machined workpiece was aluminium. Figure 5 and 6 show results of material removal rate and electrode wear rate at various material compositions.

Table 2. EDM Die-Sinking parameters.

Machining Time (sec)	Electrode Polarity	Ip (A)	Voltage (V)	Pulse On Time (μs)	Pulse Off Time (μs)
300	+ ve	5.4	60	86	43

 DOI: 10.1051/matecconf/201713500011

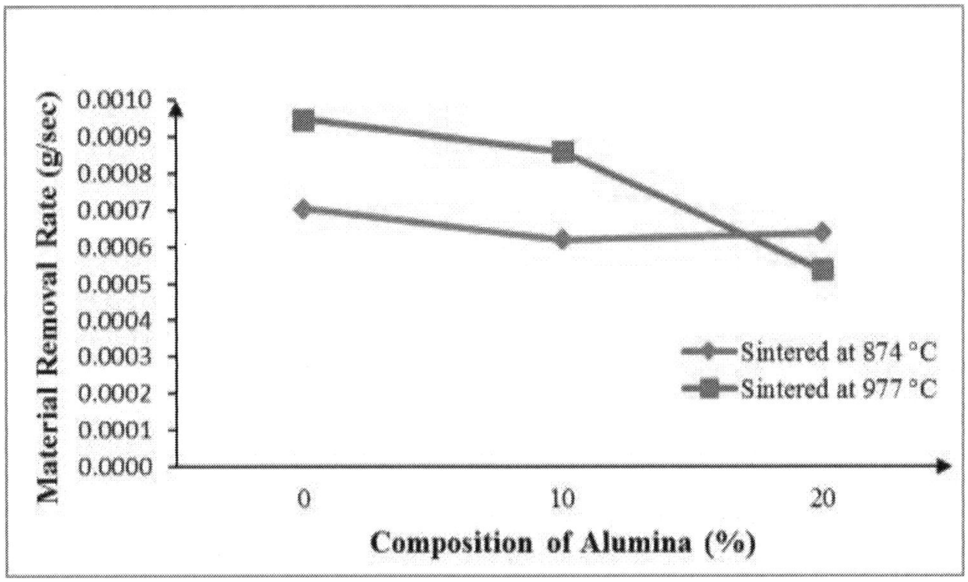

Fig. 5. Effect of different alumina compositions on material removal rate (MRR)

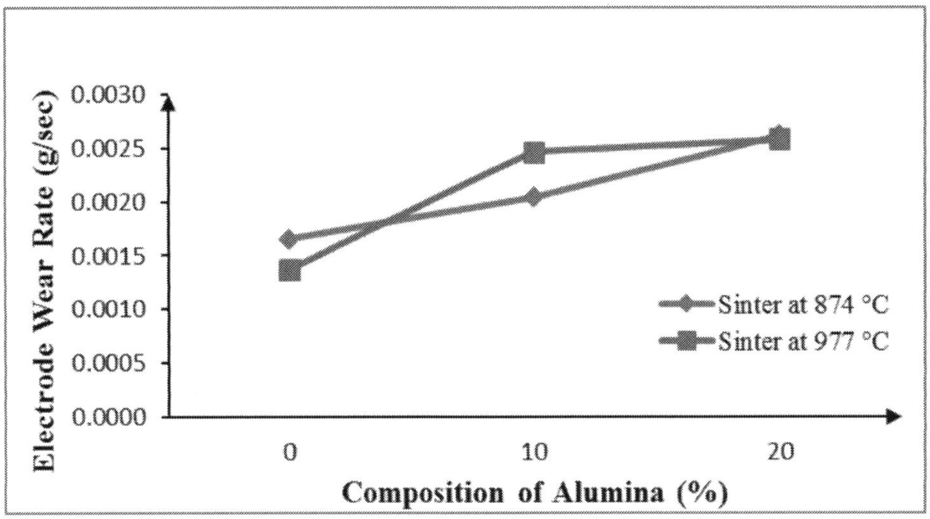

Fig. 6. Effect of various alumina compositions on electrode wear rate (EWR)

It is found that electrode with Cu – 0 vol. % Al_2O_3 composite and is sintered at temperature 977 °C shows better metal removal rate (MRR) with less electrode wear rate (EWR), but it is beyond the acceptable level as EDM electrodes if compared to conventional electrode. This is due to poor inter-particle bonding between Cu and Al_2O_3. During the EDM process, the alumina is held poorly by the copper powders and the bonding among the metal powders loose easily. The composite unable to withstand high thermal stresses that counterattacked the bonding of particles before it's gradually melted by the spark erosion. Disruption on copper powders bonding by the existence of alumina powder contributed to the low thermal conductivity of the electrode.

According to [3], for pure copper sample, most interparticle necks are well developed although their size is not uniform due to the random packing structure. For the composite with the lowest alumina content, the alumina particles appear to be isolated and less likely surrounded by the copper particles. It seems that many necks between copper particles in the close neighbourhood of an inclusion are less developed than the necks in pure copper specimen or the necks in the regions of the composites containing only copper particles, probably because of the constraint created by the presence of a rigid inclusion (alumina). Additionally, pores in the neighbourhood of an inclusion are noticeably larger than those inside a group of copper particles.

Besides, alumina is an electrical insulator but has a relatively high thermal conductivity. This will cause the MRR in EDM worsen when more alumina powder was mixed with the copper powder. The alumina will disturb the electrical and heat conductivity during the EDM process. The MRR proportional decreases with the accumulation of alumina compound in the electrode. According to [4], mechanically alloyed powder shows a progressive hardness increase with alumina addition better than powder that only underwent mixing and blending. Since hardening due to mechanical alloying imparts more scattering of electrons, thus it decreases the electrical conductivity and this is something not preferable in EDM. An increase in sintering temperature enhances thermal conductivity effectiveness in the respective composites. Figure 7 shows the effect of alumina on surface roughness.

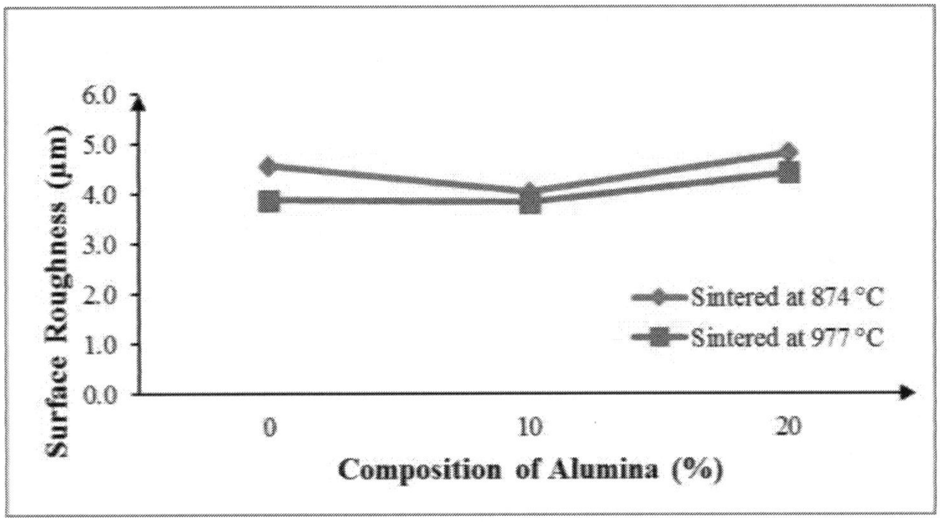

Fig. 7. Effect of Alumina on surface roughness, (Ra).

It was found that electrode with Cu – 10 vol. % Al_2O_3 composite and sintered at temperature 977 °C shows better surface finish than other samples. From the findings, surface roughness increased with increasing the alumina powder. However, if the compound of alumina powder is mixed up in small quantity, the surface roughness of workpiece is better than pure copper tool. Alumina is the most widely used oxide ceramic, either in pure form or as raw material to be mixed with other oxides. If the percentage of alumina powder is too much, the surface roughness will increase due to intense spark formation between the tool and workpiece. According to [8], the application of voltage pulses causes electrical breakdown of the dielectric in a channel of radius 10 µm. The breakdown arises from the acceleration toward the anode of both electrons emitted from the cathode by the applied field and the stray electrons present in the gap. With an increase alumina content, more electrons collide with neutral atoms of the dielectric, thereby creating more positives ions and further electrons, which in turn, are accelerated towards and bombards the workpiece at larger capacity to cause a rougher surface creation.

4 Conclusion

The 3D printing technique introduced to fabricate the EDM tools in this study offers a great flexibility that makes the production of complex and intricate shape tools become much easier and faster. Based on the findings, it can be concluded that the electrodes produced by the 3D printing technique could have potential to be used in EDM die sinking, but further improvement is still required. It is due to the machining related performances known as the MRR is lower while the EWR and the surface roughness are higher compared to conventional electrodes. The results also show the electrode produced by 100% copper powder and sintered at 977 °C revealed the highest MRR and lowest EWR compared to other electrodes. For surface roughness, electrode with 90% copper powder and 10% of alumina and also sintered at 977°C was lower than other samples. It shows that 977°C is the best sintering temperature cycle and recommended for the Cu-Al_2O_3 composite. High sintering temperature augments more inter-particle necking within the adjacent boundary of the copper particles. On top of that, in order to get a better machined surface, the alumina compound in the copper powder must be controlled. Improvement on the tool strength and EWR can be done by utilizing powder with spherical shape, reduce the powder size and embedding infiltration process with bronze during the post-processing stage.

Ackowledgement

The authors would like to express the deepest appreciation to the Office for Research, Innovation, Commercialization, Consultancy Management (ORICC), UTHM for funding this project and also partly supported by Sustainable Manufacturing and Recycling Technology, Advanced Manufacturing and Materials Center (SMART-AMMC), Universiti Tun Hussein Onn Malaysia (UTHM).

References

1. J. Kechagias, V. Iakovakis, M. Katsanos, S. Maropoulos, EDM electrode manufacture using rapid tooling: a review, Journal of Materials Science, **43(8)**, pp.2522-2535 (2008)
2. S. Singh, S. Maheshwari, P.C. Pandey, Some investigations into the electric discharge machining of hardened tool steel using different electrode materials, Journal of Materials Processing Technology, **149(1)**, pp.272-277 (2004)
3. A. Arthur, P.M. Dickens, R.C. Cobb, Using rapid prototyping to produce electrical discharge machining electrodes, Rapid Prototyping Journal, **2(1)**, pp.4-12 (1996)
4. H.M. Zaw, J.Y.H. Fuh, A.Y.C. Nee, L. Lu, Formation of a new EDM electrode material using sintering techniques, Journal of Materials Processing Technology, **89**, pp.182-186 (1999)
5. L. Olmos, C.L. Martin, D. Bouvard, Sintering of mixtures of powders: Experiments and modelling, Powder Technology, **190(1)**, pp.134-140 (2009)

6. A. Upadhyaya, G.S. Upadhyaya, Sintering of copper-alumina composites through blending and mechanical alloying powder metallurgy routes, Materials & Design, **16(1)**, pp.41-45 (1995)
7. P.K. Yarlagadda, P. Christodoulou, V.S. Subramanian, Feasibility studies on the production of electro-discharge machining electrodes with rapid prototyping and the electroforming process, Journal of Materials Processing Technology, **89**, pp.231-237 (1999)
8. Hasan El-Hofy, *Fundamentals of Machining Processes*, Taylor & Francis, New York, pp.371-383 (2007)
9. J.A McGeough, P. Regius, *Advanced Method of Machining*, London New York: Chapman and Hall, pp133-144 (1998)
10. S. Kalpakjian, S.R. Schmid, *Manufacturing Engineering and Technology 4th.ed.*, New Jersey: Prentice Hall, Inc., pp.148-149, 873-875 (2001)

Flame Spread Behaviour over Combustible Solid of Paper, Bagasse and Paper/Bagasse

Afifah Ramli[1,*] *Mohd Azahari* Razali[2], *Azwan* Sapit[2], *Mohd Faisal* Hushim[2], *Normayati* Nordin[2], *Amir* Khalid[2] and *Md Nor Anuar* Mohammad[2]

[1]Faculty of Mechanical & Manufacturing Engineering, UniversitiTun Hussein Onn Malaysia, 86400 Pt. Raja, Batu Pahat Johor, Malaysia
[2]Centre for Energy and IndustrialEnvironmentStudies (CEIES), Faculty of Mechanical & Manufacturing Engineering, Universiti Tun Hussein Onn Malaysia, 86400 Pt. Raja, Batu Pahat, Johor, Malaysia

Abstract. Flame spread over combustible solid must be precisely identify in fire hazard and this is the basic problem in fire safety engineering field. To explore the flame spread mechanism over combustible solid, the downward flame spread behaviour over paper and bagasse has been investigated in this experiment. It found that the combustion pattern for paper was slightly different from the bagasse on the front view and this was analysed through the observation. The result also showed that the flame spread rate for paper is higher compared to the flame spread rate for bagasse. It seem that, although bagasse is the combustible solid the rate of flame spread is lower and the usage of bagasse can help to control the flame spread.

1 Introduction

Flame spread can described as the rate of movement crosswise a fuel surface and because the flame spread happen over a surface, enormous complications ascend which are not establish in simple gas-phase combustion problems [1]. So, flame spread behavior over combustible solid identified as a fundamental and an extremely dangerous occurrence in fire [2]. This is the purpose why it turn out to be important to understanding the flame spread behavior of this material. The examination that correlated to flame spread over combustible solid was the elementary problem in fire safety engineering field [3]. Previously, there are many researcher that concern to study the flame spread over combustible solid permitted to increase the capability of fire prevention.

The downward flame spread over polymethylmethacrylate (PMMA) [4] was scrutinized based on experimentally, computationally and analytically in an oxygen/nitrogen environment in normal gravity. It originate that the influence of ambient pressure on the spread rate and the alteration thickness seem only qualitatively, perhaps because of finite-rate effects at low pressure. The flame spread over PMMA also examined experimentally in the quiescent air in an open space founded on the different thickness which is from 1.5 to 10 mm to discover flame spread mechanism over combustible materials. From this enquiry it

*Correspondingauthor: hd160083@siswa.uthm.edu.my

seem that the flame spread rate growths with the reducing thickness of sheet and it tends to a continual value for thick sample. Then, it also scrutinized the width and thickness effects on flame spread over thermally-thin horizontal sample of PMMA [5]. The result indicated, as the width and thickness increase, convective heat transfer from the flame to the fuel surface decrease as additional air flow around the flame surface moves away from the pyrolysis area. In radiation condition, it increases as the size of flames increases while through conduction the fuel sample was measured and determined to have a negligible impact on flame spread.

Additionally, upward flame spread behavior over PMMA also examines by various number of width 100, 200, 300, 400, 500, and 700 mm with the same thick and tall which is 18 mm and 1000 mm respectively [6]. The lateral diffusion throughout the flame width was proposed to grounds thicker flame along its centerline for wider flames and improve combustion competence. It indicated the flame spread rate was correlated to the width with a power of 0.35. The downward flame spread behavior also forecast over thermally thin cellulose in a normal gravity quiescent environment [7]. It states that, one of the vital element to apprehension when it emanates to flame spread is the orientation of flame spread. It is because it be able to ensue in many ways which is by horizontally, vertically, upward or downward, and upside or underside. While the spread rate alteration radically from one orientation to another, it rests steady nonetheless.

Then, flame spread behavior over fabric also investigated based on the diverse weft thread angle as it is eminent that the natural fiber reveals different flame spread behavior compared to the synthetic fiber [8-9]. This study had been used combined fabric of kenaf/polyester, cotton/polyester and cotton/nylon. The result indicates that the flame spread rate for the combined fabric of kenaf and polyester enslavement on the angle as the angle is increasing from 0°, 45° and 90° the flame spread rate are decreasing from 1.4702 mm/s, 1.4579 mm/s and 1.4122 mm/s respectively. Not only for this material, the outcome for combined fabric of cotton/polyester and cotton/nylon also shown that both fabrics have comparable angle reliance where the flame spread rate decreases as the angle increases.

Next, it also had an investigated on downward flame spread towards paper based on the thickness from 0.4 to 10 mm [10]. Paper is one of the combustible solid and statistic data from Malaysia Fire and Rescue Department, about 5817 cases that involved in a building fire and one of the factors that contributed to this fire is paper [11]. The flame spread found to take place for sample thickness less than 8.4 mm under the condition of natural convection and the flame spread becomes unstable when the thickness is more than 8.4 mm. It also investigated the flame spread behavior towards thin paper disk in narrow space and it was measured both upwards and downward flame spread [12]. This experiment agree that flame spread decrease when leading edge of the flame reaches 15-20 mm from the center of paper disk. The upward flame spread over corrugated cardboard [13] also explore and the result support that the non-homogeneity of the cardboard helped to reduce the acceleration of the upward spread rates by physically disorderly flow in the boundary layer close to the vertical surface and thereby modifying heating rates of the solid fuel above the pyrolysis region.

It seem that it important to investigated more about the flame spread behavior over paper but it is lack attention to study the flame spread towards paper. As known paper was produce from the pulp that derived from the wood and this production causes deforestation and the subsequent possibility of a limited timber resources crisis. Due to this awareness, many research had been conducted to produce paper by using natural fiber and the common natural fiber that use is banana fiber [14], palm fiber [15] and pineapple leaf fiber [16]s. Unfortunately, these studies have been conducted to study the properties of paper and focus on the mechanical properties of the paper only.

Instead of examine the flame spread behaviour over paper it is also significant to study the effect of natural fiber likes bagasse in combustion process as bagasse is one of the natural fiber that used in papermaking process in order to less the usage of wood pulp. The choose of bagasse in this experiment due to the number of cellulose, hemicellulose, lignin and small amount of extractives and mineral salts that compose from the bagasse [17]. In this context, the aim of this work is to test the flame spread behavior towards paper, bagasse and also the mixture of both materials as a main material used in papermaking process. At the end of this experiment the flame spread behavior also flame spread rate headed for both material will acknowledged. This study also important to know the effects of natural fibres which is bagasse influencing the behaviour of fire.

2 Experimental Setup

As mention above, this experiment has explored about the flame spread behavior towards combustible solid as in this experiment it used paper, bagasse and the mixture of this two materials. So, there are 3 specimen with pure paper, mixture of paper with bagasse and pure bagasse. The size for each specimen were same which is 10 cm x 10 cm as the aim is to investigated the flame spread behavior based on this different material.

The specimen sheet were made from the A4 office recycled paper, bagasse and the mixture between this two materials. The bagasse has to be cleaned first to remove any dirt before bagasse will be soaked in hot water that has mixed with Sodium Hydroxide (NaOH) for one day (24 hour) and this step also applied on recycle paper. Then, the mixture between this two materials will be blend at the same specified period rate which is 20 minutes. After that, the blended ingredient will placed on the net that are made from wooden frame. This step was seek to remove liquid contained in the material and let the sample sheet to dry at room temperature.

Then, all the specimen need to be examine by using Scanning Electron Microscope JEOL (SEM), to determine the structure of each specimen before the combustion testing. This specimen has testing with combustion testing in Vertical Flame Chamber and the schematics drawing for experiment arrangement as shown in Figure 1 below where the specimen was clamped to the holder. The camera was used to videotape the combustion process that occur during the experiment for all specimen until the end of the combustion process and it is videotape from the front view. The flame spread behavior that happens along the experiment had been analyzing through the videotape.

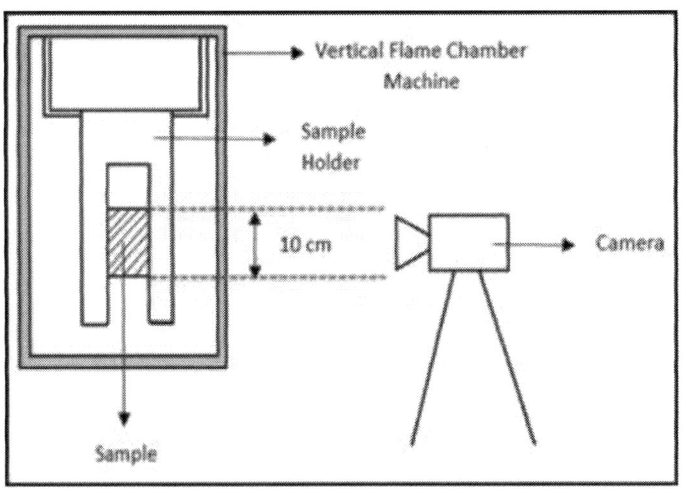

Figure 1: Schematics drawing for experimental setup

DOI: 10.1051/matecconf/201713500012

3 Result and Discussion

3.1 Flame Spread Behaviour

In this experiment, the specimen is ignited at a point on its top edge where the flame spreads is in the downward direction. The comparison between the flames spread behavior that occur for each specimen can be seen in Figure 2 (a) to (c), that attached below as this result was videotaped to observe the flame spread behavior. The detail propagation of flame shape during the experiment can be seen in Figure 3 (a) to (c) below as this figure shows the shape of flame front for every 15s. The black line represents the combustion process while the red line represents the extinguished of flame.

From the result, it is known that flame spread for pure paper was the most stable and nearly consistent flame spread shape and combustion process occurred at the overall of the specimen. It started with "U" shape as showed in Fig. 3(a) below then slowly turned to "V" shaped. As the combustion process continued the flame spread behavior back to "U" shaped. As can be refer to Fig. 3(a), the propagation of the flame spread is limited to the central region at the beginning but after 45s the flame spread take place at the whole of the specimen and behavior remained until the end of the specimen.

The result was significant different for the other type of specimen in term of flame spread behavior although at the beginning of ignition the flame spread started with "U" pattern and this occurrence take place at each of specimen as showed in Fig. 3(a) to (c) below. For paper/bagasse specimen, the flame spread also take place at the whole of the specimen as shown in Fig. 3(b) below, but the flame spread shape for this specimen does not remained until the end of specimen. The flame spread shape become unstable after 150s and extinguished while the experiment is still carried out.

For the case of pure bagasse, it found that the flame spread shape becomes unstable and the flame is retarded after a few second in the very short period compared to paper/bagasse specimen, thus it is extinguished while the experiment was carried out as shown in Fig. 3 (c). The result seem that the usage of bagasse may influence on the flame spread shape.

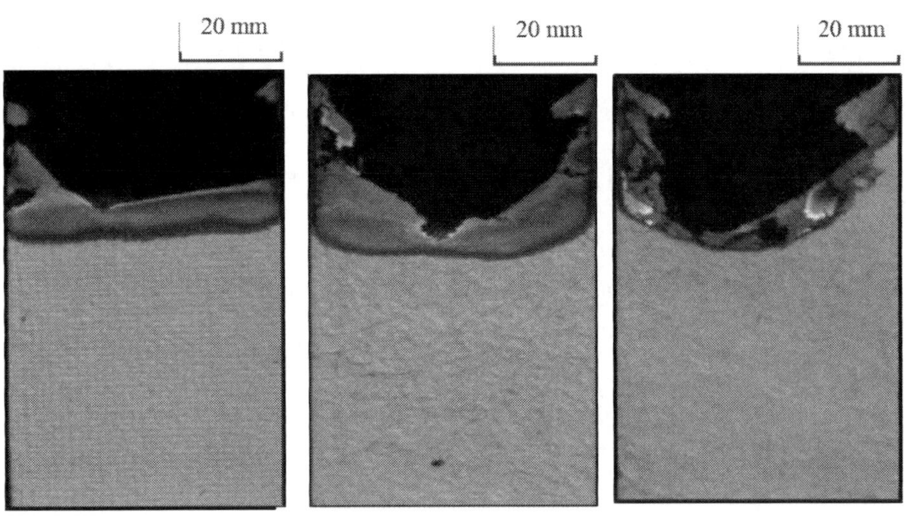

(a) Pure Paper (b) Paper / Bagasse (c) Pure Bagasse

Figure 2: Flame spread behavior towards specimen

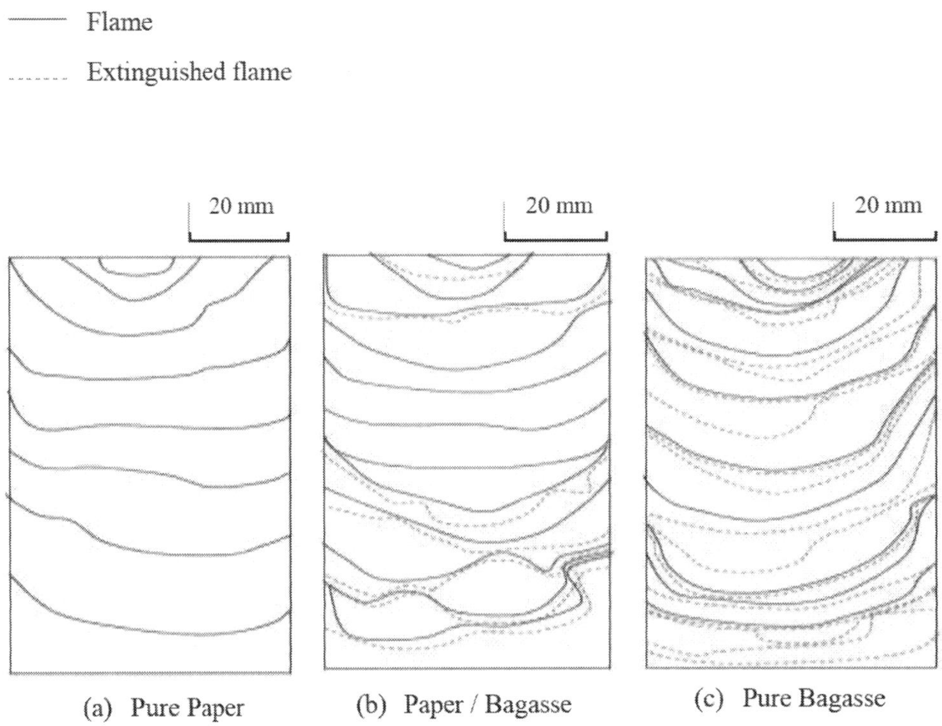

Figure 3: Overall flame spread behavior for specimen

3.2 Flame Spread Rate

Flame spread rates are obtained by measuring the position of the most preceding point of the burning front at each time. The flame spread rate for each specimen is different for each sample where the pure bagasse has the lower flame spread rate which is 0.4814 mm/s, follow by paper/bagasse specimen 0.5513 mm/s and the higher flame spread rate goes to pure paper with 0.8133 mm/s as shows in Figure 4 below. From this research, it is inferred that bagasse has significant impacts on the flame spread shape and also on the flame spread rate. However, the additional results are needed in order to examine about the influence of the material composition. This result will be useful to validate simulation works in the future.

DOI: 10.1051/matecconf/201713500012

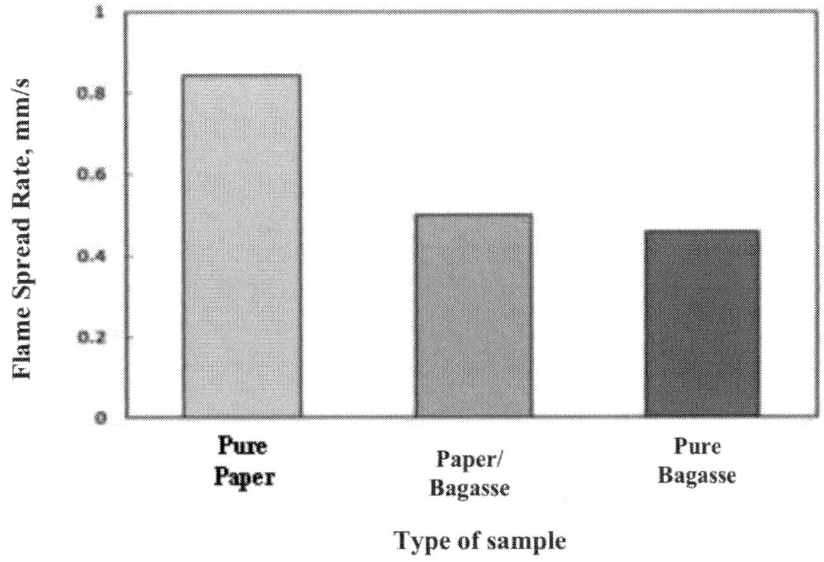

Figure 4: Flame spread rate for different type of specimen

Besides that, flame spread rate also influence by the size of the material fibre as shown in the Figure 5 below, it can see that the size of the paper fibre is trifling compare to bagasse fibre as the previous research specifies that the good distributive dispersion and the small particle size tend to result in a better flame retardancy [18]. From the experiment result it indicates that small size of fibre increase the flame spread rate as the flame spread rate for pure paper is the highest flame spread compared to specimen paper/bagasse and pure bagasse where the pure bagasse has the lower flame spread rate due to the hefty size of fibre.

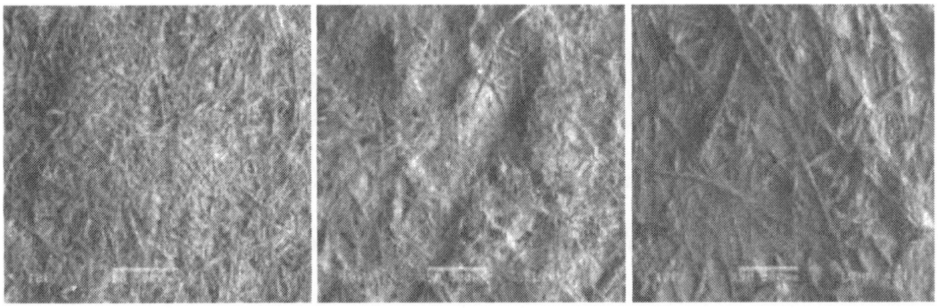

(a) Pure Paper (b) Paper / Bagasse (c) Pure Bagasse

Figure 5: Microstructure image for different type of specimen

4 Conclusion

Flame spread behavior for each specimen are different and the flame spread started to become unstable with the usage of bagasse. The usage of bagasse also made the flame often to extinguish during the combustion process while the flame of pure paper was stable and does not extinguish until the end of the combustion process. The usage of bagasse in this experiment found that this substance can be used in order to slow down the flame spread. As mentioned early, the flame spread rate for pure paper had the highest value of flame spread rate compared to pure bagasse that had the lowest flame spread rate. From the analysis, it found that the density value of the material also affected flame spread behavior and flame spread rate as the highest value of density the lower flame spread rate. From this experiment, it can say that substance with high density can use in order to control the flame spread.

Acknowledgment

The authors would like to thank Universiti Tun Hussein Onn Malaysia (UTHM) and Ministry of Higher Education Malaysia (MOHE) for their financial support of the present work through Fundamental Research Grant Scheme (FRGS-1465).

References

1. M. B. Ayani, J. A. Esfahani, and R. Mehrabian, "Downward flame spread over PMMA sheets in quiescent air: Experimental and theoretical studies," *Fire Saf. J.*, **41**, 164–169, (2006)
2. S. Charuchinda, M. Suzuki, and R. Dobashi, "Behavior of flames spreading downward over napped fabrics,"**36**, 313–325, (2001)
3. M. Azahari, A. Sapit, M. Faisal, and A. Khalid, "Flame Spread Behavior Over Combined Fabric Of Cotton / Polyester," 1–5, (1999)
4. K. K. Bhaskaran, M. D. King, S. Takahashi, T. Nagumo, and K. Wakai, "Downward Flame Spread Over Poly (Methyl) Methacrylate,"**28**, 2891–2897, (2000)
5. L. Jiang, C. H. Miller, M. J. Gollner, and J. Sun, "Sample width and thickness effects on horizontal flame spread over a thin PMMA surface," *Proc. Combust. Inst.*, **36**, 2987–2994, (2017)
6. K. Tsai, "Width effect on upward flame spread," *Fire Saf. J.*,**44**, 962–967, (2009)
7. S. Bhattacharjee, C. Paolini, W. Tran, J. Ray, and S. Takahashi, "ScienceDirect Temperature and CO 2 fields of a downward spreading flame over thin cellulose : A comparison of experimental and computational results," *Proc. Combust. Inst.*, **35**, 2665–2672, (2015)
8. M. Azahari, A. Sapit, A. N. Mohammed, M. Faisal,and H. Peraman, "Polyester Effect of Thread Angle on Flame Spread Behaviour over Combined Fabric of Kenaf / Polyester," **020016**, (2017)
9. M. Azahari, K. Amir, M. Suzuki, W. Masuda, S. Hamidon, and S. Azwan, "Effect of Synthetic Material on Angle Dependency of Flame Spread Behavior over Combined Fabric," **020014**, (2017)
10. M. Suzuki, R. Dobashi, and T. Hirano, "Behavior Of Fires Spreading Downward Over Thick Paper," 1439–1446, (1994)
11. JBPMWebsite (2013). JBPM Website Search. Retrieved on Disember 13, 2013 from:Searchhttp://www.bomba.gov.my/index.php/pages/view/542 (2013)
12. T. Takahashi, T. Daitoku, and T. Tsuruda, "Behavior of the flame spread along a thin paper-disk in a narrow space," *Proc. Combust. Inst.*, **36**, 3011–3017, (2017)

13. M. J. Gollner, F. A. Williams, and A. S. Rangwala, "Upward flame spread over corrugated cardboard," *Combust. Flame*,**158**, 1404–1412, (2011)
14. A. Yosephine, V. Gala, A. Ayucitra, and E. S. Retnoningtyas, "Dalam Pembuatan Kertas Serat Campuran," **11**, 94–100, (2012)
15. K. Syamsu, L. Haditjaroko, G. I. Pradikta, and H. Roliadi, "Campuran Pulp Tandan Kosong Kelapa Sawit dan Selulosa Mikrobial Nata de Cassava dalam Pembuatan Kertas," *J. Ilmu Pertan. Indones.*, **19**, 14–21, (2014)
16. S. Wahab, "Producing Paper using Pineapple Leaf Fiber'", **390,**3382–3386, (2012)
17. F. Mandelli, L. B. Brenelli, R. F. Almeida, R. Goldbeck, L. D. Wolf, Z. B. Hoffmam, R. Ruller, G. J. M. Rocha, A. Z. Mercadante, and F. M. Squina, "Simultaneous production of xylooligosaccharides and antioxidant compounds from sugarcane bagasse via enzymatic hydrolysis," *Ind. Crop. Prod.*,**52**, 770–775 (2014)
18. H. Huang, M. Tian, L. Liu, W. Liang, and L. Zhang, "Effect of Particle Size on Flame Retardancy of Mg (OH) 2 -Filled Ethylene Vinyl Acetate Copolymer Composites," *Journal of Applied Polymer Science.*,**100**, 4461-4469 (2006)

DOI: 10.1051/matecconf/201713500013

Solid-State Recycling of Light Metal Reinforced Inclusions by Equal Channel Angular Pressing: A Review

*Sami. Al-Alimi, M. A. Lajis, S. Shamsudin**

Sustainable Manufacturing and Recycling Technology, Advanced Manufacturing and Materials Centre (SMART-AMMC), Universiti Tun Hussein Onn Malaysia, 86400 Parit Raja, Batu Pahat, Johor, Malaysia.

Abstract. Solid-state recycling of light metals reinforced inclusions through hot Equal Channel Angular Pressing (ECAP) is performed to directly recycle metal scraps and reduce cost of material in engineering applications. The ECAP is one of the most important method in severe plastic deformation (SPD) that can convert light metals into finished products. This paper reviews several experimental and numerical works that have been done to investigate the effects of the ECAP parameters such as die angles, material properties, outer corner angle, friction coefficient, temperature, size of chips, pressing force, ram speed and direct effects of number of passes on the strain distributions. It also includes the performance enhancement of aluminium matrix composite reinforced ceramic-based particles that derived from direct recycled aluminium chips for sustainable manufacturing practices.

1 Introduction

Recycling aluminium and its alloys by solid-state recycling method is relatively simple, consume lower energy, and not to have harmful impact on the environment. It is significant to study the direct aluminium recycling in order to reduce the energy consumption and emissions level that causing global warming [1]. Chips can be directly recycled and formed by ECAP, one of the most promising severe plastic deformation (SPD) techniques for fabrication of bulk ultrafine grain materials [2]. The common steps associated with this method are pulverizing the chips, cleaning, drying, cold compaction and hot/cold pressing the chip-based billet. The process depends on the design methods and the factors affecting the mechanical properties of aluminium chips such as size of chips, cold pressing

*Corresponding author : shazarel@uthm.edu.my

parameters, extrusion temperature, extrusion rate and ratio [3]. Enhancing mechanical properties of the recycled light metal alloys and decreasing the recycled part porosities are the primary aimed. The ECAP can apply design with an internal angle of 90° and 20° of the outer angle. The ECAP dies and billet can be heated to decrease porosity during the process. The decrease in porosity was due to the compressive force applied through the interaction of the billet material with die and container [4]. In metal matrix composites (MMCs), silicon carbide, alumina oxide and boron carbide are the most commonly used particulates in fibre reinforced metal matrix composite. The boron carbide (B_4C) particulate is a superior ceramic reinforced material for MMCs than silicon carbide (SiC) and alumina (Al_2O_3) due to its high hardness, low density, high strength, high wear and impact resistance, high melting point, low coefficient of thermal expansion and good chemical stability [5, 6]. However, MMCs showed the superior plastic properties in the SPD [7]. While boron carbide (B_4C) is widely common reinforced metal matrix composites that has a low density value, good hardness, excellent thermal stability and remarkable chemical inertness [8]. This paper presents a review on solid-state recycling of aluminium alloys reinforced ceramic particles in metals matrix composites (MMCs). It also covers design variations of the ECAP dies and the processing parameters. The review considers how ultra-fine grained material through the ECAP technique can affect the microstructure, mechanical and physical properties as well as the overall performance of the consolidated chips.

2 Experimental Factors Influencing ECAP Material

2.1 Inner Angle

The inner angle Φ, is one of the important factor that enhance the strain in each pass. The factor of inner angle presented is directed to have influences on the microstructure development. The researchers proposed that angles should be selected between 90° to 120° and this supported the evidence from the results of finite element analysis and microstructure investigations [6, 10]. Die angle, Φ contributed to direct effect on strain distribution of ECAP under ideal conditions. An inner angle of 90° is the most significant angle because of its effectiveness of achieving an ultrafine-grained microstructure development [2]. The influence of processing routes is an alternative way to consider between consecutive numbers of passes and angle [6].

2.2 Outer Angle

A direct effect of an outer angle, ψ on the extruded billet through the ECAP process can become important. The outer angle proposed to be $\psi = 20°$ to cause an impact on the better strain homogeneity that is very effective on the grain refinement. If $\psi = 0$ was taken, the arc curvature leads to be dead zone where the billet is no longer contacted to the die wall and the problem could be eliminated or removed by having movable ECAP dies. Increasing in ψ will lead to less strain as it is obtained from the derived equation [12, 13]. Smaller corner angles maximize the homogenous of the billet while the friction reduced at the corner gap. ECAP channel and angles have the influence on generating the average equivalent plastics strain. The friction causes an important effects to the required punch load to extrude the billet and maximum strain is obtained by the following parameters with $\varphi = 90°$, $\psi = 15°$ and $\mu = 0.3$ [2, 13].

2.3 Pressing Speed

Pressing speed is a significant factor of ECAP process. Pressing force could be validated by using the FEM software and to be compared with the experimental work. Furthermore, the punch force of experimental and simulation have the same conditions so that, the results of FEM assumed as agreeable [14]. However, when the temperature is low the need for high straining is required to avoid the problems associated with ductility and durability of the material and die set of ECAP processing [3]. High capacity hydraulic press is required to generate high ram speed between the ranges from 1 to 20 mm/s. The investigations confirmed that the pressing of pure Al have not influences on the equilibrium of ultra-fine grain formed by ECAP process [6].

2.4 Pressing Force

Pressing force is an important factor of ECAP die. However, the pressing force could be validated by using the FEM software and to be compared with experimental work or theory relations. The punch force of experimental and simulation have the same conditions so that the results of FEM simulation assumed to be agreeable as the pressing force simulated values are 121.5kN and 113kN. On the other hand, for the first pass when $\psi = 15^\circ$, $\Phi = 60^\circ$ where the effective strain values obtained between the whole and the plan of the ECAP metal. In improvement of the ECAP tool was noted throughout the mechanical and physical properties and the grain size was reduced about 8 times [15].

2.5 Pressing Temperature

Temperature effect is one of the most important parameter in SPD processing methods such as ECAP processing. Metallic powder of aluminium and steel were investigated. The process of was proposed that the consolidation was performed at room temperature. Furthermore, the tensile strength decreases by increasing the temperature of ECAP processing and the total elongation is also increased [17]. The researcher proposed that when the temperature is increased from 120°C to 180°C, it can cause decreases of mechanical properties while the precipitate and grain size are increased. On the other hand, temperature is effective to cause decreasing on high angle boundaries and transformation on metal phase. The die was heated up until 15 minutes to desire temperature furnace [16]. Higher temperatures of 200°C, 250°C, 450°C result in decreasing the size of grains and the microstructure during continues ECAP process [18]. Lower temperature may cause new systematic recovery, recrystallization, twining, and mechanism of dislocation climb become impossible. By lowering the temperature values the need for straining force is required avoid problems emerged that associated with ductility of the material and durability of the die-set of the ECAP process [19, 20].

2.6 Processing Routes of ECAP

The theory of deformed billet routes were introduced by rotating billets throughout 0°, 90° and 180° degree along each ECAP extrusion pass [5]. The process was proceeded repetitively to obtain the strain distribution along crossing area and there are four main processing routes that have been used to study the microstructural development during ECAP. It is recognized that after pressing, the characteristics within material crystalline would be changed due to the pressing and possible to carry the main processing routes definition [21].

Route A: The sample is pressed with no rotation.

DOI: 10.1051/matecconf/201713500013

Route B_A:The sample is rotated with 90◦ alternatively, clockwise or counter clockwise
Route B_C:The sample is rotated 90◦ clockwise between the passes.
Rout C: The sample to be rotated with an angle of 180◦.
The principle of ECAP processing routes influence on the light metal compression including mechanical properties, microstructure, deformation and strain distribution [55]. Figure 1 illustrated the basics routes of ECAP process.

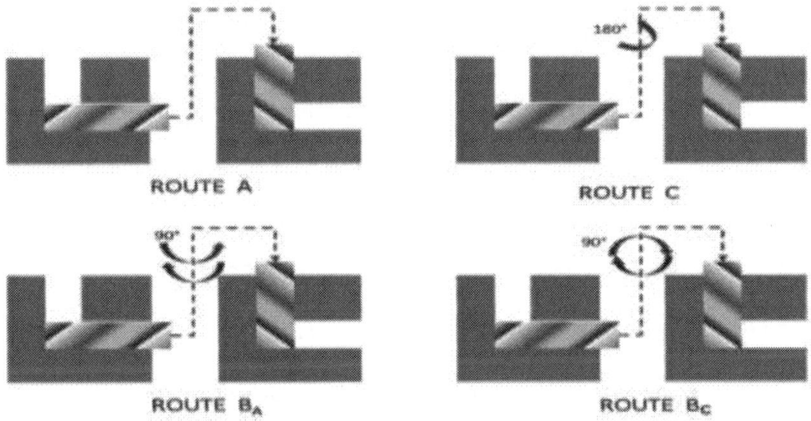

Fig. 1. Illustrated The Basics of ECAP Processing Routes [22]

However, the strain for one pass could be calculated by the equation below that developed by Iwahashi [39]

$$\epsilon = \frac{1}{\sqrt{3}}\left(2 \cot\left(\frac{\Phi+\psi}{2}\right) + \frac{1}{\sqrt{3}}\left(\psi \csc\left(\frac{\Phi+\psi}{2}\right)\right)\right)$$

$$\epsilon = \frac{1}{\sqrt{3}}\left(2 \cot\left(\frac{\Phi+\psi}{2}\right) + \frac{1}{\sqrt{3}}\left(\psi \csc\left(\frac{\Phi+\psi}{2}\right)\right)\right)$$

And for multiple no of pass the magnitude effective strain is proposed by [15]

$$\varepsilon_{eq} = \frac{N}{\sqrt{3}}\left[2 \cot\left(\frac{\Phi+\psi}{2}\right) + \psi \operatorname{cosec}\left(\frac{\Phi+\psi}{2}\right)\right]$$

$$\varepsilon_{eq} = \frac{N}{\sqrt{3}}\left[2 \cot\left(\frac{\Phi+\psi}{2}\right) + \psi \operatorname{cosec}\left(\frac{\Phi+\psi}{2}\right)\right]$$

3 Reinforced AA6061 Alloy with Metal Matrix Composites

Metal matrix composites (MMCs), especially the aluminium based matrix, have become structural materials that possess light weight, high strength and modulus, low thermal expansion, and good wear resistance. MMCs are promising materials for number of specific applications in aerospace, defence, and automobile industries [23]. The parameters such as ECAP routes, inner angle Φ, outer angle ψ. At the first pass the material deformed to rhombohedral shape and the straining shear plane x, y, z, depends on what a way of the material to be inserted into the ECAP die. The route A yields to increase distortion of in the coordinates of x, y but not with z , the route BA yields for increasing distortion at all sides of orthogonal planes , the route B and C the element is restored after for an every four pressings [24, 20, 9]. Table 1 is showing the particulate properties of the reinforced particles.

Table 1: Properties of Particulate Composites [25]

Properties	WC	SiC	Al_2O_3	B_4C	Si_3N_4
Density g/cm2	13-15.3	3.21	3.7-3.97	2.51	3.31
Compressive Strength MPa	3100-5860	1725-2500	2070-2620	2900	689-2760
Modulus of Rigidity GPa	483-641	476	393	445	317

The corresponding volume friction could be calculated from the relation below:

$$Vp = ((M\rho)/(P\rho))/(\frac{M\rho}{P\rho} + \frac{Mm}{Pm})Vp = ((M\rho)/(P\rho))/(\frac{M\rho}{P\rho} + \frac{Mm}{Pm})$$

Where, Vp volume fraction of ceramic particles. mρ, Pρ mass and density of the particles mρ, Pρ mass and density of matrix [26].

3.1 Silicon Carbide Reinforced AMC

Saheb investigated Metal Matrix Composites (MMCs) aluminium based particles with SiC, alumina (5%, 10%, 15%, 20%, 25%, and 30%), and graphite of weight fraction 2%, 4%, 6%, 8% and 10%. The findings indicated that the method is successfully developed MMCs materials properties. The observations proposed an increasing of hardness with weight volume friction and the results of maximum hardness is 25% of SiC silicon carbide [27].

Weight friction and 4% graphite. Investigation on the microstructure, wear resistance and mechanical properties of silicon carbide (SiC) reinforced particles reinforcing aluminium matrix composites was also reported. The variations of silicon percentage (0, 5, 10 and 20 wt. %) were proceeded with stir casting process. The findings showed that the maximum hardness obtained was at 20% of SiC addition as long as porosities and wear test were also increased by increasing the volume friction of silicon carbide particles [28]. Other researchers investigated the mechanical properties of hybrid reinforced with groundnut ash and SiC of the volume friction of (10:0, 7.5:2.5, 5.0:5.0, 2.5:7.5 and 0:10) by stir casting process. The results showed with increasing of Groundnut Shell Ash (GSA), the fracture toughness and elongation percentage were improved at 6%wt Al–Mg–Si/SiC–GSA for 15.7% but with 10%wt the fracture toughness reported as 7.33% improvement. Hardness and tensile decreased with the increasing of the GSA volume friction of reinforcing phase [29].

3.2 Aluminium Oxide Al_2O_3 Reinforced AMC

Fogagnolo demonstrated cold and hot press followed by hot extrusion to recycle AA6061 aluminium reinforced with Al_2O_3. The findings proved that cold press and hot extrusion contributed to higher profit compared with primary material produced by conventional casting method [61]. Al-7075 chips reinforced with 10 vol% Al_2O_3 particles by high pressure torsion with 20 turns was investigated by Fogagnolo et al. [30]. The finding showed a grain refinement from 8 μm to 300 nm after the process of HTP has been achieved. The microhardness also improved from 167Hv at initial value to 260Hv after the HTP process and super plasticity of MMC with maximum elongation of 670% was achieved in tensile test when the strain rate reported at 1.0 x 10^{-2} s^{-1}. Besides, investigation on utilizing SiC and Al_2O_3 as reinforced particles in aluminium metal matrix composite have been developed with stir casting process route. The effect of reinforcement volume fraction on microstructure, physical and mechanical properties of the composite have been increased with increasing the volume friction of ceramic particulates. The yield and tensile strengths showed an increases of 19% and 19% with the 20 vol% SiC and 20vol% Al2O3 respectively [31].

3.3 Boron Carbide (B_4C) Reinforced AMC

Boron carbide is a significant hard element that has high elastic modulus and fracture toughness. However it is superior reinforced ceramic particles for enhancing aluminium material composites (AMC) than SiC or Al_2O_3 due to its low density, higher wear resistance, higher strength, higher melting point, higher impact resistance and lower coefficient of thermal expansion and stable chemical stability [32, 9]. The effects of adding micro – nano B_4C particles by using casting process was investigated and the findings proved that 6% vol. nano particles indicate better mechanical properties particularly in tensile, ductility and impact energy than micro B4C reinforcement particles. While the wear resistant gradually increased up to 8% B_4C addition [9]. Ramnath et al. [33] investigated three types of MMC reinforced samples. Sample 1: aluminium alloy 95%, alumina 3% and boron carbide 2%; Sample 2: aluminium alloy 95%, alumina 2%, and boron carbide3 %. ; Sample 3: aluminium alloy only. The findings revealed that the flexural and tensile strength of sample 3 was higher than other two samples because of its aluminium content. The Table 2 is illustrated the amount the type and volume friction of MMCs B_4C particles.

Table 2: Volume Friction of B4C Particles

no	MMCs	Volume Friction % B_4C	Method	Author
1	AA6061/B4C	16 , 30 vol.%	stir casting route	[65]
2	AA6061/B4C	3%, 6% and 9%	stir casting method.	[64]
3	AA6061/B4C	0%,10%,15%,20%	DRAs followed by squeeze casting	[66]
4	AA6061/B4C	0%,2%,4%,6%,8%,10%	ultrasonic cavitation	[18]
5	2024 / B4C	10 vol%B4Cp/2024A	Hot extrusion	[73]
6	Al / B4C + Al2O3	95% AL + 3%Al2O3 + 2%B4C 95%Al + 2% Al2O3+ 3% B4C	stir casting method	[74]

3.4 Zircon ZrO_2 Reinforced AMC

Zircon is a hybrid reinforced particles that has superior characteristics of material matric bonding which results better wear resistance compared to other particles such as alumina Al_2O_3 [8]. Hossein et al. [34] investigated the combination of AMCs composites and the addition amount of zircon particles in different percentage (5%, 10 %, 15 %, 20 %, and 25 %). The findings showed an increases of hardness to the maximum value of 75 BHN and density increases up to 92%. By increasing temperature from 600 °C to 650 °C, the mechanical properties such as yield strength and elongation are improved. The researcher also proposed that the optimum value obtained on compressive strength is 248 MPa with 5% vol. zircon at 650 °C. The effects of adding ZrO_3 particles to aluminium at 750 °C by stir casting process was studied. The mechanical properties were enhanced with the addition of the ZrO_2 until 2% vol., the strength reported as higher as 197 MPa in this

DOI: 10.1051/matecconf/201713500013

experiment with 64% of the pure Al. The reason of increasing yield strength at the desired temperature 750 °C is because of the dislocation density of composites pile-up beyond the zirconium particles, the dislocation occurred during forming metal processing. The mechanical and physical properties were increased as proposed by Ch.V.M.Prasad, when 3 vol% ZrO3. The Table 3 is showing the amount of volume particles proposed by pervious researchers.

Table 3: Volume Friction of ZrO2 Reinforced Particles

No	MMCs	Volume Friction % ZrO_2	Method	Author
1	Al_2O_3/ ZrO_3	5%,10%,15%,20%	Surface modification with aqueous solutions	[67]
2	Al7075/ZrO_3/Fly Ash	0% , 3% ZrO_2 , 6% fly ash 4th 3% ZrO_2+ 6% flyash	stir casting	[11]
3	Al/ ZrO_2	5 – 80 %	Powder technology	[68]
4	Al / ZrO_2	1%, 2% , 3%	Liquid Routing	[69]
5	Al/ZrO2/Al_2O_3/TiO_2	0% - 5 %	stir casting method	[70]

3.5 Fly Ash Reinforced AMC

Mindivan et al. [36] aimed to conduct study on aluminium reinforced composite of aluminium 6082 alloy chips mixed with fly ash by direct conversion. Within the addition of the fly ash results in considerable increase of hardness to superior level and wear properties than those conventionally produced. Aluminium with 10 wt% added fly ash/lead in metal matrix composites was developed using powder metallurgy method. The composites reinforced powder consist of 0, 5, 10, 15 and 20 wt % fly Ash. The experiment was carried out using the mixing chamber component with fixed and rotated at 32 rpm for approximately one hour with changing in direction for every five minutes to ensure homogeneous mixing. The density is directly increased with increasing the percentage of added material content and reached up to 10 wt % [37]. The reinforcement fly ash particles from thermal power plant was also conducted at nano level of reinforcement to form Al matrix materials. The investigation of the nano fly ash particles (1 wt.% to 3 wt.%) causes in increase of AA2024 hardness of which reported as 75 Hv for 1 wt.% to 114 Hv for 3 wt.% [38]. Mechanical property such as compressive strength of nano fly ash/Al composite enhanced with increase in wt. % of nano fly ash. The strength of AA2024 alloy is 289 MPa and it was increased to 345 MPa for 3 wt. % fly ash/AA2024 composite.

4 Conclusion

In this review, direct conversion of aluminium chips reinforced with multiple ceramic particles through the ECAP tool into final products have been presented intensively. An overview of this paper is presented to propose a new approach of improving the performance of aluminium composites made of chips with addition of boron carbide (B_4C)

or zirconium dioxide (ZrO_3). The chip based composite reinforced materials offers an alternative to manufacturing automotive industries of as a way of recycle, reuse the machined materials as a secondary source of metal and to prevent our earth from green house gas for sustainable life. However, B_4C and ZrO_2 are superior ceramic materials that improve the tensile strength, wear resistance, stiffness and microstructure of ultra. Currently ECAP is the most potential tool of all SPD techniques. Due to the mechanical behaviour the ECAP deals with metal flow and microstructure evaluation. More investigations were carried out recently on using ECAP die in direct recycling. The factors that influence the final results such as geometries, die inner and outer angles, backpressure, pressing speed, pressing temperature, and die related accessories were also important to encompass in the ECAP method. Due to the importance of achieving objective to confirm that the aluminium alloy Al6061 reinforced ceramic inclusions can be utilized in engineering applications such as automotive manufacturing.

Acknowledgement

The authors would like to express the deepest appreciation to the Centre for Graduate Studies, Universiti Tun Hussein Onn Malaysia (UTHM). Additional support was also provided by Sustainable Manufacturing and Recycling Technology, Advanced Manufacturing and Materials Center (SMART-AMMC), Universiti Tun Hussein Onn Malaysia (UTHM).

References

1 J.J. Rino, D. Chandramohan, D, K.S. Sucitharan, V.D. Jebin, (2012). An overview on development of aluminium metal matrix composites with hybrid reinforcement. IJSR, India online ISSN, 2319-7064.
2 Medvedev, A. E., Neumann, A., Ng, H. P., Lapovok, R., Kasper, C., Lowe, T. C., & Estrin, Y. (2017). Combined effect of grain refinement and surface modification of pure titanium on the attachment of mesenchymal stem cells and osteoblast-like SaOS-2 cells. Materials Science and Engineering: C, 71, 483-497.
3 Gronostajski, J., Marciniak, H., & Matuszak, A. (2000). New methods of aluminium and aluminium-alloy chips recycling. Journal of materials processing technology, 106(1), 34-39.
4 Härtel, M., Wagner, S., Frint, P., & Wagner, M. F. (2014). Effects of particle reinforcement and ECAP on the precipitation kinetics of an Al-Cu alloy. In IOP Conference Series: Materials Science and Engineering (Vol. 63, No. 1, p. 012080). IOP Publishing.
5 Harichandran, R., & Selvakumar, N. (2016). Effect of nano/micro B 4 C particles on the mechanical properties of aluminium metal matrix composites fabricated by ultrasonic cavitation-assisted solidification process. Archives of Civil and Mechanical Engineering, 16(1), 147-158.
6 Samuel, M. (2003). Reinforcement of recycled aluminum-alloy scrap with Saffil ceramic fibers. Journal of Materials Processing Technology, 142(2), 295-306.
7 Sabbaghianrad, S., & Langdon, T. G. (2016). Developing superplasticity in an aluminum matrix composite processed by high-pressure torsion. Materials Science and Engineering: A, 655, 36-43.

8 Onoro, J., Salvador, M. D., & Cambronero, L. E. G. (2009). High-temperature mechanical properties of aluminium alloys reinforced with boron carbide particles. Materials Science and Engineering: A, 499(1), 421-426.
9 Sanusi, O., Makinde, D., & Oliver, J. (2012). Equal channel angular pressing technique for the formation of ultra-fine grained structures. South African Journal of Science, 108(9-10), 1-7.
10 Siddique, T. A., Islam, M. T., Kabir, M. S., & Haque, M. N. (2014). Effect of SiCp addition on the indentation hardness of as-cast Al metal matrix composites. Int. J. Innov. Sci. Res, 11, 433-438.
11 Jithin Jose, 2016 studies on mechanical and wear properties of Al7075/ Zircon/ Fly ash hybrid metal matrix composites.
12 Haghighi, R. D., Jahromi, S. J., Moresedgh, A., & Khorshid, M. T. (2012). A comparison between ECAP and conventional extrusion for consolidation of aluminum metal matrix composite. Journal of materials engineering and performance, 21(9), 1885-1892.
13 Alo, O. (2015). Mechanical Properties of Al-Si-SiCp Composites. International Journal of Scientific & Engineering Research, 6(9), 1602-1609.
14 Roshan, M. R., Mousavian, T. R., Ebrahimkhani, H., & Mosleh, A. (2013). Fabrication of Al-based composites reinforced with Al2O3-TiB2 ceramic composite particulates using vortex-casting method. Journal of Mining and Metallurgy, Section B: Metallurgy, 49(3), 299-305.
15 Djavanroodi, F., & Ebrahimi, M. (2010). Effect of die channel angle, friction and back pressure in the equal channel angular pressing using 3D finite element simulation. Materials Science and Engineering: A, 527(4), 1230-1235.
16 Shaeri, M. H., Shaeri, M., Ebrahimi, M., Salehi, M. T., & Seyyedein, S. H. (2016). Effect of ECAP temperature on microstructure and mechanical properties of Al–Zn–Mg–Cu alloy. Progress in Natural Science: Materials International, 26(2), 182-191.
17 Zare, H., Jahedi, M., Toroghinejad, M. R., Meratian, M., & Knezevic, M. (2016). Compressive, shear, and fracture behavior of CNT reinforced Al matrix composites manufactured by severe plastic deformation. Materials & Design, 106, 112-119.
18 Sitdikov, O. S. (2016). Comparative analysis of microstructures formed in highly alloyed aluminum alloy during high-temperature equal-channel angular pressing and multidirectional forging. Inorganic Materials: Applied Research, 7(2), 149-157.
19 Valiev, R. Z., & Langdon, T. G. (2006). Principles of equal-channel angular pressing as a processing tool for grain refinement. Progress in materials science, 51(7), 881-981.
20 Furukawa, M., Horita, Z., Nemoto, M., & Langdon, T. G. (2001). Review: processing of metals by equal-channel angular pressing. Journal of materials science, 36(12), 2835-2843.
21 Prasad, C. V., & Rao, K. M. (2016). A Study On Effect Of Mechanical Properties Of Al-Zro2 Composite By Liquid Routing. IJSEAT, 4(4), 189-192.
22 Šnajdar-Musa, M., & Schauperl, Z. (2013). ECAP-new consolidation method for production of aluminium matrix composites with ceramic reinforcement. Processing and Application of Ceramics, 7(2), 63-68
23 Abbas, A. T., Taha, M. A., Ragab, A. E., El-Danaf, E. A., & Abd El Aal, M. I. (2017). Effect of Equal Channel Angular Pressing on the Surface Roughness of Solid-state Recycled Aluminum Alloy 6061 Chips. Advances in Materials Science and Engineering, 2017.
24 Xu, C., & Langdon, T. G. (2007). The development of hardness homogeneity in aluminum and an aluminum alloy processed by ECAP. Journal of materials science, 42(5), 1542-1550.

25 Chmura, W., & Gronostajski, Z. (2006). Bearing composites made from aluminium and aluminium bronze chips. Journal of materials processing technology, 178(1), 188-193.
26 Ramu, G., & Bauri, R. (2009). Effect of equal channel angular pressing (ECAP) on microstructure and properties of Al–SiC p composites. Materials & Design, 30(9), 3554-3559.
27 Saheb, D. A. (2011). Aluminium silicon carbide and aluminium graphite particulate composites. ARPN Journal of Engineering and Applied Sciences, 6(10), 41-46.
28 Rahman, M. H., & Al Rashed, H. M. (2014). Characterization of silicon carbide reinforced aluminium matrix composites. Procedia Engineering, 90, 103-109.
29 Alaneme, K. K., Bodunrin, M. O., & Awe, A. A. (2016). Microstructure, mechanical and fracture properties of groundnut shell ash and silicon carbide dispersion strengthened aluminium matrix composites. Journal of King Saud University-Engineering Sciences.
30 Fogagnolo, J. B., Ruiz-Navas, E. M., Simón, M. A., & Martinez, M. A. (2003). Recycling of aluminium alloy and aluminium matrix composite chips by pressing and hot extrusion. Journal of Materials Processing Technology, 143, 792-795.
31 Dhanasekaran, S., et al. "SiC and Al2O3 Reinforced Aluminum Metal Matrix Composites for Heavy Vehicle Clutch Applications." Transactions of the Indian Institute of Metals 69.3 (2016): 699-703.
32 Kačmarčik, I., Pepelnjak, T., & Plančak, M. (2012). Solid-state recycling by cold compression of al-alloy chips. Journal for Technology of Plasticity, 37(1), 35-47.
33 Ramnath, B. V., Elanchezhian, C., Jaivignesh, M., Rajesh, S., Parswajinan, C., & Ghias, A. S. A. (2014). Evaluation of mechanical properties of aluminium alloy–alumina–boron carbide metal matrix composites. Materials & Design, 58, 332-338.
34 Abdizadeh, H., Ashuri, M., Moghadam, P. T., Nouribahadory, A., & Baharvandi, H. R. (2011). Improvement in physical and mechanical properties of aluminum/zircon composites fabricated by powder metallurgy method. Materials & Design, 32(8), 4417-4423.
35 Xiong, B. (2015). Optimization of recycling process of die cast aluminum A380 machining chips.
36 Mindivan, H., Cimenoglu, H., & Kayali, E. S. (2009). Production of the composite from 6082 Al alloy chips and fly ash particles by hot pressing. In TMS annual meeting.
37 Reddy, S. P., Ramana, B., & Reddy, A. C. (2010). Compacting Characteristics of Aluminum-10 wt% Fly Ash-Lead Metal Matrix Composites. International Journal of Materials Science, 5(6), 777-783.
38 Muley, A. V., Aravindan, S., & Singh, I. P. (2015). Nano and hybrid aluminum based metal matrix composites: an overview. Manufacturing Review, 2, 15.
39 Patil, B. V., Chakkingal, U., & Kumar, T. P. (2010). Influence of Outer Corner Radius in Equal Channel Angular Pressing. *World Academy of Science, Engineering and Technology*, *62*, 714-720.

Improvement of Solid-Gas Interaction in Fluidized Bed Systems via Secondary Air-Injection

Mohd Faizal Mohideen Batcha[*], *Sulastri Sabudin and Jamal Hazri Zakaria*

Centre for Energy and Industrial Environment Studies (CEIES), Faculty of Mechanical and Manufacturing Engineering, Universiti Tun Hussein Onn Malaysia, 86400 Parit Raja, Batu Pahat, Johor, Malaysia

Abstract. Since the invention of fluidized bed systems to augment chemical reactions, numerous innovations were introduced to the system. These innovations result-in substantial improvement to the overall efficiency apart from resulting the fluidized bed system to become more compact. One of the innovations were the introducing of secondary air inside the bed so that a secondary motion in lateral direction can increase the amount of solid-gas interaction. This paper reviews various techniques proposed by researchers in providing secondary air inside the fluidized bed system. In general, four techniques were common in which the bed was named after: vortexing bed, cyclonic bed, rotating bed and swirling bed. Though resemble similarity, each of these beds were unique and comes with different advantageous. In conclusion, a fluidized bed system can be improved by the secondary motion in the bed which provides opportunities for further innovations.

1 Introduction

Fluidization is the operation by which solid particles are transformed into fluid-like state through suspension in a gas or liquid. This method of contacting between solid and fluid has some unusual characteristics, and fluidization engineering puts them in to good use [1]. The basic mechanism of a fluidized bed can be seen simply as fluid percolation through particle interstices, in which particles begin to exhibit fluid-like characteristics upon experiencing sufficient drag force by the fluid.

Since the last 80 years, fluidized bed technology has served many industries involving solid – fluid contact due to their apparent advantages such as excellent mixing and homogeneity in heat and mass transfer. Today, fluidized bed systems are widely used in processes such as combustion and gasification of solid fuels, biomass processing, heat recovery, drying of particles, particle heating, oxidation of metal ores, metal surface treatments and catalytic and thermal cracking. The system has made substantial technological advancements ever since and become mature. Nevertheless, the energy crisis and growing industrial needs have motivated continuous initiatives to further improve the existing

[*] Corresponding author: mfaizal@uthm.edu.my

fluidized bed technology. Different designs of the system components, particularly the distributor, and their effect on bed performance are some of the examples.

Among the well-established variants in fluidized bed systems, such as the centrifugal fluidized bed, tapered fluidized bed, spouted bed, rotating bed and vibrating bed, there is a group of fluidized beds which operate based on swirling (or vortexing or cyclonic) principles. Although the terms swirling, vortexing and cyclonic may seem to address the same system, there are several salient features distinguishing them from each other.

2 Fluidized bed distributor

In a fluidized bed system, the distributor may be regarded as the 'heart' of the system. A good design of gas distributors is vital to ensure satisfactory performance of the fluidized bed system as well as success in industrial applications which adopt the fluidization technique. The large numbers of publications on the distributor design since the last three decades prove that there still are unresolved questions in gas distributor designs in actual operation. According to Geldart and Baeyens [2], some of the basic requirements of a gas distributor are uniform distribution of gas into the fluidized bed, operation at as low pressure drop as possible to minimize power consumption, sufficient strength to withstand both thermal and mechanical stresses, prevention of particle drainage, reduction in particle attrition and prevention of plate erosion. These requirements may be contradictory and their relative importance depends on the process needs.

Although distributors with low pressure drop are preferred to reduce energy consumption during processing, it should be noted that distributors having too low pressure drops may result-in poor fluidization. This is due to more flow of process gas taking place at the zone with lowest flow resistance. As a result, the bed may be permanently defluidized, whilst in other parts semi-permanent spouts or channels may be occurring [2]. Therefore, the distributor pressure drop has to be large enough to overcome the small local pressure disturbances of the fluidized bed.

Many early researchers proposed a ratio between distributor pressure drop and bed pressure drop ($\Delta P_d/\Delta P_b$) as design guidelines. Agarwal et al. [3] proposed that ΔP_d should be approximately 10% of the bed pressure drop and never less than about 3500 Pa (350 mmH_2O), while Siegel [4] proposed a minimum value for $\Delta P_d/ \Delta P_b$ ranging from 12% to 24%. In more recent industrial designs by Gupta and Sathiyamoorthy [5] and Mujumdar [6], pressure drop ratio values of not less than 25% to 50% have been found successful. In line with the rapid progress in this area over the years, a large variety of distributors has been studied and reported.

Investigation on the influence of various distributor designs, namely perforated plate, punched plate and Dutch weave mesh on fluidized bed dryer hydrodynamics were carried out by Wormsbecker et al. [7] and Wormsbecker and Pugsley [8]. Particles in the transition region between B and A (according to Geldart's classification) were tested at various bed loadings and superficial velocities. The distributors were assessed in terms of drying kinetics and pressure fluctuations. The authors found that the punched plate has the best performance as a result of lateral mixing inside the bed, due to swirling flow of fluidizing air. This swirling flow was generated by the horizontal gas jets from the hooded openings of the punched plate.

The effect of air distributor on fluidization hydrodynamics was also studied by Son et al. [9] and Ciesielczyk [10] for large particles. The former conducted experimental work using perforated plate distributors with five different fractions of open area, ranging from 0.9% to 3.7% using 1 mm glass beads as bed material, while the latter proposed a novel distributor consisting of a perforated grate with conical plenum with a solid cone at the bottom for fluidizing various biomass chunks (equivalent diameter ranging from 6.6 mm to 9.2 mm).

Both authors plotted the pressure-velocity curves and concluded successful fluidization though they reported different regimes of operation.

The distributor's influence on the uniformity of fluidization and selection of the aspect ratio at the critical fluidization quality was reported by Sathiyamoorthy and Horio [11]. The authors conducted experiments on two types of multi-orifice distributor with three bed materials having size distribution between 70 – 161 μm and proposed a model equation for predicting critical $\Delta P_d / \Delta P_b$ ratio. The authors claimed that the aspect ratio values were important in selecting a distributor and operating velocity for shallow beds.

Apart from the stationary distributors as mentioned above, Sobrino et al. [12] proposed a rotating distributor with the aim to improve solid mixing as well as achieving fluidization for a shallow bed. The authors tested Geldart B silica sand with various bed heights on a perforated plate distributor with only 1% open area ratio. With the increase of rotational speed of the distributor, the authors reported a decrease in minimum fluidization velocity while the fluidization quality was left unaffected. Perhaps due to the complicated design, the authors discontinued the study and later published their findings on the effect of perforated and bubble-cap distributor on turbulent fluidized bed [13].

Recently, Aworinde et al. [14] proposed a novel gas distributor plate with many upward-facing nozzles incorporated with removable helical coils. Their aim was to swirl the fluidizing gas prior entering the bed. They've conducted experiments using magnetic resonance imaging and concluded that the swirling flow induced by the helical coils significantly improved the bed's fluidization quality despite some increase in the distributor pressure drop.

First patented by Dodson [15] and commercialized by Torftech Ltd since the 1980's, the Torbed® reactor (acronym from Toroidal Bed), consists of angled blades in an annular form at the reactor bottom. When a process gas stream is forced to pass through openings of the blades, the pressure head is converted in the blade array to a velocity head. The resultant high velocity jets keep the bed particles in suspension and rotating toroidally [16].

3 Vortexing/cyclonic/rotating/swirling beds

As fluidized bed systems gained popularity for various industrial applications, the amount of research work also steadily increased to further improve its performance. One of the aspects that has been given particular attention is the lateral mixing inside the bed. These resulted in various innovations to the existing beds, such as introducing secondary air injection at the freeboard of the bed, modifications of the bed geometry (such as the tapered bed and conical bed), utilization of mechanical agitation or stirrer inside the bed apart from introducing various types of distributors. This review addresses some of these efforts but limits its discussion to the fluidized bed systems which resemble the swirling beds. While the terms 'swirling', 'vortexing', 'cyclonic' and 'rotating' indicate a similarity, there are some fundamental differences between them which distinguishes their hydrodynamic characteristics.

3.1 Vortexing fluidized beds

The vortexing fluidized bed was first patented by Sowards [17] in 1977. This system is generally designed as a cylindrical chamber with significant height-to-diameter ratio, with a combination of conventional fluidized bed (in the bottom part of the reactor) with vortexing gas-solid flow in its freeboard [18-22]. The latter is achieved by injecting secondary air tangentially through nozzles, with the aim of increasing the residence time of both gas and solid in the freeboard while reducing elutriation of fine particle from the fluidized bed. The flow rate of the secondary air in vortexing beds is usually a significant fraction of the primary air fed from the bottom of the bed.

3.2 Cyclonic fluidized beds

Cyclonic fluidized beds usually have moderate height-to-diameter ratio, typically 2.5 - 3.0 [23]. This bed possesses intricate hydrodynamics where the primary air enters the bed tangentially from the top while the secondary air enters from the bottom [24]. Vortex rings are employed to avoid large solids from ascending with secondary air from the bottom as well as dividing the bed into upper and lower parts. However, the authors reported the presence of bed agglomeration due the formation of char during combustion and hence introduced stirring blades at the bed distributor as a countermeasure [25] apart from carrying the combustion study without inert particles [26].

3.3 Rotating fluidized beds

Unlike both types of fluidized bed mentioned above, the rotating fluidized bed actually fluidizes the bed via generating a high centrifugal force field, which was reported by Nakamura et al. [27] to enable fluidizing fine particles which have high inter-particle cohesive forces. The system consists of a porous cylindrical gas distributor inside a plenum chamber, both axisymmetric, in which the former rotates inside the latter. This rotational motion forces the bed particles towards the porous wall by centrifugal force while the fluidizing gas flowing inwards balances this force. In a more recent design, Wilde and Broqueville [28] and Wilde [29] proposed a rotating fluidized bed but with tangential injection of fluidizing air via multiple gas inlet slots. While being successful in reducing particle entrainment, the authors reported channeling and slugging of the bed due to low bed loading. The system was further improved by Eliaers et al. [30] for drying of biomass.

3.4 Swirling fluidized beds

As the name implies, the swirling fluidized bed creates swirling motion of the bed, which is achieved either by injecting air using multiple tangential inlets at the bottom of the bed [31] or by using annular-blade type air distributors [32-35]. While the tangential injection of air creates swirling in the former, the inclined distributor blades create fluidization and swirling of the bed in the latter [33-35]. The bed is also known as the toroidal bed (Torbed®) owing to the toroidal motion of the particles superimposed on the swirling motion. Recently, the annular-blade distributor was successfully integrated with conical bed and developed as biomass combustors by [36-38]. Although the bed was conical, the authors designated their bed also as swirling fluidized bed.

The swirling fluidized bed is however limited only to shallow bed operations as the bed experiences rapid lateral momentum attenuation. Apart from this, the bed requires the use of a centre body to eliminate possible creation of a dead zone in the center of the bed. As a result, the effective bed area is an annular region. At higher operating velocities, due to the high centrifugal force, the bed is thrown outwards and some valuable amount of fluidizing gas bypasses the bed at the inner part of the annular region.

4 Conclusion

Several conclusions can be drawn from this review:

a) Numerous distributors have been designed, developed and studied for utilization in specific applications of fluidized bed systems. The efforts were focused mainly on achieving good air distribution inside the bed without proportionate increase in pressure drop due to the distributor.

b) Although similar, the vortexing, cyclonic, rotating and swirling fluidized bed have fundamental differences and unique hydrodynamics. The existence of these beds is proof of their advantage over conventional fluidized beds.

c) Several important parameters of interest among researchers in characterizing the improvements of the fluidized bed system were U_{mf}, ΔP_d, ΔP_b and the regime of operation which are normally presented in the form of pressure-velocity curves.

Acknowledgement

The authors would like to extend their deepest gratitude to Universiti Tun Hussein Onn Malaysia, especially the Office of Research, Innovation, Commercialization, Consultancy Management (ORICC) for funding and facilitating this research via contract research fund, vote no. U525.

References

1. D. Kunii, O.Levenspiel, *Fluidization Engineering*, 2nd ed., Butterworth-Heinemann, (1991)
2. D. Geldart, J. Baeyens, The design of distributors for gas-fluidized beds, Powder Technology, **42(1)**, pp. 67-78 (1985)
3. J.C. Agarwal, W.L. Davis, D.T. King, Fluidized-bed coal dryer, Chem. Eng. Progr. **58(11)**, pp. 85-90 (1962)
4. R. Siegel, Effect of distributor plate-to-bed resistance ratio on onset of fluidized-bed channeling, AIChE Journal, **22(3)**, pp. 590-592 (1976)
5. C.K. Gupta, D. Sathiyamoorthy, *Fluid Bed Technology in Materials Processing*, CRC Press, New York (1998)
6. A.S. Mujumdar, *Handbook of Industrial Drying*, 3rd ed., CRC Press: Boca Raton (2007)
7. M. Wormsbecker, T.S. Pugsley, H. Tanfara, The influence of distributor design on fluidized bed dryer hydrodynamics, 12th Int. Conf. on Fluidization, paper 100 (2007)
8. M. Wormsbecker, T.S. Pugsley, Distributor induced hydrodynamics in a conical fluidized bed dryer, Drying Technology, **27(6)**, pp. 797-804 (2009)
9. S.Y. Son, D.H. Lee, G.Y. Han, D.J. Kim S.J. Sim, S.D. Kim, Effect of air distributor on the fluidization characteristics in conical gas fluidized beds, Korean Journal of Chemical Engineering, **22(2)**, pp. 315-320 (2005)
10. W. Ciesielczyk, Novel gas distributor for fluidized bed drying of biomass, Drying Technology, **27(12)**, pp. 1309-1305 (2009)
11. D. Sathiyamoorthy, M. Horio, On the influence of aspect ratio and distributor in gas solid fluidized beds, Chemical Engineering Journal, **93(2)**, pp. 151-161 (2003)
12. C. Sobrino, J.A. Almendros-Ibáñez, D. Santana, M. De Vega, Fluidization of Group B particles with a rotating distributor, Powder Technology, **181(3)**, pp. 273-280 (2008)

13. C. Sobrino, N. Ellis, M. de Vega, Distributor effect near the bottom region of turbulent fluidized bed, Powder Technology, **189(1)**, pp. 25-33 (2009)
14. S.M. Aworinde, D.J. Holland, J.F. Davidson, Investigation of a swirling flow nozzle for a fluidised bed gas distributor, Chemical Engineering Science, **132**, pp. 22-31 (2015)
15. C.E. Dodson, Torftech Limited, *Apparatus for processing matter in a turbulent mass of particulate material*, US Patent 4,479,920, (1984)
16. J. Shu, V.I. Lakshmanan, C.E. Dodson, Hydrodynamic study of a toroidal fluidized bed reactor, Chemical Engineering and Processing, **39(6)**, pp. 499-506 (2000)
17. N.K. Sowards, Energy Products of Idaho, *Low pollution incineration of solid waste*, US Patent No. 4,060,041, (1977)
18. S. Nieh, G. Yang, Particle flow pattern in the freeboard of a vortexing fluidized bed, Powder Technology, **50(2)**, pp. 121-131 (1987)
19. J. Zhang, S. Nieh, Swirling, reacting, turbulent gas-particle flow in a vortex combustor, Powder Technology, **112(1)**, pp. 70-78 (2000)
20. C.S. Chyang, K.C. Lo, K.L. Wang, Performance evaluation of a pilot scale vortexing fluidized bed combustor, Korean Journal of Chemical Engineering, **22 (5)**, pp.774-782 (2005)
21. F. Duan, C.S. Chyang, S.H. Hsu, J. Tso, Combustion behavior and pollutant emissions of batch fluidized bed combustion, Journal of the Taiwan Institute of Chemical Engineers, **44(6)**, pp. 1034-1038 (2013)
22. F. Duan, C. Chyang, Y. Chin, J.Tso, Pollutant emission characteristics of rice husk combustion in a vortexing fluidized bed incinerator, Journal of Environmental Sciences, **25(2)**, pp. 335-339 (2013)
23. R. Kaewklum, V.I. Kuprianov, Experimental studies on a novel swirling fluidized-bed combustor using an annular spiral air distributor, Fuel, **89(1)**, pp. 43-52 (2010)
24. T. Madhiyanon, N. Piriyarungroj, S. Soponronnarit, Cold flow behaviour study in novel cyclonic fluidized bed combustor (Ψ-FBC), Energy Conversion and Management, **49(5)**, pp. 1202–1210 (2008)
25. T. Madhiyanon, A. Lapirattanakun, P. Sathiruangsak, S. Soponronnarit, A novel cyclonic fluidized bed combustor (Ψ-FBC): Combustion and thermal efficiency, temperature distributions, combustion intensity and emission of pollutants, Combustion and Flame, **146(1)**, pp. 232-245 (2006)
26. T. Madhiyanon, P. Sathitruangsak, S. Soponronnarit, Co-combustion of rice husk with coal in a cyclonic fluidized-bed combustor (Ψ-FBC), Fuel, **88(1)**, pp. 132-138 (2009)
27. H. Nakamura, T. Tokuda, T. Iwasaki, S. Watano, Numerical analysis of particle mixing in a rotating fluidized bed, Chemical Eng. Science, **62(11)**, pp. 3043-3056 (2007)
28. J. De Wilde, A. de Broqueville, Experimental investigation of a rotating fluidized bed in a static geometry, Powder Technology, **183(3)**, pp. 426-435 (2008)
29. J. De Wilde, Gas–solid fluidized beds in vortex chambers, Chemical Engineering and Processing: Process Intensification, **85**, pp. 256-290 (2014)
30. P. Eliaers, J.R. Pati, S. Dutta, J. De Wilde, Modeling and simulation of biomass drying in vortex chambers, Chemical Engineering Science, **123**, pp.648-664 (2015)
31. S.H. Kumar, D.V.R. Murthy, Minimum superficial fluid velocity in a gas–solid swirled fluidized bed, Chemical Engineering and Processing: Process Intensification, **49(10)**, pp. 1095-1100 (2010)

32. F. Ouyang, O. Levenspiel, Spiral distributor for fluidized beds, Industrial & Engineering Chemistry Process Design and Development, **25(2)**, pp. 504-507 (1986)
33. M.F. Mohideen, S. Md Seri, V. Kumar, V.R. and Raghavan, Experimental studies on a swirling fluidized bed with annular distributor, Journal of Applied Sciences, **Vol. 11, No. 11**, pp. 1980-1986 (2011)
34. M.F. Mohideen, S.M. Seri, V.R. Raghavan, Fluidization of Geldart Type-D Particles in a Swirling Fluidized Bed, In Applied Mechanics and Materials, **Vol. 110**, pp. 3720-3727, Trans Tech Publications (2012)
35. M.F. Mohideen, M.F., S.A. Sulaiman, V.R. Raghavan, Hydrodynamics of oil palm frond in a swirling fluidized bed dryer, In Applied Mechanics and Materials, **Vol. 117**, pp. 1829-1833, Trans Tech Publications (2012)
36. V.I. Kuprianov, R. Kaewklum, K. Sirisomboon, P. Arromdee, S. Chakritthakul, Combustion and emission characteristics of a swirling fluidized-bed combustor burning moisturized rice husk, Applied Energy, **87(9)**, pp. 2899-2906 (2010)
37. P. Arromdee, V.I. Kuprianov, A comparative study on combustion of sunflower shells in bubbling and swirling fluidized-bed combustors with a cone-shaped bed, Chemical Engineering and Processing: Process Intensification, **62**, pp. 26-38 (2012)
38. V.I. Kuprianov, R. Kaewklum, S. Chakritthakul, Effects of operating conditions and fuel properties on emission performance and combustion efficiency of a swirling fluidized-bed combustor fired with a biomass fuel, Energy, **36(4)**, pp.2038-2048 (2011)

DOI: 10.1051/matecconf/201713500015

Performance measurement of a new concept reciprocating piston expander (RPE) using a newly developed small-scale dynamometer unit

MNA Mohamad[1],*, MF Hushim[2], WNAW Muhammad[1], A Razali[1], A Khalid[2], and MF Zakaria[3]

[1] Faculty of Mechanical and Manufacturing Engineering, Universiti Tun Hussein Onn Malaysia, 86400 Parit Raja, Batu Pahat, Johor, Malaysia.

[2] Faculty of Engineering Technology, Universiti Tun Hussein Onn Malaysia, 86400 Parit Raja, Batu Pahat, Johor, Malaysia.

[3] Faculty of Electronic and Electric Engineering, Universiti Tun Hussein Onn Malaysia, 86400 Parit Raja, Batu Pahat, Johor, Malaysia.

E-mail: anuarm@uthm.edu.my

Abstract. This paper presents the progress of a small-scale dynamometer prototype development for performance measurement of a reciprocating piston expander (RPE). Since the available dynamometer systems in the market are limited to specific applications that require for the customization, their price normally very expensive. Since the current study on the RPE required a dynamometer unit, therefore, a new and cheaper dynamometer prototype that was suitable for RPE application has been developed. Using air as RPE working fluid, a case study has been carried out to measure its performance at different inlet fluid conditions, i.e., within 20°C–140°C and 3–5 bars. The results observed that the performance of RPE was proportionally increased to the increased of inlet fluid pressure and temperature. The maximum brake power produced was 27 Watt when the RPE operated at 140°C, 5 bars and the speed of 820 rpm. It also revealed that the changes in the pressure of inlet fluid can give significant change on the performance of the RPE due to its direct relation to the RPE actual rotating force. Although the RPE and dynamometer seem good being adapted to each other, both of them require some improvements to ensure both systems well operated and reliable.

1. Introduction

Most of the environmental problems faced by the world today caused by vast consumption of fossil fuel in order to meet the current energy demands [1]. Although there are efficient energy conversion systems have investigated and introduced, the

*Corresponding author: anuarm@uthm.edu.my

fuel consumptions can no longer be reduced when the design points have been met [2], there are still the methods to increase the overall efficiency of the system with fewer fuels and lessen the pollutions. Besides, the introduction of cleaner fuels and replacement or modification of the energy conversion systems being a solution to meet the above objectives, recovering the leftover energy especially in terms of wasted heat into a useful energy has been considered and proposed as the best solution [3],[4]. Among the heat recovery systems, recovering through an Organic Rankine Cycle (ORC) system is a viable method since the ability of this system to recover the most abundant amount of heat at low to medium temperature heat [5] and for small-scale applications [6].

Although the ORC system seems feasible, it has a limitation on its critical component, i.e., its expander as it is normally designed for certain operating conditions. In order to evaluate the performance of the ORC systems, especially during the development of its expander unit, there is a system that is used to measure the power output delivered by the expander at a predetermined condition. Most of the power driving unit such as engines, motors, transmissions, generators, turbines, and expanders require a dynamometer system to measure the power, torque, and speed from the rotating force of the driving unit by introducing variable loading conditions on the driving unit tested across the range of engine speeds and durations [7]. When a driving unit operates, it generates the potential and kinetic energy thus rotates its drive shaft. As a load absorbing unit, the dynamometer absorbs the energy through a braking action, while at the same time it measures the performance of the driving unit. Although the dynamometer system is commercially available equipment, its price is very expensive since the system has been customized to suit the requirement for the overall test system. Some of the ready-made dynamometers such as vehicle or engine dynamometers are designed for a specific purpose, i.e., to measure the performance of the engines that operate at the specific operational range, and of course, these dynamometers are bulky and not suitable for small scale driving unit measurement.

Since the dynamometer is a crucial system while developing the expander for the performance measurement, with the limited resources and funds, it is wise to develop own dynamometer system that suitable and can operate at acceptable conditions. Therefore, this paper presents the progress on developing a prototype of a new small-scale type dynamometer for performance measurement of the RPE unit. To ensure the operation of this prototype met the requirements, a case study has been carried out by measuring the actual RPE performance at different operating conditions. The weaknesses and improvements ideas for the dynamometer were also discussed at the end of this paper.

2. Design and development of a small-scale dynamometer prototype

Nowadays, there are many types of dynamometers used in the industries, research institutes, service agencies, etc. with some few advantages and disadvantages for each of them based on its counterparts. Although they look different, their designs are originated from the commonly classified dynamometer types such as fluid (water or air) brake, eddy current, prony brake, and cradled generator type dynamometers [8]. Here, the prony brake dynamometer is considered as the simplest design for measuring power output (brake power) and could be potentially developed by anyone at low cost compared to the other type dynamometers. Based on prony brake dynamometer, in this study, the prototype of a small-scale dynamometer that has been developed and its design is shown in figure 1.

Based on prony brake principle, the rotating force (brake power) of a driving equipment is measured by applying the frictional load in attempt to stop the rotation of the shaft equipment by mean of braking action by the shaft supporting arm that is attached to the load. The supporting arm is normally fixed with the braking mechanism (normally at one end) such as brake block or pad that is attached to the shaft. When the load or force gradually applied on the other end side of the arm, the brake mechanism gives the friction to the rotating shaft until as it tries to stop the shaft. During this condition, the brake power is calculated as

$$P_b = 2\pi NT \tag{1}$$

where, N is angular velocity in revolution per second (rps), and
T is the torque applied to the shaft (Nm).

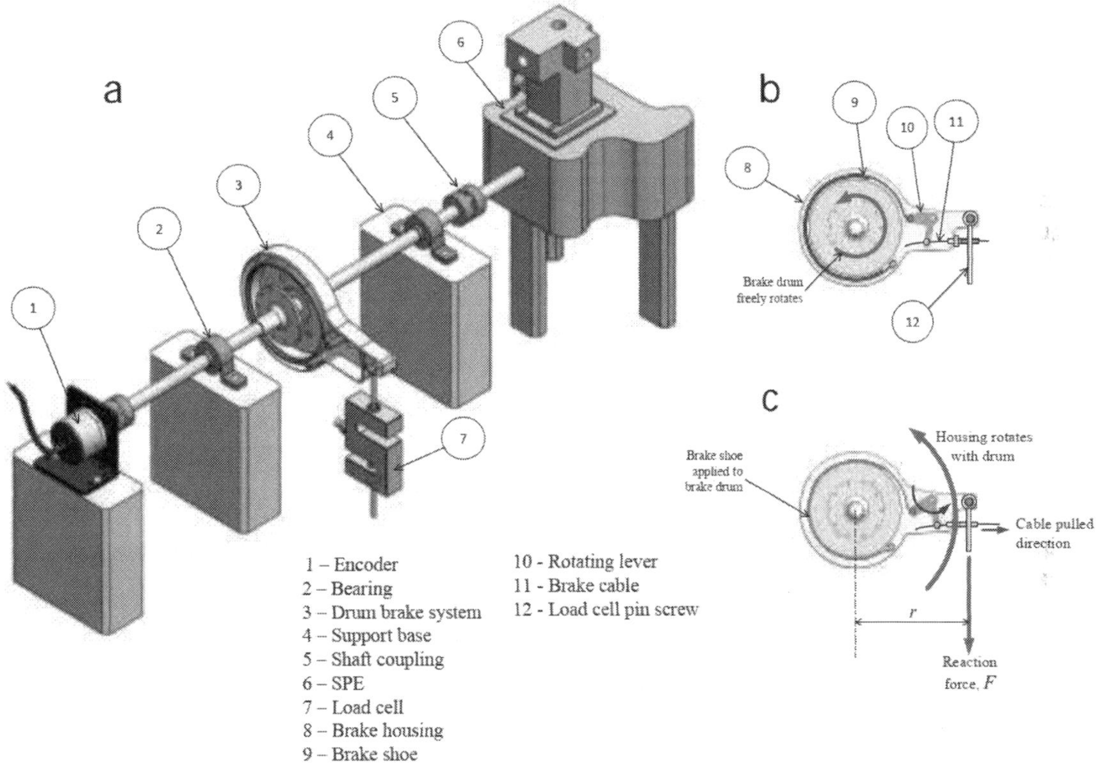

Figure 1. (a) Design layout of a small-scale dynamometer prototype, (b) drum brake condition before load applied and (d) drum brake after load applied.

This type dynamometer was utilized to measure the performance by absorbing the brake power produced by the RPE. From the figure 1, the RPE rotating shaft transferred its power to the dynamometer through the shaft coupling to the drum brake system, i.e., the critical component that enables whether the load has been applied (or not) to the shaft. If the load was not applied, then the brake drum, i.e., the component that has been fixed to the shaft was freely rotates as shown in figure 1(b). The condition if the load applied to the shaft happen when the brake shoe has been forced to rub the drum which caused the entire drum brake system especially its housing to rotate together with the drum. The more force applied to the drum, the more load was given to the RPE shaft, thus the more torque can be measured by multiplying the reaction force, F to the distance, r of the force and the center of the shaft. The torque, T is written as

 DOI: 10.1051/matecconf/201713500015

$$T = Fr \quad (2)$$

3. Case study: Performance measurement of the RPE

3.1. The reciprocating piston expander (RPE)

The expander that has been investigated in this study was a newly developed expander by modifying a few parts of a small two-stroke grass cutter engine. The spark plug port has been replaced with the inlet port of the RPE while the inlet and scavenging ports of the engine have been closed by replacing the original cylinder block of the engine with the new block for the RPE. Therefore, the operational characteristics of the RPE were different from the engine with high pressure working fluid entered the RPE through the spark plug port from the top and the expansion process occurred not by the combustion process but due to high-pressure fluid pushed the piston downward. The high-pressure fluid entrance to the RPE was controlled by the valve control switches which was also introduced in [9] to ensure the opening and closing of the solenoid valves happened at the right timing. Figure 2 shows the conversion of the engine into the RPE unit.

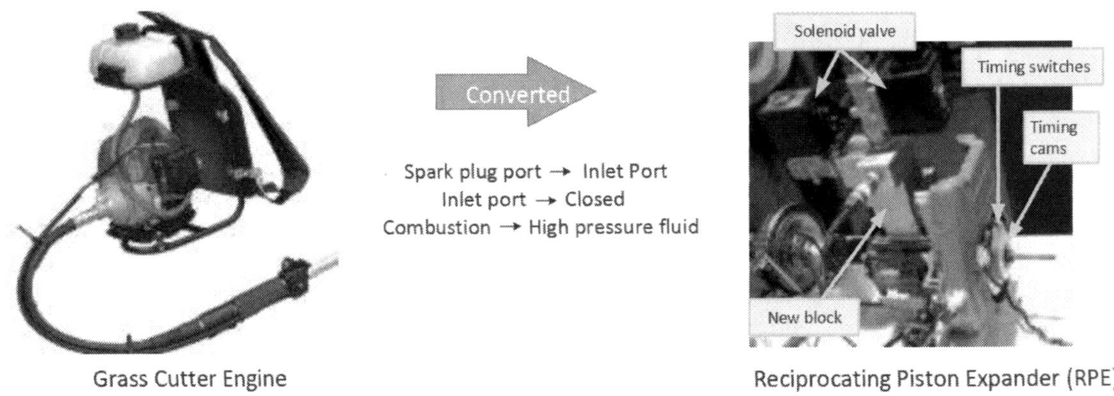

Figure 2. Conversion of the small two-stroke engine into RPE.

3.2. Performance test of the RPE

The experimental test system for the RPE performance measurement is shown in figure 3. In this study, there were five fluid inlet pressures, (i.e., 3, 3.5, 4, 4.5 and 5 bars) with five temperature conditions (i.e., 20°C, 80°C, 100°C, 120°C and 140°C) have been tested for the RPE performance. Here, the compressed air was used as working fluid to operate the expander. The air was supplied from the air compressor unit through a boiler before entering the RPE.

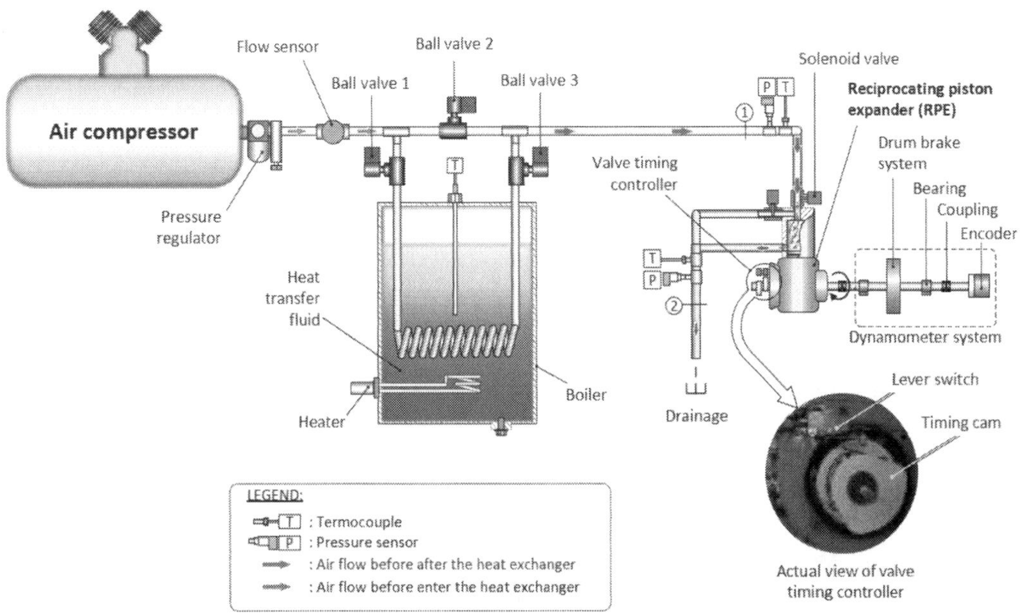

Figure 3. Experimental test system for RPE performance measurement.

In the only case where the RPE operation condition was at room temperature (20°C), the air bypassed the boiler when the ball valves 1 and 3 closed, but the ball valve 2 was kept open. For the other cases (80°C – 140°C), the air flowed through ball valve 1, the boiler, the ball valve 3 and the RPE, when the valve 3 was closed but both valves 1 and 3 were opened. The thermodynamics properties of the air at RPE inlet and outlet were measured by thermocouples and pressure sensors, while the performance (i.e., rotational speed, force, torque, and power) were measured through the dynamometer system. This figure also showed the valve control system that was used in this study by adapting the lever switches and timing cams in order to control the right opening and closing angle of the inlet and exhaust valves. In this study, the best timing for inlet and exhaust valves have been determined as 80° after top dead center (TDC) and 100° after bottom dead center (BDC), respectively.

4. Results and discussions

The study observed that the performance of the RPE was increased with the increase of the pressure and temperature of the inlet fluid. The rotational speed varied at the lowest range within 440–586 rpm when the RPE operated at room temperature (20°C) and 3 bars of pressure, while at the highest temperature (140°C) and pressure of 5 bars, the speed varied at the highest range within 820–1075 rpm. Figure 4 shows the rotational speed of the RPE when it was tested at no load and full load conditions with different inlet pressure and temperature.

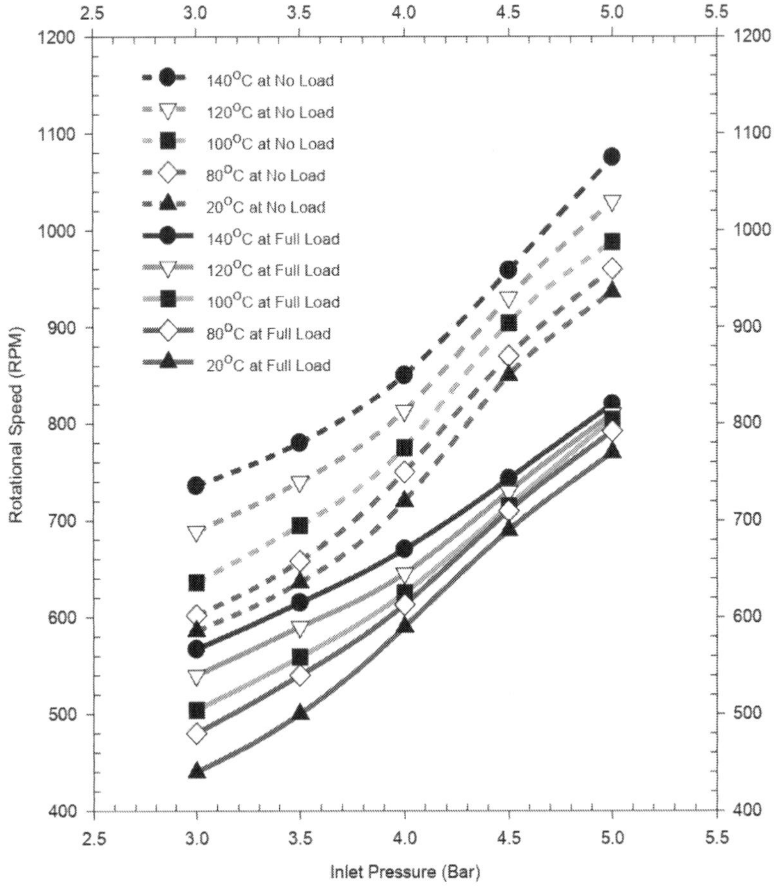

Figure 4. The rotational speed of the RPE at no load and full load conditions with different inlet pressure and temperature conditions.

The same phenomenon has shown by the brake power of the RPE as it was also increased with the increase of the fluid inlet pressure and temperature. The result for the brake power of the RPE shown in figure 5 as it was RPE tested at full load conditions. At the lowest brake power recorded was 6 Watt at 3 bars and 20°C, and the highest power delivered by the RPE was 27 Watt, recorded at 5 bars and 140°C. The RPE speed at both conditions was 440 rpm and 820 rpm, respectively. It was also observed here that the effect of pressure variations gave significant changes in the RPE brake power conditions compared to the temperature variations due to the direct effect of the pressure to the force applied to produce the brake power.

DOI: 10.1051/matecconf/201713500015

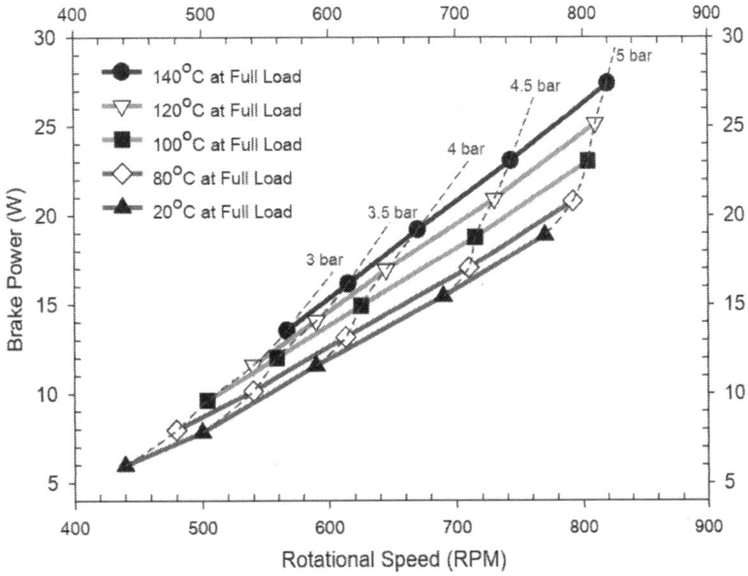

Figure 5. RPE brake power against rotational speed at full load conditions with different inlet temperature and pressure.

Fluid temperature is known to play important roles on the condition of heat capacity and the enthalpy of the fluid, therefore, when the fluid temperature has been increased, the heat capacity of the fluid and its enthalpy also increased, thus, increased the internal work of RPE. In the case of the fluid inlet pressure, the increased in the pressure directly increased the force acted on the RPE piston (Force, $F = \text{Pressure}, P \times \text{Area}, A$) and therefore increased the brake power delivered by the RPE. These reaction mechanisms were closely related to the RPE speed since the internal work and brake power were directly proportional to the rotational speed, and this explains the phenomenon of the increased in fluid inlet temperature and pressure caused the increased in the rotational speed of the RPE as being highlighted above. The study also proved that the developed small-scale dynamometer prototype was able to measure the performance of the RPE although it has been tested up to the maximum conditions, i.e., at full load, and at the highest temperature (140°C) and pressure (5 bars).

5. Conclusion

This study revealed that the performance of RPE was directly proportional to the working fluid inlet pressure and temperature, and proved that the developed small-scale dynamometer prototype was successfully measured the performance of the RPE for the entire testing conditions, even it is still under development. From the study, it was observed that the highest brake power produced by the RPE was 27 Watt when it operated at 140°C and 5 bars of inlet fluid, at a rotational speed of 820 rpm. The pressure of the inlet fluid can give significant changes on the performance of the RPE due to its direct relation to the RPE actual rotating force.

For the future work, it is strongly suggested that the current valve timing control system should be replaced by the electronic control system that is fast enough to react in order to ease the study especially in terms of valve timing (intake and exhaust). The dynamometer prototype that has been developed in this study requires a comprehensive calibration work to ensure to measurement can be done more precisely according to the published standard for the force and torque measurement. Although the dynamometer prototype investigated here was mainly used for RPE performance measurement, this not the only limitation of this dynamometer application. The dynamometer can also be used to measure the performance of any other power generation devices, provided that the power generated is in the form of rotating shaft.

6. Acknowledgements

Authors would like to acknowledge the Ministry of Higher Education Malaysia (MOHE) for the financial support given under the Research Acculturation Grant Scheme (RAGS: R056) and Universiti Tun Hussein Onn Malaysia for the facilities and place provided for this research project. Authors also would like to express thank to all individuals and groups for ideas, guidance and assistance contributed during the completion of this research.

7. References

[1] Othman M F, Adam A, Naja G, and Mamat R, 2017, "Green fuel as alternative fuel for diesel engine : A review," *Renew. Sustain. Energy Rev.*, vol. **80**, no. May, pp. 694–709.

[2] Penkuhn M, and Tsatsaronis G, 2017, "A decomposition method for the evaluation of component interactions in energy conversion systems for application to advanced exergy- based analyses," *Energy*, vol. **133**, pp. 388–403.

[3] Quoilin S, Declaye S, Tchanche B F, and Lemort V, 2011, "Thermo-economic optimization of waste heat recovery Organic Rankine Cycles," *Appl. Therm. Eng.*, vol. **31**, no. 14–15, pp. 2885–2893.

[4] Cavazzini G, Bari S, Pavesi G, and Ardizzon G, 2017, "A multi- fluid PSO-based algorithm for the search of the best performance of sub-critical Organic Rankine Cycles," *Energy*, vol. **129**, pp. 42–58.

[5] Zhang X, Wu L, Wang X, and Ju G, 2016, "Comparative study of waste heat steam SRC , ORC and S-ORC power generation systems in medium-low temperature Organic Rankine cycle," *Appl. Therm. Eng.*, vol. **106**, pp. 1427–1439.

[6] Chang J, Hung T, He Y, and Zhang W, 2015, "Experimental study on low-temperature organic Rankine cycle utilizing scroll type expander," *Appl. Energy*, vol. **155**, pp. 150–159.

[7] Batzer R E, and D D Frey, 2011, "Design and Construction of a Dynamometer," Massachusetts Institute of Technology.

[8] Menon S K, 2006, "Performance Measurement and Scaling in Small Internal Combustion Engines," University of Maryland.

[9] Mohamad M N A, Manshoor B, Hushim M F, Amirnordin S H, and

Muhammad W N A W, 2016, "Development of twin piston expander with solenoid valve for organic rankine cycle," *ARPN J. Eng. Appl. Sci.*, vol. **11**, no. 18, pp. 11154–11158.

Prelimenary Investigation of Parametric Optimization on Turning Process Using UVAT Tool Holder Based on Taguchi Method

Haris Rachmat[1,2], *Mohd Rasidi* Ibrahim[1], and *Sulaiman* Hasan[1]

[1]Premach FKMP University Tun Husein Onn Malaysia, Batu Pahat 8400 Malaysia.
[2]School of Industrial Engineering Telkom University, Bandung, Indonesia.

Abstract. Determining of cutting parameters on cutting forces and cutting temperatures is critical during the turning operation. Most of the research is focused on cutting using special toolholder for conventional turning. In this study, the main objective is to study effect of cutting speed, feed rate and depth of cut on cutting force and cutting temperature of alumunium alloy (Al6061-T6) in turning operation. Two different design of UVAT toolholders (D1 and D2) were conducted. Experiments have been applied using Taguchi's experimental design technique. An orthogonal array, signal to noise ratio, and analysis of variance are employed to investigate cutting characteristics of each design. Experimental results show that cutting speed is the most significant machining parameter on cutting forces and cutting temperatures. Continously were followed by feed rate and depth of cut. The lowest cutting force and cutting temperature were obtained when the setting of cutting parameters are cutting speed is 50 m/min, feed rate is 0.05 mm/nin and depth of cut is 0.25 mm.

1 Introduction

In modern era, the adoption of new technologies in various sectors is highly required. Ultrasonic vibration is one of the new technologies implemented in various manufacturing areas such as ultrasonic welding, ultrasonic assisted turning, ultrasonic assisted drilling, ultrasonic assisted milling etc., and ultrasonics are also used in medical applications.

Ultrasonic vibration assisted turning (UVAT) is proposed to provide an alternative towards reduction of excessive plastic deformation during the machining process of advanced materials [1]. Ultrasonic assisted turning is a cutting technique in which a certain frequency (in ultrasonic range) of vibration is applied to the cutting tool or the work-piece (besides the original relative motion between these two) to achieve better cutting performance [2]. Ductile mode of cutting can also be achieved which induces good compressive residual stresses in the workpiece during dry cutting process. High frequency and low amplitude motion during UVAT which imparted on the tool insert against the rotating work piece has benefitted the cutting tool life with great cutting force reduction and production of good surface finish [3].

Corresponding author: rasidi@uthm.edu.my

The fundamental difference between conventional turning (CT) and UVAT in terms of tool holder is the presence of a place for the installation of piezoelectric actuator (PZT) on UVAT, as long as there is no in CT. Ultrasonic vibrations in the tool holder are generated by piezoelectric actuators. Vibration with frequencies above 20kHz is propagated through the tool holder to the cutting tool. The tool will vibrate in an ultrasonic frequency during the turning process is run .

Many research works [4,5,6,7] have proposed the use of UVAT in order to cut various advanced materials (Titanium, Inconel, Aluminium matrix composite and others). However, at present the ultrasonic assisted turning process applies customized cutting tool assemblies. This includes different type of tool holders, dissimilar PZT and uncommon position of the PZT embedment. The former provides that each researcher has their own unique proposals on the application of UVAT process.

Traditionally, the selection of cutting conditions for turning process is left to the machine operator. In such cases, the experience of the operator plays a major role, but even for a skilled operator it is very difficult to attain the optimum values each time. There are various machining parameters affecting the output of turning process. Among these parameters, cutting speed, feed rate and depth of cut are the primary parameters that should be selected properly for machining process to ensure good surface quality and to prolong tool life. The effect of cutting parameters on cutting force, cutting temperature, surface roughness and tool life have been modelled by various researchers [8,9,10].

The present study proposes prelimenary investigation of machining parametric optimization on conventinal turning process using UVAT tool holder with the aim of Taguchi method. The machining parameters were used to result of cutting force and cutting temperature. Two different UVAT tool holder design (D1 and D2) were conducted to perform the machining process and each of it used same cutting tool. In this study three levels of speed, feed and depth of cut are evaluated.

2 Taguchi Method

The method presented in this study is an experimental design process called the Taguchi design method. The method is an efficient and complex technique for optimization primarily to improve quality, cost and performance design. This method is widely used in various fields of engineering for the design of experimental planning. The approach of this method is to reduce the number of experiments using a special orthogonal circuit developed by Taguchi. With this method, it is possible to obtain process optimization while simultaneously minimizing the sensitivity of experimental external variation [11]

The result of this method is expected that the optimal process parameters obtained from the parameter design are not sensitive to the variation of environmental conditions and other noise factors. Therefore, parameter design is a key step in Taguchi's method of achieving high quality without increasing cost [12].

In Taguchi method, the main parameters have influence on process results, which are positioned at different rows in a designed orthogonal array. The difference between the functional value and objective value is recognized as the loss function that can be expressed by signal-to-noise (S/N) ratio.

Regardless of the category of the performance characteristic, the larger S/N ratio corresponds to the better performance characteristic. Therefore, the optimal level of the process parameters is the level with the highest S/N ratio. Furthermore, a larger S / N ratio is associated with better performance characteristics. Therefore, the level with the highest S/N ratio indicates the optimal level of process parameters. Furthermore, an analysis of variance (ANOVA) is performed to see which process parameters are statistically significant. According to S/N and ANOVA analysis, optimal combinations of process parameters can be predicted.

3 Methodology

3.1 Design of Uvat Tool Hoder

There are 2 designs used in this study. Both of of these designs have similarities in geometry and the type of cutting tool. There is a fundamental difference in the number of parts used. It lies in the place to put the PZT as a generator of ultrasonic vibration. Ultrasonic vibration in the tool holder propagates to the insert part. Kyocera's VBMT is used as a cutting insert that is specified for the finishing process.

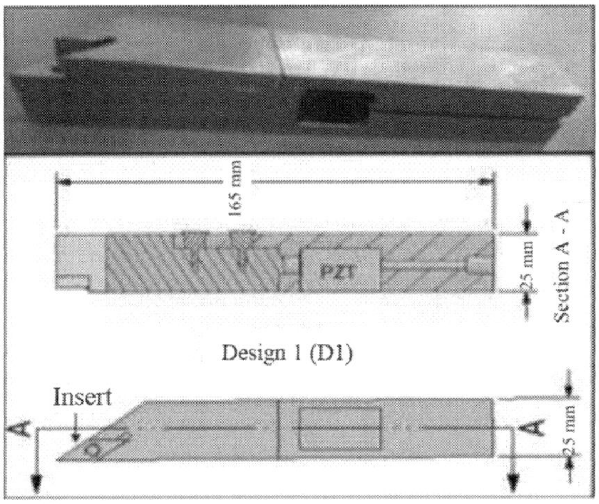

Fig. 1. Tool holder design 1 (D1)

The first design (D1), the design is divided into 2 parts, namely the head and body as shown in Figure 1. Bolts are made of the same material such as AISI 1045. Both are separated and put together using a 4 mm bolt with a depth of 10 mm. Vibration which generate from PZT is propagated on the cutting tool past the pin mounted on the head that will move linearly.

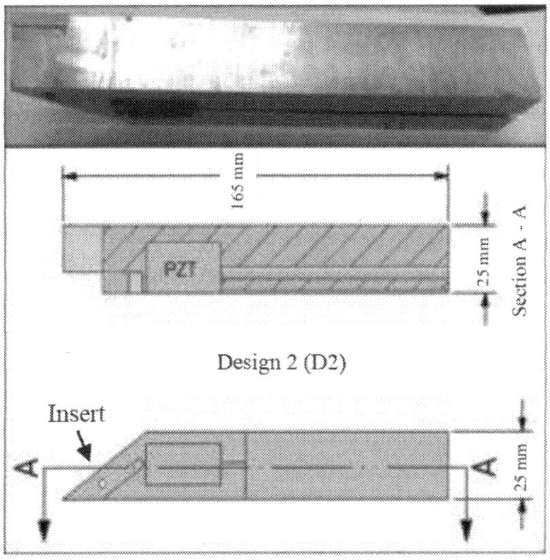

Fig. 2. Tool holder design 2 (D2)

The second design (D2), there is no separated parts in this tool holder's design. It is made from stainless steel 201 as shown in Figure 2. The ultrasonic vibration is propagated towards the tool insert directly through the tool's structure.

3.2 Machining experiment

The turning experiments were performed using Harrison Alpha 400 lathe machine. In this study, the value of cutting speed, cutting depth and feed speed were varied. There are nine variations where the cutting speed, the feed rate is the same, but the depth of cut is varies.

Experiment is done by lathe surface of workpiece made of allumunium AL 6061 T6. The inserted TNT-mounted insert tool holder is installed on the lathe and the turning operation is performed 10 mm long.

During the dry-cutting process is carried out, a measurement of cutting force using the 9257 B Kistler Dynamometer model. Dynamometer was attached to the tool holder and was supported by charge amplifier and signal conditioning to measure cutting forces. Dynoware software was connected to the dynamometer for recording purposes. The charge generated at the dynamometer was amplified using multi channel charge amplifier (type 5070A). The sensitivities of the multi-channel charge amplifiers were set for -7.939 pC/N for normal force (Fx), −7.960 pC/N for feed force (Fy), and -3.722 pC/N for axial force (Fz). The sampling frequency of data was set at 7,000 Hz. This experimental setup can be seen in Figure 3 below.

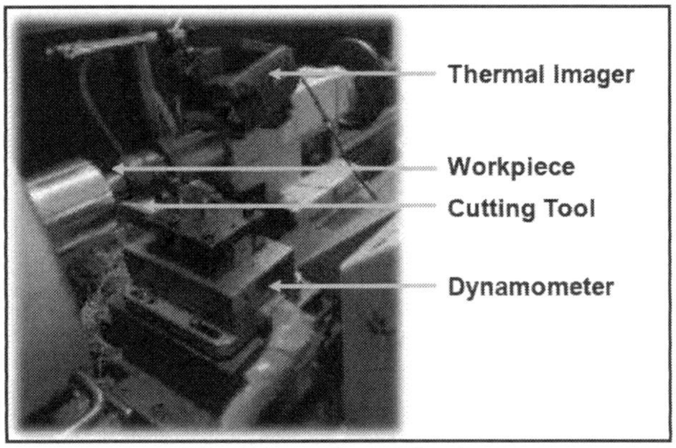

Fig. 3. Experimental setup

The temperature has been measured using infrared thermal camera FLIR T640. FLIR tool+ software was used to record temperature and to analyse the thermal image. This thermal camera has high thermal sensitivity of ≤ 60 mk at 30°C. The temperature range is from -20°C to 2000°C with high accuracy. The result of the measurement of these two machining outputs is then used as object for analysis.

3.3 Experimental design and parameter selection

There are three cutting parameters: cutting speed, feed rate and depth of cut has been considered for three levels (low, medium, high). Three variables are studied for three levels and hence nine experiments were designed. L9 orthogonal array had been used in this study as it has the capability to perform parametric study with three factors and three levels.

As shown in Table 1, three cutting parameters have been selected to conduct the experiment. The number of experiments and combination of cutting parameters and level for L9 orthogonal array is shown in Table 2.

Table 1. Experimental parameters and their levels.

Factors	Level		
	1 (Low)	2 (Medium)	3 (High)
A: Cutting speed (m/min)	50	75	100
B: Feed rate (mm/min)	0.05	0.08	0.1
C: Depth of cut (mm)	0.25	0.375	0.5

Table 2. L9 orthogonal array.

Trial number	Control factor		
	A:Cutting speed	B: Feed rate	C:Depth of cut
1	1	1	1
2	1	2	2
3	1	3	3
4	2	1	1
5	2	2	2
6	2	3	3
7	3	1	1
8	3	2	2
9	3	3	3

4 Results and Discussion

The response table, which contains the sums of the S/N ratios for each level and for each factor is shown in the Table 3. The category the smaller the better was used to calculate S/N ratio for cutting force and cutting temperature according to the equation:

$$S/N_s = -10*\log\left(\frac{1}{n}\sum_{i=1}^{n} y_i^2\right) \quad (1)$$

where y, is the average of observed data, S_y^2 is the variance of y, n is the number of observations and y is the observed data. Notice that these S/N ratios are expressed on a decibel scale. We would use S/N_T if the objective is to reduce variability around a specific target, S/N_L if the system is optimized when the response is as large as possible, and S/N_S if the system is optimized when the response is as small as possible. Factor levels that maximize the appropriate S/N ratio are optimal.

Table 3. Experimental result

Trial no	**Machining parameters**			**Output response**		**S/N Ratio**	
	A (m/min)	**B** (mm/min)	**C** (mm)	**CF** (N)	**CT** (°C)	**CF**	**CT**
1	50	0,05	0,25	21,99	47,00	-26,84	-33,44
2	50	0,05	0,375	30,39	49,87	-29,65	-33,96
3	50	0,05	0,5	42,17	53,11	-32,50	-34,50
4	50	0,08	0,25	31,79	54,21	-30,05	-34,68
5	50	0,08	0,375	44,35	56,84	-32,94	-35,09
6	50	0,08	0,5	68,54	81,31	-36,72	-38,20
7	50	0,1	0,25	35,09	60,21	-30,90	-35,59

8	50	0,1	0,375	49,02	67,18	-33,81	-36,55
9	50	0,1	0,5	62,5	78,06	-35,92	-37,85
10	75	0,05	0,25	39,08	60,01	-31,84	-35,57
11	75	0,05	0,375	43,37	64,39	-32,74	-36,18
12	75	0,05	0,5	45,47	65,32	-33,15	-36,30
13	75	0,08	0,25	44,4	66,82	-32,95	-36,50
14	75	0,08	0,375	58,02	74,12	-35,27	-37,40
15	75	0,08	0,5	64,22	89,38	-36,15	-39,02
16	75	0,1	0,25	47,93	58,82	-33,61	-35,39
17	75	0,1	0,375	58,6	71,92	-35,36	-37,14
18	75	0,1	0,5	78,36	82,28	-37,88	-38,31
19	100	0,05	0,25	35,35	65,39	-30,97	-36,31
20	100	0,05	0,375	44,71	67,79	-33,01	-36,62
21	100	0,05	0,5	54,82	76,36	-34,78	-37,66
22	100	0,08	0,25	52,39	57,21	-34,38	-35,15
23	100	0,08	0,375	68,83	63,42	-36,76	-36,05
24	100	0,08	0,5	71,31	71,63	-37,06	-37,10
25	100	0,1	0,25	42,03	61,11	-32,47	-35,72
26	100	0,1	0,375	63,69	68,47	-36,08	-36,71
27	100	0,1	0,5	80,27	70,80	-38,09	-37,00

4.1 Cutting Force

The effect of cutting parameters on cutting forces can be described using the response graph of S/N ratio. However, the percentage contribution for each factor cannot be determine using the graph. Pareto ANOVA is used to show the percentage contribution for the factors. Its simplified ANOVA by providing simple and faster step to determine the contribution of selected factor on experimental response.

The effect of changing parameters for different design (D1 and D2) is shown in Figure 4 (a) and (b) respectively. The response graph and pareto ANOVA (Figure 5 (a) and (b)) showed that the combinations of cutting speeds and feedrates have significantly effected on cutting force for both design. However, cutting speed had the strongest effect at 47.1 % for D1 while feed rate contributes 32.26% and depth of cut 20.73 %. Figure 5(a) indicates to get the lowest cutting force in machining when used D1, low cutting speed (A1), feedrate(B1) and depth of cut (C1) is required.

The result for D2 as shown in Figure 5 (b) agreed with D1, which showed that the most significant effect is the cutting speed and followed by feed rate. The percentage contribution of feedrate and cutting speed is 22.80% and 61.13% respectively. However, to get lower cutting forces A1 and B1 are preferred.

The percentage contribution of depth of cut shown that the influence of this parameter on cutting forces for both design is the lowest.

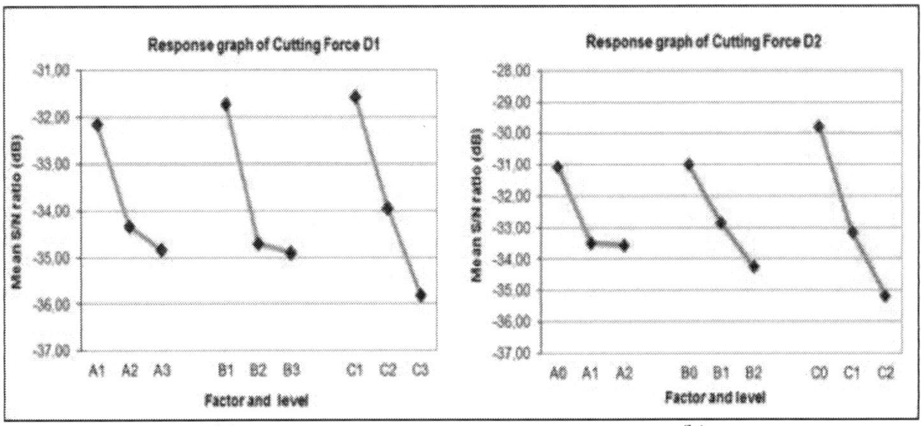

(a) (b)

Fig. 4. Effect of cutting parameters on cutting force :
(a) Design 1, (b) Design 2

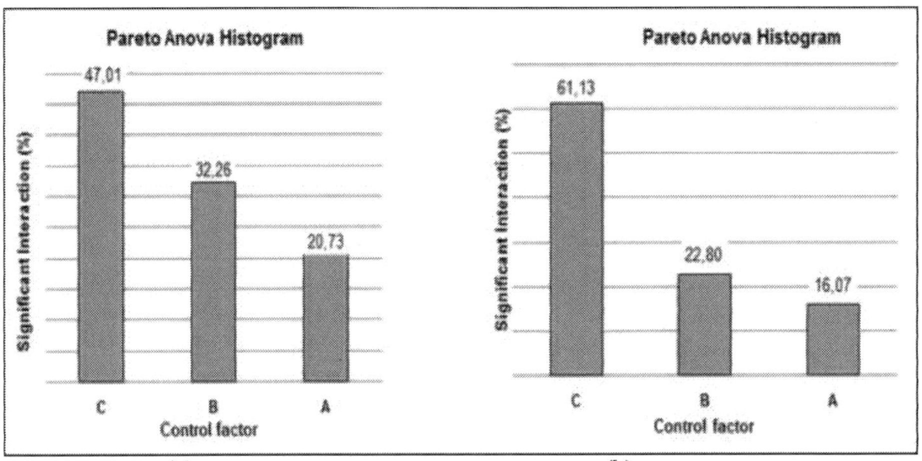

(a) (b)

Fig. 5. Percentage contribution of cutting parameters on cutting force :
(a) Design 1, (b) Design 2

4.2 Cutting temperature

Many factors affect cutting temperature during machining process. The factors include the properties of cutting material and cutting tool as well as geometry of the cutting tool. According [8], these factors and cutting parameters such as cutting speed feedrate and depth of cut have significant contribution to the temperature of the cutting area.

Response graph in Figure 6 (a) and (b) and Pareto ANOVA in Figure 7 (a) and (b) shows the effect of selected cutting parameters on cutting temperature. The graph 7 (a) shows that cutting speed is the most influential cutting parameter on cutting temperature for D1 with 45.76% contribution. Similar result was obtained in Figure 7 (b) for D2 with 53.99% contribution on cutting temperature came from cutting speed. This is due to the adiabatic effect at high cutting speed caused by the trapped heat in the shear deformation zone. The heat cannot escape in the very short time during the process causing highly localized temperatures in the chip[8].

Correspondingly, there is a similarity of feed rate effect and depth of cut for both designs. Feed rate has a greater influence on the cutting temperature for D1 and D2.

DOI: 10.1051/matecconf/201713500016

Whereas, the depth of cut has the least influence of the three parameters tested. The results show that the depth of cut has a weak effect on cutting temperature. This is influenced by the thinness of the cutting depth so it did not required a large amount of cutting force.

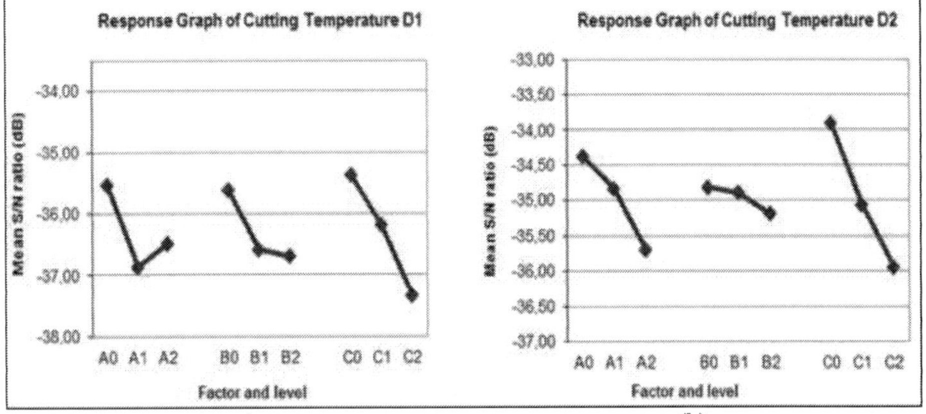

a) (b)

Fig. 6. Effect of cutting parameters on cutting temperature :
(a) Design 1, (b) Design 2

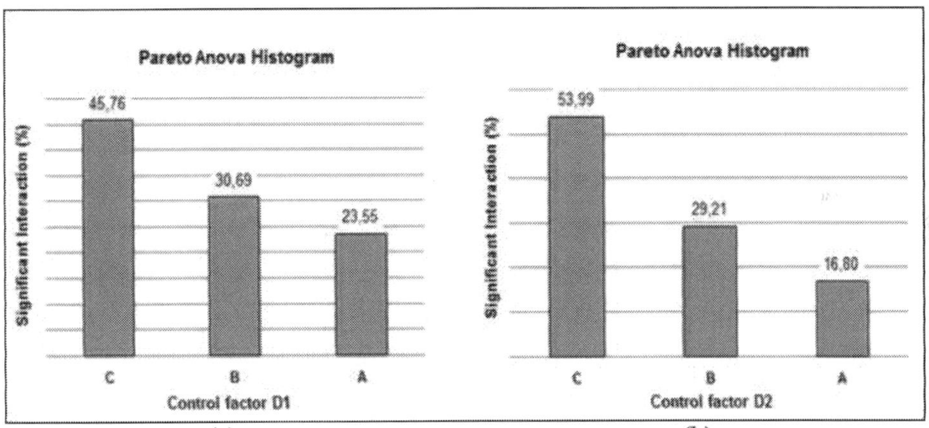

(a) (b)

Fig. 7. Percentage contribution of cutting parameters on cutting force :
(a) Design 1, (b) Design 2

5 Conclusion

The experiment has demonstrated an application of the Taguchi method for investing the effects of cutting parameters on cutting force and cutting temperature in turning aluminum metal. With analysis of results in this study using S/N ratio approach and ANOVA provides a systematic and efficient methodology for the optimization of cutting parameters.

The cutting force and cutting temperature on D1 and D2 are mainly affected by cutting speed, feed rate and depth of cut, by decreasing any one the force and temperature is decreased. The parameters considered in this experiment are optimized to attain minimize force and temperature when machining run. The best parameters on cutting force for D1 and D2 have similarity. The parameters are cutting speed at 50 m/min, feed rate at 0,05 mm/min and depth of cut at 0,025 mm. According to Anova analysis on cutting forcr for D1 and D2, the cutting speed has highly influence and followed by feed rate and depth of cut.

However with cutting force, the best parameters on cutting temperatures D1 and D2 also have similarity. Cutting speed parameters at 50 m / min, feed rate at 0.05 mm / min and cutting depth at 0.025 mm are parameters that can be used to perform the minimum cutting temperature. Similarly, based on Anova's analysis of cutting temperature for D1 and D2, cutting speed is very influential and followed by feed rate and cutting depth.

Acknowledgement

This paper was supported by UTHM GIPS grant monitored by Office for Research, Innovation, Commercialization and Consultancy (ORICC) UTHM Malaysia. Additional support was received from Precission Machining Research Center (FKMP - UTHM) and FRI - Telkom University Indonesia.

References

[1] Sharma & Pandey, *Recent advances in ultrasonic assisted turning: A step towards sustainability*, Cogent Engineering, **3**: 1-20 (2016).

[2] H. Rachmat, M. R. Ibrahim, and S. Hasan, *Design selection of an innovative tool holder for ultrasonic vibration assisted turning (IN-UVAT) using finite element analysis simulation*, AIP Conference Proceedings **1831**, 020029 (2017).

[3] D. E. Brehl and T. A. Dow, *Review of vibration-assisted machining*, Precision Engineering, **32**, 153–172 (2008).

[4] Y. Xing, , J. Deng, and J. Zhao, *Cutting performance and wear mechanism of nanoscale and microscale textured Al2O3/TiC ceramic tools in dry cutting of hardened steel*, International Journal of Refractory Metals and Hard Materials, **43**, 46 – 58 (2014).

[5] A.Maurotto, R. Muhammad, A. Roy, and V. V. Silberschmidt, *Enhanced ultrasonically assisted turning of a β-titanium alloy*, Ultrasonics, *53*, 1242–1250 (2013).

[6] M. R. Ibrahim, *Experimental Analysis on Ultrasonic Assisted Turning (UAT) Based on Innovated Tool Holder in the Scope of Dry & Wet Machining*, Applied Mechanics and Materials, Vol: **660, 8**, pp. 104-108, (2014).

[7] H. Rachmat, T. Mulyana, S. Hasan, M. R Ibrahim, *Design Selection of In-UVAT Using MATLAB Fuzzy Logic Toolbox*, Recent Advances on Soft Computing and Data Mining book series (AISC, volume **549**), pp 538-545, (2017).

[8] M. Bacci da Silva and J. Wallbank, *Cutting temperature: prediction and measurement methods-a review*, J. Matel Process Technology, vol **8 no 1- 3**, pp. 195-202, (1999).

[9] W.H Yang and Y.S Tarng, *Design Optimization of Cutting Parameters for Turning Operations based on the Taguchi Method*, Journal of Materials Processing Technology, Vol. **84**, pp. 122 – 129, 1998.

[10] A. Manna, S. Salodkar, *Optimization of machining conditions for effective turning of E0300 alloy steel. Journal of Materials Processing Technology*, vol. *203*, pp. 147-153, (2008).

[11] M. Kaladhari and K.V Subbaiah, *Application of Taguchi approach and Utility Concept in solving the Multi-objective Problem when turning AISI 202 Austenitic*

Stainless Steel, Journal of Engineering Science and Technology Review **4 (1)** : pp. 55-61 (2011).

[12] M. Sanyilmaz, *Application Taguchi method on experimental design and quality dvelopment*, Master Thesis, Applied Natural Science Institute, Dumlupinar University, Kutahya. (2006).

Designing Train Passenger Seat by Kansei Engineering in Indonesia

Sakya Nabila Hapsari[1,*], *Teddy Sjafrizal*[1], and *Rino Andias Anugraha*[1]

[1]Telkom University, School of Industrial Engineering, Industrial Engineering Department, Jl.Telekomunikasi Terusan Buah Batu, 40247 Bandung, Indonesia

Abstract. Public transport adoption is still major problem in Indonesia. Integrating impression and emotion to service may change the image of public transport, leading to fast adoption rate. As product development technique, Kansei Engineering is well known to facilitate that integration. This paper communicates the implementation of Kansei Engineering (KE) in designing train passenger seat in Indonesia. The train between Bandung-Solo (8 hours journey) was chosen to be the investigated object. The train was selected to be refurbished by introducing a new passenger seat design, emphasizing on the improvement of passenger experience. As the first step of KE type I technique, travel experience of the existing passengers were taken into the consideration for selecting Kansei words. It was found that the relevant Kansei words to design passenger train seat in Indonesia were: comfort, people-friendly, multifunction, safe, soft, spacious, relaxing, superior, maintainable, strong, producible, minimalist, attractive, modern, unique, and durable. Upon understanding the semantic meaning of those words, physical properties and technical specification were produced by consulting to the related members. Finally, a passenger seat with synthetic leather cover, modular design, foldable table, individual power point, blue-silver color, and adjusted to Indonesian's anthropometry data was proposed as the final design

1 Introduction

Congested traffic is one of the chronic issues of major cities with high density population. This phenomenon is getting serious at the emerging countries. Idea of pushing people to use public transport is believed to be the solution for reducing the road congestion [1]. However, asking people to use public transport involves converting the habit.

Habit to use public transport is built up by the society and flourished by the perception towards the travelling mode [2]. Social push and the good image of public transport accelerate the conversion. Thus, designing the service to the society requires people-centred perspective from different groups of users [3]. In system-level design, some values are considered as the importance to form the consumer perceptions and preferences through product appearance communication: aesthetic, symbolic, functional, ergonomic, attention-drawing and categorization value [4].

* Corresponding author: sakyahapsari@gmail.com

Appearance of product is the first thing attracts customer attentions. Having a product with attractive esthetical value results in high likelihood of buying. Esthetical and impression gained by the customers at the first sight drive their attention to buy the product/service. Functional value may come in the reliability and durability of the product/service. Ergonomic value ensures the product/service to deliver the comfort, effective, efficient, safe, and healthy design. Packaging condition is the example of attention-drawing value. Unique food packaging draws more attention. Finally, customers have the ability to categorize the product based on those values above. All these values determine the likelihood of potential customers to own the product/services.

In public transport, people-centered perspective in designing the service could be derived from travel experience. Travel experience determines the loyalty of passenger in taking public transport. Study on bus passengers perspectives suggest that bus provider or travel agent shall considering travel experience prior to launching the service [5]. Study by [6] identifies seven dimensions of experience factor: individual space, information provision, staff's skills, social environment, vehicle maintenance, off-board facilities, and ticket line service. In addition, the attitudes toward the standard quality level of public transport fluctuate across the passenger socio-economics groups, including gender [7]. Therefore, the product/service development process of public transport shall have the capability to process experience factor. It can be fulfilled by deploying Kansei Engineering techniques.

Established by the Japanese, Kansei Engineering (KE) has become one of the powerful approach in developing ergonomics consumer-oriented products by considering the interaction between human-machine system, environment and the society [8]. KE combines both functional and emotional aspect of the user needs as the quality level to be achieved [9]. Successful KE stories are best delivered by the product of Mazda, the Miata; the good-up Bra made by Wacoal and the Sharps's liquid Crystal viewCAM [8]. Other success stories were reported on the house design[10] & costume [11], packaging industry[12], aviation interior[13], e-commerce [14] office furniture [15] and many more.

KE unites the aspect of emotion and feeling through engineering process. KE has widely used for many kinds of product development and it becomes more complex. By the technology advancement, there are at least 4 types of KE techniques [4]. KE type I focuses on identification and development of product strategy and market segment into a tree structure that represents the manual way connection of customer's affective needs and product properties. KE Type II frequently uses computer aided system in a form of interference engines and Kansei databases, therefore Kansei and product properties are connected using mathematical statistical tools. Type III utilizes mathematical modelling in proposing design solution. Lastly, type IV has the ability to visualize and interact with the design by adopting virtual reality technology. Among those types, KE type I is believed to be the most versatile technique since it requires less resources.

This communication describes the adoption of KE type I in public transport, especially train. Previous study on the experience factor requirements of public transport (e.g. bus) finds that safety, waiting time and body comfort are the most important and requires to perform in high performance [5]. Unlike body comfort, safety and waiting time are not purely decided by the design. Since sitting on the passenger seat is the most frequent activities during travelling by bus or train, body comfort is heavily determined by seat design. Therefore, seat design was taken as the object to be investigated for the study of KE in public transport especially in the train. Additionally, outcomes of KE for public transport of one region may result in different outcomes for other regions. It is due to the fact that the needs, impression, experience and expectation of users are different towards public transport. Thus, study on deploying KE type I for designing train passenger seat in a region is unique and reasonable. Additionally, to the knowledge of authors, there is research gap on the study of KE type I for passenger train in Indonesia. Hence, this study aimed to explore and describe the deployment of KE type I at Indonesia's train service especially for designing the passenger seat.

2 Kansei engineering techniques

The implementation of KE follows the structural method that has been proven in several studies [5,10,12,14,16–19]. Generally, the method may be modified or replaced based on certain purposes. Here is the generic structure of KE [4]:

- Choice of Participant Groups

The selection of participant group members may come from different people with similar behaviour in using the certain product. Since the KE needs the homogenous subjects to perform the implementation, the information of this criteria can be gained from the company or experts who have studied about market segmentation. This information is collected not in context of KE, but it may be used anytime as the input for KE implementation.

- Collection of Kansei Words

Kansei Words are the group of words that describes the product properties. In common, these words are adjective or grammatical form. It can be collected from any reliable sources, such as literatures, magazines, experts, manuals, experienced users, or related Kansei studies. The Kansei Words should be collected in a range of 50-600 words until no new words occur, therefore it has to be checked by dictionary.

- Data reduction methods

Reduction of Kansei Words is useful to find the proper number of Kansei Words, as it may have poor quality because of the error from the participants or it does not suitable to the number of participants. In order to solve this problem, there are two empirical tests that commonly used for data reduction, which are validity test and factor analysis. The validity test comes from the questions of design importance degrees, while the factor analysis comes from semantic differential evaluation that measuring how the Kansei Words corresponding to the product.

- Rating-scale Types used

There are two types of rating-scale types that commonly used in KE, which are Visual Analogue Scale (VAS) and Likert scale. Likert scale is easier and more common to use. It has three types of scales, which are five-Likert scales, seven-Likert scales, and nine-Likert scales.

- Connecting the Kansei Words to product properties

Every Kansei Word can generate the product properties, therefore it becomes the most important step of Kansei Engineering. The ways of connecting the Kansei Words to product properties are executed qualitatively or quantitatively. The qualitative way performs the Kansei Words connection based on the intuition that cannot be communicated easily as it is done in Quality Function Deployment (QFD), furthermore this way is commonly used in Kansei Engineering Type 1. Meanwhile, the quantitative way utilizes statistical tools, such as neural networks algorithm [20], fuzzy logics[21], rough set analysis [22] and some other tools.

KE type I has been applied in several industries [4]. The legendary example of implementation of KE type I was performed by Mazda for its vehicle interior design. Intimate and lifestyle product such the brassiere design of Wacoal was designed by utilizing KE type I. Extended beyond Japanese industries and products, it is recorded that KE type I was also implemented in packaging industry in UK [12] and sunglasses e-commerce in Malaysia [14]. Having said so, the idea of taking KE type I to be implemented at the designing process of Indonesian's passenger train seat is very reasonable.

The framework covered in Kansei Engineering Type 1 are divided into several steps, which are as follows [4] :

Table 1. Steps of Kansei Engineering type I

Steps	Characteristics
Step 1: Identification of Target	- Determine targeted customer - Gather customer's Kansei words - Extensively apply marketing research and expert judgment
Step 2: Determination of Product Concept	- Deciding product concept to be further developed. - Considering customer interaction with the concept
Step 3: Breaking Down The Product Concept	- Assigning physical characteristics. - KJ method (affinity diagram) could be applied
Step 4: Deployment to Physical Design Characteristics	- Able to relate concept to technical words - May end up in deciding the physical characteristics such as size, appearance design, color and function.
Step 5: Translation to Technical Specifications	- Specification is defined, consultation with the expert or lead users are likely.

3 Procedures

This descriptive study was implemented on the train passenger seat design of PT.KAI (Indonesian Railway Company). As many as 96 respondents were questioned by interviewing techniques for refining the Kansei Words. Validity and reliability testing were performed on the gathered Kansei Words. They were performed by distributing questionnaires to the respondents to measure the degree of importance for each Kansei Words by five Likert scales. The results were developed as the input for Spearman-Rho validity test and Cronbach Alpha reliability test. Subsequently, semantic differential evaluation was performed by pairing the Kansei Words with the antonyms and questioned those pairs by the questionnaire using five Likert scales. Factor analysis is helpful to direct the process of concept mapping by clustering the Kansei Words into the smaller groups [4]. The analysis was done with the support of Kaiser-Meyer-Olkin (KMO) and Bartlett's Test and Principal Component Analysis (PCA), then the results were further analysed by identifying the clusters or factors. KJ method was performed by using affinity diagram, in which the result of factor analysis was proceeded by giving the name of each cluster or factors based on the relationship to Kansei Words. Upon clustered, the Kansei words were processed further to realize the product concept. The concept mapping was accomplished using the tree structure that contains several levels of Kansei Words. It started from zero-level Kansei Word, which is defined as the name of cluster in this research. The following levels contain the Kansei Words, semantic meaning, metrics, physical properties, and technical specification. Thus, seat technical specification could be derived upon discussion with the reliable source (company). Finally, the proposed passenger seat design was visualized.

4 Results and Discussion

4.1 Identification of target market

The development of passenger train seat was conducted by Rolling Stock Engineering, a division of PT KAI. This division is responsible for developing the transportation of goods and passengers. The role of this division aligns with PT KAI product development strategy. The company has decided to perform convergence of three passenger train classes (Executive-Business-Economy) into two classes (Executive-New Economy). Specifically, this direction has consequences for the change of the new train class interior concept which is expected to increase the quality compared to the previous one. Change at the old class interior concept is expected to be represented by a change in the concept of the seat that can create new impression of travel experience during the train journey. This strategy is aimed to increase efficiency and competitive advantage of PT KAI in the next few years.

In order to facilitate the discussion of seat design for new train class development, the constraint of trip duration was determined. The case study concerned at long distance train, such as Kereta Api Bandung-Solo which has 8 hours-long trip. This selection was done based on the data of passenger number that exceeds other trains for each semester, which indicates the high needs of using Bandung-Solo service line.

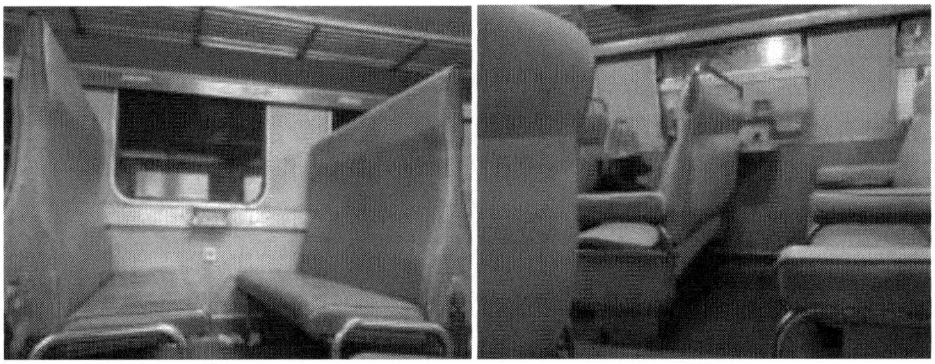

(A) **(B)**

Fig. 1. (A) Current design of economy class (B) Current design of business class.

4.2 Determination of product concept

Below steps are critical and to be considered as the main process of KE. The steps were started from collecting Kansei Words, ensuring the validity and reliability of the words by appropriate statistical tools, applying semantic differentiation technique to understand the impression of the respondents on each Kansei Words upon reviewing seat design, and finally clustering the Kansei Words to the product matrices.

4.2.1 Collection of Kansei words

Kansei Words collected in this study are related to the general impression or functions represented by a train seat, either through magazines, literatures, corporate strategies, and train users. Kansei Words amount collected was 50-60 words. Words that are synonymous, then it would only have one word that is most relevant to the semantic context of the need

DOI: 10.1051/matecconf/201713500017

for train seats. This resulted in a reduction process into 19 Kansei Words, as shown in the following table.

Upon gathering the Kansei Words, feedback from respondents were compiled to understand the validity and reliability. After 96 people were available (confidence level: 95%, variance: 1.25 and error margin: 5%) as respondents, feedback in the form of questionnaires of importance degree were developed. The questions were mainly to ask the importance level of each words in designing passenger train seat from the perspective of users. There were 19 closed questions contained Kansei Words, then the answers were represented by five levels of Likert Scale. This result was then processed further by validity and reliability test.

Spearman Rank Validity test was selected as the validity test. It is necessary to check the relevancy of Kansei Words for the purpose of designing the new passenger seat. The relevancy is expressed by the correlation between each value of Kansei Words with the total value of all variables. By using the alpha value (α) of 0.05 and number of 96 respondents, the correlation coefficient of each Kansei Words shall be more than 0.201 (critical values of Spearman's Rank table). Result in Table 2 showed that there was no correlation coefficient that less than 0.201, therefore all Kansei Words was decided as valid and appropriate to be synthesized in the following steps of Kansei Engineering.

Once deciding all Kansei Words as valid, reliability test were conducted to ensure the consistency of those words. Following the common cut off criteria for measuring reliability, results of Cronbach Alpha for each Kansei Words must exceed or equal to 0.7 [23]. Fortunately, all the Kansei Words were believed to be reliable based on the value of Cronbach Alpha (refer to table 2 for more details). Finally, the selected Kansei Words were valid and reliable as the words representing the impression/feeling/emotion of passengers in passenger seat design.

Table 2. Spearman rank validity and Cronbach alpha

No	Kansei Words	Correlation Coefficient	Cronbach Alpha
1	Comfort	0.261	0.747
2	People-friendly	0.492	0.738
3	Multifunction	0.579	0.736
4	Safe	0.452	0.742
5	Viable	0.663	0.732
6	Soft	0.338	0.741
7	Spacious	0.578	0.736
8	Relaxing	0.555	0.736
9	Superior	0.711	0.727
10	Maintainable	0.673	0.734

No	Kansei Words	Correlation Coefficient	Cronbach Alpha
11	Light	0.728	0.724
12	Strong	0.691	0.733
13	Producible	0.784	0.727
14	Minimalist	0.768	0.723
15	Attractive	0.728	0.729
16	Colorful	0.466	0.734
17	Modern	0.588	0.734
18	Unique	0.573	0.729
19	Durable	0.617	0.736

4.2.2 Semantic differentiation

KE was invented by recognizing the semantic meaning of mentioned words as the actualization of the hidden feelings and emotions[24]. The previous steps were focused on identifying the impression/feeling/emotion of user towards the passenger seat, whereas semantic differentiation stages were taken to understand the realization of the impression/feeling/emotion on product design [25,26]. According to [4], SD evaluation is useful to study words as a scale in term of semantic space by preparing the antonyms of each Kansei Words. The antonyms are necessary to be checked first by help of dictionary or thesaurus. The evaluation was constructed through a questionnaire with the same number of respondents in the previous test. In order to help the respondents to answer the questionnaire, product samples of similar train passenger seat were attached. The answers were represented by the scale and they would be analyzed using factor analysis in the next step.

Table 3. Questionnaire form for semantic differentiation analysis

Comfort	1---2---3---4---5	Discomfort
People-friendly	1---2---3---4---5	People-unfriendly
Multifunction	1---2---3---4---5	Single function
Safe	1---2---3---4---5	Dangerous
Viable	1---2---3---4---5	Impractical
Soft	1---2---3---4---5	Hard
Spacious	1---2---3---4---5	Narrow
Relaxing	1---2---3---4---5	Distressing
Superior	1---2---3---4---5	Inferior
Maintainable	1---2---3---4---5	Not maintainable
Light	1---2---3---4---5	Heavy
Strong	1---2---3---4---5	Brittle
Producible	1---2---3---4---5	Not producible
Minimalist	1---2---3---4---5	Lavish
Attractive	1---2---3---4---5	Boring
Colorful	1---2---3---4---5	Plain
Modern	1---2---3---4---5	Conventional
Unique	1---2---3---4---5	Common
Durable	1---2---3---4---5	Temporary

To be reliable for further analysis, the answers of 96 respondents were further evaluated for reliability by Kaiser-Meyer-Olkin (KMO) test and Barlett's test. KMO value states the adequacy of sample size, whereas the Barlett's test of sphericity determines the significant level of interrelationship between variables.

Table 4. Reliability result by Kaiser-Meyer-Olkin (KMO) test and Barlett's test

Kaiser-Meyer-Olkin test of sampling adequacy	0.772
Barlett's test of sphericity	0.000

The common acceptable cut-off criterion for KMO test is 0.50 [27,28]. Test result showed that the KMO value fulfilled the condition that the sample set was adequate and might proceed to factor analysis process. By having value of 0.772 this adequacy of the data is considered as moderate [27]. Subsequently, the data were then tested by Barlett's test for understanding the interrelationship with the concept of passenger seat design. Bartlett test sphericity result was 0.000 (significant level of $p < .05$). Thus, it was decided to the decision of rejecting significant interrelationship between variables affecting the concept of passenger seat design. At last, Both KMO test result and Barlett's test result permitted for further PCA process.

KMO and Bartlett's Test was proceeded by Principal Component Analysis (PCA). It may lead to clustering the Kansei Words into smaller groups. It was started from understanding the communalities value. The extraction value of communalities becomes the interpretation of how Kansei Word explain the factors. For example, the Kansei Word "comfort" had communalities value of 0.725, meaning that the word "comfort" had an ability to explain the factor as much as 72.5%. If the value exceeds 0.5 or 50%, the Kansei Words can strongly explain the factors or exist within specific factors [28].

Table 3. Communalities Analysis Results

Kansei Words	Extraction Value	Kansei Words	Extraction Value
Comfort	0.725	Light	0.497
People-friendly	0.637	Strong	0.649
Multifunction	0.606	Producible	0.536
Safe	0.606	Minimalist	0.682
Viable	0.391	Attractive	0.554
Soft	0.644	Colorful	0.403
Spacious	0.559	Modern	0.555
Relaxing	0.601	Unique	0.584
Superior	0.556	Durable	0.613
Maintainable	0.617		

*items highlighted should not be taken into account (< 0.50 cut-off criterion).

Result of communalities analysis is shown by table X. Since the majority of Kansei Words reached the value of > 0.5, it indicated that the respondents were able to correlate the seat design impression with the Kansei Words. Nevertheless, It was found that the word "viable", "light", and "colorful" became the Kansei Words which had extraction value <0.5, thus they were not necessary to be taken into account those words in clustering and further steps in KE. Hence, there were 16 Kansei words that were believed to be necessary as design consideration of train passenger seat.

As these adjective were gathered based on the customers experience as train passenger, it is believed that this result may be correlate to the experience factor of bus passenger. Correlation between the findings of study by [29] on experience factors of bus passengers with the Kansei words of train passenger seat is shown by figure 1. Majority of the identified Kansei Words are related to the experience factor, especially the one that related to design. However, out of 16 Kansei Words, there are 4 words that are not related to the experience factors. "Multi-function", "minimalist" and "attractive" are believed to be unique to product design, whereas "producible" has both users and producer interest.

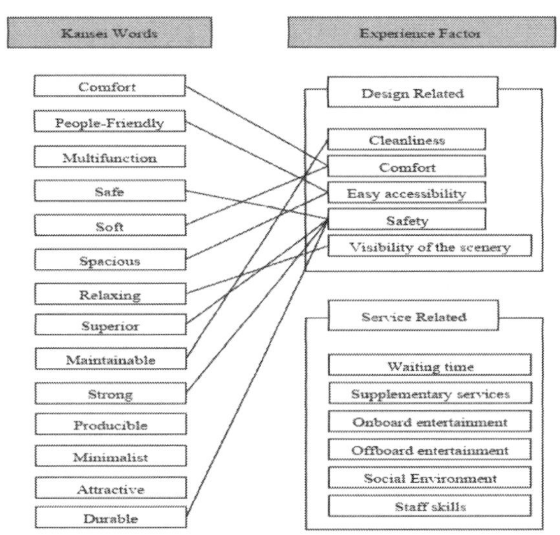

Fig. 3. Correlation between Kansei Words and Experience Factor of [31]

After considering the communalities, the PCA was expanded to orthogonal rotation by varimax technique. Varimax is considered as the most widely method for orthogonal rotation [27,28] In term of identifying the exact cluster for each Kansei Words, it can be indicated by the value of factor loading of each cluster or component. For example, if one Kansei Word has highest loading factor in component 1, it means that this Kansei Word is categorized in component 1, and so the other Kansei Words.

Table 4. Correlation Matrix& Loading Factor
(rotation method: Orthogonal, Varimax)

Kansei Words	Clusters/Components/Factors				
	1	2	3	4	5
Comfort	0.038	-0.04	0.065	0.847	-0.019
People-friendly	-0.026	0.085	0.715	0.039	0.342
Multifunction	0.231	0.204	0.223	-0.134	0.665
Safe	0.087	0.451	-0.12	0.402	0.468
Soft	-0.217	0.152	0.46	0.602	-0.023
Spacious	0.267	-0.047	0.681	0.133	-0.066
Relaxing	0.287	0.248	0.549	0.019	-0.079
Superior	0.298	0.425	0.517	-0.069	-0.117
Maintainable	0.117	0.767	0.106	0.05	0.012
Strong	0.358	0.617	-0.125	0.341	0.089
Producible	0.448	0.536	0.211	0.03	0.056
Minimalist	0.735	0.287	0.184	0.002	-0.156
Attractive	0.715	0.156	0.061	-0.027	0.12
Modern	0.651	0.096	0.209	-0.058	-0.273
Unique	0.688	-0.185	0.155	-0.086	0.211
Durable	0.314	0.703	0.098	0.091	-0.045

*The highest loading factor on each Kansei Words is highlighted

Table 4 shows the correlation matrix of the Kansei Words and the components/clusters. There were 5 components identified in this analysis. Each cluster had at least two Kansei Words that may represent the expected impression of passengers towards that cluster.

4.3 Determination of product concept

In the previous step, factor analysis has been developed in order to lead the step of making product concept. Since the analyzed Kansei Words were classified into five groups, it was necessary to give the name for each group. According to [4] , KJ method is the best method to collect the amount of data into groupings based on natural relationships. It uses affinity diagram as the tool to group each Kansei Words into specific categories, called affinity cluster. In this context, the relationship among the Kansei Words is shown by the score of loading factor in factor analysis. Study by [29] proposes the affinity diagram of Kansei Words in designing a product. Referring to that finding and with some adjustment, the identified clusters were named accordingly.

Table 5. Clusters naming by KJ method

Clusters/Factors	Degree	Impression	Spatial	Easy	Pleasure
Kansei Words	Maintainable	Minimalist	People-friendly	Multifunction	Comfort
	Producible	Attractive	Spacious	Safe	Soft
	Strong	Modern	Relaxing		
	Curable	Unique	Superior		

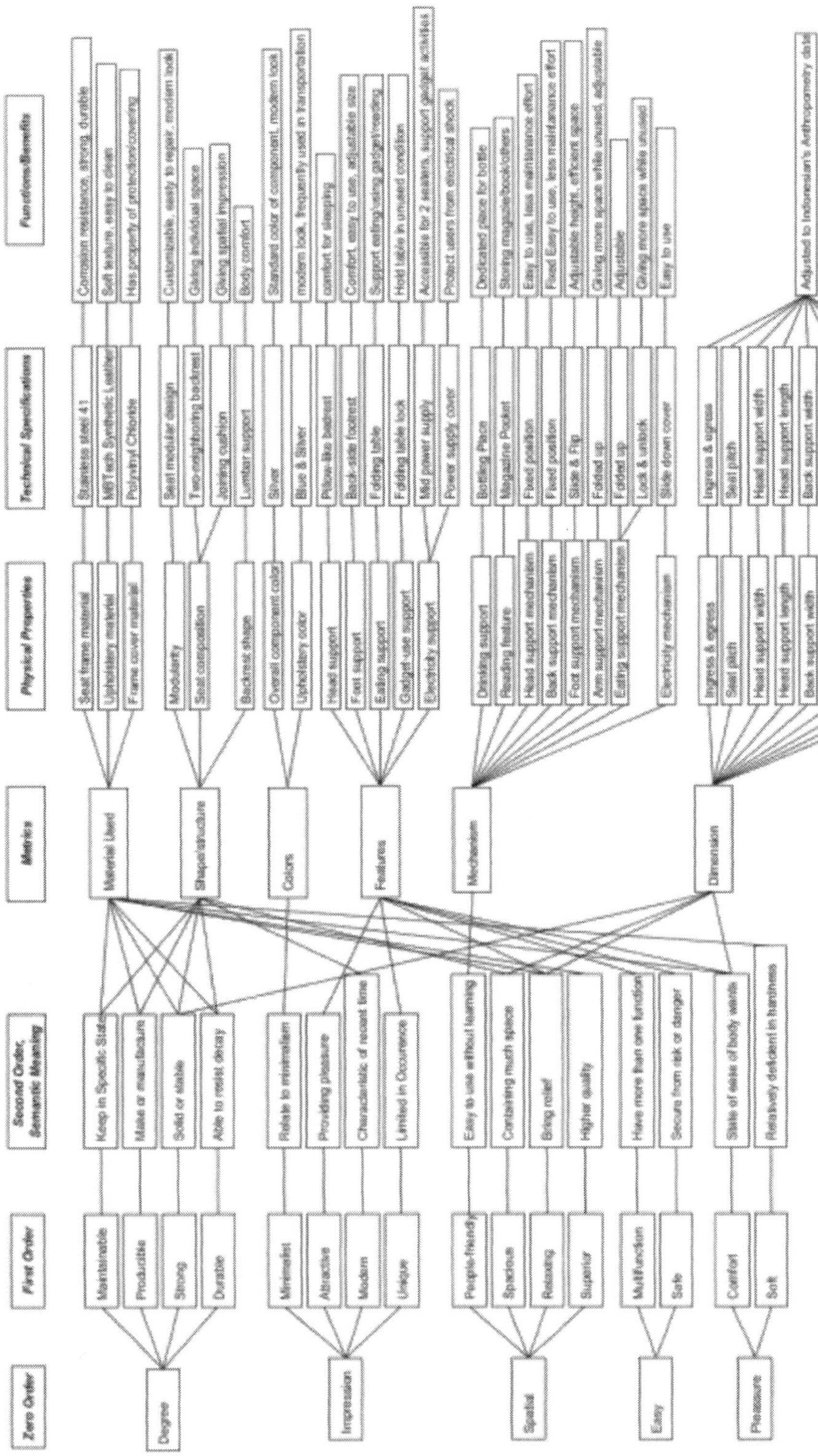

Fig. 4. Design concept mapping

Once clustered Kansei Words were formed, the design concept mapping could be executed. Fig. 4 describes the design concept mapping result. It was started from defining the semantic definitions of each Kansei Words. Subsequently, each Kansei Word was connected to its physical product attributes. Material, shape or structure, colors, features, mechanism and dimension are considered as the common physical product attribute to be addressed for physical product development. One Kansei Words with one defined semantic meaning had at least one relationship to the product attribute.

4.4 Deployment of physical design characteristics

The adjective of Kansei Words and the product attributes were linked to the related design components of the passenger seat. This process was performed together with the company representatives, since they could act as the expert as well as the lead users. The translating process is described as below.

From the physical product attributes of "colors", it had "minimalist" as the related Kansei Word. Hence, for defining the physical design characteristic, any part that had function to portray minimalist impression from its color projection were designed accordingly. For this case, there were 2 items identified, "overall component color" and "upholstery color". The colors of these 2 items were selected from any color with impression of "minimalist". If there were more than one related Kansei Words, all words were considered in deciding the physical design characteristics. Similar processes were implemented on each physical design characteristic.

4.5 Translation into technical specifications

At last, in order to come up with proposed design, the physical design characteristics were required to be translated to technical specifications. In other words, technical specification are the understanding of product attributes from engineering perspective. The functions, mechanism, needs and condition as defined in the physical design characteristics were translated to the relevant engineering terms. The translating process is described as below.

Previously, "minimalist" was defined to be applicable to the color of "overall component color" and "upholstery color". At this stage, the color was defined accordingly. Upon discussion and considering some information in the public domain, it was concluded to select blue and silver as the color to represent "minimalist". Each of the physical design characteristics had at least one technical specification. Finally, since the technical specifications have been defined in the concept mapping, it should be visualized as the concept design.

Once the design was visualized, it was be examined for design validation. In many cases, the validation is done by questioning potential customers for acceptance. Fortunately, the company representatives acted as the expert and lead users, hence their decision were use as design acceptance decision

5 Summary and recommendations

Designing passenger train seat by applying Kansei engineering type I has been achieved. The proposed design was succeed in translating consumers' impressions/feeling/emotions towards passenger train set. A passenger seat with synthetic leather cover, modular design, foldable table, individual power point, blue-silver colour, and adjusted to Indonesian's anthropometry data was concluded as the proposed design. Those specifications were gathered upon considering the Kansei Words. The identified Kansei Words are comfort, people-friendly, multifunction, safe, soft, spacious, relaxing, superior, maintainable, strong, producible, minimalist, attractive, modern, unique, and durable. Statistically, the 16 identified Kansei Words are valid and reliable.

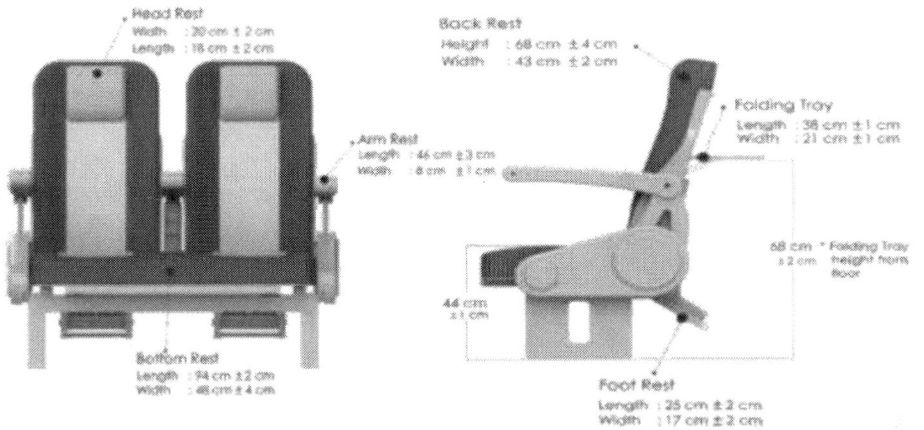

Fig. 5. Visualization of the proposed train passenger seat

The outcomes of this study may become the reference for similar studies in the future. Since, perception in travel experience may be developed by value and norms of the regions, user emotions toward public transport design may differ by location [2]. The study may be extended to other public services provided by the local government. It is also recommended to broaden similar studies to e-government service. The experience of users in using previous e-system influences their adoption to the new e-service [30]. Moreover, particularly in a multi-cultural and multi-racial society, the expectation and needs that are developed by the values and norms could be captured by deploying Kansei Engineering technique.

References

1. Federal Highway Administration, *Traffic Congestion and Reliability: Linking Solutions to Problems* (2017)
2. N. Klein, Urban Stud. 1 (2016)
3. L. Li and B. P. Y. Loo, Cities **58**, 50 (2016)
4. M. Nagamachi and A. M. Lokman, *Innovations of Kansei Engineering* (2011)
5. R. Carreira, L. Patrício, R. N. Jorge, and C. L. Magee, J. Eng. Des. **24**, 738 (2013)
6. R. Carreira, L. Patrício, R. Natal Jorge, and C. Magee, Transp. Policy **31**, 35 (2014)
7. C. Morton, B. Caulfield, and J. Anable, Case Stud. Transp. Policy **4**, 199 (2016)
8. M. Nagamachi, Ergon. Des. **10**, 5 (2002)
9. A. Lanzotti, P. Tarantino, A. Lanzotti, and P. Tarantino, TQM J. **20**, 324 (2008).
10. S. Ishihara, T. Matsubara, M. Nagamachi, and Y. Matsubara, Adv. Human-Computer Interact. **2011**, (2011)
11. M. Nagamachi, 1999 IEEE International Conference on Systems, Man, and Cybernetics

6, 273 (1999)

12. C. Barnes, T. Childs, B. Henson, and S. Lillford, TQM J. **20**, 372 (2008)
13. J. Guerin, Sixt. Symp. Qual. Funct. Deploy. 8 (2004)
14. N. K. Chuan, A. Sivaji, M. M. Shahimin, and N. Saad, Procedia - Soc. Behav. Sci. **97**, 707 (2013)
15. J. Park and S. H. Han, Int. J. Ind. Ergon. **34**, 31 (2004)
16. M. Nagamachi, Int. J. Ind. Ergon. **15**, 3 (1995)
17. B. Razza and L. C. Paschoarelli, Procedia Manuf. **3**, 6228 (2015)
18. S. Schütte and J. Eklund, Appl. Ergon. **36**, 557 (2005)
19. C. Tanoue, K. Ishizaka, and M. Nagamachi, Int. J. Ind. Ergon. **19**, 115 (1997)
20. S. Ishihara, K. Ishihara, M. Nagamachi, and Y. Matsubara, Int. J. Ind. Ergon. **15**, 13 (1995)
21. Y. Shimizu and T. Jindo, Int. J. Ind. Ergon. **15**, 39 (1995)
22. M. Nagamachi, *Rough Sets and Current Trends in Computing*, 27 (2006)
23. C. E. Lance, M. M. Butts, and L. C. Michels, Organ. Res. Methods **9**, 202 (2006)
24. S. Schütte, *Designing Feelings into Products* (2002)
25. S. H. Hsu, M. C. Chuang, and C. C. Chang, Int. J. Ind. Ergon. **25**, 375 (2000)
26. C. Llinares and A. F. Page, Landsc. Urban Plan. **87**, 247 (2008)
27. A. S. Beavers, J. W. Lounsbury, J. K. Richards, S. W. Huck, G. J. Skolits, and S. L. Esquivel, Prat. Assesment, Res. Eval. **18**, (2013)
28. A. Yong and S. Pearce, Tutorials Quant. Methods **9**, 79 (2013)
29. A. M. Lokman and K. A. Kamaruddin, *2010 International Conference on User Science Engineering, 38* (2010)
30. D. Witarsyah, T. Sjafrizal, M. F. M. Fudzee, and M. A. Salamat, Int. J. Adv. Sci. Eng. Inf. Technol. **7**, (2017)
31. R. Carreira, L. Patrıcio, R. Natal Jorge, C. Magee, and Q. Van Eikema Hommes, Transp. Policy **25**, 233 (2013)

Design and Development of a Portable Metal Chip Baler using A System Design Approach

Mohd Fahrul Hassan[1,*], *Mohd Faiz* Afham[1], *Ahmad Mubarak Tajul* Arifin[1], *Al Emran* Ismail[1], *Mohd Rasidi* Ibrahim[1], *Mohammad Zulafif* Rahim[1], *Md Fauzi* Ahmad[2]

[1]Faculty of Mechanical and Manufacturing Engineering, Universiti Tun Hussein Onn Malaysia, 86400 Parit Raja, Batu Pahat, Johor, Malaysia
[2]Faculty of Technology Management, Universiti Tun Hussein Onn Malaysia, 86400 Parit Raja, Batu Pahat, Johor, Malaysia

Abstract. A large amount of metal chips at workplace will result in untidy and unsafe condition thus measurements of safety are needed in some industries, where the metal chips will be collected and put into a container until the volume is sufficient to be recycled. Due to that reason, the metal chips require a lot of spaces for storage before going to recycle. In this study, a portable metal chip baler as a device for compacting those metal chips is presented based on a system approach of engineering design. Basically, the system design evolves through four phases of development that are started from conceptual design, preliminary system design, detail design and development to system test and evaluation. The portable metal chip baler uses current technology such as pneumatic cylinder to compress the metal chips so that the system capable to operate efficiently. The output from this system is the metal chips are compacted into a block shape and a working prototype was developed to prove the concept of the system. As a summary, the conceptual design of portable metal chip baler was proven and was presented using the philosophy of the systems design approach. This tool may assists workers especially in the Small-Medium Enterprise (SME) manufacturing industries, school or universities' workshops for managing metal chips easily and systematically.

1 Introduction

Waste is a type of substance that are unwanted or unused. The examples of waste are municipal waste (household trash), hazardous waste, wastewater (sewage), radioactive waste and others [1]. This waste can cause negative effect to the environment, social and economy. Exposure to hazardous wastes, particularly when they are burned, can cause various other diseases including cancers. Toxic waste materials can contaminate water, soil, and air which causes more problems for humans, other species, and ecosystems [2]. The issue of metal waste associated with environmental management has been faced by manufacturing industry sectors, and immediately has to complement innovating technologies to productive processes in order to reach sustainability [3].

In this study, the waste that involved are metal waste that are produced from machining process in industry. Usually the process that involves in machining process of metal are

[1]Corresponding author: fahrul@uthm.edu.my

drilling, milling, lathe and others. The main purpose of these processes are mainly to achieve a single objective that is to obtain a desired geometry on the metal workpiece. These processes produce metal waste by removing the unwanted metal on the workpiece and the type of metal waste that are produced are metal chips. Some industries have taken a good initiative to overcome those issues where it gives benefit to the industries as well in terms of end-of-life profit by recycling the metal chips [4]. In order to recycle the metal chips, a larger space is needed to store these metal chips before the metal chips can be recycled. Fig. 1 shows the problem that has been faced by the manufacturing industries.

Fig. 1. Metal Chip formation from machining process.

Due to this problem, this paper presents a portable metal chip baler as a device for compacting those metal chips using a system design approach. The system design is used in this study to assist the design and development process of the portable metal chip baler systematically. The detail of the system design process is described in the next section, continued with the design and development of the portable metal chip baler and end up with the conclusion section.

2 Methodology

2.1 A System Design Process

The system design has been introduced with in great detail ever since around 1940 [5]. The purpose of this scientific approach is to assist in the planning, design, development, manufacture and evaluation of engineering systems. In this section, a part of the system design process from Misra [5] as a reference of this study has been described, as follows:

2.1.1 Phases of System Design

Basically, the system design evolves through several phases of development that started from (1) Conceptual Design, (2) Preliminary System Design, (3) Detail Design and Development to (4) System Test and Evaluation.

Phase 1: Conceptual Design. Conceptual design is the basis on which the lifecycle phases of the remaining stages of system design. Conceptual design changes from:

- Functional definition of the system based on identified needs of the system and the customer requirements.
- Establishment of design criteria.

Phase 2: Preliminary System Design. The system level requirements obtained from the conceptual design phase will be translated into subsystem level requirements in this phase.

It also extends functional analysis and requirements allocation from the baseline, to the depth that is needed to identify specific requirements for specific components and other related resources. These resources may be in the form of software, hardware, people, data, facilities, or their combinations. A review on system design is undertaken to ensure that the needs are achieved and the results of the functional analysis and allocation process, the design approach selected, the trade-off studies, etc., are reviewed for compliance with the initially set requirements. All information are documented, and the necessary corrective measures as considered appropriate are initiated. This phase provides results that are supporting detail design and development.

Phase 3: Detail Design and Development. In this phase, the design requirements are derived from the system specifications from the previous phase. The specifications include appropriate design dependent parameters, technical performance measures and associated design to criteria for characteristics that must be incorporated into the system design, subsystems and components. Design requirements for each system element are specified through the allocation process and the identification of detailed performance and effectiveness parameters for each element in the functional analysis. Detail design documentation is an essential part of detail design phase and generates a database for the purpose of information processing, storage and retrieval so that it can be used during the testing and is also available for future designs. At this stage, the design system may be evaluated through the fabrication of a prototype model or using a physical working model. Detail design review is undertaken generally after the detail design has been completed, but before the release of firm design data to initiate production and/or fabrication.

Phase 4: System Test and Evaluation. The idea of evaluation is that the functions that the system must perform to satisfy a specific user need should be assessed along with the expectations in terms of costs, time, effectiveness, frequency and any other factors. However, the functional requirements that start at the system level are expected to determine the characteristics that should be incorporated within the system design and its subsystems and components. System evaluation is a continuous process and is undertaken starting with the conceptual design, and extends to the operational use and support phase, and concludes only when the system is retired. The objective of system evaluation is to determine (through a combination of prediction, analysis and measurement activities) true system characteristics and to ensure that the system successfully fulfils its intended purpose or mission.

In the meantime, testing is done at each design stage to ensure that the design is evolving in the intended direction and goal. For example, feasibility testing is done by the designer to prove the design concept and to choose the most promising concept from several possible design concepts. Evaluation testing is done to test early hardware in the operating and environmental conditions for which it was designed. Test procedures and test results are recorded.

3 The System Design Process for a Portable Metal Chip Baler

3.1 Conceptual Design

The solution of a problem is the main aspect of designing. The purpose of design is to make something new that is something that is not existed yet or modifying to an existing product to fulfil customers' needs and to improve the design to ease human daily task.

Function analysis is a method used for analysing a function structure which is an abstract model of a new product, without material features such as shape, dimensions and materials of the parts. It describes the functions of the product and its parts and indicates the mutual relations which underlays ideas that a function structure may be built up to from a number of functions. Fig. 2 shows the black box model for portable metal chip baler. The input of the black box for the portable metal chip baler is electricity, air, signal and also the metal waste. These inputs are the necessary items to perform the design objective. Meanwhile, the output of the black box is compacted metal chip.

Function structure analysis explains about the process that occurs starting from the supplied input until an output is obtained. In the portable metal chip baler system, there are several processes have been carried out before the complete product or processed product is obtained. For initial process, the electricity is used to run the compressor so air can be sucked and compressed into a storage air tank to obtain high air pressure. When high compressed air are obtained, the metal chip will be poured into the product design feed. Then it is fill up the compaction chamber that are design for the metal chip. Next process is the compaction process that occurs after the metal chip fills up the compaction chamber. The high air pressure are then released into the pneumatic cylinder and pushed the metal plate thus compacting the metal chip and this process are controlled by solenoid valve as it is using pneumatic system. A sensor will detect when the metal chip are fully compacted thus sends a signal to the dropping mechanism. The dropping mechanism will drop the compacted metal chip into the storage container thus the final processed product are obtained. Fig. 3 shows the function structure for the portable metal chip baler.

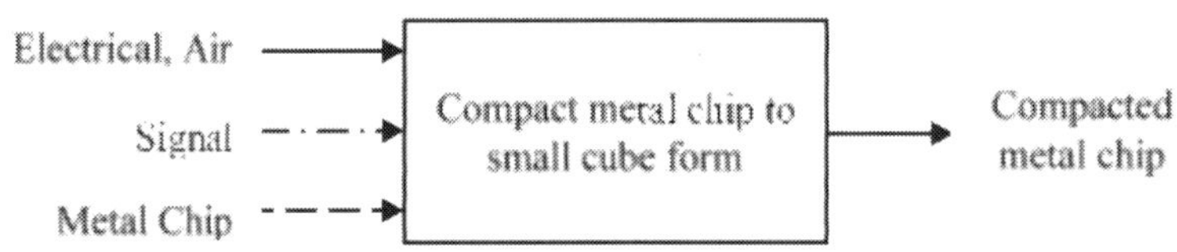

Fig. 2. Function structure black box for a portable metal chip baler.

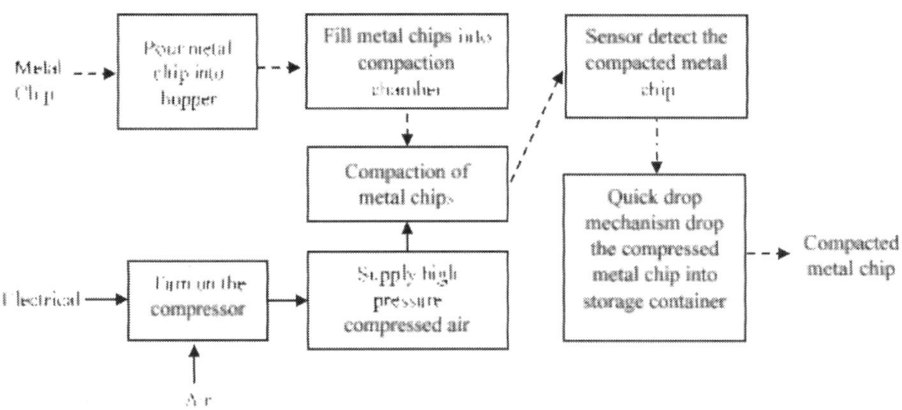

Fig. 3. Function structure for a portable metal chip baler.

3.2 Preliminary System Design

Due to achieve that solution, a new or modifying an existing product is designed based on concept generation and interpreting ideas that are obtained from the previous phase. In this study, the method of generating concepts is using morphological chart method. The morphological chart encourages the designers to widen the search of possible new solution regarding to the present problem, to identify the combination and elements or components that can be implemented to the design concept [6].

Table 1 shows the morphological chart for portable metal chip baler system that has been structured to search several possible combinations. For the complete combinations, eight sub-systems have been combined where only one alternative for each sub-system will be selected. Fig. 4 shows the three combinations that have been structured based on the morphological analysis.

After generated with several combinations, an evaluation is needed to select the best combination. The evaluation can be done by considering the objectives that the design supposed to be achieved. The tool that has been used to evaluate the combinations is weighted rating method that is carried out by comparing the selected alternative design and evaluated using rating scales 0 to 5 where the highest rating is better. Then the assigned rating is multiplied with the importance factor and the total points are summed up and the weightage of each features or specs are obtained.

Table 1. Morphological Chart for Portable Metal Chip Baler System

No	Sub-system	Alternatives		
		1	2	3
1	Power Source	Electricity	Petrol Engine	Solar
2	Metal Chip Detection Mechanism	Touch Sensor	Distance Sensor	Infrared Sensor
3	Control System	Pneumatic Control Valve	Solenoid Valve	Electric Switch
4	Air Supply	Piston System	Air Supply Device	Compressor
5	Vibration Insulator	Rubber	Insulation Pad	Spring
6	Portability	3 inch Caster Wheel	5 inch Caster Wheel	7 inch Caster Wheel
7	Metal Chip Compaction Mechanism	Hydraulic	Flat plate with pneumatic cylinder	Motorized
8	Product Material	Aluminium	Stainless Steel	Galvanized Steel

 DOI: 10.1051/matecconf/201713500018

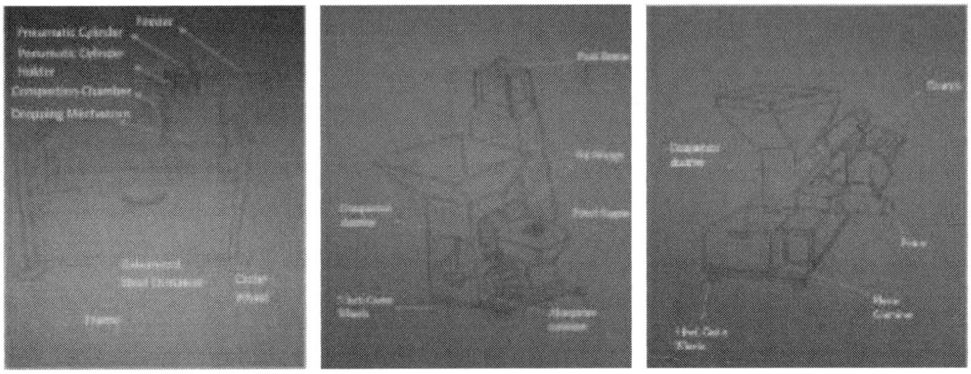

(a) Concept 1 (b) Concept 2 (c) Concept 3

Fig. 4. Sketches of Portable Metal Chip Baler concepts.

Table 2 shows the weighted rating method for evaluating the combinations of portable metal chip baler. For the evaluation, a set of 12 criteria that was based on the design objectives has been considered. Based on the evaluation, combination 1 has the highest total score among the other combinations. It is believed, the combination 1 has the highest score due to the simplest design, low maintenance cost as the spare parts are cheap and portable. However, the dimension has an average score. The safety and green friendly aspects of each alternative is equal among the alternatives.

Table 2. Weighted rating method for Portable Metal Chip Baler

Design Criterion	Weight Factor	Concepts					
		1		2		3	
		R	W	R	W	R	W
Interesting Design	0.16	4	0.64	3	0.48	2	0.32
Machine cost	0.14	3	0.42	2	0.28	3	0.42
Spare parts price	0.02	3	0.06	1	0.02	2	0.06
Labour cost	0.04	4	0.16	3	0.12	3	0.12
Easy to use	0.04	4	0.16	3	0.12	2	0.08
Number of operator	0.03	4	0.12	2	0.06	3	0.09
Portable	0.04	3	0.12	2	0.08	3	0.12
Size	0.06	3	0.18	3	0.18	4	0.24
Weight	0.05	3	0.15	2	0.10	3	0.15
Low maintenance cost	0.14	3	0.42	2	0.28	2	0.28
Safety	0.17	4	0.68	4	0.68	4	0.68
Green friendly	0.11	4	0.44	3	0.33	4	0.44
Total			3.49		2.55		3

* R=Rating; W=Weightage

Table 3 shows the sub-systems and the selected components with the function that they are required to perform on the portable metal chip baler. In this study, the power supply that suits for this product is from electricity because it is easy to obtain and it does not product any greenhouse effect as this study is to reduce metal waste. For the metal chip detection function, a touch sensor is chosen to be used as it more effective as when the plate that compacts the metal chip touches the sensor. It will operate the dropping mechanism so the compacted metal chip will fall into the container.

Table 3. Assigned functions for Portable Metal Chip Baler System.

Sub-system	Component	Function
Power Source	Electricity	Act as a power source for the project
Metal Chip Detector	Touch Sensor	To detect the metal chip after compaction process
Control system	Solenoid Valve	To control the pneumatic system
Air Supply	Air Compressor	To generate and store air pressure.
Vibration	Rubber Absorber	To absorb vibration from the compaction process
Portable	Caster Wheel	To support and make the project portable
Metal Chip Compaction Mechanism	Flat plate with pneumatic cylinder	Used to compact the metal chip in the compaction chamber
Storage	Metal Container	To store the metal chip that has been compacted.

Pneumatic system using solenoid valve is used to control the air flowing into the cylinder thus control the compaction mechanism. Air compressor is used to generate air pressure and store it into a storage tank before it is used to compact the metal chip. The product may produces vibration as it compacts the metal chip so rubber insulator is selected to reduce the vibration from the product to the floor surface. Caster wheels are selected to be used due to its availability as it is easy to find. The storage container will be made from galvanized steel because it is much stronger ad cheaper thus it is suitable to get the job done.

3.3 Detail Design and Development

In this section, the proposed design concept of a portable metal chip baler is modelled using SolidWorks software. In order to assemble a complete 3D CAD model of the portable metal chip baler (as shown in Fig. 5a), all components that has been identified in the previous phase have been drawn in the first step. After complete in the design stage, the proposed design of portable metal chip baler (as shown in Fig. 5b) is fabricated for prototype testing in the next phase.

DOI: 10.1051/matecconf/201713500018

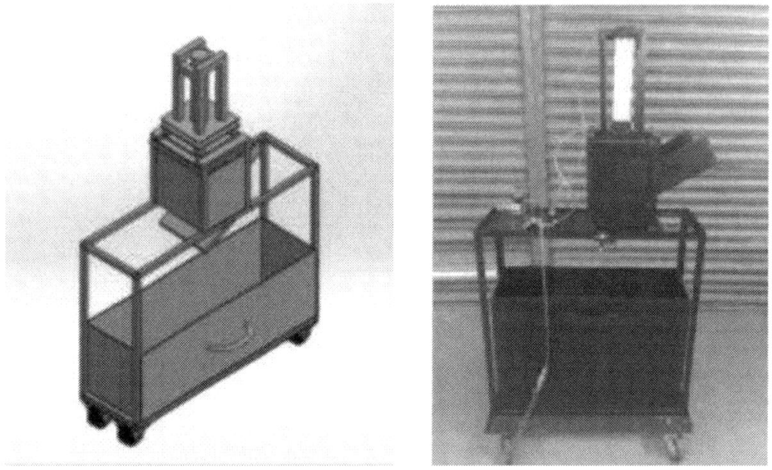

(a) 3D CAD model. (b) Prototype development.

Fig. 5. Design and development of Portable Metal Chip Baler.

3.4 System Test and Evaluation

After the portable metal chip baler is fabricated, prototype testing is conducted in order to verify whether the design system satisfies the objective of this study or not. During the prototype testing, a pile of metal chip is gathered and feed into the portable metal chip baler and the compacting process is carried out. The testing is conducted three times to ensure the result is satisfied. Fig. 6 shows the process of testing from a barrel of metal chips in a workshop and compaction process using the prototype to several compacted metal chips.

From the testing, the prototype successfully compact the metal chip into a compact square block shape thus reducing the initial size of the metal chips. The prototype had been succeeded and the objective of this study has been achieved.

(a) Metal chips ; (b) Developed metal chip baler ; (c) Several compacted metal chips

Fig. 6. Prototype testing

4 Conclusion

In conclusion, a portable metal chip baler that can compact metal chips using a system design approach is presented. Based on the system design approach, several phases have been performed in order to fulfil the required objectives. The phases include conceptual design, preliminary system design, detail design and development, and system test and evaluation. The process is started from the conceptual design where the function analysis is performed. This phase identified the important system that should be implemented to support the portable metal chip baler. In this phase, the input and output of the system, also the detail system inside of the black box are determined. Then, in the preliminary system design, the process is continued with generating three possible alternatives using morphological chart, evaluating the generated combinations using weighted rating method, and selection process where combination 1 is preferred concept since the score obtained is higher than the others with regard to the assigned criteria. Next, the 3D CAD model of the proposed concept for the portable metal chip baler is designed and fabricated for prototype testing as suggested in the detail design and development phase. In the last phase, the developed prototype has been tested three times for verification purpose. This testing has been conducted for proof-of-concept test so that the proposed concept of portable metal chip baler is working as suggested. The output from this prototype is compacted metal chips where the result means the successful of this study. This result is a preliminary work where further improvement will be made for better result. At the end of this study, the portable metal chip baler can be used by industries for proper metal waste management.

Acknowledgement

This research was funded by Universiti Tun Hussein Onn Malaysia (UTHM) under Incentive Grant Scheme for Publication (IGSP) with Vot no. U411.

References

1. G. Esenduran, E. Kemahlioglu-Ziya, J.M. Swaminathan, *Product Take Back Legislation and Its Impact on Recycling and Remanufacturing Industries. In Sustainable Supply Chains*. (Springer, New York, 2012).
2. Skanska, *4R Guide: Reduce, Reuse, Recycle, Recover.* (Stockholm: Skanska Group, 2012).
3. L. Simon, C.A.M. Moraes, R.C.E. Modolo, M. Vargas, D. Calheiro, F.A. Brehm, Journal of Cleaner Production, **153** (2017).
4. L. Wzorek, M. Wedrychowicz, T. Skrzekut, P. Noga, M. Wiewiora, J. Wiewiora, W. Sajdak, M. Richert, Metallurgy and Foundry Engineering, **42**, 2 (2016).
5. K.B. Misra, *Engineering Design: A Systems Approach. In Handbook of Performability Engineering*. (Springer, London, 2008).
6. M.F. Hassan, M.Z.M. Saman, S. Sharif, B. Omar, Int. J. Sustainable Manufacturing, **2**, 4 (2012).
7. M.F. Hassan, M.K. Keat, M.Z. Yunos, S. Adzila, A.M.T. Arifin, M.N.A. Rahman, R.H.A. Haq, ARPN Journal of Engineering and Applied Sciences, **11**, 12 (2016).

Shape Optimization of High-Speed Rail by Biomimetic

Chin Toong Foo[1], *Badrul* Omar[1,*], *Ishkrizat* Taib[2,*]

[1]Universiti Tun Hussein Onn Malaysia, Engineering Design Research Group (EDRG), 86400 Batu Pahat Johor, Malaysia

[2]Universiti Tun Hussein Onn Malaysia, Flow Analysis, Simulation, and Turbulence Research Group (FAST), 86400 Batu Pahat Johor, Malaysia

Abstract. The Southern Corridor High-Speed Rail (HSR) has the potential to connect Kuala Lumpur and Singapore with only 90 minutes of travel time. The HSR needs to travel at minimum drag in order to promote ride comfort and environmental sustainability. Eight combinations of train head bio-models are generated from a morphological chart of three key concepts formed by biomimetic method, referring to the peregrine falcon, belted kingfisher, sailfish and dolpine. The three key concepts are the type of train nose, cross-section shape, and train nose shape. The bio-models are drawn by computer-aided drawing software and the air flow around the bio-models are simulated in order to obtain the drag coefficients in counterflow condition. The drag coefficient of each bio-models are compared and validated with published drag coefficient of Harmony CRH380A, which is 0.1971. The bio-models with lowest drag coefficient is the bio-model with sharp-streamlined nose and round-based cross-section, which is 0.1559. The drag coefficient of Harmony CRH380A is reduced by 20.9% if the windscreen is removed and the paraboloid nose is replaced with sharp nose. These bio-models can be a guideline for future studies in developing a more sustainable train shape in the railway industry.

1 Introduction

High-speed rail (HSR) is one of the rail-type transports that enables mass passengers to travel from places to places within a short time. HSR is generally defined as rail transport that could operate at a high speed, and the European Union specifically defines high-speed rail as lines specially built for train of speeds equal or greater than 250 km/h. In countries or cities with high population such as China and Japan, high-speed rail is a necessity to accommodate the movement of the citizens. [1]

Malaysian government has listed Southern Corridor HSR which has the potential to connect Kuala Lumpur and Singapore as one of the Entry Point Projects for the Economic

*Corresponding author: badrul@uthm.edu.my

Transformation Programme (ETP). High-speed rails are important where demand for travel between large cities is high and will continue to increase [2]. There is a need to improve connectivity between the two countries, as currently daily traffic congestion is critical and far exceeds the capacity of the Causeway by 33%, with the current infrastructure provisions [3][4]. Travelling with car is currently the most preferred mode of transportation between the two cities [5]. Traffic of the Causeway will continue to grow at a rate comparable to the Gross Domestic Product (GDP) growth of Malaysia-Singapore, at an average 3 ~ 5%. Over longer term, the growth rate is expected to reach an average of 3.2% per annum from 2011-2060 with a market of 251 million passenger trips by 2060 [3][4].

From the Davis formula, the aerodynamic drag is known to be the main contributor to the overall drag force acting on the HSR, thus becoming one of the main limiting factor of the speed. The poor design of external shape of a HSR is one of the factors of high aerodynamic drag. The aerodynamic drag will cause the increment in the drag force which might reduce the speed of the train drastically. Furthermore, improper aerodynamic drag will induce loud noise in the train, increase structural vibrations, and trigger impulse pressure waves between two trains passing by each other. All these will lead to the discomfort of passengers on board. Thus, the aim of reducing aerodynamic drag is very important in designing a high-speed rail. [6]

In this paper, biomimetic method is used to optimize the shape of high-speed rail, where CRH380A is the scope of this research. The world fastest animals were used to extract the three key concepts. The efficiency of these key concepts which is the type of train nose, shape of train nose and the cross-sectional shape of the train, is compared in this paper. From these key concepts, eight bio-models of HSR are formed from a morphological chart and are compared to find the best combinations of key concepts. Finally, the shape of CRH380A is then optimized based on the best bio-models in terms of lowest drag coefficient.

2 Aerodynamics of Train Nose

According to Schetz [7], the nose and tail only contribute 8 ~ 13% of the aerodynamic resistance. The significance of nose and tail pressure reduction is low over a large variety of slender shapes, as long as there is no sharp edges. A poorly designed train nose will cause the train to exit a tunnel with an explosive noise because the train nose is subjected to micro-pressure wave. Micro-pressure waves are affected by train speed, train-to-tunnel area ratio, slenderness of the nose and *et cetera* [8].

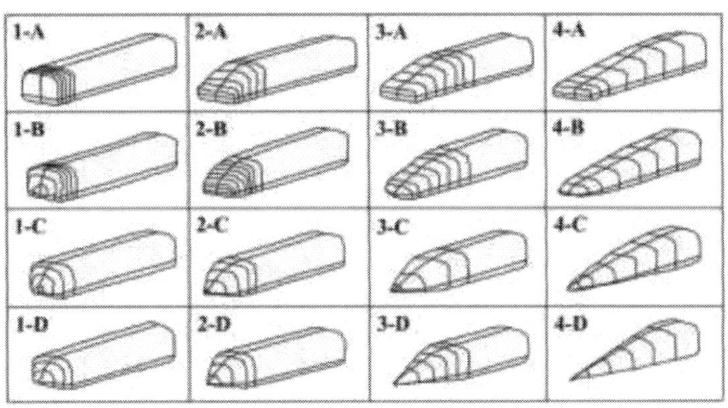

Fig. 1. Model train configuration. [9]

Figure 1 shows a typical specimens of wind tunnel test which yields 16 results. The alphabets A, B, C, and D represents the different cross-sectional area of the nose while the numbers represents the length of change of cross-sectional area. The train head series-1 and series-A has the highest drag coefficient compared to the other series of train head. The reduction in drag coefficient is insignificant after the length-to-width ratio reaches 1.0. However, the length-to-width ratio of the train tail significantly influence the drag coefficient. The higher the length-to-width ratio, the lower the drag coefficient. [9]

The nose shape of the G7 Project of Korean Government is an ellipsoid. Lee & Kim [8] had discovered that paraboloid nose shape is the most efficient shape compared to ellipsoid and circular cone nose shape. The paraboloid nose shape gives the lowest maximum micro-pressure waves. However, an optimized nose shape designed by Ku *et. al.* [10] is compared to the paraboloid nose shape of Lee at the length range of 5 ~ 15 m. The optimal shape has a negative gradient along the length of the nose in the middle section as shown in Figure 2. The maximum micro-pressure waves of the optimal shape compared to the paraboloid shape is reduced by 18 ~ 27%.

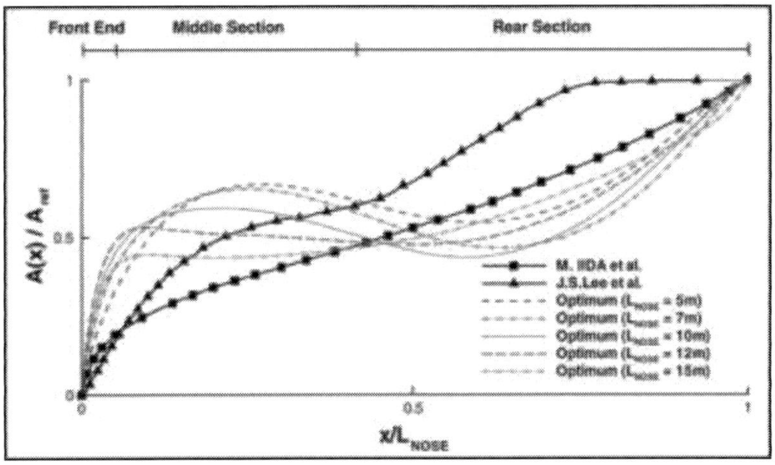

Fig. 2. Comparison of paraboloid shape (J. S. Lee) and optimal shape (Optimum). [10]

The CRH380A has eight cars with the pantographs located at the fourth-car and sixth-car. From Figure 3, we are able to know that the drag force of the fourth-car and the sixth-car has the largest drag coefficient. Relatively, the train-head and train-tail have the largest drag.

Yao *et. al.* [11] optimized the streamline of the head of CRH380A by using Kriging method, CFD analysis, and multi-objective optimization method. Table 1 shows the result of the force coefficient of the optimized shape and original shape. Figure 4 shows that the pressure contour of the optimized shape at the train nose is reduced and the shape of the train head is more streamlined by smoothing out the edge marked with "1".

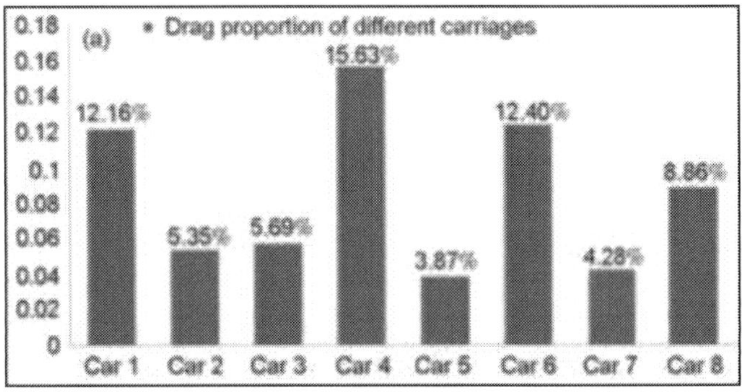

Fig. 3. Drag coefficient distribution of CRH380A. [6]

Table 1. Aerodynamic force coefficients. [11]

Model type	Total C_D	Reduction
Original shape	0.1971	3.20%
Optimal shape	0.1908	

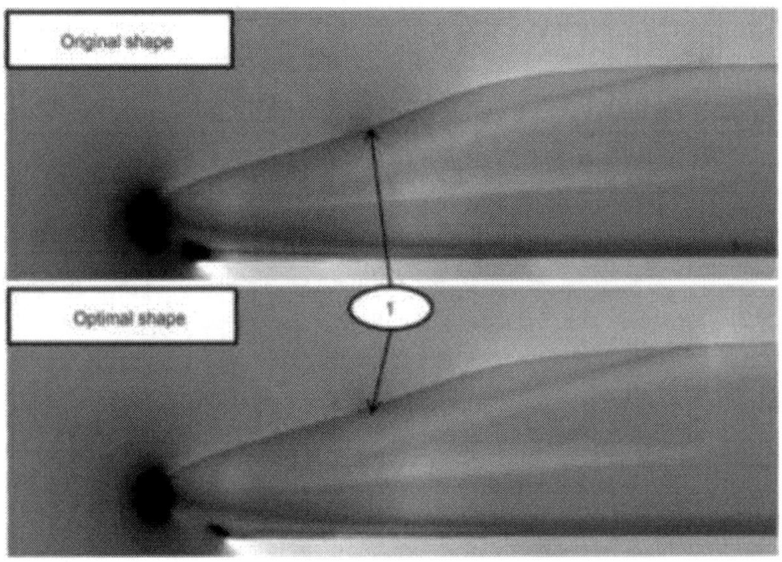

Fig. 4. Pressure contour around the streamline of the optimal shape and original shape. [11]

3 Biomimetics

Biomimetic is widely used in designing streamlined train nose shape [12]. As discussed in section 2.3.2, Kim & Lee [12] developed the BTM with drag reduction by 21%. The chief engineer of the *Shinkansen*, Eiji Nakatsu, used biomimetic to mimic the bill of a belted kingfisher as the basic shape of the *Shinkansen* 500 Series train head. The explosive sound caused by the impact of micro-pressure wave on the train nose as the train enter or exit a tunnel is greatly reduced [13, 14].

It is not the best way to achieve an optimal design by mimicking directly. Verification and validation of a solution-shape is crucial in developing bio-model. This can be done by evaluating and modifying every sub-features in relation to design constraints. Adapting the parametric modelling from evolutionary biology could provide a complementary methodology by which to validate and verify the original bio-model shape, allowing designers to seek a solution shape closer to optimal for each specific design challenge. Some previous effort ignored the verification and validation of a solution-shape because they were too focused on mimicking the original shape and delivering its original characteristics [12]. In order to overcome the issues of credibility of designs by biomimetic, a set of proper procedures and morphological evaluation is included in this research in order to choose the best combination of bio-model generated.

Aerodynamics and hydrodynamics characteristics of different animals are the subject of interest in this research. Head shapes of different fast-moving organisms such as birds, fishes, and eagles, are studied and observed for developing the train nose. Marine animals is included in this biomimicry perspective report. The animals chosen for this research is peregrine falcon, belted kingfisher, sailfish, and common dolphin.

An adult peregrine falcon is the fastest animal in the world that dives at a speed over 386 km/h (230 mph). The subject of interest on the peregrine falcon is the shape profile of its head. A belted kingfisher could fly at a speed ranged between 58 ~ 72 km/h. However, speed is not the subject of interest. A kingfisher is able to dive into water, separating the water with its bill to capture its prey and flew back out of the water within seconds. The subject of interest on the belted kingfisher is its ability to penetrate into water with its long bill at high-speed which other birds could not. Sailfish is known as the fastest fish in the ocean. It could swim at a speed up to 110 km/h. It has highly streamlined body enabling it to swim fast. The subject of interest in the sailfish is its long mouth and streamlined body. Its fins are not considered in this research because trains do not need fins to balance while travelling on a rail track. Common dolphin could swim up to a speed of 40 km/h, which is faster than the world fastest Olympic swimmer. The subject of interest in the dolphin is its highly streamlined shape because it has strong muscles which allows it to sprint. In order for it to sprint at high-speed, streamlined is important for it. These four animals are chosen as they seems to possess the characteristics required to solve the problem of aerodynamic drag faced by high-speed rail.

The key concepts that characterizes each of these organisms also characterize the strategies adopted by each organism and could be used to extend our understanding of such strategies, increasing the chances of finding answers for the challenges proposed. These key concepts and new functions are then used to produce different combinations of train nose shape by morphological evaluation. The three key concepts obtained from the animal model test are type of nose, cross-section shape, and nose shape. Bio-models are obtained by combining different key concepts by morphological evaluation. The bio-models are then compared with the relative design of CRH380A as the benchmark of this research.

The three key concepts are first tabulated in a morphological chart, shown in Figure 5, yielding eight different combinations of bio-model train head as shown in Figure 6. Model drawings are prepared with SOLIDWORKS 2016 software based on the combinations from the morphological chart. The parameters of the bio-model train heads are scaled down to $1/25^{th}$ of the original size as well as the CRH380A model so that the result can be compared at a common standard.

3 CFD Modelling

3.1 Fluid flow characteristics

DOI: 10.1051/matecconf/201713500019

In this research, only two of the governing equations are used; the continuity equation:

$$\dot{m}_{in} - \dot{m}_{out} = \iint V \cdot \bar{n} dA = 0 \quad (1)$$

and the momentum equation:

$$\rho \frac{\partial u}{\partial t} + \rho u \frac{\partial u}{\partial x} + \rho v \frac{\partial u}{\partial y} + \rho w \frac{\partial u}{\partial z} = \left(-\frac{\partial P}{\partial t} + \frac{\partial \sigma_{xx}}{\partial x} + \frac{\partial \tau_{yx}}{\partial y} + \frac{\partial \tau_{zx}}{\partial z} \right) + \rho g \quad (2)$$

The energy equation is not used because the system of this research is adiabatic, therefore no heat energy is transferred.

Key Concept	**Option 1**	**Option 2**
Nose type	*Protruding Nose* 	*Streamlined Nose* 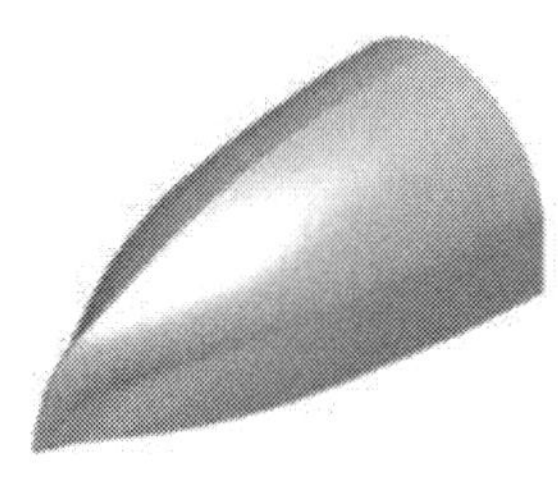
Cross-section	*Round-based* 	*Square-based* 
Nose Shape	*Sharp* 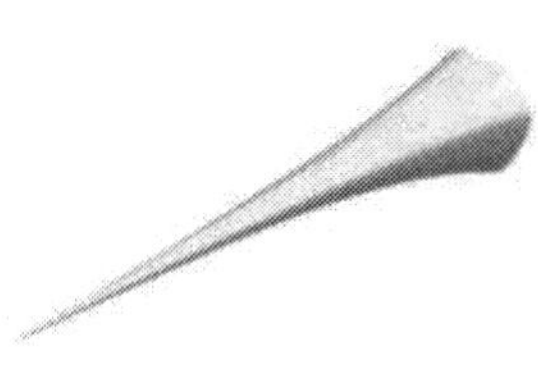	*Rounded* 

Fig. 5. Morphological chart.

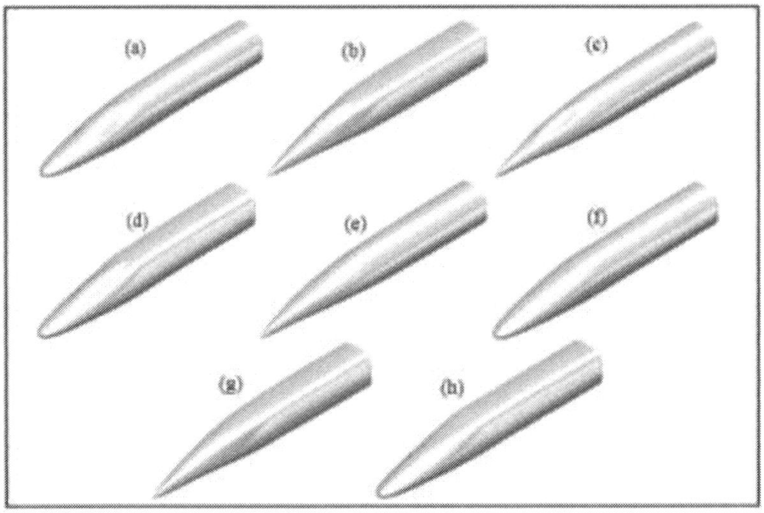

Fig. 6. Bio-models (a) 111, (b) 112, (c) 121, (d) 122, (e) 211, (f) 212, (g) 221, (h) 222.

The momentum equation is then expressed in Reynolds Averaged Navier-Stokes equation:

$$\rho\left(\frac{\partial u'}{\partial t}+\frac{\partial}{\partial x}\left(\bar{u}^2+u'^2\right)+\frac{\partial}{\partial x}\left(\overline{uv}+u'v'\right)+\frac{\partial}{\partial x}\left(\overline{uw}+u'w'\right)\right)=\rho g_x-\frac{\partial\bar{P}}{\partial x}+\mu\left(\frac{\partial^2\bar{u}}{\partial x^2}+\frac{\partial^2\bar{u}}{\partial y^2}+\frac{\partial^2\bar{u}}{\partial z^2}\right) \quad (3)$$

Since the flow is inviscid and the flow is time independent, therefore the body force and the time-dependent variable are equal to zero. The RANS Equation is then reduced to:

$$\rho\left(\frac{\partial}{\partial x}\left(\bar{u}^2+u'^2\right)+\frac{\partial}{\partial x}\left(\overline{uv}+u'v'\right)+\frac{\partial}{\partial x}\left(\overline{uw}+u'w'\right)\right)=-\frac{\partial\bar{P}}{\partial x}+\mu\left(\frac{\partial^2\bar{u}}{\partial x^2}+\frac{\partial^2\bar{u}}{\partial y^2}+\frac{\partial^2\bar{u}}{\partial z^2}\right) \quad (4)$$

The turbulent model *k-ε* realizable model is used in this research because the *k-ε* model focuses on the mechanisms that affect the turbulent kinetic energy. It is used for far field model. This model is made up of two model equations; the transport equation for the mean turbulence kinetic energy, *k*:

$$\frac{\partial}{\partial t}(\rho k)+\frac{\partial}{\partial x_i}(\rho k u_i)=\frac{\partial}{\partial x_j}\left[\left(\mu+\frac{\mu_t}{\sigma_k}\right)\frac{\partial k}{\partial x_j}\right]+G_k+G_b-\rho\varepsilon-Y_M+S_k \quad (5)$$

and the transport equation for the dissipation of mean kinetic energy, ε that occurs in the smallest eddies:

$$\frac{\partial}{\partial t}(\rho\varepsilon)+\frac{\partial}{\partial x_i}(\rho\varepsilon u_i)=\frac{\partial}{\partial x_j}\left[\left(\mu+\frac{\mu_t}{\sigma_\varepsilon}\right)\frac{\partial\varepsilon}{\partial x_j}\right]+C_1\frac{\varepsilon}{k}\left(G_k+C_3G_b\right)-C_2\rho\frac{\varepsilon^2}{k}+S_\varepsilon \quad (6)$$

3.2 CFD simulation

The speed of airflow is 30 m/s. Only counter flow of wind is tested in this simulations. The size of the fluid domain is as shown in Figure 7. Taking the length, L of the train head as a reference, the length of inflow direction is 0.5, the total length of the domain is 4L, the

 DOI: 10.1051/matecconf/201713500019

width and height is 0.5L, and the clearance between the underbody and the rail is 0.00235L. In common practices the fluid domain is usually bigger almost twice the size of the fluid domain used in this research. However, this smaller domain size is used because the focus of this research is the fluid flowing near the surface of the train model and the differences of characteristics of fluid flow between the train head and train tail instead of the characteristics of the entire flow in the domain.

The model mesh is made up of 1.1 million nodes. The element size set is as default; minimum size of 5.1162e-04 *m* and maximum size of 0.102320 *m*. Proximity and curvature is turned on for advanced size factor, fine relevance centre, and high smoothing. The train wall surface is covered by a five layer prism with a stretching factor of 1.2. Grid independent test is not done due to limited time and limited computing resources. Therefore, the best setting available due to constraints are chosen and verified with reference result.

Additional body sizing is also added to refine discretized parts of the model which are the nose tip, protruding head region, and the wake region. The element size for these sizing is set as 0.01 *m* and 0.02 *m* for the wake region. The face sizing for the train wall is set as 0.02 *m*. Smooth transition is set as the inflation option at the train wall with transition ratio of 0.272, maximum layer of five and growth rate of 1.2.

The solution method used is Pressure-Based Simple Solution. The flow is steady and the velocity formation is absolute. The turbulent model used is k-ε realizable model. The setting of the turbulent model is as default.

The inlet boundary layer, the density and dynamic viscosity of air at 303.5 K atmospheric pressure is given by 1.164 kg/m^3 and 1.875X10^{-5} kg/ms respectively. The fluid velocity component is -30 m/s at the z-direction. The turbulent intensity is 0.3%, turbulent viscosity ratio is 10, and the backflow turbulent length scale is 0.012 m. The gauge pressure box is set as zero Pascal because the atmospheric pressure is not taken into consideration for gauge pressure. The outlet boundary conditions used is pressure outlet. The turbulent intensity is set as 5% while the remaining setting is the same as the inlet. The surrounding boundary condition is set as moving frictionless wall at a speed of 30 m/s to the opposite of *z*-direction because the wall is a part of the flowing air. However, the rail is a stationary frictionless wall.

The turbulent kinetic energy, k applies at the first order upwind while the turbulent dissipation rate, ε is second order upwind for the solution method. The solver method is simple scheme. The surface area that is used to calculate the drag coefficient is the cross-sectional area of the frontal side.

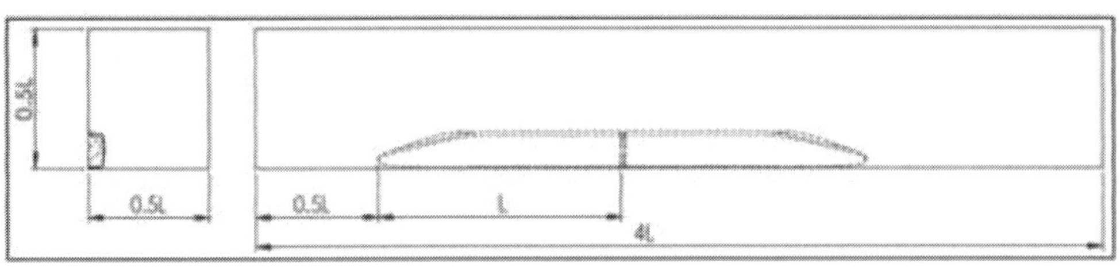

Fig. 7. Fluid domain size.

 DOI: 10.1051/matecconf/201713500019

4 Results

The drag coefficient of CRH380A model is 0.1973, which is very near to the reference drag coefficient. Among the eight bio-models, bio-model 211 has the lowest drag coefficient, which is 0.1559. The drag coefficient of each respective models are tabulated in Table 2 and compared in Figure 8. The drag coefficient for model CRH380A is 0.1973 and will be used as the benchmark in this research. The bio-models' drag coefficient obtained is lower than that of CRH380A model, ranged between 0.1559 ~ 0.1922.

Figure 9 and Figure 10 are the pressure distribution along CRH380A model and bio-model 211. The pressure at the nose tip of CRH380A model is almost doubled of that of bio-model 211. The occurrence of high pressure region at the train head of CRH380A model is also more frequent compared to bio-model 211.

The drag coefficient of bio-models with protruding nose is higher because the occurrence of stagnation point is more due to the presence of windscreen. These high pressure regions around the windscreen that are shown in Figure 11 for model 111 is not found in model 211 and any other bio-models without windscreen. These stagnation points in the head increased the total pressure of the head which leaves the tail with a relatively lower pressure coefficient, therefore increasing the drag.

Table 2. Drag coefficient result.

Train model	Published C_D	Simulated C_D
CRH380A	**0.1971**	**0.1973**

Bio-model	C_D	Bio-model	C_D
111	0.1582	*211*	0.1559
112	0.1922	*212*	0.1826
121	0.1655	*221*	0.1651
122	0.1845	*222*	0.1785

DOI: 10.1051/matecconf/201713500019

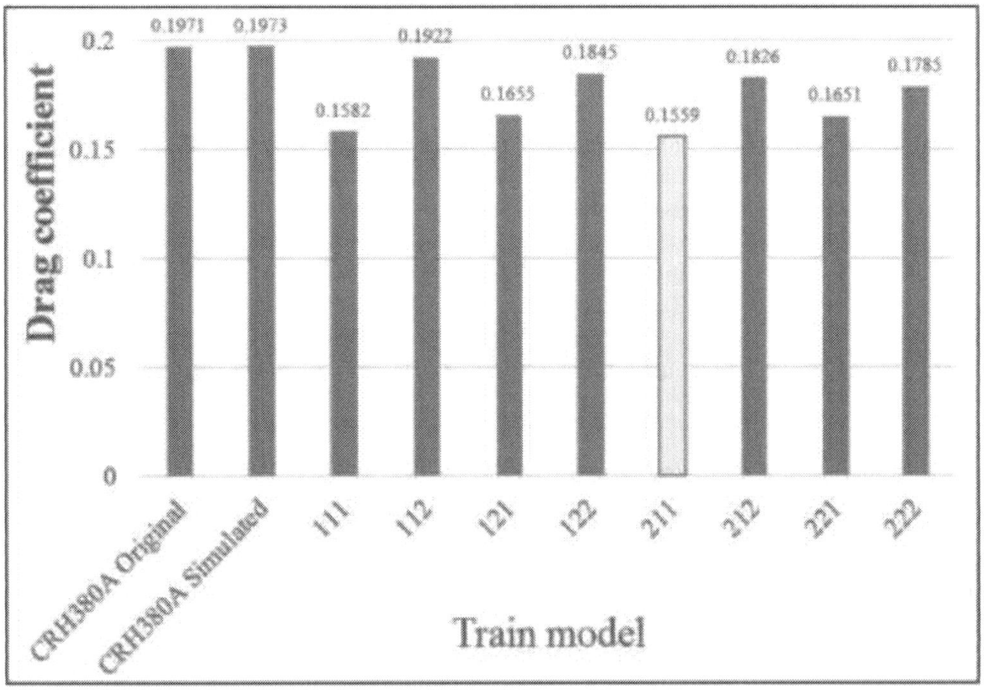

Fig. 8. Drag coefficient of different train models.

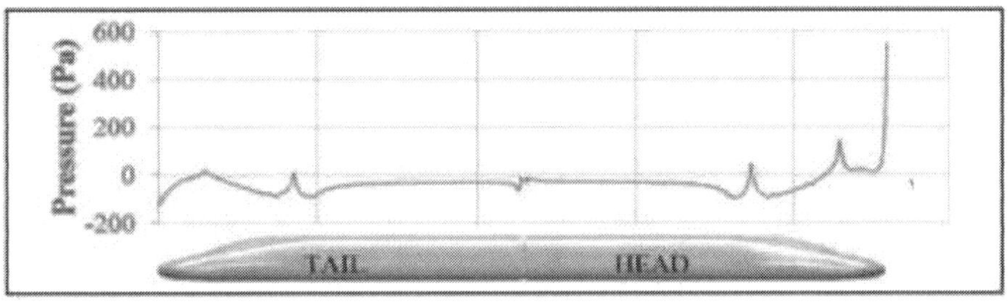

Fig. 9: Pressure distribution of Harmony CRH380A.

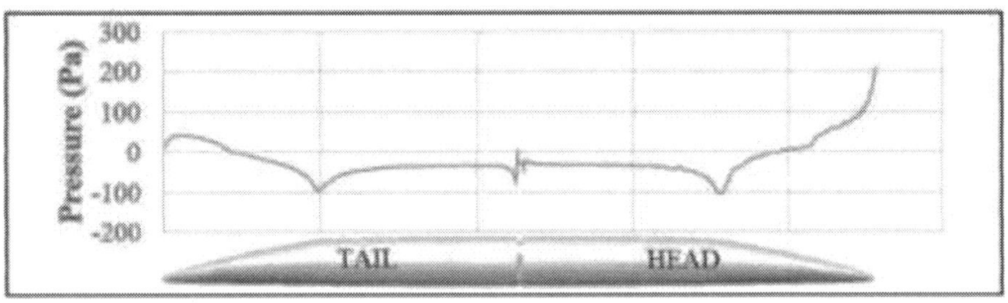

Fig. 10. Pressure distribution of Bio-model 211.

The drag coefficient of the HSR is not dependent to the cross-section shape of the train body, neither round-based (prefix 1) nor square-based (prefix 2). Figure 12 shows the comparison of drag coefficient for different groups of bio-models; round-based cross-section and square-based cross-section. The drag coefficient of 111 and 211 is lower than that of 121 and 221 by 5% while the drag coefficient of 112 and 212 is higher than that of 122 and 222 by 3%.

DOI: 10.1051/matecconf/201713500019

The drag coefficient for rounded nose shape models are higher because the area of stagnation point for rounded nose shape is larger and the impact is greater compared to that of sharp nose shape as shown in Figure 13. The maximum pressure at the nose tip of model 211 is 209 Pa. However, the maximum pressure at the nose tip of model 212 is 555 Pa. There is a huge difference in pressure between the two types of nose shape. Therefore, sharp nose shape is more efficient than rounded nose shape.

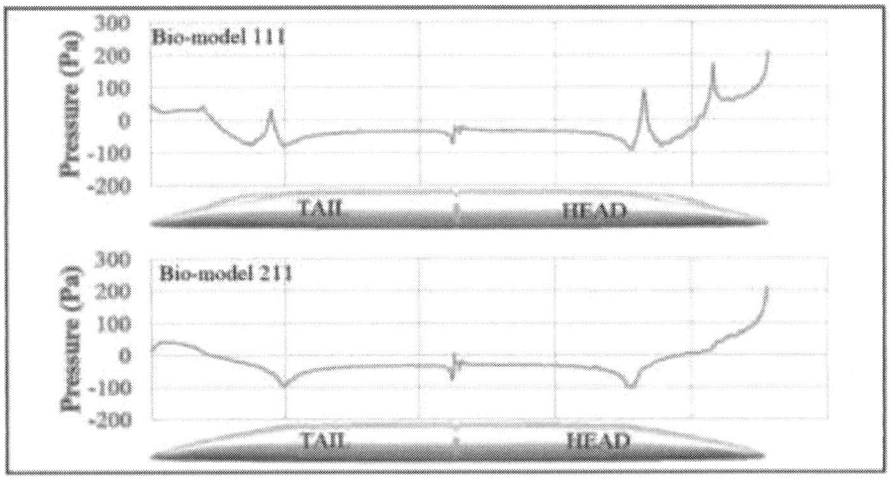

Fig. 11. Pressure distribution comparison of different nose type.

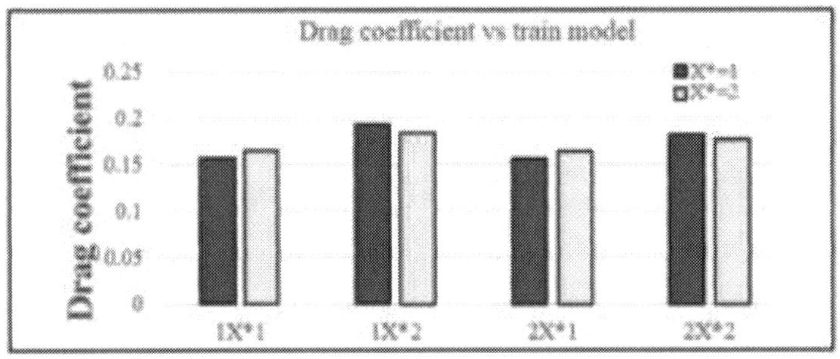

Fig. 12. Drag coefficient of train models with round-based and square-based cross-section.

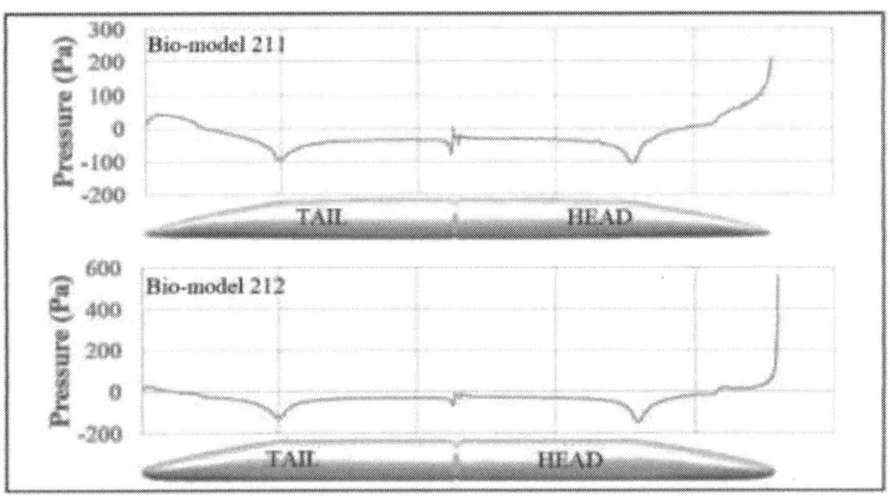

Fig. 13. Pressure distribution comparison of different nose shape.

5 Conclusion

By comparing eight bio-models, it is found that bio-model 211, Figure 6 (e), is the best model with a drag coefficient of 0.1559. Bio-model 211 has sharp-streamlined nose and round-based cross-section. Though the cross-section shape doesn't affect the drag force much. However, the simulated result shows that bio-model 211 is the model with lowest drag coefficient. Train heads are best without the dome-shape windscreen in order to be more streamlined, because the windscreen will increase the frequency of micro-pressure waves at the train head. Sharp train nose is best in reduction of micro-pressure waves because the area of stagnation point is smaller.

By comparing the drag coefficient of the generated bio-models with CRH380A, it is found that all of the bio-models have lower drag coefficient. The model of CRH380A resembles bio-model 112 and 122 because both of these two models have round-protruding nose. In fact, the morphological chart of this researched is a basic guideline for building any train head model because what differs one HSR's shape from another is the shape of the nose, cross-section shape, and the presence of the windscreen. The drag coefficient of bio-model 112 and 122 is nearest to that of CRH380A, which is 0.1922 and 0.1845 respectively while the drag coefficient of CRH380A is 0.1971.

Since CRH380 resembles bio-model 112 and 122, the optimization criteria should be the key concepts used to build bio-model 211 where the windscreen of CRH380A should be removed and be made streamlined in order to reduce the micro-pressure waves at the train head. The original nose shape of CRH380A is paraboloid. But if the nose shape is replaced by sharp nose shape, the micro-pressure waves at the train nose can be reduced dramatically. By implementing such modifications on the current model of CRH380A, the new model is expected to have a drag coefficient of 0.1559, which is the drag coefficient of bio-model 211, with a percentage reduction of 20.9%.

6 Acknowledgement

This paper was partly funded by the Centre for Graduate Studies UTHM and Postgraduate Research Grant (GPPS/U799). The authors are thankful to Universiti Tun Hussein Onn Malaysia (UTHM) for providing facilities support.

References

1. UIC, The High-speed Definition of the European Union, Retrieved March 26, 2016, from http://www.uic.org/highspeed (2012).
2. T. Gourvish, The High Speed Rail Revolution: History and Prospects, *High Speed Two Ltd, United Kingdom*. United Kingdom: High Speed Two Ltd, (2010).
3. MyHSR Corp, Kuala Lumpur – Singapore High Speed Rail Project. Retrieved March 26, 2016, (2015).
4. SPAD, Suruhanjaya Pengangkutan Awam Darat, High Speed Rail. Retrieved March 26, 2016, from http://www.spad.gov.my/projects/high-speed-rail Links, (2014).
5. H. H. Ling, *A Comparison Study on the Carbon Footprint between the Proposed Kuala Lumpur - Singapore High Speed Rail and Other Transportation Modes*. Universiti Tun Hussein Onn Malaysia, (2016).
6. G. Yang, D. Guo, S. Yao, & C. Liu, Aerodynamic Design for China New High-Speed Trains. *Science China Technological Sciences*, *55*(7), 1923–1928, (2012).

7. Y. C., Rho, J. H., Yun, S. H., Kwak, M. H., Kim, K. H., Kwon, H. Bin, & Lee, D. H. Optimal Cross-Sectional Area Distribution of a High-Speed Train Nose to Minimize the Tunnel Micro-Pressure Wave. *Structural and Multidisciplinary Optimization*, *42*(6), 965–976, (2010).
8. J. Lee & J. Kim, Approximate Optimization of High-Speed Train Nose Shape for Reducing Micropressure Wave. *Structural and Multidisciplinary Optimization*, *35*(1), 79–87, (2008).
9. R. S. Raghunathan, H. D. Kim, & T. Setoguchi, *Aerodynamics of High-Speed Railway Train. Progress in Aerospace Sciences* (Vol. 38), (2002)
10. Y. C. Ku, J. H. Rho, S. H. Yun, M. H. Kwak, K. H. Kim, H. B. Kwon, & D. H. Lee, Optimal Cross-Sectional Area Distribution of a High-Speed Train Nose to Minimize the Tunnel Micro-Pressure Wave. *Structural and Multidisciplinary Optimization*, *42*(6), 965–976, (2010).
11. S. Yao, D. Guo, Z. Sun, G. Yang, & D. Chen, Multi-Objective Optimization of the Streamlined Head of High-Speed Trains Based on the Kriging Model. *Science China Technological Sciences*, *55*(12), (2012).
12. S. J. Kim & J. H. Lee, Parametric Shape Modification and Application in a Morphological Biomimetic Design. *Advanced Engineering Informatics*, *29*(1), 76–86, (2015).
13. S. J. Kim & J. H. Lee, How Biomimetic Approach Enlarges Morphological Solution Space in A Streamlined High-Speed Train Design ? *Proceedings of the 16th SIGraDi Conference*, 538–542, (2012).
14. A. Deshpande, A. L. Goh, A. Goossens & S. Javdani, *Biomimicry for Coastal Eco-fcities: Towards a Carbon Neutra Dover, UK.* (R. A. Shenoi, P. A. Wilson, & S. S. Bennett, Eds.) (Vol. 4). Southhampton, England: University of Southampton, (2013).

Review on Design for Medical Device

Naqibah Azman[1], *Nik Hisyamudin* Muhd Nor[1], *Mohd Fahrul* Hassan[1], *Saifulnizan* Jamian[1], *Muhammad Hanif* Abd Rashid[1], *Badrul* Omar[1], *Siti Zarina* Mohd Muji[2], *Fariza* Mohamad[2], *Mohd Fauzi* Ahmad[3], *Faiz* Mohd Turan[4], *Elmi* Abu Bakar[5]

[1]Department of Design and Material Engineering, Faculty of Mechanical Engineering and Manufacturing, Universiti Tun Hussein Onn Malaysia (UTHM), Batu Pahat, Johor, Malaysia.
[2]Faculty of Electric and Electronic Engineering, Universiti Tun Hussein Onn Malaysia (UTHM), Batu Pahat, Johor, Malaysia.
[3]Faculty of Technology Management and Business, Universiti Tun Hussein Onn Malaysia (UTHM), Batu Pahat, Johor, Malaysia.
[4]Faculty of Manufacturing Engineering, Universiti Malaysia Pahang, Pekan, Pahang, Malaysia.
[5]School of Aerospace Engineering, Universiti Sains Malaysia, Nibong Tebal, Seberang Perai Selatan, Pulau Pinang Malaysia.

Abstract. Medical devices are one of the uprising star in product development due to high demand among users. Developing safe, effective and usable design is indeed a challenging work which faced by designer nowadays. There are comparative methods in designing medical device to achieve desired outcome. Weather designing a "User-friendly", "Safe to use" or "Eco-friendly" product requires an exclusive method. This review paper might illuminate some light to the designers and researchers who choose to integrate those aspect in their design work in healthcare sector. A review on current method used on user-friendly approach, ensuring safety, regulations on safety of medical device and suggested standards to ensure safety in less complicated way. Thus, there is no solid method or standard yet to establish on safety for medical device.

1 Introduction

Reported by the Department of Health United Kingdom, in an article "An Organization with a Memory" which record repeated cases which occur not only in medical sector but also in industry and research area. However, the patient safety is not generally secured in this particular aspect [1]. Reported by The National Patient Safety Agency's (NPSA), 24,207 patient safety incident was reported which involve medical device since April 2006 until March 2007 which equivalent to 3% of overall incident that been reported in the time period [2]. From these reported incident in 2008, 303 incidents were classified as a result of fatal or severe harm to patient [3]. Investigations shows that such incident experience a classic system failure thus amenable for prevention [4][5]. Record have found weakness at every stage process which include poor labeling, confusing procedure, lack of standardization across units and other cause [6]. Further analysis have been made throughout the year have found that medical device design was one of the issue.

*Corresponding author: hd160039@uthm.edu.my, nhisyam@uthm.edu.my

For example, incorrect connection, identification issue and lack of feedback from user have led to patient safety incident. The awareness of the role played by the user in recognizing user error by the staffs and manufacturer have contributes in improving procurement decisions, scopes and selecting well-designed device. Thus, suggested by NPSA was to demonstrate safe design philosophies and guidelines for a more standardized device which suits user requirements, usage context and potential usage environment need to be taken into measures [7].

Table 1 shows the classification of incident according to their groups that is reported by NPSA in 2008. The collected data have shown that incident cause by medical device is in the top-five causes of incident reported although the safety of medical device is often neglected [8].

Table 1. Classification of incident that is reported in 2008 by NPSA [8]

	Number in group in incident reports	Number remaining in initial group after reclassification	Percent remaining in initial group after reclassification, %	Number in group reallocated by investigators
Access admission, transfer and discharge	362	312	86	1032
Clinical assessment	266	217	82	49
Consent, communication, confidentiality	232	172	74	549
Documentation	247	227	92	552
Implementation of care	412	336	82	1047
Infection control	174	161	93	257
Infrastructure, staff	814	714	88	1289
Medical device/ equipment	647	586	91	1003
Medication	1145	1096	91	1450
Patient accident	330	273	83	286
Treatment procedure	786	366	43	576
Other	200	57	23	365
Total	5615	4517	80	8905

Reported by the European Pressure Sore Advisory Panel grades that 119 cases of pressure sore are associated with medical device. In 87 cases of pressure sore which are often associated with medical device are being presented on admission, suggested that implementation of care and regular monitoring should be implemented [9][10][11]. There is no 'accurate' or 'safe' number of patient safety to be considered as safe however a 'low' reporting rate can be interpreted as 'safe' and a 'high' report rate can be describe as 'unsafe' which represent culture of greater openness [12].

2 Medical Device

2.1 What Is Medical Device?

Primarily define by Medical Device Authority of Malaysia, medical device is any instrument, apparatus, implement, machine, appliance, implant , calibrator or in vitro reagent , material or software produce by the manufacturer to be used alone or combined for humans in various purpose [13]. Generally explained by Top Market report in 2016 produce by the US government, a medical device also defined as any piece of equipment or apparatus used to treat or diagnose diesease and involves direct contact with patients. US medical device companies have highly respected ranks due to their innovations and high-tech product. In recent years, the investment in medical device research and development have ben doubled than previous years [14]. The medical device universe, largely encompasses specifically emitting special technological innovation, which includes hundreds of different technologies and thousands of types of products. Six major technology themes were identified and elaborated as highly to prefigure medical device innovation over the next decade: (1) electronics technology; (2) detection, diagnosis, and monitoring technologies; (3) decentralized care technologies; (4) minimally invasive technologies; (5) synthetic organs, tissues, and combination device/biological and device/drug technologies; and (6) demographically oriented technologies [15]. More than billions patients across the globe depend on medical device and its technologies. Most of them are used in diagnosis, prevention and disease treatment. Medical technologies such as wheelchairs, contact lenses and pregnancy test are frequently used by humans [16].

2.2 User-Centered Design Approach

User-Centered approach are the most common approach that been used in designing medical device aside from data driven design due to organized and more reliable in designing process [17]. It also adds to customer satisfaction when using the product because it fits the needs of user.

2.2.1 Medical Devices and User-Centered Design

In previous years, the focus on designing medical device have increased due to high demand in such area. In relation of a user-device design embodiment, most designers have focus more on device design, human error, usability and patient safety in their conceptual design. The aim of improving such aspect, a numbers of initiative have been organized. The role of medical device in patient safety incident have been investigated in 2007 by the United Kingdom's National Patient Safety Agency. Looking at all death and severe incidents reported by them, the cause of the incident is mainly the design of the medical device that is not safe to be used [18]. In addition, the incident includes developer who does not solely understood the context of the device and do not predicted hazard scenario. They also do not consider user practicality first before manufacturing resulting the device did not reached customer's satisfaction [19][20][21]. Ergonomic or widely known as human factor have contribute significantly in medical device safety have highlighted contribution which will improving safety in health care discipline especially on design [7][22][23][24].

Reported by the Institute of Medicine , that medical error report have significantly increased awareness on the magnitude, frequency, complexity and seriousness of error. Top eight leading cause of death in US is motor vehicle accidents, breast cancer, or AIDS, and medical error [25]. There are technologies such as Health Information Technology (HIT) which reduced the risk on getting a serious injury for hospitalized patients [26]. Therefore, it proves validates that the usage of technology itself can saves life. A study was conducted in the field of shoulder surgery where "A brief fatigue inventory of shoulder health developed by QFD technique". The usage of QFD techniques is to develop an instrument to diagnose the severity of symptoms on neck and shoulder thus trace the cause of the symptoms [27].

Previous research made by Kianfar have used QFD methodology to Reliability-Centered Maintenance (RCM) to improve RCM capability in preserving the functions of the plants. Their objective is mainly to preserve the function of plant with least resource. By adding such method to the RCM have regain more efficiency [28]. Self-management initiatives increasingly rely on the use of technologies to facilitate the process of care in the home. These technologies range from medical devices such as glucose monitors to comprehensive computer-mediated telemedicine systems that provide interactive support as well as World Wide Web access. Although such devices are required to meet certain standards, very little is known about their usability [25].

3 Medical Device Safety Assurance Method

3.1 Review on method in ensuring safety of medical device

Table 2 shows review on previous researchers in order to ensuring the safety of medical device and method chosen to cater certain case study.

Table 2. Review on method in ensuring safety of medical device.

No	Case Study / Medical device	Objective	Methodology	Reference
1	General	✓ Meeting design requirement ✓ Reduce medication error ✓ Ergonomic methods	a. System based user control approach b. Mapping workshop c. Designers workshop ✓ Ideas for solution ✓ Design concept ✓ Improve design ✓ Patient-centred	[29] [30]
2	General	✓ Reduce medication error ✓ Increase safety ✓ Reorganized design system	✓ User centred design ✓ Standardization ✓ Information sharing ✓ Effective monitoring ✓ Risk management ✓ Action plan	[31] [32]

3	Laparoendoscopic single-site (LESS)	✓ Evaluate new medical device usability ✓ User-centred ✓ Improving ergonomic posture	✓ Clinical simulation lab ✓ ETI training ✓ EMG sensor and DataLINK™ software	[33]
4	General	✓ Reduce device related errors ✓ Increase efficiency ✓ Functionality ✓ Performance ✓ Increase safety	✓ Human machine interface ✓ Use machine ✓ User training	[34][35] [24] [36]
5	Patient-controlled analgesia	✓ Decrease patient injuries ✓ Increase safety ✓ Introduce to new interface ✓ Decrease human error ✓ Faster performance	✓ Continuous of representative during design ✓ Field observation ✓ Task analysis (bottle neck) ✓ Technology compatibility with human limitations ✓ Iterative design process ✓ Experiment evaluation ✓ Market study ✓ Programming sequence	[2], [18], [20], [26], [34], [35]
6	Infusion pumps	✓ Identify risk ✓ More usable by end users	✓ Heuristic analysis ✓ Nielsen-Schneiderman ✓ Rules	[11] [37]
7	Glucose meter	✓ Framework that introduces concentric layers to dicot	✓ DiCoT-CL	[38]

4 Medical Device and Regulations

The medical device industry has undergone significant changes in recent years due to the passing of the Affordable Care Act (ACA) and the expansion of government regulations, including the adoption of Unique Device Identification (UDI) [39]. The regulatory procedures for medical devices will vary according to their class. In general, higher-risk devices will require more regulations and a more stringent conformity assessment process. The regulatory procedure for medical devices varies according to their class [40].

Nowadays, medical device regulations differ across the globe which makes the compliance gets complex and difficult process. The main scope of these medical device regulations around the world is mainly on risk management to ensure that the device do not compromised with either clinical condition or safety of patient and its users. Medical device developers need to eliminate nor reduce as much as possible the risk associate with the device [7]. Taking decrease in non-renewable resource into measures, manufacturing sustainability is indeed a crucial issue. Stricter regulation is needed on environment and occupational safety as customer preference for more eco-friendly product [41].

4.1 Review on existing standardized medical device

There are a few methods that have already achieved certain standard fixed by ISO. Table 3 shows review on method used to standardized current medical device according to regulations.

Table 3. Review on method use to standardized medical device

ISO/IEC	Case study	Method	Reference
ISO 14117	Electromagnetic Interference (EMI)	Risk-based analysis : Computational Human Phantom (CHP)	[42]
ISO 9241-11	General	User-centred design in development	[43]
ISO 13407	Laparoendoscopic Single-Site (LESS)	User-centered design	[33]
ISO 13485:2003	General	Design for Reliability : S&E assessment	[44]
ISO 14971	Fault Injection	FMEA	[45]

4.2 Preference standard for medical device developer

One of most crucial issue when it comes to designing or development of medical device is safety. There are no certain standards put to every medical device in ensuring safety. Therefore, IEC 60601 covers mainly in electrical medical device which more safety measures need to be taken.

4.2.1 IEC 60601-1

One standard that commonly used by medical device developers are likely to encounter is IEC 60601-1 (2004): general requirements for safety of electrical medical equipment. This standard is identical to the equivalent European standard (EN) and the British standard (BS) [7]. IEC 60601-1 originally have represent a major issue on medical electrical equipment on its safety standard in 1977 [46] [47]. Initially the scope was not covered on a few important devices such as automatic external defibrillator (AED) which is used commonly in airports and other handling issue. IEC 60601-1 only covers basic safety standards that contribute to basic safety and performance of a device. After IEC 60601-1 have been revised in 2006, the general requirements involved basic safety and essential performance with collateral standard in general requirements, tests and guidance for alarm systems in medical electrical equipment and medical electrical systems. The manufacturer company need to have ISO 14971 certificate for risk management process in order to obtain the standard that is required by IEC 60601-1 [46][48]–[50]. As suggested by J. Martin, the best approach in achieving IEC 60601-1 is by using user-centred design approach because this method has specific aims of improving medical device usability in order to reduce medical error [18]. Usability and other user-centred issues are covered in the collateral standard: 60601-1-6, which specifies the usability requirements for safety of electrical medical devices. This collateral standard requires developers to adopt a usability engineering process to ensure medical electrical equipment safety and function accordingly [7]. Usability engineering should begin early and continue through the equipment design and development life cycle [51].

5 Conclusion

This review paper has confirmed that there is no solid standard procedure in designing medical device to ensure the safety inclusively. There are comparative methods suggested by other researcher in achieving safety in design for several type of medical device which cause arguments among designers. This review is mainly highlighted the procedure and design method used currently in designing medical device. Medical device industries have put abundance of effort in ensuring safety of the product in line with the increasing demand in this area. For further research, standardized procedure and design method on ensuring safety both need to be performed before the product is released to the market so the safety of its user is guaranteed, thus decrease number of incident in healthcare sector. Therefore this paper suggested the implementation of IEC 60601-1 in medical device might be the holy grail of endless arguments among researchers by looking at its simple yet effective method in ensuring safety in a universal range of electronic medical device.

Acknowledgement

The authors would like to thank the financial support from Centre for Graduates Studies and Research Management Centre, Universiti Tun Hussein Onn Malaysia (UTHM).

References

1. Department of Health London (2000). "An organisation with a memory - Report of an expert group on learning from adverse events in the NHS," Clin. Med., vol. 2, no. 5, pp. 452–457.
2. Patient Safety Observatory (2007). "Safety in doses: medication safety incidents in the NHS," pp. 9–12.
3. National Patient Safety Agency (2008) "Device-related incidents : reports of serious harm," vol. 9, no. 9, pp. 2007–2009.
4. B. Toft (2001) "External inquiry into the adverse incident that occurred at queen's medical centre, Nottingham, 4th January 2001," Inquiry, no. January, p. 76.
5. S. I. Panel (2007) "Report on a Medication Incident of Intrathecal Administration of Vincristine in Prince of Wales Hospital," no. August.

6. B. D. Franklin, S. S. Panesar, C. Vincent, and L. J. Donaldson (2014). "Identifying systems failures in the pathway to a catastrophic event: An analysis of national incident report data relating to vinca alkaloids," BMJ Qual. Saf., vol. 23, no. 9, pp. 765–772.
7. J. L. Martin, B. J. Norris, E. Murphy, and J. a. Crowe (2008). "Medical device development: The challenge for ergonomics," Appl. Ergon., vol. 39, no. 3, pp. 271–283.
8. A. N. Thomas, U. Panchagnula, and R. J. Taylor (2009). "Review of patient safety incidents submitted from Critical Care Units in England & Wales to the UK National Patient Safety Agency," Anaesthesia, vol. 64, no. 11, pp. 1178–1185.
9. A. N. Thomas and B. A. McGrath (2009). "Patient safety incidents associated with airway devices in critical care: A review of reports to the UK National Patient Safety Agency," Anaesthesia, vol. 64, no. 4, pp. 358–365.
10. A. N. Thomas and U. Panchagnula (2008) "Medication-related patient safety incidents in critical care: A review of reports to the UK National Patient Safety Agency," Anaesthesia, vol. 63, no. 7, pp. 726–733, 2008.
11. A. N. Thomas and I. Galvin (2008). "Patient safety incidents associated with equipment in critical care: A review of reports to the UK National Patient Safety Agency," Anaesthesia, vol. 63, no. 11, pp. 1193–1197.
12. National Patient Safety Agency (2016). "Guidance notes on National Reporting and Learning System quarterly data summary publications," no. September.
13. European Commission and D. G. Health (2001). "Medical Devices : Guidance document medical device software," no. July 2001, pp. 1–21.
14. T. M. Report (2016). "2016 Top Markets Report Medical Devices Overview and Key Findings," pp. 1–9.
15. W. a. Herman and G. B. Devey (2007). "Future Trends in Medical Device Technologies," FDA Natl. Sci. Found.
16. J. Cunningham, B. Dolan, D. Kelly, and C. Young (2015). "Medical Device Sectoral Overview," Galw. City Cty. Econ. Ind. Baseline Study.
17. O. M. F. De Troyer and C. J. Leune (1998). "WSDM: a user centered design method for Web sites," Comput. Networks ISDN Syst., vol. 30, no. 1–7, pp. 85–94.
18. J. L. Martin, D. J. Clark, S. P. Morgan, J. A. Crowe, and E. Murphy (2012) "A user-centred approach to requirements elicitation in medical device development: A case study from an industry perspective," Appl. Ergon., vol. 43, no. 1, pp. 184–190.
19. D. R. Kaufman, V. L. Patel, C. Hilliman, P. C. Morin, J. Pevzner, R. S. Weinstock, R. Goland, S. Shea, and J. Starren (2003). "Usability in the real world: Assessing medical information technologies in patients' homes," J. Biomed. Inform., vol. 36, no. 1–2, pp. 45–60.
20. J. N. Amoore (2014). "A Structured Approach for Investigating the Causes of Medical Device Adverse Events," vol. 2014, no. Table 1.
21. M. Aljazzazi, M. Rawashdeh, T. Alshawaheen, and A. Malkawi (2013). "A User - Requirements Approach in Medical Devices Maintenance System Development : A Case Study from an Industry Perspective," World Acad. Sci. Eng. Technol., vol. 75, no. 3, pp. 780–790.
22. O. Safety (2000). "Ergonomics : The Study of Work," vol. 2000.
23. N. Moray (2000) "Culture, politics and ergonomics.," Ergonomics, vol. 43, no. 7, pp. 858–868.
24. L. Fennigkoh and H. Diego (2009). "Human Factors and the Control of Medical Devi|Freescale," pp. 39–43.
25. J. Zhang, T. R. Johnson, V. L. Patel, D. L. Paige, and T. Kubose (2003). "Using usability heuristics to evaluate patient safety of medical devices," J. Biomed. Inform., vol. 36, no. 1–2, pp. 23–30.

26. J. Horsky, J. Zhang, and V. L. Patel (2005). "To err is not entirely human:Complex technology and user cognition," J. Biomed. Inform., vol. 38, no. 4, pp. 264–266.
27. S. F. Liu, Y. Lee, and Y. Huang (2009). "A brief fatigue inventory of shoulder health developed by quality function deployment technique," J. Shoulder Elb. Surg., vol. 18, no. 3, pp. 418–423.
28. A. Kianfar and F. Kianfar (2010) "Plant function deployment via RCM and QFD," J. Qual. Maint. Eng., vol. 16, no. 4, pp. 354–366.
29. P. Buckle, P. J. Clarkson, R. Coleman, J. Bound, J. Ward, and J. Brown (2010) "Systems mapping workshops and their role in understanding medication errors in healthcare," Appl. Ergon., vol. 41, no. 5, pp. 645–656.
30. P. Buckle, P. J. Clarkson, R. Coleman, J. Ward, and J. Anderson (2006). "Patient safety, systems design and ergonomics," Appl. Ergon., vol. 37, no. 4 SPEC. ISS., pp. 491–500.
31. N. Khambete and A. Murray (2012). "National efforts to improve healthcare technology management and medical device safety in india," pp. 1–5,.
32. Department of Health (2003). "A System-wide Design-led Approach (Design for Patient Safety)," Design, p. 45.
33. de Laveaga and Adam E. (2012) "An Investigation Of Medical Device Design And Physical Ergonomics In Healthcare," Industrial and Management Systems Engineering - Dissertations and Student Research., Paper 32.
34. P. Carayon and K. E. Wood (2010). "Patient safety: The role of human factors and systems engineering," Stud. Health Technol. Inform., vol. 153, pp. 23–46.
35. L. Lin, K. J. Vicente, and D. J. J. Doyle (2001). "Patient Safety, Potential Adverse Drug Events, and Medical Device Design: A Human Factors Engineering Approach," J. Biomed. Inform., vol. 34, no. 4, pp. 274–84.
36. A. Çetin (2004). "Applying Product Design Methods to Medical Device Design With a Case Study on Home Care Devices," Strategy.
37. L. Duque-Muñoz, C. A. Aguirre-Echeverry, and G. Castellanos- Domínguez (2014). "EEG rhythm analysis using stochastic relevance," IFMBE Proc., vol. 41, pp. 658–661.
38. H. Care (2012). "2012 Symposium on Human Factors and Ergonomics,".
39. The Association for Packaging and Processing Technologies (2016). "Pharmaceutical & Medical Devices: Trends and Opportunities in Packaging Operations,".
40. N. Radhadevi, V. Balamuralidhara, T. M. P. Kumar, and V. Ravi (2012). "Regulatory guidelines for medical devices in India: An overview," Asian J. Pharm., vol. 6, no. 1, pp. 10–17.
41. M. Mutingi, P. Dube, and C. Mbohwa (2017) "A Modular Product Design Approach for Sustainable Manufacturing in A Fuzzy Environment," Procedia Manuf., vol. 8, no. October 2016, pp. 471–478.
42. J. E. Brown, R. Qiang, P. J. Stadnik, S. Member, L. J. Stotts, S. Member, and J. A. Von Arx (2016)."MR Conditional Safety Assessment of Implanted Medical Devices : Advantages of Computational Human Phantoms," pp. 6465–6468.
43. National Patient Safety Agency (2010). "Design for Patient Safety User testing in the development of medical devices," Patient Safety, 2nd Ed., p. 68.
44. S. Weininger and D. Ph (2009). "and Its Relationship to Safety and Effectiveness," Pace Pacing Clin. Electrophysiol..
45. J. Park, C. Yi, K. Kwon, and J. W. Jeon (2014). "Method Of Fault Injection for Medical Device Based on ISO 26262," pp. 1–2.
46. L. Eisner (2006) "Iec 60601-1 – the Third Edition," Regulation, pp. 8–17,.
47. Taylor A. (2013) "Standards : Medical Devices," pp. 1–3.
48. Mitch Gaudyn (2012). "The third edition of IEC 60601-1-A manufacturer's perspective," no. February.

49. J. O. Hare (2006) "Today's Moves Towards Eco-innovation in the Medical Electronics Sector," pp. 1–10.
50. P. G. Rajendra K. Songara, Ganesh N. Sharma, Vipul K. Gupta (2010). "Need For Harmonization Of Labeling Of Medical Devices: A Review," J. Adv. Pharm. Teachnology, vol. 1, no. 2, pp. 1–36.
51. Rigel Medical (2007). "A Practical Guide to IEC 60601-1,".

MATEC Web of Conferences **135**, 00021 (2017) DOI: 10.1051/matecconf/201713500021
ICME'17

An Analysis of Iron Flux Bridges Topologies for 12Slot-14Pole HE-FSM

Siti Khalidah Rahimi[1,*], *Md. Zarafi Ahmad*[2] , *Erwan Sulaiman*[2] and *S.M.N.S.Othman*[1]

[1]Research Center for Applied Electromagnetics (EMCenter), Universiti Tun Hussein Onn Malaysia, Batu Pahat, Malaysia
[2]Dept. of Electrical Power Engineering, FKEE Universiti Tun Hussein Onn Malaysia, Batu Pahat, Malaysia

Abstract. In this paper, a novel Hybrid Excitation Flux Switching Machine (HE-FSM) with DC-field excitation coil (DC-FEC) wounded in radial direction is proposed. The design is able to generate high torque and power performance because of a few points of interest such as robust rotor structure, non-overlapping windings and reduction of flux cancellation in stator yoke. However, the design with FEC in radial direction required no space between upper FEC and outer stator of machine which is lead to flux loss to surrounding area and flux saturation in stator yoke. As a solution, iron flux bridges have been introduced on the stator core to produce high torque execution. By using JMAG Designer ver. 14, numerous configuration of iron flux bridges have been presented and overviewed. Then, performances of torque at numerous flux bridges is analysed and compared. A design with highest torque is chosen to investigate the other performances of the machine.

1 Introduction

Principal idea of Flux Switching Machine (FSM) has been introduced and reported in the year 1950. [1-2]. Over the last decade, there are numerous novelty and several of the design have been applied for different applications, running from minimal effort residential machines, car, wind power, aviation, and etc. Generally, a typical FSM such as Permanent Magnet-FSM (PM-FSM), Field Excitation- FSM (FE-FSM), and Hybrid Excitation- FSM (HE-FSM) are categorised as shown in Fig. 1.

PM-FSM and FE-FSM only have a single main flux source for the motor to operate, one of the source comes from the permanent magnet (PM) itself and the other is the field excitation coil (FEC). As for HE-FSM, it require flux source from the PM and FEC, respectively as their main flux sources [3]. PM-FSM exhibit a high torque density and high efficiency meanwhile the price of PM material is high. In addition, the operating temperature of PM itself may restrict its application and performance. At high speed operation, the flux-weakening operation affected the PM machines due to fixed excitation

*Corresponding author : sitikhalidah17@gmail.com

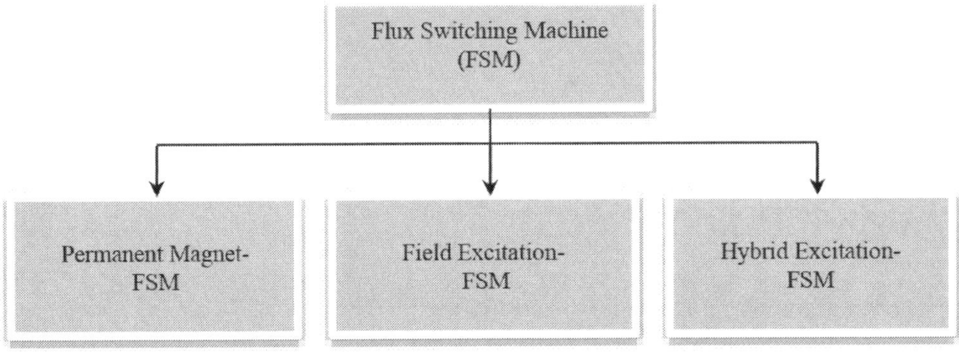

Fig. 1. Three categories of Flux Switching Machine (FSM)

from PM only. To overcome the high cost, PM excitation can be replaced with the FEC to form a FE-FSM.

FE-FSM is a model quite similar to salient-rotor reluctance machine with its own novel topology, by combining the principles of the Switch Reluctance Motor (SRMs) and inductor generator [4]. Theoretically, FE-FSM involve switching the flux vector with the armature coil winding, which relate to the rotor tooth position respectively. This type of machine has the potential of low cost and variable flux capabilities suitable for different performances [5]. Furthermore, the design structure is simple, easy to construct and no PM required. However, single-phase FE-FSM has issues of high torque ripple, low pull-up torque and one way rotating direction only. Then, armature coil and FEC windings are overlapped to each other resulting high copper loss [6].

Furthermore, HE-FSM is introduced as another option by bringing together the advantages of both PM machines and DC FEC synchronous machines. Therefore, the machine can possibly improve the output torque and power, enhance flux weakening capability, flux control capability and increase its efficiency [7-9]. Numerous combinations of rotor pole and stator slot for HE-FSM have been studied and promote for high-speed application such as 12S-10P HE-FSM configuration is proposed with a separated PM and C-type stator core, however this design is difficult to manufacture [10-11]. In order to decrease the supply frequency of the inverter, a design with 6S–5P HE-FSM has been proposed.

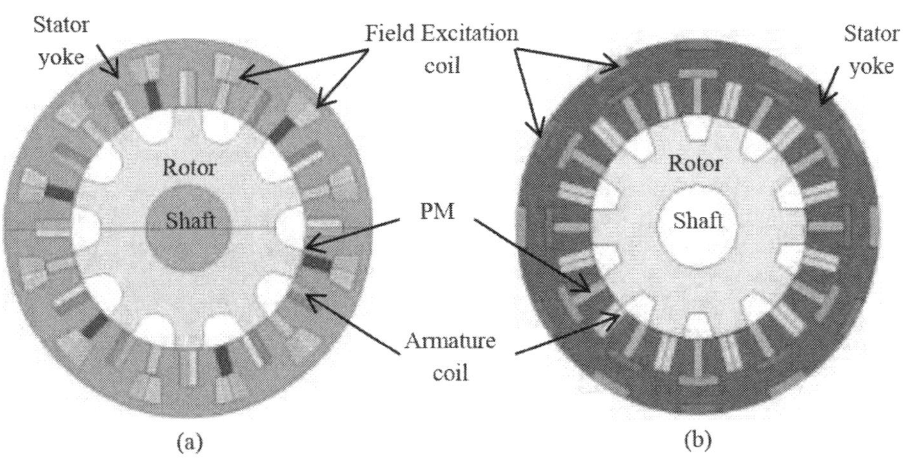

Fig. 2. The design comparison of 12Slot HE-FSM (a) Theta direction (b) Radial direction

Previously, all HE-FSM design have arranged their armature coil and FEC in thetha direction which create an issue of flux cancellation between PM flux and FEC flux. Thus, a new design of HE-FSM with arrangement of DC FEC in radial direction is introduced to overcome high flux cancellation. The design comparison of 12Slot HE-FSM with FEC in theta direction and radial direction is shown in Fig. 2(a) and Fig. 2(b), respectively [3]. In this paper, performance analysis is carried out for numerous iron flux bridges topologies of 12Slot-14Pole HE-FSM with DC FEC in radial polarity. The performance of selected design is also discussed.

2 Design restrictions, specifications and parameter of 12Slot 14Pole HE-FSM

Parameter specifications and dimensions of the proposed 12Slot-14Pole HE-FSM are listed in Table 1 and Table 2, respectively. Electrical restrictions related with the inverter has a maximum 650V DC bus voltage and maximum of 30A/mm^2 and 30A$_{rms}$/mm^2 for DC FEC winding and armature winding, respectively. Additionally, the dimensions of the proposed HE-FSM such as outer diameter of the stator and shaft radius are set to 269mm, 30mm respectively where air gap is set at 0.7mm. Besides, the weight of PM has been kept constant at 1.3kg.

For the motor dimension in Table 2, rotor parameters specified as outer rotor radius, L1, rotor tooth depth, L2 and rotor tooth width, L3 fixed at 80.25mm, 20.2mm and 9.5mm, respectively. Moreover, the PM depth, L4 is firmed at 26.775mm. For DC FEC parameters, DC FEC width and depth, L5 and L6 are fixed at 29.98mm and 6.67mm, respectively with 60 turns of DC FEC coil. Lastly, the armature coil parameter such as armature coil width and armature coil depth, L7 and L8, are set to 6.46mm and 26. 775mm, respectively, this gives the 7 number of turns for armature coil [3].

Table 1. HE-FSM design restriction and specification for HEV

Description	HE-FSM
Max. DC-bus voltage inverter (V)	650
Max. inverter current (A$_{rms}$)	360
Max. current density in armature winding, J$_a$ (A$_{rms}$/mm^2)	30
Max. current density in excitation winding, J$_e$ (A/mm^2)	30
Stator outer diameter (mm)	269
Motor stack length (mm)	84
Shaft radius (mm)	30
Air gap length (mm)	0.7
PM weight (kg)	1.3
Maximum speed (r/min)	20000
Maximum torque (Nm)	303
Maximum power (kW)	123

Table 2. Dimension of HE-FSM

Parameter	Details	HE-FSM
	PM volume (kg)	1.3
P1	Rotor radius(mm)	80.25
P2	Rotor pole depth (mm)	20.2
P3	Rotor pole width (mm)	9.33
P4	PM height (mm)	26.775

P5	FEC width (mm)	29.98
P6	FEC height (mm)	6.67
P7	Armature coil width(mm)	6.46
P8	Armature coil height(mm)	26.775
Na	No. of turns of AC	7
Ne	No. of turns of FEC	60

Besides, in 2D-Finite Element Analysis (FEA) analysis, JMAG-Designer ver.14.1, released by Japanese Research Institute (JRI) is used for this design. In case of PM, Neodiyum-Ferrite-Boron based on Neomax-35AH material is used, its residual flux density and coercive force at 20C°, is 1.2T and 932kA/m. The chosen electrical steel that is utilized for rotor and stator is 35H210. Based on Fig. 3, it demonstrates the initial design of DC FEC in radial direction for 12Slot-14Pole HE-FSM.

3 Numerous iron flux bridges topologies

The new HE-FSM is known as HE-FSM with radial direction since the FEC positioned is below and upper to each other in the stator. Meanwhile, the flux generated by FEC cannot flow through upper FEC because no space provide at this part. This phenomena leads to less flux generated in the machine. By adding iron flux bridges at upper FEC, the flux can flow easily and flux generation can be increased as compare to the design without iron flux bridges. Moreover, it can embellish the performance of the FEC with several flux excitation level. Several arrangements of 1mm thickness of iron flux bridges are designed to the proposed in HE-FSM as highlight in Fig. 4. The presence of FEC winding, PMs and iron flux bridges in stator can vary the machine performance in terms of PM magnetic flux at no load condition, excitation current utilization, toque capability and so on.

From the figure, A1 shows the initial configuration of the proposed design in the absence of iron flux bridge, which partially affects flux leakage from both outer and inner stator core to the air gap. Meanwhile, A2, A3, and A4 demonstrate different arrangements of a 1mm iron flux bridge placed between FEC slot-top layer of stator core, FEC slot and PM, finally between PM and rotor air gap, respectively. It is clear from the figure, that for A2 and A3 condition, a flux drainage occurs from PM to air gap. For A4 condition, a flux leakage occur and results in flux loss to outer region of the stator. Furthermore, there is two more possible combination of a 1mm iron flux bridges situated between PM, FEC slot and stator, known as A5 and A6.

Although, at this configuration, overall flux has been reduce since the flux oppose from one bridge to another, which causes reduction of total performance of the machine. Torque performances of all iron flux bridge arrangement are analyzed and the maximum torque obtained are recorded as shown in Fig. 5. It is clear that, in case of A2 configuration, highest torque has been achieved which is approximately 221Nm. For the lowest torque, 194.06Nm is obtained for A4 configuration. It is because there exist a flux drainage and elimination occur at the air gap as well as flux flows to outer most part of the stator core. Since A2 gives the highest maximum torque output, the performances of A2 configuration in load condition in term of torque and power with respect to various armature current density, J_a and field excitation current density, J_e conditions are analyzed and discussed.

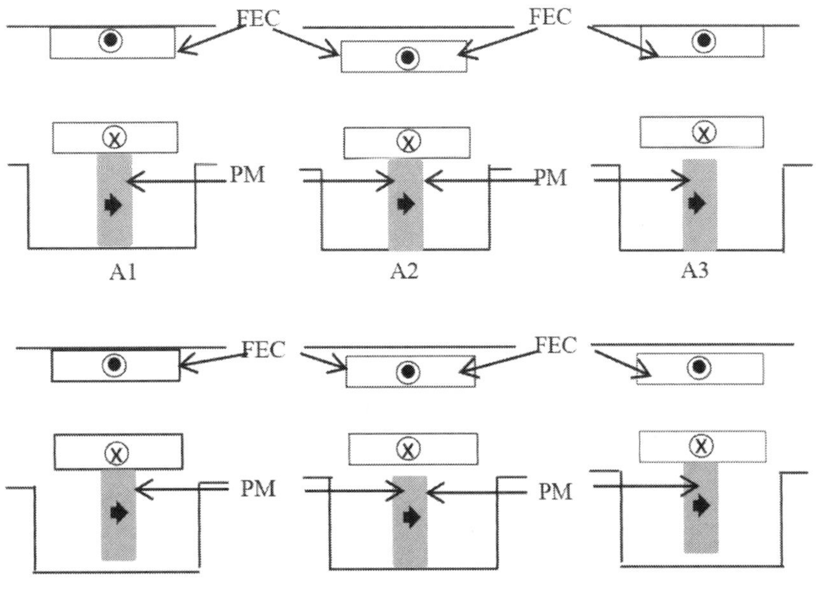

Fig. 3. The arrangement of iron flux bridges

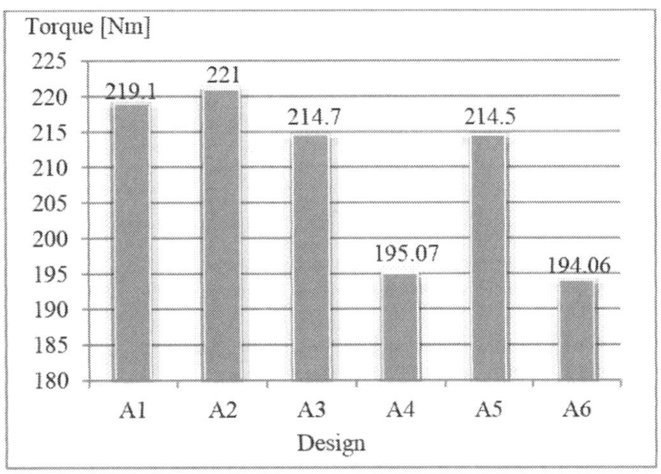

Fig. 4. The maximum torque for all configuration

4 Performance of A2 design

The instantaneous torque waveform of 12Slot-14Pole HE-FSM with A2 configuration are plotted in Fig. 5, based on 2D-FEA at maximum current density condition, of 30A/mm^2 and 30A$_{rms}$/mm^2 for DC FEC and armature coil, respectively. From the figure, it is clear that the average torque reaches 221.4Nm with peak-to-peak of 20Nm. Since the peak-to-peak torque of A2 design is quite low, it is expected that the machine produce low vibration and noise in practical applications.

Furthermore, Fig. 6 represents the torque versus J_e at various J_a characteristics. Both J_e and J_a are varied from 0 to 30A/mm^2 and 30A$_{rms}$/mm^2. It is examined that the torque increased as J_a and J_e increased. Additionally, the increment of the torque shows the machine possible to be applied for higher current density condition. This feature gives huge advantages of the HE-FSM which is suitable for wide range applications [12]. However, the torque remains constant after a certain limit even it is at higher J_a. From the graph, the maximum torque obtained 221Nm when J_a and J_e are set to maximum.

Graph of power versus maximum armature current density, J_a at various J_e is analyzed and depicted in Fig. 7. From the figure, it shows that the power increased linearly with increasing of DC FEC current density, Je. The maximum power output is approximately 95.14kW at maximum armature and FE current density.

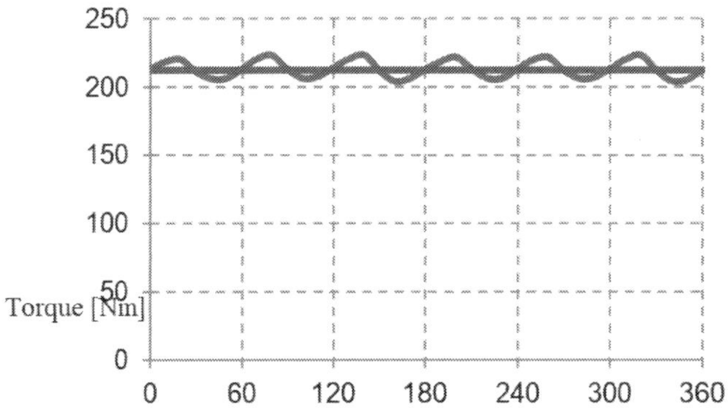

Fig. 5. Instantenous torque for A2

Electric cycle [°]

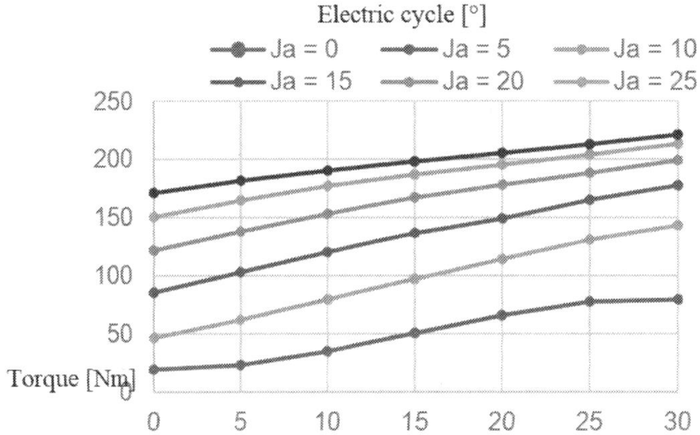

Fig. 6. Torque versus J_e at various J_a characteristics for A2

Je [A/mm^2]

 DOI: 10.1051/matecconf/201713500021

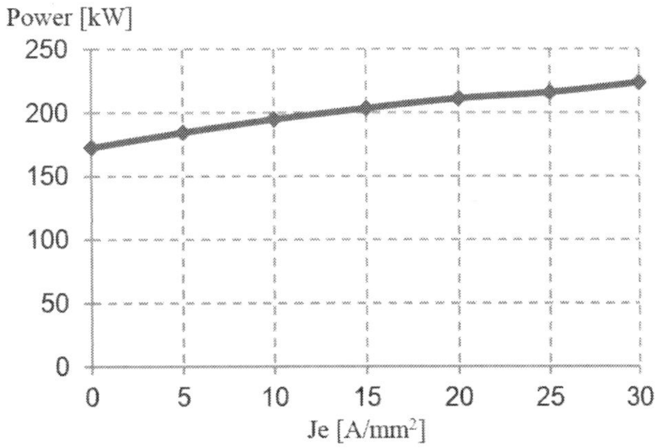

Fig. 7. Power versus J_e at maximum J_a for A2

5 Conclusions

As a conclusion, numerous design or topologies of the iron flux bridges for 12Slot-14Pole HE-FSM have been overviewed and discussed. The iron flux bridges with 1mm thickness are introduced on the initial design in order to solve some drawbacks of the previous design and improve the capability of the FEC with various excitation levels. Moreover, configuration A2 with a single iron flux bridge located between upper FEC slot and outer stator produces higher torque compared with other configuration, which is approximately 221Nm. Finally, the performance of the instantaneous torque, torque and power have been analysed.

Acknowledgements

This work was supported by GPPS (Vot. U714) and IGSP (Vot. U683) under Ministry of Education Malaysia, and Universiti Tun Hussein Onn Malaysia (UTHM), Batu Pahat, Johor, Malaysia.

References

1. Z. A. Husin, E. Sulaiman, F. Khan and M. F. Omar, *Performances comparison of 12S-14P field excitation flux switching motor with overlap and non-overlap windings for hybrid electric vehicles*, IEEE Student Conference on Research and Development, *pp. 1-6 (2014)*
2. E. Sulaiman, S. N. U. Zakaria and T. Kosaka, *Parameter sensitivity study for optimization of single phase E-Core hybrid excitation flux switching machine*, IEEE International Conference on Mechatronics (ICM),pp. 697-702 (2015).
3. S. K. Rahimi and E. Sulaiman, *Design investigation of hybrid excitation flux switching machine for high-speed electric vehicles*, IEEE 8th International Power Engineering and Optimization Conference (PEOCO), pp. 303-307 (2014).
4. E. Sulaiman, T. Kosaka and N. Matsui, *A novel hybrid excitation flux switching synchronous machine for a high-speed hybrid electric vehicle applications*, International Conference on Electrical Machines and Systems, pp. 1-6 (2011).
5. E. Sulaiman, T. Kosaka and N. Matsui, *A novel hybrid excitation flux switching synchronous machine for a high-speed hybrid electric vehicle application*, International Conference on Electrical Machines and Systems, pp. 1-6 (2011).

DOI: 10.1051/matecconf/201713500021

6. M. Z. Ahmad, E. Sulaiman, F. Khan and Z. A. Haron,, *Design and performance analysis of 12Slot-14Pole HE-FSM with outer-rotor configuration*, IEEE 8th International Power Engineering and Optimization Conference (PEOCO), pp. 369-374 (2014).
7. M. M. A. Mazlan, E. Sulaiman, M. Z. Ahmad and S. M. N. S. Othman, *Design optimization of single-phase outer-rotor hybrid excitation flux switching motor for electric vehicles*, IEEE Student Conference on Research and Development, pp. 1-6 (2014).
8. H. Ali, E. bin Sulaiman, M. Jenal and M. F. Omar, *Three phase segmental rotor hybrid excitation flux switching motors for various applications*, IEEE Student Conference on Research and Development (SCOReD), pp. 373-377 (2015).
9. E. Sulaiman, T. Kosaka and N. Matsui, *A novel hybrid excitation flux switching synchronous machine for a high-speed hybrid electric vehicle applications,* International Conference on Electrical Machines and Systems, pp. 1-6 (2011).
10. M. K. Hassan, E. Sulaiman, G. M. Romalan, M. F. Omar and M. Jenal, *Load analysis of dual Rotor Hybrid Excitation Flux Switching Machine (DR-HE-FSM)*, IEEE Conference on Energy Conversion (CENCON), pp. 155-159 (2015).
11. N. A. Jafar and E. Sulaiman, *Design analysis of 12S-10P hybrid-excitation flux-switching permanent-magnet machines for hybrid electric vehicle*, IEEE 8th International Power Engineering and Optimization Conference (PEOCO2014), pp. 308-312 (2014).
12. M. M. A. Mazlan, E. Sulaiman, M. Z. Ahmad and S. M. N. S. Othman, *Topologies of single-phase outer-rotor hybrid excitation flux switching motor for in wheel drive applications*, IEEE Conference on Energy Conversion (CENCON), pp. 112-116 (2014).

MATEC Web of Conferences **135**, 00022 (2017) DOI: 10.1051/matecconf/201713500022
ICME'17

Combination of Skid Control and Direct Yaw Moment Control to Improve the Safety and Stability of the Small Electric Vehicle with Two In-Wheel Motors

P.M. Heerwan[1,*], *S.M. Ashraf*[2], *M.I. Ishak*[1]

[1]Automotive Engineering Research Group (AERG), Faculty of Mechanical Engineering, University Malaysia Pahang, 26600 Pekan, Pahang

[2] Faculty of Mechanical Engineering, Universiti Malaysia Pahang, 26600 Pekan, Pahang

Abstract. For a small electric vehicle (EV) with the rear two in-wheel motors, the hydraulic brake system and the mechanical brake system are installed at the front and rear tire respectively. The mechanical brake system is used at the rear tire because there is no enough space for the hydraulic brake system. In a braking condition, the in-wheel motor at the rear tire will generate the regenerative braking force and it can improve the braking performance of the vehicle. However, during braking on the low adhesion road surface, anti-lock brake system (ABS) is very crucial to prevent the tire from lock-up. To improve the safety and stability of the vehicle, the combination of anti-skid control system and direct yaw moment control system is proposed. The anti-skid control system contains a hydraulic unit of ABS at the front tires and regenerative brake timing control at the rear tires. The control method of the regenerative brake timing control is same as ABS and it will turn on and off to prevent the tire from lock-up. On the other hand, the direct yaw moment control system is developed to increase the steer performance of the vehicle. The optimal control is used as the control strategy method to control the yaw moment. The simulation is developed in MATLAB Simulink and the result shows that the proposed model can improve the stopping distance from 9 seconds to 8.2 seconds. In addition, the combination of skid control and yaw moment control also improved the steer performance of the vehicle.

1 Introduction

Recently, research of the electrical vehicles (EVs) is become trending among automotive researchers [1]. It is because many advantages of EVs that can achieve than conventional internal combustion engine vehicles (ICEVs). One of the advantages of EV is the driving system, which is an electric motor is environmentally friendly and more easy to control than a combustion engine. There are several advantages of electric motor such as fast

* Corresponding author: mheerwan@ump.edu.my

torque response, motor torque can be measured easily and the motor can be installed at each tire [2].

Although small EVs have many advantages, one of the challenges is the safety system. The small EVs only have the basic components of a safety system which is a seat belt and hydraulic-mechanical hybrid brake system (HMHBS). In the HMHBS, the hydraulic brake system is installed at the front tire and the mechanical brake system at the rear tire. The mechanical brake system is used at the rear tire because the space at the rear tire is not enough for the hydraulic brake system. Compared with the hydraulic brake system, the response performance of the mechanical brake system is lower than hydraulic brake system. The other challenges in the study to improve the safety of the small EVs are to engage the HMHBS with the anti-lock brake system (ABS). ABS is a basic skid control system to prevent the tire from lock-up and also can avoid the vehicle from skidding. The hydraulic unit of ABS cannot be installed at the rear tire because the space at the rear tire is very limited [3]. From the braking system of small EVs, it can be considered that the safety system needs to be improved.

S. Kobayashi et al have developed the simulation model of the hydraulic-mechanical hybrid brake system (HMBS) with ABS to improve the braking performance of small EV [4]. In that study, the 3 hydraulic unit of ABS is engage with the HMHBS ; 2 hydraulic unit at each front tire, and one hydraulic unit is in between master cylinder and rear power cylinder. The result shows that during braking on an icy road, ABS only can prevent the front tire from lock-up. Although the hydraulic unit of ABS is install in between master cylinder and rear power cylinder, due to the low response performance of the mechanical brake system, the rear tire is lock-up.

M.Heerwan et al have improved the braking system of small EVs by using the in-wheel motor. During braking, the in-wheel motor will generate the regenerative braking force. By using the same control method of ABS, the regenerative braking force timing control is developed [5]. The control method of the regenerative braking force timing control is similar with the ABS, which is based on the slip ratio of the tire. The result shows that by controlling the regenerative braking force, it can prevent the driving tire from lock-up.

From the study of ABS and regenerative braking force timing control, the braking performance of small EVs with in-wheel motors can be improved. However, the steer performance of the small EVs also needs to be improved. During braking on the slippery road and at the same time the driver turns the steering wheel, the turning direction of the vehicle is not same as desired turning radius. This circumstance is occurred due to the low friction between the tire and road. M. Heerwan et al. have developed the HMBS with ABS at the front tire and regenerative brake timing control at the rear tire to improve the braking and steer performance on the cornering [6]. The result shows that HMBS with ABS and regenerative brake timing control also can improve the steer performance of the small EVs. Although that model can improve the steer performance, the stability of the vehicle is still need to be improved because the yaw rate is unstable.

In this paper, the combination of anti-skid control system and direct yaw moment control is proposed to increase the safety and stability of the small EVs. Anti-skid control system contains ABS and regenerative brake control. The hydraulic unit of ABS is installed at the front hydraulic brake system, while the regenerative brake timing control is used to control the braking force at the rear tire. In addition, for the direct yaw moment control, the optimal control method is used to improve the steer performance of the vehicle. The yaw moment from the linear model is compared to the non-linear model and feedback to the input, which is steer angle. The simulation model is developed in the MATLAB Simulink and the simulation result shows that the combination of anti-skid control system and direct yaw moment control system can increase the safety and stability of the small EVs.

2 Vehicle Model

The Toyota Coms AK10E-PC is used as the vehicle analysis model. This vehicle employed in-wheel motors as a driving system. The capacity of the passenger is one person and the maximum speed is 50 km/h. Figure 1 shows the vehicle model and the specifications of the vehicle is shown in Table 1.

Fig 1. Analysis Vehicle Model, Toyota COMS, AK10E-PC

Table 1. Specifications of Toyota COMS

Vehicle mass	361.9 (kg)
Maximum speed	50 (km/h)
Height of center of gravity (CoG)	0.105 (m)
Tread front	0.840 (m)
Tread rear	0.815 (m)
Wheel base	1.280 (m)
Inertia of front tire	0.430 (kgm^2)
Inertia of rear tire	2.530 (kgm^2)
Driving system	2 in-wheel motors
Motor power output	0.290 (kW)
Lead acid batteries	6 sets, 12 (V), 33 (Ah)

2.1 Braking system of the analysis vehicle model

2.1.1 Hydraulic-Mechanical Hybrid Brake System (HMHBS)

The braking system of the analysis vehicle model is hydraulic-mechanical hybrid brake system (HMHBS) [3,4]. The hydraulic brake system is installed at the front tires, while the mechanical brake system is installed at the rear tires. Compared with the hydraulic brake system, due to the rigidness of the mechanical brake system, the response performance and the braking force of the mechanical braking system are lower than hydraulic brake system.

Fig 2 shows the schematic diagram of the HMHBS. The components of HMHBS are master cylinder, front and rear brake tube, brake hose, front and rear wheel cylinder, power cylinder, link and wire. When the driver presses a brake pedal, the pressure of the brake fluid from the master cylinder is directed to the front wheel cylinder. However, for the mechanical brake system, the pressurised fluid from the master cylinder is directed to the rear power cylinder. Spring and damper in the rear power cylinder will generate the force and this force is transferred to the rear wheel cylinder by the link and wire.

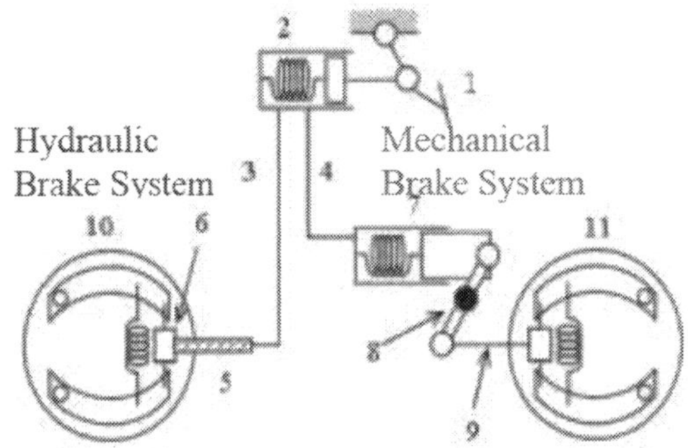

1:Brake Pedal; 2:Master Cylinder; 3:Front Brake Tube; 4:Rear Brake Tube; 5:Brake Hose; 6:Wheel Cylinder; 7:Power Cylinder; 8:Link ; 9:Wire; 10:Front Drum Brake; 11:Rear Drum Brake

Fig 2 Schematic diagram of hydraulic-mechanical hybrid brake system

2.1.2 Regenerative Brake System

Fig 3 (a) and (b) shows how in-wheel motor works in a driving and braking condition respectively [10]. In the driving condition, the electrical energy from the battery will rotate the electric motor. In contrast, in the braking condition, the rotation of the tire will convert the kinetic energy into electrical energy and fed back to the battery. This mechanism is known as regenerative braking. Based on the mechanism, the regenerative braking force is proportional to the wheel speed. As a consequence, during braking the regenerative braking force will decrease according to the speed of the tire. Therefore, the regenerative brake system must work together with the mechanical brake system [5,6].

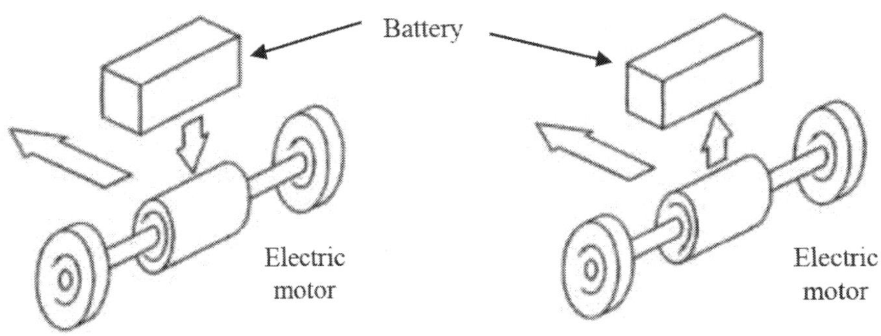

(a) In a driving condition (b) In a braking condition

Fig 3 In-wheel motor at the rear tire

3 Control system method

In this study, there are two control strategy methods; skid control to prevent the tire from lock-up and yaw moment control to improve the cornering performance. The effectiveness of the control model is developed in the MATLAB Simulink.

3.1 Skid control

Slip ratio is defined as the difference between the vehicle actual longitudinal velocity, u and the rotational velocity of the tire, $r\omega$ [7]. In this study, the slip ratio is used as the control parameter to prevent the tire from lock-up. The equation of slip ratio is:

$$\rho = u / r\omega \tag{1}$$

Fig 5 shows the relation between friction coefficient and slip ratio at the dry asphalt and icy road [5-6]. From Fig 4, the optimum slip ratio at the icy road is 0.2 to 0.3. In this optimum value, the friction between tire and road is maximum and the tire can grip at the road.

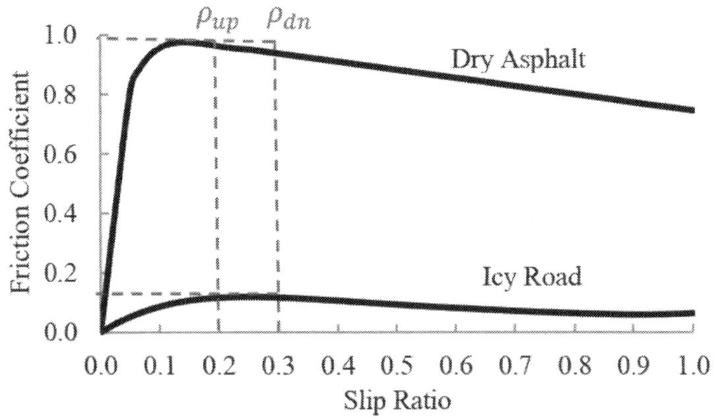

Fig 5 Relation between friction coefficient and slip ratio

3.2 Direct yaw control

In the direct yaw moment control (DYC), the two wheels of the linear vehicle model is used as an observer. The linear equations of motion are as below [8] :

$$mv\dot{\beta} + 2(K_f + K_r)\beta + \left[mv + \frac{2}{v}(l_f K_f - l_r K_r)\right] r = 2K_f\delta \tag{2}$$

$$2(l_f K_f - l_r K_r)\beta + I\dot{\gamma} + \frac{2(l_f{}^2 K_f + l_r{}^2 K_r)}{v} r = 2l_f K_f \delta \tag{3}$$

The cornering stiffness for the front and rear tire, K_f and K_r in equation (2) and equation (3) are the lateral force of the front and rear tire, Y_f and Y_r per unit side-slip angle of the front and the rear tire, β_f and β_r. The equations of K_f and K_r are:

 DOI: 10.1051/matecconf/201713500022

$$K_f = \frac{\partial Y_f}{\partial \beta_f} \quad (4)$$

$$K_r = \frac{\partial Y_r}{\partial \beta_r} \quad (5)$$

The state-space equation is used to describe a system in a linear motion.

$$\dot{\mathrm{x}} = \mathrm{Ax} + \mathrm{b}u \quad (6)$$

where,

$$\mathrm{A} = \begin{bmatrix} -\frac{2(K_f+K_r)}{mv} & -1-\frac{2(l_fK_f-l_rK_r)}{mv^2} \\ \frac{2}{I}(l_fK_f-l_rK_r) & -\frac{2(l_f{}^2K_f+l_r{}^2K_r)}{Iv} \end{bmatrix} \quad (7)$$

$$\mathrm{x} = [\beta \quad \gamma]^T \quad (8)$$

$$\mathrm{b} = \begin{bmatrix} \frac{2K_f}{mv} \\ \frac{2l_fK_f}{I} \end{bmatrix} \quad (9)$$

$$u = \delta \quad (10)$$

Here, equation (7) and equation (9) can be simplified as:

$$\mathrm{A} = \begin{bmatrix} a & b \\ c & d \end{bmatrix} \quad (11)$$

$$\mathrm{b} = \begin{bmatrix} e \\ f \end{bmatrix} \quad (12)$$

In the state-space equation, x is a system state, u is a scalar representing the input of the system and y is a scalar representing the output of the system. The matrixes A, b and c determine the relationships between state and input and output variables [9]. In this system, the input of the system is steering angle, δ.

To increase the steer performance of the vehicle, the feedback gain is added to the input, u and the equation for input is:

$$u = -\mathrm{g}^T\mathrm{x} + \delta \quad (13)$$

where,

$$\mathrm{g} = [\mathrm{g}_1 \quad \mathrm{g}_2]^T \quad (14)$$

Substituting equation (11) into equation (4), gives:

$$\dot{x} = Ax + b(-g^Tx + \delta) \quad (15)$$

where δ is a steering angle of the front tire. The block diagram of the state-space form in equation (15) is shown in Fig 6.

DOI: 10.1051/matecconf/201713500022

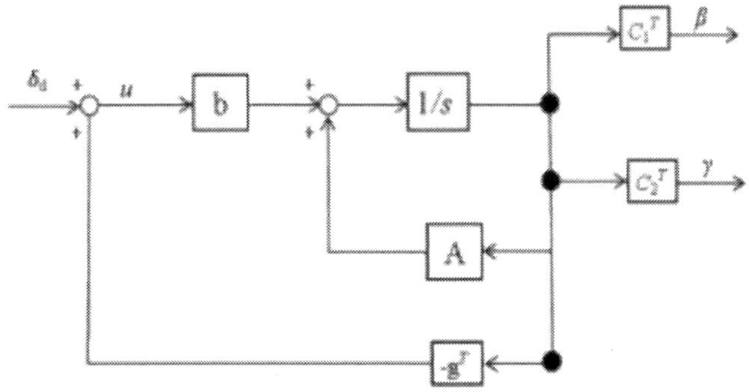

Fig 6 Block diagram of the state feedback control with the gain feedback to the input

In this study, the optimal control is used as a control strategy method in the observer. From equation (15), the evaluation function, *J* is [11]:

$$J = \int_0^{\infty}(x^T Q x + R{\delta_d}^2)dt \tag{16}$$

where,

$$Q = \begin{bmatrix} q_{11} & 0 \\ 0 & q_{22} \end{bmatrix} \tag{17}$$

Substitute equation (17) into equation (16) gives:

$$J = \int_0^{\infty}(q_{11}\beta^2 + q_{22}\gamma^2 + R{\delta_d}^2)dt \tag{18}$$

where q_{11} , q_{22} and *R* denote the state and input weighing matrixes.

Fig 7 shows block diagram of the state feedback control and gain with the non-linear system. The difference between yaw moment of the non-linear system with the yaw moment of the linear system is called as error. To eliminate the error, the weighting *H* is introduced into the system. The state equation for the DYC with optimal control is as below:

$$\dot{x} = Ax + b(-g^T x + \delta) + H(\gamma_{non-linear} - \gamma) \tag{19}$$

From equation (19), the gain is added to the input system to improve the steer angle, and the yaw angle can be improved by multiplying the error with the weighting, *H*.

DOI: 10.1051/matecconf/201713500022

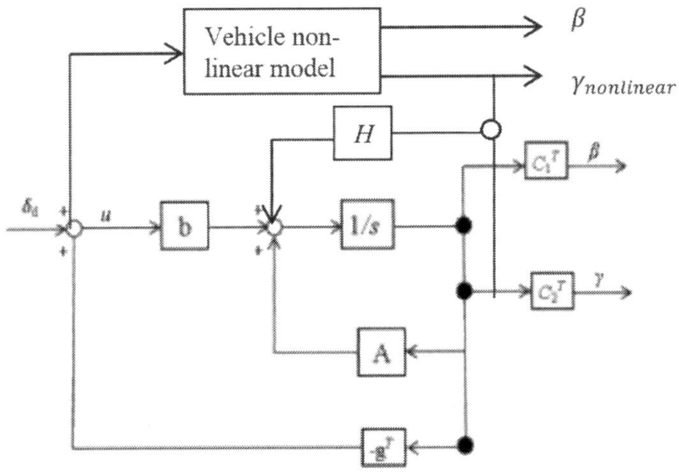

Fig 7 Block diagram of direct yaw moment control of non-linear model with linear model

4 Results and discussion

In the simulation, the road condition is set as an icy road and the steer angle is 15°. The initial speed of the vehicle is 30 km/h and the braking pressure at the master cylinder is 1.8 MPa.

4.1 Speed of the vehicle and tires during braking without ABS and regenerative brake control

Fig 8 shows the velocity of the vehicle and the tires during braking on an icy road. The y-axis is speed in m/s and the x-axis is the time in *ms*. From this result, the stopping time of the vehicle, *u* is 9.5 s. The result also shows that the speed of the front tires, vFL and vFR are rapidly decreased to 0 m/s in 0.3 s. Without ABS control, the hydraulic braking force at the front tire is very large, and as a result, the tire is decreased rapidly. This phenomenon is called tire lock-up [3]. However, for the rear tires, vRL and vRR are decreased from 8.3 m/s to 1.5 m/s in 2 s. From 2 s to 7 s, vRL and vRR are constant at 1.5 m/s. Due to the inertia of the vehicle and slippery road condition, the vehicle will slide on the surface. In this situation, the rear tire will still rotate at a low speed. From this result, it is approved that during braking on a slippery road condition, without ABS the tire will lock up and the vehicle will skid.

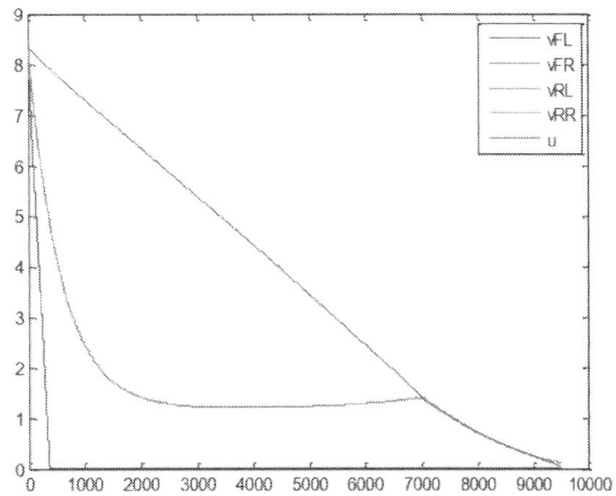

Fig 8 Speed of the vehicle and tires during braking without ABS, regenerative brake control and DYC

4.2 Speed of the vehicle and tires during braking with combination control of ABS, regenerative brake control and DYC

Figure 9 shows the speed of the vehicle, u, and the speed of the tires, vFL, vFR, vRL and vRR during braking on an icy road with ABS, regenerative brake control and DYC. From the result, the stopping time of the vehicle is improved to 8.2 s. By using ABS and regenerative brake control, the speed of the front and rear tires are fluctuated and approach to the speed of the vehicle. When ABS is operational, the IN valve and OUT will open and close rapidly. Same mechanism as regenerative brake control, the in-wheel motor will on and off rapidly based on the slip ratio of the tire. From this mechanism, the speed of the tire are fluctuated and the tire lock-up can be prevented.

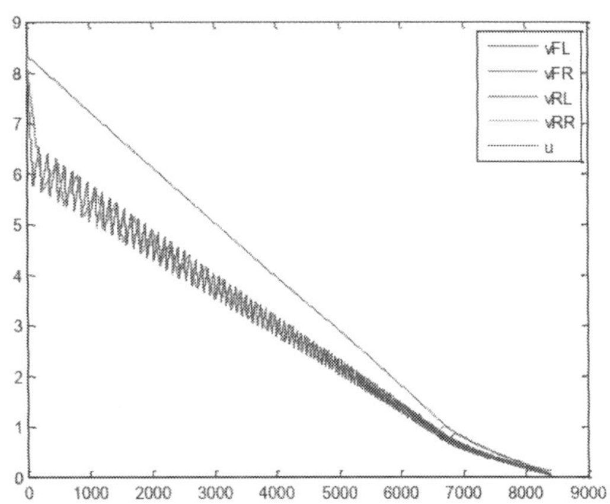

Fig 9 Speed of the vehicle and tires during braking with ABS, regenerative brake control and DYC

4.3 Steer performance of the vehicle during braking with combination control of ABS, regenerative brake control and DYC

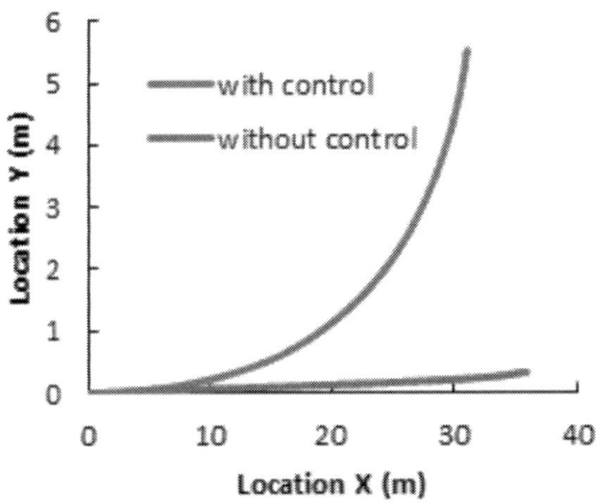

Fig 10 Trajectory of the vehicle during braking on an icy road

Fig 10 shows the trajectory of the vehicle during braking on an icy road. During braking without ABS, regenerative brake control and DYC, due to the low friction force between tire and road surface, the vehicle cannot be steered as driver desired. However, when the ABS, regenerative brake control and DYC are applied, the vehicle can be steered smoothly. ABS can regenerative brake control can increase the friction between the tire and road, and as result the tire will have more grip on the road surface. In addition, in the DYC, the yaw rate error between linear model and actual model is fed back to the steer angle to improve the steer performance of the vehicle. This result approved that the combination of ABS, regenerative brake control and DYC can prevent the vehicle from skidding and improve the steer performance of the vehicle.

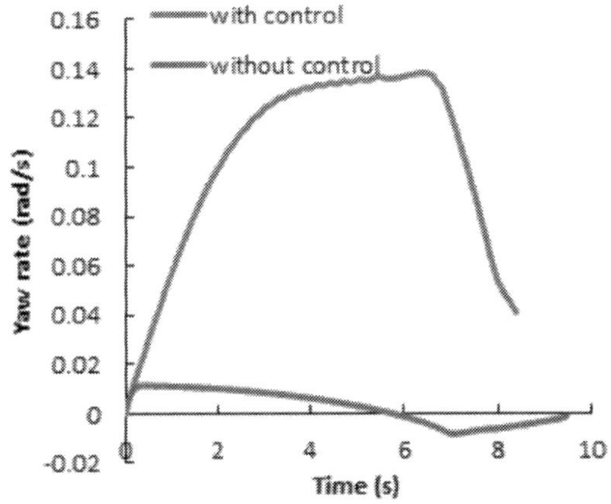

Fig 10 Yaw moment of the vehicle during braking on icy road

The yaw moment of the vehicle during braking is shown in Fig 11. From Fig 11, the yaw moment of the vehicle when ABS, regenerative brake control and DYC applied is increased from 0 to rad/s to 0.14 rad/s. When the yaw rate increases, the turning radius of the vehicle also will increase. However, without ABS, regenerative brake control and DYC, the yaw moment of the vehicle is very small. This result approved that the proposed control system can improve the yaw rate of the vehicle.

5 Conclusion

In this paper, the combination of the anti-skid control system and direct yaw moment control are introduced to improve the safety and stability of the vehicle. The anti-skid control system contains ABS at the front tire and regenerative brake control at the rear tire. The direct yaw moment control is combined with the anti-skid control system to improve the stability and steer performance of the vehicle. From the result, the proposed control system can prevent the vehicle from skidding and also can increase the stability of the vehicle.

Acknowledgments

The authors would like to thank Universiti Malaysia Pahang (UMP) for providing laboratory facilities and financial assistance under project no. RDU160312. Special thanks to Ministry of Higher Education Malaysia for providing Research Acculturation Grant Scheme (RAGS) under project no RDU151403.

References

1. H. Mirzaeinejad and M. Mirzaei, Control Engineering Practice A novel method for non-linear control of wheel slip in anti-lock braking systems, Control Eng. Pract., Vol. 18, no. 8, pp. 918–926, (2010)
2. K. Fujii and H. Fujimoto, Traction control based on slip ratio estimation without detecting vehicle speed for electric vehicle, Proceedings of the 2007 Power Conversion Conference, pp.688-693, (2007)
3. S. Hasegawa and Ogino, H., Research on skid control of small electric vehicle with hydraulic-mechanical hybrid brake system (1st Report: Simulation of 2 wheel model of anti-lock braking system), Proceedings of the School of Engineering, Tokai University, Vol.49, No.2, pp.89-94 (2009) (in Japanese)
4. S. Kobayashi and H. Ogino, Research on skid control of small electric vehicle with hydraulic-mechanical hybrid brake system (Simulation of 4 wheel brake model), Proceeding of the 2010 JSME Annual Congress (2010) (in Japanese)
5. M. Heerwan, H. Ogino, Y. Oshinoya, Skid control of a small electric vehicle with two in-wheel motors: Simulation model of ABS and regenerative brake control, International Journal of Crashworthiness, Vol 21, (2016)
6. M. Heerwan, H. Ogino, Y. Oshinoya, Skid control of a small electric vehicle with hydraulic-mechanical hybrid brake system (Effect of ABS and regenerative brake on an icy road), Mechanical Engineering Journal, Vol 5, (2014)
7. S. Sakai and Y. Hori, New skid avoidance method for electric vehicle with independently controlled 4 in-wheel motors, pp. 934–939 (1999)
8. W.Wang, L.Yuan, S.Tao,W.Zhang and T.Su, Estimation of vehicle side slip angle in

nonlinear condition based on the state feedback observer, Proceedings of the 2010 IEEE International Conference on Automation and Logistics, pp. 632-636, (2010)
9. M. Heerwan, H. Ogino, Y. Oshinoya, Skid control of small electric vehicles (Direct yaw moment control using tire steer angle), Proceedings of School of Engineering, Tokai University, Vol. 39, (2014)
10. M.Ehsani, Y.Gao, S.E.Gay and A.Emadi, Modern electric, hybrid electric and fuel cell vehicles, Fundamentals, Theory and Design, CRC Press, Ltd., Boca Raton, Florida, (2003).
11. M.I.Ishak, H.Ogino, Y.Yamamoto, Numerical simulation analysis of an oversteer in-wheel small electric vehicle integrated with four-wheel drive and independent steering, International Journal of Vehicular Technology, Vol. 2016, pp 1-12, (2016)

 DOI: 10.1051/matecconf/201713500023

A Review on Methods for Assessing Risk Factors of the Upper Limb Disorders among Cashiers in Grocery Retail Industries

Muhammad Fareez Ahmad Zuhaidi and *Mohd Nasrull* Abdol Rahman*

Department of Manufacturing and Industrial Engineering, Faculty of Mechanical Engineering and Manufacturing, Universiti Tun Hussein Onn Malaysia (UTHM), 86400 Batu Pahat, Johor, Malaysia

Abstract. This paper provides an overview of the various methods that have been developed for the assessment of risk factors for upper limb disorders among cashiers in grocery retail industries. This paper is essential as upper limb disorders have been known as one of the prime cause of work-related disability in various countries. The methods used for the assessment were the Disabilities of the Arm, Shoulder and Hand (DASH) questionnaire, the Assessment of Repetitive Tasks (ART), the Job Strain Index (JSI) and the Occupational Repetitive Actions (OCRA) Checklist. DASH questionnaire is a structured interview in determining the prevalence of the upper limb disorders, while ART, the JSI and OCRA Checklist act as a direct observation in examining the involvement of risk factors. Basically, these tools dealt with the measurements of symptoms, posture observation, and workplace risk assessment, as they are the most relevant tools for assessing the risk factors faced by the workers in grocery industries. The use of these tools assessed the risk factors and consequently reduces the risk of injuries among the workers.

1 Introduction

Cashiers have been declared as one of top ten professions having the risk of developing musculoskeletal disorders (MSDs) concerning the upper limb [1]. Most grocery store managers and employees have carried out all sorts of efforts in reducing occupational injuries and illnesses [2]. Unfortunately, most of them still are exposed to high-risk injury due to improper workstations and insufficient training in basic biomechanical principles provided to the employees [3].

Upper extremity musculoskeletal disorders (UEMSDs), also known as ULDs have been regularly described as elevated risks for the workers where they are exposed to psychosocial work factors [4]. In fact, ULDs are widely known as the leading cause of work-related disability within various countries [1][5]. Generally, supermarket cashiers are the common workers having the risk of experiencing musculoskeletal disorders of the upper limb [6], commonly the carpal tunnel syndrome (CTS) [7]. They tend to perform repetitive actions and light manual material handlings involving scanning and dealing with products,

[1]Corresponding author: mnasrull@uthm.edu.my

where these actions are reported as a risk factors contributed to the upper limb musculoskeletal disorders [8]. These include body posture and movements (sitting, standing, lifting, pulling, and pushing) and environmental factors (noise, lighting, temperature, and humidity) [9]. Besides that, repetition, overload, awkward positions or the combination of any of these are also some factors that contributed to the disease [10]. Other than grocery industries, workers in mould manufacturing industries also are highly exposed to upper limb disorders due to similar risk factors such as repetitive movements, uncomfortable work postures and excessive work without breaks [11]. The same applies for workers dealing with wall plastering job in construction industry [12].

There were various methods used for the assessment of risk factors for the upper limb disorders. The Disabilities of the Arm, Shoulder and Hand (DASH) questionnaire [13] acts as a structured interview in determining the prevalence of ULDs. On the contrary, the Assessment of Repetitive Tasks (ART) [14], Job Strain Index (JSI) [15] and Occupational Repetitive Actions (OCRA) Checklist [16] act as a direct observation in examining the involvement of risk factors. The objective of this research is to review the methods used in assessing risk factors for the upper limb disorders among cashiers in grocery retail industries.

2 Methods used in assessing risk factors for upper limb disorders among cashiers in grocery retail industries

In this paper, the questionnaire and tools used in assessing risk factors among cashiers in grocery retail industries were discussed. The usage of these methods is important to assess the risk factors for the upper limb disorders. The tools dealt with the measurement of symptoms, posture observation, and workplace risk assessment. These tools are beneficial in obtaining crucial surveillance data for the management to determine the disorders and reducing risk injury in a workplace. The relevant questionnaire and tools in the context of grocery industries are discussed.

2.1 The Disabilities of the Arm, Shoulder and Hand (DASH) questionnaire

The Disabilities of the Arm, Shoulder and Hand (DASH) Questionnaire is an upper extremity specific outcome measurement that has been introduced back in 1996 by the American Academy of Orthopedic Surgeons in collaboration with other certain organisations [13]. Basically, DASH is a self-report questionnaire aimed to measure physical function and symptoms in subjects with any musculoskeletal disorders of the upper limb.

DASH Questionnaire has additional optional modules. These modules were used to measure symptoms and functions among athletes, artists, and workers who require a high level of function. The module inquires about the impact of the arm, shoulder, and hand problems on the ability to perform tasks during work. Another additional optional module was the sports/art module. The module inquires about the impact of the arm, shoulder, and hand problems on activities such as playing musical instruments or doing sports, or both. The questionnaire in the additional optional module is the same as DASH Questionnaire having five different responses. The scale score was calculated ranging from 0 (no difficulty) to 100 (unable).

The DASH was separated into two sections, the disability/symptom questions which consist of 30 items, the optional work module and performance sport/music module, each consists of 4 items. The values from the completed response were summed and averaged to produce a score out of five. These values were converted according to Equation 1, to be compared to other measurement scaled on 0 to 100. The higher the score the greater the disability.

DASH Disability Score = [(Sum of response)-1] / (n (where n is the number of completed response)) x 25 (1)

2.2 Assessment of Repetitive Tasks (ART)

The Assessment of Repetitive Tasks (ART) was developed by Health and Safety Executive in the United Kingdom back in 2009. ART tool was developed to assess risk on tasks that require repetitive movements of the upper limbs [14].

ART tool was developed based on the Manual Handling Assessment Charts [17] which uses a logical flowchart process and 'traffic light' system to grade the risks. The ART tool was composed of some factors in assessing the repetitive movements of the upper limbs. These factors consist of four stages, namely frequency and repetition of movements (A), force (B), awkward postures (C), and additional factors (D). The duration stated in additional factors represents the duration of the task performed by the worker. The duration of the task was used to determine the duration multiplier.

Table 1. Duration Multiplier of the Duration Task [14]

Duration Task	Duration Multiplier
≤ 2 hours	x 0.5
2 ≤ hours ≤ 4	x 0.75
4 ≤ hours ≤ 8	x 1
≥8 hours	x 1.5

In order to calculate the task score, the scores of all factors from Stage A to D was added. The task score was then multiplied by duration multiplier. Finally, the exposure score was determined. Low exposure level should consider each individual circumstance if the score recorded was in the range of 0 to 11. On the other hand, medium requires further investigation if the score recorded was in the range of 12 to 21. Lastly, further investigation is required for high exposure level if the score recorded was above 22.

$$\text{Task Score}=A+B+C+D \quad (2)$$

$$\text{Task Score x Duration Multiplier} = \text{Exposure Score} \quad (3)$$

Table 2. Exposure level for ART risk factors [14]

Exposure Score	Exposure Level	
0-11	Low	Consider individual circumstances
12-21	Medium	Further investigation required
22 or more	High	Further investigation required urgently

2.3 Job Strain Index (JSI)

In 1995, the Job Strain Index (JSI) was developed by J. Steven Moore and Arun Garg. This tool was developed to analyse the single task jobs in which there are similar motions or only one repeated motion with the same effort level needed throughout the task [15]. The JSI needs both the qualitative and quantitative measurements of six variables, namely intensity of exertion (IE), duration of exertion per cycle (DE), efforts per minute (EM), hand/wrist posture (HWP), speed of work (SW), and duration of task per day (DD) [18]. Duration of exertion per cycle, efforts per minute and duration of task per day are the three quantitative variables while the intensity of exertion, hand/wrist posture and speed of work are categorised as qualitative variables. Each variable has its own rating score ranging from one to five.

Table 3. Rating values for each variable [15]

Rating Values	IE	DE	EM	HWP	SW	DD
1	Light	< 10	< 4	Very good	Very slow	≤ 1 hour
2	Somewhat hard	10 – 29	4 – 8	Good	Slow	1 – 2 hours
3	Hard	30 – 49	9 – 14	Fair	Fair	2 – 4 hours
4	Very hard	50 – 79	15 – 19	Bad	Fast	4 – 8 hours
5	Near maximal	≥ 80	≥ 20	Very bad	Very fast	≥ 8 hours

In order to calculate the JSI score, the multipliers should be determined first. The multipliers for each variable were based on the rating values.

Table 4. Multipliers for each variable [15]

Rating Values	IE	DE	EM	HWP	SW	DD
1	1	0.5	0.5	1.0	1.0	0.25
2	3	1.0	1.0	1.0	1.0	0.50
3	6	1.5	1.5	1.5	1.0	0.75
4	9	2.0	2.0	2.0	1.5	1.0
5	13	3.0	3.0	3.0	2.0	1.5

The JSI scores were calculated by multiplying all the variables (intensity of exertion, duration of exertion, hand/wrist posture, and speed of work and duration per day) multipliers as described in Equation 4. The JSI exposure was then determined using the JSI score. The job was considered safe if the JSI score recorded was lower than 3. If the JSI score recorded was in the range of 3 to 7, the job may increase the risk for distal upper extremities disorders. Furthermore, if the JSI score recorded was higher than 7, then the job was considered hazardous to the workers.

$$JSI = IE \times DE \times EM \times HWP \times SW \times DD \quad (4)$$

Table 5. Exposure level for JSI risk factors [15]

JSI Score	JSI Exposure Level
JSI ≤ 3	The job is probably safe
3 < JSI < 7	The job may increase the risk for distal upper extremity disorders
JSI ≥ 7	The job is probably hazardous

2.4 Occupational Repetitive Actions (OCRA) Checklist

The Occupational Repetitive Actions (OCRA) Checklist was developed by Occhipinti & Colombini back in 1996. Basically, this method was developed to analyse the workers' exposure to various upper limb risk factors such as repetitiveness, force, awkward postures and movements, lack of recovery periods, and other additional features [16].

OCRA Checklist consists five parts where each devoted to the analysis of different risk factor. Originally, there is four main risk factors namely lack of recovery time, movement frequency, force, and awkward postures with stereotyped movements. The other additional risk factors considered were vibration transmitted to the hand-arm system, ambient temperatures below 0°C, precision work, and the usage of inadequate gloves.

OCRA score can be calculated once the assessment of each part has been completed. OCRA score is the sum of frequency, force, recovery, posture, and additional risk factors, which then multiplied with the multiplier for the net duration of the repetitive task as described in Equation 5. The multiplier was based on the total net duration of the repetitive tasks recorded previously.

Table 6. Multiplier of the net duration repetitive tasks [16]

Duration Multiplier of the Net Duration Repetitive Tasks		
60 – 120 min = 0.5	241 – 300 min = 0.85	421 – 480 min = 1
121 – 180 min = 0.65	301 – 360 min = 0.925	sup 480 = 1.5
181 – 240 min = 0.75	361 – 420 min = 0.95	

The final score was then evaluated and the exposure level was determined. The final score below 7.5 was considered as an acceptable risk of exposure level. The final score ranging from 7.6 to 11.0 and 11.1 to 14.0 were considered as very low and light risk of exposure level, respectively. Furthermore, the final score ranging from 14.1 to 22.5 was considered medium exposure level and final score above 22.6 was considered as high-risk exposure level.

$$\text{OCRA Score} = (\text{Frequency} + \text{Force} + \text{Recovery} + \text{Posture} + \text{Additional}) \times \text{Duration Multiplier} \quad (5)$$

Table 7. Exposure level for OCRA Checklist risk factors [16]

OCRA Checklist Score	**Exposure Level**	
Up to A 7.5	Green	No risk (acceptable)
7.6 – 11.0	Yellow	Very low risk
11.1 – 14.0	Red light	Light risk
14.1 – 22.5	Red medium	Medium risk
≥ 22.6	Red high	High risk

3 Conclusion

To sum up, appropriate tools used are an important part of an ergonomic assessment. This research provides an overview of various methods used for the assessment of risk factors for the upper limb disorders among cashiers in grocery retail industries. These methods are appropriate to be used as they focused on respondents experiencing repetitive movements. Apart from the mentioned tools, there are various tools able to analyse tasks, determining symptoms, examine risk level, design work, and collection of data. There are some significant problems still experienced by the developers and ergonomic users when using these methods and tools, such as in developing methods integrating with other methods, methods related to ergonomic theory, to ease the applications, and also in proving the reliability and validity of the methods used.

This paper is funded by Office for Research, Innovation, Commercialization and Consultancy Management (ORICC) and Universiti Tun Hussein Onn Malaysia (UTHM) under Geran Penyelidikan Pascasiswazah Scheme (GPPS, Vot U800).

References

[1] US Department of Labor. (2000). Ergonomics : The Study of Work. OSHA 3125, 2000. Retrieved from www.osha.gov.

[2] OSHA, Occupational Safety and Health Act. (2004). Guidelines for Retail Grocery Stores : Ergonomics for the Prevention of Musculoskeletal Disorders. OSHA, (1), 1–19.

[3] A. Shinnar, J. Indelicato, M. Altimari, and S. Shinnar. (2004). Survey of ergonomic features of supermarket cash registers. *International Journal of Industrial Ergonomics*, **34**(6), 535–541.

[4] J. K. Park and S. H. Jang. (2010). Association between Upper Extremity Musculoskeletal Disorders and Psychosocial Factors at Work: A Review on the Job DCS Model's Perspective. *Safety and Health at Work*, **1**(1), 37–42.

[5] M. N. A. Rahman and N. S. A. Razak. Review on Pen and Paper Based Observational Methods for Assessing Work-related Upper Limb Disorders. *Indian Journal of Science and Technology*, **9**, 1-11 (Special Issue) (2016).

[6] L. Forcier, C. Lapointe, M. Lortie, P. Buckle, I. Kuorinka, J. Lemaire, et al. (2008). Supermarket workers: their work and their health, particularly their self-reported musculoskeletal problems and compensable injuries. *Work (Reading, Mass)*, **30**(4), 493–510.

[7] D. Roman-liu, J. Bugajska, and T. Tokarski. (2014). Comparative Study of Upper Limb Load Assessment and Occurrence of Musculoskeletal Disorders at Repetitive Task Workstations. *Industrial Health*, **52**, 461–470.

[8] K. R. Lehman, J. P. Psihogios, and R. G. Meulenbroek. (2001). Effects of sitting versus standing and scanner type on cashiers. *Ergonomics*, **44**(7), 719–738.

[9] J. Dul, B. Weerdmeester. (2008). Ergonomics for Beginners: A Quick Reference Guide. CRC Press.

[10] M. N. A. Rahman, F. A. Aziz, and R. M. Yusuff. (2010). Survey of body part symptoms among workers in a car tyre service centre. *Journal of Human Ergology*, **39**(1), 53–56.

[11] M. N. A. Rahman, H. R. Haq, M. F. Hassan, M. A. Arifin, M. Z. Yunos, S. A. Bakar. et al. Musculoskeletal discomfort among workers in mould manufacturing industry. *ARPJ Journal of Ergineering and Applied Sciences*, **10**(15), 6269-6273 (2015).

[12] M, N. A. Rahman, M. R. A. Rani and M. J. Rohani. (2012). Investigation of work-related musculoskeletal disorders in wall plastering jobs within construction industry. WORK: A Journal of Prevention, Assessment and Rehabilitation, **43** (4), 507-514.

[13] P. L. Hudak, P. C. Amadio, and C. Bombardier. (1996). Development of an upper extremity outcome measure: The DASH (disabilities of the arm, shoulder, and head). *American Journal of Industrial Medicine*, **29**(6), 602–608.

[14] HSE. Health & Safety Executive. (2010). Assessment of Repetitive Tasks of the upper limbs (ART tool), 1–16.

[15] J. S. Moore and A. Garg. (1995) 'The strain index: a proposed method to analyze jobs for risk of distal upper extremity disorders', *American Industrial Hygiene Association Journal*, **56** (5), 443-458.

[16] E. Occhipinti and D. Colombini. (1996). Alterazioni muscolo-scheletriche degli arti superiori da sovraccarico biomeccanico: metodi e criteri per l'inquadrament o dell'esposizione lavorativa. *La Medicina del Lavoro,* **87**(6), 491-525.
[17] S. C. Monnington, A. D. Pinder and C. Quarrie. (2002). Development of an inspection tool for manual handling risk assessment. (Sheffield, Health and Safety Laboratory), HSL. Internal Report ERG/02/20.
[18] P. Drinkaus, D. S. Bloswick, R. Sesek, C. Mann, and T. Bernard. (2005). Job level risk assessment using task level strain index scores: a pilot study. *International Journal of Occupational Safety and Ergonomics : JOSE*, **11**(2), 141–152.

Inter-Rater Reliability of the New Observational Method for Assessing an Exposure to Risk Factors Related to Work-Related Musculoskeletal Disorders (WMSDS)

Mohd Nasrull Abdol Rahman*, *Syahrul* Aziana Abdul Rahman, *Al Emran* Ismail, and *Azmahani* Sadikin

Department of Manufacturing and Industrial Engineering, Faculty of Mechanical and Manufacturing Engineering, Universiti Tun Hussein Onn Malaysia (UTHM), 86400 Batu Pahat, Johor, Malaysia

Abstract. The Entire Body Risk Assessment (ENBORA) method is a new observational method used to evaluate risk factors linked to Work-related Musculoskeletal Disorders (WMSDs). Hence, this study aims to verify the inter-rater reliability of the ENBORA method. 16 raters were involved in a one day ENBORA training session. All of the participants needed to assess three different tasks using the ENBORA paper checklist. The tasks involved were; lifting from pallet to conveyor for Task A, creels of wire onto spindles for Task B, and bottling line inspection for Task C. The Cohen's Kappa (K) coefficient and percentage of agreement were calculated. As a result, the inter-rater reliability for Task A has a Cohen's Kappa value (K) of 0.88, while Task B and Task C achieved K=0.91 which indicated very good agreement. This study found that trained practitioners can reliably use the ENBORA method to assess exposure to risk factors related to WMSDs.

1 Introduction

Work-related musculoskeletal disorders (WMSDs) fall under the category of musculoskeletal disorders (MSDs) that are caused by occupational exposure which could be the reason for work restriction, work-time loss, and at times work leave [1]. It has been reported that WMSDs can affect a person's quality of life and career with frequent absenteeism at work, increasing work restriction, job hopping, or even incapability compared to the other groups of diseases with a huge economic toll on oneself, the organisation, and the public all together [2]. It is a condition that affects the nerves, tendons, muscles and the body's supporting structures which are triggered by cumulative trauma associated with non-neutral body postures, sustained static postures, repetitive motions, as well as psychosocial factors and individual characteristics [3]. Many job sectors which have contributed to WMSDs include car tyre service centres [4], mould making manufacturing industry [5] and the construction industry [6]. Burdorf and Beek explained

*Corresponding author: mnasrull@uthm.edu.my

that, there are three types of methods to assess exposure to WMSD risk factors namely; subjective judgment (e.g., questionnaire and measurement scale), systematic observation, and direct measurement [7]. Takala et al, claimed that the most common method used by practitioners is the observational method [8]. Checklists which act as assessment toolsfor observational methods are applicable to determine the existence and acuteness of WMSDs risk factors in the workplace. Other than that, the checklists also provide ergonomic specialists that have a stable structure for the purpose of analysing of risk factors in different occupations [9]. Several observational methods for evaluating the exposure of risk factors linked to WMSDs are; Ovako Working Posture Assessment System (OWAS)[10], Posture, Activity, Tools and Handling (PATH)[11], Quick Exposure Check (QEC)[12], Rapid Entire Body Assessment (REBA)[13], Workplace Ergonomic Risk Assessment (WERA)[14], and Individual Risk Assessment (ERIN)[15]. However, the majority of the tools only emphasise on evaluating the physical risk factors of WMSDs specifically in terms of different body postures which do not involve other factors of WMSDs. Other risk factors of WMSDs include psychosocial risk factor, work organisation risk factor, and individual risk factors. The Entire Body Risk Assessment (ENBORA) method is a new observational method for assessing exposure to risk factors related to WMSDs that assesses physical risk factors, psychosocial risk factors, work organisational risk factors and individual risk factors. Reliability is required to prove validity as it ought to be operated on any kind of new measurement method [16]. Reliability is also known as the concord between the object's recurring measurements and the one that has been tested to support the propositioned method [17]. A measurement's reliability implies a thorough and scrupulous measurement that will guarantee consistency through numerous items in the instrument [18]. Reliability of a measure also depicts the stability and regularity of the instrument that determines the concept while assisting in determination of its quality at the same time [18]. The inter-rater reliability has been analysed for the ENBORA method. Inter-rater reliability is defined as the results' level of agreement derived by numbers in monitoring similar work conditions [19]. Streiner et al, acknowledged that inter-rater reliability resembles a tool's capability to come out with consistent outcomes. However, it also depends on the person in charge of the measurement [20]. Hence, this research aims to analyse the inter-rater reliability of the ENBORA method.

2 Methods

2.1 Study design

Researchers from Company A conducted the inter-rater reliability analysis which involved 16 participants in a-day ENBORA workshop. This research concentrated on the assessment and investigation of inputs delivered from 16 of the observers. Individually, the observers were also needed to measure distinct job task using the ENBORA method. Among 16 of the observers from Company A, they possess different expertise where ten observers were categorized as executive level group including process engineer, production executive, health, safety and environment (HSE) executive, quality management executive, service executive, parts executive and warehouse assistant manager, while other six observers were categorized as non-executive level including mechanic, administration officer, production operator, production officer, production line leader, and health, safety and environment supervisor. Six of them had been staying in the same position within the company for two to six years. Nevertheless, four of them had been in employment in the same job scope for 11 to 18 years and six observers had been employed in their current position for longer than 20 years.

2.2 Data collection

Prior to the execution of the ENBORA method, all participants were required to undergo risk assessment. They were also provided with proper training to execute the method. A 90 minute briefing was given to introduce the and step by step procedure of the method. Additionally, observers were presented with examples of the assessment to familiarise them with the method. There were three different tasks which were assessed by the observers. These tasks were shown through individual video clips which include lifting from pallet to conveyor for Task A, creels of wire onto spindles for Task B, and bottling line inspection for Task C. 30 minutes were spent by the observers to evaluateeach case study. Observers were given a brief talk regarding the purpose of the tool, instructions, scoring system, and terms used in the ENBORA method before the videos were played. After that, the videos were played as the observers assessed every recorded task using the ENBORA method. The total time used to evaluate the three tasks was one and half hours. In the case where any subject failed to finish the evaluation within the given period of time, the video films were shown again to make certain that all observers managed to complete the evaluation on the specific task.

2.3 Data analysis

The reliability of the ENBORA method has been assessed using inter-rater reliability. This test aimed to determine the applicability of the ENBORA method in assessing the workplace risks for all sectors and job scopes that involve the entire body. For the analysis of the inter-rater reliability of the ENBORA method, Cohen's Kappa coefficients were calculated for the assessment by 16 observers including the executive and non-executive level team, who evaluated three recorded tasks. Many researchers have used Kappa statistics in posture reliability studies[10, 12, 14, 21]. Knowing that the subject's occupation may have some influence in defining the results, further investigation of the results was performed. If the Cohen's Kappa (K) value are less than 0, it indicates lesser value than change agreement, K=0.01-0.20 is poor agreement, K=0.21-0.40 is fair agreement, while K=0.41- 0.60 is moderate agreement, K=0.61-0.80 is a good agreement and if K=0.81-1, it is very good agreement [22]. In order to obtain Cohen's Kappa value, the differences between the values of observed agreement and expected agreement were compared. Observed agreement is the existing agreement, while expected agreement is the agreement that is supposed to occur only by chance [23]. The formula below was used for the inter-rater reliability analysis [22].

$$Kappa,\ K = (po-pe)/(1-pe) \tag{1}$$

Where, po = observed agreement,
pe = expected agreement

The percentage of agreement was also calculated for the observers group, which included the executive and non-executive level team. The mean, standard deviation (SD) were used to evaluate and analyse the demographic data of the 16 participants in ENBORA training session.

DOI: 10.1051/matecconf/201713500024

3 Results

3.1 Demographic item

Table 1 shows the demographics data of the observers in ENBORA training sessions. The total mean age of the observers (N=16) for the age range between 27 to 49 years old is 37.06 years and (SD=8.27). Meanwhile, the total mean for working experience for the range between the 2 to 27 years are 14.19 years and (SD=9.53).

Table 1. Demographics data of the observers in ENBORA training session.

Group of observer	Age (Year)			WorkingExperience (Year)		
	Mean	SD	Range	Mean	SD	Range
Executivelevel[a] (n=10)	36.80	8.35	22	14.20	10.51	25
Non-executivelevel[b](n=6)	37.50	8.89	22	14.17	8.59	22
Total (N=16)	37.06	8.27	22	14.19	9.53	25

[a]*Executivelevelincludeprocessengineer, production executive, health, safety and environment (HSE) executive, quality management executive, service executive, parts executive and warehouse assistant manager*
[b]*Non-executivelevelincludemechanic, administration officer, production operator, production officer, production line leader, and health, safety and environmentsupervisor.*

3.2 Inter–rater reliability

The results of inter-rater reliability test for Task A until C are shown in Table 2. For Task A, the range of Cohen's Kappa coefficient for all ENBORA items was 0.55-0.93 and the range of the percentage of agreementwas 68.8-93.8%. The total agreement of Cohen's Kappa coefficient for Task A was 0.88, which indicated that very good agreement existed among observers who assessed Task A using the ENBORA method. The total of percentage of agreement was 90%. Furthermore, for Task B the range of Cohen's Kappa coefficient for all ENBORA items was 0.55-0.93 and the range of the percentage of agreement was 68.8-93.8%. The total agreement of Cohen's Kappa coefficient for Task B was 0.91, which indicated that very good agreement existed among observers who assessed Task B using the ENBORA method and the total percentage of agreement was 92.3%. Additionally, the range of Cohen's Kappa coefficient for all ENBORA items for Task C was 0.77-0.93 and the range of the percentage of agreement was 81.3-93.8%. The total agreement of Cohen's Kappa coefficient for Task C was 0.91, which indicated that very good agreement existed among observers who assessed Task C using the ENBORA method and the total of percentage of agreement was 91.8%.

Table 2. Result of inter-rater reliability analysis

ENBORA Items	Sub-item	Inter-rater Agreement (16 Observers)					
		Task A		Task B		Task c	
		*K**	%	*K**	%	*K**	%
A. Neck posture	A1. Flexion	0.93	93.8	0.86	87.5	0.93	93.8
	A2. Extension	0.93	93.8	0.86	87.5	0.93	93.8
	A3. LateralBending	0.93	93.8	0.93	93.8	0.86	87.5
	A4. Rotation	0.86	87.5	0.93	93.8	0.93	93.8
B. Shoulders posture	B1. Flexion	0.93	93.8	0.93	93.8	0.93	93.8
	B2. Extension	0.93	93.8	0.93	93.8	0.86	87.5
	B3. Abduction	0.86	87.5	0.93	93.8	0.86	87.5
	B4. Adduction	0.86	87.5	0.93	93.8	0.93	93.8
	B5. Medial Rotation	0.93	93.8	0.93	93.8	0.86	87.5
	B6. Lateral Rotation	0.86	87.5	0.86	87.5	0.86	87.5
C. Elbows posture	C1. Flexion	0.93	93.8	0.93	93.8	0.93	93.8
	C2. Pronation	0.86	87.5	0.93	93.8	0.86	87.5
	C3. Supination	0.93	93.8	0.93	93.8	0.93	93.8
D. Hands/wrists posture	D1. Flexion	0.93	93.8	0.93	93.8	0.93	93.8
	D2. Extension	0.93	93.8	0.93	93.8	0.93	93.8
	D3. Radial Deviation	0.77	81.3	0.86	87.5	0.86	87.5
	D4. UlnarDeviation	0.77	81.3	0.86	87.5	0.86	87.5
E. Back posture	E1. Flexion	0.86	87.5	0.93	93.8	0.93	93.8
	E2. Extension	0.93	93.8	0.93	93.8	0.93	93.8
	E3. LateralBending	0.93	93.8	0.93	93.8	0.93	93.8
	E4. Rotation	0.86	87.5	0.86	87.5	0.86	87.5
F. Legs flexion		0.93	93.8	0.93	93.8	0.93	93.8
G. Shouldersrepetition		0.77	81.3	0.93	93.8	0.93	93.8
H. Elbowsrepetition		0.93	93.8	0.86	87.5	0.86	87.5
I. Hands/wristsrepetition		0.93	93.8	0.86	87.5	0.86	87.5
J. Back repetition		0.93	93.8	0.93	93.8	0.93	93.8
K. Forcefulexertion	K1. Force	0.77	81.3	0.93	93.8	0.93	93.8
	K2. Repetition	0.93	93.8	0.93	93.8	0.93	93.8
L. Contact stress		0.77	81.3	0.77	81.3	0.77	81.3
M. Vibration		0.93	93.8	0.93	93.8	0.93	93.8
N. Task duration		0.86	87.5	0.93	93.8	0.93	93.8
O. Work stress		0.93	93.8	0.93	93.8	0.93	93.8
P. Workload		0.93	93.8	0.93	93.8	0.93	93.8
Q. Work pace	Q1. Dynamic Motion	0.93	93.8	0.93	93.8	0.93	93.8
	Q2.Static Posture	0.93	93.8	0.93	93.8	0.93	93.8
R. Technology changes		0.93	93.8	0.93	93.8	0.93	93.8
S. Social environment		0.93	93.8	0.93	93.8	0.93	93.8
T. Monotonytask		0.86	87.5	0.93	93.8	0.93	93.8
U. Workschedule		0.86	87.5	0.93	93.8	0.93	93.8
V. BMI		0.55	68.8	0.93	93.8	0.93	93.8
W. Smoking		0.86	87.5	0.93	93.8	0.93	93.8
	Total agreement	0.88	90.0	0.91	92.3	0.91	91.8

**K*, Cohen's Kappa value : <0 (lessthan change agreement), 0.01-0.20 (Poor agreement), 0.21-0.40 (Fair agreement), 0.41- 0.60(Moderate agreement), 0.61-0.80(Good agreement) and 0.81-1(Very good agreement)[22]

4 Discussion

Table 2 indicates the result of the inter-rater reliability analysis for Task A, Task B and Task C. According to the results for Task A's ENBORA method on six main body regions, the Cohen's Kappa (K) ranged between 0.77 to 0.93 with good to very good agreement. The inter-rater reliability results for Task B show thatthe K values slightly increase compared to Task A in terms of the body region analysis. The K values for Task B ranged between 0.86 to 0.93 with very good agreement, whereas the K values ranged between 0.86 to 0.93 for Task C's six main body postures with very good agreement.

In general, the observers obtained the highest K value for the assessment of flexion and extension posture compared to bending and rotation posture. Task A's Cohen's Kappa value for neck flexion and extension was 0.93, and K=0.86 for the neck rotation. On the other hand, the K value for shoulder flexion and shoulder extension was 0.93, and K=0.86 for shoulder abduction, adduction and lateral bending. Additionally, the back flexion and extension for Task B recorded the highest value where K=0.93 compared to the K value for back rotation which was 0.86. The K value for bending and rotation posture for Task C was 0.86. However, the K value for flexion posture for the six main body regions was 0.93.

The other physical items including repetition, forceful exertion, contact stress, vibration and task duration show good to very good agreement with K values ranging from 0.77 to 0.93 for Task A until Task C. For psychosocial risk factors, for the K values for Task A ranged between 0.86 to 0.93, and 0.93 for each item for Task B and Task C.For work schedule items, Task A scored a K value of 0.86 while Task B and Task C both obtained a K value of 0.93. The individual items resulted in a moderate to very good agreement of K=0.55-0.86 for Task A whereas Task B and C obtained very good agreement because the value of K was 0.93. The total agreement of the inter-observer reliability for Task A resulted in a high Kappa value of 0.88 with a total agreement of 90%. The total agreement for Task B was K=0.91 with a total agreement of 92.3%. Nevertheless, Task C's total agreement also resulted in a high Kappa value of K=0.91 with a total agreement of 91.8%. The results indicated very good agreement.

Based on the feedback obtained from the observers in the reliability analysis, the observers mentioned that they had difficulty understanding the neck rotation and bending angles due to the ambiguous illustration of the posture while the head position was also a bit confusing. The same problem occurred during the evaluation of the medial rotation and lateral rotation of the shoulders. The observers found the illustrations to be confusing because the view from the head position was difficult to understand. In addition, there are similarities between the inter-rater reliability obtained in this research and the results presented by other researchers which were used to examine body posture. For example, Coenen et al.,had obtained very good agreement for trunk flexion angle, trunk lateral flexion, elbow flexion and shoulder flexion. However, poor agreement was obtained for trunk rotation and shoulder abduction [24]. Meanwhile, Burt and Punnett had discovered that gross body motions including back flexion were easier to observe which will then develop a better agreement compared to smaller motions such as wrist deviation [21].

5 Conclusion

The main purpose of the Entire Body Risk Assessment (ENBORA) is to provide a quick way to identify exposure to physical risk factors, psychosocial risk factors, work organisational risk factors and individual risk factors associated with work-related musculoskeletal disorders (WMSDs). The results for inter-rater reliability were (K=0.88) for Task A, (K=0.91) for Task B, and (K=0.91) for Task C which showed very good agreement. To conclude, the ENBORA method is a reliable tool which can be applied by well-guided practitioners to assess the exposure to risk factors related to WMSDs.With this tool, it will be easier for employers to recognise and identify the improvements required in their work stations in order to reduce the high occurrence of musculoskeletal disorders and improve employee productivity.

Acknowledgements

This research is funded by the Ministry of Higher Education of Malaysia (MOHE) and Universiti Tun Hussein Onn Malaysia (UTHM) under Fundamental Research Grant Scheme (FRGS, Vot 1495)

References

[1] M. S. Forde, L. Punnett, D. H. Wegman, "Pathomechanisms of work-related musculoskeletal disorders: conceptual issues," Ergonomics**45**, 9, pp 619-630 (2002)
[2] K. Kemmlert, "Labour inspectorate investigation for the prevention of occupational musculo-skeletal injuries," Natl. Inst. Occup. Heal. Solna, Sweden (1994)
[3] K. Macdonald, D. King, "Work-related musculoskeletal disorders in veterinary echocardiographers: A cross-sectional study on prevalence and risk factors," J. Vet. Cardiol**16**, 1, pp. 27–37 (2014)
[4] M.N.A. Rahman, F. A. Aziz, R. M. Yusuff, "Survey of body part symptoms among workers in a car tyre service centre.," J. Hum. Ergol**39**, 1, pp. 53–56 (2010)
[5] M.N.A. Rahman, U. J. Hui, R. H. A. Haq, M. F. Hassan, A. M. T. Arifin, M. Z. Yunos, S. Adzila, "Musculoskeletal discomfort among workers in mould making manufacturing industry," ARPN J. Eng. Appl. Sci**10**, 15, pp. 6269–6273 (2015)
[6] M.N.A. Rahman, M.R.A. Rani, J. M. Rohani, "Investigation of work-related musculoskeletal disorders in wall plastering jobs within the construction industry." Work **43**, pp 507-514 (2012)
[7] A. Burdorf, A. Van Der Beek, "Exposure assessment strategies for work-related risk factors for musculoskeletal disorders," Scand J. W. Env. Health **4**, pp 25-30 (1999)
[8] E.P. Takala, I. Pehkonen, M. Forsman, G. A. Hansson, S. E. Mathiassen, W. P. Neumann, G.Sjøgaard, K. B. Veiersted, R. H. Westgaard, J. Winkel. "Systematic evaluation of observational methods assessing biomechanical exposures at work." Scand J. W. Env. Health **36**, 1, pp 3-24 (2010)
[9] T. J. Albin, "Measuring the validity and reliability of ergonomic checklists," Work **43**, pp. 381–385 (2012)
[10] O. Karhu, P. Kansi, I. Kuorinka, "Correcting working postures in industry : A practical method for analysis," Appl. Ergon **8**, 4, pp. 199–201 (1977)
[11] B. Buchholz, V. Paquet, L. Punnett, D. Lee, and S. Moir, "PATH: A work sampling-based approach to ergonomic job analysis for construction and other non-repetitive work," Appl. Ergon **27**, 3, pp. 177-187 (1996)
[12] G. David, V. Woods, G. Li, P. Buckle, "The development of the Quick Exposure Check (QEC) for assessing exposure to risk factors for work-related musculoskeletal disorders," Appl. Ergon**39**,1, pp. 57–69 (2008)
[13] S. Hignett, L. McAtamney, "Rapid entire body assessment (REBA)," Appl. Ergon **31**, 2, pp. 201–205 (2000)
[14] M.N.A. Rahman, M.R.A. Rani, J. M. Rohani, "WERA: An observational tool develop to investigate the physical risk factor associated with WMSDs," J. Hum.

Ergol. (Tokyo) **40**, pp. 19–36 (2011)
[15] Y. Rodríguez, S. Viña, R. Montero, "A Method for Non-experts in Assessing Exposure to Risk Factors for Work-related Musculoskeletal Disorders — ERIN," Industrial Health **51**, pp. 622–626 (2013)
[16] J. C. Nunnally, I. H. Bernstein, "Psychometric Theory." McGraw-Hill, New York (1994)
[17] I. De Bruijn, J. A. Engels, J. W. J. Van Der Gulden, "A simple method to evaluate the reliability of OWAS observations," Appl. Ergon **29**, 4, pp. 281–283 (1998)
[18] U. Sekaran, Research Methods For Business: A Skill Building Approach, 4Th Ed. Wiley India Pvt. Limited (2006)
[19] J. Park, J. Boyer, J. Tessler, G. Perez, L. Punnett, "Exposure assessment of musculoskeletal disorder risk factors in hospital work: Inter-rater reliability of PATH observations," In Proceedings of the Human Factors and Ergonomics Society Annual Meeting **49**, 14,pp. 1385–1389 (2005)
[20] D. L. Streiner, G. R. Norman, J. Cairney, Health measurement scales: a practical guide to their development and use. Oxford University Press, USA (2014)
[21] S. Burt, L. Punnett, "Evaluation of interrater reliability for posture observations in a field study," Appl. Ergon **30**, 2, pp 121-135 (1999)
[22] D. G. Altman, Practical statistics for medical research. CRC press (1990)
[23] A. J. Viera, J. M. Garrett, "Understanding Interobserver Agreement: The Kappa Statistic," Fam Med **37**, 5, pp. 360–363 (2005)
[24] P. Coenen, "Inter-rater reliability of a video-analysis method measuring low-back load in a field situation," Appl. Ergon **44**, 5, pp. 828–834 (2013)

MATEC Web of Conferences **135**, 00025 (2017) DOI: 10.1051/matecconf/201713500025
ICME'17

Feedback Survey on the Usability of the OFFERA Method for Assessing an Exposure Risks of Computer Work Related to WMSDs

Mohd Nasrull Abdol Rahman*, *Siti Shafika* Mohamad, *Salihatun* Md Salleh, *Shahruddin* Mahzan and *Sufizar* Ahmad

Department of Manufacturing and Industrial Engineering Faculty of Mechanical and Manufacturing Engineering, Universiti Tun Hussein Onn Malaysia (UTHM), 86400 Parit Raja, Batu Pahat, Johor, Malaysia

Abstract. The OFFERA method was designed to assess the exposure of the office workstation risk factors associated with WMSDs. This method involved six domains which include chair, desk, input device, monitor, accessories, and the environment. The purpose of this study was to assess the usability of the OFFERA method for assessing exposure risks of computer work related to WMSDs. The participants were trained to conduct the OFFERA method. Trial assessments on three different office jobs were conducted. The usability of the OFFERA method was identified based on the feedback questionnaire survey obtained from 44 practitioners (undergraduate students) after the training. From the usability test, the OFFERA method was found be easy to use (mean 4.48 ± 0.698) and quick to use (mean 4.48 ± 0.821). However, the observers found that the font used was too small hence it was difficult to read the instruction (mean 3.93 ± 1.096). The pictures or illustrations in the OFFERA tool were also recorded as unclear based on the relatively low score for 18 items (mean 3.73± 1.128). Besides that, all participants agreed that OFFERA method is user-friendly, cost effective and applicable to a wide range of office-related activities.

1 Introduction

Work-related musculoskeletal disorders (WMSDs) common health problems related to work. [1-2]. Besides the lower-back region, the neck and upper extremities are the most affected regions for industries such as construction, mould manufacturing, car service centre, and occupational computer use [2-5]. Some of the risk factors associated with MSDs among of the computer users are maintaining a static sitting posture for a long time, awkward posture of the head, neck and upper limb, wrist, and forearm and repetitive movement of the fingers [5-10]. There are a variety of methods which have been developed to assess posture in order to assess risk factors for WMSDS. An ergonomic risk assessment tool is the one of the methods that can be used [11]. Seven ergonomic risk assessment tools

* Corresponding author: mnasrull@uthm.edu.my

which have been widely used for WMSDs related to office workstations include Rapid Upper Limb Assessment [12], Rapid Entire Body Assessment [13], Computer Workstation e-Tool [14], Quick Exposure Checklist [15], Assessment of Repetitive Task [16], Office Ergonomic Assessment [6], and Rapid Office Strain Assessment [5]. The Office Ergonomic Risk Assessment (OFFERA) method was developed to assess the exposure of office risk factors associated with work-related musculoskeletal disorders (WMSDs). This method involves the assessment of office components (chairs, desks, input devices, monitors, and accessories) and the office environment. The usability test is a method that shows the strengths and weakness in the usability of the tool or system for a measure of comparing the usability [17]. However, most of the eligible tools were not tested in terms of the usability [12-14]. Therefore, this study was carried out to assess the usability of the OFFERA method for assessing the exposure risk of computer work related to WMSDs.

2 Method

2.1 Study design

Forty-four mechanical and manufacturing engineering undergraduates' students were involved in a workshop conducted in Universiti Tun Hussein Onn Malaysia (UTHM), Johor. This training session was conducted by the researchers. All the participants who had undertaken risk assessment were trained to conduct the OFFERA method. This involved a trial assessment of three jobs (admin counter, accountant, and research assistant) captured in videos to familiarise the participants with the OFFERA process. After the training session (intra-observer reliability) the participant assesses the usability of OFFERA method based on the three jobs via questionnaire survey which took 30 minutes to complete.

2.2 Data Collection

The usability of the OFFERA assessment was assessed using feedback questionnaires survey after the participants had completed the prototype assessment for three jobs during a training session (intra-observer reliability test). The survey consisted of 13 question items about OFFERA (prototype) assessment which include; 1) easy to use and straightforward, 2) quick to use, 3) instruction is clearly to read and understand, 4) wording is clearly to read and understand, 5) pictures/illustrations is clearly to see and understand, 6) scoring system is easy to interpret the results, 7) score for each item is easy to understand, 8) final score is easy to understand, 9) action level is easy to understand, 10) applicable to workplace assessment, 11) applicable to wide range of job/tasks, 12) valuable to works (cost effective), and 13) valuable at works (provide a good basis for intervention proposal). The survey consisted of close-ended questionnaire that used the five points of the Likert Scale rating (1=strongly disagree, 5=strongly agree). The questionnaire survey also included demographic items such as the subject's age, gender, and education level. Besides that, the survey consisted of general comments and suggestions for part B whereas Part C consisted of specific comment and suggestion for each item. The completed prototype assessment and feedback questionnaires were collected and stored securely. Data from the prototype tool and feedback questionnaires were analysed using Microsoft Excel 2007.

2.3 Data Analysis

The usability of the OFFERA tool was assessed using descriptive statistics. The mean (X) and standard deviation (SD) were used to evaluate and analyse each item of the questionnaire answered by all participants (N=44). Statistical analysis was performed using Statistical Package for Social Sciences (SPSS) software Version 20 and Microsoft Office Excel (Version 2007).

3 Results

3.1 Feedback Survey for Usability of the OFFERA method

The results from the feedback surveys used to assess the usability of the OFFERA method are shown in Table 1 and Figure 1. All observers' (N=44) rating (1= strongly disagree, 5=strongly agree) indicated that the prototype of the OFFERA method was easy to use and straightforward (mean 4.48 ± 0.698), quick to use (mean 4.48 ± 0.821), instruction is clearly to read and understand (mean 3.93 ± 1.149), wording is clearly to read and understand (mean 3.93 ± 1.096), pictures / illustrations which are clearly to see and easy to understand (mean 3.73± 1.128), scoring system is easy to see and understand (mean 4.59 ± 0.622), score for each item is easy to understand (mean 4.50 ± 0.792), final score is easy to understand (mean 4.68 ± 0.639), action level is easy to understand (mean 4.52 ± 0.549), applicable to workplace assessment (mean 4.52 ± 0.549), applicable to wide range of job / tasks (mean 4.02 ± 0.952), cost effective (mean 4.36 ± 0.685), and provide a good basis for intervention proposal (mean 4.41 ± 0.583).

Table 1. Feedback Survey on the usability of the OFFERA method

No	Questions	Rating Score					Total score
		1	2	3	4	5	Mean (SD)
1	Easy to use and straightforward	1	-	2	16	25	4.48 (0.698)
2	Quick to use	1	-	3	14	26	4.48 (0.731)
3	Instruction is clearly to read and understand	2	4	6	15	17	3.93 (1.149)
4	Wording is clearly to read and understand	2	3	7	17	15	3.91 (1.096)
5	Pictures / Illustrations is clearly to see and understand	2	5	8	17	12	3.73 (1.128)
6	Scoring system is easy to see and understand	-	-	3	12	29	4.59 (0.622)
7	Score for each items is easy to understand	1	-	2	14	27	4.50 (0.792)
8	Final score is easy to understand	1	-	1	9	33	4.68 (0.639)
9	Action level is easy to understand	-	-	1	19	24	4.52 (0.549)
10	Applicable to workplace assessment	-	-	1	19	24	4.52 (0.549)
11	Applicable to wide range of job / tasks	2	-	7	21	14	4.02 (0.952)
12	Valuable to works (cost effective)	-	-	5	18	21	4.36 (0.685)
13	Valuable at works (provide a good basis for intervention proposal)	-	-	2	22	20	4.41 (0.583)

Notes: rating scale range from 1=strongly disagree to 5=strongly agree

Figure 1 shows the results in terms of percentage on the feedback survey of usability of the OFFERA method by 44 observers (undergraduate's student). All of the observers strongly agree and agree that the prototype of the OFFERA method was easy to use and straightforward (93%), quick to use (91%), has a clear instructions and easy to understand (73%), clear wording and easy to understand (73%), clear pictures/illustrations which are easy to understand (66%), scoring system is easy to see and understand (93%), score for each item is easy to understand (93%), final score which is easy to understand (95%), action levels which are easy to understand (98%), applicable to workplace assessment (98%) applicable to a wide range of job/tasks (80%), cost effective (89%), and provide a good basis for intervention proposal (95%). From the results, more than 60% of the observers agree and strongly agree with the questionnaire items.

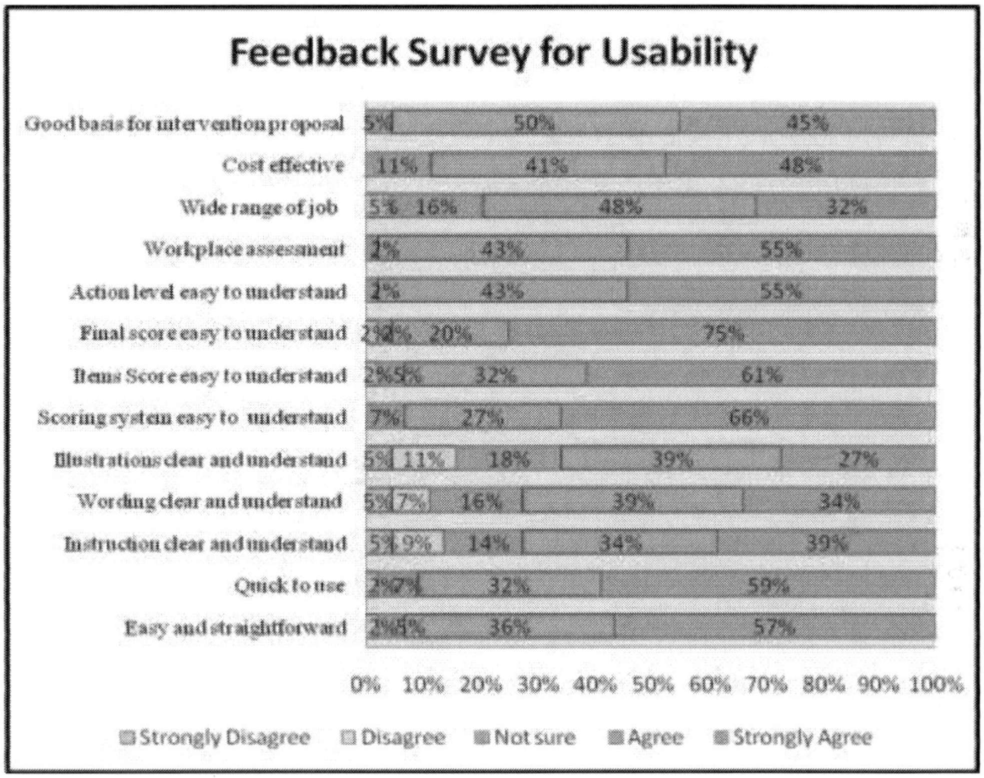

Fig 1. Feedback Survey on the usability of the OFFERA method

3.2 General comments and suggestions for the OFFERA method

All of the participants gave their general comments and suggestions on the OFFERA method. Table 2 shows the summary of feedback given by the observers. The observers commented that the assessment tool was very good for assessing ergonomic risks (good for ergonomic users) and that the tool was very useful as it can be used for a wide range of office work (useful for assessing the workplace environment). This is because the assessment tool was easy to use and understand (easy to judge). However, some of the observers commented that the tool was too difficult to read because the font used was too small. The observers also found the tool confusing and difficult to assess when the instructions on how to use the assessment were not provided especially when at the part where more than one action have shown, hence the final sub score is a combination of two actions. Observers also commented that the section about environment part, they state the range of the environment especially for lighting and noise was not flexible enough.

Table 2. Feedback Comments from observer

No	Comment
1	Very good tool to access the ergonomic risk (good for ergonomic users)
2	Easy to understand and easy to use (easy to judge)
3	This OFFERA Tool is a good to ensure the person always in good posture that can be avoid from hazard
4	Suitable for learning session because it easy to understand
5	OFFERA tool is very useful to office workers survey and can be used for all range of office works (useful to access workplace environment).
6	Easy to get final score and make the action based on final score
7	Font of the tool too small – difficult to read
8	For mouse positioned, put box left and right make it more confusing
9	C1 and E1 make a confused because need give the score for two part
10	Range for environment part especially for lighting and noise not enough
11	For some part, more than one action is shown; hence the final sub score is combination of two action. Make it confused because not provided instruction how to use this assessment
12	The probability for employees to work more than 4-5 hours a day is high because it multiplication
13	Use of colour and illustration/ diagram not clear and hard to understand.

In addition, Table 3 shows a summary of the suggestions from practitioners. Most of the practitioners found the wording or font of the sentences too small making them difficult to read. Other than that, the observers also suggested that the illustrations and the dimension for each diagram should be enlarged to make them clearer and more understandable. When the wordings, illustration, and dimension clear, it easier to observer access the level of office workers risks using this tool. The observers suggested enlarging the font size of the sentences. Observers suggested putting one box if the item only for one user. An example mouse positioned which are only uses one side either left or right hand to clicking the mouse. The rigid range of the section on the environment part made it difficult for the observers to assess. Therefore, improvement needs to make by adding a wider range for lighting and noise.

Table 3. General comments and suggestions from the observers

No	Suggestions / recommendations
1	Wording need to be clear and enlarge the font
2	If only one user, no need put box left and right, just put one box only.
3	Need to bold the keyword , C1 and E1 need have distance
4	Need improvement for office environment part with added more range for lighting and noise
5	Suggested a simple instruction should be provided before proceed
6	Adjust work hours are more suited to the actual working time for Step 7
7	Use basic colour or contrast between one section ratings. Enlarge the diagram and dimension to make it more clear and understandable.

3.3 Specific comments and suggestions for the OFFERA method

In this section, the comments and suggestions from 44 observers for each part of the OFFERA Method which consisted of six part of risk factors which include Part A (chair), Part B (desk), Part C (device input), Part D (monitor), Part E (accessories), and Part F (environment) are presented. Table 4 shows the summary of the specific comments and suggestions for all of the risk factors items.

Table 4. Comments and suggestions for Part A, B, C, D, E, and F

Items	Comments	Suggestions
Part A: Chair Risk Factor A1 – Seat Pan Height A2 – Seat Pan Depth A3 – Backrest A4 – Armrest	- Dimension for item A2 not cleared - Dimension too small - Clear and interesting - For A4 section- how do you differentiate between the hands placed on a desk with a chair - Using "OR" word in each evaluation make confused - Item A2 not valid (is it sure A2 effect ergonomic). - Need considered leg position for both side - Have armrest but not used. The risk is high - Font too small - No rating for comfort seat - Too many words - Seat pan height – if legs sticking behind what score rating - A1 – at third illustration it confusing	- Make dimension more bigger and clear - Bold the keyword - Make another diagram which shows the conditions laid hands on the table - Range no of seat pan depth - Make the diagram clear - Make separation for hand / leg for right and left - Added types of chair (span type or not) for rating comfort seat - Have armrest but not used – must put the high score. - Change the font size - Use simple words to explain - Fix the illustration – A1
Part B :Desk Risk Factor B1 – Desk Height B2 – Desk Depth	- Illustration for item B2 confusing / not clear - Font and box small - Instruction not clear - Keyword for that sentence not clear - Desk depth difficult to understand / to assess - B2- more bigger desk, more comfortable - Need more detail for item B2 - B1 – not understand for sizing (confused because don't have maximum height) - B2 – illustration for second and third are confusing	- Can put range - Change the fornt size (enlarge) - Bold the keyword - Added range / score for >720mm (too high) - Put more range for desk wide - Repaired B2 rating - Make range for a few sizing – B2 - Perhaps the dimension for B2 in range and be helpful (clear to understand for evaluation) - Try too add symbol > for easy to understand – B2

Continued table

Items	Comments	Suggestion
Part C : Input Device Risk Factor C1 – Keyboard C2 – Mouse Positioned C3 – Mouse Size	- Good - C1 need have distance - C1 why don't have score for keyboard position in line with elbow - CI difficult to determine the angle of wrist - Illustration too small - C1- only have too high or too low keyboard positioned - C1 – sentence not clear/ confusing - C1 – combine both side, it is related - C2 need add instruction – only for one hand - C2 confusing	- Need bold the keyword - Add keyboard for another level (keyboard position) - Enlarge the illustration - C1- added neutral keyboard positioned (to avoid assessor confused) - C1- consisted two evaluated section. It should be do separately - C1-Fix the sentence to more clear. - C2- need more detailed explanation - Just put one side for mouse position (more easy) – no need put left and right side
Part D: Monitor Risk Factor D1 – Angle View D2 – Distance View	- Good - Keyword for that sentence not clear - Difficult to assess the neck angular range for monitor range	- Need bold the keyword - Add screen brightness factor - Angle 15° still ok, need make range from 0°-30°
Part E : Accessories Risk Factor E1 – Telephone E2 – Document Holder E3 – Keyboard Wrist rest E4 – Mouse Wrist rest	- Good - The Illustration E2 not clear (confused) but the sentence clear. - Dimension too small - Font and Keyword for that sentence not clear - E1 need have distance - E1 not easy to completed (illustration too confusing) - What if: document holder is in front of monitor but beck flexion is not > 15°	- Make dimension more bigger - Need bold the keyword - Illustration need more specific - Give instruction for E2 more clear - E1- consisted two evaluated section. It should be do separately
Part F: Office Environment Risk Factor F1 – Lighting F2 – Temperature F3 – Noise	- Scoring for item F1 confusing - Font too small - Too much types of office (confused). - Not colourful - This part make confused - Keyword for that sentence not clear - Range of score limited (rigid) - The recommended temperature rating may not be appropriate	- Make it more colourful like other part - Need enlarge the font size - Need bold the keyword - Need to add more range - Better if there is no double range. - Suitable temperature at 20°C – 30°C - Make more research for lighting rating - Better if part for F1, F2, and F3 separated for easy to understand but still in office environment part

4 Discussion

4.1 Feedback Survey on the Usability of the OFFERA method

All the participants (N=44) gave their feedback on the usability of the OFFERA Tool (Table 1 and Figure 1) is the results indicated that the OFFERA tool is easy to use (mean 4.48 ± 0.698) and quick to use (mean 4.48 ± 0.821) with the percentages of 93% (agree) and 91% (strongly agree) respectively. The observers found that, the OFFERA method was easy to understand and suitable for evaluating ergonomic risks. Besides that, they stated that the OFFERA method will be very handy for determining office ergonomic factors. The observations also showed that the prototype tool is applicable in workplace assessment (mean 4.52 ± 0.549), and a wide range of jobs (mean 4.02 ± 0.952) with percentages of 93% (agree) and 80% (strongly agree) respectively. This is because the observers found that the OFFERA tool useful for assessing the workplace environment. Other than that, observer found, this prototype is a good ensure the person always in a good posture that can be avoided from the hazard.

The feedback from the observers during the training session on the score for the item 'clear instruction and easy to understand' (mean 3.93 ± 1.149) and 'clear wording and easy to understand (mean 3.91 ± 1.096) showed that the assessment was not clear. In addition, the percentage of the observers who responded either 'agree' or 'strongly agree' was less than 80%. The observers found that the font used was too small and hence made it difficult for them to read the instructions. The pictures or illustrations in the OFFERA tool ware also said to be unclear based on the relatively low score (mean 3.73 ± 1.128) as the percentage of the observers who either 'strongly agree' or 'agree' was only 66%. This is because the observers found it difficult to observe and define the range of angular values for monitors angle, keyboard, telephone and desk depth/ width. Besides that, the observers found it difficult to observe and define the range of angular values during the assessment of items related to neck posture. In fact, other studies revealed that it is difficult for observers to determine the worker's posture angle based on observation only [8, 15]

A study by Dartt et al., [18] found that the assessment of wrist posture to be more difficult to assess compared to other joints because of its dynamic posture. In the OFFERA tool, the participants confirmed that it was easy to understand the scoring system (mean 4.59 ± 0.622), score for each items (mean 4.50 ± 0.792), the final score (mean 4.68 ± 0.639), and the action level (mean 4.52 ± 0.549) are useful. This is because the observers concluded that based on the final score of the tool, it was easy to obtain the final score and determine the action level. The feedback from the practitioners on the score for items such as 'cost effective' (mean 4.36 ± 0.685), and 'provide a good basis for intervention proposal' (mean 4.41 ± 0.583) indicated that the tool is also practical and useful for the workplace.

4.2 General comments and suggestions for the OFFERA method

Most of the practitioner commented that the wording or font used was too small and made it difficult to read the sentences. Meanwhile, the percentage of observers who agree and strongly agree with the 'wording clear and understand' (mean 3.91 ± 1.096) only 73% (below 80%). This shows that their responses in the questionnaire are also reflected in their comments. Other than that, the observers also suggested that the illustrations and the dimension for each diagram should be enlarged to make them clearer and more understandable. In terms of reliability, the observers mentioned that it was difficult to assess the monitor angle (K=0.58) because the illustration of the angle was not clear. They also commented that it was difficult and confusing to assess the neck posture. According to other studies, the observers found it difficult to observe and determine the angle during the assessment of body posture [8]. This is because only 66% of the observers agree and strongly agree that the illustrations are clear and easy to understand. Other than that, during the training session, the observers stated that it was difficult and confusing to assess the keyboard position item. This happened because the illustrations of that item were not clear. It was difficult to determine whether the wrist posture was below or above especially for typing job. However, the questionnaire revealed that the observers agree (43%) and strongly agree (55%) that the tool is 'applicable to workplace assessment'. The observer found that the OFFERA method can be used for a wide range of office work (useful for assessing the workplace environment). Besides that, the observer found that the OFFERA method is a good way to ensure that office workers maintain a good posture at work.

4.3 Specific comment and suggestion for OFFERA method

Table 4 shows that most of the observers commented that the font and dimension for all items were too small and made it difficult for them to read. In this case, they suggested enlarging the font size and dimension for the items. Besides that, some of the observers were confused with item A4 (armrest) about how to differentiate between the hands placed on the desk with a chair and have an armrest but not used it. The observers suggested that a high score should for the risk factor "armrest but not used". Other than that, observers also suggested to put the keywords for each item in bold. It can make the observer more understand and clearer about the sentences. Besides that, the observer stated using the simple words to explain each items will make the assessment easier because using the word "OR" in each evaluation would confuse the observers. Other than that, the observers found Part A to be good, clear and interesting to assess. Besides that, Dockrell et al., [19] stated that the sitting posture or static posture is easier to assess compared to dynamic postures.

For part B, most of the observers commented about item B2 (desk depth/width). Some of the observers commented that they did not understand how to assess the desk depth/width because they were confused about the range and the illustration of the desk. Therefore, the percentage of agreement for the item 'illustration is clear and easy to understand' was only 66%. Besides that, they opined that a bigger desk would be more comfortable. The observers suggested that a range for B2 should be a provided in a few size and perhaps the dimension of B2 in a range. For B1 (desk height), the observers found that there was no maximum height provided only have a low height and suitable height. Because of that, they suggested adding the range/score for 'a desk too high' at B1.

For Part C (input device), most of the observers commented about item CI (keyboard). They found this item very confusing to assess because they needed to combine two scores in one item. Therefore they suggested that should be a distance between the score and that clearer instruction should be provided. Besides that, it was difficult for them to determine the wrist angle for item C1 because the illustration was too small. In addition, the observers had difficulty in determining the position of the wrist posture and the angle when assessing the wrist posture during typing activity. Besides that, another researcher also found that it is was difficult for an observer to assess whether the wrist angle was 15° or 20° from its neutral position during task performance [20]. Other than that, assessing the wrist posture for typing and clicking activities was more difficult to assess compared to other joints because it is dynamic posture (repetitive movement occur) [18]. The sentence for the item was also not clear so the keywords need to be bold.

For Part D (monitor), the observers found it too difficult to assess the range of angles of the neck for monitor range. Because of that, they suggested that a suitable angle range for the neck be adjusted to 0°-30° because an angle of 15° was still acceptable for them. According to David et al., [15], observers found it difficult to determine the worker's posture angle based on observation only. Besides that, Li and Buckle [20] found the observation easier to assess when using the terms such as 'bend up' (extension) and 'bend down' (flexion) compared to use of angular angle. Other than that, they suggested the use of 'bold' keywords in the sentence to avoid observer confused during assess the tool.

In Part E (accessories), the observers found it difficult to assess the telephone and document holder. They stated that items (telephone) need to be separated or have a distance because there was a combined score. Besides that, they found it hard to use the tool because the illustrations were too confusing. This happened, because the evaluation on the telephone consists of two sections. The observers suggested separating the scores. Besides that, the illustrations of the telephone and document holder need to be more specific and clear. For the document holder, the illustration was not very clear but the description was easy to understand. The observers asked about neck flexion below 15° but the document holder is in front of the monitor. They also suggested the need to have scores for that situation.

For part F (environment), the observers found that the range of the scores for lighting and noise was limited (rigid) and made assessment difficult. The observers suggested that the range should be extended but a double range should be employed because it can be confusing. Besides that, the observers suggested that items F1, F2, and F3 should be separated but still in the office environment part. It makes the observers easier to assess.

5 Conclusion

To conclude, from the feedback pertaining to the usability of the OFFERA method gathered via questionnaires, all participants agreed that the OFFERA method is both user-friendly and cost-effective. The observers found that, the prototype was suitable to evaluating ergonomic risks and easy to understand. Besides that, they stated that the tool will be very handy to determining the office ergonomic factors. The observations show that the OFFERA method is applicable in workplace assessments and in a wide range of jobs. The observers also noted that, it can be useful for assessing the workplace environment. In addition, the participants confirmed that the OFFERA method had scoring systems, scores for each item, final score, and action levels which were clear and easy to comprehend. However, the observers lamented that the font use was too small and thus made it difficult to read the instructions. The picture or illustrations in the OFFERA method were also recorded as unclear based on the relatively low score for 18 items. This is because the observers found it difficult to observe and define the ranges of angular values for monitors angle, keyboards, telephones, and desks depth/width. After the usability test, the OFFERA method has been revised. Besides that, OFFERA tool needs to be reassessed in terms of usability in future studies.

This research is funded by the Ministry of Higher Education of Malaysia (MOHE) and Universiti Tun Hussein Onn Malaysia (UTHM) under the Fundamental Research Grant Scheme (FRGS, Vot 1495)

References

1. B.P. Bernard, *Musculoskeletal disorders and workplace factors: critical review of epidemiologic evidence for work-related musculoskeletal disorders of the neck, upper extremity, and low back.* National Institute for Occupational Safety and Health (NIOSH). Cincinnati, OH (1997).

2. M.N.A. Rahman, M.R.A. Rani, and J.M. Rohani, Investigation of work-related musculoskeletal disorders in wall plastering jobs within the construction industry. *Work*, **43(4),** 507-514 (2012).
3. M.N.A. Rahman, F.A. Aziz, and R.M. Yusuff, Survey of body part symptoms among workers in a car tyre service centre. *Journal of human ergology*, **39(1),** 53-56 (2010).
4. M.N.A. Rahman, H.R. Haq, M.F. Hassan, M.A. Arifin, M.Z. Yunos, S.A. Bakar, and S. Adzila, Musculoskeletal discomfort among workers in mould making manufacturing industry. *ARPN Journal of Engineering and Applied Sciences*, **10(15),** 6269-6273 (2015).
5. M. Sonne, D. L. Villalta, and D. M. Andrews, Development and evaluation of an office ergonomic risk checklist: ROSA–Rapid office strain assessment. *Applied ergonomics*, **43,** 98-108 (2012).
6. M. Robertson, B. C. Amick, K. DeRango, T. Rooney, L. Bazzani, R. Harrist, A. Moore, The effects of an office ergonomics training and chair intervention on worker knowledge, behavior and musculoskeletal risk. *Applied Ergonomics*, **40(1),** 124-135 (2009)
7. J. Village, D. Rempel, K. Teschke, Musculoskeletal disorders of the upper extremity associated with computer work: a systematic review. Occup. Ergon. **5**, 205-218 (2005).
8. M.N.A. Rahman, M.R.A. Rani, J.M. Rohani, WERA: an observational tool develop to investigate the physical risk factor associated with WMSDs. *J. Hum. Ergol*, **40**,19-36 (2011)
9. A. Loghmani, P. Golshiri, A. Zamani, M. Kheirmand, N. Jafari, Musculoskeletal symptoms and job satisfaction among office-workers: A Cross-sectional study from Iran. *Acta Medica Academica*, **42**, 46-54 (2013).
10. M.N.A. Rahman, N.H. Zakaria, I. Masood, S. Adzila, and N.F. Nasir, Risk Assessment for Assessing the Subjective Occupant Seating Discomfort related Office Works. *International Information Institute (Tokyo). Information*, **19(7B),** 3025-3030 (2016).
11. L. Eyal, J. Ribak, Y. Badihi, Remote online ergonomic assessment in the office environment as compared to face-to-face ergonomic assessment. Work: A Journal of Prevention, Assessment and Rehabilition, **41**,516-523 (2012)
12. L. McAtamney, and E.N. Corlett, RULA: A survey method for investigation of work-related upper limb disorders. *Applied Ergonomics*, **24(2),** 91 – 99 (1993)
13. S. Hignett, & L. McAtamney, Rapid entire body assessment (REBA). *Applied ergonomics*, **31(2)**, 201-205 (2000).
14. Occupation Safety & Health Administration, *Computer Workstation e-Tool Checlist,* United State Department of Labor (2003).. Retrived from https://www.osha.gov.
15. G. David, V. Woods, G. Li, P. Buckle, he development of the Quick Exposure Check (QEC) for assessing exposure to risk factors for work-related musculoskeletal disorders. *Applied Ergonomics*, **39(1)**, 57-69 (2008).
16. J.G. Ferreira, M. Hunter, L.M. Birtles, & D. Riley, Development of an assessment tool for repetitive tasks of the upper limbs (ART) (2009). *Derbyshire: Health and Safety Executive.*
17. J.S. Dumas, and M.C. Salzman, Usability assessment methods. *Reviews of human factors and ergonomics*, *2*(**1**), 109-140 (2006).
18. A. Dartt, J. Rosecrance, F. Gerr, P. Chen, D. Anton, L. Merlino, Reliability of assessing upper limb postures among workers performing manufacturing tasks. *Applied Ergonomics*, **40(3)**, 371-378 (2009).

19. S. Dockrell, E. O'Grady, K. Bennett, C. Mullarkey, R. Mc Connell, R. Ruddy, C. Flannery, An investigation of the reliability of Rapid Upper Limb Assessment (RULA) as a method of assessment of children's computing posture. *Applied Ergonomics*, **43(3)**, 632-636 (2012).
20. G. Li, & P. Buckle, Current techniques for assessing physical exposure to work-related musculoskeletal risk, with emphasis on posture-based methods. *Ergonomics,* **42(5)**:674-695 (1999).

MATEC Web of Conferences **135**, 00026 (2017) DOI: 10.1051/matecconf/201713500026
ICME'17

The Balance Improvement of Final Assembly Water Pump Business Unit Tracks Section : Case Study

Ikhsan Siregar[1*], *Gretty* Margaretha [1] , *Tigor* Hamonangan Nasution[2] and Ulfi Andayani [3]

[1] Department of Industrial Engineering, Faculty of Engineering, University of Sumatera Utara, Almamater Rd., Medan 20155, Indonesia.
[2] Department of Electrical Engineering, Faculty of Engineering, University of Sumatera Utara, Almamater Rd., Medan 20155, Indonesia.
[3] Department of Information Technology, Faculty of Engineering, University of Sumatera Utara, Almamater Rd., Medan 20155, Indonesia.

Abstract. This research was conducted at an electronic field assembly company. The company implements a production system that starts from semi-finished goods to finished products. The problem of the company is the imbalance of the production trajectory. The track balance in the final assembly will affect the amount of product that can be produced. The unbalanced track conditions resulted a bottleneck condition in the material flow from one work center to the other, resulting in delay time on the production floor. The problems are caused by large cycle times, thus resulting in different production capacities for each workcenter. The implementation of the theory of constraints is to identify the existing constraints and to obtain the optimum amount of workcenter as well as to allocate elements of activity in the workcenter, thereby improving the efficiency of the production line.. In actual trajectory there are 7 work centers with track path efficiency 59,80%, balance delay 40,20%, and smoothing index 74,47. The result of balancing in the production line using theory of constraint method obtained the repair solution by number of work center as many as 5 work centers with track efficiency value 88,65%, balance delay 11,35% and smoothing index 17,07.

1 Introduction

The balance of the production process assembly line in the mass manufacturing industry is an important part of the production process. This is related to the existence of several jobs that are grouped in several WC (work center). The operator's ability to do assembly activities will affect the number of products in each process. The assembly time of each work element is determined by the length of time required in the assembly. Track balance in the assembly process will affect the achievement of production targets. Unfulfilled production targets will cause losses for the company.

---

* Corresponding author: ikhsan.siregar@usu.ac.id

Previous research has suggested that line balancing issues to determine the optimal allocation of assembly operations to a set of workstations, with related constraints and choose which precedence. This study based on case study that is ultimately discussed to demonstrate the capabilities of the proposed method in finding the optimal solution [1] . Another research on line balancing suggests that multi-objective hybrid optimization algorithms aimed at simultaneously resolving line balancing, equipment and buffer selection, measuring problems with capacity-oriented capacity and objectives, in other words looking to optimize in the production path [2]

Research for line balancing is widely applied to the manufacturing industry as in research conducted on the world of airplanes [3]. In other hand, there was also a line balancing study conducted in the automotive industry such as done in the europe automotive industry [4]. Line balancing is very influential on workers, such as previous research that discusses the Heuristic Algorithm associated with the assignor to the worker [5]. The purpose of line balancing is clear, ie optimizing production and looking for better benefits than before as in the research conducted about the production line with maximal stretch and get the optimal profit [6].

Previous research has also been studied with many methods similar to this research method, namely the Theory of Constraint (TOC) method. In a previous study it was mentioned that analyzing whether the Theory of Constraints (TOC) could be beneficial to hospital record logistics, the study proves that this method can be applied anywhere [7]. In another study stated that the method is also very good in its use in project management, where the study discusses the system of supervision and monitoring in project management [8]. Some even apply to the furniture industry to look for optimal products [9]. This method is categorized in the category of artificial intelligence, where the artificial intelligence is much studied by the field of Computer Science before, but has now been applied in all fields such as application on the production of light bulbs [10]. Another study of artificial intelligence is the comparison of the theory of forecasting in sood production [11].

Trajectory imbalances are also experienced at the site of this research, where the company where the research is conducted is engaged in electronics and is a manufacturing company. One of the company's results is a water pump. The initial naming done to the assembly of the company in the water pump product is symbolized by X, where the constraints are due to the presence of idle operators, another constraint is to wait for the material to be processed. The difference in production speed between the workcenter causes an imbalance in the production line and the production target is not achieved. Thus it can be concluded that the problem is caused by the allocation of work elements are not balanced and cause congestion. This can be overcome by using Theory of Constraints (ToC) method. This study is limited by not taking into account the cost of losses due to imbalances that occur and does not analyze the achievement of production targets.

2 Research Method

2.1 The Aim of This Study

The general purpose of this study is to obtain improved production lines through a more efficient equilibrium with Theory of Constraint (TOC) method so as to facilitate the flow of production and production targets achieved, to identify bottleneck work centers, to allocate elements of activity, and to compare initial track conditions with proposed trajectories better.

2.2 Data Processing Method

The steps in the process of data processing are:

A. Identify a System Constraint
B. Calculate the standart time of the final assembly process
C. of the five steps of the theory of constraint
   a. Identify system constraints with CRT (Current Reality Tree)
   b. Determine how to know the constraints with Identify the workstation bottleneck with CRD (Conflict Resolution Diagram) and Calculation of time required and available time
   c. Repairs with Line Balancing Solutions to Reduce Constraints. Determination of the best solution based on Line balancing criteria, namely Line efficiency, Balance Delay, and Smoothing Index.

Trajectory efficiency

$$E = \frac{Tproduct}{Tmax \ X \ WC} \ X \ 100\% \quad \text{............. (1)}$$

Balance Delay

$$D = \frac{nSm-\Sigma Si}{(n)(Sm)} \ x \ 100\% \ \text{...........(2)}$$

Smoothing Index

$$SI=\sqrt{\Sigma_{i=1}^{K}\left(STi_{max}-STi\right)^{2}} \quad \text{............... (3)}$$

3 Results and Discussion

3.1 Application of Theory of Constraints on Final Assembly Process

The theoretical steps of constraints consist of five steps:

1. Identify a System Constraint

This stage requires observation of the problem on the final assembly process to be solved. This requires cause-and-effect to identify the basic causes of the imbalance of the assembly line path to the core problem. Based on the observations that have been done, then the equilibrium of the path that has been found is arranged in the form of the current reality tree diagram (CRT) to see the causal relationship. Figure diagram of CRT can be seen in Figure 1.

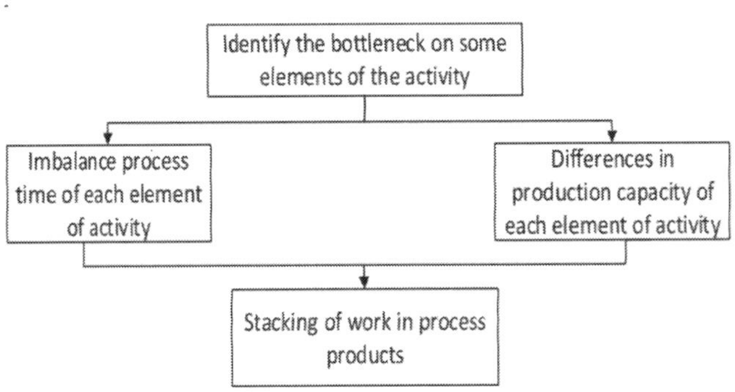

Figure 1. Current Reality Tree (CRT)

Figure 1 explains the main cause of track balance problem and the conclusion that can be drawn from the CRT diagram that the main factor causing constraints on the final assembly process in the company during the research is the existence of bottlenecks.

2. Exploit Existing Obstacles

This second phase aims to build a simple and practical solution to overcome bottleneck constraints on the final assembly process of the CRT diagram. If the CRT sought is the main cause of the problem, then in the constraint exploitation phase used the Conflict Resolution Diagram (CRD) tool which prioritized the goal, ie reducing or eliminating the bottleneck and seeing the conflicts arising from the various solutions. The solution focuses on the cause of the path prose imbalance in the final prose assembly with respect to processing time. The CRD image can be seen in Figure 2.

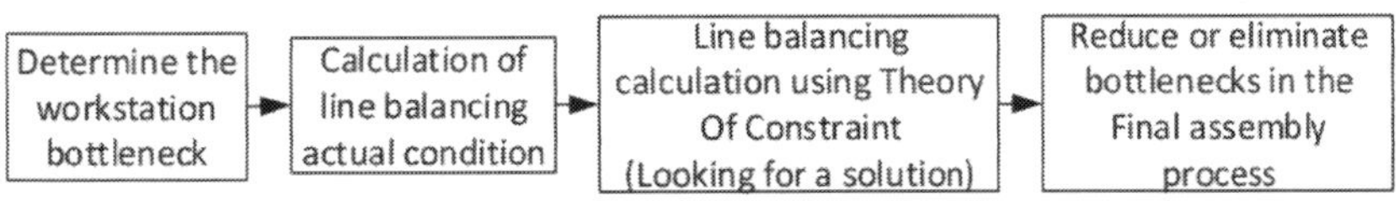

Figure 2. Conflict Resolution Diagram

Figure 2 describes the steps that must be done in solving the path equilibrium problem. Once the solution is determined to overcome the existing constraints by using CRD, then the next stage is to deploy solutions to the Work center bottleneck.
Bottleneck occurs when the time required is greater than the time available in the sense that the work centre cannot produce in accordance with predetermined production targets and consequently the existence of raw materials that accumulate. Work center bottleneck can be seen in Table 1.

From Table 1 can be seen there are three elements of work that experience the bottleneck of the work elements of checking pumps and water remover and attachment of barcodes, warranty paper, and manuals (work elements 16 and 17). From the results of this calculation clearly shows that the existence of bottleneck constraints and the imbalance of production time of each work element.

Table 1. Bottleneck Working Element on Final Assembly Process

Working Element	Time Available (sec)	Time Taken (sec)	Difference (sec)
1	5748,91	5607,54	141,37
2	7724,38	7534,44	189,94
3	4365,57	4258,22	107,35
4	5678,42	5538,79	139,63
5	3684,67	3594,07	90,61
6	14646,03	14165,84	480,20
7	4571,65	4421,76	149,89
8	16115,92	15851,72	264,20
9	5553,74	5462,70	91,04
10	6586,64	6478,67	107,98
11	15801,84	15542,79	259,05
12	5856,00	5760,00	96,00

13	17083,43	16943,40	140,03
14	7836,10	7771,87	64,23
15	8217,87	8150,51	67,36
16	39837,69	40164,23	-326,54
17	6919,81	6976,53	-56,72
18	14105,19	13989,57	115,62
19	3799,43	3768,29	31,14
20	6586,64	6532,66	53,99
21	3310,45	3283,31	27,13

3. Repairs with Line Balancing Solutions to Reduce Constraints

The track balancing steps can be seen as follows.

a. Actual Condition

Shows the actual condition of the company according to the precedence sequence diagram. Based on precedence diagrams that have been sorted can be seen idle time each work center that can be seen in Table 2.

Table 2. Task Determination of Each Work Station According to the Actual Condition

Work Center	Working Element	Standard Time (second)	Cycle Time (second)	Idle (second)
I	1	6.98	40.28	19.50
	2	11.80		
	3	7.16		
	4	8.43		
	5	5.91		
II	6	11.84	19.33	40.44
	7	7.49		
III	8	15.05	34.95	24.83
	9	9.10		
	10	10.80		
IV	11	15.50	25.44	34.34
	12	9.93		
V	13	13.71	40.02	19.75
	14	12.85		
	15	13.47		
VI	16	49.76	59.78	0.00
	17	10.01		
VII	18	13.12	34.82	24.95
	19	6.43		
	20	10.23		
	21	5.04		
	TOTAL		254.62	163.80

Trajectory efficiency

$$E = \frac{258.08}{7 \times 61,65} \times 100\% = 59,80\% \text{ ....(1)}$$

Balance Delay

$$D = \frac{7 \times 61,65 - \sum 258.08}{7 \times 61,65} \times 100\% = 40,20\% \text{...........(2)}$$

Smoothing Index

$$SI = \sqrt{(41,29^2)+(19,20^2)+\ldots+(35,58)^2} \text{.............(3)}$$

 DOI: 10.1051/mateccconf/201713500026

=59,80

The solutions in Table 2 are based on actual conditions. The approach solution is determined based on the number of workcenter, number of work elements on each workcenter, and the standard time of each work center. Iteration 0 is done by following the number of workcenter from initial that is as much as 6 workcenter, and work element is sorted according to actual situation at company. The optimal number of workcenter can be searched by using the following formula.

$$WC = \frac{T_{product}}{T_{max}} = \frac{258{,}08}{50{,}31} = 5{,}13 \approx 5$$

So on the iteration done the number of workcenter reduction one by one approaching the optimal number of workcenter. At 0 iteration, the division of work elements in each work center is done by trial and error which can be seen in Table 3.

Table 3. Zero Iteration of Final Assembly Process

Work Center	Working Element	Standart Time (second)	Cycle Time (second)	Idle (second)
I	1	6,12	41,29	9,01
	2	12,66		
	3	7,16		
	4	9,31		
	5	6,04		
II	6	11,71	43,63	6,68
	7	7,49		
	8	15,32		
	9	9,10		
III	10	10,80	35,90	14,41
	11	15,50		
	12	9,60		
IV	13	13,71	40,02	10,28
	14	12,85		
	15	13,47		
V	16	50,31	50,31	0,00
VI	17	11,34	46,92	3,39
	18	13,12		
	19	6,23		
	20	10,80		
	21	5,43		
		TOTAL		**43,77**

Trajectory efficiency

$$E = \frac{258{,}08}{6 \times 50{,}31} \text{ x } 100\% = 85{,}28\ \%$$

Balance Delay

$$D = \frac{6 \times 50{,}31 - \sum 258{,}08}{6 \times 50{,}31} \text{ x } 100\% = 14{,}50\%$$

Smoothing Index

$$SI = \sqrt{(9{,}01)^2+(6{,}68)^2+....+(3{,}39)^2}$$
$$=15{,}59$$

In Table 3. which is the 0 iteration steps, it occurs that the number of work elements in WC II-WC V. Items affects the value of track balance criteria. Better track efficiency is available than actual condition, and smoothing index is smaller than actual (15.59 <59,80). Therefore iterated back. The next iteration is done by subtracting 1st work center, and re-balancing the trajectory by trial and error ie replacing all working elements on the WC contained in iteration 0. The result of 1st iteration can be seen in the following calculation.

Trajectory efficiency

$$E = \frac{258{,}08}{5 \times 58{,}22} \times 100\% = 88{,}65\%$$

Balance Delay

$$D = \frac{5 \times 58{,}22 - \sum 258{,}08}{5 \times 58{,}22} \times 100\% = 11{,}35\%$$

Smoothing Index

$$SI = \sqrt{(5{,}22)^2+(0{,}00)^2+....+(11{,}30)^2}$$
$$= 31{,}48$$

In the 1st iteration the path's efficiency is higher than 0 iteration (1st iteration > Iteration 0 = 88,65> 85,50) and the optimal number of work center has been reached that is 5 work center. Re-do trial and error to see if the track efficiency value will be better if the work center is reduced once again to 4. Thus, the method on iteration 1 can be continued to iteration 2. The iteration is done until it is filled with the stop rule ie the amount of idle time increases and Reduced efficiency at iteration 2, Working elements modified at all work centers by trial and error. Here are the calculations:

Trajectory efficiency

$$E = \frac{258{,}08}{4 \times 76{,}63} = 84{,}20\%$$

Balance Delay

$$D = \frac{4 \times 76{,}63 - \sum 258{,}08}{4 \times 76{,}63} \times 100\% = 15{,}80\%$$

Smoothing Index

$$SI = \sqrt{(16{,}13)^2+(2{,}59)^2+(0)^2+(29{,}71)^2}$$
$$= 35{,}37$$

In the 2nd iteration, the path's efficiency is lower than 1st iteratio. The smoothing index value at 2nd iteration is 33.90. This indicates that the SI value is greater than 1st iteration (33.90> 17.07), as well as the idle value of iteration 2 which is greater than iteration 1 (48,43> 33,03).

Each iteration yields a percent difference in track efficiency and balance delay, and the indicator values of the smoothing index are also different, as well as the number of idle times and different cycle times at each work center. So, its necessary for us to make a comparison of the three iterations to choose the best. Recapitulation of iterative comparison can be seen in Table 4 below.

Table 4. Comparative Results of Final Assembly Processes Iteration

Iteration	Total WC	Efficiency (%)	Balance Delay (%)	Smoo-thing Index	Idle Time (s)
0	6	85,50	14,50	15,59	43,77
I	5	88,65	11,35	17,07	33,03
II	4	84,20	15,80	33,90	48,43

Reviewing from Table 4, the comparison of the above three iterations is based on the aspect of efficiency value, balance delay, smoothing index, and idle time, the best iteration chosen is 1st iteration. Illustration of balance condition of 1st Iteration can be seen in Figure 3.

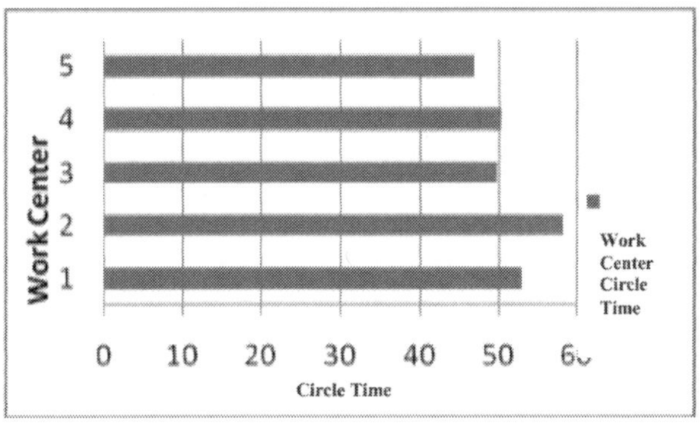

Figure 3. Track Balance Condition with 1st Iteration

The final assembly trajectory in Figure 3 is the result of the solution splitting using TOC. Can be seen the cycle time of each work center does not differ too much (balanced) compared with actual conditions. It can also be seen from the smoothing index value of the selected solution ie 1st iteration which is smaller than the actual condition.

In the processing steps of theory of constraints has been done identification constraints that occur in the company by using tools current reality tree. The results of identification indicate the existence of bottleneck in some work center that is WC 5 and WC 6 because of the cycle time between different operators far. Illustration of trajectory of the actual final assembly process can be seen in Figure 4.

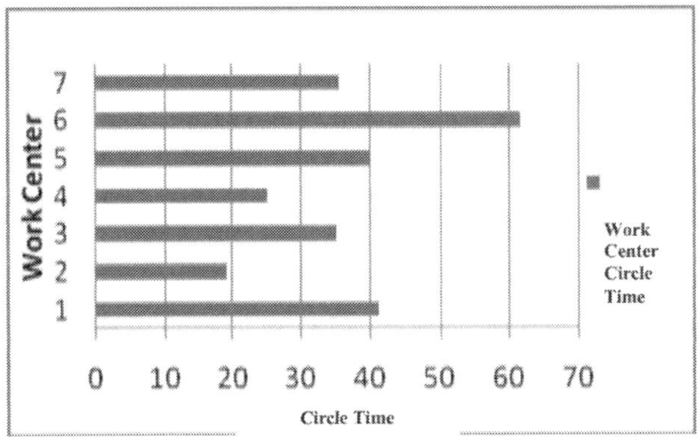

Figure 4. Actual Trajectory Condition of Final Assembly Process

The actual trajectory condition shows in Figure 4, can be seen that the path with 7 work centers has an efficiency that is 59.80%, balance delay is 40.20%, and smoothing index is 73.47. These values indicate the need for track balancing, so that the Theory of Constraints (TOC) principle is used.

To determine the best method between the actual condition and after the improvement design analysis by using performance parameters to be applied to the company. Comparison of result of balancing parameter of production line used is efficiency index (EI). The TOC method shows that the conditions in Iteration I are better than the actual condition with the increase of efficiency as follows:

$$(IE) = \frac{E\ (TOC)}{E\ (Actual)} = \frac{88{,}65}{59{,}80} = 1{,}48$$

An efficiency index of 1.48 indicates that the final assembly path with TOC method is better than the actual condition so that the company can apply the proposed trajectory design to obtain the optimum production trajectory.

The number of actual work centers is 7 and after balancing the TOC method, the number of work centers is 5. The result will not interfere with the production process due to the increased efficiency, balance delay down, and smoothing index down compared to the actual condition.

4 Conclusions

The conclusions of the observation and data processing are as follows:

1. The final assembly process has a bottleneck on WC 6.
2. Implementation of constraints theory allocates 21 elements of actual work with 7 work centers into 5 work centers.
3. Production path with TOC heuristic approach method produces efficiency about 88,65%, balance delay 11,35%, and smoothing index 17,07.
4. Efficiency index of TOC usage increased by 1.48 compared to actual condition.

References

1. Dalle Mura, M., & Dini, G. (2016). Worker Skills and Equipment Optimization in Assembly Line Balancing by a Genetic Approach. *Procedia CIRP*, *44*, 102-107.

2. Oesterle, J., Bauernhansl, T., & Amodeo, L. (2016). Hybrid Multi-objective Optimization Method for Solving Simultaneously the line Balancing, Equipment and Buffer Sizing Problems for Hybrid Assembly Systems. Procedia CIRP, 57, 416-421.
3. Borreguero, T., Mas, F., Menéndez, J. L., & Barreda, M. A. (2015). Enhanced Assembly Line Balancing and Scheduling Methodology for the Aeronautical Industry. Procedia Engineering, 132, 990-997.
4. Pröpster, M., März, L., Reinhart, G., & Intra, C. (2015). Validation of Line Balancing by Simulation of Workforce Flexibility. *Procedia CIRP*, *33*, 93-98.
5. Borba, L., & Ritt, M. (2014). A heuristic and a branch-and-bound algorithm for the Assembly Line Worker Assignment and Balancing Problem. *Computers & Operations Research*, *45*, 87-96.
6. Bentaha, M. L., Battaïa, O., & Dolgui, A. (2014). Disassembly line balancing and sequencing under uncertainty. *Procedia CIRP*, *15*, 239-244.
7. Aguilar-Escobar, V. G., & Garrido-Vega, P. (2016). Applying the theory of constraints to the logistics service of medical records of a hospital. *European Research on Management and Business Economics*, *22*(3), 139-146.
8. Izmailov, A., Korneva, D., & Kozhemiakin, A. (2016). Effective Project Management with Theory of Constraints. *Procedia-Social and Behavioral Sciences*, *229*, 96-103.
9. Okutmuş, E., Kahveci, A., & Kartašova, J. (2015). Using theory of constraints for reaching optimal product mix: an application in the furniture sector. *Intellectual Economics*, *9*(2), 138-149.
10. Siregar, I., Suantio, H., Hanifiah, Y., Muchtar, M. A., & Nasution, T. H. (2017, January). Fuzzy linear programming for bulb production. In *Journal of Physics: Conference Series* (Vol. 801, No. 1, p. 012001). IOP Publishing.
11. Siregar, B., Butar-Butar, I. A., Rahmat, R. F., Andayani, U., & Fahmi, F. (2017, January). Comparison of Exponential Smoothing Methods in Forecasting Palm Oil Real Production. In *Journal of Physics: Conference Series* (Vol. 801, No. 1, p. 012004). IOP Publishing.

 DOI: 10.1051/matecconf/201713500027

Factors to Consider in Selecting an Organisational Improvement Initiative: Survey Results

Musli Mohammad[1,*], *Maher* Ali Hussein[2] and *Mohd* Shahir Yahya[3]

[1]Industrial and Systems Engineering Focus Group, Department of Manufacturing and Industrial Engineering, Faculty of Mechanical and Manufacturing Engineering, Universiti Tun Hussein Onn Malaysia (UTHM), 86400, Batu Pahat, Johor, Malaysia

[2]Al-Furat Al-Awsat Technical University - Technical Institute of Al-Mussaib, 54003, Kufa, Iraq

[3]Department of Mechanical Engineering, Centre for Diploma Studies, Universiti Tun Hussein Onn Malaysia (UTHM), 86400, Batu Pahat, Johor, Malaysia

Abstract. Organisations should select the appropriate improvement initiative that will fit with the context of organisation and provide value to the organisation. This paper presents 18 factors to be considered when selecting an organisational improvement initiative. Organisational improvement initiatives are approaches, management systems, tools and/or techniques that can be used for managing and improving organisations, such as Lean, ISO9001, Six Sigma and Improvement Team. A survey was conducted to identify the level of importance of these 18 factors as criteria for selecting an improvement initiative. Purposive sampling was used for this survey involving practitioners, managers, engineers, executives, consultants and/or academicians, who have been involved in the selection and/or implementation of organisational improvement initiatives in Malaysia. Two factors were rated as 'very high importance', which involve: (1) The ability to gain top management commitment and support to introduce and implement the initiative successfully, and (2) The initiative is aligned to the vision, mission and/or purpose of the organisation. All these factors can be adopted by the organisations as decision criteria to assist in the selection of the most appropriate improvement initiative based on rational decision making.

1 Introduction

There are many factors to consider in selecting an appropriate organisational improvement initiative, such as the current maturity level of an organisation, areas in which the initiatives are adopted, type or size of an organisation and the capabilities of its workforce [1-3]. 'Improvement initiatives' refers herein to approaches, management systems, tools and/or techniques that can be used for improving organisation and include, for example: Six

* Corresponding author: mmusli@uthm.edu.my

Sigma, Lean, ISO9001, Business Continuity Management, and Environmental Management System [4-6]. Since every organisation is unique, there is no one best initiative that can solve all organisational problems. Each initiative also has its own purpose, strengths and limitations. The most appropriate initiative depends on the context in which it is adopted, rather than assumed to be universally applicable. Managers and/or leaders in the organisations should therefore consider all related factors before selecting the right initiative for the situation [4, 7-8]. Therefore, the objectives of this study are to identify factors that should be considered when selecting organisational improvement initiative and assess the level of importance of each factor for selecting the improvement initiative.

2 Literature review

Many existing literature have highlighted the importance of using contingency approach in managing organisation and understanding complex and dynamic situations, for instance Donaldson [9], Koontz and O'Donnell [10], Lawrence and Lorsch [11], and Sousa and Voss [12]. According to Kast and Rosenzweig [13], "the essence of contingency approach is that there is no one best way and that there is a middle ground between 'universal principles' and 'it all depends'. This approach recognises the complexity involved in managing modern organizations but uses the existing body of knowledge to relate environment and design, to match structure and technology, to integrate strategy and tactics, or to determine the appropriate degree of subordinate participation in decision making, given a specific situation".

Based on literature review, several contingency factors have been identified for the selection of improvement initiative which are as follow:

- Capability of the workforce to introduce and implement the initiative successfully [3, 14].
- Ability to gain top management commitment and support [1, 3, 14].
- Ability to allocate necessary resources to introduce and implement the initiatives successfully, such as funding and equipment [8, 15].
- Suit with the level of organisational excellence maturity [2, 16].
- The initiative is aligned to the vision, mission and/or purpose of the organisation [3, 14].
- The initiative is aligned to the direction, strategic plan and/or goals of the organisation [3, 14].
- The initiative will satisfy the requirements / expectations of the customers and other stakeholders [16].
- Suit with the organisational culture [3, 12, 14].
- Suit with the external environment in which the organisations operate, which includes: political, economic, social, technological, legal and environmental factors [16-17].
- The initiative fits the sector / industry in which the organisation operates (e.g. private, public or non-profit) [1-3, 14].
- The initiative fits the size of the organisation (e.g. small, medium or large) [1-3, 14].
- The ability of the initiative to fit in with complement, integrate and/or to support other initiatives already in place, and might be used in the future [8, 15].
- Ability to implement the initiative easily in the organisation [4]
- Value for implementing the initiative (based on the cost benefit analysis, Return of Investment (ROI) analysis and/or payback analysis) [3, 8, 14, 18].
- Duration taken to obtain the value of implementing the initiative [16].

- Whether the initiative has been proven to provide value / benefit in other similar organisations [16].
- Popularity of the initiative (e.g. many other organisations are implementing the initiative) [3].
- Areas in which the initiative will be implemented (e.g. leadership and social responsibilities, planning and strategy, information, customer, process, and/or people) [8, 19].

3 Research design and methodology

Survey was used to collect data for this research. Self-administered questionnaire was selected as the survey instrument. The questionnaire was adapted from [4]. The questionnaire consists of an introductory page and four main sections. The introductory page provides the background to the researchers, describes the potential respondents, purposes of the survey, expected time taken to complete the survey, benefits of participating in the survey and the researchers contact information. The first section of the questionnaire covers general information about the respondent and respondents' organisation. The second section of the questionnaire obtains the respondents opinion on the level of importance of the factors that should be considered when selecting an appropriate improvement initiative. The third section requests respondents to evaluate the computer aided decision support tools for selecting organisational improvement initiative. The last section is to obtain the respondents opinions on strengths and weaknesses of the proposed computer aided decision support tool for selecting the organisational improvement initiative. This paper only presents results from Section 2 of the questionnaire.

Purposive sampling was used for this survey involving practitioners, managers, engineers, executives, consultants and/or academicians, who have involved in the selection and/or implementation of the organisational improvement initiatives. The reason for choosing this sample was due to the respondents' knowledge and/or experience in the selection and/or implementation of organisational improvement initiatives. Twenty three (23) respondents completed the questionnaire. 78.26% of total respondents have seven years or more than seven years of experience, which indicate that most of the respondents have vast experience in the area of quality, productivity and/or operational improvement. The respondents consist of managers, assistant managers, academicians, consultants, engineers, and vice president. 39% of respondents are managers and assistant managers. Four percent of the total respondents were from organisations that had won Business Excellence or Quality Awards.

In section 2 of the questionnaire, the respondents were requested to indicate the level of importance of each factor that should be considered when selecting the improvement initiative. Eighteen factors were evaluated by the respondents. A six point likert scale was used to represent the levels of importance for each factor:

- Very high importance = 5
- High importance = 4
- Moderate importance = 3
- Low importance = 2
- Very low importance = 1
- Not important = 0

The returning of the completed questionnaire was taken as participants' consent to participate in this survey. Median was used to analyse the likert scale data because it is suitable for ordinal data and non-parametric analysis [20-21].

4 Results and discussion

The results were summarised in Table 1. Two factors were rated as 'very high importance' (Median = 5). The factors are: (1) The ability to gain top management commitment and support to introduce and implement the initiative successfully, and (2) The initiative is aligned to the vision, mission and/or purpose of the organisation. Survey conducted by Mohammad [4] also found that 'the ability to gain top management commitment and support to introduce and implement the initiative successfully' was rated as 'very high importance'. Fourteen factors were rated as 'high importance' (Median = 4). The factors are:

- Capability of the workforce to introduce and implement the initiative successfully.
- Ability to allocate necessary resources to introduce and implement the initiative successfully (e.g. funding and equipment).
- The initiative is suitable for the level of the organisational excellence maturity.
- The initiative is aligned to the direction, strategic plan and/or goals of the organisation.
- The initiative will satisfy the requirements / expectations of the customers and other stakeholders.
- The initiative fits the organisation's culture.
- The initiative fits the sector/industry in which the organisation operates (e.g. private, public or non-profit).
- The initiative fits the size of the organisation (e.g. small, medium or large).
- The ability of the initiative to fit in with complement, to integrate and/or to support other initiatives already in place, and might be used in the future.
- Ability to implement the initiative easily in the organisation.
- Value for implementing the initiative (based on the cost benefit analysis, Return of Investment (ROI) analysis and/or payback analysis).
- Duration taken to obtain the value of implementing the initiative.
- Whether the initiative has been proven to provide value/benefit in other similar organisations.
- Areas in which the initiative will be implemented (e.g. leadership and social responsibilities, planning and strategy, information, customer, process, and/or people).

Subsequently, two factors were rated as 'moderate importance' (Median = 3). The factors are: (1) The initiative is suitable for the external environment in which the organisation operates, which includes: political, economic, social, technological, legal and environmental factors, and (2) popularity of the initiative (e.g. many other organisations are implementing the initiative).

Table 1. Survey results on the levels of importance of each factor to be considered when selecting improvement initiative

Factors	Valid	Missing	Median	Min	Max
Q2.2 Ability to gain top management commitment and	23	0	5	3	5

DOI: 10.1051/matecconf/201713500027

support to introduce and implement the initiative successfully.					
Q2.5 The initiative is aligned to the vision, mission and/or purpose of the organisation	23	0	5	3	5
Q2.1 Capability of the workforce to introduce and implement the initiative successfully	23	0	4	3	5
Q2.3 Ability to allocate the necessary resources to introduce and implement the initiative successfully (e.g. funding and equipment).	23	0	4	3	5
Q2.4 The initiative is suitable for the level of organisational excellence maturity.	23	0	4	3	5
Q2.6 The initiative is aligned to the direction, strategic plan and/or goals of the organisation.	23	0	4	3	5
Q2.7 The initiative will satisfy the requirements / expectations of the customers and other stakeholders.	22	1	4	3	5
Q2.8 The initiative fits the organisation's culture.	23	0	4	2	5
Q2.10 The initiative fits the sector / industry in which the organisation operates (e.g. private, public or non-profit).	23	0	4	2	5
Q2.11 The initiative fits the size of the organisation (e.g. small, medium or large).	23	0	4	2	5
Q2.12 The ability of the initiative to fit in with complement, integrates and/or to support other initiatives already in place, and might be used in the future.	23	0	4	2	5
Q2.13 Ability to implement the initiative easily in the organisation	23	0	4	3	5
Q2.14 Value for implementing the initiative (based on the cost benefit analysis, Return of Investment (ROI) analysis and/or payback analysis).	23	0	4	2	5
Q2.15 Duration taken to obtain the value of implementing the initiative.	23	0	4	3	5
Q2.16 Whether the initiative has been proven to provide value / benefit in other similar organisations	23	0	4	3	5
Q2.18 Areas in which the initiative will be implemented (e.g. leadership and social responsibilities, planning and strategy, information, customer, process, and/or people).	23	0	4	3	5
Q2.9 The initiative is suitable for the external environment in which the organisation operates, which includes: political, economic, social, technological, legal and environmental factors.	23	0	3	2	5
Q2.17 Popularity of the initiative (e.g. many other organisations are implementing the initiative).	23	0	3	1	4

Remark:

5.0 : Very high importance
4.0 : High importance
3.0 : Moderate importance
2.0 : Low importance
1.0 : Very low importance

5 Conclusions

This paper has presented 18 factors that should be considered when selecting organisational improvement initiative. Two factors were rated as 'very high importance' (Median = 5), which involve: (1) The ability to gain top management commitment and support to introduce and implement the initiative successfully, and (2) The initiative is aligned to the vision, mission and/or purpose of the organisation. All the factors discussed in this paper can be adopted by the organisations as decision criteria to assist in the selection of the most appropriate improvement initiative based on rational decision making. The adoption of improvement initiatives requires time, resources, financial and knowledge. To avoid unnecessary waste and frustration as well as enhance confidence in implementation, the organisations should select the right initiative that will fit with the context of organisation and provide value to the organisation [4, 14]. Further research has been conducted to develop a decision support tool for selecting improvement initiative using the factors discussed in this paper.

Acknowledgement

The authors would like to thank Universiti Tun Hussein Onn Malaysia (UTHM) for funding this project under Multi-Disciplinary Research Grant (Vot. U092).

References

1. Benson, P.G., J.V. Saraph, and R.G. Schroeder, The effects of organisational context on quality management: An empirical investigation. Management Science, **37**(9): p. 1107-1124, (1991)
2. Dahlgaard, J.J. and S.M. Dahlgaard-Park, The 4P quality strategy for breakthrough and sustainable development. European Quality, **10**(4): p. 6-20, (2004)
3. Thawesaengskulthai, N., *Selecting quality management and improvement initiatives: Case studies of industries in Thailand*, Nottingham University Business School, University of Nottingham, (2007)
4. Mohammad, M., *Development of a guidance model for the selection of organisational improvement initiatives*, School of Engineering and Advanced Technology, Massey University: New Zealand, (2012)
5. Mohammad, M., Mann, R., Grigg, N., & Wagner, J. P., Business Excellence Model: An overarching framework for managing and aligning multiple organisational improvement initiatives. Total Quality Management & Business Excellence, **22**(11): p. 1213-1236, (2011)
6. Van der Wiele, A., Van Iwaarden, J. D., Dale, B. G., & Williams, A. R. T., Improvement approaches, in Managing Quality, B.G. Dale, A. Van der Weile, and J. Van Iwaarden, Editors, Blackwell Publishing: Malden, MA. p. 559-575, (2007)
7. Yahya, M.S., Mohammad, M., Omar, B., & Ramly, E. F., A review on the selection of lean production tools and techniques. ARPN Journal of Engineering and Applied Sciences, **11**(12): p. 7721-7727, (2016)
8. Mohammad, M., Mann, R., Grigg, N., & Wagner, J. P., *The right improvement initiative for the right situation: A contextual and systems approach*, Proceedings of the 40th International Conference on Computers and Industrial Engineering, H. Ishii,

et al., Editors., Institute of Electrical and Electronics Engineers (IEEE): Hyogo. p. 1-6, (2010)
9. Donaldson, L., *The contingency theory of organizations*, California, CA: Sage Publications, (2001)
10. Koontz, H. and C. O'Donnell, *Management: A systems and contingency analysis of managerial functions,* 6th ed., Tokyo: McGraw-Hill, (1976)
11. Lawrence, P.R. and J.W. Lorsch, *Organization and environment: Managing differentiation and integration*, Massachusetts: Harvard University Press, (1967)
12. Sousa, R. and C.A. Voss, Contingency research in operations management practices, Journal of Operations Management, **26**: p. 697-713, (2008)
13. Kast, F.E. and J.E. Rosenzweig, *Organization and management: A systems and contingency approach*, 4th ed., New York, NY: McGraw-Hill, (1985)
14. Thawesaengskulthai, N. and J.D.T. Tannock, A decision aid for selecting improvement methodologies. International Journal of Production Research, **46**(23): p. 6721–6737, (2008)
15. Dale, B.G., *Tools and techniques: An overview, in Managing quality*, B.G. Dale, T. Van der Wiele, and J. Van Iwaarden, Editors, Blackwell Publishing: Malden, MA. p. 336-381, (2007)
16. Mohammad, M., Mann, R., Grigg, N., & Wagner, J. P., *Selecting appropriate organisational improvement initiatives: A five steps approach*, in Proceedings of the 15th International Conference on ISO and TQM S.K.M. Ho, Editor, Asia Pacific Business Excellence Standard Academy: Hong Kong. p. 1-6, (2011)
17. Capon, C., *Understanding organisational context: Inside and outside organisations*, 2nd ed., Harlow: Prentice Hall Financial Times, (2004)
18. Thawesaengskulthai, N. and J.D.T., Tannock, Pay-off selection criteria for quality and improvement initiatives. International Journal of Quality & Reliability Management, **25**(4): p. 366-382, (2008)
19. Saunders, M. and R. Mann, Business excellence tools: *The tools used by companies at different stages of business excellence maturity*, in Unpublished consultancy report, Centre for Organisational Excellence Research: Palmerston North, (2007)
20. Allen, I.E. and C.A. Seaman, Likert scales and data analyses. Quality Progress, **40**(7): p. 64-65, (2007)
21. Jamieson, S., Likert scales: how to (ab) use them. Medical Education, **38**(12): p. 1217-1218, (2004)

Implementation of Total Productive Maintenance (TPM) to Improve Sheeter Machine Performance

Nofri Eka Candra[1,*], *Anita* Susilawati[2], *Herisiswanto*[3], and *Wahyu* Setiady[4]

[1]Mechanical Engineering Student, Faculty of Engineering, Universitas Riau, Pekanbaru, Indonesia
[2,3]Mechanical Engineering Lecturers, Faculty of Engineering, Universitas Riau, Pekanbaru, Indonesia
[4]Manager of Finishing Department, PT Riau Andalan Pulp and Paper, Riau, Indonesia

Abstract. This paper purpose is an evaluation of TPM implementation, as a case study at sheeter machine cut size line 5 finishing department, PT RAPP, Indonesia. Research methodology collected the Overall Equipment Effectiveness (OEE) data of sheeter machine and computed its scores. Then, OEE analysis big losses, statistical analysis using SPSS 20 and focused maintenance evaluation of TPM were performed. The data collected to machine sheeter's production for 10 months (January-October 2016). The data analyses was resulted the OEE average score of 82.75%. This score was still below the world class OEE (85%) and the company target (90%). Based the big losses of OEE analysis was obtained the reduce speed losses, which most significant losses of OEE scores. The reduce speed losses value was 44.79% of total losses during the research period. The high score of these losses due to decreasing of machine production speed by operators, which intended to improve the quality of resulting products. The OEE scores statistical analysis was found breakdown losses and reduces speed losses, which significantly affected to OEE scores. Implementations of focused maintenance of TPM in the case study may need to improve because there were still occurred un-expecting losses during the research period.

1 Introduction

Companies have implemented the TPM for their production activities, which minimize losses incurred in the company. The losses incurred in the company usually originate from the production floor caused by the operator, personal care, the process, the tooling problem and the unavailability of components [1]. Other forms of losses that occur in the production floor is the idle machine condition due to the absence of operators, broken machines, rejected parts and so forth [1]. Waste is very necessary to be eliminated because it involves the sustainability of the company and the profits earned by the company. The concept of zero tolerance for waste, defect, breakdown, and zero accident is a prerequisite in the company. To implement this concept, many companies in the world adopted Total Productive Maintenance (TPM) method as their equipment management method [1]. TPM

*Corresponding authors: nofricandra234@gmail.com

is an equipment management method from Japan that has been developed based on the concept of productive maintenance. TPM is an innovative approach for equipment maintenance to improve equipment effectiveness, eliminate equipment damage and promote autonomous activities for operators to maintain daily machines that involve employees. Characteristic of the TPM is autonomous maintenance that operator understands how to care on the equipment so as to reduce the breakdown on the equipment [2]. The objective of TPM is to maximize the value of Overall Equipment Effectiveness (OEE) on the equipment by eliminating the equipment failures, defects on product and accident. Improvements in the quality, productivity and creation of a more positive atmosphere in the work area are critical to implementation of the TPM concept. This improvement would affect the effectiveness of a production activity [3]. To achieve the objectives of the TPM program that can be included in 5 main areas: 1) raising awareness for all employees ranging from top management to production line workers to jointly engage in TPM implementation to improve effectiveness in the work area, 2) carry out all activities based on teamwork, 3) introducing 5S activities in the work area, 4) creating an impartial assessment system and comparison of the assessment results for each work area, 5) develop and implement sustainable management awareness of the health and safety system for employees [4].

The OEE can be defined as a measurement of the effectiveness value of equipments in term of availability equipments to be used, the performance efficiency of equipments and the ability the equipments to produce the quality products. According Ahuja & Khamba [2] stated the OEE measurement can help the production teams to improve equipment performance on actual activities of the production processes. The OEE measurement is focused on the analysis of six big losses that contribute to decreasing the effectiveness of the equipment. The six big losses are equipment failure or breakdown losses, setup and adjustment losses, idling and minor stoppage, reduce speed, defect in process and reduce yield. Therefore, based on the OEE measurement can provide information to all management and workers to determine the critical losses that may be occurred in the production process activities and take action to remove the most critical losses [2].

The OEE is one of the many benchmarks of TPM success. The OEE is the most effective measurement to encourage improvement in the work environment by continues to focus on the concept of zero waste [3]. Besides, the OEE measurement can be benchmarks of TPM success's program to improve productivity, quality, cost, delivery, safety and morale [2].

The purpose of this paper is to analyze implementation of TPM on sheeter machine cut size line 5 finishing department PT Riau Andalan Pulp and Paper. The implementation of TPM was analyzed from OEE measurement on the sheeter machine in term of availability, performance efficiency and capability to produce quality products. Furthermore, the OEE scores of the sheeter machine were computed and determined the big losses factors. The causes of the big losses can be identified by using a fishbone diagram, then, determination of action to minimize the critical losses. Finally, from the history of the monthly losses in company's case study can be analyzed the benchmarking of the TPM implementation of focused maintenance that achieved the company's target or not.

2 Literature review

2.1 Total Productive Maintenance

The TPM represents a new culture, philosophy and attitude toward for equipment or machine maintenance. Concept of TPM is a maintenance, which involves all employees, ranging from top management to production line personnel [4]. According [5] the TPM includes participation of all sections in planning and maintaining of equipments. Based on [2] stated that the TPM encompass eight pillar i.e. autonomous maintenance, focused maintenance, planned maintenance, quality maintenance, education and training, TPM office, development management, safety, health and environment. Autonomous maintenance activities of TPM give an awareness of operators and workers to maintain the equipments or machines in respectable manner. The essence of this activity is to keep machine work and prevent equipment or machine failure during production activities. By applying TPM system, the machine on each production line can keep running which no failure occurs during the production process activities [4].

2.2 Overall Equipment Effectiveness

The Overall Equipment Effectiveness (OEE) is a method to measure the effectiveness of the use of equipment. The OEE is one of key success parameters of TPM implementation. The OEE measurement is required for continuous improvement in TPM implementation. Based on the OEE measurement is revealed the OEE scores that is used for performance measurement activities of TPM pillar implementation such as focused maintenance [5]. The OEE scores indicate whether equipment that is optimally work or not to produce a quality product by reducing losses during the production process [6]. World class standard of OEE score is 85%, which availability score rate of 90%, performance efficiency score rate of 95% and rate of quality score of 99.9% [5]. The formula of OEE is [2]:

$$OEE = Availability \ x \ Performance \ efficiency \ x \ Rate \ of \ quality \quad (1)$$

2.3 Big Losses

The key to success of TPM implementation is focus in minimizing losses which occur in production activities thus increasing overall production effectiveness. TPM focus to minimize six big losses that is equipment failures or breakdown losses, setup and adjustment losses, idling and minor stoppage losses, reduce speed losses, defect in process losses and reduce yield losses [2]. With focus to minimize big losses in production process, the company will get high availability of the equipment used, the equipment is able to achieve the ideal performance and product of production activities will be good [5].

2.4 Correlation and Regression

Correlation test is as statistical methods to test the strength of relation of two variables interval or ratio scales. The relation between two variables was usually denoted r, or called r Pearson. According Lind et al [7], the reason was called r Pearson, because this method was introduced Karl Pearson in 1900. The coefficient correlation has a range values between -1.0 to 1.0. Two variables have perfect correlation when the values of coefficient correlation of -1.0 or 1.0. Positive and negative symbols represent the relation between two variables has tested. And regression test as statistical method that aims to reveal the linear relation between dependent variable and independent variables. From the regression test obtained coefficient regression then can be formed into linear equation. The regression equation function is to estimate change of dependent variable value in the future to affect on the independent variables [7].

The general form of multiple regression equation is [7]:

$$Y = a + b_1X_1 + b_2X_2 + ... + b_kX_k \quad (2)$$

Where: "a" is the intersection of Y value or Y value when all X values equal to zero. And "b" is the slope of line or average change of Y value for one unit change of X values. This regression test can be done using SPSS software.

3 Research Methods

A quantitative research method was adopted to evaluate the implementation TPM in the case study company. The OEE data was collected in sheeter machine cut size line 5 finishing department PT RAPP Indonesia, from January 2016 to October 2016. The stages of research conducted to analyze the OEE and evaluate focused maintenance of TPM that can be seen in Figure 1. Research begins by collecting data that is in the form of primary data and secondary data. The primary data was obtained from direct observation of TPM implementation in the company. The secondary data was obtained from the collection production data on sheeter machine such as time availability, plan shutdown, external shutdown, un-plan shutdown, no roll or no pack, operation speed, average speed, net production and roll used. After all the data was obtained, then the next performed processing data that was calculation of availability, performance efficiency, rate of quality, OEE scores, big losses identification, then correlation and regression test of OEE scores and big losses were performed. The problem solving of this research conducted some analysis that was analysis of OEE scores, analysis of big losses, analysis of correlation and regression of OEE scores with big losses, cause-effect analysis to identify the causing of losses from the using of sheeter machine, and evaluation of implementation of focused maintenance pillar of TPM.

The calculation of OEE used the formula (1). The availability, performance efficiency and rate of quality used formula (3), (4) and (5) below. The availability is a parameter to measure the equipment to be used in production activities in accordance with the schedule that has been planned [8]. The formula to compute the availability is [2]:

$$Availability\,(\%) = \frac{Loading\ Time - Downtime}{Loading\ time}\ x\ 100 \tag{3}$$

The performance efficiency is the ratio of actual production speed of the equipment to ideal speed based on design capacity. Thus, performance efficiency will be good if the production speed can be maintained in accordance with the ideal speed of equipment [8]. Based on the calculation of performance efficiency can be seen the percentage of wasting time, which contribute to decrease the production speed on equipment [9]. The formula to compute the performance efficiency is [2]:

$$Performance\ efficiency\,(\%) = \frac{Processed\ amount\ x\ Ideal\ cycle\ time}{Operating\ time}\ x\ 100 \tag{4}$$

The rate of quality is a parameter of ability equipment to produce product in accordance with the standards [8]. Reworking of a reject product is considered to be one of the quality parameters. If more reject product was produced that will waste time and resource to produce the product [9]. The formula to compute the rate of quality of sheeter machine is:

$$Rate\ of\ quality\,(\%) = \frac{Net\ Production}{Rollused}\ x\ 100 \tag{5}$$

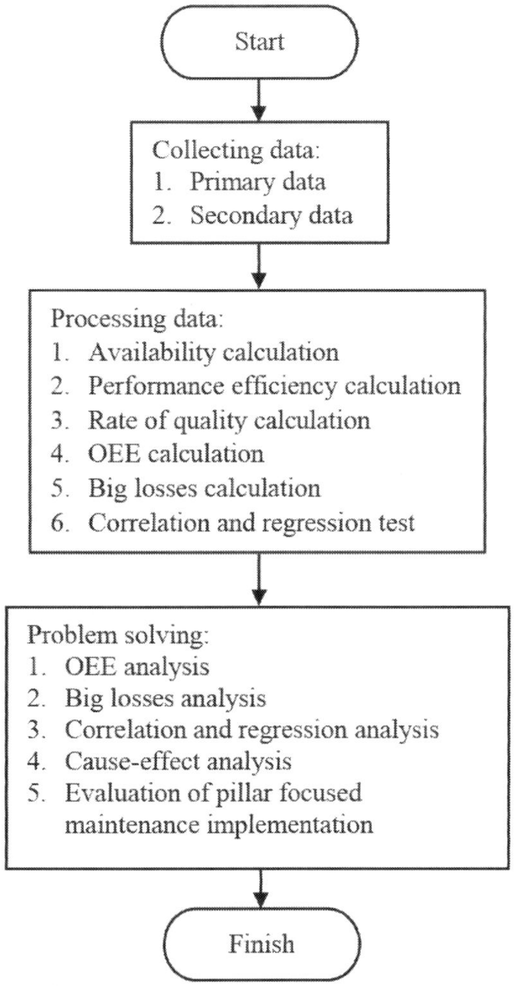

Fig. 1. The Stages of Research Method

4 Results and Discussion

4.1 Result

4.1.1 Production data

The production data of sheeter machine cut size line 5 from January 2016 to October 2016 is as depicted in Table 1.

Table 1. Production Data of Sheeter Machine Cut Size line 5

Month	Time availability (%)	Plan shutdown (%)	External shutdown (%)	Un plan shutdown (%)	No roll or no pack (%)	Operation speed (ream/min)	Average speed (ream/min)	Net production (tons)	Roll used (tons)
January	100	2.02	0	5.11	4.17	170	156	8528	8864
February	100	2.16	0	4.45	0	170	152	8492	8740
March	100	2.02	0	4.44	0.13	170	161	9839	10188
April	100	2.08	0	8.32	0.13	170	158	9022	9378
May	100	1.48	0	2.82	0.27	170	159	10172	10550
June	100	12.36	0	5.83	0.14	170	146	7780	8075
July	100	2.02	0	8.33	0	170	155	8038	8433
August	100	2.02	0	3.23	4.17	170	161	9220	9622
September	100	1.94	0	5.69	1.39	170	156	8770	9107
October	100	2.42	0	1.48	0.13	170	155	9466	9886

4.1.2 OEE Scores

From processing data production of sheeter machine cut size line 5 was obtained the OEE scores of the machine from January 2016 to October 2016 that can be seen in Table 2.

Table 2. The OEE Scores of Sheeter Machine Cut Size Line 5 from January 2016 to October 2016

Month	Availability (%)	Performance Efficiency (%)	Quality (%)	OEE (%)
January	90.53	91.76	96.21	79.92
February	95.45	89.41	97.16	82.92
March	95.34	94.71	96.57	87.20
April	91.37	92.94	96.20	81.70
May	96.86	93.53	96.42	87.35
June	93.19	85.88	96.35	77.11
July	91.50	91.18	95.32	79.52
August	92.45	94.71	95.82	83.90
September	92.78	91.76	96.30	81.99
October	98.35	91.18	95.75	85.86

4.1.3 Big Losses Scores

Total big losses in the sheeter machine during the period of January 2016 to October 2016 can be seen in Table 3.

Table 3. The Big Losses in Sheeter Machine Cut Size Line 5 From January 2016 to October 2016

Number	Factors Losses	Total Time Loss (minutes)	Percentage (%)
1	Reduce Speed Losses	32756.68	44.79
2	Breakdown Losses	21774.88	29.78
3	Quality Losses	13920.01	19.04
4	Setup and Adjustment Losses	4675.84	6.39
	Total	73127.41	100

4.1.4 Correlation and Regression

The OEE data correlation test results and big losses sheeter machine cut size line 5 is shown in Table 4. Whilst, the results of regression test of OEE data and big losses of sheeter machine cut size line 5 is shown in Table 5.

DOI: 10.1051/matecconf/201713500028

Table 4. Results of Correlation Test Using SPSS 20

		Breakdown losses	Setup and adjustment losses	Reduce speed losses	Quality losses	OEE
Breakdown losses	Pearson correlation	1	-.198	.201	-.071	-.715*
	Sig. (2-tailed)		.584	.578	.846	.020
	N	10	10	10	10	10
Setup and adjustment losses	Pearson correlation	-.198	1	-.358	.158	-.133
	Sig. (2-tailed)	.584		.309	.663	.714
	N	10	10	10	10	10
Reduce speed losses	Pearson correlation	.201	-.358	1	-.373	-.635*
	Sig. (2-tailed)	.578	.309		.289	.049
	N	10	10	10	10	10
Quality losses	Pearson correlation	-.071	.158	-.373	1	.110
	Sig. (2-tailed)	.846	.663	.289		.762
	N	10	10	10	10	10
OEE	Pearson correlation	-715*	-.133	-.635*	.110	1
	Sig. (2-tailed)	.020	.714	.049	.762	
	N	10	10	10	10	10

Table 5. Results of regression test using SPSS 20

Model	*Coefficients unstandardized*		*Standardized coefficients*	T	Sig.
	B	Std. error	Beta		
(Constant)	99.973	.026		3895.709	.000
Breakdown losses	-1.000	.001	-.678	-876.567	.000
Setup and adjustment losses	-.998	.002	-.508	-625.230	.000
Reduce speed losses	-.999	.001	-.729	-843.728	.000
Quality losses	-.994	.006	-.129	-159.926	.000

4.2 Discussion

4.2.1 OEE Scores

From OEE calculation was obtained the average score of OEE from January 2016 to October 2016 on sheeter machine cut size line 5 of 82.75% with average score of availability was 93.78%, average score of performance efficiency was 91.71% and average score of quality product was 96.21%. The OEE score was still below the world class OEE (85%) and the company target (90%). However, the OEE scores in March, May and October obtained the higher score than word class OEE, that was 87.20%, 87.53% and 85.86% respectively. This is indicated the company has an opportunity to achieve more than word class OEE along time in the future.

4.2.2 Big Losses

From big losses calculation that most contributed to the low achievement of OEE scores was reduce speed loss, which time losses of 32576.68 minutes or 44.79 % of total time losses (73127.41 minutes), from January 2016 to October 2016. The percentage of losses contribution to OEE is shown in Pareto Diagram at Figure 2.

 DOI: 10.1051/matecconf/201713500028

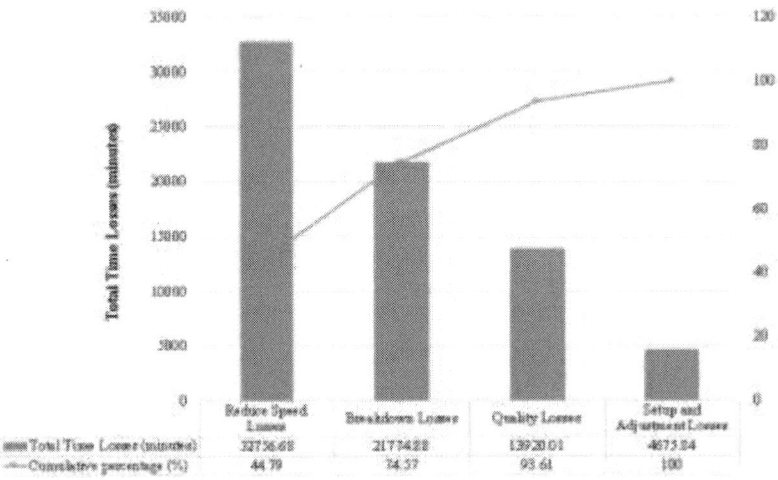

Fig. 2. Pareto Diagram Big Losses Sheeter Machine Cut Size Line 5

4.2.3 Correlation and Regression Analysis

From the correlation test was found the losses factors that closely affects the low scores of OEE of reduce speed losses (Pearson correlation = -0.635) and breakdown losses (Pearson correlation = -0.715). And to predict the change of OEE scores in the future, in this paper has developed a regression equation as follow:

$$Y = 99.973 - 1.000X1 - 0.998X2 - 0.999X3 - 0.994X4 \tag{6}$$

As the dependent variable is OEE (Y) and the independent variables are breakdown losses (X1), setup and adjustment losses (X2), reduce speed losses (X3) and quality losses (X4).

4.2.4 Fishbone Diagram

The critical losses on the sheeter machine during research period were "reduce speed losses". For future improvement, based a cause-effect analysis can help to minimize the "reduce speed losses" during the research period, it is shown in Figure 3. The fishbone diagram is known effect of several factors influences to take action such as man, machine, method, environment and material. From man factor, "reduce speed losses" influenced by (1) lack of operator concentration on the production activities and (2) achieving production targets. From machines factor, "reduce speed losses" influenced by (1) damage the slitter and cross cutter, (2) resulting products were not according to quality products standard of company. From method factor, "reduce speed losses" influenced by decreasing the speed that was intended by operators to reduce the number of products reject. From environment factor, "reduce speed losses" was influenced by (1) the dust contained in the raw material and (2) wear on engine components. And from material factor, "reduce speed losses" was caused by perishable materials.

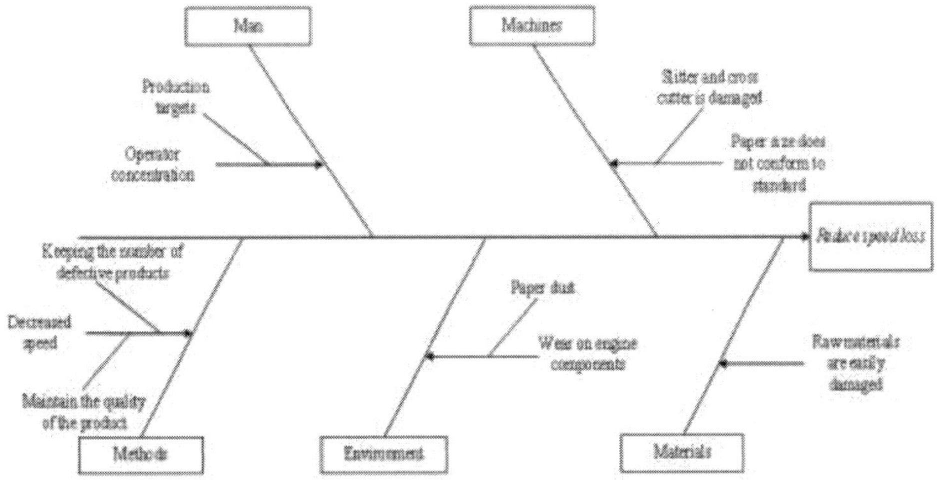

Fig. 3. Fishbone Diagram of Reduce Speed Losses

4.2.5 Evaluation of Implementation Focused Maintenance

Implementation of focused maintenance of TPM on the sheeter machine during the research period can be analyzed from the OEE scores. The OEE average score of 82.75% was less than the world class OEE (85%) and the company target (90%). This is indicated that the TPM implementations in the case study company may need to improve due to still occurred un-expecting losses during the research period. This can be seen in Figure 4, based the calculation of big losses of each month during the period of research indicated frequency occurring losses. Reduce speed losses awarded the highest losses value and followed by breakdown losses for almost each month of research period.

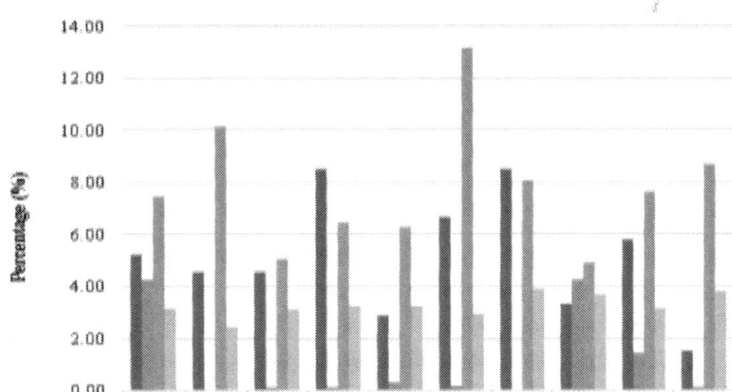

	Jan	Feb	Mar	Apr	May	Jun	Jul	Aug	Sep	Oct
Breakdown Losses	5.22	4.55	4.53	8.50	2.86	6.65	8.50	3.30	5.80	1.52
Setup and Adjustment Losses	4.26	0.00	0.13	0.13	0.27	0.16	0.00	4.26	1.42	0.13
Reduce Speed Losses	7.46	10.11	5.05	6.45	6.27	13.16	8.07	4.89	7.64	8.68
Quality losses	3.15	2.42	3.09	3.22	3.25	2.92	3.91	3.66	3.15	3.81

Fig. 4. Percentage of Big Losses of the Sheeter Machine

5 Conclusions and Recommendations

5.1 Conclusion

From the results of research on the sheeter machine cut size line 5 during the period January to October 2016 was obtained the following conclusions:

- The average of OEE score during research period was 82.75%, the OEE score still below world class OEE (85%) and the company target (90%).
- The most significant losses decreasing the OEE scores during research period was reduce speed losses with a percentage of 44.79% of the total losses and a total losses time was 32756.68 minutes.
- Statistical test results from OEE scores show significant losses was the breakdown losses and reduce speed losses. The result of the regression test was constant on 99.973 and the regression coefficient of each independent variables of breakdown losses (X1 = -1,000), setup and adjustment losses (X2 = -0.998), reduce speed losses (X3 = -0.999) and quality losses (X4 = -0.994).
- From the history of losses data each month during research period was known that implementation of focused maintenance of TPM on sheeter machine may need to improve, because there were still losses occurred during research period.

5.2 Recommendations

From this case study results, it would be recommended that company pay attention the method of decreasing speed of sheeter machine to get a good product quality in optimize manner. An optimization decision making alternative between 'reduce speed losses" and "quality losses" needs to be analyzed breakeven point of this alternative whether it is feasible to be implemented. Decrease sheeter machine speed will be able to reduce the quality losses and otherwise. This is related to the role of the operator in operating the machine and implementation autonomous maintenance of TPM.

Finally, recommending for the company may focus more on machine maintenance such as implementation of autonomous maintenance and conduct multi-skill training for operators to improve the ability and expertise of them to tackle the existing problems to reduce OEE losses value on the machine for the future.

References

[1] R. Singh, A.M. Gohil, D.B. Shahand, S. Desai. *Total Productive Maintenance (TPM) Implementation in a Machine Shop: A Case Study: Proc. Chemical, Civil and Mechanical Engineering Track of 3rd Nirma University International Conference on Engineering (NUiCONE 2012)* (Nirma University: India),592-599 (2013)

[2] I.P.S. Ahuja, J.S. Khamba. Total Productive Maintenance: Literature Review and Directions. *International Journal of Quality and Reliability Management,* **25**, 709-756 (2008)

[3] T.R. Pomorski. *Total Productive Maintenance (TPM) Concepts and Literature Review* (Principal Consulting Engineer Brook Automation Inc) (2004)

[4] B. Gajdzik. Introduction of Total Productive Maintenance in Steelworks Plants *Metalurgija 48,* **2,** 137-140 (2009)

[5] A.K. Gupta A K, R.K. Garg. OEE Improvement by TPM Implementation: A Case Study *International Journal of IT Engineering and Applied Sciences Research (IJIEASR),* **1,** 115-124 (2012)
[6] I. Madanhire, C. Mbohwa. *Implementing Successful Total Productive Maintenance (TPM) in a Manufacturing Plant: Proc. World Congress on Engineering 2015* (London, UK, 1-3 July 2015) (2015)
[7] D.A. Lind, W.G. Marchal, S.A. Wathen. *Statistical Techniques in Business and Economics* (United State of America: McGraw-Hill Companies Inc) (2012)
[8] A. Mansur, R. Rayendra, M.I. Mastur. Performance Acceleration on Production Machines Using the Overall Equipment Effectiveness (OEE) Approach *IOP Conf. Series: Material Science and Engineering,* **105,** 1-7 (2016)
[9] V. Lahri, P. Pathak. A Case Study of Implementation of Overall Equipment Effectiveness on CNC Table Type Boring & Milling Machine of a Heavy Machinery Manufacturing Industry *IOSR Journal of Mechanical and Civil Engineering (IOSR-JMCE),* **12,** 63-70 (2015)

Usability of the Upper Limb Risk Assessment (UPLIRA) Method for Assessing the Risk Factors of Upper Limb Disorders

Noor Syafiqa Abd Razak and *Mohd Nasrull* Abdol Rahman*

Department of Manufacturing and Industrial Engineering, Faculty of Mechanical and Manufacturing Engineering, Universiti Tun Hussein Onn Malaysia (UTHM), 86400 Batu Pahat, Johor, Malaysia.

Abstract. It is important to design ergonomic methods or assessment with a high usability and good ergonomic features, so the method is easily adaptable to the task and the workplace environment. Usability is a measurement on how well the user can use that functionality. The study converged on testing the usability of the new tools for assessing ULDs namely as Upper Limb Risk Assessment (UPLIRA). The UPLIRA method was evaluated by 6 students (3 undergraduate students and 3 postgraduates students) from University Tun Hussein Onn Malaysia (UTHM) who are conducting research in ergonomic. From the feedback survey of UPLIRA method, the observers agree that the prototype of UPLIRA tool was applicable to workplace assessment for the wide range of jobs/task (mean 4.00, SD:0.632) with 83% percentage of agreement . They also indicate that UPLIRA method is quick to use (mean 3.67, SD:1.033) with percentage agreement of 67%. The scoring system and action level was rated as easy to understand (mean 3.67, SD:1.211) and (mean 4.17, SD:0.408). Conclusively, the UPLIRA method was rated as straightforward to use, applicable to wide range of tasks, and time saving as assessment can be completed within 10 minutes. In addition, the UPLIRA assessment covers an extensive range of physical, psychosocial, work organizational, and individual risk factors.

1 Introduction

Upper limb disorders (ULDs) are injury that affected the upper part of the body including neck, shoulders, elbows, wrists, and upper back [1]. ULDs correspondingly cause disability, absenteeism, pain, heavy financial costs among workers worldwide, and reduced productivity [2-3]. Prolonged exposure to risk factors at workplace may contribute to the occurrence of occupational health disorders and development of occupational diseases such as repetitive movements, awkward posture, forceful exertion, awkward posture, and vibration [3-4]. In addition, the previous study reported that psychological and individual factors such as gender, age, BMI, smoking, and anthropometry also play a role in the development of ULDs [5]. The psychosocial factors such as workload, job dissatisfaction, limited social support, time pressure and limited job control further increase the potential for the worker to get an upper limb disorders [5]. However, only a few studies have considered the multifactor association of job, individual, and psychosocial factors in assessing the prevalence of ULDs [6].

* Corresponding author : mnasrull@uthm.edu.my

Pen and paper based observational methods are probably the most frequently used to evaluate physical workloads in order to identify hazards at workplace. The number of available methods are large, but no single method is suitable for all purposes. Since, different methods often focus on different exposures, it may be worth to combine several methods to perform a more comprehensive risk assessment [7]. The current techniques for assessing exposure to risk factors associated with ULDs includes Rapid Upper Limb Assessment (RULA) [8], Job Strain Index (JSI) [9], Occupational Repetitive Actions (OCRA) [10], Cumulative Trauma Disorders (CTD) [11], Loading on the Upper Body Assessment (LUBA) [12], Upper Limb Disorders (ULDs) Health and Safety Guide (HSG-60) [13], and Assessment of Repetitive Task (ART) [14]. From the objective for each tool, it can be concluded that the aim was to not entirely fixate on assessing the work-related upper limb disorders. Most of them only focused on general ergonomic risks [15].

It is substantial to design ergonomic methods or assessment with a high usability and good ergonomic features, so the method is adaptable to the task and the workplace environment. Usability measured the ability of a user to use that functionality [16]. In the study of [17-18] found that the usability of ART has the highest mean score. Meanwhile, the lowest rated method was the OCRA checklist. The OCRA checklist also required a relatively extensive period in average for assessment. This is probably because the ART tool is simpler as compared to OCRA checklist. Therefore, the main objective of this study is to test the usability of the new tools for assessing ULDs namely as Upper Limb Risk Assessment (UPLIRA).

2 Method

2.1 Study Design Data Collected

The prototype was evaluated by 6 students (3 under graduate students and 3 postgraduates students) from University Tun Hussein Onn Malaysia (UTHM) who are conducting research in ergonomic. Firstly, the participants were given a brief discussion about the purpose of the tool, the applications of UPLIRA, the instructions, the scoring system, action level system, and exposure level. Then, all the participants were briefed on how to conduct an UPLIRA assessment for about 1 hour during the day. The video of 3 different tasks being performed such an operator in packaging section, a cashier, and an accountant was presented. After that, the participants made their assessment on each recorded task. The video was re-played to ensure that all observers could complete their assessment on that particular task. About 10 minutes were taken by the observers to evaluate each task.

After the observers completed all their assessment, they fill out the feedback survey questions to evaluate the usability of the UPLIRA method. The questionnaire comprises 3 parts; part 1 consists of 13 questions about the UPLIRA method such as , simple and quick operation, the clarity and conciseness of the instructions, the readability of the wordings, the clarity and conciseness of the pictures/illustrations , level of results interpretation of scoring system, understandability of the score for each items, understandability of the final score and action level applicability towards workplace assessment, applicability in wide range of jobs/tasks, valuable at works (cost effective) and valuable at works (provide a good basis for intervention proposals) as shown in Figure 1. These closed-ended questions were rate with five point Likert scale (1 = strongly disagree, 5 = strongly agree). Part 2 on the other hand consist of overall comments and suggestions for UPLIRA method and Part 3 consist of comments and suggestions for each part as shown in Figure 1 and 2.

DOI: 10.1051/matecconf/201713500029

Feedback Survey for Usability of Upper Limb Risk Assessment (UPLIRA) Method

The UPLIRA method is designed to assess the risk factors of Work-related Upper Limb Disorders (WRULDs) at work place that including the upper body regions such as neck, shoulders, elbows, wrists, and back that contribute to the development of Work-related Upper Limb Disorders (WRULDs). Risk factors cover for this tool are physical, psychosocial, work organizational and individual risk factors. This objective of this survey are developed to identify the usability of the UPLIRA method for user and practitioners.

Name: .... Age (Years): ....

Job Position: .... Working Experienced (Years): ....

Education Level: Date: ....

☐ SPM / Certificate ☐ Degree

☐ Diploma ☐ Master / PhD

Please circle the answer that best represents your rating for each statement and add comments in order to help us understand better your opinion.

1	2	3	4	5
Strongly Disagree	Disagree	Not Sure	Agree	Strongly Agree

No.	Question	Rating				
1	Easy to use and straightforward	1	2	3	4	5
2	Quick to use	1	2	3	4	5
3	Instructions is clearly to read and understand	1	2	3	4	5
4	Wordings is clearly to read and understand	1	2	3	4	5
5	Pictures/illustrations is clearly to see and understand	1	2	3	4	5
6	Scoring system is easy to interpret the results	1	2	3	4	5
7	Score for each items is easy to understand	1	2	3	4	5
8	Final score is easy to understand	1	2	3	4	5
9	Action level is easy to understand	1	2	3	4	5
10	Applicable to workplace assessment	1	2	3	4	5
11	Applicable to wide range of jobs/tasks	1	2	3	4	5
12	Valuable at works (cost effective)	1	2	3	4	5
13	Valuable at works (provide a good basis for intervention proposals	1	2	3	4	5

a) Overall comments for UPLIRA method (if any)

....

....

....

b) Suggestions or recommendations for UPLIRA method (if any)

....

....

....

Fig 1. Part 1 and Part 2 of UPLIRA usability feedback questions.

DOI: 10.1051/matecconf/201713500029

c) **Comments and suggestions for each part (if any)**

Items	Comments	Suggestions
Part 1: Physical Risk Factors A – Awkward Posture: A1 – Neck A2 – Shoulders A3 – Elbows A4 – Wrists A5 – Back B – Force / Load B1 – Lifting / Lowering B2 – Pushing / Pulling B3 – Gripping / Pinching C – Vibration D – Temperature E – Noise Level F – Lighting Level		
Part 2: Psychosocial Risk Factors G1 – Lack of support from supervisors of co-workers G2 – Difficult to keep up with their work G3 – Frequent tight deadlines to meet G4 – Task requires high levels of attention and concentration G5 – Overtime/shift work that is unplanned G6 – Sufficient training and information G7 – Sudden changes in workload G8 – Job/role is lacking in importance to your organization G9 – If your friends interested in working in your job, what would you tell him/her G10 – How satisfied would you say you are with your job		
Part 3: Work Organizational Risk Factors H1 – Job require you to work hard H2 – How often can you use the skills from your previous experience and training H3 – Relationship between others employees is harmonious in attaining the overall organizational goals H4 – Does your company give wage incentives?		
Part 4: Individual Risk Factors I1 – How any of each of the following do you smoke (Cigarette per day) I2 – During the past 6 months, have you had any on the job accidents? I3 – Within the past 12 months, has the kind or amount of work you can do been limited by any disabilities I4 – Was this disability caused by your job? I5 – Was this disability a result of an accident at work? I6 – Have you applied for financial benefits as a result of this disability I7 – Do you currently receive financial benefits as a result of this disability?		

Fig. 2. Part 3 of UPLIRA usability feedback questions

2.2 Data Analysis

Statistical analysis was performed using Statistical Package for Social Sciences (SPSS) version 22. The feedback survey was presented as mean, standard deviation (SD) and percentage.

3 Results and Discussion

This section discussed the feedback for usability of UPLIRA tool, general comment, and suggestion for UPLIRA tool, and specific comments and suggestion for UPLIRA tool. From the result, the observers (n = 6) indicates that the prototype of UPLIRA tool was easy to use and straightforward (mean 3.60 SD:1.140), quick to use (mean 3.67 SD:1.033), instructions is clearly to read and understand (mean 3.17 SD:0.753), wordings is clearly to read and understand (mean 3.33 SD:0.816), pictures or illustrations is clearly to see and understand (mean 3.83 SD:0.753), scoring system is easy to interpret the results (mean 3.67 SD:1.211), score for each items is easy to interpret the results (mean 3.50 SD:1.049), final score is easy to understand (mean 3.67 SD:1.506), action level is easy to understand (mean 4.17 SD:0.408), applicable to workplace assessment (mean 4.00 SD:0.632), applicable to wide range of jobs/tasks (mean 4.17 SD:0.753), valuable at works which is cost effective (mean 4.17 SD:0.753), and finally valuable at works which provide a good basis for intervention proposals (mean 4.17 SD:0.753). Figure 1 tabulated the results of percentage feedback survey of usability from 6 observers (3 undergraduate students and 3 postgraduate students). Table 1 shows the observers rating of feedback surveys for usability of UPLIRA method.

Table 1. Observers ratings of feedback survey of UPLIRA method (n = 6 observers)

No	Questions	Rating Score					Total Score
		1	2	3	4	5	Mean (SD)
1	Easy to use and straightforward	-	1	2	2	1	3.60 (1.140)
2	Quick to use	-	1	1	3	1	3.67 (1.033)
3	Instructions is clearly to read and understand	-	2	2	2	-	3.17 (0.753)
4	Wordings is clearly to read and understand	-	-	3	3	-	3.33 (0.816)
5	Pictures/illustrations is clearly to see and understand	-	-	2	3	1	3.83 (0.753)
6	Scoring system is easy to interpret the results	-	1	2	1	2	3.67 (1.211)
7	Score for each item is easy to understand	-	1	2	2	1	3.50 (1.049)
8	Final score is easy to understand	1	-	-	3	2	3.67 (1.506)
9	Action level is easy to understand	-	-	1	4	1	4.17 (0.408)
10	Applicable to workplace assessment	-	-	1	4	1	4.00 (0.632)
11	Applicable to wide range of jobs/tasks	-	-	1	3	2	4.17 (0.753)
12	Valuable at works (cost effective)	-	-	1	3	2	4.17 (0.753)
13	Valuable at works (provide a good basis for intervention proposals)	-	-	1	3	2	4.17 (0.753)

Notes: Rating scale range from 1 = strongly disagree to 5 = strongly disagree)

Figure 1 shows the percentage for feedback survey on usability of UPLIRA method. All of the observers agreed and strongly agreed that UPLIRA method is easy to use and straightforward (60%), quick to use (67%), instructions is clearly to read and understand

 DOI: 10.1051/matecconf/201713500029

(83%), wordings is clearly to read and understand (50%), picture or illustrations is clearly to see and understand (67%), scoring system is easy to interpret the results (50%), score for each items is easy to understand (50%), final score is easy to understand (67%), action level is easy to understand (100%), applicable to workplace assessment (83%), applicable to wide range of jobs/tasks (83%), valuable at works which is cost effective (83%), and lastly valuable at works which provide a good basis for intervention proposals (83%). From the results, it is deduced that more than 50% of observers has either agreed or strongly agreed with the questions of the survey.

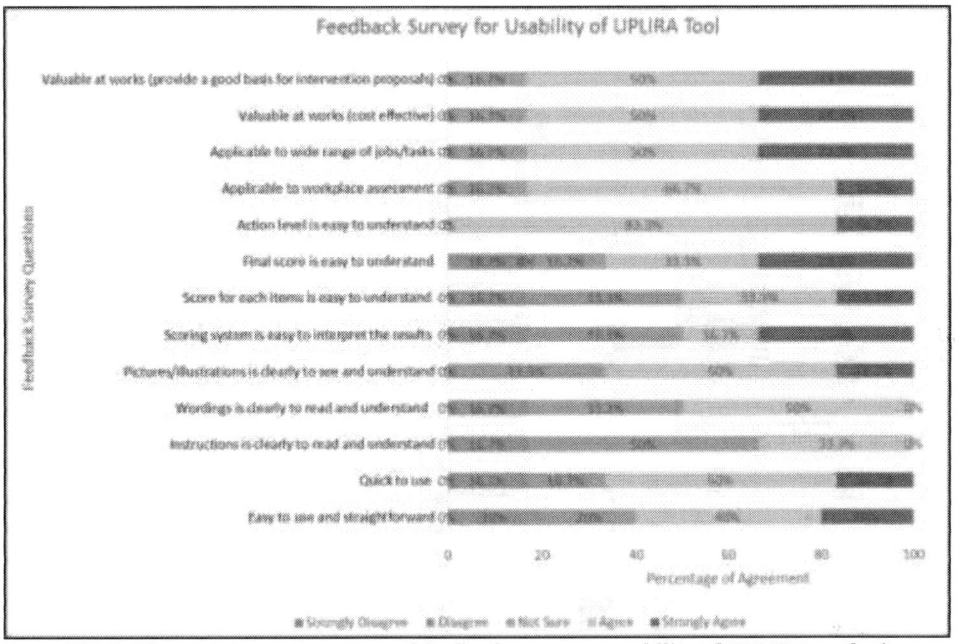

Fig. 3. The percentage agreement of feedback survey on usability of UPLIRA method

From the feedback questionnaire about usability of the UPLIRA method for all participants (n = 6) using ratings (1 = strongly disagree to 5 = strongly agree), found that the action level is easy to understand with mean (mean 4.17 ± 0.408) with percentage of agree and strongly agree was 100%. This is because the final score is easy to understand that lead to choosing the suitable action level. The observers considered that these prototype is applicable to wide range of job/task (mean 4.17 ± 0.753), valuable at works (mean 4.17 ± 0.753), and provide a good basis intervention proposals (mean 4.17 ± 0.753) with percentage agreement of each was 83%. This is because the observers believe that the UPLIRA method can be used at different workstations. Unfortunately, the observers found that the instructions and wordings is not readable and understandable based on the relatively low score (mean 3.17 ± 0.753) and (mean 3.33 ± 0.816) with percentage of agreement of 83% and 50% respectively. This is because the observer found that the font is small and the color of figure and box of risk level is confusing. In addition, the observers found that the pictures or illustration is vague (mean 3.83 ± 0.753) and percentage agreement was 67%. The observers found that the picture and the motion is confusing. In other study found that it was difficult for observers to determine the specific angle solely by observation [19]. In addition, the observers found that the UPLIRA method is not really easy to understand, time consuming with the results (mean 3.60 ± 1.140) and (mean 3.67 ± 1.033) with percentage agreement 60% and 67% respectively. While for the scoring system and final score, the observers take more time to understand before they properly decide based on the score (mean 3.67 ± 1.211) and (mean 3.67 ± 1.506) with the percentage agreement was 50% and 67%. The observers stated that, the UPLIRA method is applicable to workplace assessment with the score (mean 4.00 ± 0.632) with percentage agreement of 83%.

Part 2 of feedback questions explored the general comments and suggestion for UPLIRA method. Most of the observers found that the left and right side is really confusing because it's not synchronized and the instructions should be explained appropriately for better understanding. Other than that, they believe that UPLIRA method can be a good reference for workplace improvement and good for ergonomic assessment especially in upper body. The UPLIRA method is confusing for the first-time user and the observers suggest that it need to be revised. Table 2 shows the summary of overall feedback comments from observers.

Table 2. The overall feedback comments from observers (n = 6)

Observers	**Comments**
1	The left and right side should be synchronized Good references for workplace improvement.
2	Slightly confusing for the first time user Good for ergonomic assessment
3	It is a good for create a new method in the ergonomic assessment tool. This will be one of the alternative to reduce the risk
4	The prototype need to be revised Useful tool to use It is difficult to use without any reference or user guide
5	The instructions is better in circle than cross (x) The left and right side should be synchronized
6	The right and left side are confusing The instruction not clear The words in the score box is not clear The angle line is not clear

Table 3 shows the summary of suggestions from the observers. Some of the suggestions mentioned about using simple and short words for instruction, reducing the questions, adding more exposure level range, improving the range of angle motion, and minimizing the use of paper. The others mentioned about the technicality of the UPLIRA tool such as the keywords should be bold, increasing the size of risk level box, putting a degree (◦) symbol in every classes of motion, and changing the color of the score box because the color of moderate box is same as the figure.

Table 3. The suggestions for UPLIRA tool (n = 6)

Observers	**Suggestions/recommendations**
1	Use simple words to better understanding Add more exposure level
2	Make it short and simple Need to improve the angle range (e.g. 0-20°, 21- 45°, > 46°)
3	Reduce the questions Add more exposure level range
4	Add more exposure level range Increase the size of risk level box
5	The keywords should be bold Add more exposure level range

	Minimize the use of paper for scoring sheet
6	Put a degree (◦) symbol in every classes of motion Make a short and simple instruction or used symbol The keywords should be bold Change the color of the score box Minimize the used of paper Synchronized the left and right side

Table 4 shows shows the summary of specific comments and suggestions for UPLIRA tool. This is the third part of feedback questions from UPLIRA tool. Most of the comments and suggestions are from Part 1 of the UPLIRA prototype which is physical risk factors items and some from psychosocial and individual risk factors. The repetition for every motion should be added in the UPLIRA tool. In addition, the illustration is not clear especially for wrist motion. Unclear definitions of postures can be a source of disagreement between the assessors [20]. Most of the questions comprises of choice (OR) options and it's confusing. Furthermore, the right (R) and left (L) side should be synchronized. In addition, the noise and lighting level should be used in numerical range instead of general value such as slow, loud, very loud. While for psychosocial risk factors, the observer believes that 'always' and 'often' have the similar meaning. They suggest using another word and reduce the questions and use only ones that are related to upper limb disorders. For individual risk factors, some of the questions is confusing. The observers also imply reducing the questions and only ask the questions related to upper limb disorders.

Table 4. Specific comments and suggestions for UPLIRA tool

Items	Comment	Suggestions
Part 1: Physical Risk Factors A – Awkward Posture A1 – Neck A2 – Shoulders A3 – Elbows A4 – Wrists A5 – Back B0 – Force/Load B1 – Lifting/Lowering B2 – Pushing/Pulling B3 – Gripping/Pinching C – Vibration D – Temperature E – Noise Level F – Lighting Level	A1 – The color of picture and column for ticks are same color (confuse) The question is a bit confuse because it has a choice (OR) A2 – The Right and Left side a bit confused	The instruction should be a circle than the a cross (x) No neck extension repetition column Make a clear wrist picture. Add repetition column for every motion D – The position D1 and D2 should be change E – Should be used a range instead of a value (e.g. slow/loud/very loud) F – Should be used a in numerical range instead of general value
Part 2: Psychosocial Risk Factors G1 – Lack of support from supervisors and workers G2 – Difficult to keep up with their work G3 – Frequent tight deadlines to meet G4 – Task requires high levels of attention	'Always' and 'Often' have the same meaning	Reduce the questions

Table 4. Continue

G5 – Overtime/shift work that is unplanned G6 – Sufficient training and Information G7 – Sudden changes in workload G8 – Job/role is lacking in importance to your organization		
G9 – If your friends interested in working in your job, what would you tell him/her? G10 – How satisfied would you say you are with your job?		
Part 3: Work Organizational Risk Factors H1 – Job require you to work hard H2 – How often can you use the skills form your previous experience and training? H3 – Relationship between others employees is harmonious in attaining the overall organizational goals H4 – Does your company give wage incentives?		
Part 4: Individual Risk Factors I1 – How any of each of the following do you smoke (Cigarette per day) I2 – During the past 6 months, have you had any on the job accidents? I3 – Within the past 12 months, has the kind or amount of work you can do been limited by any disabilities? I4 – Was this disability caused by your job? I5 – Was this disability a result of an accident at work? I6 – Have you applied for financial benefits as a result of this disability? I7 – Do you currently receive Financial benefits as a result of this disability?	The question is somewhat confusing	Reduce the questions and use only related to upper limb disorders

DOI: 10.1051/matecconf/201713500029

4 Conclusion

As the conclusion, the UPLIRA method was rated as it is straightforward to use, widely applicable, and time saving. In addition, the UPLIRA assessment covers an extensive range of physical, psychosocial, work organizational, and individual risk factors. Meanwhile, the disadvantages of UPLIRA assessment is that it is undoubtedly a complex method. Therefore, further refinements are required to render the UPLIRA to be a better assessment in evaluating work related ULDs

This research is funded by Office for Research, Innovation, Commercialization and Consultancy Management (ORICC) and Universiti Tun Hussein Onn Malaysia (UTHM) under *'Geran Penyelidikan Pasca Siswazah'* (GPPS), Vot U594.

References

[1] A. Garg, J. M. Kapellusch. Job analysis techniques for distal upper extremity disorders. *Reviews of Human Factors and Ergonomics,* **7(1)**, 149-196 (2011).
[2] L. Punnett, D. H. Wegman. Work-related musculoskeletal disorders: the epidemiologic evidence and the debate. *Journal of Electromyography and Kinesiology,* **14(1)**, 13-23 (2004).
[3] B. M. Tinubu, C. E Mbada, A. L Oyeyemi, A. A Fabunmi. Work-related musculoskeletal disorders among nurses in Ibadan, South-west Nigeria: a cross-sectional survey. *BMC musculoskeletal disorders,* **11(1),** 12. (2010).
[4] B. R. da Costa, E. R. Vieira. Risk factors for work- related musculoskeletal disorders: a systematic review of recent longitudinal studies. *American Journal of Industrial Medicine,* **53(3)**, 285-323 (2010).
[5] B. P. Bernard. Musculoskeletal disorders and workplace factors: a critical review of epidemiologic evidence for work-related musculoskeletal disorders of the neck, upper extremity, and low back. Cincinnati, OH: U.S. Department of Health and Human Services, Centers for Disease Control and Prevention, National Institute for Occupational Safety and Health, DHHS (NIOSH) Publication No. **97-141**. (1997)
[6] P. R. Mosaly. Multifactor association of job, individual and psychosocial factors in prevalence of distal upper extremity disorders and quantification of job physical exposure. *International Journal of Industrial Ergonomics,* **55**, 40-45, (2016).
[7] E. P. Takala, I. Pehkonen, M. Forsman, G. A. Hansson, S. E. Mathiassen, W. P. Neumann, J. Winkel Systematic evaluation of observational methods assessing biomechanical exposures at work. *Scandinavian journal of work, environment & health,* **3-24**, (2010).
[8] L. McAtamney, E. N. Corlett. RULA: a survey method for the investigation of work-related upper limb disorders. *Applied Ergonomics,* **24(2)**, 91-99, (1993).
[9] J. S. Moore, A. Garg, The strain index: a proposed method to analyze jobs for risk of distal upper extremity disorders. *American Industrial Hygiene Association,* **56(5)**, 443-458, (1995).
[10] E. Occhipinti. OCRA: a concise index for the assessment of exposure to repetitive movements of the upper limbs. *Ergonomics,* **41(9)**, 1290-1311, (1998).
[11] V. Seth, R. L. Weston, A. Freivalds, Development of a cumulative trauma disorder risk assessment model for the upper extremities. *International Journal of Industrial Ergonomics,* **23(4)**, 281-291, (1999).
[12] D. Kee, W. Karwowski. LUBA: an assessment technique for postural loading on the upper body based on joint motion discomfort and maximum holding time. *Applied Ergonomics,* **32(4)**, 357-366, (2001).

[13] R. J. Graves, K. r. Way, D. Riley, C. Lawton, L. Morris. Development of risk filter and risk assessment worksheets for HSE guidance—'Upper Limb Disorders in the Workplace'2002. *Applied Ergonomics,* **35(5)**, 475-484. (2004).
[14] J. Ferreira, M. Gray, L. Hunter, M. Birtles, D. Riley. Development of an assessment tool for repetitive tasks of the upper limbs (ART). *Derbyshire: Health and Safety Executive*. (2009).
[15] M. N. A. Rahman, N. S. A. Razak. Review on Pen and Paper Based Observational Methods for Assessing Work-related Upper Limb Disorders. *Indian Journal of Science and Technology,* **9(S1)**, (2016).
[16] J. Nielsen. Usability engineering, Elsevier, (1994).
[17] K. Eliasson, T. Nyman, M. Forsman. Usability of six observational risk assessment methods. *Paper presented at the Proceedings 19th Triennial Congress of the IEA*. (2015)
[18] M. E. Chiasson, D. Imbeau, K. Aubry, A. Delisle. Comparing the results of eight methods used to evaluate risk factors associated with musculoskeletal disorders. *International Journal of Industrial Ergonomics,* **42(5)**, 478-488. (2012).
[19] G. David, V. Woods, G. Li, P. Buckle. The development of the Quick Exposure Check (QEC) for assessing exposure to risk factors for work-related musculoskeletal disorders. *Applied Ergonomics,* **39(1)**, 57-69, (2008).
[20] S. Burt, L. Punnett. Evaluation of interrater reliability for posture observations in a field study. *Applied Ergonomics,* **30(2)**, 121-135, (1999).

MATEC Web of Conferences 135, 00030 (2017) DOI: 10.1051/matecconf/201713500030
ICME'17

Review on Item Selection for Development of the Self-Reported Questionnaire Related to Lower Back Disorders at Workplace

Nur Atifah Jaffar and *Mohd Nasrull* Abdol Rahman*

Department of Manufacturing and Industrial Engineering, Faculty of Mechanical Engineering and Manufacturing, Universiti Tun Hussein Onn Malaysia (UTHM), 86400 Batu Pahat, Johor, Malaysia

Abstract. This paper reviewed the current method in item selection available for developing self-reported questionnaire related to low back disorders at workplace. A systematic review of the literature was performed by searching using database and the searching strategy employed combined keyword for self-reported questionnaire, item selection, back disorders, and selection of items. Fourteen self-reported questionnaires related to low back disorders were review, there are different techniques implemented in selection of item in developing questionnaire. There are various techniques in selecting items in developing the questionnaire such as by surveying, correlation of item, factor analysis and by means of expert. Most of the self-reported questionnaire selected the items by survey technique based on risk factors related, where the previous self-reported questionnaires were reviewed and taken based on daily activities related to low back disorders. Thus, from the paper reviewed surveying technique were mostly used in selecting the item for the development self-reported technique with eight out of fourteen questionnaires. There are three self-reported questionnaires used in factor analysis to select the item and two self-reported questionnaires use item correlation technique to select the item for development of questionnaire. Only one self-reported questionnaire makes use of expert for selecting item in questionnaire development.

1 Introduction

Self-reported questionnaire is used to identify the status of functional and pain to subsequently evaluate the patients before and after treatment besides being used detect changes of systems and disabilities [1]. Self-reported functional disability is one of the most important outcome measurements in back pain research [2]. Those whom work in automotive industries commonly may be affected on various areas on the body such as back, wrists, arms, knees, hand, and neck [3]. Workers in wall plastering job were reported to experience discomfort in back regions [4]. People that always operate with computers or do other task in sitting posture commonly have experienced low back pain and

* Corresponding author: mnasrull@uthm.edu.my

musculoskeletal injury [5]. The step in developing the questionnaire is based on international literature, stages of item selection, domain constructions and psychometric properties [6-9]. The committee of bilingual subject's matter experts (SMEs) [6] can be employed to determine the percentage of agreement score calculated for each questionnaire item [10]. Method of principal components analysis (PCA) is used to identify the items loaded in developing the questionnaire [11]. Criteria for the selection of the items in developing the questionnaire consist of conceptual and psychological [12]. Phase for item reduction is done by surveying with open-ended questionnaires selected from previous studies and from the important factor related to the case studies [13].

2 Methods

There are three stages of method used in reviewing the self-reported questionnaires which include selection of literatures from database, developing the evaluation criteria and evaluation process.

2.1 Selection literatures from database.

The literatures were explored using several of databases such as Science Direct, Google Scholar, Pub Med, Scopus, Seagate, and e-Journal to search and select a suitable references and related articles. The materials related were covered from 1980 until 2015.The combination of several keywords, terms, tittle, and abstract were screened and used in identifying journal related to Self-Reported Questionnaires for assessing the exposure of low back disorders.

2.2 Developing the evaluation criteria.

The evaluation criteria for these review methods contained the, items, questionnaire development and selection of item technique were evaluated.

2.3 Evaluation process.

The structure and content of evaluation process for selected publications were conducted by two researchers. The evaluation process will focus on the identifying the published questionnaires based on self-report method related to low back disorders from the previous case studies. The evaluation process involved finding any dissimilarities of information or gap to do documentations.

3 Results

According to the techniques for selecting the items in developing self-reported questionnaire that reviewed in ten self-reported questionnaires related to low back disorders is showed in Table 1. This table similarly describes the result in list of risk factors related to low back disorders for survey techniques, correlation value for item correlation technique, results for factor analysis technique and person whom observed as the expert in selection of items. Table 1 presents an overview of the item selection technique in developing questionnaire related to low back disorders.

Table 1. Item selection techniques in self-reported questionnaire related to low back disorders

Self-reported questionnaire	Items	Items selection techniques	Result
Oswestry Disability Scale (ODI) [14]	10 items	survey	Survey on the most relevant problems in daily activities related to low back disorders
Revised Oswestry Disability Questionnaire [24]			
Modified Version [25]			
The Million Visual Analogue Scale (MVAS) [15]	15 items	Item correlation	Intra and inter observer (0.966 and 0.918)
Roland-Morris disability questionnaire (RDQ) [16]	24 items	survey	Survey in aspect of daily living
RDQ-23 [28]	23 items	Survey based on Sickness impact profile	
RDQ-18 [29]	18 items	Item correlation	6 items were deleted
The Waddell disability Index (WDI) [17]	9 items	Factor analysis	Total internal consistency is 0.76
The Low Back Outcome Score (LBOS) [18]	13 items	Survey	Survey on factor related to low back disorders
The Low Back Pain Rating Scales (LBPRS) [19]	25 items	Survey	Based on Nordic Questionnaire an daily task activities
The Clinical Back Pain Questionnaire (CBPQ) [20]	19 items	Factor analysis	0.94 (correlated well)
The Quebec back pain disability scale (QBPDS) [21]	20 items	Factor analysis and statistical program (TESTGRAF)	46 items from 48 items by using factor analysis and 20 items reducing by using statistical program (TESTGRAF)
The NASS Lumbar Spine Outcome Assessment Instrument (NASS LSO) [22]	62 main question	Expertise person	Group members from Task Force evaluated the questionnaire.
Risk Factor Questionnaire (RFQ) [23]	25 items	Survey	Survey based on risk factors related to lower back disorders in form of four categories.

3.1 Oswestry Disability Index (ODI)

Oswestry Disability Index [14] questionnaire is an original paper with 10 items by surveying technique for the selection of items. The items are chosen by the most relevant problems that experienced by people with low back pain. This technique remains stagnant to two other versions of self-reported questionnaire which are Revised Oswestry Disability Questionnaire [24] and Modified Version [25] questionnaire and both of these questionnaires correspondingly haves same number of items which is 10 items. The entire self-reported questionnaire selected the items that are most relevant to daily activities that are related to low back disorders such as pain intensity, personal care, lifting, walking, sitting, standing, sleeping, sex life, social life and travelling.

3.2 The Million Visual Analogue Scale (MVAS)

The Million Visual Analogue Scale [15] is utilized to assess the disability and pain intensity in patients with LBP and consists of 15 items by using 100mm Visual Analogue Scale (VAS). This questionnaire makes use of item correlations to select the item while developing the self-reported questionnaire. Item correlation was tested by using intra and inter observer. Intra-observer is observed by one person during collecting the data and the result for first trail is found to have high correlation of 0.966. The inter-observer method is employed by using more than one person to observe the activities and give two trials to gain a high correlation of 0.918. First trial of inter-observer failed due to the negative value for correlation in items number 3, thus the items was replaced in second trial for improvement.

3.3 Roland-Morris disability questionnaire (RDQ)

Roland-Morris disability questionnaire [16] is an original paper and modified sporadically and its newest version is Roland Disability Questionnaire-12 [26]. First version consists of 24 items by surveying technique from Sickness Impact Profile [27]. The items were selected based on daily living activities that used physical and mental function. The physical functions include walking, bending over, sitting, lying down, dressing, sleeping, self-care and daily activities. Another self-reported questionnaire was revised in Roland Disability Questionnaire-23 [28] by using similar techniques that reviews from the SIP [27] and has 23 items. Then, the previous questionnaire was revised again and forms the Roland Disability Questionnaire-18 [29] with 18 items by using item correlation techniques and consequently suggested that 6 items were deleted.

3.4 The Waddell disability Index (WDI)

The Waddell disability Index (WDI) [17] consists 9 items and items were selected based on factor analysis technique. Result from factor analysis showed that all the nine activities were well correlated. Nine activities in items listed are heavy lifting (0.44), sitting one-half hour (0.44), travelling one-half hour (0.51), standing one-half hour (0.51), walking one-half hour (0.65), sleep disturbance (0.61), social life restriction (0.62), sex life restriction (0.62) and help with footwear (0.54). All the nine activities have a total of internal consistency with high correlation which is 0.76 and (r=0.70).

3.5 The Low Back Outcome Score (LBOS)

The Low Back Outcome Score [18] to assess the patients with low back pain and consist of 13 items with 3, 6 and 9 point scoring system with survey techniques to select the items. The survey was based on factors related to low back disorders such as current pain, employment domestic chores, sport activities, resting, medical treatments or consultations, drug use, sex life and daily activities (sleeping, walking, sitting, travelling and dressing).

3.6 The Low Back Pain Rating Scales (LBPRS)

The Low Back Pain Rating Scales [19] was used to evaluate the clinical outcome of low back pain patients and consists of 25 items with VAS, (Yes/No) and NSAID scoring system. It's contains of three different components which are pain, disability, and physical impairment. These self-reported questionnaires used survey technique to select the items for develop the questionnaire. For pain section, the items was selected from the Nordic Questionnaire [30] based on a few parameters such as pain medication reduction, physical function, estimation and psychological adjustment [31]. For disability and physical impairment both of these section were surveyed based on daily task activities such as sleeping, ability to perform housework, walking, sitting, lifting, working, dressing, driving, running, getting up from the chair, climbing stairs, contact with people and expectations of future pain.

3.7 The Clinical Back Pain Questionnaire (CBPQ)

The Clinical Back Pain Questionnaire (CBPQ) [20] or Aberdeen LBP scale is used to measure of health outcome for patients with low back pain and it consists of 19 items with multiple choice and single choice answer. By employing the factor analysis techniques the item was selected for development this questionnaire. The item was reviewed from clinical literature item and it was devised from commonly used in the clinical assessment for patients with low back pain. Two measurements were used which are an internal consistency measured by Cronbah's alpha to measure the correlation between item and item-total correlation is to compare individual question with total score. Both of these measurements were considered as rejection as the total item-correlation is below 0.2. Thus, factor analysis was used to identify the health factors within the questionnaire. The result showed well correlation of 0.94.

3.8 The Quebec back pain disability scale (QBPDS)

The Quebec back pain disability scales [21] is used to assess the degree of disability in patient with back pain and has 6-point difficulties scale as scoring system method used. This questionnaire consists about 20 items and the item was selected by using factor analysis technique and statistical program (TESTGRAF). 46 items were factor analysed from 48 items and decreased to 20 items by using statistical program (TESTGRAF) that was developed by Dr. J. Ramsay, Department of Psychology, McGill University, Montreal, Canada. The items were based on si domains which are bed/rest, sitting/standing, ambulation, movement, bending/stooping and handling large/heavy objects.

3.9 The NASS Lumbar Spine Outcome Assessment Instrument (NASS LSO)

The NASS Lumbar Spine Outcome Assessment Instrument [22] aims to assess the instrument's reliability, validity and acceptability to the patients and consists of 62 main questions and was grouped into 5 categories which are group 1: Demographic data, Group 2: Medical history, Group 3: Pain, neurogenic symptoms & function, Group 4: Employment history and Group 5: Data about outcomes of treatment. This questionnaire was derived from a consensus of the NASS and all the 62 questions obtained from 3 different existing questionnaires which are SF 36, modified ODI and modified employment assessment published by BIGOS. Within 18 months, expert in questionnaire which are Task Force group members were recruited to review the instrument and critique. Thus, it is observed the questionnaire became shorter, increase in readability, concise and simpler to follow.

3.10 Risk Factor Questionnaire (RFQ)

Risk Factor Questionnaire [23] focus to prevent extended disability, treat the acute and sub-acute stages of low back pain and employed 25 items of occupational risk factor questionnaires with Yes/No selection scoring system. The items were selected based on survey techniques. The selection of items from survey based on risk factors related to low back disorders such as load features (bulk or grip), handling activities (pushing, pulling, carrying), body position (sitting, climbing) and environmental demands (vibration, slippery surfaces, or work on elevated surfaces).

4 Discussions

There are 14 self-reported questionnaires related to low back disorders were reviewed in selection of items step. There are various techniques employed to select the item in developing the self-reported questionnaire which are by surveying either by revision from the previous study or by relating the risk factor of the case studies (low back disorders), item correlation, factor analysis and by using an expert to select the item. Basically, there are ten self-reported questionnaires related to low back disorders including Oswestry Disability Index (ODI) Version 1.0 [14], the Million Visual Analogue Scale (MVAS) [15], Roland-Morris Disability Questionnaire (RDQ) [16], the Waddell Disability Index (WDI) [17], the Low Back Outcome Score (LBOS) [18], the Low Back Pain Rating Scales (LBPRS) [19], the Clinical Back Pain Questionnaire (CBPQ) [20] or Aberdeen LBP scale, the Quebec Back Pain Disability Scale (QBPDS) [21], the NASS Lumbar Spine Outcome Assessment Instrument (NASS LSO) [22] and Risk Factor Questionnaire (RFQ) [23]. For Oswestry Disability Index (ODI) there are another two versions which are Revised Oswestry Disability Questionnaire [24] and Modified Version [25]. Both of the versions implemented similar techniques in selecting the items for developing the self-reported questionnaire. Other than that, Roland Disability Questionnaire had other versions a well which are RDQ-23 [28] and RDQ-18 [29] and both of these self-reported questionnaires also had same techniques for selection item in development of self-reported questionnaires which is by surveying technique based on SIP [27]. Most of the self-reported questionnaires use survey technique to select the item in development of self-reported questionnaire. The survey was taken based various characteristics which are established on daily activities, Sickness Impact Profile (SIP) [27] and based on risk factor related to low back disorders. There are 8 self-reported questionnaires that utilizing this method which are ODI [14],

Revised ODI [24], Modified ODI [25], RDQ [16], RDQ-23 [28], LBOS [18], LBPRS [19] and RFQ [23]. 3 self-reported questionnaires which are WDI [17], CBPQ [20] and QBPDS [21] have used factor analysis method to select the item in development of self-reported questionnaire and 2 self- reported questionnaires which are MVAS [15] and RDQ-18 [29] had item correlation technique as the method to select the item for the development of self-reported questionnaires. There only one self-reported questionnaire that utilize expert to select the item in developing the self-reported questionnaire which is NASS LSO [22].

5 Conclusions

In this study, fourteen self-reported questionnaires related to lower back disorders were reviewed. There are four current techniques for selecting item in developing self-reported questionnaire related to low back disorders were reviewed and covered from 1980 until 2001, which are by survey, factor analysis, item correlation and by using expert. Most of the self-reported questionnaires selected the items by using surveying techniques which is ODI, RDQ, LBOS, LBPRS and RFQ. The most used survey techniques based on risk factors related to low back disorders which are ODI, LBOS and RFQ. Besides that, there are several self-reported questionnaires using techniques of surveying based on daily activities such as sitting, walking, standing, sitting, and bending which are RDQ and LBPRS. The least technique used among self-reported questionnaires is by using expert which is NASS LSO. 3 out of 14 self-reported questionnaires were found to use technique of factor analysis to select the item for development questionnaires. 2 out of 14 self-reported questionnaires were using item correlation method which are are MVAS and RDQ-18.

Acknowledgement

This research is funded by Office for Research, Innovation, Commercialization and Consultation (ORICC) and Universiti Tun Hussein Onn Malaysia (UTHM) under Geran Penyelidikan Pasca Siswazah (GPPS, Vot U745).

References

1. Grotle M, Brox JI, Vøllestad NK. Functional status and disability questionnaires: What do they assess? A systematic review of back-specific outcome questionnaires. *Spine*, **30**, 130–140 (2004)
2. Deyo RA. Measuring the functional status of patients with low back pain. *Arch Phys Med Rehabil* **69**,1044-53 (1988)
3. Rahman, M.N.A., Aziz, F.A. and Yusuff, R.M: Survey of body part symptoms among workers in a car tyre service centre. *Journal of Human Ergology*, **39** (1), 53-56 (2010)
4. Rahman, M.N.A., M.R.A. Rani, and M.J. Rohani: Investigation of work-related musculoskeletal disorders in wall plastering jobs within the construction industry. WORK: *A Journal of Prevention, Assessment and Rehabilitation*, **43** (4), 507-514 (2012).
5. Rahman, M. N. A., Zakaria, N. H., Masood, I., Adzila, S., & Nasir, N. F: Risk assessment for assessing the subjective occupant seating discomfort related office works. Information, **19** (7B), 3025-3030 (2016)

6. Lynn, M.R., Determination and quantification of content validity. *Nurs. Res.* **35** (6) 382-385 (1986)
7. Streiner, D.L., Norman, G.R. Health Measurement Scales: a Practical Guide to Their Development and Use, second ed. Oxford University Press, New York (1995)
8. Turner, R., Quittner, A.L., Parasuraman, B.M., Kallich, J.D., Cleeland, C.S., Mayo/FDA Patient-Reported Outcomes Consensus Meeting Group, Patient-reported Outcomes: Instrument Development and Selection Issues, *Value Health* **10**(Suppl. 2) 86-93 (2007)
9. Snyder, C.F., Watson, M.E., Jackson, J.D., Cella, D., Halyard, M.Y., Mayo/FDA Patient-Reported Outcomes Consensus Meeting Group. Patient-reported Outcome Instrument Selection: Designing a Measurement Strategy. Value Health 10(Suppl. 2), 76-85 (2007).
10. Tilden, V.P., Nelson, C.A., May, B.A. Use of qualitative methods to enhance content validity. Nurs. Res. **39** (3), 172-175 (1990)
11. Intermediate Care National Evaluation Team. A national evaluation of the costs and outcomes of intermediate care for older people. Final report. University of Leciester, (2006)
12. Leal, Rossana B., et al. "Development of a questionnaire for measuring health-related quality of life among children and adolescents with mouth breathing." American journal of rhinology 7 allergy **29** (6) 212-215 (2015)
13. Saab, Sammy et al. " Development of a disease-specific questionnaire to measure health-related quality of life in liver transplant recipients." Liver Transplantation **17** (5) 567-579 (2011)
14. Fairbank J, Couper J, Davies J et al. The Oswestry low back pain questionnaire. Physiotherapy **66,** 271- 273 (1980)
15. Million R, Hall W, Nilsen KH, Baker RD, Jayson MI. Assessment of the progress of the back -pain patient- Volvo Award in Clinical Science. Spine **7,** 204- 212 (1981)
16. Roland M, Morris R. A study of the natural history of low back pain- Part 1. Development of a reliable and sensitive measure of disability in low back-pain-. Spine **30,** 130- 140 (1983)
17. Waddell G, Main CJ. Assessment of severity in low-back disorders. Spine **9,** 204- 208 (1984)
18. Greenough CG, Fraser RD. Assessment of outcome in patients with low-back pain. Spine **17,** 36-41 (1992)
19. Manniche C, Asmussen K, Lauritsen B, Vinterberg H, Kreiner S, Jordan A. Low Back Pain Rating scale- validation of a tool for assessment of low back pain-. Elsevier Science **57** (3), 317–326 (1994)
20. Ruta DA, Garrant AM, Wardlaw D et al. Developing a valid and reliable measure of health outcome for patients with low back pain. Spine **19**, 1887-1896 (1994)
21. Kopec JA, Esdaile JM, Abrahamowicz M et al. The Quebec back pain disability scale- measurement properties-. Spine **20**, 341- 352 (1995)
22. Daltroy LH, Cats Baril WL, Katz JN, Fossel AH, Liang MH. The North American spine society lumbar spine outcome assessment instrument: reliability and validity tests. Spine **21**, 741-749 (1996)
23. Halpern N, Hiebert R, Nordin M, Crane M. The test – retest reliability of a new occupational risk factor questionnaire for outcome studies. *Applied Ergonomics*, 32(**1**), 39- 46 (2001)

24. Hudson-Cook N, Tomes-Nicholson K, Breen A A revised Oswestry disability questionnaire. In Roland M, Jenner JR (eds) Back Pain: New Approaches to Rehabilitation and Education. Manchester: *Manchester University Press*, 187–204 (1989)
25. Fritz JM, Irrgang JJ. A Comparison of a modified Oswestry low back pain disability questionnaire and the Quebec back pain disability scale. *Phys Ther*, **81**, 776–788 (2001)
26. Atlas SJ, Deyo RA, van den Ancker M, Singer DE, Keller RB, Patrick DL. The Maine-Seattle Back Questionnaire: a 12-item disability questionnaire for evaluating patients with lumbar sciatica or stenosis. Results of a derivation and validation cohort analysis. Spine **28**, 1869–1876 (2003)
27. Bergner, Marilyn et al. "The sickness impact profile: conceptual formulation and methodology for the development of a health status measure." *International Journal of Health Services* **6** (3), 393-415 (1976)
28. Patrick DL, Deyo RA, Atlas SJ et al. Assessing health-related quality of life in patients with sciatica. Spine **20**, 1899–1909 (1995)
29. Stratford PW, Binkley JM Measurement properties of the RM-18: a modified version of the Roland-Morris Disability Scale. Spine, **22**, 2416–2421 (1997)
30. Kuorinka, I., Jonsson, B., Kilbom, A., Vinterberg, H.. Biering-Sorensen, F., Andersson, G. and Jorgensen, K., Standardised Nordic questionnaires for the analysis of musculoskeletal symptoms. Appl. Ergonom., **18**, 233-237 (1987)
31. Evans, James H and I. I Abbott Kagan. "The development of a functional rating scale to measure the treatment outcome of chronic spinal patients." Spine **11** (3) 277-281 (1986)

MATEC Web of Conferences **135**, 00031 (2017) DOI: 10.1051/matecconf/201713500031
ICME'17

Safety Culture and Issue in the Malaysian Manufacturing Sector

Danish Ali*[a], *Yusri* Yusof[b] and *Anbia* Adam[c]

Faculty of Mechanical and Manufacturing Engineering, Universiti Tun Hussein Onn Malaysia

Abstract. . This paper highlights the Safety culture and issue in the Malaysian Manufacturing Sector and emphasis the high occupational accidents due to lack of safety culture and non-compliance of the requirements of Occupational Safety and Health Act 1994. The aim of this study is to review the occupational accidents occurrence in the Malaysia workplace since 2012-2016. Malaysia aimed to reduce the occupational accidents, the results show by DOSH increase that Occupational Noise Induced Hearing Loss 83.7%, occupational musculoskeletal diseases, 4.4% and occupational lung diseases 2.3%. But the as per the record from DOSH that in last 5-Years, the increment in the fatal accidents by Average 26%, Permanent Disability by Average 71% and Non-Permanent Disability by 64 % are investigated only in Manufacturing Industries. The government must show their high interest on such a vulnerable employees to accomplish the above aim. This step will be helpful for planning to reduce the accidents in workplaces and it will also detect the prevention for the future accidents.

1 Introduction

The Occupational Safety and Health Act 1994 (OSHA 1994) [1]was made considering the way that the Factories and Machinery Act (FMA) 1967 just secured word related safety and health in the construction industries, mining and manufacturing industry, though alternate businesses were not considered. The extent of OSHA 1994 covers all people that works in public and private sectors with the exception of military forces. The Occupational Health and Safety directions force the punishment to the employers, which they present of unsafe working conditions for their employees noncompliance with the standards [2-3].

To ensure the occupational health and safety of the persons at working in danger places, the objectives of OSHA 1994 are to protect workers at the workplace, to encourage an occupational safe environment for the employee at their workplace adjusted to their physical

*Corresponding author: yusri@uthm.edu.my

and psychological needs, and to offer the means whereby the associated occupational safety and health legislations can be gradually changed by a system of regulations and endorsed industry codes of practice.

Nowadays, work related accident became well-known as major concerned to the industries involved. This kind of problem also closely faced by the manufacturing industry's workers that involve in the different operation of handing machine and different production equipment like the use of molding machine, fabrication equipment, cranes, prime movers, and forklifts and truck that is very much associated with the issue of safety and health awareness. In an effort to promote the development of safety and healthy workplace, as it is one of the important functions between the organization and workers to conduct their daily operation in a safe and correct manner. Workers have the high level of health and safety awareness that are able to conduct their work efficiently and professionally, where it possibly might to prevent accidents that occurred at the work related. In the work environment mishap happening during the everyday operations brings about hopeless scenario. So, when accident happens, it should always reported to the health and safety officer to ensure that instant action could be taken by the organization. Encourage assessment could be led to recognize the factors that reason of accident and so in the future to prevent the accident. A worker additionally earnest to report any near miss accident or injury either supervisor or health and safety officer. This paper helps to identify potential hazards happen in different sector, special focus on manufacturing sector and also address the different occupational accident and diseases lungs, skin and muscular disorder and in the future so that avoid accident and disease for the betterment. [4-5].

This review anticipated that would raise the level of awareness towards wellbeing of health and safety needs at workplace. Furthermore employers or supervisors for most part can recognize what are the causes that play important role accident among their workers and can take the protection measures in controlling them. [6].

2 Safety culture and Problems in the Malaysian Manufacturing Sector

These days, safety culture has turned into the key factor determining the association's wellbeing execution safety performance. Safety culture plays an important role to deciding the organization's success or failure. The role of safety culture also play among the others targeted organization towards outstanding safety performance. In Malaysia, there are different categories of manufacturing industry. There are the different kinds of industry such as textile, petrochemical automotive, plastic, wood based, electronics and many others. As most of accident were occur by the unsafe performances at their workplace, so it's very important that good attitude s should be possessed by individuals in order to exhibit safety behaviors [7-8].

For this situation, safety culture gives direction to the workers to understanding the importance of working environment with safe workplace. Safety culture organization promote belongingness, trust and participation among the worker through their cohesiveness and membership. They remind each other on the significance of safety and encourage communication through teamwork. In additional safety culture offer them for the guidance of safety practices in the organization and also provides the share values and beliefs that assist workers to realize organizational safety. Furthermore to propagate safety information for launching communication networks and safety activities, it emphasizes on broad participation by employees. It keeps up a fundamental concentrate on the interior authoritative safety standard. In an association with wellbeing safety trends, working

 DOI: 10.1051/matecconf/201713500031

environment safety has never been traded off with efficiency. Pioneers lead the laborers by showing safety is an incentive to them and to the association. They convince that the working environment is safe for the staff to perform their responsibilities. In this way, workers will have more sureness to do their duties and they can also enhance the productivity. As far as molding individual wellbeing practices, safety culture supports the display of wellbeing consistence and safety interest conduct. It advances wellbeing in the feeling of strengthening and self-belongingness.

Table # 1 respectively present the total number of accidents and by types of accident in Malaysia during 2012 to 2016. There was significant increase in the total number of accidents reported for all sectors, increase 24.90% from 2780 in 2012 to 3702 in 2016. There was also significant increase in the total number death and non-permanent Disability respectively reported for all sectors, increase 20.4% from 191 in 2012 to 240 in 2016 and increase 28.9% from 2382 in 2012 to 3348 in 2016. There was significant decrease in the total number of permanent Disability reported for all sectors, Decrease 43.7% from 207 in 2012 to 144 in 2016[9-13].

Table 1. Number of Report accident 2012-2016

Years	Total Accident Reported	Death	Permanent Disability	Non -Permanent Disability
2012	2,780	191	207	2,382
2013	2,826	185	165	2,476
2014	2,805	204	145	2,456
2015	3,344	213	122	3,009
2016	3,702	240	144	3,348

Among all sectors, Table #2 the number of accidents reported for the manufacturing sector has been the highest throughout the period since increase 26.2% from 1722 in 2012 to 2333 in 2016 . This reflects specialists in the manufacturing area are presented to high accidental dangers. It can be watched that the pattern of accident reports varies from one sector to another sector, reflecting the difference of hazard crosswise over all sectors.

Table 2. Industrial accidents reported by sectors, Malaysia, (2012 to 2016)

Sector	2012	2013	2014	2015	2016
Manufacturing	1,722	1,655	1,667	2,040	2,333
Mining and Quarrying	42	35	62	39	25
Construction	177	164	172	237	233
Agriculture, Forestry, Logging and Fishery	446	535	492	480	471
Utility	94	108	70	96	75
Transport, Storage and Communication	95	93	102	131	130
Wholesale and Retail Trade	73	78	83	108	109
Hotel and Restaurant	15	20	57	62	90
Financial, Insurance, Real Estate and Business Services	62	71	74	119	126
Public Services and Statutory Bodies	54	67	26	32	110
Grand Total	2780	2826	2805	3344	3702

 DOI: 10.1051/matecconf/201713500031

During 2011 to 2016 Table # 3, the statistic of occupational accident in Malaysia illustrated that the first highest non-permanent and permanent disabilities were recorded in manufacturing industry (DOSH Malaysia, 2012 to 2016) and second highest Death/Fetal accident were also recorded in manufacturing Industry (DOSH Malaysia, 2011 to 2016).

There is significant increase in the total number of non-permanent Disability in manufacturing industry reported for all sectors and increase from 1535 in 2012 to 2187 in 2016 with an increment of 42.5%. Also there is a significant increment in the total number permanent Disability in manufacturing industry as compare to other sector.

While the total number of permanent Disability in manufacturing industry reported for all sectors having a decrease from 147 in 2012 to 74 in 2016 with a decrement of 49.7%. But an increment in the total number Death in manufacturing industry has been reported with numbers from 40 in 2012 to 72 in 2016 by 80% increment.

Table 3. Industrial occupational accidents categories (D, PD & NPD) reported by sectors, Malaysia, (2012 to 2016)

Sector	2012			2013			2014			2015			2016		
	D	NPD	PD	D	NPD	PD	D	NPD	PD	D	NPD	PD	D	NPD	PD
Manufacturing	40	1,535	147	58	1,469	128	45	1,510	112	45	1,906	89	72	2,187	74
Mining and Quarrying	7	32	3	5	30	0	15	43	4	4	32	3	4	21	0
Construction	67	98	12	69	83	12	72	94	6	88	138	11	99	129	5
Agriculture, Forestry, Logging and Fishery	38	382	26	33	488	14	42	441	9	31	440	9	25	437	9
Utility	5	86	3	7	100	1	0	69	1	6	86	4	2	69	4
Transport, Storage and Communication	22	68	5	8	84	1	15	84	3	22	107	2	13	115	2
Wholesale and Retail Trade	4	63	6	5	66	7	6	74	3	3	102	3	0	105	4
Hotel and Restaurant	0	14	1	0	19	1	0	56	1	0	62	0	3	85	2
Financial, Insurance, Real Estate and Business Service	4	55	3	0	70	1	4	65	5	14	105	0	16	99	11
Public Services and Statutory Bodies	4	49	1	0	67	0	5	20	1	0	31	1	6	101	3
Grand Total	191	2,382	207	185	2,476	165	204	2,456	145	213	3,009	122	240	3,348	114

As per the record from DOSH Table # 4 show that in last 5-Years, the increment in the fatal accidents by Average 26% , Permanent Disability by Average 71% and Non-Permanent Disability by 64 % are investigated only in Manufacturing Industries.

Table 4. Industrial accidents of Manufacturing sector last five years (2012 to 2016)

Year	Death (D)			Permanent Disability (PD)			Non-Permanent Disability (NPD)		
	Total	MFG	%	Total	MFG	%	Total	MFG	%
2012	191	40	21	207	147	71	2,382	1,535	64
2013	185	58	31	165	128	78	2,476	1,469	59
2014	204	45	22	145	112	77	2,456	1,510	61
2015	213	46	22	122	89	73	3,009	1,906	63
2016	240	72	30	114	74	65	3,348	2,187	65
Average % of 5 Years (D)			25%	**Average % of 5 Years (PD)**		75%	**Average % of 5 Years (NPD)**		64%

In this way, accident and their results keep on being a noteworthy general for public health concern. Moreover, the Occupational Safety and Health Act (OSHA) 1994 expressed that the self-direction idea was declared in view of the essential duty of ensuring safety, wellbeing and welfare of all people at all work environments. In this way, a presentation of safety culture can be viewed as an orderly arrangement towards the foundation of zero accident in the working environment [14].

3 Occupational disease

Manufacturing industries in Malaysia give most extreme consumer loyalty ceaselessly experience advancements. These developments in new developing innovation that has prompting greater progression in modern assembling forms with a subsequent increment the field of industrial manufacturing with resulting new growing technology has leading to new type of disease impact on human health [15]. The more up to date sorts of work related diseases, illnesses and injuries are also developing without the labors been realizing that new innovation brings new health issues [16-17].

Depicted occupational health as the progression and safety of the most elevated level of mentally, physical and social prosperity of specialists in all occupations. Manufacturing industries are emphases on the raw materials and that raw material convert into finish product that goods will be fulfil customer satisfaction according to their needs. So diseases occur during the process manufacturing of raw material convert into finish product also generate term as industrial disease. As occupational health and safety focus on two expect Health and safety, health associated free from diseases and safety associated free accident or injury happen during the manufacturing.

Epidemiologic reviews have credited overabundance work related deaths, sicknesses, diseases and injuries due to related manufacturing industry [18].The expanding rate of fatalities because of work related occupational disease and sickness is a reasonable sign that worker are not completely aware of the potential source occupational disease and sickness related with poor health practices in their work related environment [19].Another work related wellbeing estimation for cases with days from work is musculoskeletal disorders (MSDs) [20].

Many of the worker working without health care hazard exiting in their work related area [21]. Industries laborers are likewise disregard of the obligations in term of health care because of the carelessness of employer and administration group. This is the poor work related health practices existing in the Malaysia manufacturing industry.

In 2015, a total of 5960 cases of occupational disease and poisoning have been reported to the Occupational Health Division as compared with 2648 cases reported in the year 2014 as Table # 5. This shows an increment of 55.57 % in occupational disease and poisoning. A sum of 4034 cases have been effectively examined for the purpose of carrying out improvements to the workplace in terms of occupational health. An aggregate of 3377 instances of work related diseases that were investigated were of noise-induced hearing loss (NIHL) and the disease is still the most common occupational disease experienced by workers (83.7%) as compared with other diseases. This was trailed by work-related musculoskeletal diseases, i.e. a total of 147 cases (4.4%) and occupational lung diseases, of 91 cases (2.3%) were investigated in 2015 as shown in Table. 6. Diseases caused by physical agents were the least investigated diseases by the Division whereby no cases were investigated in 2015.[22] [23]

Table 5. Statistic of Occupational Poisoning and Diseases from the Year 2012 – 2015

Years	Number of Occupational Diseases Cases
2011	1,198
2012	1,792
2013	2,588
2014	2,648
2015	5,960

Table 6. Statistics of Occupational Poisoning and Diseases by Type of Diseases for the Year 2015

Sr.No	Type of Diseases	Number of Cases Received	Number of Cases Received %	Number of Cases Investigated	Number of Cases Investigated %
1	Occupational Lung Diseases (OLD)	126	2	91	2.3
2	Occupational Skin Diseases (OSD)	46	1	27	0.7
3	Occupational Noise Induced Hearing Loss (NIHL)	5,366	90	3,377	83.7
4	Occupational Muscular - Skeletal Disorders (OMD)	249	4	147	4.4
5	Occupational Poisoning	32	1	27	0.7

Number of Cases Investigated %	0.0	1.3	0.0	0.0	0.1	7.5	4,034
Number of Cases Investigated	0	52	2	1	6	304	
Number of Cases Received %	0	1	0	0	0	1	5,960
Number of Cases Received	5	73	3	2	19	39	
Type of Diseases	Disease cause by Physical Agent	Disease cause by Biological Agent	Occupational Cancer	Psychosocial Disorders	Other Types of Occupational Diseases	Non Occupational Diseases	Total Number of Cases
Sr.No	6	7	8	9	10	11	

Investigation of work related diseases and poisoning by sector revealed that the manufacturing sector noted that most number of cases of work-related disease and poisoning in 2015, a total of 3487 cases (86.4%). This was followed by the public service sectors, of 147 cases (4.4%) and the agricultural sector, a total of 122 cases (3.0%) As shown in table # 7. The Occupational Health Division also investigated ex-gratia compensation cases for occupational disease confirmation purpose for the government employees and a total of 38 cases have been successfully investigated for the year 2015 as shown in Table # 8.

Table 7. Occupational Disease and Poisoning By Sector Dosh 2015

Sr.No.	Sector	No. of Cass	%
1	Manufacturing	3,487	86.4
2	Mining and Quarrying	109	2.7
3	Construction	5	0.1
4	Agriculture, Forestry, Logging and Fishery	122	3.0
5	Utility	77	1.9
6	Transport, Storage and Communication	17	0.4
7	Wholesale and Retail Trade	4	0.1
8	Hotel and Restaurant	14	0.3
9	Financial, Insurance, Real Estate and Business Services	50	1.2
10	Public Services and Statutory Bodies	149	3.7

Table 8. Occupational Disease and Poisoning (Ex Gratia) By Type of Diseases DOSH 2015

Types of Occupational Diseases	No.
Occupational Lung Diseases	22
Occupational Muscular - Skeletal Disorders (OMD)	2
Occupational Skin Diseases	1
Diseases Caused Biological Agents	3
Other Occupational Diseases	1
Non Occupational Diseases	9

A decent national OSH framework is basic for the powerful execution of national strategies and developers to strengthen the avoidance of work related diseases; it should be incorporate [24]:

1. Laws and regulation and, where proper, aggregate understandings joining the counteractive action of work related diseases;
2. Law obedience instruments, including viable OSH investigation frameworks;
3. Participation amongst administration and laborers and their delegates in the usage of OSH measures;
4. Providing of occupational health facilities;
5. Satisfactory components for the accumulation and investigation of information on occupational diseases;
6. Occupational health and Safety information and training;
7. Collaboration between departments of Labour, department of health and social security schemes covering work-related injuries and diseases.

4 Conclusion

In conclusion, research on occupational accident in Malaysia is generally lacking and needs to be strengthened. Future research should focus on high-risk industries (manufacturing, construction, agriculture and transportation). Future occupational management systems and the role of a safer design of the environment including equipment and machinery. The effectiveness of OSHA should be evaluated and measures to further implement it should be developed. Comparative research into the effectiveness of occupational safety practices in our country with other developed nations will be helpful.

Despite more than 18 years have passed since OSHA was first implemented in Malaysia, there has been no revision done on OSHA. Since Malaysia is a fast growing country since the last three decades, there are certain provisions that may no longer be compatible with current situation. For example, in the manufacturing industries, new machineries with new technologies have been introduced which consequentially requires some of OSHA's provision to be revised.

For government, industry administration and workers an improvement in occupational safety and health is needed. Mishaps at the job environment ought to be kept with the participation from both managers and representatives on top of customary authorization from the legislature. Occupational accidents are avertable and occupational accidents can be omitted through proper safety measures. Compared to disabilities in terms of non-permanent are less likely as compared to permanent disabilities and deaths.

The administration, staffs and DOSH must consolidate strengths and attempt to forestall future occupational accidents by tending to the underlying reason of accidents. To develop a good safety trend at workplace, the rules from government, employer and workers need to combine together.

Malaysia aimed to reduce the occupational accidents, however; the results show by DOSH increase that Occupational Noise Induced Hearing Loss 83.7%, occupational musculoskeletal diseases, 4.4% and occupational lung diseases 2.3%. But the as per the record from DOSH that in last 5-Years, the increment in the fatal accidents by Average 26%, Permanent Disability by Average 71% and Non-Permanent Disability by 64 % are investigated only in Manufacturing Industries. The government must show their high interest on such a vulnerable employees to accomplish the above aim. This step will be helpful for planning to reduce the accidents in workplaces and it will also detect the prevention for the future accidents.

5 Acknowledgement

This research work is fully funded by office for research, innovation, commercialization, consultancy management (ORICC), Universiti Tun Hussein Onn Malaysia (UTHM) Parit Raja, Johor under vote number U595.

Reference

1. Government of Malaysia. Occupational Safety and Health Act (Act 514) 1994
2. Knowledge Management to Promote Occupational Safety and Health at the Malaysian Manufacturing Workplace: Reposed in Occupational Safety and Health Committees? Ummu Kolsome Farouk, Stanley Richardson and Arul Jeganathan Solucis Santhapparaj Multimedia University, Malaysia
3. Soehod, K. (2007). 'Law on Safety and Health in Malaysia,' Unpublished manuscript, Universiti Teknologi Malaysia.
4. Government of Malaysia. Occupational Safety and Health Act (Act 514) 1994.
5. Safety Culture in Malaysian Workplace: An Analysis of Occupational Accidents Hui-Nee A* Hui-Nee 2014 Health and the Environment Journal, 2014, Vol. 5, No. 3.Workplace Accident in Malaysia: Most Common Causes and Solutions Noorul Huda Zakaria, Norudin Mansor, Zalinawati Abdullah Business and Management Review Vol.

2(5) pp. 75 – 88 July, 2012 ISSN: 2047 - 0398 Available online at http://www.businessjournalz.org/bmr.

6. Factors Contribute to Safety Culture in the Manufacturing Industry in Malaysia, Ong Choon Hee Universiti Utara Malaysia, Sintok, Kedah, Malaysia , DOI: 10.6007/IJARBSS/v4-i4/753 URL: http://dx.doi.org/10.6007/IJARBSS/v4-i4/753 , International Journal of Academic Research in Business and Social Sciences April 2014, Vol. 4, No. 4 ISSN: 2222-6990.
7. Mullen, J. (2004). Investigating factors that influence individual safety behavior at work. Journal of Safety Research, 35, 275-285.
8. Preliminary Study of the Safety Culture in a Manufacturing Industry ,Ezrin Hani Sukadarin , Nurud Suria Suhaimi ,Norhidayah Abdull , Department of Technology Management Faculty of Manufacturing Engineering and Technology Management University Malaysia Pahang 26300, Pahang, Malaysia, International Journal of Humanities and Social Science Vol. 2 No. 4 [Special Issue – February 2012].
9. http://www.dosh.gov.my/index.php/en/archive-statistics/2012-archive-stat/795-occupational-accidents-statistics-2012
10. http://www.dosh.gov.my/index.php/en/archive-statistics/2013/843-occupational-accidents-statistics-by-sector
11. http://www.dosh.gov.my/index.php/en/archive-statistics/2014/1225-occupational-accidents-statistics-by-sector-2
12. 12http://www.dosh.gov.my/index.php/en/archive-statistics/2015/1713-occupational-accidents-statistics-by-sector-until-december-2015
13. http://www.dosh.gov.my/index.php/en/archive-statistics/2016/1846-occupational-accidents-statistics-by-sector-until-december-2016
14. Belanger, P.L. (1981). Health hazard evaluation report. Center for disease control, NIOSH report no. HETA 81 – 107
15. David A.G.(2006).Occupational contribution to chronic cough. Journal of occupational medicine and toxicology. Retrieved 27th December, 2012 from www.occup/med.com
16. Eilbert, A. (1996). Occupational health hazards. Retrieved 27th December, 2012 from www.occup/med.com.
17. Wikipedia (2011). Work-related diseases, industries. Retrieved 27th December, 2012 from www.wikibooks.com

DOI: 10.1051/matecconf/201713500032

Value Stream Mapping to Improve Workplace to support Lean Environment

Ahmad Nur Aizat *Ahmad*.[1,*], Te Chuan *Lee*.[1] Rohaizan *Ramlan*.[1] , Md. Fauzi *Ahmad*. [1], Norafifah *Husin* [2] and Mustaqqim *Abdul Rahim*.[3]

[1]Department of Production and Operation Management, Faculty of Technology Management and Business, Universiti Tun Hussein Onn Malaysia (UTHM), Malaysia
[2]Department of Industrial Engineering, Faculty of Mechanical Engineering, Universiti Teknologi Malaysia (UTM), Malaysia
[3]Department of Civil Engineering, School of Environmental Engineering, Universiti Malaysia Perlis (UniMAP), Malaysia

Abstract. In recent years, lean manufacturing is being followed by various sectors in order to keep their competitiveness in the global markets. Lean manufacturing plays a vital role in improving the efficiency of operation by eliminating or reducing wastes. Nonetheless, most of small and medium enterprises (SMEs) lack sufficient knowledge or information on the benefits of implementing lean manufacturing. The main objective of this study is to apply value stream mapping, one of lean manufacturing tools, for improving the productivity in a SME by eliminating non-value added activities. In this study, lean manufacturing was adopted at a SME, particularly a food industry. Value stream mapping was served as main tool to identify the wastes and improvement opportunities in production line. Subsequently, different lean manufacturing tools such as Kaizen Burst, one piece flow, and 5S were applied to eliminate or reduce identified wastes. Based on the future state value stream mapping, final results showed that the total operation time and non-added value activities time were successfully decreased from 1993 seconds to 1719 seconds, and 234 seconds to 104 seconds, respectively. The findings of this study indicate that value stream mapping is an effective approach to eliminate the wastes and improve the productivity.

1.0 Introduction

Despite an abundance of obstacles, the biggest challenge faced by manufacturer is how to deliver their products or services quickly at low cost but high quality. A number of approaches have been undertaken using computer simulation, statistical analysis, and lean manufacturing for improving the productivity and efficiency of production line [1]. Among these, lean manufacturing is one of most popular approaches and has been widely implemented in many manufacturing industries around the world.

* Corresponding author: aizat@uthm.edu.my

Lean manufacturing is currently becoming more popular in manufacturing industries. It can be seen that lean manufacturing has been developed, growth up, and adopted progressively in many major manufacturing industries in order to remain their competitive in an increasingly global market [2]. The main focus of this approach is on cost reduction by eliminating waste or non-value added activities. In other words, the perfect strive of the manufacturing system can be achieved through successful implementation of lean elements [3].

Lean manufacturing has been proven its capability in improving workplace effectiveness, most notably in waste reduction. In general, workplace is defined as physical structure and location where work is undertaken and organized. It is necessary to ensure the processes in the workplace are run smoothly and absence of non-value added activities such as waiting time, overproduction, and unnecessary movement. Nowadays, almost all the manufacturing industries trying to improve the productivity performance by minimizing the production lead time and waste. However, most of SMEs facing obstacle such as production waste management. This is attributed to the fact that SMEs lack sufficient knowledge or information on the benefits of implementing lean manufacturing [4,5].

In order to address these issues, a case study was conducted in a domestic SME, particularly food industry, to reduce the total production lead time through elimination of non-value added activities. Value stream mapping (VSM), one of lean manufacturing tools, was applied to eliminate non-added value activities in production line. There are three objectives for this study: (1) To apply VSM for identifying non-value added activities and highlighting the improved area, (2) To analyse the current state of VSM to improve process flow of the operation, and (3) To propose future state of VSM for the company.

2.0 Literature Review

2.1 Lean Manufacturing

Lean manufacturing concept was originated from Japan after the World War II owing to the vast shortage of materials, financial, and human resource. Lean manufacturing or known as Toyota Production System (TPS) is a production strategy that aims to eliminate waste and deliver increased value to the product and customer. Generally, lean manufacturing focuses on pinpointing the major source of wastes and eliminate the wastes using lean manufacturing tools [6–10]. There are several lean manufacturing tools have been extensively used to identify and eliminate wastes, for example, include but not limited to Kanban, 5S, Kaizen, Value Stream Mapping, Total Productive Maintenance, and Poka-Yoke [11]. Apparently, lean manufacturing attract a great of attention in manufacturing industries owing to its exclusive advantages, i.e. reduce lead time, reduce inventory, require less space, less process waste, increase process understanding, and financial saving [10].

It is often suggested that lean should be understood on two levels: (1) the strategic level of how to understand value, and (2) the operational level (tools) of how to eliminate waste [12]. It is noteworthy to mention that lean manufacturing strives for perfection through continuous improvement to eliminate waste by identifying value added activities (VA) and non-value added activities (NVA).

Lean means manufacturing without waste. Waste, or known as "Muda" in Japanese term, is defined as any activity in a process which does not add value to the customer. However, sometime the waste cannot be eliminated as it is a necessary part of the process and adds value to the company, for instance, financial control [10]. There are seven sources of the NVA wastes commonly found in manufacturing industries, which are, overproduction, waiting, transportation, unnecessary motion, unnecessary inventory, defects, and inappropriate processing. Elimination of these wastes is achieved through the successful implementation of lean elements. Continuous improvement is the core of lean thinking, therefore, all the processes will continually improve to reduce the waste and ultimately achieve a waste-free process [6–8,20].

2.2 Value Stream Mapping

Value stream mapping (VSM) is one of the lean manufacturing tools to identify wastage of resource and eliminate or reduce them for improving productivity [14]. Nowadays. VSM has been widely used in many sector such as healthcare industry, manufacturing industry, finance firm, office environment, etc [15]. It has been recognized that VSM is a powerful lean tool to improve the productivity of manufacturing industries.

Generally, VSM is a collection of all actions such as value added and non-value added activity that are required to bring a product through the main flows, starting with raw material and ending with customer [16]. According to Rother & Shook [16], VSM is a drawing map using a predetermined set of standardized icons to virtualize a product's production path from start to end. There are two different in the method of VSM, which are future state of VSM and current state of VSM.

• *Current state map*: A current state map (CSM) is a snapshot of how a process is currently done. It is also known as current state process flowchart, or a current state value stream map, but the principle is the same. The CSM shows the current methodology of how the products being produced or service being performed to customer. The CSM enables process flow to be identified, analysis data, as well as identity and prioritise areas for improvement [7,9].

• *Future state map*: Future state map (FSM) illustrates how things should be done based on takt time requirements [17]. FSM presents the flow of product after eliminating all inefficiency activities. There are three phases for mapping the future state which have to be performed together, namely (1) Customer demand phase: Understanding the customer needs for service, quality, and lead time, (2) Continuous flow phase: Implementation of continuous flow by make sure customer receive the right item, in the right quality, and at the right time, and (3) Levelling phase: Distribute the works evenly to reduce queue times [18].

Basically, there are 4 steps to improve efficiency of manufacturing industries using VSM, which are: (1) choose a particular product or product family as the target for improvement, (2) illustrate a current state map after walking and observing along the actual process, (3) analyse the current system and identify the weakness and waste, (4) create the future state map to virtualize a picture of how the production line should look after eliminating wastes [9].

In the practical situation, Ventakaraman et al. [19] applied the VSM for reduction of cycle time in a machining process in order to increase the export sales. The results showed that the manufacturing time successfully reduced by 40 %, defects were reduced, higher process capability and quick response to the customer demand in small lots were achieved.

Meanwhile, Rohani and Zahraee [1] conducted a case study to improve the production line using VSM. A leading manufacturer of industrial building paint was selected as a case study for their research. The FSM showed that production lead time decreased from 8.5 days to 6 days, and the value added time decreased from 68 minutes to 37 minutes after implementing some lean manufacturing tools such as 5S, Kanban method, and Kaizen.

Similarly, Ahmad, A.N.A [20] applied VSM in a manufacturing firm producing aircraft parts. He mentioned that application of VSM uncovered number of problems and bottlenecks in company production process. Surprisingly, huge improvements had been made after applying VSM. Total lead time successfully reduced by 24.8%. On the other hand, the VA index increased around 32%.

Recently, Stadnicka and Ratnayake [21] demonstrated the use of VSM in aircraft maintenance processes for minimizing the lead-time of maintenance services and subsequently minimizing the costs of maintenance services. As expected, production lead-time decreased by 63% after eliminating NVA. Additionally, they also revealed that the process cycle efficiency has increased about four times.

3.0 Research Methodology

A SME company in Batu Pahat, Johor was selected as a case study for this paper. This company is a popular manufacturer of various fried chips. The total number of operators is 19. All the operators in the production line were involved in this study. The average daily demand of customer is approximately to 462 kg. Meanwhile, the working time is 8 hours per day including 1 hour break.

Qualitative approach was used in this study for collecting data. Observation and time study were carried out for obtaining accurate data. The cycle time of each process was recorded using a stopwatch. As mentioned previously in literature review, there are seven types of wastes in production line. However, underutilizing talent was added in this study as the eighth waste because this factor is directly affected the production system and create unwanted waste, as discussed by other author [22].

Initially, the observation was made for collecting data such as layout and facilities of company, number of operators, number of processes and cycle time for each process. Next, the current state map of the company is prepared in order to identify the potential wastes and improvement opportunities in the production line. Subsequently, several lean manufacturing tools such as Kaizen Burst, one piece flow, and 5s were implemented for enhancing the efficiency of production line. Takt time also was calculated to set the pace of production process. In the end of the study, a future state map was illustrated to examine the effectiveness of proposed improvements by comparing the total operation time between the current state and future state map.

 DOI: 10.1051/matecconf/201713500032

4.0 Results and Discussion

4.1 Current State Value Stream Mapping (CSVSM)

The ultimate goal of CSVSM is to illustrate how a product flow through the value stream, starting from raw materials to finished product. In this study, 9 processes were identified as follows:
Process 1: Materials preparation
Process 2: Peeling process
Process 3: Cutting process
Process 4: Wash and transfer into basket
Process 5: Frying process and put on the tray
Process 6: Packaging for 5 kg
Process 7: Arrange on the rack
Process 8: Packaging for 500 g
Process 9: Warehouse

4.1.1 Takt Time

Takt time is defined as the time needed to produce a product to meet customers' demand. Basically, if the demand increases, the Takt time decrease, and vice versa. On the other hand, if bottleneck cycle time less than Takt time, then customers' demand met, and vice versa [7]. Hence, Take time is served as a management tool to determine whether the production line is ahead or behind the production schedule. Additionally, it also plays a vital role as an alignment tool to calculate the real-time production needs. In this case, Takt time was calculated as below:

Customer demand per day	*: 462 kg*
Working time	*: 8 hours (one shift per day)*
Break time	*: 1 hour*
Available time	*: 7 hours per day*

Takt time = Net Operating Time / Customer Demand
= (8 hours – 1 hours) x 60 min / 462 kg
= 0.9 min per kg
= 54 sec

Fig. 1 demonstrates the CSVSM. Meanwhile, a summary of NA and NVA data is tabulated in Table 1. Based on Fig. 1 and Table 1, it can be clearly seen that the total lead time for manufacturing process is 2227 seconds (approximately to 37.12 minutes). Specifically, the time of VA and NVA are 1993 seconds (approximately to 33.22 minutes), and 234 seconds (approximately to 3.9 minutes), respectively, Hence, it can be inferred that the time of NA is higher than NVA, and the flow of process is stable. However, there's still have room for improvement.

DOI: 10.1051/matecconf/201713500032

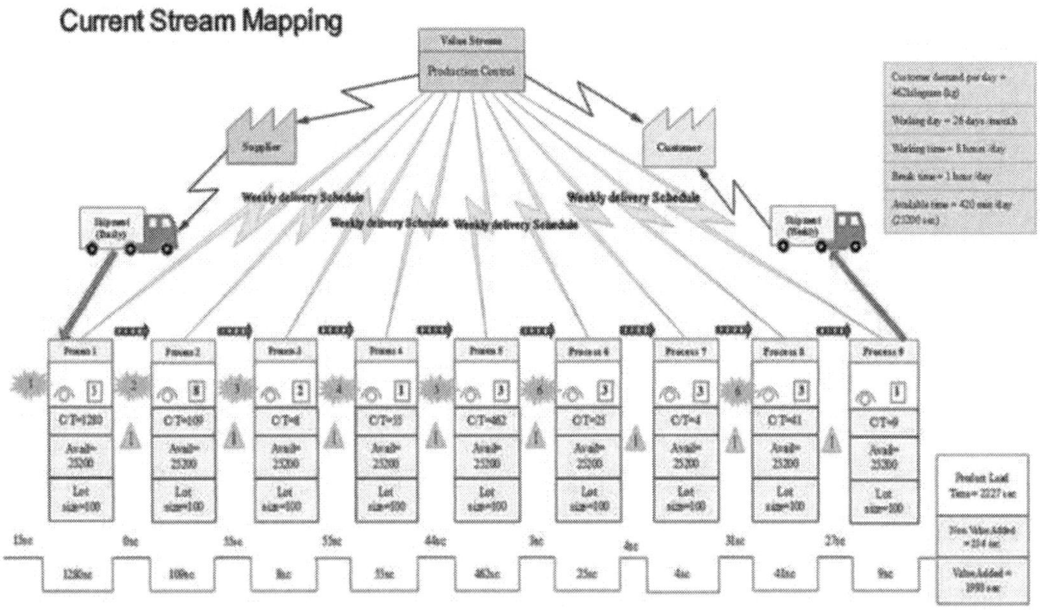

Fig. 1: The current state value stream mapping.

Table 1: Summary of VA and NVA

Operation Number	*Processes*	*Total Operation Time (sec)*			*% NVA (to be eliminated)*
		VA	*NVA*	*Lead Time*	
1	Materials preparation	1280	15	1295	1.17
2	Peeling process	109	0	109	0
3	Cutting process	8	55	63	87.3
4	Wash and transfer into basket	55	55	110	50.00
5	Frying process and put on the tray	462	44	506	8.7
6	Packaging for 5 kg	25	3	28	10.71
7	Arrange on the rack	4	4	8	50
8	Packaging for 500 g	41	31	72	43.06
9	Warehouse	9	27	36	75
	Totals	*1993*	*234*	*2227*	

4.2 Problem Identification and Improvement Proposal

To improve the process flow of the production, identifying the waste in the process is the first step to be carried out. The problems can be solved when the wastes are identified as well as where the problems are coming from. Based on observations and CSVSM, a summary of improvement proposal is presented in Table 2.

Table 2: The improvement proposal

No.	Waste	Observations	Proposed Improvement
1	*Unnecessary motion*	Operators needed to walk further to get the materials	• New floor layout for the production area • Apply one-piece flow in the production line
2	*Space*	Moving space of forklift was limited owing to a portion of finished product and raw materials inventory were placed outside of the warehouse	• Make the packaging room become the storage room • Apply the 5S concept in the production area
3	*Underutilizing talent*	Creative and talented operators were instructed to do only one task	• Give the operators an opportunity to express their opinions for improving the current production line flow
4	*Defects*	The materials is overcooked or undercooked	• Set the timer to avoid overcooked or undercooked
5	*Excess Processing*	Packaging for 5 kg and 500 g products were done in different workstations.	• Combine the packaging process for 5 kg and 500 g
6	*Waiting*	Operators cannot proceed their task if the previous processes were delayed	• Reallocate the number of operator at each workstation
7	*Transportation*	Lack of wheelbarrow and trolley cart and consequently affect the daily output	• Rearrange the tool cart's schedule to maximize its usage

4.2.1 Kaizen Burst

Kaizen Burst is a lean manufacturing tool to generate the improvement ideas during pahse of developing current and future state of VSM. In this study, there are 6 Kaizen Burst had been applied to reduce the NVA and lead time. As shown in Fig. 1, the 6 Kaizen Burst applied in this study are as follows:

1st : Reallocated the operators
2nd : Reduced the operators in the workstation
3rd & 5th : Transferred the operator to other workstation
4th : Place the basket in the basin during washing process
6th : Combine the process 6 and 8

4.2.3 One-piece Flow Concept in the Production Line

Continuous flow, or sometimes to be called as single piece flow or one-piece flow, refer to the concept of the movement of work in progress (WIP) /work piece at 1 time between production process in a work cells. It keeps WIP at the lowest possible inits and encourage balance in work as well as improve the quality of work. [23]. As mentioned previously, unnecessary movement of operators resulted increasing of lead time. Therefore, one piece flow process was suggested in this case to avoid spending time on walking from one station to another.

4.2.4 Application of the 5S in the Production Area

5S is a method on how to organize workplace in the proper and tidy environment. This method is applied at a production area either others areas which to entrust the areas are in proper arrangement f material and in proper flow of the material movement. In this case, the floor in production area was dirty, dusty, and slippery. This occurred mainly due to the fallen soil from the raw materials, spillage of water during the washing process, and spillage of cooking oil during the frying process. Therefore, 5S was implemented to eliminate wastes that resulted from a poorly organized production area and avoid the unwanted accident. Few improvements were made in this case, for example, the floor of production line were swapped and mopped frequently, the rack was arranged properly to maximize the usage of the space for other purposes, and tools were cleaned and placed properly after being used.

4.2.2 New Floor Layout

The biggest waste occurred in the production line was unnecessary movement, especially operators from the cutting and frying station. Operators of both stations needed to walk further to get the materials and return with heavy loads. Consequently, the cycle time was increased. Fig. 2 and 3 illustrates the movement of operators in the production area before and after implementing proposed improvement, respectively.

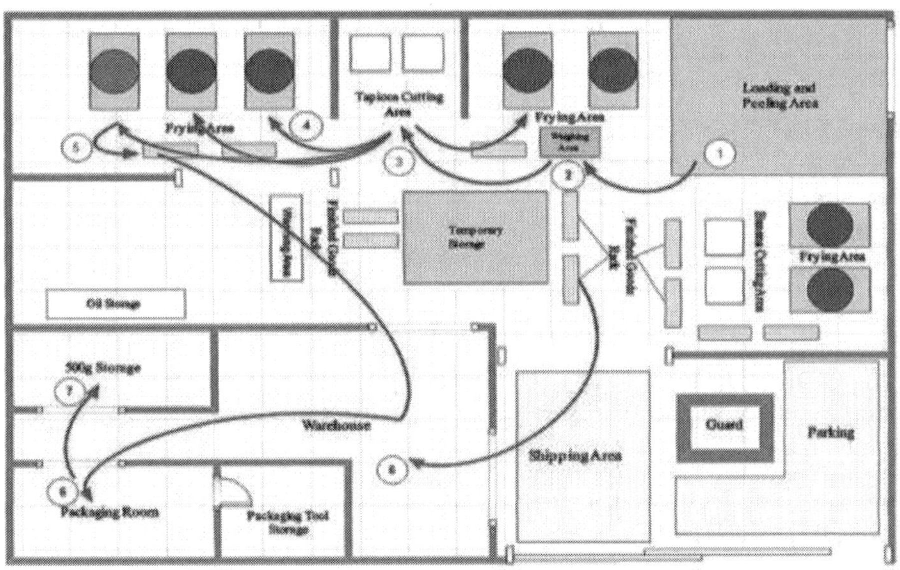

Fig. 2: Floor layout before implementing proposed improvement.

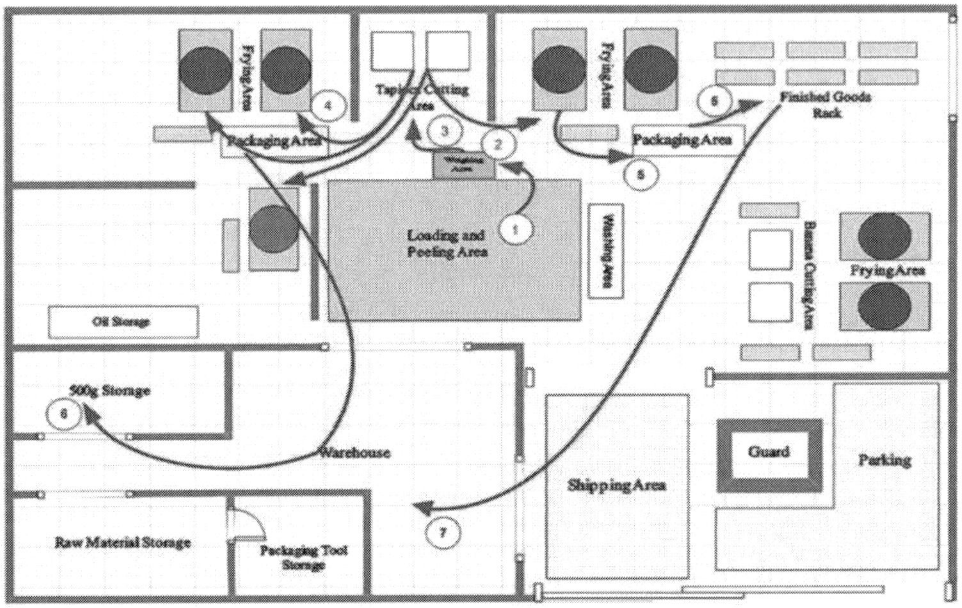

Fig. 3: Floor layout after implementing proposed improvement.

4.3 Future State of Value Stream Mapping (FSVSM)

FSVSM is a tool to illustrate the improvement plan after implementing lean manufacturing tools. In this case, the existing processes had been combined and reduced to 8 processes, as follows:

Process 1: Materials Preparation
Process 2: Peeling process
Process 3: Cutting process
Process 4: Wash and transfer into basket
Process 5: Frying process and put on the tray
Process 6: Packaging for 5kg and 500g
Process 7: Arrange on the rack
Process 8: Warehouse

The FSVSM of this study is shown in Fig. 4. Meanwhile, comparison of cycle time between current state and future state is presented in Table 3. It is noteworthy to mention that the quality of finished product remain unchanged even though the 2 packaging processes were combined. As shown in Table 2, it can be inferred that the total operation time reduced from 1993 seconds to 1719 seconds, or 13.75 % reduction. On top of that, the NVA was successfully decreased from 234 seconds to 104 seconds. Additionally, negative values of total reduced can be noticed from Table 2. This occurred mainly due to the increasing number of operators at that particular process. However, it resulted in a reduction of waiting time, bottleneck at workstation, and production flow became smoother. It is critical to note that increasing of cycle time in this study does not lead to increasing of waste, and conversely increased the value added time to the products.

DOI: 10.1051/matecconf/201713500032

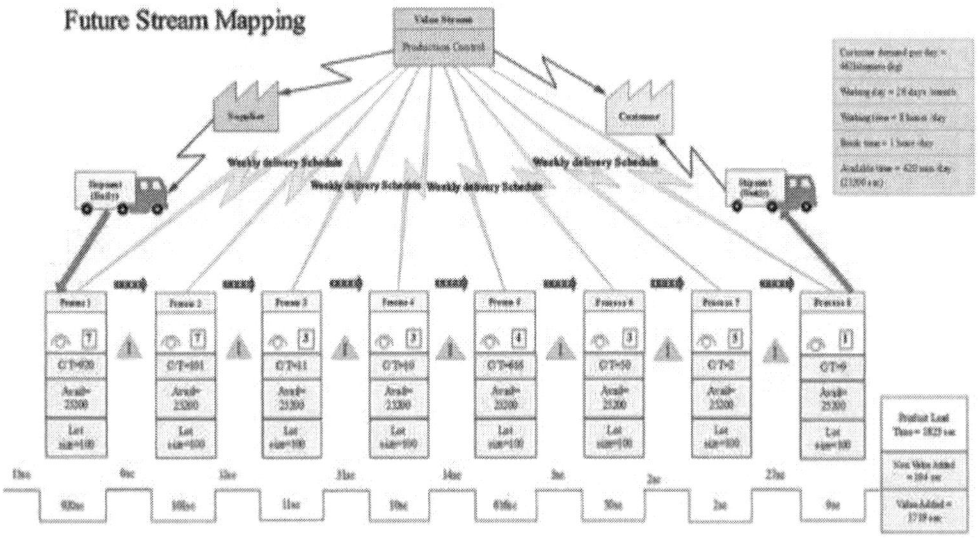

Fig. 4: Future state value stream mapping.

Table 3: Comparison between current and future value stream mapping

Processes	*Total Operation Time (sec)*			*% reduced or increased*
	Current State	*Future State*	*Total reduced*	
Operation Number	9 processes	8 processes	1	12.5
Peeling process	109	0	109	0
Process 1	1280	920	360	39.13
Process 2	109	101	8	7.92
Process 3	8	11	-3	-27.27
Process 4	55	10	45	450
Process 5	462	616	-154	-25
Process 6	25	50	16	32
Process 8	41			
Process 7	4	2	2	100
Process 9	9	9	0	0
Total	***1993***	***1719***	***274***	***13.75***

5.0 Conclusions

In summary, this study has shown that VSM is an effective tool to eliminate waste in production operation of industry. The application of VSM associated with different lean manufacturing tools had successfully enhance the efficiency of production line. Moreover, the new floor layout and combination of two packaging processes were effectively reduced the unnecessary movement and NVA. The total operation time and NVA time were successfully reduced from 1993 seconds to 1719 seconds, and 234 seconds to 104 seconds, respectively. Consequently, it can reduce the company's operation cost, enhance the operating performance of production line, and lastly meet the customers' demand.

6.0 Acknowledgment

Appreciation to UTHM and ORICC, Universiti Tun Hussein Onn for supporting this research. Appreciation also to Manufacturing Technology Management (MTM) focus group, Faculty of Technology Management, UTHM.

References

1. J.M. Rohani, S.M. Zahraee, Procedia Manuf. **2,** 6–10 (2015)
2. A. Azizi, T. T. Manoharan, Procedia Manuf. **2**, 153–158 (2015)
3. A. Agarwal, A. Kumar, A.K. Madan, Int. J. Ind. Eng. Technol. **7,** 31–38 (2015)
4. A. Raghunath, R.V. Jayathirtha, J. Adv. in Res. & Technol **2**, 1–7 (2013)
5. O. Bakås, T. Givaert, H. Van Landeghem, *Challenges and success factors for implementation of lean manufacturing in European SMES* (Tapir Academic Press, Norway, 2011)
6. E.W. Dickson, S. Singh, D.S. Cheung, C.C. Wyatt, A.S. Nugent, J. Emerg. Med. **37**, 177–182 (2009)
7. R. Sundar, A.N. Balaji, R.M. Satheesh Kumar, Procedia Eng. **97**, 1875–1885 (2014)
8. M.G. Yang, P. Hong, S.B. Modi, Int. J. Prod. Econ. **129**, 251–261 (2011)
9. F.A. Abdulmalek, J. Rajgopal, Int. J. Prod. Econ. 107, 223–236 (2007)
10. T. Melton, Chem. Eng. Res. Des. **83**, 662–673 (2005)
11. A. R.R. Rahani, M. Al-Ashraf, Procedia Eng. **41**, 1727–1734 (2012)
12. P. Hines, N.L. Rich, Int. J. Oper. &Production Manag. **17**, 46–64 (1997)
13. R. Shah, P.T. Ward, J. Oper. Manag. **25**, 785–805 (2007)
14. S.K.P.N. Silva, Int. J. Lean Think. **3**, 36–56 (2012)
15. P. Senna, A. Gomes, A. Monteiro, D. Pinha, Brazilian J. Oper. Prod. Manag. **13**, 406–420 (2016)
16. J. Rother, M. & Shook, *Learning to see: value stream mapping to create value and eliminate muda* (The Lean Enterprise Institute, Massachusetts, 1999)
17. D. Seth., V. Gupta, Prod. Plan. Control. **16**, 44–59 (2005)
18. T. Tapping, D. Shuker, *Value stream management for the lean office* (Productivity Press, New York, 2003)
19. K. Venkataraman, B.V. Ramnath, V.M. Kumar, C. Elanchezhian, Procedia Mater. Sci. **6**, 1187–1196 (2014)
20. A.N.A. Ahmad, H. Sulaiman, Int. J. Mech. Eng. And Adv. Mat. **1**, 13-18 (2015)
21. D. Stadnicka, R.M.C. Ratnayake, Procedia Eng. **182**, 665–672 (2017)
22. A. Muse Ouma, D.A. W. Njeru, J. Dennis, Int. J. Acad. Res. Bus. Soc. Sci. **4**, 361–368 (2014)
23. A. C. Yamagar; P. M. Ravanan, *Proc. 2nd Int. Conf. Manuf. Eng. Qual. Prod. Syst.* 104–112 (2010)

A Novel Approach for Automatic Machining Feature Recognition with Edge Blend Feature

Chen Wong Keong[1], *Yusri* Yusof[2*]

[1]Politeknik Kuching Sarawak, Department of Mechanical Engineering, 93050 Kuching, Sarawak, Malaysia
[2]Universiti Tun Hussein Onn Malaysia, Faculty of Mechanical and Manufacturing Engineering, 86400 Parit Raja, Johor, Malaysia

Abstract. This paper presents an algorithm for efficiently recognizing and determining the convexity of an edge blend feature. The algorithm first recognizes all of the edge blend features from the Boundary Representation of a part; then a series of convexity test have been run on the recognized edge blend features. The novelty of the presented algorithm lies in, instead of each recognized blend feature is suppressed as most of researchers did, the recognized blend features of this research are gone through a series of convexity test before this blend features are used in automatic machining features recognition. A new graph-based method is also introduced with taking account of this edge blend features used in automatic machining feature recognition. This study has contributed to CAD/CAPP integration based on STEP standard in the life cycle of product development.

1 Introduction

Most of the solid modelling is represented in B-rep. Point, edge, and face are the most basic elements in the B-rep model which could provide effective topological data of the solid model for feature recognition analysis. Most of researchers prefer taking the topological relationship of adjacent entities in feature recognition. A thorough review on feature recognition methodologies has been done by several researchers [1-4]. Ansaldi et al.[5] proposed a face adjacency graph structure with a rule-based algorithm for generating graphs of the topological relationship of adjacent faces for several types of features. Although, the algorithm was valid for recognizing several types of features, but it failed to recognize complex features. Joshi et al.[6] proposed the attributed adjacency graph (AAG) method, which is based on the face adjacency graph structure and adds rules in accordance with edges and face conditions for recognizing features types. Tian et al.[7] defined three types of edges and developed rules for identify holes. Recently, a new algorithms is proposed to tool approach direction identification, dimensional and geometric tolerances determination, reference faces identification have been studied by [8]. However, these aforementioned methods consider only those machining features without fillets or blend faces.

*Corresponding author : yusri@uthm.edu.my

DOI: 10.1051/matecconf/201713500033

Fillets (or blend faces, BFs) always created in CAD models to increase machinability, but their presence changes the geometric relationship of adjacent faces, which can complicate the topological structure of faces used in machining feature recognition. Therefore, blend faces must first be recognized, before the topological relationship between the edges and faces of a target feature to be examined. Cui et al. classified three types of blend face: edge blend faces (EBFs), vertex BFs (VBFs), and mixed BFs. They proposed an algorithm using smooth edges, support faces, and span angle, and the area of target feature for automatically recognizing BFs. Venkataraman et al. explained that edge blend faces as face blend faces, and cliff blend faces as face-edge blend faces. They developed an algorithm to evaluate the curvature of smooth edges and automatically detect edge blend faces.

In this research, a new mechanism is proposed for recognizing edge blend face (or specifically called fillet feature) in the solid models. Furthermore, instead of suppressing the recognized edge blend faces as mostly done by researchers[9] in machining feature recognition, attributes of recognized edge blend face is determined as convex/ concave to the adjacent planes, through convexity test. Finally, this edge blend face feature is added to graph-based method for automatic machining feature recognition with edge blend features.

2 Basic concepts

Before describing the algorithm for recognizing edge blend face, we first introduce some basic concepts as described by [9].

(1) Blend face: the face generated by blending operation, such as f_3 in figure 1. The surface of a blend may be a NURBS surface of a complex quadric surface.

(2) Support face: the face that is not a blend face but adjacent with a blend face at its smooth edge. Taking blend face f_3 as an example, f_1 and f_2 are two support faces of it.

(3) Edge blend face: the blend face that replaces a sharp edge between two faces as shown in figure 2.

(4) Vertex blend face: the blend face that is not edge blend face, it connects smoothly several edge blend faces that meet at a common vertex.

(5) Blend feature: a set of connected blend faces corresponding to a vertex or an edge of the product model.

(6) Edge blend feature: the blend feature corresponding to an edge of the suppressed model, such as the grayer face in figure that forms an edge blend feature.

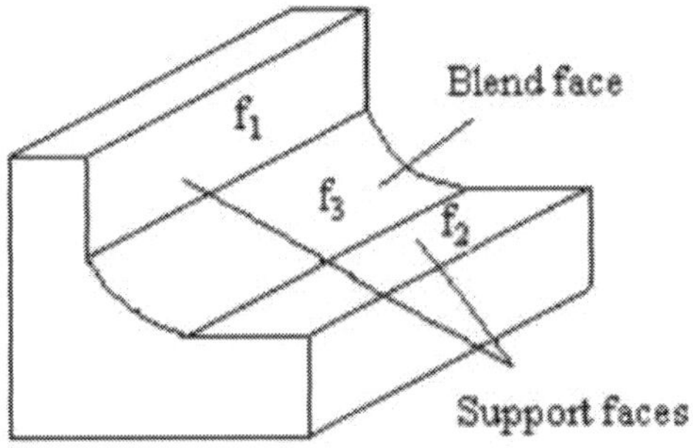

Fig 1. Blend face and its support faces [9]

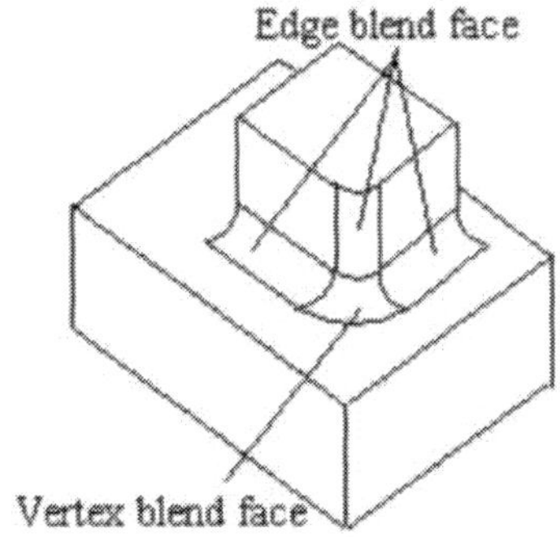

Fig 2. Vertex blend face and edge blend face[9]

DOI: 10.1051/matecconf/201713500033

3 Algorithm for recognizing edge blend face with convexity characteristics

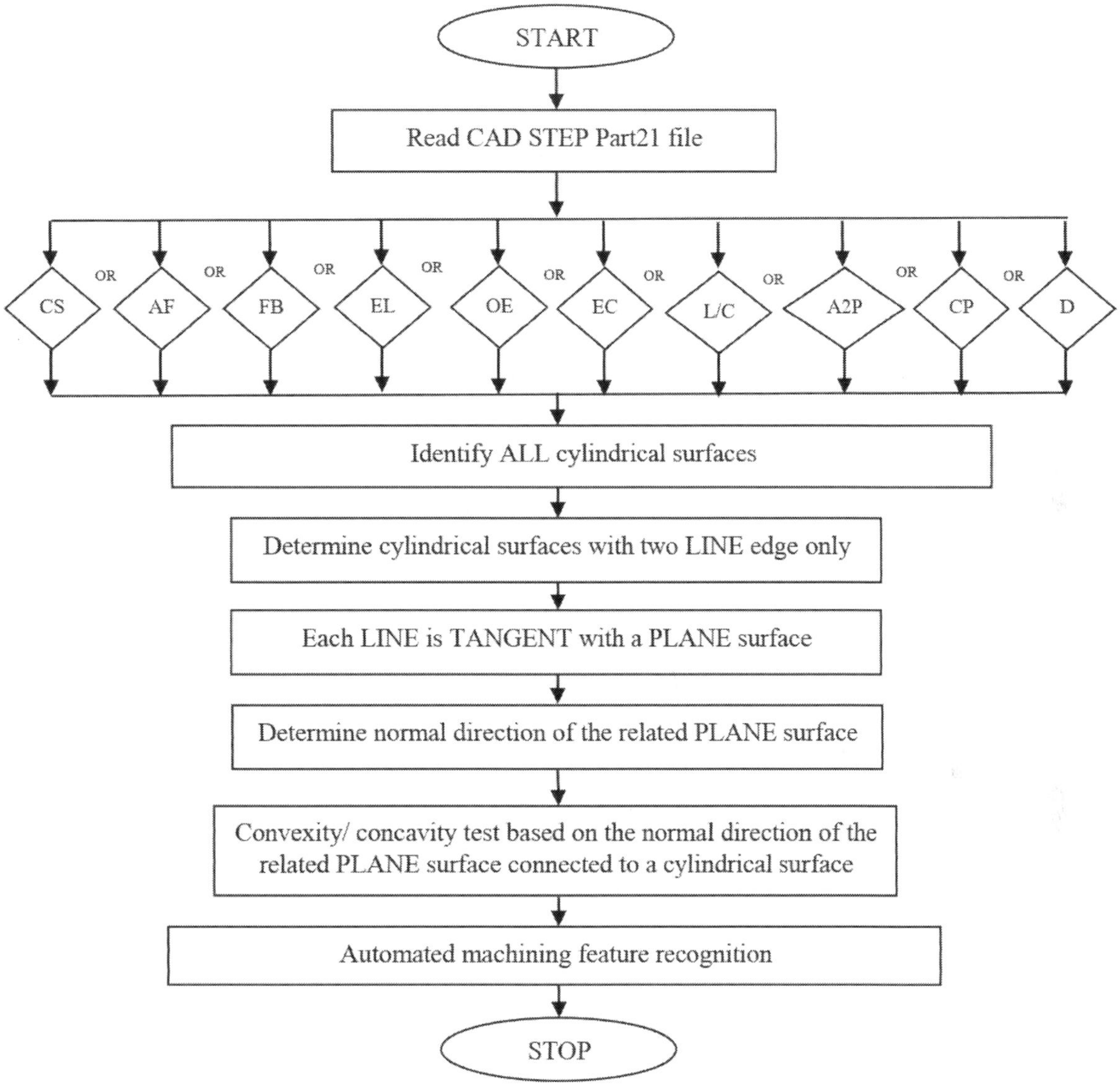

Basic Entity of CAD STEP Part21 file
CS: closed shell
FB: face bound
EL: edge loop
OE: oriented edge
EC: edge curve
L/C: line or circle
A2P: axis2 placement
CP: Cartesian point
D: direction

Fig 3. Algorithm for recognizing edge blend face with convexity test

Figure 3 shows a new proposed algorithm on how to recognize an edge blend face with convexity attribute and used in automatic feature recognition. First, all CYLINDRICAL surface entities are identified from B-rep of solid models in CAD STEP Part21 file. Then, only CYLINDRICAL surface entities with two LINE edges which is tangent with adjacent PLANE are selected for following analysis. The normal direction of the adjacent PLANE is recognized. Convexity test is done based on the direction of the related PLANE surface which is connected a CYLINDRICAL surface. Then, edge blend face is assigned with the attribute as convex or concave based on the result of convexity test. The recognized edge blend faces with attributes will be used in automated machining feature recognition processes.

3.1 Determination of convexity/concavity of an edge or edge blend faces

Convexity of every edge of the object must be determined before other procedures taken for automatic machining feature recognition. Figure 4 shows some of the important parameters used for determining the convexity of an edge.

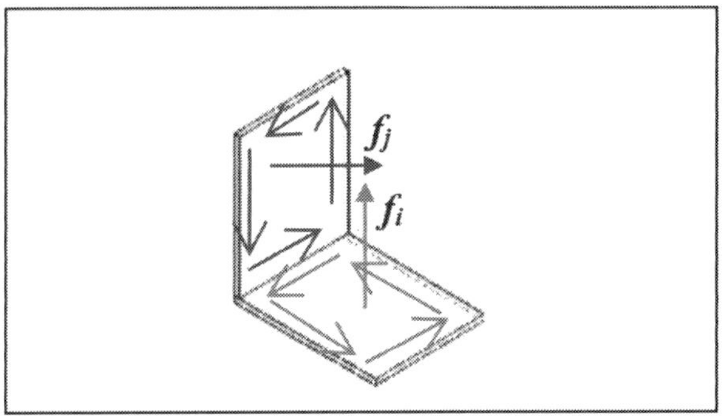

Fig 4. Important parameters for determining the convexity/concavity of an edge or edge blend face

The direction c of the edge shared by two faces can be determined by finding the result of cross product operation between two normal vectors of two faces as shown below.

$$\mathbf{c} = f_i \times f_j \tag{3.1}$$

Taking the edge direction of f_i as a reference,
If **c** direction is same with oriented edge f_i direction,
Then the edge is **CONVEX**
If **c** direction is **NOT** same with oriented edge f_i direction,
Then the edge is **CONCAVE**

The usage of the concavity index for the effective recognition of the concave, open manufacturing features like open pockets, through slots and steps have been presented by [10]. The determination of this index using B-Rep representation is presented in detail by [10]. Besides that, algorithm for the recognition of manufacturing features defined in the STEP-NC standard. However, the recognition of the concave for edge blend faces have been neglected in previous study. As a result, this proposed research has adopted the convexity test for edge, used in determining the concave of edge blend faces.

 DOI: 10.1051/matecconf/201713500033

3.2 A Novel Graph-Based Method For Machining Feature Recognition With Edge Blend Faces

At this stage, a novel method for machining feature recognition have been proposed with considering the edge blend faces in rectangular closed pocket recognition. However, graph based method seldom discussed the concavity or convexity edge blend feature present in the feature recognition process. In this case, concave edge blend face will be assign as **-1** and convex edge blend face as **1**. Lastly, a new graph method which includes the edge blend feature which is mark as darken edge in the graph is presented as shown in figure 5.

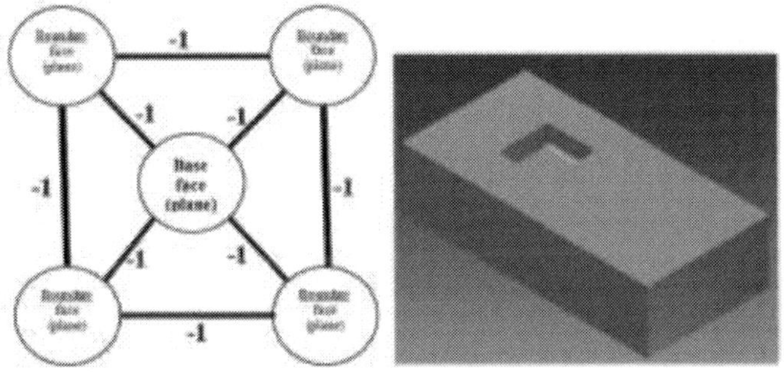

Fig 5. New graph-based method for pocket feature recognition with edge blend face

4 Implementation

The Microsoft Window based MS Access is used to develop the product model database system, and this system able to extract the geometry and topological data from EXPESS entities of CAD STEP Part21 for mapping to relational model. Each extracted entity is mapped to a table with columns for attributes. A column with a unique identifier is present for each instance in each of the table. Attributes with primitive values are stored in place, and composite values like entity instances, selects, and aggregates are stored as foreign keys containing the unique instance identifier. Structured Query Language (SQL) is used for both data definition and data manipulation in this system. One of the advantage of this database is it can be easily migrated into other advanced RDBMS [11]. Figure 6 shows the developed database system used for recognize pocket feature with edge blend face successfully.

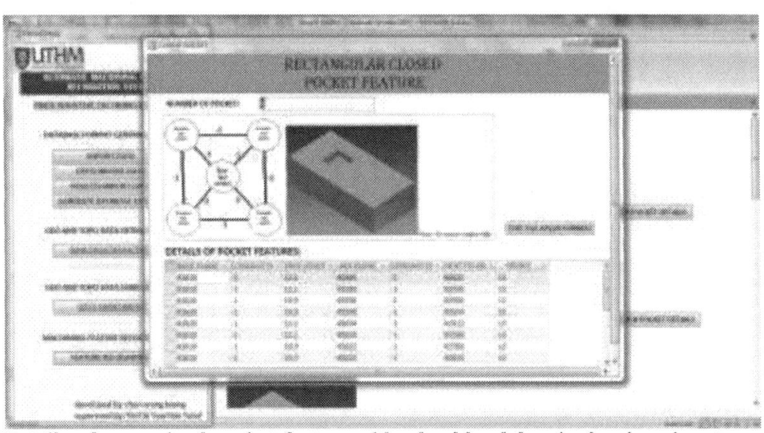

Fig 6. Details of recognized pocket feature with edge blend face in developed system

5 Conclusion and recommendation

In conclusion, this research concentrates on automatic machining feature recognition with edge blend face of geometry and topological data from CAD STEP Part21 file using EXPRESS schema entities are in the backend. Edge blend face (e.g. fillet feature) is one of the important elements have to be considered in machining feature recognition. However, it is seldom discussed in graph based method. In this research, a novel graph based method with edge blend face has been proposed with new table mapping used for machining feature recognition. This implementation provides flexible environment to the people, who are using STEP data and manage the EXPRESS entity data. This recognized machining feature with edge blend face is useful for subsequent manufacturing process, especially in computer-aided process planning (CAPP) and computer-aided manufacturing (CAM).

6 Acknowledgements

This work was supported by the Malaysian Government under Science Fund of Ministry of Science, Technology and Innovation (MOSTI) with vote No S021, Malaysian International Scholarship (MIS) under Ministry of Education Malaysia (MOE), and Universiti Tun Hussein Onn Malaysia (UTHM).

References

1. Grabowik, C., K. Kalinowski, I. Paprocka, and W. Kempa. *A survey on methods of design features identification*. in *IOP Conference Series: Materials Science and Engineering*: IOP Publishing (2015)
2. Verma, A.K. and S. Rajotia, A review of machining feature recognition methodologies. Int. J. Comput. Integr. Manuf. **23**(4): p. 353-368 (2010)
3. Babic, B., N. Nesic, and Z. Miljkovic, A review of automated feature recognition with rule-based pattern recognition. Computers in Industry. **59**(4): p. 321-337 (2008)
4. JungHyun, H., M. Pratt, and W.C. Regli, Manufacturing feature recognition from solid models: a status report. Robotics and Automation, IEEE Transactions on. **16**(6): p. 782-796 (2000)
5. Ansaldi, S., L.D. Floriani, and B. Falcidieno, Geometric modeling of solid objects by using a face adjacency graph representation. SIGGRAPH Comput. Graph. **19**(3): p. 131-139 (1985)
6. Joshi, S. and T.C. Chang, Graph-based heuristics for recognition of machined features from a 3D solid model. Computer-Aided Design. **20**(2): p. 58-66 (1988)
7. Tian, F., X. Tian, J. Geng, Z. Li, and Z. Zhang. *A Hybrid Interactive Feature Recognition Method Based on Lightweight Model*. in *2010 International Conference on Measuring Technology and Mechatronics Automation* (2010)
8. Manafi, D., M. Nategh, and H. Parvaz, Extracting the manufacturing information of machining features for computer-aided process planning systems. Proceedings of the Institution of Mechanical Engineers, Part B: Journal of Engineering Manufacture, (2016)
9. Cui, X., S. Gao, and G. Zhou, An Efficient Algorithm for Recognizing and Suppressing Blend Features. Computer-Aided Design and Applications. **1**(1-4): p. 421-428 (2004)
10. Pobożniak, J., Using Concavity Index For Manufacturing Feature Recognition. Technical Transactions,(7) (2015)
11. Babu, K.S., D.D.N. Rao, D.A. Balakrishna, and C.S. Rao, Development Of A Manufacturing Database System For Step-Nc Data From Express Entities. International Journal of Engineering Science and Technology **2** (11): p. 6819-6828 (2010)

The Dispositional Attribution of Customer Satisfaction through the Juxtaposition of QFD Aand Servqual in Service Industry Design

*Khairul Salleh Mohamed Sahari[1], Hasril Hasini[1], A.M. Megat Hamdan[2] and A.Z Syahmi[1]
[1] Department of Mechanical Engineering, Universiti Tenaga Nasional, (UNITEN)
[2] Department of Mechanical Engineering, Universiti Pertahanan Nasional Malaysia (UPNM)

Abstract. This study has been carried out to analyze the dispositional source of satisfaction through the juxtaposition of Quality Function Deployment (QFD) and Service Quality (ServQual) in the service industry. ServQual is one of the famous tools to measure the customer satisfaction. The customer satisfaction is measure through five dimensions, which are reliability, assurance, tangible, empathy and responsiveness. QFD is one of the mathematical approach to transform the customer needs into technical requirements. The difference between these two models is, ServQual evaluates the priorities of the basic customer needs, which within these five dimensions, which one is the most important and which one is the least important, according to responds obtained from the customer. Meanwhile, for QFD evaluate the priorities of technical requirement of service that will be able to satisfy the customer. The study has been carried out at one customer service centre in Bangi. The study applies ServQual methodology to measure the customer satisfaction after the service has been delivered through its five dimensions. The QFD methodology is used as one of the mathematical approach to transform the customer needs into technical requirement and evaluate the priorities of technical requirement of service that will be able to satisfy the customer. The results show ServQual and QFD ought to be consolidated to get alternate points of view regarding the behavior of the customers. Acted as variables to quantify in terms of the contentment felt by the customers in the service quality. Besides, it has a distinctive methodology to help service industry being able to gauge the satisfaction of customers. Together they give an intense instrument that is not exclusively will reveal whether the customer satisfaction is fulfilled or not, but rather additionally how great the distinction of the service is as well as the level of competence the organization functions. In addition to that, it will let the companies in the service industries know that they should take action in order to enhance the service quality and along these lines to make the customers satisfied. Through the theoretical base questionnaires, as both of the juxtaposition approaches' analysis, the main priorities need to be taken into consideration is 'facilities' in service centre. The service centre needs to ease the customer in whatever their purpose to come to the service centre and make them comfortable enough. While, the factor 'Ease of contact' is become the vital consideration in ServQual before integrating towards QFD. These elements are the improvement needed by the service centre to look into details in order to increase and grab more attention towards customer satisfaction.

1 Introduction

Disposition has been defined in many different ways. Some of them are the state of mind regarding something, mood and feelings. In this manufacturing service case study, the dispositional source of satisfaction is defined as the tendency of individual to react or respond to the situation. There are two traits to express the tendency, which are to view things in either positive or negative way. In service industry as well, the dispositional is relate to the job satisfaction. Equally, important, dispositional approach indicates that individuals differ in their propensity to be enthusiastic about their job or task, the degree of individual trait in the job fulfillment [1]. The dispositional of job satisfaction give impact to the emotions and actions of individual and most of them happen unconsciously. Past analysis has shown there is a connection between dispositional effect, psychology and science. The connection such as the character or behavior, lifestyle, decision making of a person is very important in most organizations or association in order to achieve stability and success especially in socioeconomic impact that will lead to others such as customer.

Nowadays, service organizations are all related to human whether the service provider or the service receiver. For the service organizations to succeed it is important for them to give high quality of service and fulfilling or meet the customer needed. The fulfillment of customer expectation plays a very significant role in service industries as it controls the up and down in sales and profits. A good quality service provided bring the customer loyalty and increase the organization's efficiency [2]. Thus, the level of getting the customer's loyalty and loyal customers depends on how the organizations work to meet the customer's needs and expectations. This study is carry out to classify and investigate the service industry's customer satisfaction by collecting data and information.

Consequently, the customer chooses the service provider and not the other way around [3]. The service organizations must realize the importance of the customers and the service quality in managing their organizations. Second, in current global economy service industry is a very competitive and rough sector. The service providers need to set their goals for customers and just by fulfilled their needs are not enough nowadays. Service providers need to put the customers as their top priority. They need to move beyond to meet customer's needs such as providing opportunities for additional profit in order to survive and thrive in this industry.

Thus, other than just to meet the customer's expectations, service providers need to improve their service quality to ensure better effectiveness and efficiency [4]. Understanding their customer is a matter of concern but how to understand them is another major concern in this industry. Third, detail understanding of the customer's situation, issues and concerns will help organizations to find solution for customer's problems in meeting their expectations and perspectives [5]. This will also improve the organization opportunities and knowledge in delivering their service quality in the future. Based on those reasons, the study is carry out by highlighting the implication of juxtaposition between ServQual method and Quality Function Deployment (QFD) method in measuring dispositional source of satisfaction in service provided. The juxtaposition of these two methods will lead to dispositional attribution of customer satisfaction in service industry design.

*Corresponding author: Fevilia@uniten.edu.my

2 Theoretical Approach

2.1 Customer Satisfaction

Customer's satisfaction has been define in many ways nowadays. Customer satisfaction is view as something that is pleasurable to fulfil [6]. While Kotler [7] defined it when the pre-consumption expectations of customer has been given the customer, satisfaction is including of post consumption assessment about the goods or service quality. In different perception, customer satisfaction is describe as the consumer's assessment that products or services meet or fall to meet customer's expectation [8][9]. There are two types of customer in service industry, which are external customer and internal customer. An external customer is the individual who does not involve in the organization directly while the internal customer is surely the one who does involve directly in the organization such as the employees, stakeholders and shareholders.

More than that, matter such as salesperson, service provider, experience and decision-making can affect the customer satisfaction towards organization. Customer satisfaction is an extremely private evaluation that is affect by the individual expectations. Without customer information and feedback, there will be no clue to design product or delivering service according to customer expectation, thus decreasing the organization effectiveness and losing of customers. The simplest way, which is customer feedback survey, really helps and gives insight about customer expectation and perception but it also has their pros and cons. Dispositional satisfaction of customer satisfaction is all about how the perceptions, views and personality traits regarding the product or service and their reaction. This is often see in situations after consuming or purchasing the product or service by getting feedbacks whether they are satisfied or dissatisfied. Thus, resulting in their action and emotion such as whether they dislike the organization and want to unsubscribe or satisfied enough and resulting loyalty, returns and increasing of customer.

2.2 ServQual

Service quality (ServQual) is an evaluation of how well a provided product or service is in accordance to the customer's objectives. Service industry organization often look at the service quality offered to their customers in order to enhance their service, to recognize problems at a fast rate as well as for better evaluation of customer service. Service quality is an approach to manage business for ensuring fulfilment of customer satisfaction and quality of service delivered to customer by prioritizing them [10].

ServQual model is an empiric model invented by Parasuraman et al. [11] to compare service quality performance with customer service quality needs. The usage of this model is for the gap analysis of service quality performance and the needs of customer's service quality for organization in service industry. It is also call the GAP model. It considers the views of customer of the comparative significance of service attributes thus, prioritizing them.

ServQual evaluates on 10 aspects of service quality, which are reliability, receptiveness, capability, access, courtesy, communication, credibility, security, comprehending the customer and tangibility. Then it narrowed down until it only measures five dimensions, which are reliability, assurance, tangible, empathy and responsiveness. The purpose of these dimensions is to measure the gap between customer expectations and experiences. The five dimensions of ServQual are as follow:

1. Reliability: the ability to achieve the promised service dependably and accurately
2. Assurance: the knowledge and courtesy of employees and their ability to express trust and confidence
3. Tangibles: the appearance of physical facilities, equipment, personnel and communication resources
4. Empathy: the provision of concerned, individualized attention to customers
5. Responsiveness: the readiness to help customers and to provide prompt service

Service quality becomes more and more popular in service industry nowadays as it involves and give big impact in expenses, profitability customer satisfaction and customer retention of organization. Most common reasons why organization implements this model is that to evaluate the company's service quality through the five dimensions of ServQual. Besides that, organization often use them to determine and check the customer expectation and perception that may be different in their evaluation of organization service performance. In this competitive era, organization needs to develop and succeed by keeping track about the customer expectation and perception to maintain the customer loyalty.

It is indeed one of the best model in closing the gap between the consumer and manufacturer by process such as communication with customers and by keeping track the customer expectation and perception and comparing the score of service quality within the industry.

2.3 Quality function Deployment (QFD)

Quality function Deployment (QFD) is not only a quality tool, but also a significant planning tool, is defined as the method to exceed customer expectation by turning and planning the voice of customer (VOC) into the quality properties of services or products [12]. Akao,[13] the founder of QFD itself stated that QFD is an approach to satisfy the customer by transforming the customer's demand into design targets and major quality assurance points to be used throughout the production phase and QFD is the way to ensure the quality of a design while in the designing stage.

QFD helps to create operational details of the requirements, which may be unclear when first expressed. It prioritizes each products or services characteristic while at the same time sets the development objectives for the products or services. Likewise, QFD is a structured process in planning and designing a product or service and it is a process of restructuring the existing product and service. It emphasized on understanding of the needs and requirements of customers. Then the needs and requirements are transform into characteristics of product or service. There are four phases of QFD process, which is product planning, part development, process planning and production planning.

In the phase of product planning, the House of Quality (HOQ) is implementing. The same phase is the process of transforming the needs and requirements of customer into the product technical requirements. It links the customer requirements with product attributes. Some said there are few stages and steps of generating HOQ. According to Becker &Associates [14], there are 12 steps of generating HOQ as depicted in Figure 1.

The usage of QFD can be apply in various kinds of industry. In the service industry, implementation of QFD can be vital to ensure the business' success and sustainability. In service industry, QFD refers to determine the customer needs and the requirements to satisfy them. After requiring the information from customer, the process continues with transforming the needs into design characteristics. QFD is a customer driven tool in executing total quality management. The main goal of QFD is to maximize the customer satisfaction thus giving the opportunities to organization to reduce costs and the production time, improving the revenues and lastly to thrive in the competitive industry by improving and produce product or services [15].

3 Methodology

The quality attributes of ServQual dimension (Table 1) is used to identify and investigate the service industry's customer satisfaction that required the concept of QFD to identify the strengths and weaknesses of quality attributes related to the service given from the customers' viewpoints [16]. The integration between ServQual dimensions (customer requirement) and technical requirement in HOQ will give an impact how the evaluation of both conceptual approaches would come out with the gaps of satisfaction related to the disconfirmation theory [16]. These asymmetries are use as the customer satisfaction improvement required against the quality of service provided.

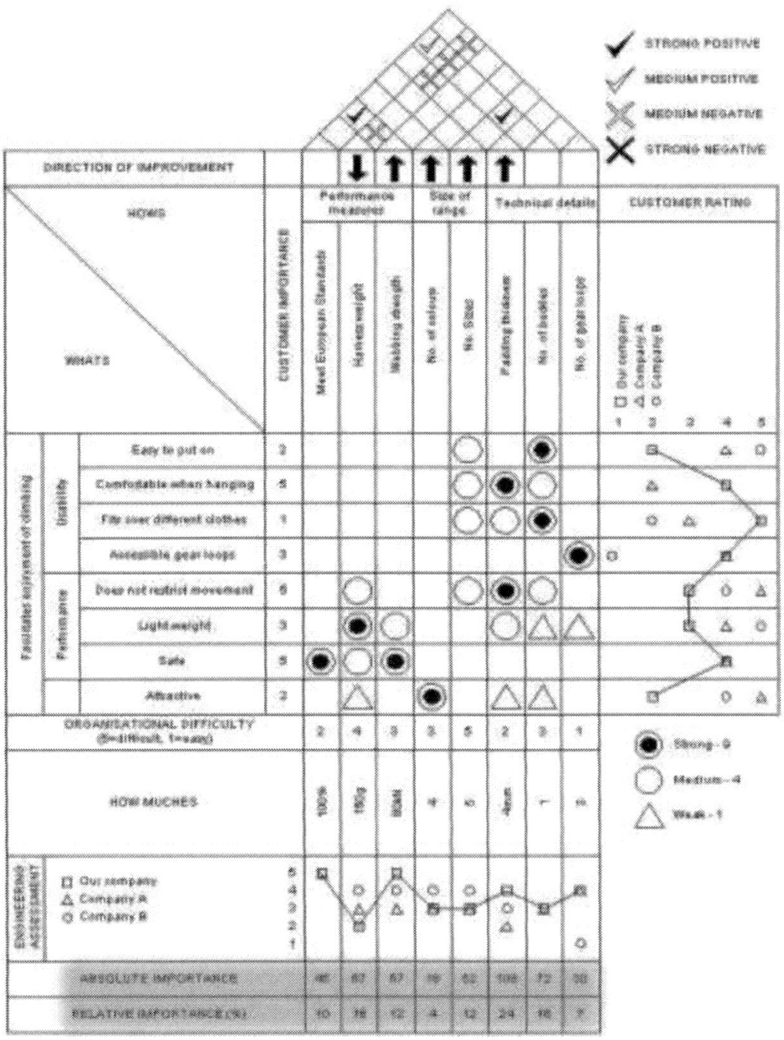

Fig. 1. House of Quality [17]

DOI: 10.1051/matecconf/201713500034

Table 1. ServQual dimensions

ServQual Elements (variables)	Operationalization Items	Question number in questionnaire
Reliability	•Consistence in quality of services given •The degree of trust in service •Attention to specifics of the service delivery by the staff •Ease of contact	1-5
Assurance	•Assurance that the rate charge is fair and facilities provided is enough for customers •Assurance that the location is suitable and parking lot is enough	6-10
Tangibles	•Visually appealing external outlook of the staff •Having appropriate places and facilities for effective service •The facilities are functioning well •Cleanliness and safety of building, places and facilities	11-15
Empathy	•Staff have knowledge, polite and necessary service skills •Approachability •Staff pay attention to individual needs	16-20
Responsiveness	•Competence •Service given tally with company's advertisement •The service system for booking, registration and payment is easy and reliable	20-25

4 Analysis & Discussion

4.1 ServQual Result

The analysis results are in the form of value, it stated the ranking of five dimensions of ServQual according to the most important until the least important dimension. Table 2 concludes the result as the sum of average of each dimensions in ServQual. The lower the rating given the higher the satisfaction. This is because the items in likert scale were gauged on a five-point scale ranging from "1 = strongly agree" to "5 = strongly disagree". Respondents have rated the importance of the dimensions as reliability as the most vital attribute. While, Table 3 and clearly depicted in Figure 2 shown the value of ServQual's dimension ranking, which is based on customers' priority point of view that considered as mostly concerned by the customers. The lower the score rated by the respondents the higher the rank of importance of question.

Table 2. Average value of each ServQual dimensions

No.	Dimension	average
1	reliability	2.3
2	assurance	2.3
3	tangible	2.5
4	empathy	2.6
5	responsiveness	2.6

Table 3. Value of ServQual's dimension ranking

Dimension	Rank	Average rating
Reliability	1. Ease of contact	1.6
	2. Service given tally with advertisement	1.8
Assurance	3. Operating hour	1.9
	4. Staff pay attention to individual needs	1.9
Tangible	5. Location of service centre	2.0
	6. Parking lot	2.1
Empathy	7. Accessibility of staff when needed	2.2
	8. Consistency of service	2.3
Responsiveness	9. Up-to-date equipment	2.4
	10. Cleanliness and safety	2.4

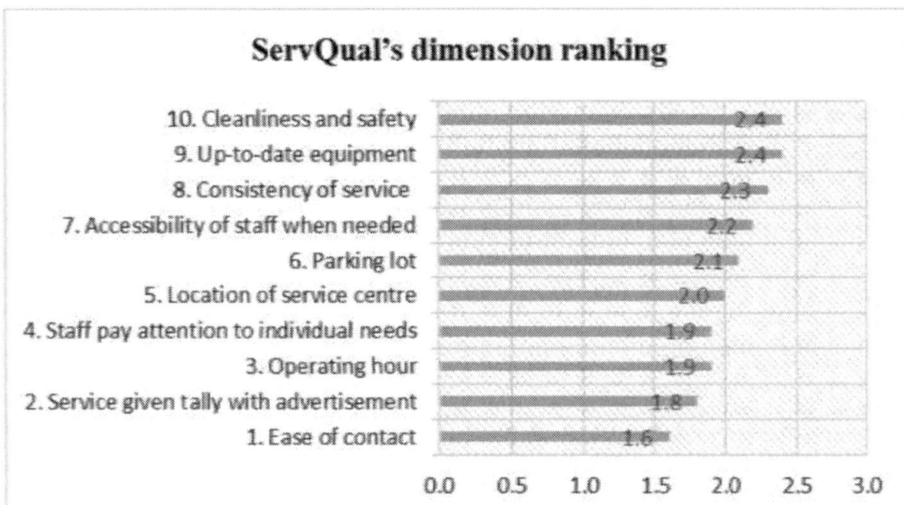

Fig. 2. ServQual's dimension ranking

Based on the ranking, the top two factors that need to be considered as the first priority is in Reliability dimension, that give the feeling of '*the ability to achieve the promised service dependably and accurately*' are:

1. Ease of contact
2. Service given tally with advertisement

4.2 QFD Results

It founds that from the previous ServQual analysis, reliability dimension has been rank as the first priority, so the Table 4 below is the ranking of importance in customer requirement (from ServQual dimension) in HOQ perspective of view.

Table 4. Importance Rating on Customer Needs

		Importance
Reliability	Ease of contact	10
	Consistency of service	8
Assurance	Location of service centre	5
	Parking lot	6
Tangible	Up-to-date equipment	9
	Cleanliness and safety	1
Empathy	Operating hours	3
	Staff pay attention to individual needs	4
Responsiveness	Service given tally with company's advertisement	2
	Accessibility of staff	7

The technical requirement part has been identified on the foundation of service elements. The key question in this step is "how" the staff would be able to provide the required service to their customer. In this study, three important issues are considered which are processes, people and physical evidence [18]. These three issues are used to define the technical requirement as a top roof of House of Quality (Figure 3).

The importance and percentage weight of service design features are measured by using the relative importance values and the relationship matrix developed. This calculation process intertwines customer's needs with service technical requirements so that resulting value gives the percentage weight of each of the requirements as compared to customer's needs (Table 5). The impact relationship between ServQual (customer requirement) and technical requirement in HOQ's perspective can be seen as depicted in Figure 4.

Courtesy	Queue system	Payment system	Self-service kiosk	Reconnection & termination	Payment & service counter	New power supply application	Complaint handling	Customer details & information	Staff behaviour	Staff knowledge	Information counter	Guideline	Facilities	Safety & security
1	2	3	4	5	6	7	8	9	10	11	12	13	14	15
Processes				**People**					**Physical**					

Fig. 3. Technical requirement as top roof evaluation in HOQ

Table 5. Percentage Weightage of Service Technical Requirements

No of attributes	1	2	3	4	5	6	7	8	9	10	11	12	13	14	15
Importance	70	27	99	69	33	25	25	101	13	14	14	12	108	108	63
%	9	3.5	13	8.8	4.2	3.2	3.2	13	1.7	1.8	1.8	1.5	14	14	8.1

		Column		1	2	3	4	5	6	7	8	9	10	11	12	13	14	15
		Functional Requirements / Customer Requirements (Explicit and Implicit)		Courtesy	Queue system	Payment system	Self service kiosk	Reconnection & termination	Payment & service counter	New power supply applicatio	Complaint handling	Customer details & informat	Staff behaviour	Staff knowledge	Information counter	Guideline	Facilities	Safety & security
			Importance	Processess							People				Physical			
1	RELIABILITY	Ease of contact	10								●							○
		Consistency of service	8	○		●		▽					▽	▽				
2	ASSURANCE	Location of service centre	5				○									●		○
		Parking lot	6														○	○
3	TANGIBLE	Up-to-date equipment	3		○	○	○				▽	▽					●	
		Cleanliness and safety	1	▽													○	
4	EMPATHY	Operating hours	3				●											
		Staff pay attention to individu	4	●				▽	▽	▽		▽			○			
5	RESPONSIVENESS	Service given tally with advert	2	▽							▽		○	○			○	
		Accessibility of staff	7	▽				○	○	○						●		
			Importance	70	27	99	69	33	25	25	101	13	14	14	12	105	105	63
			%	9	3.5	13	8.8	4.2	3.2	3.2	13	1.7	1.8	1.8	1.5	14	14	8.1

Fig. 4. The relationship of ServQual (customer requirement) and technical requirement in HOQ perspective

Based on the percentage ranking, the top two factors that need to be considered as the first priority are:
1. Facilities (14%)
2. Guideline (14%)

5 Conclusion

ServQual and QFD ought to be consolidated to get alternate points of view regarding the customer behaviour and distinctive methodologies of how a service industry can gauge the customer satisfaction. Grounded on the theoretic background portrayed, the observational discoveries assembled and the later analysis, these models consolidated provide another extremely intriguing perspective. They display the customer satisfaction model and the variables that customer discover more significant to enhance the nature of service quality. ServQual gives this estimation and great information regarding the quality of service and the characteristics, customer needs and desires, after QFD gives the connection between these needs and the service technical requirements that results in an exact of focuses to enhance.

Joining the best of every model it gets to be distinctly simpler to recover the essential data and after that applying all the data accumulated into a technique to quantify the satisfaction. As both of the approaches' analysis, the main priorities need to be taken into consideration are facilities and guideline in service centre. The service centre needs to ease the customer in 'warm welcoming' in whatever their purpose to come to the service centre and make them comfortable enough. The expansion of waiting area, expansion of parking lot and place signboard for every facilities can be considered and need more attention to satisfy them.

While before integrating, the priority for ServQual itself rated that, the factor 'Ease of contact' is become the vital. These required the further improvement action by the company to fulfil those requirements in satisfaction towards services provided. The previous objective to propose the improvement of the customer satisfaction by using the juxtapostion of ServQual approach and QFD is achieved.

6 Acknowledgement

A special tribute to Universiti Tenaga Nasional for funding this research under the UNITEN Internal Research Grant.

References

[1] A. Erez, & T. A. Judge. Dispositional source of job satisfaction: The role of self-deception (CAHRS Working Paper #94-14). Ithaca, NY: Cornell University, School of Industrial and Labor Relations, Center for Advanced Human Resource Studies (1994).

[2] M. Ramees Rahman, & Safeena, P. K. Customer Needs and Customer Satisfaction. In *Theeranaipunya – A Capacity Building Training Programme Equipping the Fisherwomen Youth for the Future* Kerala, India: Central Marine Fisheries Research Institute. Pp. 259-262 (2016).

[3] M. Colgate & M. Norris. Why Customers Leave - or Decide Not To Leave - Their Bank. *University Of Auckland: Business Review*, Vol. *2(*2), Pp. 40-51(2000).

[4] N. Nabi. Customer Expectations of Service Quality: A Study on Private Banks of Bangladesh. *World Review of Business Research*, Vol. 2(2), Pp.172-186 (2012).

[5] A. Bollen & Emes, C. Understanding Customer Relationships: How important is the personal touch? London: Ipsos MORI. (2008).

[6] R. L. Oliver. *Satisfaction: A Behavioural Perspective on the Consumer*. New York, NY: McGraw-Hill (1997).

[7] P. Kotler. *Marketing Management: Analysis, Planning, Implementation and Control* (1st ed.). Englewood Cliffs, N.J.: Prentice-Hall (1991).

[8] R. L. Oliver & J. E. Swan. Consumer Perceptions of Interpersonal Equity and Satisfaction in Transactions: A Field Survey Approach. *Journal of Marketing*, Vol.53 (2) (1989).

[9] Y. A. Yi. Critical Review of Customer Satisfaction. In V. Zeithaml (Ed.), *Review of Marketing,* Chicago: American Marketing Association. Pp 68-123 (1991).

[10] A. Shahin. SERVQUAL and model of service quality gaps: A framework for determining and prioritizing critical factors in delivering quality services. *Service quality – An introduction* Andhra Pradesh: ICFAI University Press. Pp. 117-131 (2006).

[11] A. Parasuraman., V. Zeithaml, & L. A Berry. Conceptual Model of Service Quality and Its Implications for Future Research. *Journal of Marketing*, Vol. 49(4) (1985).

[12] L. Bernal., T. Byrnes., U. Dornberger., & A. Suvelza. *Quality Function Deployment (QFD) for services*. Leipzig: University of Leipzig (2009).
[13] Y. Akao. Quality function deployment (1st ed.). Cambridge, *Mass.: Productivity Press*.(1990).
[14]Becker and Associates. Steps to the House of Quality (2000).
[15]M. Moradi & S. A. Raissi. Quality Function Deployment Based Approach in Service Quality Analysis to Improve Customer Satisfaction. *International Journal of Applied Operational Research*, Vol 5(1), pp. 41-49(2014).
[16] F. Syaifoelida., P. Azman., I. Ismail., M. R. Eqwan. The Quantification of Two Grading Elements in Satisfaction based on ServQual Approach by Using IPA and Kano Method: Automotive Industry. *International Journal of Business and Economic Affairs (IJBEA)*, Vol. 1(1), pp. 60-66 (2016).
[17] A. J. Lowe & K. Ridgway. Quality Function Deployment. University of Sheffield. (2000).
[18] N. Seth., S.G. Deshmukh & P. Vrat. Service Quality Models: A Review. *International Journal of Quality Reliability Management*. Vol. 22(9), pp. 413-449 (2005).

Evaluation of Business Excellence among Halal Certified Food Manufacturers in Malaysia

Nor Ratna binti Masrom[1], *Raja Zuraidah binti Raja* Mohd Rasi[2], *Badru At Tamam bin* Daut[3]

[1]Faculty of Technology Management and Technoprenuership, Universiti Teknikal Malaysia Melaka, Melaka, Malaysia
[2, 3] Faculty of Technology Management and Entrepreneurship, Universiti Tun Hussein Onn Malaysia, Johor, Malaysia

Abstract. Business excellence is a key tool to improve the efficiency of the company. Through the business excellence model, the organization can execute their strategies to determine, forecast and survive in highly competitive market. This paper presents on identifying those elements in business excellence model (leadership, strategy and planning, customer focus, measurement, analysis, and knowledge management, operation focus) this paper provides empirical support for halal food certified companies reviewing and improvement. Survey has been done and analyzed among halal food certified manufacturers in Malaysia. Thus, seven categories with 35 key elements regarding American MBNQA and eight equations are examined. The main conclusions are that, strategy and measurement, analysis and knowledge management are the main weakness of the halal food certified manufacturers in Malaysia. Overall, halal food certified manufacturers in Malaysia is at the silver level of excellence based on the calculation.

1 Introduction

Worldwide, the halal product industry is highly in demand because Islam is the world's second largest religion (Hill, 2008). In the year 2010, the world's Muslim population is approximately around 1.6 billion, and this number is expected to increase by 19.8% to 20% in 2050 (DeSilver, D and Masci D,2017). The global Halal food market is currently one of the fastest growing market segments. Halal food refers to items that are allowed for consumption by Sharia law. Currently, halal food and beverages on the market are worth US$1.2 trillion, and this is expected to rise to US$1.9 trillion in 2021 (Asia Pacific Food Industry, 2016). Although numerous halal food manufacturers are available worldwide, the current supply of Halal food is still insufficient to cater the ever-increasing demand (HDC, 2014).

* Corresponding author: norratna@utem.edu.my

This scenario of high demand and lack of supply provides prospective opportunities to venture into this industry, thus, resulting in countries all over the world including the non-Muslim countries showing their interest for the halal industry. The promising profit in this industry has resulted in many food companies venturing into halal food.

MS1500 is the standard developed to enable businesses to have a clear view of halal food production and quality maintenance. Positive image of halal food has been increased as to product safety, quality, and hygiene because of the spread of contagious infections such as birds' flu or prion disease (Muhammad et al. 2009). Systems and processes of an organization, which further leads to the improvement of performance, thus, creating value for investors.

Besides the increasing of awareness for wellness and hygiene, the awareness about animal welfare and taste has contributed to the increasing attention given towards halal and kosher products (Treena, 2008). Business Excellence (BE) focuses on developing and strengthening the management Apart from setting up a quality system, BE emphasizes on attaining excellence in all aspects of an organization which includes leadership, strategy, customer focus, information management, people, and processes. Additionally, achieving superior business results is the ultimate goal.

The Malcolm Baldrige National Quality Award (MBNQA) functioned to recognise companies that show outstanding performances. Its excellence framework is the most comprehensive quality framework. The MBNQA model which is the basis of US quality award now functions as a foundation for quality awards in many countries including Malaysia. It continuously concentrates on the improvement of all components of the organization, which includes tasks and activities at any level (Nabitz and Klazinga, 1999).

Productivity in Malaysia is low when compared with other countries like Singapore, United States, Taiwan, and South Korea. Reports by the Malaysia Productivity Corporation (2016) identified huge disparities between productivity and business efficiency in Malaysia. A great deal of work is required to overcome the existing barrier between business efficiency and productivity in Malaysia

Our main aim is to provide empirical support for the Quality Management review and, where possible, suggest ways to improve it. To do so, two questions were formulated as follows:

- To find the strongest and weakest of quality management in halal certified food manufacturers in Malaysia

- To find the level of excellence in halal certified food manufacturers in Malaysia

By answering these questions, we also sought to test the validity of a business self-assessment questionnaire. This paper will discuss the importance of the research, survey design, data collection, method used to carry out the assessment, some key results and critical lessons learned, suggestions, conclusions and implications for future research.

2 Literature Review

This section discusses three important topics: halal food industry, halal food certification (MS1500) and quality management.

2.1 Halal Food Industry

Muslims represent an estimated 23% of the global population or about 1.8 billion consumers with an average growth rate of 3% per annum. If the trend continues, Muslims will be 26% of the world's total projected population of 2.2 billion in 2030. Halal market is not limit to Muslim only but it has been accepting by the non-Muslim. The demand for the halal certified products has gain popularity among them due to high quality, safe and ethical products. Halal food market has been estimated around USD667 million and represents 20 % of the entire global food industry. Many economists believe that based on four prevalent trends. Firstly, Islam is now the fastest growing religion in the world thus it will make big global demand for halal products. The annual growth in consumption of halal food is estimated at 16%. Secondly, the growing trend of consuming. Next the halal consumer power has been increasing because of the increasing number of Muslim and rising of their disposable income. Lastly, the increasing of awareness of consuming halal food among Muslims Halal food is becoming an increasing part of the Western diet and has become a multi-billion-dollar global industry involving multinationals like Tesco, Unilever and Nestle who have aggressively expanded their halal-certified product lines. Halal is originated from Arabic word which means lawful and permissible according to Islamic Sharia law which refers to the teachings of Al-Quran and exemplary conduct of Prophet Muhammad (Peace and blessings of Allah be upon Him) as quoted in Hadith. The opposite of Halal is Haram which means forbidden, Roland (1996).

2.2 Halal Food Certification (MS1500)

According to Department of Standard Malaysia, 'standards are required by industry, government and consumers to facilitate both domestic and international trade; enhance industrial efficiency and technological development; enforce regulations for public safety, health, environment protection and prevention of deceptive practices.'

This means that halal certified food manufacturers have to follow the standard of MS 1500 in order to produce their products. In light of Clause (g) of Item 8, Federal List, Ninth Schedule of the Federal Constitution expresses that trade, commerce and industry additionally include foundation of measures of quality of products made in or sent out from the Federation. MS 1500 was developed which The Minister of Science, Technology and Innovation is responsible for the approval of the draft Malaysian Standards (MS). Malaysian Standards (MS) are consensus documents developed by Standards Development Committees (SDC) within the Malaysian Standards Development System and approved by the Minister of Science, Technology and Innovation in accordance with Act 549. Malaysian Standards (MS) specifies the optimum requirements of quality and safety for voluntary use by the public. A standard becomes mandatory when a regulatory agency enforces its use through the relevant Acts or Regulations.

2.3 Business Excellence

MBNQA is a system that manage the organization and key activities processes to get results. There are seven criteria categories that are designed to endow organizations with an integrated approach to performance excellence. These criteria categories are (1) leadership, (2) strategic planning, (3) customer and market focus, (4) measurement, analysis, and knowledge management, (5) workforce focus, (6) operation focus, and (7) results.

MBNQA is a useful indicator of successful TQM implementation (Bailey, 2011). Lau *et al.* (2004) found out that firms that implemented quality systems according to the MBNQA criteria actually practiced TQM. In addition, Prajogo (2005) used the MBNQA criteria to evaluate the relationship between QM and quality performance and agreed that the MBNQA content was a valid and good representation of TQM constructs. Sila and Ebrahimpour (2003) further added that the MBNQA criteria served as helpful guidelines for quality management implementation. National productivity organization of Malaysia introduce Prime Minister Quality Award (PMQA) based on MBNQA. Since the 1930s, QM has gotten the consideration of scientists and experts (Abrunhosa and Moura E Sa, 2008). QM and business excellence turned out to be exceptionally well known point and has been utilized for picking up and managing an aggressive edge in the worldwide market (Quazi and Padibjo, 1998; Yusof and Aspinwall, 2000; Lakhal et al., 2006; Sharma and Kodali, 2008; Pinho, 2008). QM as key device to enhance hierarchical execution in both substantial and private companies in any piece of the world as said in Fenning et al. (2008).

3 Methodology

The survey was done by distributing it among 200 halal certified food manufacturers. a response rate of 132 is required. However, due to few constraints, the useable response was 100. Although it is not 100% representative of the population, it is acceptable for the purpose of exploratory research.

The degree of each indictor was determined by the respondents using a 5-point Likert scale as a self-assessment survey. The respondents assessed the components of their organizations concerning each questionnaire item in five level with 1 being very bad and 5 very good. The key elements are focused on seven major categories leadership, strategic planning, customer focus, measurement, analysis and knowledge management (MAKM), operation focus, workforce focus and operational performance of MBNQA 2015 criteria.

This research is using IBM® SPSS® Statistics 22 (SPSS) to find the weakness and strengths of the companies and to find the level of excellence, the calculation developed by Gomes d SJ, et al (2002) as follows:

$$\text{Leadership} = (L_1+\ldots+L_4)*CV_L/5* \Delta_{ke_L} \quad (1)$$

$$\text{Strategic} = (S_1+\ldots+S_5)*CV_s/5* \Delta_{ke_S} \quad (2)$$

$$\text{Customer} = (C_1+\ldots+C_5)*CV_c/5* \Delta_{ke_C} \quad (3)$$

$$\text{MAKM} = (M_1+\ldots+M_5)*CV_M/5* \Delta_{ke_M} \quad (4)$$

$$\text{Workforce} = (W_1+\ldots+W_6)*CV_W/5* \Delta_{ke_W} \quad (5)$$

$$\text{Operation} = (O_1+\ldots+O_4)*CV_O/5* \Delta_{ke_O} \quad (6)$$

$$\text{Operational Performance} = (OP_1+\ldots+OP_6)*CV_{OP}/5* \Delta_{ke_{OP}} \quad (7)$$

Above equation is to find each categories achievement, L_1 and S_1 are the key elements of leadership that the total of number 5 that the respondent answered at question L_1 and elements of strategic planning that the total of number 5 that the respondent answered at question S_1 respectively. They were the independent variables and these principles are applicable to all the equations. The scoring of criteria is based on MBNQA, CV_L=120,

CV_S=85, CV_C=85, CV_m=90, CVw=85, CV_O=85, CV_{op}=450, which will bring total of 1000.

For normalization,5* ΔkeL, is the maximum value can be allowed in leadership, where 5 represent excellent. (1 is being very bad and 5 is excellent), keL reprensent the leadership`s number of key elements (which is all the independent variables with 5 multiply by independent variables) and this formula is also applicable to all the criteria. For performance of the halal certified food manufacturers, f(Pi(c)) f(Pi(c)) = Leadership + Strategy +Customer +MAKM+Operation +Operational Performance.

4 Results and Discussions

Among the seven criteria, respondents of the study have the propensity in emphasizing operation focus criteria, followed by leadership. Thus, this indicates that operation focus criteria and leadership criteria are the two areas which were perceived by the respondents to be of the greatest importance in the company. Conversely, strategy and MAKM criteria has the lowest mean.

For performance of the halal certified food manufacturers is 823.7 Halal certified food manufacturers performance is at the silver level of excellence based on AT&T Co with permission. The Level of excellence applied from George, S. and Weimerskirch, A. (1998) as following table.

Table 1. Level of excellence

Award/point score	Description
Gold (876-1000 points)	Outstanding effort result in all categories. Effective integration and sustained results. National and world leaders
Silver (751-875 points)	Effective efforts in all categories and outstanding in many. Good integration and good to excellent results in all areas. Many industry leaders
Bronze (600-750 points)	Evidence of effective efforts in most categories, and outstanding in several. Deployment and result show strength, but some efforts may lack maturity. Clear areas for further attention
Crystal (500-600 points)	Evidence of effective in many categories and outstanding in some. A good prevention-based process. Further deployment and results needed to demonstrate continuity

5 Conclusions

Strategy is a vital element towards an organization's success. By holding people accountable, insisting that they deal with tough issues and helping them manage dissent, leaders can help make leadership happens. Organizations and teams that learn to harness conflict and use it constructively come up with more creative ideas and innovative solutions to fulfil the customer needs. Healthy conflict and debate are essential precursors for organizational learning and growth. Sadly, in most organizations, this adaptive work is usually avoided, more dirt is swept under the carpet, and the organization suffers.

6 Acknowledgement

The author is grateful to Malaysian Ministry of Higher Education (MoHE) under Fundamental Research Grant Scheme (FRGS) to the Office of Research, Innovation, Commercialization and consultancy (ORICC) Universiti Tun Hussein Onn Malaysia (UTHM) under Vot 1430.

References

Ab Rahman, L. (2008, January). Shariah and Malaysia Halal certification system. In Paperwork presented in Halal Food Seminar.

Ab Talib, M. S., & Johan, M. R. M. (2012). Issues in halal packaging: a conceptual paper. International Business and Management, 5(2), 94-98

Da Silva Jonas, G., Kikuo, N., & Tadashi, O. (2002). Evaluation of TQM in Japanese industries and validation of a self-assessment questionnaire. The TQM magazine, 14(5), 318-328.

Desilver, D and Masci D., World's Muslim population more widespread than you might think. (Jan 23, 2017) retrieved from http://www.pewresearch.org/fact-tank/2017/01/31/worlds-muslim-population-more-widespread-than-you- might-think/

Economic Developments in 2015. (2016, March 23). Retrieved from http://www.bnm.gov.my/files/publication/ar/en/2015/ar2015_book.pdf

George, S. and Weimerskirch, A. (1998) Total quality management: Strategies and Techniques proven at today's most successful companies. Place: New Yourk.Chichester, Wiley.

Direct Investment Casting For Pattern Developed By Desktop 3D Printer

O.M.F. Marwah[1,3,*], *M.S.* Shukri[1,3] , *E.J.* Mohamad[2] , *M.A.* Johar [1], *R.H.A.* Haq[1,3] , *R.K.* Khirotdin[1,3]

[1]Department of Manufacturing and Industrial Engineering, Faculty of Mechanical and Manufacturing Engineering, Universiti Tun Hussein Onn Malaysia, 86400 Johor, Malaysia

[2]Department of Mechatronics and Robotics Engineering, Faculty of Electrical and Electronic Engineering, Universiti Tun Hussein Onn Malaysia, 86400 Johor, Malaysia

[3]Additive Manufacturing Research Group, Faculty of Mechanical and Manufacturing Engineering, Universiti Tun Hussein Onn Malaysia, 86400 Johor, Malaysia

Abstract. Development of RP technologies has encouraged rapid study on portable 3D Printer in which there are varieties of portable 3D printer machines in market. Nevertheless, less reports regarding its consumption of fabricated pattern to be used in direct investment casting. This study has focused on development of ABS P400 pattern in terms of collapsibility behaviour which has capability to be used as sacrificial pattern in direct investment casting. In addition, the internal built structures have been built into two designs such as square and polygon patterns respectively. The patterns were constructed in semi cylindrical geometry which comes with one side opened and one side closed together wrapped with 4 mm of ceramic shell. This experimental were conducted with variation temperature starting from 30°C until 150°C with increment of 5°C per minutes while for the numerical simulation, the temperature selected was between 30 °C to 120 °C with 10 °C increment per minutes. Moreover, the observation was made that glass transition temperature, Tg happened near 110°C. It was observed that the shell cracking happened on the ceramic shell. Other than that, the polygon pattern tends to collapsed inwardly rather than square pattern during the burnout process. A result also shows that, there is significant amount of stress reduction on both square and polygon which was 45 % respectively. Besides that, the amount of strain on pattern itself has shown about 9% reduction. Moreover, there is greater difference in terms of ceramic shell strain reduction which was 38% for square and polygon patterns respectively. Lastly, there is 11 % reduction of strain when compared square and polygon in terms of axial strain on ceramic shell

* Corresponding author: mdfaizan@uthm.edu.my

1 Introduction

The perception of RP to be implemented as sacrificial master pattern in Investment Casting (IC) was dated back in 1989 [1]. Initially, the block mould was used in IC, and the concept is to build the RP parts that required no additional finishing. With the innovation of RP to be implemented in IC process, the hard tooling can be eliminated. Concurrent phenomenon, the application of RP in IC was driven by prospects of reducing costs and manufacturing times as well as assuring high quality products [2].

Several RP machine have been developed since last 10 years such as Stereolithography (SL), Selective Laser Sintering (SLS), Laminated Object Manufacturing (LOM), Three Dimensional Printing (3DP), Fused Deposition Modeling (FDM) and Multijetmodeling (MJM) [3-4, 11] . Due to its capability, RP techniques have emerged at the leading edge in the casting process for manufacturing of intricate parts whereby better dimensional accuracy and fine surface were requisite. Intensification of RP technique currently pays attention to desktop 3D Printer machine which similar to FDM since considered as the cheapest in the market. Nevertheless, less reports regarding the pattern printed by the desktop 3D printer as well as its capability to be used as a sacrificial pattern in IC to generate metal cast parts. This study focused on the stress developed by the portable FDM pattern which act as a direct sacrificial pattern in the IC process for fabrication of metal parts whereby, temperature and stress are the main keys to be watched.

During the burnout process, the shell will acts as stress towards pattern due to the expansion of the pattern when heat is applied. Therefore, the shell cracking occurs when the stress-strain of pattern is greater than shell resistant [5]. The consequence to this phenomenon is thicker shell induced greater shell stresses towards pattern. In addition, another criteria which is vital in determining the collapsibility of the pattern itself is the web build internal pattern structure. If the shell rupture temperature is lower than the pattern buckling temperature and the resin glass transition temperature, the cracking of the ceramic shell will occurs [6-7]. Besides that, the glass transition temperature (Tg) and Modulus of Rupture (MOR) is a relationship that significant for every polymer material when heat is presence. By using polymer materials as sacrificial pattern, the Tg is a reference to understand how the shell cracking occurs during the burnout process [6]. There will be a certain point of temperature when the pattern expands vigorously and reached above Tg with stress is above than MOR of ceramic shell, the shell will cracks and failed. This occurrence is related to prediction of cracking in the ceramic shell.

Acrylonitrile butadiene styrene (ABS) is a thermoplastic group in a combination of monomers of Acrylonitrile, Butadiene and Styrene. Besides that, ABS basic molecule consists of three plastic to form composite group characteristics. The wide range of using ABS materials as sacrificial pattern has been attracted reseacher due to the high impact strength, good dimensional stability, good chemical resistance, ability to prevent the flow of electricity and easily to process [10]. Besides that, ABS has a mechanical properties that are strong and hard even at temperatures as low as -40 ° C. Moreover, it also has a high thermal resistance even at fluctuated temperatures as well as having the glass transition temperature (Tg) which is 110 °C. Therefore, further investigations were undertaken by using ABS materials as a sacrificial patterns in IC process by using desktop Odyssey X2 3D Printer.

2 Internal RP Pattern Structure Development

A new internal structure for investment casting(IC) for desktop 3D Printer was designed and compared between one open and one closed half cylindrical shape. Firstly, the new internal structure pattern was design with square 90° and polygon styles. Fig. 1 (a) & (b) shown the square and polygon model structures open ended. This built styles has been chosen for easier the examination of collapsibility behavior during the burnout processes. The model built with two different web structures to ensure which of both structures has tendency to collapse inwardly during the burnout process if and only if the ABS P400 does not expanding further after the glass region temperature (Tg). The internal structures represented for the ABS P400 (blue) and ceramic shell (yellow) were constructed with diameter thickness of the internal structure was 1 mm, having height of the cylinder which was 60 mm and the inner diameter which was 50 mm.

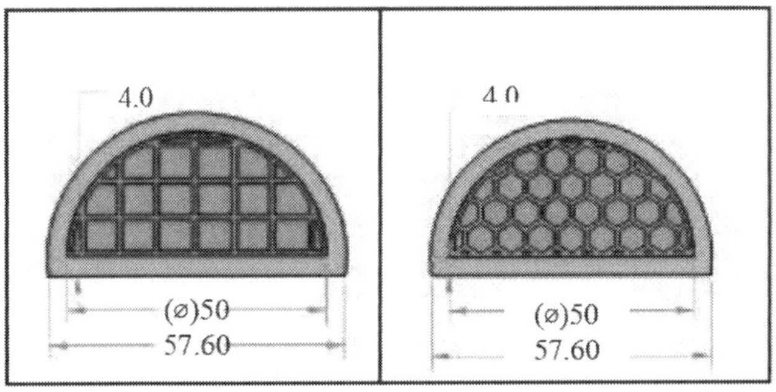

a) Square b) Polygon

Fig. 1. (a) Square and (b) polygon interanal web pattern structure.

This two types of internal patterns in which square and polygon built style that have one opened ended and one closed together wrapped with the 4 mm of ceramic shell thickness. This constructed style was different from the previous Quickcast 2.0 for external layout in which in this paper it was built with half cylinder compared to the full cylinder style. The simpler built of the pattern is to ensure that flatness of the pattern can be observed on the plain surface rather than curved surface. This constructed style was different from the previous Quickcast 2.0 for external layout in which in this paper it was built with half cylinder compared to the full cylinder style. The simpler built of the pattern is to ensure that flatness of the pattern can be observed on the plain surface rather than curved surface.

2.1 Application of strain gauge

For the experimental parts, the usage of strain gauge is vital due to it's ability to detect the presence of strain changed due to external pressure and temperature in the furnace. The cylindrical geometry of ceramic shell makes it possible to assume it as a thick cylinder. The difference between a thin cylinder and a thick cylinder is that a thick cylinder has a stress in the radial direction as well as a circumferential stress and longitudinal stress [8]. Fig. 2 shown that the location of strain gauge was installed on the pattern itself which are square and polygon internal web structure.

DOI: 10.1051/matecconf/201713500036

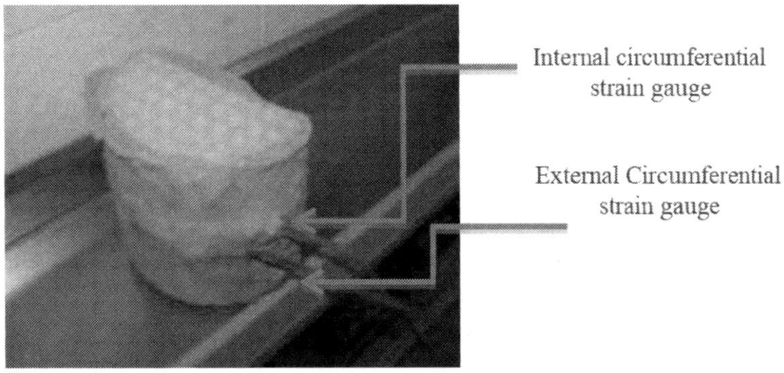

Fig. 3. The location of strain gauge attached on pattern for both square and polygon internal web structure.

In addition, radial stress has only a radial crippling effect on the wall of the ceramic shell and it is not the main criteria for defining the failure of a thick cylinder. For square and polygon models, strain gauges were installed in longitudinal and circumferential directions. Therefore, the magnitude of the strains in both directions is the comparative criteria for two structures. Besides that, the strain gauge also were attached on the ceramic shell for analyzing the strain changed on the shell surface. Furthermore, it is important to detect the strain changed on the shell due to cracking behavior of the ceramic shell when subjected to high stress and temperature.. The testing sample was placed in the controlled temperature furnace for ensuring that the temperature can be adjusted within time. The strain gauge was used with the help of data logger for collecting the responded data. Due to the sensivity of the strain gauge towards pressure and temperature, the data was taken every for 5°C per minutes increment starting from 30°C until 150°C.

3 Experimental Results

3.1 Pattern buckling of ABS P400

To understand better about the collapsibility behaviour of ABS pattern which cause the ceramic shell to cracks, many study have been conducted. Most of the previous study have been achieved through Finite Element (FE) analysis. Nevertheless, due to the 3D geometry of the square and polygon network it was able to see the weaker point inside structures [9]. Fig. 4 shown that the patterns were undergoes burnout process and buckling happened.

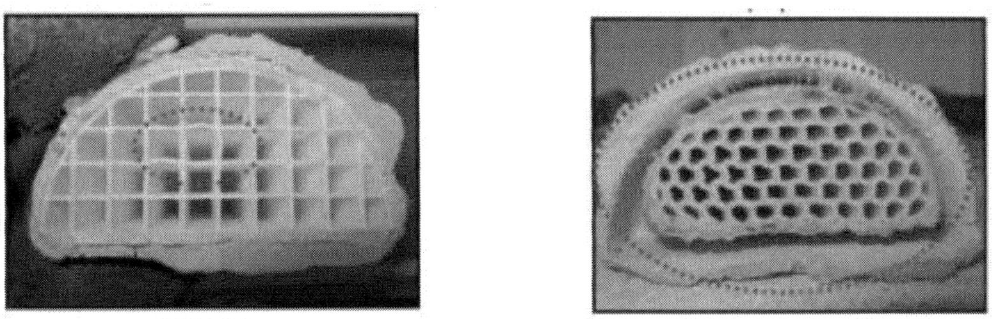

a) Square

b) Polygon

 DOI: 10.1051/matecconf/201713500036

Fig. 4. Buckling of pattern during burnout process.

Half cylindrical models were used to compare the behavior of the structures during the burnout stage. The shell tends to have more strength against outer pressures; therefore, the results of this model could be applicable for other forms of 3D Printer patterns. Since the material properties of ABS P400 changes with temperature during the burnout stage, it is essential to have the diagrams of changing the mechanical properties during the heating.
This occurrence is to review the collapsibility behavior of patterns when exposed to high temperature and pressure in the furnace. It can be seen that the polygon pattern is collapsed inward rather than outward for the open ended highlighted in the red section. While for the square internal web pattern, the buckling process does not fully takes place. It can be seen that, the red section buckling only at the middle part whereby, at the side none buckling happened. When an investment casting shell with a 3D Printer epoxy pattern inside furnace oven during the burnout process, it is subjected to high temperature rise, thermal expansion, and large strains. Since the difference of the CTEs of pattern and investment ceramic is more than one order of magnitude, the pattern can exert considerable stress on the ceramic shell.

3.2 Maximum Stress of ABS P400 Structures

As shown at Fig. 5, the square pattern intends to received more stress rather than polygon pattern during the burnout process due to the larger difference. For the square pattern, the maximum stress induced was 38.46 MPa while for the polygon pattern was 21.23 MPa. There was about 45 % of reduction stress when compared polygon and square patterns. Results also shows the glass transition temperature for square pattern was 100°C while for the polygon stated that the glass trasition temperature was 90 °C. Hence, based on the selected Modulus of Rupture (MOR) which was 30 Mpa [8], square polygon undergoes cracking of shell when subjected to high thermal stress and temperature. This phenomenon happened due to the higher stability of square pattern which was not collapsing during burnout process.

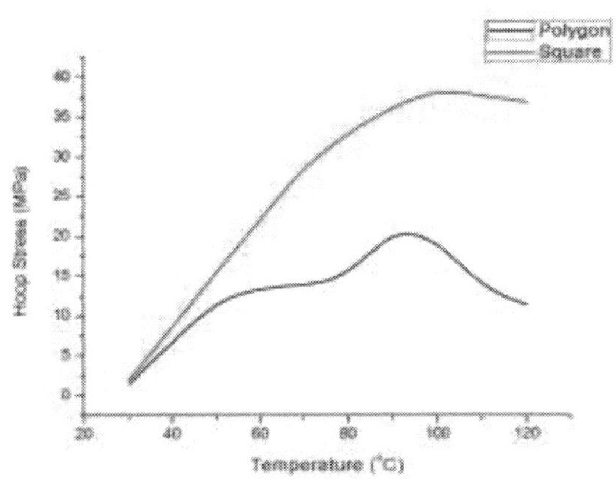

Fig. 5. Maximum Stress against Temperature

3.3 Equivalent Strain Analysis

According to Fig. 6(a) result shows that collapsibility behavior, polygon patterns tends to collapse easier compared to square pattern. This phenomenon shows that, square pattern tends to experience more strain changes compared to polygon pattern. This is due to the inner built structure of both types of ABS pattern. For the square, the design built was 90^0 internal shape along together with supports which enable it to collapsed easier. Nevertheless, during the experimental process, the square pattern less collapsed eventhough it passed glass region temperature (Tg). It continued to gain stress from the expansion of pattern from the rise of temperature of 30^0 C until 130^0 C. The strain only started to reduce when passed the temperature of 130^0 C until 150^0 C. The maximum strain induced for the square pattern was 2806 microstrain.This shows that, the square pattern is hardly collapsed due to its strong build structure. According to the principle of CTE, during this phase, the pattern is most likely undergoes phase changed from solid to rubbery like. There is a limits when the maximum of temperature can be achieved by the materials pattern before it collapsed [10]. For the polygon pattern, the maximum strain induced was 2552 microstrain. There is about 9 % reduction of strain when compared both square and polygon patterns. Meanwhile the polygon pattern tends to expands from beginning of 30 ^{0} C before it reached 110 ^{0} C which is before the glass transition temperature (Tg). This phenomenon can be acceptable for most of the resin or polymer material whereby, the pattern should be collapsed before the Tg.

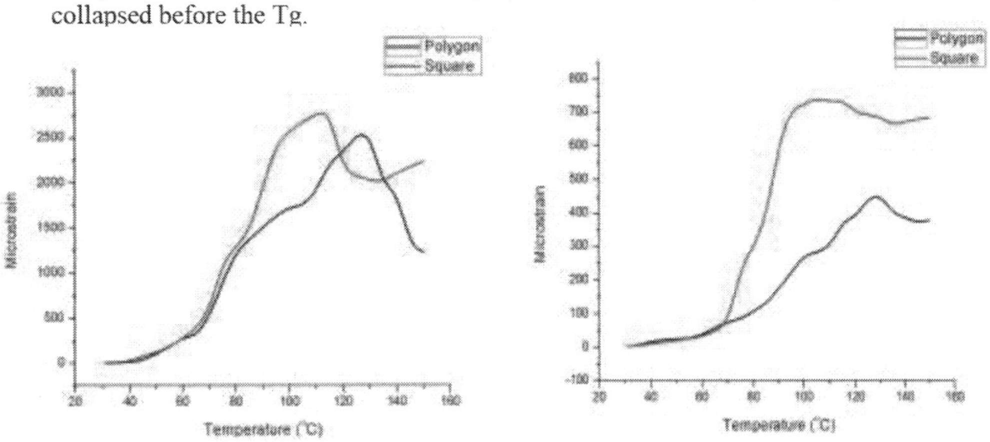

(a) Internal (b)External

Fig. 6. Circumferential Internal and External Circumferential Strain

From the Fig. 6(b) shown, it was stated that, square shell received more amout of strain when temperature was increased. The highest increased of temperature for square shell happened between temperature of 60 ^{0} C until 100 ^{0} C. During this phase, the shell tends to achieve it's ultimate stress due to pressure contact between the pattern and shell in which the pattern expanded as well as shell too. Apart from that, however, for polygon shell it tends to yield a little bit stress on ceramic shell. Most of all, the strain was slowly increased within time. Neverthelss, it remaining to received more strain during the burnout process. Up reaching 130 °C, the shell still received strain, but still less compared to the square pattern shell. This occurrence is happened due to the stronger shell to withstand the pressure expanding from the polygon pattern. It was said that, the Modulus of Rupture (MOR) of the shell must be greater than the expandeable material pattern to avoid the shell cracking during the burnout process [8].

DOI: 10.1051/matecconf/201713500036

4 Conclusion

Desktop 3D Printer ABS P400 material has validated it's ability to be used in IC process. In terms of strain on patterns there was 9 % strain reduction when compared for square and polygon respectively. This is due to different orientation of internal built structure which is also effects the collapsibility of patterns. Besides that, it was found that, there was a great difference in percentages of circumferential shell strain reduction which was 38 %. This is due to higher strain induced on the ceramic shell impact on the cracking behaviour of shell. However, in terms of axial shell strain, the reduction of strain found was 11% for square and polygon respectively. This behaviour can be understandable in terms of shell cracking, whereby, less strain happened to the shell considered to reduce the stress on the shell. By doing so, the shell can withstand the high pressure expansion of the pattern and will collapsed inwardly. Furthermore, the glass transition temperature was happened near 110 ^{0} C whereby it was an important characteristic of collapsibilty compared to MOR and CTE for determining the cracking behaviour of shell.

5 Acknowledgement

The authors wish to thank Ministry of Education Malaysia for funded Fundamental Research Grant Scheme(FRGS) Vot1423 and Additive Manufacturing Research Group(ADDMAN) at Faculty of Mechanical & Manufacturing Engineering in Universiti Tun Hussein Onn Malaysia(UTHM) for facilitating and financial support to this research programme.

References

1.C. K. Chua, C. Feng, C. W. Lee, and G. Q. Ang, 'Rapid investment casting: Direct and indirect approaches via model maker II', Int. J. Adv. Manuf. Technol., vol. 25, no. 1–2, pp. 26–32, (2005)
2.Y. Ding, H. Lan, J. Hong, and D. Wu, 'An integrated manufacturing system for rapid tooling based on rapid prototyping', vol. 20, pp. 281–288, (2004).
3.C. W. Lee, C. K. Chua, C. M. Cheah, L. H. Tan, and C. Feng, 'Rapid investment casting: Direct and indirect approaches via fused deposition modelling', Int. J. Adv. Manuf. Technol., vol. 23, no. 1–2, pp. 93–101, (2004)
4.A. Azari and S. Nikzad, 'The evolution of rapid prototyping in dentistry: a review', *Rapid Prototyp. J.*, vol. 15, no. 3, pp. 216–225, (2009)
5.S. Jones and C. Yuan, 'Advances in shell moulding for investment casting', *J. Mater. Process. Technol.*, vol. 135, no. 2–3 SPEC., pp. 258–265, (2003)
6.W. L. Yao and M. C. Leu, 'Analysis of shell cracking in investment casting with laser stereolithography patterns', *Rapid Prototyp. J.*, vol. 5, no. 1, pp. 12–20, (1999)
7.W. L. Yao and M. C. Leu, 'Analysis and design of internal web structure of laser stereolithography patterns for investment casting', *Mater. Des.*, vol. 21, no. 2, pp. 101–109, (2000)
8.Y. Norouzi, S. Rahmati, and Y. Hojjat, 'A novel lattice structure for SL investment casting patterns', *Rapid Prototyp. J.*, vol. 15, no. 4, pp. 255–263, (2009)
9.O. M. F. Marwah, S. Sharif, S. Sulaiman, M. Ibrahim, and E. J. Mohamad, 'Direct Investment Casting Numerical Study for ABS P400 FDM Materials', *Appl. Mech. Mater.*, vol. 660, pp. 99–103, (2014)
10.W. S. W. Harun, S. Safian, and M. H. Idris, 'Evaluation of ABS patterns produced from FDM for investment casting process', *WIT Trans. Eng. Sci.*, vol. 64, no. 3, pp. 319–328, (2009)

11. O. Mohd, F. Marwah, S. Sharif, and M. Ibrahim, 'Direct Fabrication of IC Sacrificial Patterns via Rapid Prototyping Approaches', vol. 6, no. 5, (2012)

Evaluation of Injection Molding Process Parameters for Manufacturing Polyethylene Terephthalate

O.M.F. Marwah[1, a)], A. Darsani[1, b)], M.A. Johar[1 c)], A.A. Rosden[2, d)], E.J. Mohamad[3, e, f)], M.F. Shaari[1, g)], R.H.A. Haq[1], A.M. Amin[1]

[1]Department of Manufacturing and Industrial Engineering, Faculty of Mechanical and Manufacturing Engineering, Universiti Tun Hussein Onn Malaysia, 86400 Johor, Malaysia.
[2]Nilai Legasi Plastik Sdn. Bhd. Jalan P/AM 3, Desaria, 71800 Nilai, Negeri Sembilan.
[3]Department of Mechatronics and Robotics Engineering, Faculty of Electrical and Electronic Engineering, Universiti Tun Hussein Onn Malaysia, 86400Johor, Malaysia

Abstract. Quality control is an important aspect in manufacturing process. The quality of product in injection moulding is influenced by injection moulding process parameter. In this study, the effect of injection moulding parameter on defects quantity of PET preform was investigated. Optimizing the parameter of injection moulding process is critical to enhance productivity where parameters must operate at an optimum level for an acceptable performance. Design of Experiment (DOE) by factorial design approach was used to find an optimum parameter setting and reduce the defects. In this case study, Minitab 17 software was used to analyses the data. The selected input parameters were mould hot runner temperature, water cooling chiller temperature 1 and water cooling chiller temperature 2. Meanwhile, the output for the process was defects quantity of the preform. The relationship between input and output of the process was analyzed using regression method and Analysis of Variance (ANOVA). In order to interpolate the experiment data, mathematical modeling was used which consists of different types of regression equation. Next, from the model, 95% confidence level (p-value) was considered and the significant parameter was figured out. This study involved a collaboration with a preform injection moulding company which was Nilai Legasi Plastik Sdn Bhd. The collaboration enabled the researchers to collect the data and also help the company to improve the quality of its production. The results of the study showed that the optimum parameter setting that could reduce the defect quantity of preform was MHR= 88°C, CT1= 24°C and CT2= 27°C. The comparison defect quantity analysis between current companies setting with the optimum setting showed improvement by 21% reduction of defect quantity at the optimum setting. Finally, from the optimization plot, the validation error between the prediction value and experiment was 1.72%. The result proved that quality of products can be improved by using the DOE approach.

*Corresponding author: mdfaizan@uthm.edu.my

1 Introduction

Plastic manufacturing has been one of the competitive markets which make producers to produce high quality parts at a low price and within the least time [1]. This study focused on the production of plastic PET preform which are produced in large volumes. A mandatory process involved in this production is injection moulding. Injection moulding is a standard method for producing parts in a large number and with high quality, value added commercial or industrial plastic parts in the market [2].

Polyethylene terephthalate (PET) belongs to a group of materials known as thermoplastic polymer of which would become softened and deformed due to heat.. PET is one of the materials suitable to be used in plastic industries especially to make bottle preform [3]. The preforms produced using this material have some advantages such as impact resistant, very lightweight and non-toxic when in contact with food [4].

In the industrial field, the quality and quantity of production are two important matters. Nilai Legasi Plastik Sdn Bhd (NLP) is one of the companies that manufactures preform products by using injection moulding process. Based on an observation, the company uses trial and error method to have set the moulding parameter to get the best quality of production. Unfortunately, there were moulding defects that have occurred in the production of the perform, Incorrect input parameters settings has caused bad defects and increased production cost. Therefore, finding an optimized parameters is highly desirable [5]. A collaboration between UTHM and NLP has opened up an opportunity to conduct this study. In order to produce good quality parts, the optimized moulding parameter based on DOE approach is suggested for the processes of plastic preform

Design of Experiment (DOE) is one of the many problem solving or improvement quality tools that can be used for various investigations such as finding the significant factor in a process, the effect of each factor on an outcome and screening a parameter. Many industries use this tool to stay competitive worldwide by designing robust products as well as improving their quality and reliability. [6]. One of the methods under DOE is factorial design method. Generally, factorial design is the most efficient method to be used in an experiment where the study of effects of two or more factors are involved [11]. The effect of factor is defined as the change in response, produced by a change in the level of the factor. The model of process and optimal condition are developed using analytic methods and heuristic algorithms [7]. Most researchers have considered mould temperature as a very important parameter. A module called Mould Temperature Controller (MTC) is used to control mould temperature. The MTC is very expensive and generally not incorporated in a basic control system. This constraints the effective control of the output of injection moulding. In the absence of MTC, optimization of process parameters can be achieved by considering the coolant flow rate along with other process parameters [8]. At NLP, there is an MTC machine which is called as Mould Hot Runner Temperature Controller. In this study, the temperature on Mould Hot Runner Temperature Controller and Water Cooling Temperature machine was strictly controlled. Moreover, the experiment and the analysis was designed by using the Minitab Software.

Finally, from this study, it was expected that the best analysis of process parameter optimization for the production of PET preform injection moulding process at the NLP by using DOE approach would be obtained.

2 Literature Review

Injection moulding is one of the popular manufacturing processes among the plastic industries. For this study, the application of injection moulding process was specifically conducted for the manufacturing of PET preforms and PET bottles. Fig.1 shows the specific injection moulding process for preforms., It started with the PET granulates are fed into the hopper (1). Then, the granules will traverse from the hopper using gravity into a heated barrel (2, 3) and crushed via a rotating screw into a molten form. The heated liquid plastic has a temperature in between 250 °C to 320 °C (PET). The screw propels the molten plastic through a nozzle (4) into a two sided mould (5) which defines the shape of the preform. When the mould (5) is filled with the pre-programmed amount of the molten plastic, it is closed by the hydraulic clamp force (6) which is generated by an electric motor and a hydraulic pump (7). The mould will be hold together by the clamp within the time period which is needed for the products to cool down. The mould is cooled with external water cooling during the process to ensure that the preform solidifies fast and becomes as homogenous as possible. After cooling down, the end products are ejected from the mould by an ejector pin (8) which pushes the preform out of the mould. The mould is then re-opened and filled using molten plastic that is fed constantly through the hopper and the traversing screw [4].

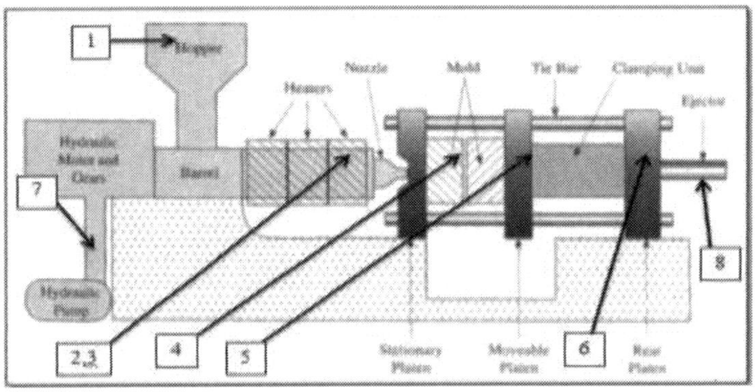

Fig. 1. Diagram of injection moulding machine process

In addition, 16 cavities preforms mould use the hot runner system rather than cold runner system. As such, the design of hot runner system plays an important role in guaranteeing mould quality. Based on foreign advanced techniques and combined with domestic clients' demand, the valve gated is designed, in which each cavity has its own heating zone and temperature is measured by an independent thermocouple. The advantage of this system is it easily avoids some problems such as white or draw bench at the bottom [8].

Lastly, there are three important things in PET preform processing which are [9]:

- Dry and dehumidify PET material which will be process by autoloader, hopper dryer & dehumidifier.
- Melting PET material into preform mould by PET injection moulding machine.
- PET preform cooled down and formed in the preform mould which can be controlled by hot runner temperature controller, water cooling chiller controller and cooling water tower.

2.1 Polyethylene Terephthalate (PET) Preforms

There is probably no subject in the PET industry shrouded in more secret problem than the design of preforms. There is no handbook, no course, and very little other material that prospective preforms designer may peruse to get prepared for the job. The main reason is that preforms design is still without a calculation or simulation that can guarantee a perfectly suitable preforms for a given bottle design and blow machine. This unpredictable factor is due to the problem that the formation of a bottle from a preform is literally explosive and even very tiny temperature variations could affect the outcome. In addition, the particular preform is manifold, still too many to make perfect predictions [3]. Figure 2.2 show the example of PET preforms product.

Based on the literature review in previous research, reference books and articles, there have been no calculation or simulation that could guarantee a perfect preform for a given bottle design and blow machine. However, the quality of the preform can be controlled, depending on the setting of the injection moulding machine used by the industry and condition of the industry. This is because, the process of making preform is very sensitive even for a simple product. PET preforms start to absorb humidity almost immediately after being manufactured by the injection moulding.

There are two methods in making the PET bottles which are "One-Step hot preform method" and "Two-Step cold preform method". One step method is a direct process from PET granule to finished bottles completed in one integrated machine. In this process, the injection moulded preform is withdrawn from the injection cavity while it still hot enough to be stretched and blown to form the bottles. On the other hand, in the two step methods, two separate machines are used. The first one is injection moulding machine that is used for making PET preforms. The other one is the blow moulding machine that is used to reheat the cold preform and then blow it to form the bottles [4]. Actually, there are some defects could happen to the preforms during their production. However, not all preforms have similar defects. Watermarks, shrinkage, burning, bubbles and scratch surface are some examples of defects to PET preforms [10].

3 Design of Experiment

Design of Experiments (DOE) is employed for screening the process parameters for the specific mould geometry. Moreover, the DOE is mainly used for checking the important level of each process parameter and quality indicator [6]. There has been a great need to seek quick and effective solutions, especially for glob- al optimal process parameter settings, and the determination of the optimal parameter setting is now recognized as a crucial step if the quality of plastic injection moulding products is to be improved [12]. The basic procedure is to perform a series of well-planned experiments with different settings of a set of variables. This is followed by an analysis of, the corresponding results to gain as much information as possible [2]. A well designed experiment is important because the result and conclusion that can be drawn from the experiment depend to a large extent on the manner how the data are collected.

Many experiments involved the study of the effects of two or more factors. In general, factorial designs are most efficient for this type of experiment. By using the factorial design, each complete trial or replication of the experiment, all possible combinations of the levels of the factors are investigated. For an example, if there are „a" levels of factor A and „b" levels of factor B, each replication contains all treatment combinations. When factors are arranged according to the factorial design, the factors are often said to be crossed. The effect of a factor is defined as the change in response to a change in the level of the factor. This is frequently called the main effect because it refers to the primary factors of interest in the experiment [11].

4 Methodology

4.1 Specification of Machine and Material

WY Preform Injection Moulding Machine was used in this study. Meanwhile, other machines including mould hot runner temperature controller, water cooling chiller controller and hopper were supplied by the other companies. The injection moulding machine has 16 cavities that produces 16 preforms in each shot. The mould hot runner temperature controller helps to control the runner for each cavity. This runner temperature is one of the factors that was controlled in this study. Next, two water cooling chiller controllers control the core and cavity of the mould directly and it is one of the other factors that affect the production. The raw material used in this company is PET which was supplied by Recron Malaysia Sdn. Bhd. The material used is of grade A quality with lot number 6868. The net weight for one packaging for this raw material is 1100 kgs.

4.2 Design of Experiment (DOE) for PET Preform

For this study, in order to identify the input factor and response, the data were collected from the NLP. As the study needed observations on the preform injection moulding process in the company, several time was spent on conducting the observations. By collaborating with NLP, this study was limited to control only the other machines such as mould hot runner temperature controller, water cooling chiller, hopper or tower tank. This was because, the owner of the company wanted to avoid any critical setting that could affect their injection moulding machine especially as the mould has a high maintenance cost. Moreover injection moulding machine is also a very sensitive machine. The company does not want to take any risk of damaging it.

A table form was prepared in order to make the analysis simpler. The data collected from the analysis were recorded in the form which was later keyed in the software. So, after collecting all the data for the parameter involved, the information is put in the Table 1.

Table 1. Details of parameter

Parameters	Temperature Level (°C)		
	Low	Center Point	High
Mould hot runner controller (MHR)	85	88	90
Water cooling chiller series 1 (CT 1)	23	24	25
Water cooling chiller series 2 (CT 2)	25	27	28

DOI: 10.1051/matecconf/201713500037

5 Results and Discussions

After the experiment was carried out, the analysis was done using in the Minitab Software. The defect quantity was added in the response column. The defect quantity was calculated based on the experiment production. The total quantity for each run order were 16 cavities x 10 shots = 160 pieces. Table 2 shows the details information of defect quantity results.

5.1 Main Effect and ANOVA Table

The graph of mean effects for defect quantity is shown in Fig. 2. The analysis shows the main effects which are Mould Hot Runner (MHR), Water Cooling Temperature (CT1) and Water Cooling Temperature 2 (CT2). The graph shows, the values of mean scores of each main effect while the low value of mean shows the best value of setting for the effects. For the MHR, CT1 and CT2, the data mean scores recorded were 88°C, 24°C and 27°C respectively were in low value of mean. The main effects were analyzed as significant where the P-value was 0.000. In the ANOVA, if the p-value is less than 0.05, the variables have statistically significant results. So this proved that the variables are important to be controlled and interaction between them also showed significant p-value.
Then, R-Sq from the table shows almost high percentage value which was 99.85%. The R-Sq indicates the percentage of variation in the response and this shows how fit the variables are in this data. The higher the percentage of R-Sq the better the variables fit in the data. The percentage indicated that all the variables were important and mostly affected the response.

Table 2. Experiment Result

Run Order	Center Pt	MHR	CT1	CT2	Defect Quantity
1	1	85	25	28	110
2	1	85	23	25	96
3	1	85	23	25	96
4	1	85	23	28	92
5	1	90	23	28	100
6	1	85	23	28	92
7	1	90	25	25	95
8	1	90	25	28	91
9	1	85	25	28	110
10	1	85	25	28	110
11	1	90	23	25	80
12	1	85	23	28	92
13	1	90	23	28	100
14	1	90	25	25	95
15	1	85	23	25	92
16	0	88	24	27	57
17	1	90	25	25	95
18	1	85	25	25	117
19	1	85	25	25	117
20	1	90	23	25	80

21	1	90	25	28	91
22	1	85	25	25	117
23	1	90	25	28	91
24	1	90	23	25	80
25	0	88	24	27	57
26	0	88	24	27	57
27	1	90	23	28	100

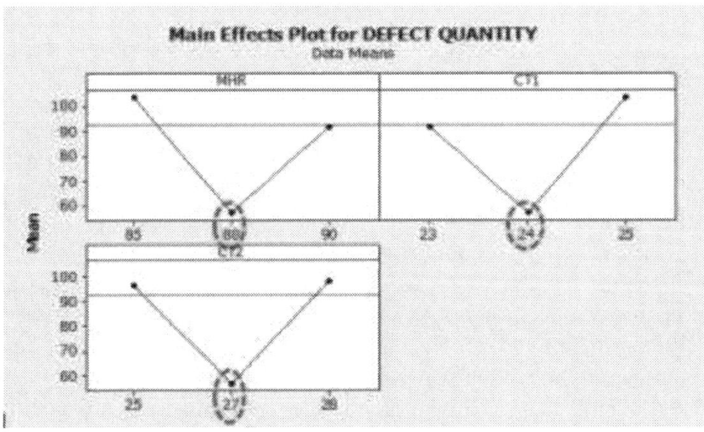

Fig. 2. Main effect plot for injection mould process parameters

5.2 Analysis of Variance (ANOVA)

The main effects are concluded as significant where the P-value was 0.000. In the ANOVA, if the p-value is less than 0.05, it shows that the variables are statistically significant result. So this proved that the variables are important to be controlled and interaction between them also showed the significant p-value. Then, R-Sq from the Fig. 3 shows almost high percentage value which was 99.85%. The R-Sq indicates the percentage of variation in the response and this is show how fit the variables are in this data. The higher the percentage of R-Sq the better the variables fit in the data. The percentage indicated that all the variables were important and mostly affected the response. Based on ANOVA table in the Minitab, the prediction value of the defect quantity could be identified. By inserting the value of the related factor in this experiment, the prediction of defect quantity could be calculated. Furthermore, from the ANOVA table analysis, the regression equation could be made based on the information of estimated coefficient using data in un-coded units.

Factorial Fit: DEFECT QUANTITY versus MHR, CT1, CT2

Estimated Effects and Coefficients for DEFECT QUANTITY (coded units)

Term	Effect	Coef	SE Coef	T	P
Constant		97.46	0.1571	620.22	0.000
MHR	-11.92	-5.96	0.1571	-37.92	0.000
CT1	11.58	5.79	0.1571	36.86	0.000
CT2	1.58	0.79	0.1571	5.04	0.000
MHR*CT1	-8.58	-4.29	0.1571	-27.31	0.000
MHR*CT2	6.42	3.21	0.1571	20.42	0.000
CT1*CT2	-7.08	-3.54	0.1571	-22.54	0.000
MHR*CT1*CT2	-4.92	-2.46	0.1571	-15.64	0.000
Ct Pt		-39.74	0.4755	-83.59	0.000

S = 0.769800 PRESS = 24
R-Sq = 99.85% R-Sq(pred) = 99.67% R-Sq(adj) = 99.79%

Analysis of Variance for DEFECT QUANTITY (coded units)

Source	DF	Seq SS	Adj SS	Adj MS	F	P
Main Effects	3	1938.83	1672.12	557.37	940.57	0.000
MHR	1	1123.58	852.04	852.04	1437.82	0.000
CT1	1	805.04	805.04	805.04	1358.51	0.000
CT2	1	10.21	15.04	15.04	25.38	0.000
2-Way Interactions	3	947.67	990.12	330.04	556.95	0.000
MHR*CT1	1	442.04	442.04	442.04	745.95	0.000
MHR*CT2	1	204.58	247.04	247.04	416.88	0.000
CT1*CT2	1	301.04	301.04	301.04	508.01	0.000
3-Way Interactions	1	145.04	145.04	145.04	244.76	0.000
MHR*CT1*CT2	1	145.04	145.04	145.04	244.76	0.000
Curvature	1	4140.75	4140.75	4140.75	6987.52	0.000
Residual Error	18	10.67	10.67	0.59		
Pure Error	18	10.67	10.67	0.59		
Total	26	7182.96				

Fig. 3. ANOVA for injection mould process

Moreover, the comparison for defect quantity was made for the three settings which were low and high value of MHR with constant value of CT1 and CT2 and the optimum setting. Fig. 4 shows the comparisons among them.

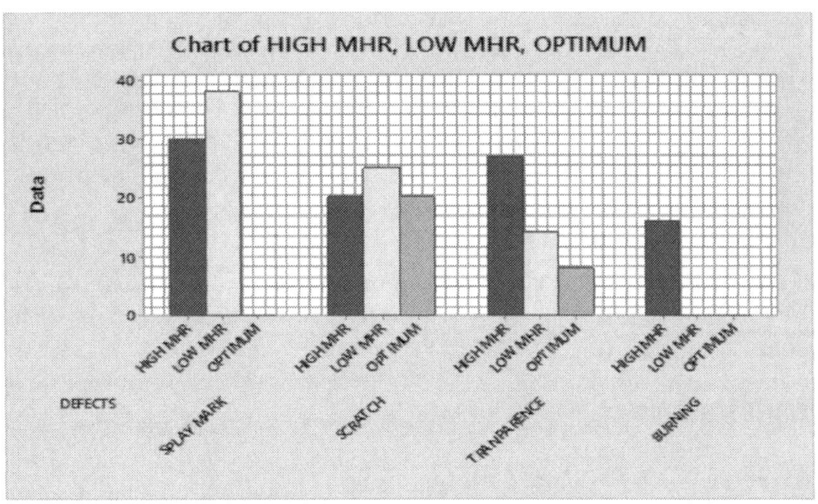

Fig.4. Defect quantity for comparison setting

Finally, based on the regression and graphical analysis, the best setting that showed the less defect quantity was 57 that came from the run order 25, 26 and 27. In this run order, the temperature of mould hot runner temperature, water cooling chiller series 1 and water cooling chiller series 2 were 88°C, 24°C and 27°C respectively. In order to show that this setting is much better than the setting that was used in the company, the experiment also run for that current setting which the mould hot runner temperature was 90°C, water cooling chiller 1 was 26°C and water cooling chiller 2 was 26°C. Thus, based on the defect quantity analysis, the comparison for the setting was be made. The experiment for both settings were ran for 10 shots. Based on the comparison, the best setting based on the experimental layout and the current setting was used in the company. Fig. 5 show that the defect quantity for optimum sets. The optimum setting graph show 21% reduction rate from 56% to 35% of defect quantity.

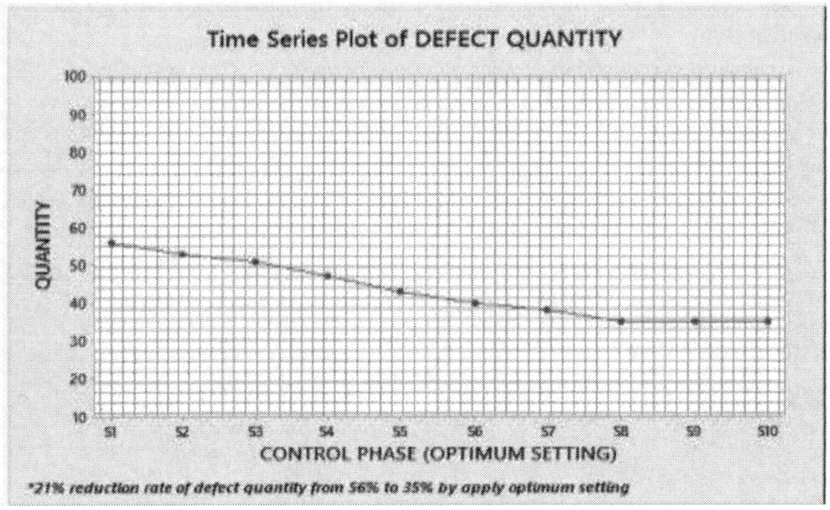

Fig. 5. Time series plot of defect quantity for optimum setting

5.3 Validation of Best Setting

For the validation of best setting predicted for the experiment, the production ran for 30 shots. For the validation, the total pieces used were 480 pieces and the analysis was also made to find the defect quantity from the validation run. The defect quantity from the validation testing was in the range of 160 per 480 pieces. The conclusion that could be made from the validation was, the best setting for the experiment gave the less quantity of defect differed from the quantity defect using the setting applied by the company.

Calculation for validation

Error % = [| prediction-experiment | / prediction] x 100 (Eq.1)

= [| 58-57 | / 58] x 100

= 1.72 %

6 Conclusions

In conclusion, the Design of Experiment by using the factorial design approach showed improvement in the reduction of defect quantity. In this study, there were several statistical analysis or regression models investigated. R_2 and p-value statistics were used to identify the best models or the optimum setting for the experiment. ANOVA was used to determine the most significant parameter in this study and the estimated coefficients for defect quantity could be calculated using data in encoded units. Moreover, from the ANOVA, all the main effects showed the three parameters which were Mould hot runner temperature and Water cooling chiller temperature 1 & 2 were significant. There was a significant improvement in the reduction of the quantity of defect in their production. As for future

 DOI: 10.1051/matecconf/201713500037

research, it is recommended that further analysis on the response using surface method and design, such analysis if conducted in a more advanced way could lead to more improvement in production quality. Furthermore, it is also recommended that the analysis includes classifying the defects quantity into types of defects. This will help to find out specifically the kind of defects caused by each cavity,

7 Acknowledgments

The authors acknowledge Ministry of Higher Education for funded PPRN Grant Vot A049 .and Faculty of Mechanical and Manufacturing Engineering (UTHM) for facilities. Special thanks to those who contributed to this project directly or indirectly.

References

1. G. Potsch and W. Michaeli "Injection Moulding : An Introduction", Munich : Hanser Publication, (1995)
2. Pei Jen Wang and Jui Ming Liang. Journal Of Injection Moulding Technology 6(1): 58 -70 , (2002)
3. Ottmar Brandau .Bottles, Preforms and Closures. 2nd Ed. USA: Plastic Design Library. (2012)
4. Kenplas Industry Limited, Hangzhou Kenplas Machinery Ltd .PET Preform Making, PET Bottle Making. Hangzhou (2016.)
5. D.V.Rosorto, M.G Rosato, "Injection Moulding Handbook" Massachusetts: Kluwer Academic Publisher, (2000)
6. R. Konda, K.P. Rajurkar, R.R. Bishu, A. Guha and M. Parson. Design Of Experiments To Study And Optimize Process Performance. International Journal Of Quality & Reliability Management. 16 (1): 56 – 71(1999)
7. F.Shi, Z.L Lou, J.G. Lu and Y.Q Zhang "Optimization Of Plastic Injection Moulding Process With Soft Computing : Int .J. Adv Manufacturing Tech. Vol 21, pp 656-661 (2003)
8. D.V.Rosorto, N.R Schott and M.G Rosato, "Plastic Engineering, Manufacturing And Data Handbook" Boston :Springer, pp 406-407 (2001)
9. P. Fletcher and D. Thesis, "Founding and requirements of starting a plastics product factory," (2013)
10. SACMI Group Article, Watermarks, Shrinkage & Surface Defects : The „Invisible" Faults On PET Preforms. (2015)
11. Douglas C. Montgomery, Design And Analysis Of Experiment. Eight Edition. Arizona State University. (2013)
12. Wen-Chin Chen, Manh-Hung Nguyen, Wen-Hsin Chiu, Te-Ning Chen, Pei-Hao Tai "Optimization of the plastic injection molding process using the Taguchi method, RSM, and hybrid GA-PSO": Int. J. Adv. Manuf. Technology, Springer-Verlag , (2015)

Green density optimization of stainless steel powder via metal injection molding by Taguchi method

AM Amin[1], *MHI* Ibrahim[1], *MY* Hashim[3], *OMF* Marwah[2], *MH* Othman[3], *Muhammad Akmal* Johar[2], *CH* Ng[3]

[1]Advanced Materials and Manufacturing Centre, Universiti Tun Hussein Onn Malaysia, 86400 Parit Raja, Batu Pahat, Johor, Malaysia
[2]Additive Manufacturing Research Group, Universiti Tun Hussein Onn Malaysia, 86400 Parit Raja, Batu Pahat, Johor, Malaysia
[3]Advanced Forming Research Group, Universiti Tun Hussein Onn Malaysia, 86400 Parit Raja, Batu Pahat, Johor, Malaysia

Abstract. Metal injection moulding (MIM) has gains much attention due to its ability in producing large amount of small part with complex geometry and intricate shape. In order to obtain better shape retention, optimum density of green part is required. This paper deals with the application of Taguchi method in optimising the green density of moulded components base on parameters setting in plastic injection moulding machine. For this purposes only 7 process parameters were considered here are injection pressure, injection temperature, cooling time, injection speed, injection time, packing time and mould temperature. An orthogonal array of L27 experimental base design was conducted. Base on the experimental results, cooling time plays significant contribution to density followed by injection pressure, time, speed, mould temperature, packing time and injection temperature. Confirmation test was done base on optimization level for each factors and shows good results in green density of the injected moulded samples.

1 Introduction

Metal injection moulding is a manufacturing process with a capability of producing complex geometry and intricate parts from mixing of metals, ceramic or cermet powder with combination of polymers, wax and surfactant [1] with a few shot as compare to other fabrication process [2]–[4]. Due to its versatility, near net shape and less materials waste, it's becomes attraction to most researchers in exploiting it into new dimensions whether in terms of its binder, powder characteristic, injection moulding conditions, debinding and sintering [5].

* **Corresponding author:** azriszul@uthm.edu.my

The traditional approach to experimental work is to vary one factor at a time, holding all other factors fixed. This method does not produce satisfactory results in a wide range of experimental settings. In order to obtain high efficiency in the planning and analysis of experimental data, various robust designs of DOE had been implemented in manufacturing area. Design of experimental (DOE) has long been applied by many researchers in terms of mixing the metal or ceramics powder with binder [6] to the sintering stage [7], [8]. Effect of powders and binders on injection moulding parameters has been analysed by Ahn et al. [3] where they indicate that the binder plays critical role in minimizing the injection pressure, clamping, temperature, cooling time and velocity. Their finding states very minimized influential by the types of powder. Application of DOE in MIM is continued by Yakimov and Coyle [9] where they use full factorial design in determining the factors that affecting the thermal debinding processes. Jorge et al. [10] using full factorial design in optimizing solvent and thermal debinding. Factorial design was also used by [11], [12] in optimizing the thermal and sintering parameters of the injected parts.

Dr. Geneichi Taguchi is founded Taguchi's robust design which functioning to reduce economically the variations of product function and optimizing the parameters involved in laboratory or manufacturing environment to meet the requirement of target quality. Taguchi parameter design applied to investigate and optimize the process involved in MIM has been implemented by Ji et al.[7] using Taguchi's method in optimizing the sintering parameters with L9 orthogonal arrays with four factors and 3-level design in their experimental data [13] of sintering.

Lee et al. [14] implementing the Taguchi's method by using L9 orthogonal arrays with 4 factors and 3 level design for studying the effect of gas the effect of processing variables on gas penetration during MIM. Ibrahim [15], [16] use Taguchi with Grey relation in optimizing the microMIM process. Chua et al.[17] optimizing the injection density and solvent process of parameters using L25 orthogonal array and found that mould temperature and powder loading are the most influential parameters for green part surface quality and injected part strength and density, respectively.

In this paper, 316 L stainless steel feedstock which contains of restaurant waste lipids (RWL) and polypropylene (PP) as binder with ratio of 40/60 respectively was moulded via injection moulding process and the parameters setting was optimized using Taguchi method for obtaining the optimum density of the green compacts. Orthogonal array of L27 with 7 factors and 3 levels factor were used in determining the optimized parameter for injected optimum green density of the moulded parts.

2 Experimental work

The 316L water atomized stainless steel powder used in this study has mean particle size of 6μm. Particles size was measured via particle analyser Fritsch analysette 22 compact. The stainless steel powder has the tap density of 8.0471 g/cm^3 and was supplied by Epson Atmix Japan. Figure 1 shows the powder particles size distribution and powder morphology of the 316L water atomized stainless steel. Physical properties and the chemical composition of the powder is shown in Table 1.

Table 1: Physical properties and chemical composition of 316L stainless steel metal powder

Powder size and density

Powder SS316L Epson Atmix Corp	Size	D10	D50	D90	Density
	6.0μm	2.87μm	5.96μm	10.65μm	8.0471 g/cm³

Powder chemical composition (% wt)

Ci	Si	Mn	P	S	Ni	Cr	Mo	Cu
0.027	0.84	0.19	0.016	0.012	12.20	16.40	2.10	0.03

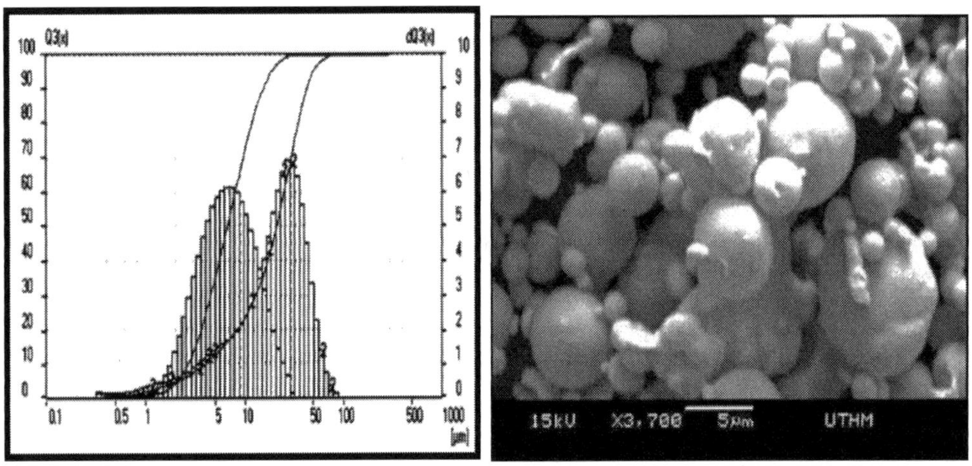

(a) (b)

Fig. 1: (a) Particle size distribution and (b) SEM image of water atomised 316L stainless steel powder (PF-10F)

Table 2: Binder composition and thermal properties of binder components

Binder	Fraction in binder	Density (gcm^{-3})	Melting Temperature (°C)	Degradation Range (°C)
Polypropylene (PP)	0.6 (8.3g)	0.90	165	350~470
RWL derivatives	0.4 (5.6g)	0.90	50	270~360

For making the feedstock, 316L stainless steel powder of 60% powder volume which is equivalent to 186.1g of powder weight along with binder fraction shown in Table. 2 were used. 60% powder loading was chosen because it is approximately 5% below the critical powder volume concentration [18] which in this case Critical Powder Volume Concentration (CPVC) of the selected stainless steel powder was found to be 64.8% [15]. The compound was blended using Brabender Plastograph EC mixer at temperature of 175°C with 30rpm for 90min. The compound was left cool at the ambient atmosphere and temperature before being crushed into smaller pallets for ease of feeding inside the injection moulding machine.

DOI: 10.1051/matecconf/201713500038

After crushing was done, the feedstock was fed into horizontal injection moulding machine modelled Nissei NP7 Real Mini to obtain the green sample as shown in Figure 2. The density optimization was conducted by implementation of Taguchi orthogonal array (Statistical Method). The injection process was run in 3 level design of experiment with 7 parameters and 3 trial of experiment. The parameters and its level is shown in Table 3. The results were analyzed by using Signal to Noise Ratio (S/N ratio) terms

In detailed explanation, S/N ratios is help in prediction the optimum result besides improve the quality via variability reduction and improved the measurement based on repetition [19]. Thus, paper are discussed the optimization of optimum green density with 60% Powder loading. The analysis is based on Signal to noise ratio or well known as S/N ratio. This analysis are based on Nominal is better characteristic base on Equation 1.

According to Table 3, injection pressure, 64.4 MPa has been used as the minimum pressure, which could force the feedstock to flow smoothly inside the cavities and avoiding from short mould. Mould temperature being implemented to aid the flow of the feedstock and avoiding the thermal shock of the green parts and also improve the ejection condition of the part. When the part was allowed to cool rapidly the part becomes so brittle, which slight bending from the ejector, could break the green part. Mould temperature was made slightly above the room temperature (40°C) for allowing the backbone becomes slightly flexible and could be ejected without any defects.

Here cool time was set for allowable shrinkage, which in turn reduce the friction between the green part and the wall cavity. Injection speed was selected to promote shear rate for the feedstock to flow inside barrel. Too high in shear rate could promotes voids and binder separation. Injection time also being considered as the factor here since it could avoid jetting, cracks, flashes, and short mould associate with improper injection speed and filling time.

Packing time plays significant effect of sink mark on the surface of the green part. As mentioned before, feedstock, which has filled up the mould cavities, could experience shrinkage due to cooling conditions. With the help of proper packing time shrinkage, flashes and voids can be prevented.

Table 3: 7 factors and 3 level for each factors involved in otimising process of injecting the green parts

Factors	Level 1	Level 2	Level 3
Injection Temperature, A (°C)	180	190	200
Injection Pressure, B (MPa)	64.4	80.5	96.6
Mould Temperature, C (°C)	40	60	80
Cool time, D (s)	5	10	15
Injection speed, E (RPM)	70	105	140
Injection time, F (s)	1	3	5
Packing time, G (s)	1	2	3

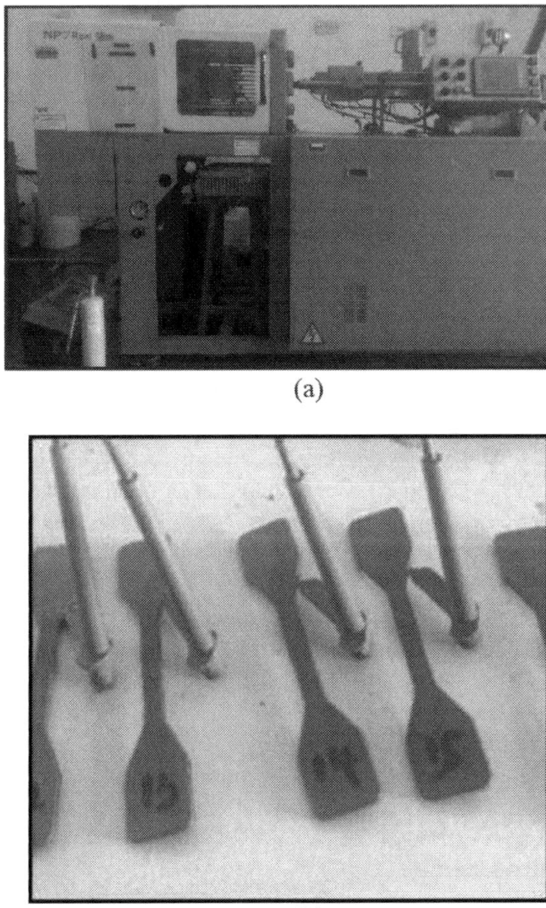

(a)

(b)

Fig. 2. (a) Injection machine used (b) injection moulded compacts

$$\frac{S}{N} = 10\log_{10}\sum\frac{\bar{y}^2}{\sigma_{N-1}^2} \tag{1}$$

Orthogonal array of L27 structure is shown in Table 4 where replication of three parts of the same tested parameters being employed. Density of the three replication injected moulded parts were taken by means of Archimedes principle and recorded for calculating Mean and S/N ratio as shown in Table 5. S/N ratio was calculated base on Nominal is better by using Equation 1. From Table 5, S/N ratio and Mean were calculated from the average density of the replication for each level structure of the L27 and the variability of the S/N ratio and Mean are shown in Table 6 and 7.

Table 4: Orthogonal array of L27 structure

A	B	C	D	E	F	G
1	1	1	1	1	1	1
1	1	1	1	2	2	2
1	1	1	1	3	3	3
1	2	2	2	1	1	1
1	2	2	2	2	2	2
1	2	2	2	3	3	3
1	3	3	3	1	1	1
1	3	3	3	2	2	2
1	3	3	3	3	3	3
2	1	2	3	1	2	3
2	1	2	3	2	3	1
2	1	2	3	3	1	2
2	2	3	1	1	2	3
2	2	3	1	2	3	1
2	2	3	1	3	1	2
2	3	1	2	1	2	3
2	3	1	2	2	3	1
2	3	1	2	3	1	2
3	1	3	2	1	3	2
3	1	3	2	2	1	3
3	1	3	2	3	2	1
3	2	1	3	1	3	2
3	2	1	3	2	1	3
3	2	1	3	3	2	1
3	3	2	1	1	3	2
3	3	2	1	2	1	3
3	3	2	1	3	2	1

Table 5: Mean and S/N ratio calculated from density replication of L27 structure experiment

Density Replication (g/cm^3)			Mean ($\bar{y}$)	$(\bar{y})^2$	$\frac{1}{(\bar{y}^2)}$	S/N Ratio
R1	R2	R3				
4.9331	5.0095	5.0366	4.993	24.93	0.0401	13.967
5.0367	4.9458	5.0373	5.007	25.07	0.0399	13.991
5.0782	5.0631	5.0719	5.071	25.72	0.0389	14.102
5.104	5.0834	5.1095	5.099	26.00	0.0385	14.150
5.0716	5.0833	5.0847	5.080	25.81	0.0388	14.117
5.1078	5.0793	5.0935	5.094	25.94	0.0385	14.140
5.0731	5.0999	5.0768	5.083	25.84	0.0387	14.123
5.1069	5.0772	5.1167	5.100	26.01	0.0384	14.152
5.0842	5.0976	5.079	5.087	25.88	0.0386	14.129
5.04	5.1017	5.1016	5.081	25.82	0.0387	14.119
5.0693	5.0458	5.0842	5.066	25.67	0.0390	14.094
5.0848	5.1002	5.06	5.082	25.82	0.0387	14.120
5.082	5.0984	5.0954	5.092	25.93	0.0386	14.138
5.0949	5.1085	5.1106	5.105	26.06	0.0384	14.159
5.0964	5.0787	5.0877	5.088	25.88	0.0386	14.130
5.0831	5.0658	5.0422	5.064	25.64	0.0390	14.089
5.0976	5.0835	5.0866	5.089	25.90	0.0386	14.133
5.1067	5.0906	5.1106	5.103	26.04	0.0384	14.156
5.0997	5.106	5.077	5.094	25.95	0.0385	14.142
5.083	5.0776	5.0562	5.072	25.73	0.0389	14.104
5.073	5.1014	5.1049	5.093	25.94	0.0386	14.140
5.1086	5.0885	5.1261	5.108	26.09	0.0383	14.165
5.0938	5.1098	5.1196	5.108	26.09	0.0383	14.165
5.0814	5.0942	5.1035	5.093	25.94	0.0386	14.140
5.0786	5.1126	5.1181	5.103	26.04	0.0384	14.157
5.0775	5.0723	5.0879	5.079	25.80	0.0388	14.116
5.0024	5.074	5.0106	5.029	25.29	0.0395	14.030

 DOI: 10.1051/matecconf/201713500038

3 Results and discussion

In this section, analysis of optimised parameters of the injection moulding parameters with its level are analysed. Variability analysis table of S/N ratio was calculated as shown in Table 6. From Table 6 average S/N ratio was calculated for each level factor and the difference between the levels of highest S/N ratio for the same factor being substituted with the lowest one. The difference was shown by the Delta value. The Delta values were then being plotted as shown in Figure 3 for clear viewing of significant factors that contributes to the green density of the moulded parts.

It was found that factor D or cooling time plays significant effect as compare to other factors (6.274) since mould temperature would assist the melt feedstock flow inside the mould cavity. The second most influencing factor was come from injection pressure (B), which has the Delta value of 5.95. Injection time (F), injection speed (E), mould temperature (C), packing time (G) and injection temperature (A) become the third, fourth, fifth, sixth and seventh important factor respectively according to delta value.

Table 6: Response table of S/N ratio variability analysis

Level	A	B	C	D	E	F	G
1	49.791	47.299	49.716	45.779	48.287	50.834	49.662
2	51.523	53.251	49.409	52.053	51.570	48.373	49.772
3	50.049	50.813	52.238	49.607	51.506	52.155	51.929
Delta Δ	1.732	5.951	2.522	6.274	3.283	3.781	2.267

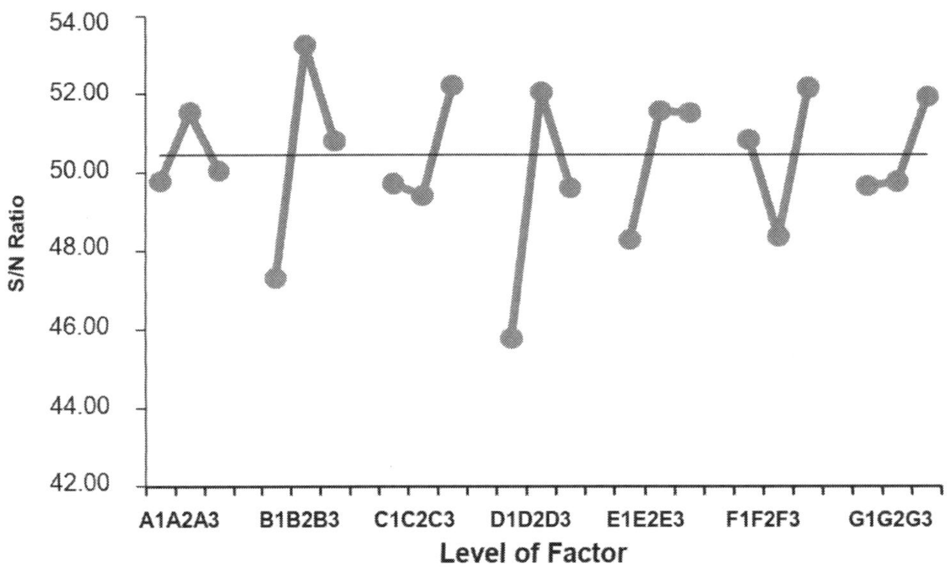

Fig. 3. S/N ratios variations of green part density at various levels of injection parameters

DOI: 10.1051/matecconf/201713500038

Response table for Mean variability analysis (Table 7) also being tabulated since this table can gives the results of any variability of green part density in response to the factors level. The delta analysis was then being plotted as shown in Figure 4.

From Figure 3 and 4, agreement of level factors of injection pressure, mould temperature and injection time were found for analysis of Mean and S/N ratio which indicates this level of parameters would give optimum and low fluctuation density of the green compact. For injection temperature, S/N ratio indicates that level 2 has the significant effect in obtaining optimum density and this result was contradicts to the Mean analysis, which indicates level 3 has significant factors in reducing the fluctuations of density obtained. Disagreement also was found for factor level for D, E and G which representing the factor of cooling time, injection speed and packing time respectively.

Although the optimum level between the four levels was not the same, if we compare the gradients of A factors for Mean between level 2 and 3 only slight change affect the optimum density variability as compare to S/N ratio gradients. Shifted the level from 1 to 2 and from 2 to 3 would results in greatest change in optimum green density. Therefore since shifted the level from 2 to 3 for factor A would results in greatest change S/N ratio as compare to Mean, therefore level 2 of factor A was selected since he Mean variability of the green density would only slightly change the density. The same goes to the factor D, E and G where level 2, 3 and 3 are chosen respectively.

Table 7: Response table of Mean variability analysis

Level	A	B	C	D	E	F	G
1	5.068	5.062	5.071	5.063	5.080	5.078	5.072
2	5.085	5.096	5.079	5.088	5.078	5.071	5.085
3	5.087	5.082	5.090	5.090	5.082	5.091	5.083
Delta Δ	0.018	0.034	0.020	0.027	0.004	0.020	0.013

Base on the analysis of factors and levels the predicted optimisation S/N ratio value is calculated using equation 2 below.

$$\frac{S}{N}\,\boldsymbol{ratio} = \overline{\boldsymbol{T}} + (\overline{\boldsymbol{A}}_2 - \overline{\boldsymbol{T}}) + (\overline{\boldsymbol{B}}_2 - \overline{\boldsymbol{T}}) + (\overline{\boldsymbol{C}}_3 - \overline{\boldsymbol{T}}) + (\overline{\boldsymbol{D}}_2 - \overline{\boldsymbol{T}}) + (\overline{\boldsymbol{E}}_2 - \overline{\boldsymbol{T}}) + (\overline{\boldsymbol{F}}_3 - \overline{\boldsymbol{T}}) + (\overline{\boldsymbol{G}}_3 - \overline{\boldsymbol{T}}) \quad (1)$$

Where T is the average S/N ratio, $\bar{A}_2$ is the injection temperature level 2, $\overline{\boldsymbol{B}}_2$ is the injection pressure level 2, $\overline{\boldsymbol{C}}_3$ is the mould temperature level 3, $\overline{\boldsymbol{D}}_2$ is cool time level 2, $\overline{\boldsymbol{E}}_2$ is injection speed level 2, $\overline{\boldsymbol{F}}_3$ is injection time level 3 and $\overline{\boldsymbol{G}}_3$ is packing time level 3.

Therefore, predicted S/N ratio is;

$$\frac{S}{N}ratio = 50.45 + (51.52 - 50.45) + (53.25 - 50.45) + (52.24 - 50.45) + (52.05 - 50.45) + (51.57 - 50.45) + (52.16 - 50.45) + (51.93 - 50.45)$$

$$\frac{S}{N}\,ratio = 60.92$$

Confirmation analysis was done base on the selected levels of factors with 10 replications of injected moulded sample. The results of S/N ratio density obtained for 10 replications are 60.65, which is 0.4% difference and below 10% as compare to the predicted S/N ratio.

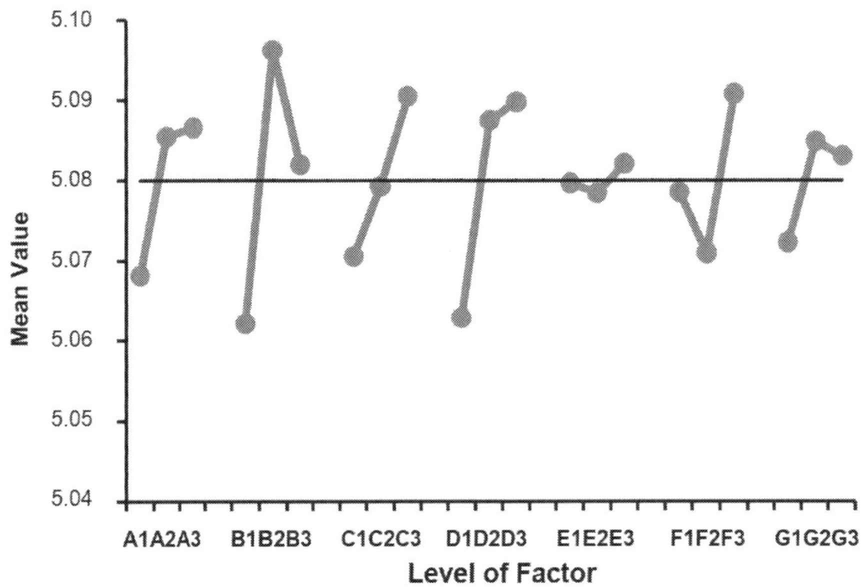

Fig. 4. Mean variations of green part density at various levels of injection parameters

Table 8: Mean and S/N ratio of confirmation test base on the optimized parameters with 10 replications sample

$\bar{y}$	$\bar{y}^2$	(δ_{N-1})	$(\delta_{N-1})^2$	S/N ratio
5.025	25.25	0.005	0.000022	60.650

4 Conclusion

From the results and discussion, Taguchi L27 was successfully being implemented in determining the optimum green density of moulded stainless steel powder. Base on the experimental results, cooling time plays significant contribution to density followed by injection pressure, time, speed, mould temperature, packing time and injection temperature. Results of confirmation test also indicate of 0.4% difference as compare to the predicted value of the S/N ratio base on the selected parameters.

5 Acknowledgement

Special thanks dedicated to Office for Research, Innovation, Commercialization, Consultancy Management (ORICC), Universiti Tun Hussein Onn Malaysia (UTHM) for their support throughout publishing this paper.

References

[1] X. Kong, T. Barriere, and J. C. C. Gelin, "Determination of critical and optimal powder loadings for 316L fine stainless steel feedstocks for micro-powder injection molding," *J. Mater. Process. Technol.*, vol. 212, no. 11, pp. 2173–2182, Nov. 2012.

[2] H. Abolhasani and N. Muhamad, "A new starch-based binder for metal injection molding," *J. Mater. Process. Technol.*, vol. 210, no. 6–7, pp. 961–968, 2010.

[3] S. Ahn, S. Jin, S. Lee, S. V. Atre, R. M. German, S. J. Park, S. Lee, S. V. Atre, R. M. German, S. Jin, S. Lee, S. V. Atre, and R. M. German, "Effect of powders and binders on material properties and molding parameters in iron and stainless steel powder injection molding process," *Powder Technol.*, vol. 193, no. 2, pp. 162–169, Jul. 2009.

[4] M. A. Omar, I. Subuki, N. Abdullah, and M. F. Ismail, "The Influence Of Palm Stearin Content On The Rheological Behaviour Of 316L Stainless Steel Mim Compact," *J. Sci. Technol.*, vol. 2, no. 2, pp. 1–14, 2010.

[5] A. M. Amin, M. H. I. Ibrahim, R. Asmawi, and N. Mustafa, "The Influence of Sewage fat Composition on Rheological Bahavior of Metal Injection Moulding," *Appl. Mech. Mater.*, vol. 660, pp. 38–42, 2014.

[6] R. Supati, N. H. Loh, K. A. Khor, and S. B. Tor, "Mixing and characterization of feedstock for powder injection molding," *Mater. Lett.*, vol. 46, no. 2–3, pp. 109–114, 2000.

[7] C. H. Ji, N. H. Loh, K. A. Khor, and S. B. Tor, "Sintering study of 316L stainless steel metal injection molding parts using Taguchi method: Final density," *Mater. Sci. Eng. A*, vol. 311, no. 1–2, pp. 74–82, 2001.

[8] D. F. Heaney, T. W. Mueller, and P. a. Davies, "Mechanical properties of metal injection moulded 316L stainless steel using both prealloy and master alloy techniques," *Powder Metall.*, vol. 47, no. 4, pp. 367–373, 2004.

[9] A. O. Yakimov and T. W. Coyle, "A statistical analysis of the factors affecting thermal debinding," *Mater. Sci. Lett.*, vol. 19, pp. 2255–2257, 2000.

[10] H. Jorge, L. Hennetier, A. Correia, and A. Cunha, "Tailoring Solvent / Thermal Debinding 316L Stainless Steel Feedstocks for PIM : An Experimental Approach," *Euro PM2005*, pp. 351–357, 2005.

[11] N. H. Loh and R. M. German, "Statistical analysis of shrinkage variation for powder injection molding," *J. Mater. Process. Technol.*, vol. 59, no. 3 SPEC. ISS., pp. 278–284, 1996.

[12] S. Butković, M. Oruč, E. Šarić, and M. Mehmedović, "Effect of sintering parameters on the density, microstructure and mechanical properties of the niobium-modified heat-resistant stainless steel GX40CrNiSi25-20 produced by MIM technology," *Mater. Tehnol.*, vol. 46, no. 2, pp. 185–190, 2012.

[13] A. T. Sidambe, I. A. Figueroa, H. G. C. Hamilton, and I. Todd, "Metal injection

moulding of CP-Ti components for biomedical applications," *J. Mater. Process. Technol.*, vol. 212, no. 7, pp. 1591–1597, 2012.

[14] K. Lee, M. De Hoyos, S. Ahn, R. Nambiar, M. A. Gonzalez, S. Jin, and R. M. German, "Gas-assisted powder injection molding : A study on the effect of processing variables on gas penetration," *Powder Technol.*, vol. 200, no. 3, pp. 128–135, 2010.

[15] M. H. . Ibrahim, "Optimization of MicroMetal Injection Moulding Parameter by Design of Experiment Method," Universiti Kebangsaan Malaysia, 2011.

[16] M. H. I. Ibrahim, N. Muhamad, A. B. Sulong, and K. R. Jamaludin, "Optimization of Micro Metal Injection Molding with Multiple Performance Characteristics using Grey Relational Grade," *J. Sci.*, vol. 38, no. 2, pp. 231–241, 2011.

[17] M. I. H. Chua, A. B. Sulong, M. F. Abdullah, and N. Muhamad, "Optimization of injection molding and solvent debinding parameters of stainless steel powder (SS316L) based feedstock for metal injection molding," *Sains Malaysiana*, vol. 42, no. 12, pp. 1743–1750, 2013.

[18] R. M. German and A. Bose, *Injection Molding of Metals and Ceramics*. Metal powders Industries Federation, 1997.

[19] N. Mustafa, M. H. I. Ibrahim, R. Asmawi, A. M. Amin, and S. R. Masrol, "Green Strength Optimization in Metal Injection Molding applicable with a Taguchi Method L9 (3)^4," *Appl. Mech. Mater.*, vol. 773–774, pp. 115–117, 2015.

 DOI: 10.1051/matecconf/201713500039

Effect of Welding Speed on Microstructure and Mechanical Properties due to The Deposition of Reinforcements on Friction Stir Welded Dissimilar Aluminium Alloys

Ravinder Reddy Baridula[1*], *Che Ku Mohammad* Faizal Bin Yahya[1], *Ramgopal Varma* Ramaraju[2] and *Abdullah Bin* Ibrahim[1]

[1]Faculty of Engineering Technology, Department of Manufacturing Engineering Technology, University Malaysia Pahang, Gambang, Malaysia [2]Faculty of Technology, Department of Mechanical Engineering, Stanford International College of Business and Technology, Toronto, Canada.

Abstract The strength of the welded joint obtained by solid state stir welding process was found to be improved as compared to fusion welding process. The deposition of reinforcements during friction stir welding process can further enhance the strength of the welded joint by locking the movement of grain boundaries. In the present study, the aluminium alloys AA2024 and AA7075 were welded effectively by depositing the multi-walled carbon nanotubes in to the stir zone. The mechanical properties and microstructures were studied by varying the traverse speed at constant rotational speed. The results show that rotating tool pin stirring action and heat input play an important role in controlling the grain size. The carbon nanotubes were found to be distributed uniformly at a welding speed (traverse speed) of 80mm/min. This enhanced the mechanical properties of the welded joint. The microstructure and Electron dispersive X-ray analysis (EDX) studies indicate that the deposition of carbon nanotubes in the stir zone was influenced by the traverse speed.

1 Introduction

Friction Stir Welding (FSW) is an efficient method for joining nonferrous metals which was invented by The Welding Institute (TWI), in the year 1991 [1]. Since, FSW utilizes a non consumable specially shaped tool, the absence of filler material is observed, which leads to less distortion tool. The tool includes a profiled pin (or probe) and a cylindrical shoulder. The probe helps to increase the frictional force between the work piece and the pin, and stirs the plastically deformed work piece material, resulting in a porosity free weld. The joining takes place at much low temperatures, in which the metal is in plastic zone. Though this method was initially applied to weld wrought aluminum alloys and later this technique applied on variety of materials with variety of joint scenarios of which metals and polymers were also included.

*Corresponding author: baridula@gmail.com

FSW has proven to be an alternative joining technique in which less distortion, less plastic deformation and low residual stresses resulted when compared with other conventional joining methods [2]. Therefore, the weld defects in FSW are less than defects in fusion welding process. The quality of the welds depends mainly on the process parameters like tool rotational speed (rpm), axial force (KN), and tool traverse speed (mm/s). The weld joints produced between pure aluminium and interstitial free steel, the positioning of the tool pin center towards the soft material which was placed on retreating side yielded good quality welds [3].

In recent years the concept of FSW was utilized to produce the surface composites by depositing different types of reinforcements (macro, micro and nano) during friction stir processing [4, 5]. The impact of particle distribution during number of pass was investigated by various researchers. It was found that the particle distribution was improved by the increasing the passes and the process parameters [6-10]. Jafari et al.[11] studied the outcome of CNT on 6mm Cu plates fabricated by friction stir processing by taking the groove size 0.3mm depth and 1.5mm width, and found that there is an enhancement of mechanical properties of the welded joint. Lim et al.[12] reported the synthesis of multi-walled CNT reinforced aluminium alloy composite with a groove of 0.3depth and 2.3mm width by friction stir processing, in which the hardness of the joint was improved. Izadi et al. [13] studied the multi-pass FSP of AA5059 alloy by inserting the carbon nanotubes of mean diameter 30-50nm in to the groove of size 2.5mm in width and 1.8mm in depth. After three passes the reinforcements were distributed uniformly in the stir zone of the nanocomposite which restricted the grain size and yielded higher microhardness. Morisada et al. [14] studied the AZ31 surface composites by dispersing Multi-walled carbon nanotubes of outer diameter 20-50nm in a groove of 1 mm × 2 mm using friction stir processing. It has been reported that the microhardness of the composite was increased due to grain refinement and reinforcement of cabon nanotubes. Morisada et al. [15] reported AA5083 composites fabricated by fullerene powder of mean grain size 25.4 μm in to a groove of 1 mm× 2 mm during friction stir processing. Sun et al. [16] reported the deposition of SiC particles in to copper plates of 2mm thickness by friction stir welding which improved the mechanical properties of the welded joint. Abnar et al. [17] investigated the friction stir welding of AA3003-H18 aluminium alloy by depositing the reinforcements in the stir zone. He reported that Cu and premixed Al-Cu powder were inserted between the two metals without making groove. The mechanical properties found to be improved more for premixed Al-Cu powder, when compared to Cu powder. Ravinder et al. [18] studied the deposition of copper nanoparticles during friction stir welding of dissimilar aluminium alloys and found that the mechanical properties of the welded joint improved in all the samples.

Thus the above literature review show that different researchers have successfully produced the surface composites by depositing the reinforcements in the stir zone. Hence, the present study focuses on the effect of traverse speed on microstructure and mechanical properties due to the deposition of multi-walled carbon nanotubes on fricton stir welded dissimilar aluminium alloys AA2024-T3 and AA7075-T6.

2 Experimental Details

In the present study the dissimilar aluminium alloys selected as base metals are AA2024-T3 and AA7075-T6 with dimensions of 100 mm× 50 mm× 4.85mm . The chemical composition and mechanical properties of the base metals are shown in Table1 and Table2. Fig. 1 shows the as received multi-walled carbon nanotubes having purity more than 90% with a nominal diameter of 10 to 20nm. The reinforcements were inserted in to the groove of 1mm in width and 2mm in depth which is being cut precisely by a wire cut machine i.e 0.5mm width and 2mm depth in each base metal. The welding is performed by a 5T NC FSW machine and the welding tool used is made of tool steel (W-Co) with dimensions- shoulder diameter of 18mm, pin height 4.65mm and cylindrical pin diameter of 5mm. The welded joints were fabricated by a double pass friction stir welding process. Initially a rotating tool with out pin is passed gently over the groove with carbon nanoparticles to compress and avoid the sputtering of nanoparticles during welding process, then the welded joints were fabricated by using a rotating tool with pin. The process parameters selected are at a contant rotational speed of 1400 rpm and axial load of 1 kN, the welding speed (traverse speed) was varied as 70,80,90mm/min. In the experimentation a total of 4 welded joints were fabricated in which thee joints are with nanoparticles and one joint with out nanoparticles. The process parameter selected for the welded joint without nanoparticles are 1400 rpm, 80 mm/min and 1kN, since with the same process parameters the strength of the welded joint with nanoparticles was found to be maximum. The tensile test specimens (ASTM-E8) were cut by using EDM wire cut machine and the tensile test was carried out on 12 samples by INSTRON universal testing machine of 30 kN. The microhardness of the samples was carried out by Wilson vickers hardness tester. The samples were prepared for microstructure examination and etched with keller's reagent. The stir zone microstructures were obtained by optical microscope and elemental identification was done by FESEM.

Table 1. The chemical composition of the alloys in this work (%wt)

Alloy Material	Cu	Mg	Mn	Zn	Si	Fe	Cr	Al
AA2024-T3	4.46	1.42	0.63	0.05	0.04	0.11	0.001	Balance
AA7075-T6	1.7	2.43	0.08	6.93	0.05	0.12	0.001	Balance

Table 2. Mechanical Properties

Base Material	Yield Strength	Tensile	Elongation	Hardness
AA2024-T3	376Mpa	446Mpa	17%	144
AA7075-T6	489Mpa	573Mpa	13%	172

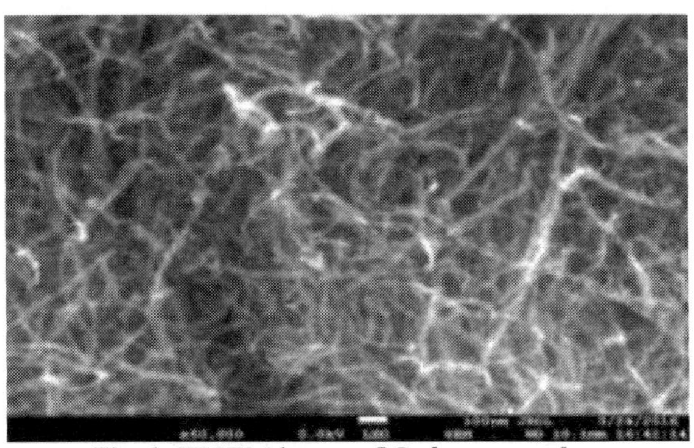

Fig. 1. FESEM image of Carbon nanotubes

3 Results and Discussion

The weld quality and weld pool geometry are observed to be effected by process parameters during friction stir welding process. During FSW, defects may be formed due to variation in the velocity field around the rotating tool. Defects may also be formed due to the material flow at high or low rotational speeds. The flow lines within the stir zone are observed to be formed at different positions of the weld as the material layers undergo different levels of plastic deformation. It has been observed that by the change in traverse speed the strength of the welded joints has changed, with nano-particles. This may be due to change in the distribution and amount of reinforcements present in the microstructure.

3.1 Microstructure

Fig. 1 and Fig. 2 show the microstructures of the welded joint at varying traverse speed. During FSW process the fine grains are formed in the material due to intense plastic deformation at high temperatures, which is referred as dynamic recrystallization[14]. Heat input is the controlling factor which determines the grain size during the FSW process, where as the conditions change due to the addition of reinforcements in to the metal matrix[6]. Also it has been reported by Humphrey et al.[19] that the migration of grain boundaries takes place at higher temperatures. Thus by increasing the rotational speed or decreasing the welding speed promotes higher temperatures in the weld zone which leads to coarse grain size [20]. The migration of grain boundaries can be controlled by the reinforcements addition which breaks the initial grains and impedes the grain growth [21]. The addition of reinforcements in to the metal matrix creates more number of misoriented low angle grain boundaries during the plastic deformation, which results more nucleating sites in the stir zone. Thus the low angle grain boundaries are transformed in to high angle grain boundaries and new grains are nucleated at favourable sites, which yielded fine grain microstructure [4]. The stir zone microstructure with carbon nanotubes at a traverse speed of 70mm/min and 80mm/min are shown in Fig. 1. The material flow in the stir zone was found to be optimum at a traverse speed of 80mm/min which improved the distribution of nanoparticles and yielded fine grains in the microstructure as shown in Fig. 1(b). At a traverse speed of 70mm/min the distribution of carbon nanoparticles were not even which yielded combination of coarse and fine grains as shown in Fig. 1(a). Similarly as the traverse speed was increased to 90mm/min the heat input decreased which influenced the material flow and the nanoparticles are distributed randomly in the microstructure as shown in the Fig. 2(a). Due to more heat input the microstructure of the welded joint without nanoparticles yielded coarse grains as shown in Fig. 2(b).

 DOI: 10.1051/matecconf/201713500039

Fig. 1. Microstructure of Stir zone at (a) 70 mm/min (b) 80 mm/min

DOI: 10.1051/matecconf/201713500039

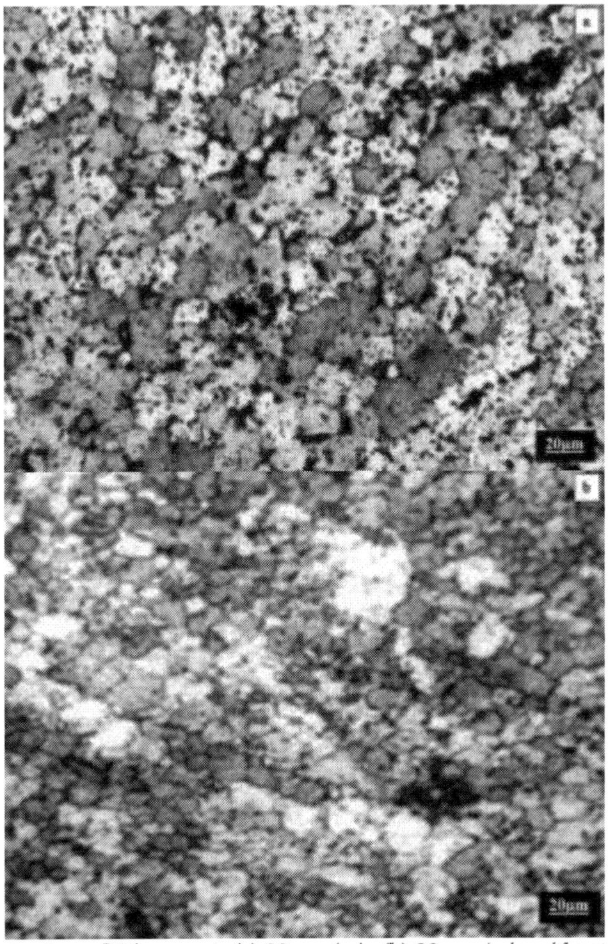

Fig. 2. Microstructure of Stir zone at (a) 90 mm/min (b) 80 mm/min without nanoparticles

3.2 EDX Results

Elemental identification was obtained by Electron dispersive X-ray analysis (EDX) as the optical microscopy does not show the amount of distribution of the elements in the stir zone. The EDX patterns of the stir zone for different traverse speeds are shown in the Fig. 3 and Fig. 4. The results show that the carbon nanotubes were deposited in all the samples, but there is a slight variation in the deposition of the reinforcements in the stir zone. This is due to the difference in material flow under various traverse speed. Thus the process parameters of 1400 rpm, 80 mm/min and 1kN are found to be optimum where the heat input is sufficient for the material flow to distribute the carbon nanotubes uniformly and maximum amount in the stir zone as shown in the Fig. 3(b). If the heat input is higher or lower than the material flow may not be proper and which influences the distribution of nanoparticles in the stir zone. Fig. 4(b) shows the EDX results for the sample without nanoparticles at a traverse speed of 80 mm/min. As the heat input was optimum at a traverse speed of 80 mm/min exhibited the maximum carbon nanotube reinforcements of 16.49 wt% and minimum amount of 12 .47 wt% at a traverse speed of 90 mm/min.

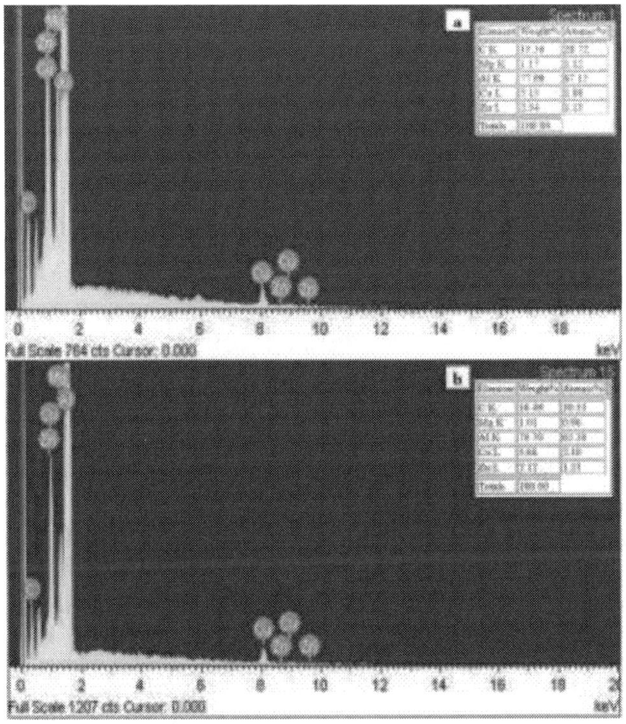

Fig. 3. EDX results of Stir zone with nanoparticles at (a) 70 mm/min (b) 80 mm/min

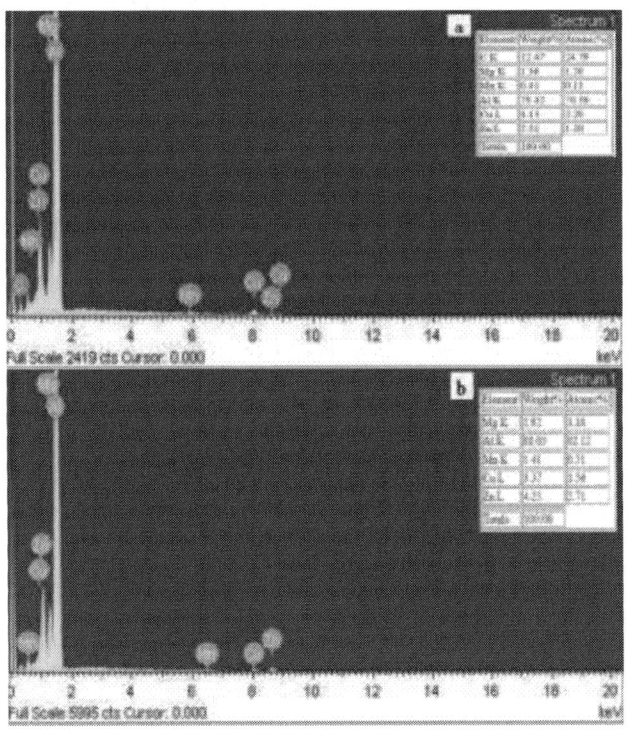

Fig. 4. EDX results of Stir zone with nanoparticles at (a) 90 mm/min (b) 80 mm/min without nanoparticles

 DOI: 10.1051/matecconf/201713500039

3.3 Tensile Strength

Tensile tests are carried out to find the effect of traverse speed on the strength of the friction stir welded joints with and without nanoparticles. The factors which influence the mechanical properties are grain size, the interaction between the reinforcements and base metals and dislocation density[4]. The generation of dislocation density was higher at the interface due to large thermal mismatch between the carbon nanotubes and aluminium matrix, which enhances the strengthening of the matrix by work hardening[22]. And the dislocations movement will be hindered by the nanosized carbon nanotubes which were distributed in the matrix. The tensile strength and microhardness of the friction stir welded joints are shown in the Table 3. Fig. 5(a) shows the effect of traverse speed on the tensile strength of welded joints. It has been observed that the strength of the welded joint was maximum at a traverse speed of 80 mm/min. This is due to the uniform distribution of carbon nanotubes in the stir zone which restricted the movement of grain boundaries and pins the movement of dislocations during elongation[14,17,18]. At 70 mm/min the heat input is more than the earlier case but the plasticity of the material is more and distribute the nanoparticles randomly and thus the strength decreases. As the welding speed is increased the heat input decreases and the material flow around the rotating tool pin influences the distribution of nanoparticles and yeilded the second heighest tensile strength of 443.2 MPa. Thus due to optimum heat input the maximum tensile strength of 449.71 Mpa was obtained for the process parameters of 1400 rpm, 80 mm/min and 1 kN. The strength of the welded joint was improved by 8.8% when compared to the welded joint without nanoparticles .

Table 3. Mechanical Properties of Friction Stir Welded Joints

Traverse Speed(mm/min)	70	80	90	80(without nanoparticles
Tensile Strength(MPa)	436.2	449.71	443.2	410.3
Microhardness(Hv)	173.5	189.3	179.7	145.2

 DOI: 10.1051/matecconf/201713500039

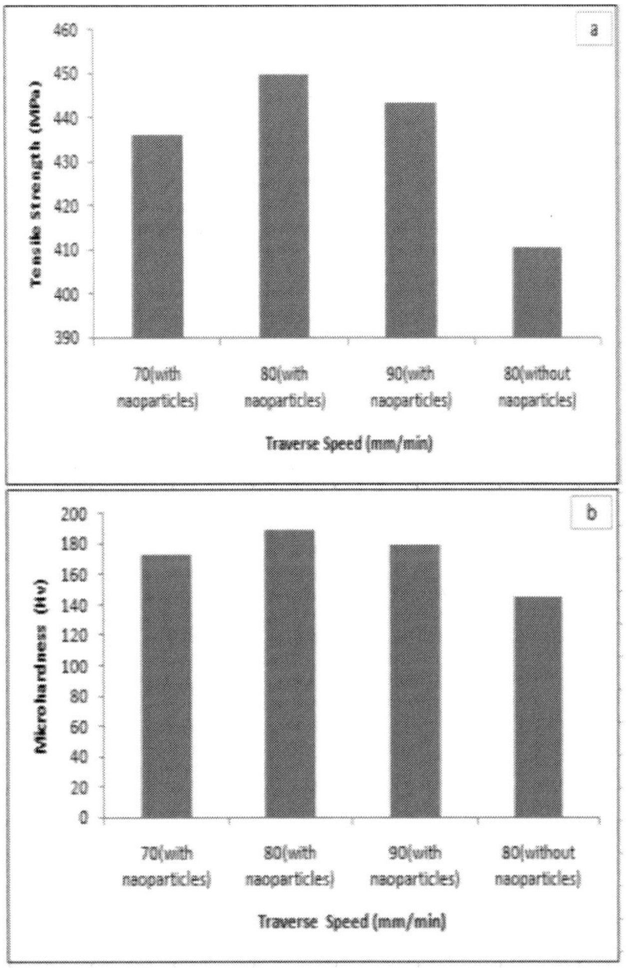

Fig. 5. Mechanical Properties of the welded joint (a) Tensile Strength (b) Microhardness

3.4 Microhardness

The important factors which influence the microhardness of the composites are amount of reinforcement distribution, heat input, grain size and dislocation density. It has been reported that the fine grains yield higher microhardness as per the Hall-Petch equation [8]. Since the reinforcing phase and the matrix have different coefficient of thermal expansion which generates the dislocations and yields high hardness [23]. The average microhardness of the stir zone is shown in the Table 3 for all welded joints. It has been observed that the microhardness of the welded joints with nanoparticles was higher than the welded joint without nanoparticles, due to the distribution of reinforcing particles in the stir zone. But the distribution is found to be varying due to change in the process parameters [16,17,18]. Fig. 5(b) shows the effect of traverse speed on the microhardness of welded joints. The maximum hardness was achieved for the welded joint fabricated at a traverse speed of 80 mm/min and minimum hardness at 70 mm/min. Even though heat input was higher at lower traverse speed the pin stirring action was found to be better at traverse speed of 80 mm/min and uniform distribution of nanoparticles was achieved. This enhanced the microhardness as fine grains

 DOI: 10.1051/matecconf/201713500039

are formed. As the traverse speed is increased the heat input decreases and non-uniform distribution of particles takes place, due to change in material flow behavior around the tool rotating pin. The microhardness of the welded joint was improved by 23.3% when compared to the welded joint without nanoparticles.

4 Conclusions

The dissimilar aluminum alloys were fabricated effectively by depositing the multi-walled carbon nanotubes during friction stir welding process. The nanoparticles were deposited in all the welded joints irrespective of process parameters and enhanced the mechanical properties. Heat input and stirring action of the tool pin are the important factors which enhances the uniform distribution of nanoparticles in the stir zone. The process parameters combination play an important role in distribution of nanoparticles in the stir zone, which enhances the mechanical properties of the welded joint. Microhardness and tensile strength were observed to be maximum at a traverse speed of 80 mm/min when compared to the welded joint obtained at 70 mm/min and 90 mm/min. The heat input was found to be optimum at 80 mm/min and uniform distribution of nanoparticles was achieved due to better stirring action of pin. Thus, the fine grain size were observed in the microstructures due to the dispersion of nanoparticles, locks the dislocations movement and enhances the strength of the welded joint. The tensile strength and microhardness of the welded joint was improved by 8.8% and 23.3% when compared to the welded joint without nanoparticles.

5 Acknowledgements

The authors are thankful to Universiti Malaysia Pahang for its constant encouragement during the course of this work. The authors are also thankful to Ministry of Higher Education, Malaysia for granting FRGS under RDU130103. The first author would like to thank University Malaysia Pahang, for awarding Doctoral Scholar Ship Scheme.

References

1. W.M.Thomas,E.D. Nicholas,J.C. Needham,M.G. Murch,P. Templesmith,C.J. Dawes, GB Patent Application No. 9125978.8, (1991).
2. M.W. Mahoney, C.G. Rhodes, J.G. Flintoff, R.A. Spurling, W.H. Bingel, Metal Mater Trans. A **29**, 1955–64 (1998)
3. S. Kundu, D. Roy, R. Bhola, D. Bhatacharjee, B. Mishra, S. Chatarjee, Materials and Design **50,**370–375 (2013)
4. M. Barmouz, P. Asadi, M.K. BesharatiGivi, M. Taherishargh, Mater Sci Eng. A **528**, 1740–9 (2011)
5. S.A.Alidokht, A.Abdollah-zadeh, S.Soleymani, H.Assadi, Materials and Design **32**, 2727-2733 (2011)
6. G.Faraji, P.Asadi, Materials Science and Engineering A **527**, 2320–30 (2011)
7. A.Shamsipur, S.F.Kashani-Bozorg, A.Zarei-Hanzaki, Surf Coat Technol. **206**, 1372–81 (2011)
8. M. Sharifitabar, A.Sarani, S.Khorshahian, M.Shafiee Afarani, Mater Des. **32**, 4164–72 (2011)
9. M. Barmouz, M.K. Besharati Givi, Composites: Part A **42**,1445–53 (2011)
10. D. Khayyamin, A. Mostafapour, A. Keshmiri, Materials Science and Engineering A **559**,217–21 (2013)
11. J.Jafari,M.K.B. Givi,M. Barmouz, Int J Adv Manuf Technol. **78**,199–209 (2015)

12. D.K.Lim,T. Shibayanagi,A.P. Gerlich, Materials Science and Engineering A, **507**(1-2), 194-199 (2009)
13. H. Izadi, A. P. Gerlich, *Carbon* **50,** 4744-4749 (2012).
14. Y.Morisada, H. Fujii, T.Nagaoka, M.Fukusumi, Materials Science and Engineering A **419**, 344–348 (2006)
15. Y.Morisada,H.Fujii,T.Nagaoka,K.Nogib,M.Fukusumi,Composites Part A: Applied Science and Manufacturing **38,** (10) 2097–2101 (2007)
16. Y.F.Sun,H. Fujii, Materials Science and Engineering A. **528** ,5470−5475 (2011)
17. B. Abnar, M. Kazeminezhad, A. H. Kokabi, Journal of Materials Engineering and Performance **24**(2) 1086-1093, (2015)
18. R. B. Ravinder, R.V. Ramaraju, A.B. Ibrahim, M.F.B. Cheku, Trans Indian Inst Met. **70**(4),1005-1017 (2017)
19. F. J. Humphreys, M. Hatherly, UK, Elsevier Ltd., University of Manchester, Institute of Science and Technology (2004)
20. M.Azizieh, A.H.Kokabi, P.Abachi, Mater Des. **32**, (4) 2034–41 (2011)
21. M. El-Rayes Magdy, A. El-Danaf Ehab, J Mater Process Technol. **212**,1157–68 (2012)
22. R.George, K.T. Kashyap, R. Rahul, S. Yamdagni , Scr Mater **53**, 1159 (2005)
23. A.Dolatkhah, P. Golbabaei, M. K. Besharati Givi, F. Molaiekiya, Materials and Design **37,** 458–464 (2012)

Designing Halal Supply Chain: Malaysia's Halal Industry Scenarios

Raja Zuraidah Rasi[1], *Nor Ratna* Masrom[2], *Siti Sarah* Omar[3], *Md Fauzi* Ahmad[4], *Rohana* Sham[5]

[1,3,4]Faculty of Technology Management and Business, Universiti Tun Hussein Onn Malaysia
[2]Faculty of Technology Management and Technopreneurship, Universiti Teknikal Malaysia Melaka
[5]Faculty of Business and Management, Universiti Teknologi Mara Pasir Gudang Campus

Abstract. This study aims to survey current halal activities in the Malaysia halal industry. The conventional supply chain focuses on one-way interconnected processes wherein raw materials are transformed into final products, and delivered to end customers. However, due to recent halal criteria requirements received from customers, the conventional supply chain has shifted to halal supply chain management. The halal supply chain is a network of supply that gives special attention and endorsement to the products and services to ensure halal integrity throughout the network. This study used a qualitative research approach. The data was gathered through in-depth interviews with five case study halal manufacturers in Malaysia. The results suggest a streamlined framework of halal supply chains that consists of (1) Halal procurement (2) Halal manufacturing (3) Halal distribution (4) Halal logistics.

1 Introduction

The demand for halal products has increased tremendously across the world, both from Muslim and non-Muslim communities. The trade of halal goods and products is estimated to be worth more than USD 2.1 trillion (Halal Journal, 2009). Based on the calculation and consumption, the global halal food market size has developed progressively from USD587.2 billion in 2004 to USD651.5 billion in 2010 (Sungkar, 2009). There were about 1.57 billion Muslims in the world in 2010, according to the studies conducted by The Pew Forum on Religion and Public Life. This figure is approximately 23.4% of the total world population.

Various countries across the world are beginning to realize the potential of the halal industry as a source of the nation's income. This attention given to halal products has led to the emergence of halal supply chain management, which focuses on halal network management to achieve halal integrity from the point of consumer purchase(Tieman, M., et

*Corresponding author: rzuraida@uthm.edu.my

al., 2012). Mohd Bahruddin, S.S. et all. (2011) mention that Halal Supply Chain Management (HSCM) is the same as Supply Chain Management (SCM), the only difference is that HSCM has been added with the element of Shariah law, providing Islamic law is the guideline for a proper halal process. Meanwhile, Omar, E.N. & Jaafar, H.S (2011) emphasizing that HSCM is a halal process from the supplier's source to the customer.

Most of Malaysian organizations that are directly involved in the halal supply chain have initiated implementation of a number of halal accredited systems due to institutional pressure, which relate to stakeholder theory; or self-determination drivers such as great halal business opportunities. While seeking to maintain a balance with this multi-faceted motivation and pressure, organizations also have to deal with unresolved halal supply chain issues and challenges. Although, a halal management system is well established, a more systematic and integrated halal supply chain is needed to reduce isolated halal practices within individual organizations. In this paper, we focus on the concept and activities of halal supply chain among Malaysian halal-certified organizations.

2 Malaysia's halal ecosystem

Halal production has seen a significant increase in Malaysia. Thus, the demands for halal material from the point of origin up until the downstream customers have become increasingly crucial to ensure the integrity of halal products. It not only covers the halalness of the material but includes all machines, tools as well as the logistic management. Ensuring halal operations throughout the supply chain is fundamental for the sustainability of halal products.

2.1 An overview of the Malaysian halal industry

The Malaysian government started to implement halal regulation in the 1960s. The regulation was initially made to protect the interest of the majority of Malaysia's Muslim people. However, due to the increasing demand of halal products worldwide, the Malaysian government has taken a further step to become a halal hub in the region and in the world. To ensure credibility of halal products, the Malaysian government has introduced a national certification system. The government has also established various agencies to facilitate the development of Malaysian halal industries. The main halal related agencies are the Department of Islamic Development Malaysia, or Jabatan Kemajuan Islam Malaysia (JAKIM), the Halal Industry Development Corporation (HDC) and the Malaysian Investment Development Authority (MIDA). Halal certification in Malaysia is organized and monitored by a government agency, JAKIM. Various halal sectors in Malaysia co-exist, supported by good policy, certification system, infrastructure and human capital development program.

2.2 Halal regulations and policy

In Malaysia, and worldwide, there is awareness of halal-certified products. In response to the growing worldwide concern for halal product integrity, new regulations and legislation related to halal products have been adopted in the Malaysia. The main elements are: (1)

Trade Description Act (2) Custom Order, (3) Animal Importation Order (4) Food Regulations and (5) Local Government Act. Table 1 below lists each of these main pieces of halal-related laws and the key provisions it contains.

Table 1. Policies related to halal products in Malaysia

Policy/ Regulation of Halal	Description
Trade Description Act 1972	Section 10 of the Act 1972 allows the minister to enact the Trade Descriptions (Use Of Expression "Halal") Order 1975.
Trade Descriptions (Marking of Food) Order, 1975 (made under s.11of the Trade Description Act 1972)	Requires food, which is halal according to the Trade Description (Use of expression "halal") Order 1975 must not be supplied unless marked with a halal label.
Custom Order (Prohibition of Imports) 1988	This order provides a control over the import of milk and milk products, poultry fat, lard, pig fat and others. Requires that all meat and livestock imported into Malaysia to be halal.
Animal Importation Order 1962	It is a requirement that all meat and livestock must be halal, safe and disease free. Enforced by the Department of Veterinary Services and the Customs and Excise Department at entry points in Malaysia.
Food Regulations 1985	Regulation 11 (1) (c) provides that where the food contains beef or pork, or its derivatives, or lard, a statement of its presence in that food should be written on the label. Regulation 11 (d): requires that food that contains alcohol to be labeled as such.
Local Government Act 1976	Section 20D (1) (a), (b) of the Licensing of Food Establishments (Federal Territory of Kuala Lumpur) (Amendment) by-Laws 1995, requires the licensee to wash utensils used for non-halal food separately from those used for halal food. It is also a requirement that the utensils used for non-halal food should not be used for the preparation and serving of halal food. Sub-section (3) states that if the establishment only serves non-halal food then the licensee must at all times display on the menu the words "Non-Halal Food Only".

Malaysia also has a halal certification system. Manufacturers use the halal certification and its associated logo as a way of informing and to reassuring their customers that the products are halal and shariah-compliant (Shafie and Othman, 2006). The Muslim halal product buyers in Malaysia look for the authentic halal certification issued by a government body, namely, JAKIM. Organizations that wish to use the halal certificate need to apply for it, then follow the JAKIM procedure. The halal certificate approved by JAKIM grants organizations the use of the standard halal logo on their products. Table 2 shows the halal certification system

Table 2. Halal Certification System in Malaysia

Certification System	Description
MS1500:2009	Halal food – Production, preparation, handling and storage Gives practical guidelines to the food industry regarding the provision, preparation, and handling of halal food (including any extra nutrition)
MS2200: Section 1: 2008,	Goods for Islamic consumers - Section 1: Cosmetics and Personal Custody – General guide
MS 2400-1: 2010	Halalan Toyyiban Assurance – Pipe – Management of transportation service systems / freight chain services
MS 2400-2: 2010	Halalan Toyyiban Assurance – Pipe – Management of warehousing services systems and related activities
MS 2400-3: 2010	Halalan Toyyiban Assurance – Retail management systems requirement

3 The Conventional Supply Chain

A basic concept of a supply chain consists of all parties involved in fulfilling a customer request (Chopra and Meindl 2013). To fulfil the customer request, organizations need to coordinate the production, inventory, locations and transportation among all parties in a supply chain (Hugos 2006). Supply chain management is the integration of suppliers, manufacturers, distributors and customers, in which raw materials source from suppliers to manufacturers who then produce them into finished products and coordinate delivery to the customers (Vonderembse, Uppal et al. 2006).

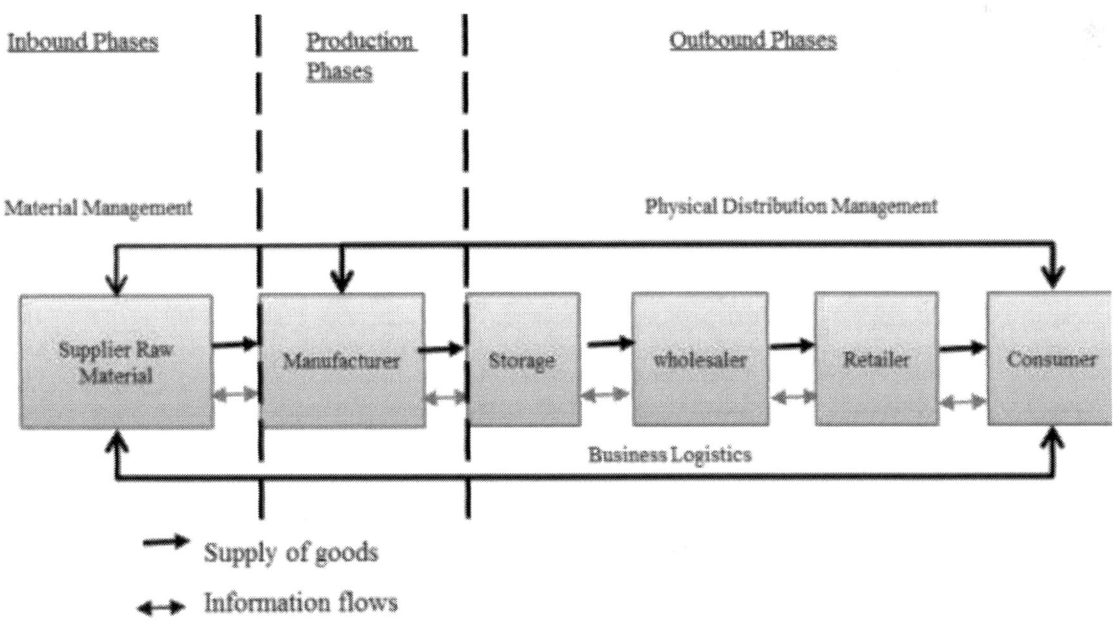

Figure 1. The flows of a conventional supply chain

Planning, modeling, designing and analysis of a conventional supply chain has mainly been driven by optimizing the procurement of materials from upstream suppliers towards the downstream distribution of products to end customers. As shown in Figure 1, basically, there are three phases of conventional supply chain which include inbound phases, production phases and outbound phases. During an inbound phase, organizations deal with material procurement and management. It involves activities of transportation, storage and delivery of the supplies coming into the plants or businesses. Meanwhile, the opposite is true of outbound logistics, which is related to similar items going out from the plant, or businesses, which include all activities of warehousing, logistics, marketing and sales to the end customer. Typically, the movement included the raw materials, any work-in-process and finished goods from upstream to downstream customers or end users.

4 Towards the Halal Supply Chain

To survey current halal manufacturing activities in Malaysia, five manufacturers were selected for case studies. An in-depth interview about halal procurement, halal manufacturing, halal distribution, and halal logistics was conducted in order to understand the halal practices of these certified organizations. According to Sekaran (2007) the in-depth interview is an effective tool for gaining a rich understanding of a new phenomenon, such as a halal supply chain. More specifically, the in-depth interview helps to understand:

- How organizations deal with halal issues in the industry
- The practices of halal supply chain among halal certified organizations in Malaysia

As stated before, to ensure the integrity of halal products and services, preventive actions should take place as early as possible, starting with the sourcing process. Figure 2 shows the flow of goods and information from suppliers to end consumers. The figure also shows activities of halal checkpoints throughout the supply chain. The concept of a halal supply chain uses similar principles to conventional supply chain management with some special attention and endorsement of the products and services related to their halalness (Mohamed et al 2016). A halal supply chain is the formulation and implementation of halal business policies and actions that simultaneously uphold market positions and generate profits. The basic notion of a halal supply chain is to ensure that the integrity and quality of the halal products are maintained throughout the entire supply chain. During each stage, it is important to ensure that organizations always consider and plan for halal quality.

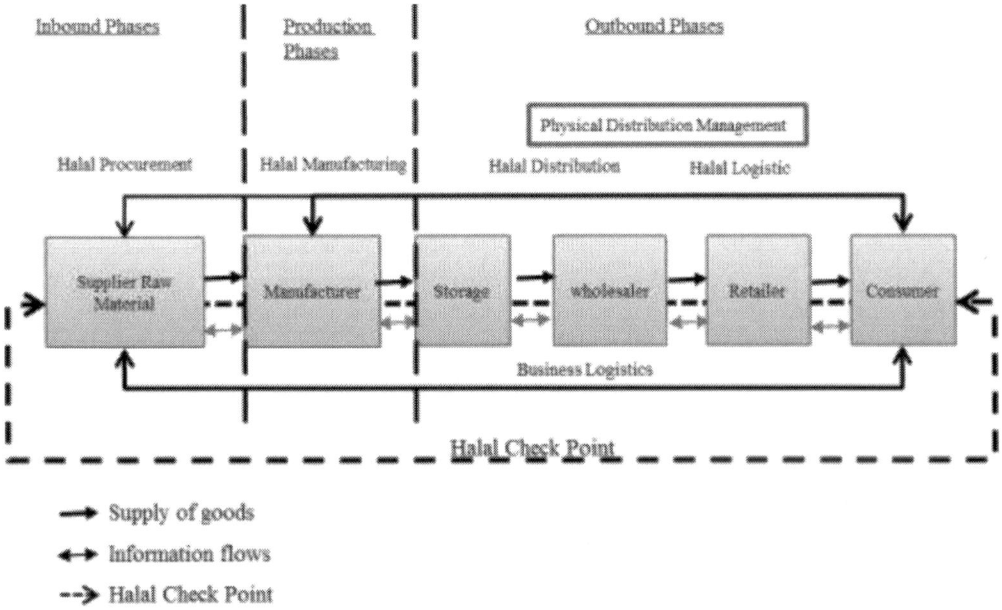

Figure 2. Conceptual Model of Halal Supply Chain Management

In general, a halal supply chain consists of four main activities: (1) Halal procurement (2) Halal manufacturing (3) Halal distribution (4) Halal logistics. In order to achieve the halal supply chain, firms must at least follow the basic principles established by MS 1500. In particular, firms must observe halal procedures in every operation involved in producing the products.

4.1 Halal procurement

Halal procurement is defined as halal purchasing, consisting of involvement in activities that focus on protecting halal integrity along the supply chain. For each sourcing activities within the halal supply chain, identification of all inputs, byproducts, and resources are crucial in order to ensure halal integrity. Given the complexity of a halal supply chain and the absence of traceability tools (Ali et al, 2013), supplier selection plays a crucial function in the sourcing process of halal material. The conformances of halal material need to be deployed at the very early stage by cautiously selecting halal certified suppliers: "Our company only chooses "halal suppliers" who satisfy all JAKIM requirements." (case D). Halal labeling is another important aspect that must be considered by the organizations when obtaining material. There is ongoing confusion surrounding the standardized halal logo, since it was produced by many agencies. Halal labeling is perceived to be more trustworthy and can actually provide market access compared to those that are labelled with International Standard Organization (ISO) certification (Shafie and Othman, 2006). The cooperation between organizations and suppliers was mentioned in each interview session. The manufacturing organizations have difficulty finding reliable suppliers that can fulfill JAKIM's requirements. Most small and medium enterprises do not have proper halal certification. On the other hand, big organizations that are halal certified often tag the price of material higher than usual market price. "(…) sometimes there are firms that can supply material with reasonable price but they don't have halal certification… We can't appoint them (...)" (case B). Therefore, a good relationship with halal certified suppliers is very important. "(…) we have our own supplier, we selected a supplier that has halal certification in order to reduce risks (…)" (case A).

4.2 Halal manufacturing

Halal manufacturing is defined as the transformation process of halal ingredients or inputs by using halal procedures on halal outputs. Halal manufacturing can lead to higher halal integrity and a reduced risk of cross contamination. Halal manufacturing consists of senior management commitment, halal certification management systems, and shariah-compliant funding. The commitment to halal manufacturing from the senior management is significant. Senior management sets up the direction of their organizations. When the organization determines to be halal producers or manufacturers, the whole ecosystem of the organization needs radical change.

4.3 Halal distribution

Halal distribution consists of halal packaging and halal containers. Packaging characteristics such as materials have an impact on the halalness of the product. Among issues that have been raised within packaging are halal certification on the packaging, halal methods of product handling and halal traceability of the packaging (Ab Talib, M.S., & Mohd Johan, M.R., 2012). In the context of halal, packaging involves certain elements such as halal certification or the halal logo, a list of ingredients and the product's origin (Rezai et all., 2012).

The other basic function of packaging in logistics is to contain and protect the goods during delivery and distribution. It is also practiced to protect workers handling the goods from hazards such as chemicals, liquids, or loose products. In the context of halal, packaging will protect goods from any cross- contamination between halal and non-halal substances. During transportation and distribution, packaging is used to segregate halal and non-halal products either by using separate carriers or separate compartments in a same carrier (Ab Talib, M.S & Mohd Johan, M.R.,2012).

4.4 Halal logistics.

Logistics include organizing, protecting and identifying products and materials before they reach the customers (Omar, E.N. & Jaafar, H.S., 2011). Ab Talib, M.S & Mohd Johan, M.R. (2012), all state that the role of packaging in logistics is to contain and protect the goods during delivery and distribution. Halal logistics capabilities are critical in ensuring the halal integrity of the supply chain from farm to fork (Tieman, 2007). The halal status does not only consider the product ingredients, it also takes into account the supply chain and the logistics aspects of halal food products. The key element of halal logistics is the segregation process between halal and non-halal food products. "(…) the container or lorries used in the delivery are used only to deliver halal products (...)" (case C).
"(…) We did not use 3PL because we do not know the cleanliness of their containers or lorries(...)" (case E)..

Furthermore, the containers used to handle halal products must not be mixed together with the ones used for non-halal products. This is done in order to avoid cross contamination (Talib et al., 2010). Rohana et al, suggest that halal authorities' bodies should be controlling the receiving of the products at the point of entry to check on the halal status of every import goods before it is allowed to be disseminating throughout the country.

All halal activities in the halal supply chain of Malaysian halal industry are shown in Figure 3.

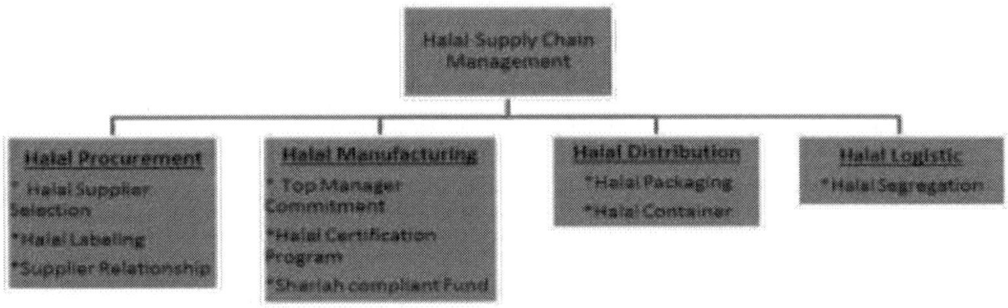

Figure 3. Halal Activities in the Halal Supply Chain of Malaysian Halal Industry

5 Conclusion

Halal supply chain management is still in its infancy. Organizations need more awareness and halal know-how, in order to enhance the halal industry's performance. This study focuses on four main categories of a halal supply chain framework: (1) Halal procurement (2) Halal manufacturing (3) Halal distribution and (4) Halal logistics. Further study needs to be done in order to develop the measurement for a halal supply chain's construction.

6 Acknowledgment

The author is grateful to Malaysian Ministry of Higher Education (MoHE) under the Fundamental Research Grant Scheme (FRGS) to the Office of Research, Innovation, Commercialization and Consultancy (ORICC) Universiti Tun Hussein Onn Malaysia (UTHM) under Vot 1430.

References

1. Ali, Mohd Helmi, Kim Hua Tan, and Zafir Mohd Makhbul. *Mitigating halal food integrity risk through supply chain integration*. Proceeding of Asia Pacific Industrial Engineering and Management System Conference. 2013.
2. Ab Talib, MS and Mohd Johan, MR 2012, *Issues in Halal Packaging: A Conceptual Paper,* International Business and Management , vol. 5, no. 2, pp. 91-5
3. Bahrudin, S.S.M., Illyas, M.I. & Desa, M.I. (2011). *Tracking and tracing technology for Halal product integrity over the supply chain*. Paper presented at the 2011

International Conference on Electrical Engineering and Informatics, Bandung, Indonesia, 17-19 July

4. Hugos, MH. 2006. Essentials of Supply Chain Management, Hoboken, NJ: John Wiley and Sons.
5. Mohamed MIKP, Raja Zuraidah Raja Mohd Rasi, Md Fauzi bin Ahmad Mohamad and Wan Fauziah Wan Yusoff, 2016. *Towards an Integrated and Streamlined Halal Supply Chain in Malaysia-Challenges, Best Practices and Framework.* The Social Sciences, 11: 2864-2870.
6. Omar, E.N. and H.S. Jaafar. *Halal supply chain in the food industry - A conceptual model.* In Business, Engineering and Industrial Applications (ISBEIA), 2011 IEEE Symposium on. 2011.
7. Rohana Sham, Raja Zuraidah Rasi, Noranita Abdamia, Suhana Mohamed and Thahira Bibi TKM, 2017 Halal Logistics Implementation in Malaysia: A Practical View IOP Conf. Ser.: Mater. Sci. Eng. 226 012040
8. Rezai, G, Mohamed, Z & Shamsudin, MN 2012, *"Non-Muslim consumers' understanding of Halal principles in Malaysia",* Journal of Islamic Marketing , vol. 3, no. 1, pp. 35-46.
9. S. Chopra, P. Meindl (2013) *Supply chain management: strategy, planning, and operations* Prentice-Hall, Upper Saddle River, NJ (2001)
10. Sekaran, U. (2007). Research methods for business (3rd ed.). New York: John Wiley.
11. Sungkar, 2009, "Rising Income and Trade Patterns of the Global Halal Meat Trade", The Halal Journal, January February 2009, p. 32.
12. Talib, Z, Zailani, S & Zainuddin, Y 2010, *"Conceptualizations on the Dimensions for Halal Orientation for Food Manufacturers: A Study in the Context of Malaysia",* Pakistan Journal of Social Sciences, vol. 7, no. 2, pp. 56-61
13. Tieman, M. (2007), "Effective Halal Supply Chains", The Halal Journal, July/August
14. Tieman, M., Jack, G.A.J., Vorst V.D. & Ghazali, M.C. (2012). *Principles in Halal Supply Chain Management.* Journal of Islamic Marketing,3(3), 217-243.
15. Shafie, S., & Othman, M.N. (2006). *Halal Certification: An International Marketing Issues and Challenges.* In Proceeding at the International IFSAM VIIIth World Congress, pp. 28-30.
16. Vonderembse, M. A., M. Uppal, et al. (2006). *"Designing supply chains: Towards theory development."* International Journal of Production Economics 100(2): 223-238.

The Impact of Business Excellence on Operational Performance among Halal Certified Food Manufacturers in Malaysia

Nor Ratna Masrom[1], *Raja Zuraidah* Raja Mohd Rasi[2] , *Badru At Tamam* Daut [3]

[1]Faculty of Technology Management and Technoprenuership, Universiti Teknikal Malaysia Melaka, Melaka, Malaysia
[2, 3]Faculty of Technology Management and Business, Universiti Tun Hussein Onn Malaysia, Johor, Malaysia

Abstract. The impact of business excellence with key elements embedded can be a tool for organization to determine opportunities and gain knowledge. This will help the organization to learn how to adopt high-performance and achieve competitive advantage. Through the business excellence model, the organization can execute their strategies to determine, forecast and survive in highly competitive market. This paper presents on identifying those elements in business excellence model (leadership, strategy and planning, customer focus, measurement, analysis, and knowledge management, operation focus) that impact on operational performance in halal certified food manufacturers in Malaysia. Data for the study were collected from 100 halal certified food manufacturers and the relationship proposed in the framework were tested using structural equation modelling (SEM). The result indicated that all the elements of Business Excellence cannot lead to operational performance except for customer focus.

1 Introduction

The challenge in today's world provides a place full of opportunities for the organizations but highly competitive. Halal product industry is highly in demand because Islam is the world's second largest religion (Hill, 2008). In the year 2010, the world's Muslim population is approximately around 1.6 billion, and this number is expected to increase by 19.8% to 20% in 2050 (DeSilver, D and Masci D,201). The global Halal food market is currently one of the fastest growing market segments. Malaysia is one of the world leader in halal food producer. In order for Malaysian to sustain as a world leader in halal food industry, they must continuously differentiate themselves from the competitors. The implementation of Business Excellence (BE) with key elements embedded can serve as a tool to find out their strengths and weaknesses. The Malcolm Baldrige National Quality Award (MBNQA) model is one of the most globally used and proven excellence models for quality awards and self-assessment (Samuelsson and Nilsson, 2002). BE focuses on developing and strengthening the management systems and processes of an organization, which further leads to the improvement of performance, thus, creating value for investors. BE emphasizes on attaining excellence in all aspects of an organization which includes leadership, strategy and planning, customer focus, measurement, analysis, and knowledge management (MAKM), workforce focus and operation focus. Additionally, achieving superior business results is the ultimate goal. MBNQA functioned to recognize companies that show outstanding performances. The MBNQA model which is the basis of US quality award now functions as a foundation for quality awards in many countries including Malaysia. It continuously concentrates on the improvement of all components of the organization, which includes tasks and activities at any level (Nabitz and Klazinga, 1999).

NIST has provided an easy-to-use questionnaire that will help the organization to learn what to improve the performance of their organization. This questionnaire is already proven as an effectiveness of BE and self-assessment is the rigor and effectiveness of the implementation process at all organizational levels.

The aim of this paper is to explore the impact of BE on operational performance in halal certified food manufacturers in Malaysia. The importance of the research lies in the placing of Malaysia as an international business hub seeking to achieve quality and excellence to gain and maintain competitive advantage and to diversify from halal food industry. One of the five guiding principles of the Malaysian strategic plan (2007-2015) is that of implementing public sector excellence.

This paper consists of four sections. Section 2 presents MBNQA model, Section 3 is. The results of a survey conducted in a number of halal certified food manufacturers in Malaysia Finally, Section 4 presents some concluding remarks, as well as future research in the context of the MBNQA model

*Corresponding author: norratna@utem.edu.my

DOI: 10.1051/matecconf/201713500041

2 Self-Assessment

Organizations need to start self-assessment process include: find the strengths and weakness of the BE embedded elements and how to improve it; provide new motivation for the improvement process and to manage the business (van der Wiele et al. (1996)).

2.1 MBNQA

The MBNQA was established in 1987 by the US government as a statement of national intention to provide quality leadership and improve the competitiveness of the USA companies. The National Institute of Standards and Technology (NIST) responsible to manage the award, with the American Society for Quality assisting with the application review process, preparation of award documents and other tasks. MBNQA perceives proceeds with change that can enhance the execution and intensity, and how it can share data adequately on execution procedures. MBNQA is a system that manage the organization and key activities processes to get results. There are seven criteria categories that are designed to endow organizations with an integrated approach to performance excellence. These criteria categories are (1) leadership, (2) strategic planning, (3) customer and market focus, (4) measurement, analysis, and knowledge management, (5) workforce focus, (6) operation focus, and (7) results. In the mid-1990s there was a change in terminology from "quality" and "TQM" to organizational (or business) excellence in the West with the previously called "Quality or TQM Models" being rebranded as business excellence models (Adebanjo, 2001).

Business excellence (BE) can be defined as "excellence in strategies, business practices, and stakeholder-related performance results that have been validated by assessments using proven business excellence models (BEMs)" (Adebanjo and Mann, 2008, p. 1).

The MBNQA BE model self-assesment questionnaire was derived from "are we making progress as a leader?" that provided in the NIST website. The criteria, description and sub-criteria of MBNQA BE model as in table 1 shown below.

Table 1: description and sub-criteria of MBNQA BE model

Criteria	description	Sub-criteria
leadership	Assesses how leaders guide and sustain their organization, how organizations view their governance system, and how organizations fulfill their ethical, legal, and community responsibilities (NIST, 2008).	1. Our workforce knows our company's mission (what we are trying to accomplish). 2. Our workforce knows our company's vision (where it is trying to go in the future). 3. Our leadership team uses our company's values to guide our company and employees. 4. Our leadership team creates a work environment that helps our employees do their jobs 5. Our leadership team shares information about the company. 6. Our leadership team asks employees what they think
Strategic planning	Determine how organizations plan their short and long term goals, and how organizations can develop relationships with customers, suppliers, and partners (NIST, 2008; Prybutok, Zhang, & Ryan, 2008).	1. As our leadership team plans for the future, we ask our employees for their ideas. 2. Our company encourages totally new ideas (innovation). 3. Our employees know the parts of our company's plans that will affect them and their work. 4. Our employees know how to tell if they are making progress on their work group's part of the plan. 5. Our company is flexible and can make changes quickly when

 DOI: 10.1051/matecconf/201713500041

		needed.
Customer focus	Examine how organizations figure out their product and service expectations, identify customer groups and market segments, and determine customer satisfaction levels to ensure the continuing relevance of product and service and to develop new business opportunities (NIST, 2008; Prybutok et al., 2008).	1. Our employees know who their most important customers are. 2. Our employees regularly ask their customers what they need and want. 3. Our employees ask if their customers are satisfied or dissatisfied with their work. 4. Our employees are allowed to make decisions to solve problems for their customers. 5. Our employees also know who our company's most important customers are.
MAKM	Figure out how the organization's selection, gathering, analysis, management, and improvement of data, information, knowledge, and information technology. Also examines how organizations use review to improve their performance (NIST, 2008).	1. Our employees know how to measure the quality of their work. 2. Our employees use this information to make changes that will improve their work. 3. Our employees know how the measures they use in their work fit into our company's overall measures of improvement. 4. Our employees get all the information they need to do their work. 5. Our employees know how our company as a whole is doing.
Workforce focus	Examines organization's ability to manage, develop, and assess its workforce. The workforce will achieve its overall mission, strategy, and action plans (NIST, 2008).	1. Our employees cooperate and work as a team. 2. Our leadership team encourages and enables our employees to develop their job skills so they can advance in their careers. 3. Our employees are recognized for their work. 4. Our company has a safe workplace. 5. Our managers and our company care about our workforce. 6. Our workforce is committed to our company's success.
Operation focus	Examines how organizations determine their core competencies and work systems, and how they design, manage and improve their key processes to implement those work systems (NIST, 2008).	1. Our employees can get everything they need to do their jobs. 2. Our company has good processes for doing its work. 3. Our employees have control over their personal work processes. 4. Our company is prepared to handle an emergency

		5. We have well established methods to measure the quality of our products. 6. We work closely with our suppliers to improve each other's processes. 7. We have site-wide standardized and documented operating procedures.

2.2 Operational Performance

The operational performance is a non-financial indicator that can measure non-financial measures can overcome the limitations of just using financial performance measures (Neely, 1999; Beamon, 1999; Merodi and Steeple, 2000). Operational performance can assess the efficiency and effectiveness of the operations within the firm. These categories of performance reflect competencies in specific areas of the supply chain, including cost, delivery speed and reliability, quality, and customer satisfaction. In measuring firm performance, much attention has been devoted to three aspects: financial, organizational, and strategic performance (Claro, 2004). Organizational theory offers three approaches to measure organizational performance, namely the goal-based, systems and multiple constituency approach. In addition, operational (non-financial) performance measures, such as product quality, customer satisfaction and market share are also frequently examined (Claro, 2004; Soboh et al., 2009). In this study, non-financial indicators are used to measure the performance of halal food operational namely as Compliance with the MS1500 standard and product standard that has been set by the companies, production efficiency and productivity, delivery and food quality.

Table 2: Description of Operational Performance

Criteria	Description	Sub-criteria
Operational performance	Examines organization's performance and improvement in all key areas including product and service outcomes, customer- focused outcomes, financial and market outcomes, workforce- focused outcomes, operation effectiveness outcomes, and leadership outcomes (NIST, 2008).	1. Maintenance of consistent compliance with legislative & regulatory requirements MS1500 2. Production efficiency 3. On time delivery 4. Increased Productivity 5. Conformity of the product to the standard 6. Food quality – Tasty, healthy and smell

3.0 Hypothesis development

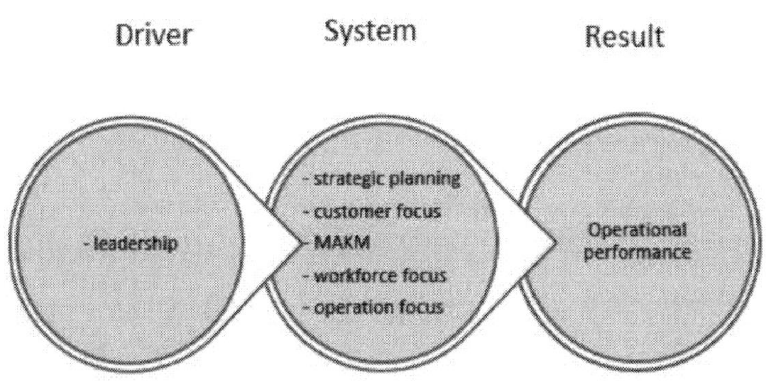

Figure 1: The elements of the general theory in the relationships of the MBNQA BE model (Flynn and Saladin, 2001)

DOI: 10.1051/matecconf/201713500041

Table 3: the elements of the general theory in the relationships of the MBNQA BE model.

MBNQA BE model criteria	Category
Leadership	Driver
Strategic planning Customer focus MAKM Workforce focus Operation focus	System
Operational performance	Result

Table 4: The causal relationships supposed by the conceptual model (hypotheses)

Hypotheses group 1 (H1): Driver criteria influences the system criteria	
H1a	Leadership has causal positive effect on Strategic planning
H1b	Leadership has causal positive effect on customer focus
H1c	Leadership has causal positive effect on MAKM
H1d	Leadership has causal positive effect on workforce focus
H1e	Leadership has causal positive effect on operation focus
Hypotheses group 2 (H2): There are interrelationships within system criteria	
H2a	Strategic planning has causal positive effect on customer focus
H2b	Strategic planning has causal positive effect on MAKM
H2c	Strategic planning has causal positive effect on workforce focus
H2d	Strategic planning has causal positive effect on operation focus
H2e	Customer focus has causal positive effect on MAKM
H2f	Customer focus has causal positive effect on workforce focus
H2g	Customer focus has causal positive effect on operation focus
H2h	MAKM has causal positive effect on workforce focus
H2i	MAKM has causal positive effect on operation focus
H2j	Workforce focus has causal positive effect on operation focus
Hypotheses group 3 (H3): Leader criteria influences the operational performance	
H3	Leadership has causal positive effect on operational performance
Hypotheses group 4 (H4): System criteria influences the operational performance	
H4a	Strategic planning has causal positive effect on operational performance
H4b	Customer focus has causal positive effect on operational performance
H4c	MAKM has causal positive effect on operational performance
H4d	Workforce focus has causal positive effect on operational performance
H4e	Operation focus has causal positive effect on operational performance

4 Methodology

4.1 Content validity

The literature reviews have been conducted. The pilot survey also has been done by giving the questionnaire to the experts from both areas academic and industry. Then, if any comments the amendment was being done to improve the content and clarity of the survey. Lastly, the final version was completed. Besides that, the Malay language version was also being translated and reviewed parallel.

4.2 Measurement Reliability

Cronbach's alpha is one of the most popular methods to assess the reliability (Rosnow and Rosenthal 1998). The value of alpha shows the relations among the measurement indicators which with higher correlations among the indicators is the higher of alpha value (Pedhazur and Schmelkin 1991). A value of alpha ranges from 0 to 1 and the higher the alpha the higher the reliability. Table 5 shows the value of Cronbach's alpha, composite reliability and average variance value for each indicator in the research model.

Table 5: Cronbach's alpha, CR and AVE value.

	Cronbach's Alpha	Composite Reliability (CR)	Average Variance Extracted (AVE)
Leadership	0.82	0.88	0.66
Strategy	0.67	0.86	0.75
Customer Focus	0.71	0.83	0.63
MAKM	0.76	0.86	0.68
Workforce Focus	0.8	0.87	0.62
Operation Focus	0.85	0.89	0.63
Operational Performance	0.81	0.88	0.64

Besides Cronbach's alpha, Werts *et al.* (1974) developed the composite reliability in assessing the reliability for a set of indicators. The composite reliability is "a closer approximation under the assumption that the parameter estimates are accurate" (Chin 1998). The reliability assessment using the composite reliability generated from SmartPLS. And the threshold value must be higher than 0.70 (Nunnally and Berbstein 1994). In table 5 it is clearly showed that composite reliabilities are above 0.70. The lowest value is strategy planning, which is 0.86. However, the values are still at acceptable range. Since both values meet the specification, thus, the measurement instrument of this study is reliable.

4.3 Measurement Validity

In this study, all constructs are reflective. Table 21 shows the factors loading, Cronbach alpha, CR and AVE values for all constructs. Fornell and Larcker (1981) suggested that AVE should be above 0.5 to achieve convergent and discriminant validities. It means that when the value at 0.5, at least 50% of more variance of the indicators should be accounted for the measured construct. Strategy planning gives the highest value of AVE at 0.75 and all other indicators are above 0.5, the suggested threshold for the validity test. Therefore, all indicators are at adequate convergent and discriminant validities.

Table 6: Value Of Outer Loading Of Initial Model

construct	Item	Outer loading	Cronbach Alpha	CR	AVE
leadership	L1	0.79	0.82	0.88	0.66
	L2	0.81			
	L3	0.83			
	L4	0.66			
	L5	0.68			
	L6	0.66			
strategy	S1	0.68	0.67	0.86	0.75
	S2	0.69			
	S3	0.71			
	S4	0.78			
	S5	0.67			
Customer focus	C1	0.75	0.71	0.83	0.63
	C2	0.57			
	C3	0.69			
	C4	0.67			
	C5	0.79			
MAKM	M1	0.7	0.76	0.86	0.68
	M2	0.81			
	M3	0.79			

construct	Item	Outer loading	Cronbach Alpha	CR	AVE
	M4	0.74			
	M5	0.74			
workforce focus	W1	0.76	0.8	0.87	0.62
	W2	0.73			
	W3	0.65			
	W4	0.6			
	W5	0.74			
	W6	0.77			
operation focus	O1	0.77	0.85	0.89	0.63
	O2	0.79			
	O3	0.68			
	O4	0.75			
	O5	0.79			
	O6	0.78			
	O7	0.69			
Operational performance	OP1	0.68	0.81	0.88	0.64
	OP2	0.83			
	OP3	0.71			
	OP4	0.82			
	OP5	0.79			

Higher outer loadings indicate that the particular indicators have much in common. The outer loading must be 0.70 or above. (Hair et al., 2013) Those below the value have to be removed. However, after removing indicators, the value of composite reliability as well as on the construct's content validity must be considered. The outer loading analysis removes weak indicators from the model.

Table 7: Value of Outer Loading of Modified Model

construct	Item	Outer loading	Cronbach Alpha	CR	AVE
leadership	L1	0.87	0.83	0.9	0.75
	L2	0.86			
	L3	0.85			
	L4	Deleted			
	L5	Deleted			
strategy	S1	Deleted	0.67	0.86	0.75
	S2	Deleted			
	S3	0.85			
	S4	0.88			
	S5	Deleted			
Customer focus	C1	0.78	0.71	0.83	0.63
	C2	Deleted			
	C3	Deleted			
	C4	0.72			
	C5	0.87			
MAKM	M1	Deleted	0.8	0.87	0.63
	M2	0.82			
	M3	0.8			
	M4	0.78			
	M5	0.77			
workforce focus	W1	0.82	0.8	0.87	0.62
	W2	Deleted			
	W3	0.73			
	W4	Deleted			
	W5	0.81			
	W6	0.79			
operation focus	O1	0.85	0.79	0.87	0.70
	O2	0.86			
	O3	0.8			

construct	Item	Outer loading	Cronbach Alpha	CR	AVE
	04	Deleted			
	05	Deleted			
Operational performance	OP1	Deleted	0.8	0.88	0.72
	OP2	0.89			
	OP3	Deleted			
	OP4	0.86			
	OP5	0.79			

To complete the convergent validity analysis, AVE and CR were revaluated. AVE and CR results in Table 22 are above 0.50 and 0.70.

4.4 Discriminant validity

The discriminant validity of the scales for constructs validity and estimated the internal consistency for reliability was done before assessing the model. In the table 8 shows below the highest value is more than 0.90 and this brings the validity has been established.

Table 8: Heterotrait-Monotrait Ratio of Correlations (**HTMT**)

	Customer	Leadership	MAKM	Operational performance	Operation	strategy
Customer						
Leadership	0.68					
MAKM	0.73	0.79				
Operational performance	0.39	0.52	0.64			
Operation	0.88	0.57	0.75	0.65		
Strategy	0.72	0.76	0.86	0.58	0.79	
Workforce	0.77	0.51	0.5	0.52	0.82	0.68

4.5 Model of fit

Model of fit in SMARTPLS was mimicking with CB-SEM. Hair et al., 2017 said that model of fits indices can judge how well a hypothesis model fits the empirical data. As a result, it can help to figure out the misspecification. This research used SRMR to identify the range of model misspecification (Dijkstra & Henseler, 2015; Henseler et al., 2014). A value less than 0.10 or of 0.08 (in more conventional criteria; see Hu and Bentler, 1999) are considered a good fit. Henseler et al. (2014) introduce the SRMR as a goodness of fit measure for PLS-SEM that can be used to avoid model misspecification.

Table 9: Model of Fit

	Saturated Model	Estimated Model
SRMR	0.08	0.10

4.6 The coefficient of determination (R-squared)

Table 10: R-squared

	R-squared	Result
strategy planning	0.33	moderate
customer	0.34	moderate
MAKM	0.55	moderate
workforce focus	0.4	moderate
Operation focus	0.62	moderate
Operational Performance	0.66	moderate

DOI: 10.1051/matecconf/201713500041

The coefficient of determination (R^2) is a measure of the model's predictive accuracy and represents the amount of variance in an endogenous construct explained by all exogenous constructs linked to it. R^2 values range from 0 to 1 with higher values indicating higher levels of predictive accuracy. The range for acceptable R^2 depends on the type of study. Henseler et al. (2009) indicates that 0.67, 0.33, and 0.19 are substantial, moderate, and weak values. Researchers seek models that are good at explaining the data with high R^2 values but also parsimonious with few exogenous constructs. Table 25 summarizes the $R^2$for model endogenous latent variables; values range between 0.33 and 0.66 reflecting a model with solid predictive accuracy.

4.7 Hypothesis testing

Path coefficients are representing the hypothesis of relationships among the constructs. The value is between -1 and +1. Estimated path coefficients close to +1 represent strong positive relationships. The closer the estimated coefficients are to 0, the weaker the relationships. Bootstrapping has been used to find out the significant of the coefficient. PLS-SEM also can figure out the relevant effects. The commonly used critical value for the two-tailed *t*-test is 1.96 for significance level of 10 percent (Hair et. al., 2013, p. 171). Table 29 summarizes the leader path coefficients, *t*-values and *p*-values. The combined analysis of path coefficients, original indicates that all path coefficients.

Table 11: Interrelation between Variables

	Leadership	Strategic Planning	Customer Focus	MAKIM	Workforce Focus	Operation Focus	Operational performance
Leadeship	-						
Strategic Planning	0.58/8.29**						
Customer Focus	0.35/3.88**	0.30/2.67*					
MAKIM	0.34/3.24**	0.33/2.61*	0.22/2.51*				
Workforce Focus	0.006/0.52	0.28/2.39*	0.44/4.86**	-0.06/0.49			
Operation Focus	-0.08/0.83	0.14/1.36	0.28/2.46*	0.27/2.11*	0.36/4.05**		
Operational Performance	0.07/0.8	0.09/1.02	0.17/2.05*	0.15/1.31	0.09/0.86	0/0.02	
R-Square	0	0.33	0.34	0.55	0.4	0.62	

Notes: R-squares are shown in bold.
[a] Represent path coefficients.
[b] Represent T-Statistics.
*Significant at the 0.05 level.
** Significant at the 0.01 level.

The strength of the correlation was assessed through the value of the correlation coefficient, which ranges from -1 to 1. In this study, given that the correlations were all positive, the range was from 0 to 1. A coefficient of 0 indicates no correlation, a coefficient of .3 indicates low correlation, a coefficient of .5 indicates medium correlation, and a coefficient of .7 or more indicates a high correlation.

Table 12: Summarization of result of hypothesis testing

Hypotheses group 1 (H1): Driver (leadership) criteria influence System Criteria		
H1a	Leadership has causal positive effect on Strategic planning	**Accepted**
H1b	Leadership has causal positive effect on customer focus	**Accepted**
H1c	Leadership has causal positive effect on MAKM	**Accepted**
H1d	Leadership has causal positive effect on workforce focus	Rejected
H1e	Leadership has causal positive effect on operation focus	Rejected
Hypotheses group 2 (H2): There are interrelationships within system criteria		
H2a	Strategic planning has causal positive effect on customer focus	**Accepted**
H2b	Strategic planning has causal positive effect on MAKM	**Accepted**
H2c	Strategic planning has causal positive effect on workforce focus	**Accepted**
H2d	Strategic planning has causal positive effect on operation focus	Rejected
H2e	Customer focus has causal positive effect on MAKM	**Accepted**
H2f	Customer focus has causal positive effect on workforce focus	**Accepted**
H2g	Customer focus has causal positive effect on operation focus	**Accepted**
H2h	MAKM has causal positive effect on workforce focus	Rejected
H2i	MAKM has causal positive effect on operation focus	Rejected
H2j	Workforce focus has causal positive effect on operation focus	**Accepted**
Hypotheses group 3 (H3): System criteria influence Results Criteria (Operational Performance)		
H3	Leadership has causal positive effect on operational performance	Rejected
Hypotheses group 4 (H4): System criteria influence Results Criteria (Operational Performance)		
H4a	Strategic planning has causal positive effect on operational performance	Rejected
H4b	Customer focus has causal positive effect on operational performance	Rejected
H4c	MAKM has causal positive effect on operational performance	Rejected
H4d	Workforce focus has causal positive effect on operational performance	Rejected
H4e	Operation focus has causal positive effect on operational performance	Rejected

Shows that customer focus has relationship with operational performance at p-values<0.05. While other criteria do not show any significant relationship. Therefore, only H4b is accepted while H4a, H4c, H4d and H4e are rejected.

5 Implications for managers

The findings of the current study provide important pointers to managers in terms of managing their organizations for superior performance. First of all, the study establishes that quality management practices are holistic in that they must be effectively linked together for organizational success. Thus, synergies must be created among the best practice factors to achieve favorable results. The study identifies strategic planning and customer focus as the two factors that act as the foundations of these synergies because they have the biggest number of direct effects on the other factors in the system. Therefore, as far as strategic planning is concerned, companies need to properly plan and develop strategies that embrace the organization's focus on quality and assume responsibility for quality and change in all aspects of management including customer relationships, MAKM and workforce focus. In terms of customer focus, organizational resources must be allocated to fulfill the need of the customer. Leadership and MAKM also has implications on a company's operational performance. The effective implementation of practices related to these factors is likely to result in improved performance. MAKM and leadership indirectly affect operational performance mediated through SCI. However, without leadership and the use of MAKM related practices operationalized in this study that affect operational performance through SCI, the impact of operation focus on operational performance may be doomed to fail. Thus, once again, managers must be aware of the significance of synergistic effects in the best practice system. The direct effect of operation focus and MAKM to SCI indicate that these factors have potential to enhance operational performance. This is due to the high impact of material costs on profits, compliance to the standard, on time delivery and operational effectiveness and increasing emphasis on just-in-time production (Sarkis et al., 2002).

6 Implication for academician

In the academic context, the current study adds to the body of literature on best practice factors and their impact on business results. This study used an extensive literature review to come up with a set of factors and indicators to measure these factors. In addition, it utilized a comprehensive literature review to formulate 21 hypotheses. Using SEM as a strong methodology to test these hypotheses, it revealed that only 10 of the 21 hypotheses were significant within the best practice system modeled. Therefore, the hypothesized relationships must be examined further in future studies to determine the reasons as to why some of the bivariate relationships that were found to be significant in other studies were empirically nonsignificant in the current study. A naive response to this would probably be that the business excellence practices of Malaysian companies are not as complex as the methods espoused by the quality researchers indicating a nonconvergence between theory and practice. However, a more in-depth look into this issue would possibly reveal that a more complex set of reasons underlie these discrepancies. For instance, although most of the quality researchers argue that quality management practices are universally applicable, there is also a group of researchers who argue that these practices are contingent upon other factors. As mentioned before, these factors may include firm type (different industries, service vs. manufacturing etc.), and country of origin of companies, which may have implications for the nature of the quality management practices implemented and so on. Therefore, it is important that future studies that test related hypotheses control for these factors in examining direct and indirect effects among factors and indicators. It is possible that these relationships occur in unique ways under different conditions. The results obtained in the current study that were similar to those of Wilson et al. (2000) and Pannirselvam et al. (2001) may suggest that these relationships are universally applicable for firms of different size and type. On the other hand, the differences in the significance of paths across three studies may indicate that the relationships under investigation are contingent upon the differing conditions in which the companies operate. One possible way to test these differences could be through the use of multiple group analyses within a SEM framework, where the significance of paths could be tested using data from different industries, from firms in different countries or even data from firms that have formal quality programs versus those that do not. In addition to testing how the best practice factors affect each other under different conditions, it would be interesting to examine how they are related to business results under such conditions, if at all. Finally, it should also be noted that it is likely for different statistical methods to yield different results regarding the relationships among variables. For instance, a simple correlation analysis (which is what many previous studies used) may indicate that two factors are highly correlated, although this relationship may turn out to be non-significant in a multiple regression or SEM analysis. Therefore, it would probably be a better idea to analyze such relationships within a multivariate framework to get more accurate results.

7 Acknowledgement

The author is grateful to Malaysian Ministry of Higher Education (MoHE) under Fundamental Research Grant Scheme (FRGS) to the Office of Research, Innovation, Commercialization and consultancy (ORICC) Universiti Tun Hussein Onn Malaysia (UTHM) under Vot 1430.

References

1. Desilver, D., 2013. World's Muslim population more widespread than you might think. *Pew Research Center, 7*.
2. U.W. Nabitz, N.S. Klazinga, (1999) "EFQM approach and the Dutch Quality Award", International Journal of Health Care Quality Assurance, Vol. 12 Issue: 2
3. Dotun Adebanjo, Robin Mann, (2000) "Identifying problems in forecasting consumer demand in the fast moving consumer goods sector", Benchmarking: An International Journal, Vol.7 Issue: 3
4. Ab Rahman, L. (2008, January). Shariah and Malaysia Halal certification system. In *Paperwork presented in Halal Food Seminar*.
5. Ab Talib, M. S., & Johan, M. R. M. (2012). Issues in halal packaging: a conceptual paper. *International Business and Management*, *5*(2), 94-98.
6. Ab Talib, M. S., Abdul Hamid, A. B., & Zulfakar, M. H. (2015). Halal supply chain critical success factors: a literature review. *Journal of Islamic Marketing*,
7. Backhouse, C. J., & Sidek, A. A. (2014). Environmental sustainability issues in Malaysian metal and fabrication SMEs: comparative analysis from a case study perspectives.
8. Bahrudin, S. S. M., Illyas, M. I., & Desa, M. I. (2011, July). Tracking and tracing technology for halal product integrity over the supply chain. In *Electrical Engineering and Informatics (ICEEI), 2011 International Conference on* (pp. 1-7). IEEE.
9. Bonne, K., & Verbeke, W. (2008). Religious values informing halal meat production and the control and delivery of halal credence quality. *Agriculture and Human Values*, *25*(1), 35-47.
10. Bovet, D., & Martha, J. (2000). Value nets: reinventing the rusty supply chain for competitive advantage. *Strategy & Leadership*, *28*(4), 21-26.
11. Bovet, D., & Martha, J. (2003). Supply chain hidden profits. *Mercer Management Consulting, available at: www. valuenets. com/book/VNPreprint. pdf (accessed 8 August 2003)*.

Effect of Extraction Process and Surface Treatment on the mechanical properties in Pineapple Leaf Fibre

Azrie Ariffin[1] *and Yusri Yusof*[2,1],

[1]Universiti Tun Hussein Onn Malaysia, (UTHM), Johor, Malaysia

Abstract. Pineapple Leaf Fibre (PALF) is a one of the natural fibre that has high potential in the industry. Natural fibres have become the main alternative source in the modern world industry. The objective of this study is to observe the effect chemical treatment using Sodium Hydroxide (NaOH) solution on the physical and mechanical properties of pineapple leaf fibre. Different concentration of NaOH solution (2%, 4%, 6%, 8%) and different treatment time (1 hour, 3 hour and 5 hour) are used for the experiment. The tensile test was conducted to obtain the mechanical properties such as tensile strength, Yong modulus, (E) and elongation at break. From the results obtained, NaOH concentration of 6% and five-hour treatment time that was used for treatment showed the higher physical and mechanical properties values. Furthermore, morphology analysis also shows the surface of the fibre at 6% NaOH after five-hour of treatment was in the better condition with good bonding arrangement of the fibre.

1 Introduction

Natural fibre has begun to pace towards becoming the main alternative source in the modern world industry [1]. Nowadays, pineapple leaf fibre (PALF) is one of the plant fibre source and good natural fibre that has the potential for the alternative source in the industry. (PALF) is used for reinforcement thermoplastic composite for application consumer goods, furniture, textile and civil structures. This type of natural fibre exhibit attractive features such as superior mechanical, physical and thermal properties, thus offer potential uses in a spectrum of applications. Pineapple Leaf Fibre contains only 2.5%-3.5% fibre, covered by a hydrophobic waxy layer which remains beneath the waxy layer [2]. In general, fibres are in form of hair-like materials which are consist of continuous filaments or in discrete elongated pieces and most likely like a thread [3]. Recent advances in natural fibre development, genetic engineering, and composites science offer significant opportunities for improved materials from renewable resources with enhanced support for global sustainability [4].

Therefore, in order to increase the mechanical properties of fibre, it may be improved by hydroxyl groups. The process is through the hydrophilization process and consequently enhancing the wetting effect of the resin on the fibres. This combination will be producing a stronger bonded [5]. The research also found that fibre surface roughness and surface area increase after alkaline treatment as shows. It possesses a smaller average diameter of about 44 μm. Each single fibre can be clearly observed because of removal of some impurities through alkaline treatment. PALF exhibits excellent mechanical properties compared to others. So, in order to improve the mechanical properties of a fibre, a treatment is applied.

[1] Corresponding author: azrieyariffin@gmail.com

There are generally consists of coupling agents or compatibilizer agents capable of inducing physical and chemical bonds between the fibre and matrix. For the physical treatments, it changes the structural and surface properties of the fibre and thereby influences the mechanical bonding with the matrix. For the chemical method, they are involving the treatment with stearic acid, coupling agents and also impregnation methods [6].

The alkaline group especially Sodium Hydroxide (NaOH) treatment is an important chemical treatment method used to enhance interfacial adhesion between natural fibre and polymer, as well as the mechanical, physical and thermal properties of the fibre. Alkali treatment is also a good alternative to these chemicals in many cases swell the cell wall of the matrix structure and give better penetration [7]. This treatment also acting to remove hemicellulose, split fibres to fibrils, and lead to a closer packing cellulose chains due to release of internal tension, which in turn increases the mechanical properties of the fibres. For the untreated fibre, it is covered with hemicellulose, lignin, pectin, waxy substances, and natural oils [8].

2 Methodology

2.1 Raw Material Preparation

Material selection in this study was a natural fibre from pineapple leaves. Pineapple leaf was chosen because of it rich in cellulose, abundantly available and easy to process and handle it. There are two most common types of pineapple planted in Malaysia, which are Josephine and Moris. For this research, Josephine is selected for testing because Josephine shows a better quality in terms of taste, leaf size and even easier to be extracted whilst maintaining a good quality of fibre [9]. Pineapple leaves (Josephine) were extracted by extractor machine called Pineapple Leaf Fibre Machine (PALF M). There were known as wet fibre, dried fibre and usable fibre. Wet fibre state was when the fibre freshly extracted from the leaf which it still contains a lot of moisture. Then, dried fibre state occurs when the fibre was dried at a certain temperature (80°C). While usable fibre state was when a total of weight of dried fibre divided by the weight of the leaf.

2.2 Pineapple Leaf Fibre (PALF) Preparation

For the preparation process, it involved weight, wash, soak, and dry process. After extraction process by using PALF M, PALF was dried under the sun for two days. This experiment required 60 pieces of fibre that have been knotted as shown in Figure 1. PALF was cut approximately twenty centimeters (20cm) and the diameter and weight were measured before through the chemical process.

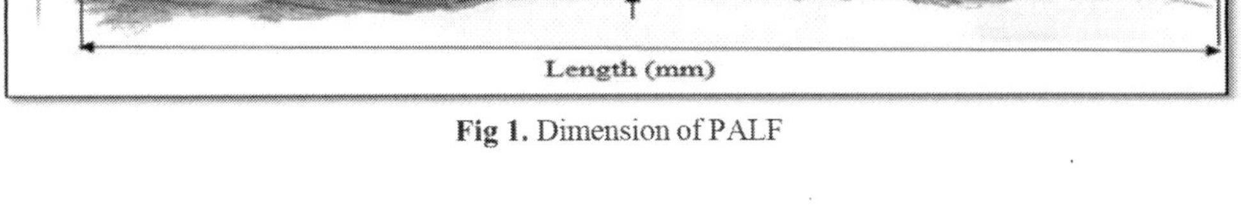

Fig 1. Dimension of PALF

2.3 Alkaline Treatment preparation

The raw material pineapple leaf fibre was undergone alkaline treatment. For this experiment, a Sodium Hydroxide (NaOH) was used as alkaline treatment. The concentration of NaOH was calculated by using equation (1). Fibres was treated with an alkaline solution to remove natural and artificial impurities which improve the adhesion between the matrix and fibres. Sodium Hydroxide solution also removes the hydrogen bonding in the network structure of the fibre cellulose. The solution also able to increased fibre surface roughness, reduce water absorption and also improving fibre-matrix adhesion.

$$\text{Concentration NaOH solution} = \frac{percentage\ needed}{100} \times total\ volume\ of\ distilled\ water \quad (1)$$

2.4 Tensile Test of PALF

The mechanical property of a material was very important in any engineering material. Strength and ability of a product produced depending on the characteristics and mechanical properties of the material. The mechanical properties were calculated by using equations (2)-(4). Besides, test results obtained were determined whether an engineering material that can be used in the application or not. For this experiment, mechanical testing methods that were carried out are based on American Standard Testing Methods (ASTM C1557-03). There were two tests performed in the experiment which is a tensile test and morphological analysis with Scanning Electron Microscope (SEM).

$$Tensile\ Strength = \frac{Force\ (load)}{Crosssectionarea} \quad (2)$$

$$Young\ Modulus, E = \frac{stress}{strain} = \frac{F/A}{\Delta L/L} \quad (3)$$

$$Strain, \varepsilon = Elongation\ \ Original\ / elongation \quad (4)$$

3 Experimental Result

3.1 Chemical Reaction between PALF and Sodium Hydroxide (NaOH)

Alkaline treatment was an important chemical treatment method used to enhance interfacial adhesion between natural fibre and polymer, as well as the mechanical, physical and thermal properties of the fibre. Figures 2 (a)-(d) show the mechanism of the reaction between the PALF and the NaOH solution. The reaction shows that the OH groups are separated from the fibre by the action of the Na+2 ions to produce a new component which is (fibre-NaO) instead of (fibre-OH). 8% of concentration NaOH was the most reactive and faster with PALF compared to other concentration because it has a high content of OH group. In addition, alkali treatment by NaOH also an alternative to these chemicals in many cases swell the cell wall of the matrix structure and give better penetration [8].

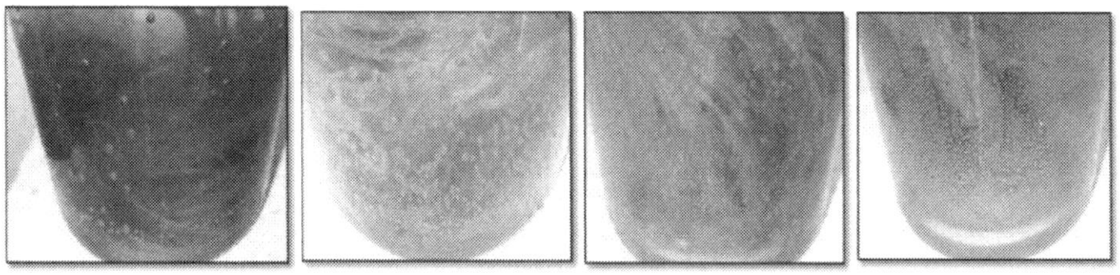

(a): 2% NaOH (b): 4% NaOH (c): 6% NaOH (d): 8% NaOH

Fig 2. The mechanism of the reaction between the PALF and the NaOH solution

3.2 Physical Properties of PALF

The data shown in Table 1 was the result of weight and diameter of PALF before the chemical treatment. All the PALF was measured and recorded in the table according to the concentration of NaOH solution. The values of weight and diameter of PALF were different depends on the sizing of pineapple leaves. All the data were taken as a reference to distinguish the result after doing the chemical treatment.

Table 1. Weight and diameter of PALF before an experiment

Concentration (%)	Weight and Diameter of Pineapple Leaf Fibre (PALF) Before an Experiment (g)	
	Weight (g)	Diameter (mm)
2	0.8672	2.22
4	0.8916	2.30
6	0.9851	2.46
8	1.0230	2.50

Figure 3 shows the higher physical reduction in weight and diameter of PALF after chemical treatment by NaOH solution for five-hour treatment. From the several conditions that used in the experiment, result for five-hour treatment stated the higher reduction physical properties of PALF. For weight reduction values of PALF, it was directly proportional to the increasing concentration of NaOH solution. Besides, the reduction value for the diameter of PALF shows a similar trend with a reduction in weight of PALF. Both of data shows that 8% NaOH was the better concentration in order to decrease of diameter and weight of PALF. This is because 8% NaOH for five-hour treatment give more reaction to the surface of fibre and removal of some hemicelluloses and lignin on the fibre surface. However, the increasing of reduction was affected the tensile strength of PALF.

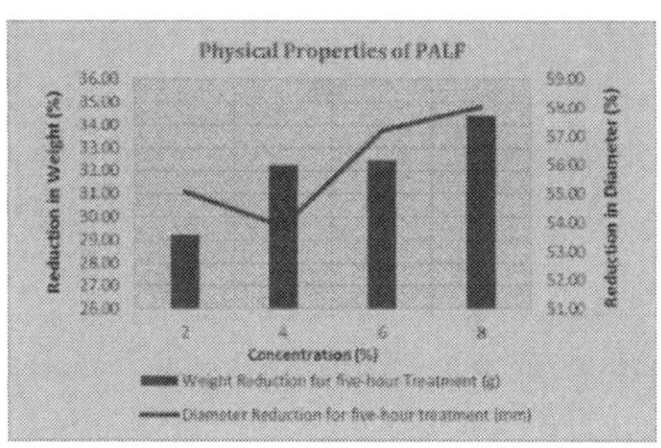

Fig 3. The higher physical properties reduction of PALF after five-hour treatment

3.3 Tensile Test of PALF

Figures 4-5 shows the relationship between the tensile strength, Young modulus, and elongation at break. From the result illustrated in Figure 7, the highest strength of fibre was occurred at 6% NaOH for five-hour treatment. At the same graph, elongation at break also gives the likely same trend for the results. The increasing of elongation of fibre was stronger the strength of the fibre. Furthermore, the maximum load required was increased to break the fibre because of strengthen of the fibre itself. Apart of that, Figure 8 shows the Young modulus and tensile strength of PALF. From the analysed, the best result for Young modulus was occur in 2% concentration of NaOH after three-hour treatment. The result of the sample was the highest compared to other samples at a different concentration of NaOH and treatment time.

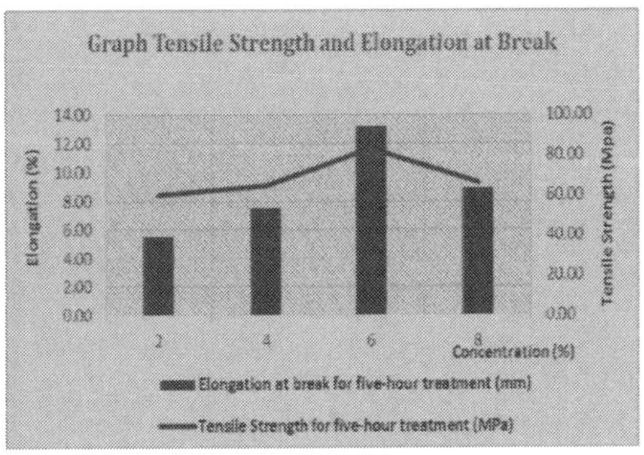

Fig 4. Graph of Tensile Strength and Elongation of PALF for five-hour treatment

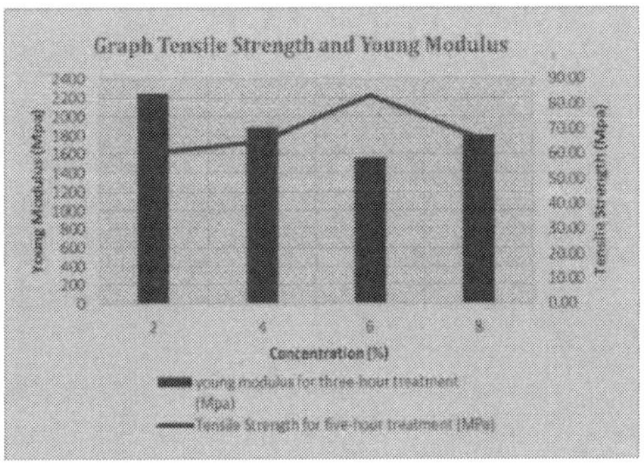

Fig 5. Graph of Tensile Strength and Young modulus, E of PALF for five-hour

For the morphological analysis, the surface structure PALF were different for each test. The results were changed due to the reaction between concentrations of chemical used with the surface of the fibre. According to the result obtained in Figures 6, the best reaction occurs between chemical and fibre at 6% NaOH after five-hour treatment. From the morphological analysis, the structure of PALF was well aligned with the matrix and give the good morphology structure. The higher strength and elongation of the fibre are produced from good structure of PALF.

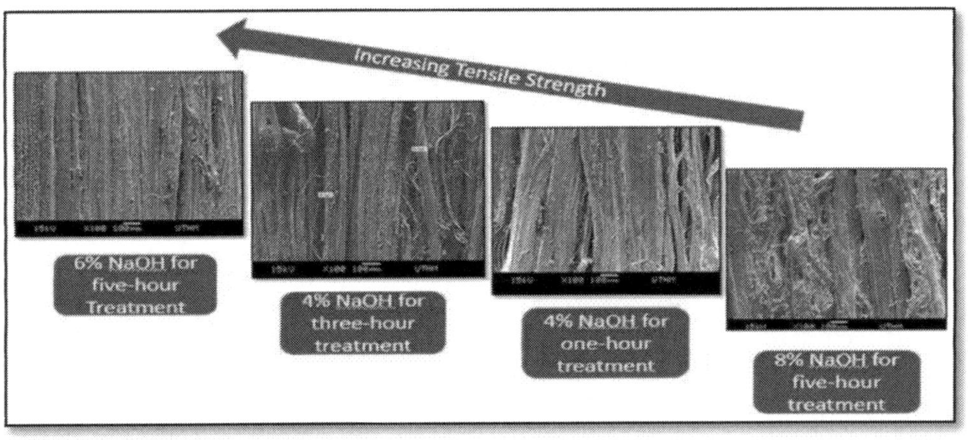

Fig 6. Surface structure of PALF with increasing the concentration of NaOH

4 Conclusion

From the results, the diameter of PALF was reduced by 38% from the diameter for untreated PALF before the experiment. The reducing of diameter recorded can be concluded that high concentration NaOH solution give more reaction to the surface of PALF and more reaction that removed the lignin and wax contained in PALF. Besides, the highest concentration of the chemical solution (NaOH) recorded the best tensile strength in the composite, giving 83.38 MPa and this condition occurs in 6% NaOH at five-hour treatment. Furthermore, the highest values of elongation for treated fibres recorded was 13.21% and it occurs in 6% NaOH for five-hour treatment. However, the Young modulus showed the maximum amount at 2% of NaOH with 2242 MPa for one-hour treatment. From the test, sodium hydroxide treatment on pineapple leaf fibre was improved the surface structure of the prepared composite material by removing waxy materials and impurities. From the result produced by SEM, PALF that treated with 6% NaOH after five-hour treatment showed the better surface structure compared to other samples.

Acknowledgements This work was supported by the Malaysian Government under Science Fund of Ministry of Science, Technology and Innovation (MOSTI) with vote No S021, Malaysian International Scholarship (MIS) under Ministry of Education Malaysia (MOE), and Universiti Tun Hussein Onn Malaysia (UTHM).

References

[1] M. Asim et al., "A Review on Pineapple Leaves Fibre and Its Composites," Int. J. Polym. Sci., vol. 2015, no. April, pp. 1–17, 2015.

[2] S. Banika,. Nag, and S. Debnath, "Utilization of pineapple leaf agro-waste for extraction of fibre and the residual biomass for vermicomposting," Indian J. Fibre Text. Res., vol. 36, no. 2, pp. 172–177, 2011.

[3] Ms. Munirah Mokhtar, Characterization and Treatments of Pineapple Leaf Fibre Thermoplastic Composite for Construction Application. 2007.

[4] A. Rijswijk, K. V., Brouwer, W.D. and Beukers, "Application of Natural Fibre Composites in the Development of Rural Societies," 2003.

[5] W. Liu, M. Misra, P. Askeland, L. T. Drzal, and A. K. Mohanty, "'Green' composites from soy based plastic and pineapple leaf fiber: Fabrication and properties evaluation," Polymer (Guildf). vol. 46, no. 8, pp. 2710–2721, 2005.

[6] J. George, S. S. Bhagawan, and S. Thomas, "Improved interactions in chemically modified pineapple leaf fiber reinforced polyethylene composites," Compos. Interfaces, vol. 5, no. 3, pp. 201–223, 1997.

[7] K. Panyasart, N. Chaiyut, and T. Amornsakchai, "Effect of surface treatment on the properties of pineapple leaf fibers reinforced polyamide 6 composites," Energy Procedia, vol. 56, pp. 406–413, 2014.

[8] J. P. S. and M. R. M. R. 1Faculty A.A. Mohammed1, 2, D. Bachtiar1,*, "Effect of sodium hydroxide on the tensile properties of sugar palm fibre reinforced thermoplastic polyurethane composites," J. Mech. Eng. Sci., vol. 10, no. 1, pp. 1765–1777, 2016.

[9] Yusri. A.Anbia Adam, "Extraction of Pineapple Leaf Fibre: Josapine and Moris," vol. 1, 2016.

Natural Pineapple Leaf Fibre Extraction On Josapine And Morris

Muhammad Firdaus Mazalan [*], *Yusri* Yusof

Faculty of Mechanical & Manufacturing, Universiti Tun Hussein Onn Malaysia

Abstract. The pineapple's leaf plant contains approximately 2.5% to 3.5% of strong white silky fibres. These fibres are useful and can be extracted from the leaves. There are a few ways to extract the fibre such as hand scrapping and by extraction machine. The objective of this research is to study the quality of fibre extraction by using different age of pineapple leaf. Next, the study aims to compare the quality of Josapine and Morris pineapple leaf with tensile test. Fibre yield percentage are calculated to determine which type of pineapple leaf produce high production of dry fibre. The mechanical properties of the fibres are analysed by Tensile Test under American Standard Testing Methods (ASTM C1577-03) and Scanning Electron Microscope (SEM). The result of the fibre yield percentage show the Josapine type on 12 month ages are the highest value fibre yield percentage which is 7.89%. Based on fibre yield percentages, it showed the Josapine type produce better dry fibre production compare to Morris type. Based on mechanical test, it showed Josapine type on 12 months ages are the strongest fibre compare to Morris type since it can withstand on 67.6 N of load.

1 Introduction

Pineapple is always use as food or desert as it is one of the famous local fruit. Mostly of the people who eat pineapple usually take the fruit and the seeds. Usually, pineapple leaf not in use and the leaves are disposed by burning or decomposed [1].

This research is focused on choose the suitable age for pineapple leaf fibre. In Malaysia, there are two types of pineapple are famous for plantation are Josapine (well-known for Johor-Sarawak-Pineapple) and Morris, which also known as Mauritius. These two types of pineapple leaf have different physical characteristic from each other. Josapine usually long and has more width than Morris [2]. Morris type are thornier and the Moris's fruit usually cheaper than Josapine. The rising of pineapple plantation around the world, human found the pineapple leaf fibre can be utilized for many purposes in different sectors of industries. The product that have be made from pineapple leaf is traditional cloth.

From the rising of pineapple plantation across the world, the pineapple leaf fibres are also being utilized for many purposes in different sectors of industries [3]. Besides, some research found the uses of pineapple leaf and there are many benefits for human use. From the research found, there are fibre in the pineapple leaf. Before this, the fibre from plant leaf

[*] Corresponding author: firdausmazalan23@gmail.com

are from banana [4]. Environmental awareness, new laws, and legislation are forcing industries to find new materials which are environmental friendly.

The objective of this paper is to study the quality of fibre extraction by using different age of pineapple leaf. Secondly, to compare the quality of Josapine and Morris pineapple leaf with tensile test.

1.1 Natural fibre

Natural fibre is a fibre that comes from the nature such as mineral fibre, animal fibre and vegetables fibre. Pineapple leaf fibre also is one of the vegetable fibres. Nowadays, the source from petroleum products is very limited and uncertain [5]. From this problem, many industry and research decided to find an alternative way with a cheap sustainable. Other than that, the alternative way also must be a source that easily available raw material.

For information, many developing countries trade lignocellulosic fibres for economic improvement. This method also can help the poor farmers as much the support from the country. Other than agricultural propose, the countries growing the plant to generate raw materials for industries. The polymer composited that containing cellulosic fibres are being under focus in the literature recently.

There are around 30 million tonnes of natural fibres that are produced every year [5]. Usually, the source of natural fibres is being used as a component in many manufacturing processes such as packaging, clothing, automobiles, sports equipment, paper making and building materials [6]. Besides that, natural fibres composites get the attention to industry due its eco-friendly nature compare traditional composites and density [7].

Basically, there are two types of retting process which is dry retting and water retting [8]. The water retting is widely used in Asian countries. The quality of dry retting usually not good as water retting. However, dry retting is most popular used in Europe. Natural fibres are divided into three major structural parts and have systematic internal cell wall structures [9].

2 Fibre yield percentage

The percentage yield of pineapple fibre is to measure the total fibre that can produce from each leaf. Both Josapine and Moris were extracted by using an extractor machine called Pinapple Leaf Fiber Machine 1 (PALF M1) and then were dried under the sun for two days. **Figure 1** show the pineapple leaf fibre machine (PALF M1). From production of fibres, there are three states in this research. The state of the fibre is wet fibre, dried fibre and usable fibre. When fibres are freshly extracted from original leaf, it contains a moisture and the name is wet fibre [2].

Figure 1: Pineapple leaf fibre machine (PALF M1)

The dried fibre is when the fibre is dried under sunlight or certain temperature after extracted from it leaf. Furthermore, the usable fibre is the total weight of dried fibre that are divided by the original weight of fibre [2]. From each types of pineapple leaf divided into three ages which is 4,8 and 12 months. For the test to be success, 5 pieces of leaf from each type of pineapple leaf are used.The percentage yield of pineapple fibre can be calculated by using equation (1).

$$fibre\ yield(\%) = \frac{Weight\ of\ dry\ palf(kg)}{Weight\ of\ fresh\ pineapple\ leaf(kg)} \times 100\% \quad (1)$$

Hence, the fibre yield projected from this equation will determine which type of pineapple leaf have the highest percentage of fibre yield.

After done with weighing process of dried fibre, the selected leaf then extracted, dried and the results for the fibre yield analysis test were shown in **Figure 2.**

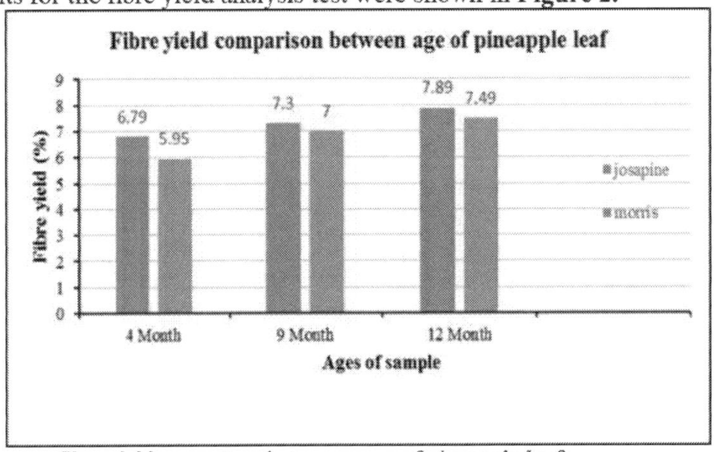

Figure 2: Average fibre yield percentage between ages of pineapple leaf

According to **Figure 2**, the highest fibre yield percentage for Josapine is 7.89% for 12 month samples and the highest fibre yield percentage for Morris is 7.49% on 12 month samples. For the Josapine type, it can give a conclusion fibre yield on 12 months with 7.89% is higher than 4 and 8 month samples which is 6.79% and 7.30%. The same result also happened on Morris type which is 12 month samples with 7.49% is higher than 4 and 8 month samples which is 5.95% and 7.00%.

For the overall, Josapine type for fibre yield percentage have higher than Morris type due to leaf size and even easier to be extracted. As conclusion, the best production of fibre between Josapine and Morris is Josapine as it gives the highest fibre yield percentage compare to Morris type. For the best ages in choosing pineapple leaf to get the highest maximum percentage yield, the most suitable age is 12 months age compare to 4 and 8 months.

3 Tensile Test Analysis

The objective of tensile test is to measure the extent to which the fibre or specimen sample can elongate to the breaking point. In other words, tensile test need to measure which force need to break the fibre or specimen sample. In this test, two types of pineapple leaf fibre are used which is Josapine and Morris type. Furthermore, each type of samples divided into different age such as 4, 8 and 12 months.

For the test to be success, 5 pieces of leaf from each type of pineapple leaf are used. The standard used for the testing was based on American Standard Testing Methods

(ASTM C1577-03). **Figure 3** show the graph for the comparison average fibre yield percentage between ages of pineapple leaf.

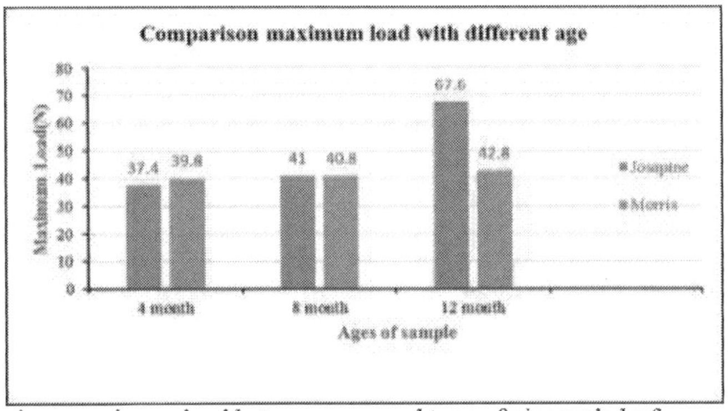

Figure 3: Comparison maximum load between ages and type of pineapple leaf

Based on **Figure 3**, the highest of the maximum load for Josapine type is 67.6 N for 12 months and the highest maximum load for Morris type is 42.8 N on 12 months. For the overall on Josapine type, it can give a conclusion maximum load value on 12 months with 67.6 N is higher than 4 and 8 months which is 37.4 N and 41 N. Meanwhile, the same result also happened on Morris type which is 12 month samples with 42.8 N is higher than 4 and 8 month samples which is 39.8 N and 40.8 N.

Next for 8 month samples between Josapine and Morris type, there are only small different which is 0.02 N between this two type of pineapple leaf. The result might be the same as the thickness of the fibre samples between Josapine and Morris type on 8 months have small difference value compare to another sample which is 4 and 12 month samples.

Furthermore, the maximum load value for Morris type higher than Josapine for 4 month samples. The result might be affected because the thickness of the fibre sample of Morris is bigger than Josapine type. Next, the lowest value of maximum load is from Josapine on 4 month samples with 37.4 N.

From the overall sample between age and type of pineapple leaf, the best and the stronger fibres since it has higher maximum load value is Josapine type with 67.6 N. Meanwhile, the stronger fibres based on age and is 12 month ages.

4 Morphological Analysis

Fibre structure can be studied by using Scanning Electron Microscope (SEM) machine. From this test, it can analyse the surface structure of the pineapple leaf fibre. In this test, there are six samples which is three sample from Josapine type and three sample from Morris type. The samples are divided based on their age which is 4, 8 and 12 month ages. Each sample from the type of pineapple leaf fibre are chosen from the highest value of fibre yield percentage. **Figure 4** show the surface structure for different ages and type of pineapple leaf fibre.

From figure **Figure 4**, it is apparent that the pineapple leaf fibre extracted using PALF M1 produces parallel fibre arrangement. Meanwhile, the surface structure for 4 month ages show Josapine type is more attractive and clean in producing dry fibre after extraction process compare to Morris type.

Other than that, the surface structure for 8 month ages show the same result as 4 month samples which is Josapine type is more clean compare to Morris. Based on 8 month Morris sample, the surface structure not clean enough after extraction process since Morris leaf surface more thicker compare to Josapine type.

Josapine	Morris
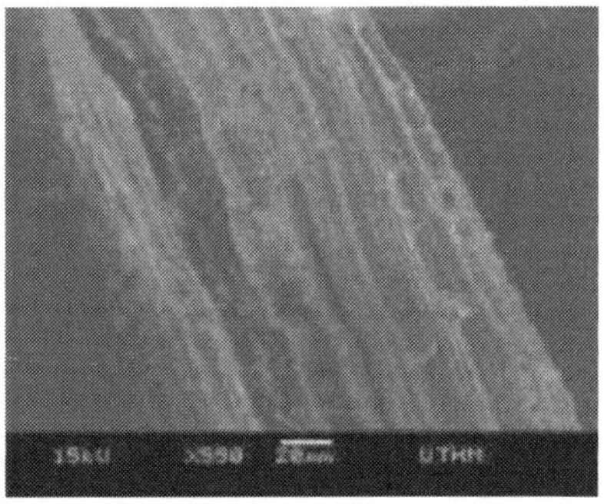 4 month	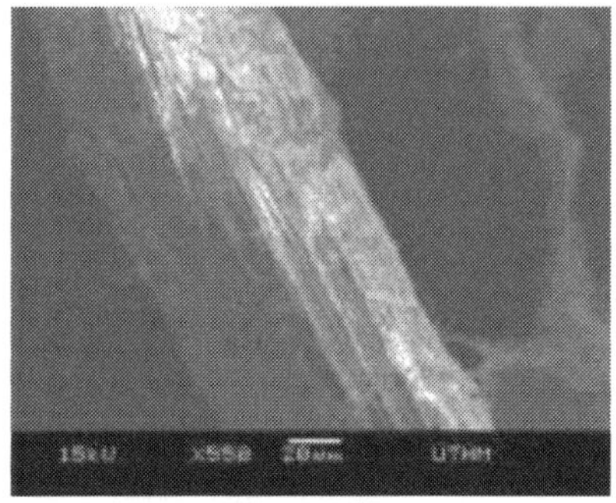 4 month
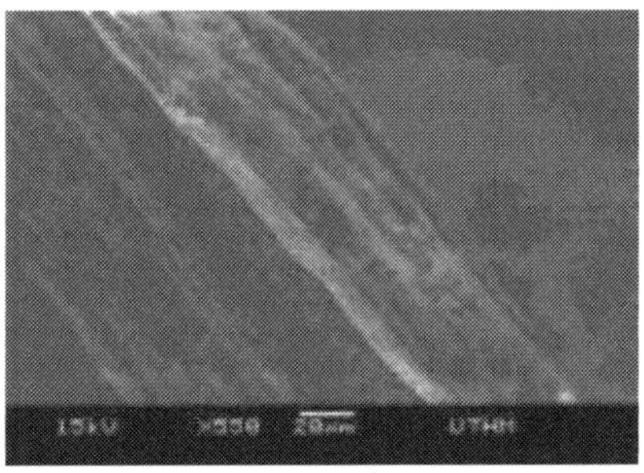 8 month	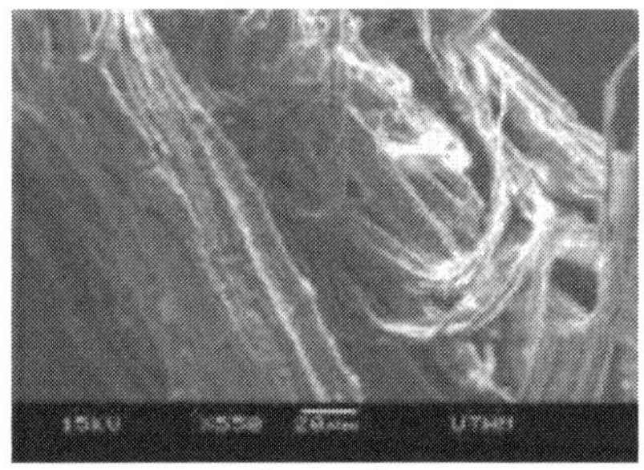 8 month
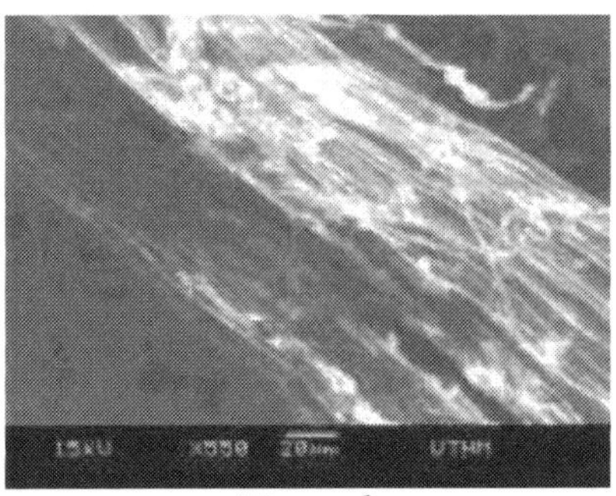 12 month	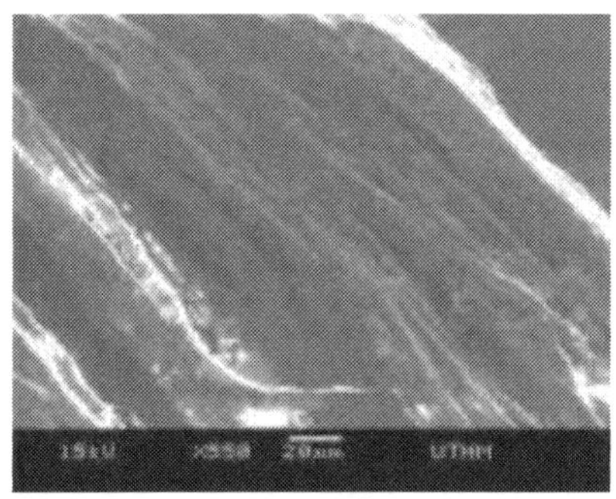 12 month

Figures 4: SEM Images of structure at 550x magnification of Pineapple leaf fibre

Finally, the surface structure for 12 month show the Josapine type produce more fibre and completely extracted compare to Morris type. For the overall, pineapple leaf fibre from Josapine type produce better quality compare to Morris as Josapine leaf easier to be extracted [2].

5 Conclusion

For the development and improvement in pineapple leaf fibre usage across the country, it is important to determine which type of pineapple leaf better. From the test and analysis, it can be said Josapine type more better than Morris type since it has the more

DOI: 10.1051/matecconf/201713500043

fibre, strong and easier to be extracted. Other than that, Josapine fruit's price higher than Morris so Josapine type more suited type to be grown in agriculture sector.

References

[1] Y. Yusof, S. A. Yahya, and A. Adam, "A New Approach for PALF Productions and Spinning System : The Role of Surface Treatments," vol. 1, no. 2, pp. 161–164, 2014.

[2] A. Adam and A. Yahya, "Extraction of Pineapple Leaf Fiber : Josapine and Moris," vol. 11, no. 1, pp. 161–165, 2016.

[3] D. B. Dittenber and H. V. S. G. Rao, "Critical review of recent publications on use of natural composites in infrastructure," in *International SAMPE Technical Conference*, 2011.

[4] R. M. N. Arib, S. M. Sapuan, M. M. H. M. Ahmad, M. T. Paridah, and H. M. D. K. Zaman, "Mechanical properties of pineapple leaf fibre reinforced polypropylene composites," *Mater. Des.*, vol. 27, no. 5, pp. 391–396, 2006.

[5] M. Asim *et al.*, "A Review on Pineapple Leaves Fibre and Its Composites," *Int. J. Polym. Sci.*, vol. 2015, pp. 1–17, 2015.

[6] M. Jawaid and H. P. S. Abdul Khalil, "Cellulosic/synthetic fibre reinforced polymer hybrid composites: A review," *Carbohydrate Polymers*, vol. 86, no. 1. pp. 1–18, 2011.

[7] M. Drzal, L. T., Mohanty, A. K., & Misra, "Bio-Composite Materials As Alternatives To Petroleum-Based Composites for Automotive Applications," *Magnesium*, vol. 40, pp. 1–3, 2001.

[8] J. Foulk, D. Akin, R. Dodd, and D. McAlister III, "Flax fiber: potential for a new crop in the Southeast," *Trends new Crop. new uses*, no. December 2015, pp. 361–370, 2002.

[9] B. Madsen, "Properties of Plant Fibre Yarn Polymer Composites," *Thesis*, p. 218, 2004.

[10] D. N. Saheb, J. P. Jog, D. Nabi Saheb, and J. P. Jog, "Natural Fiber Polymer Composites : A Review," *Adv. Polym. Technol.*, vol. 18, no. 4, pp. 351–363, 1999.

[11] A. K. Bledzki and J. Gassan, "Composites reinforced with cellulose based fibres," *Prog. Polym. Sci.*, vol. 24, no. 2, pp. 221–274, 1999.

[12] M. J. John and R. D. Anandjiwala, "Recent developments in chemical modification and characterization of natural fiber-reinforced composites," *Polym. Compos.*, vol. 29, pp. 187–202, 2008.

The Relationship between TQM Practices with TQM Tools and Techniques in Small and Medium Enterprise (SMEs)

Md Fauzi Ahmad[1,*] ,Jessica Chan Siew Yin[1] ,*Chan Shiau Wei* [1] ,*Nor Aida Abdul Rahman* [2] ,*Nik Hisyamudin Muhd Nor* [3],*Mohd Fahrul Hassan* [3] ,*Fatan Adibah Hashim* [4]

[1]Faculty of Technology Management, Universiti Tun Hussein Onn Malaysia, Malaysia
[2]Faculty of Aviation Management Department, Universiti Kuala Lumpur Malaysia Institute of Aviation technology (UniKL MIAT), Malaysia
[3]Faculty of Mechanical, Universiti Tun Hussein Onn Malaysia, Malaysia
[4]Sharp Manufacturing Corporation Malaysia, Malaysia

Abstract. Total Quality Management (TQM) is an approach which enforcing continuous improvement in all aspects in an organization for long term achievement. However, small and medium enterprises (SMEs) have limited resources for implementing TQM tools and techniques in their organisation. Hence, the purpose of this research is to identify the extent level of TQM practices and TQM tools and techniques and to examine their relationship. Research hypotheses were examined while considering six dimension of TQM practices which considered important in TQM tools and techniques implementation. Random sampling and survey have been used by distributing questionnaire to 200 respondents for final study. The study result shows that the extent level in TQM tools and techniques is in high and moderate level. Furthermore, there is a significant relationship between TQM practices and TQM tools and techniques in SMEs. The study is important for SMEs to gain insight on the effect of TQM practices towards TQM tools and techniques in organizational performance and identify the most important tools and techniques for their companies in future implementation.

1 Introduction

Small and medium enterprises (SMEs) are playing important role in developing a country [1]. In the developing countries, SMEs constitute as a main sources of national income and create an important area for business activity [2]. In recent decades, SMEs in all sectors have done breakthrough by creating and ensuring their product quality or service quality in every complexion of daily operations because quality is a key to gain competitive advantage. Hence, total quality management (TQM) had become a widespread practice in SMEs. According to Ahmad [3], a successful TQM implementation can create sustainable quality and productivity in long term by developing continuous improvement. TQM

* Corresponding author: mohdfauzi@uthm.edu.my

DOI: 10.1051/matecconf/201713500044

practices included top management, customer satisfaction, employee improvement, continuous improvement, supplier partnership and performance measures. Moreover, continuous improvement is important when an organisations begin to start implement TQM in organizations [4][5]. TQM tools and techniques are important to maintain SMEs' quality and productivity performance. Furthermore, TQM tools and techniques defined by the Deming Prize Committee of the Union of Japanese Scientists and Engineers [6] as: "a set of systematic activities carried out by the entire organization to effectively and efficiently achieve the organization's objectives so as to provide products and services with a level of quality that satisfies customers at the appropriate time and price. Both TQM practices and TQM tools and techniques are playing important role in maintaining a good quality and productivity performance of SMEs.

In fast-speeding global era, TQM tools and techniques are commonly implementing in SMEs. TQM is a multidimensional approach [7][8]. It integrates fundamental approach can lead to a dynamic capability of SMEs. The purpose of implementing TQM is to provide quality products or services to customers, which will, in turn, increase productivity and decrease costs. SMEs use the basic tools like check sheet, flow chart and control chart which is simple and easy to carry out in order to operate their daily process for the past decade. SMEs are using most basic tools because implement new tools to evaluate their daily processing need some professional knowledge, skills and training nowadays [9]. However, SMEs required high investment such as man power, skill and financial to implement TQM tools and techniques [10]. In this situation, limited TQM tools and techniques usage had affected poor quality and productivity performance of SMEs [5]. Hence, this research helps to investigate the most related tools and techniques for TQM implementation by analysing the extent level and the relationship to TQM. SMEs implement TQM tools and techniques to enhance their quality and productivity performance and maintain close relationship with supplier and customer [7][3]. Hence, SMEs should focus more on TQM practices, as this may support both knowledge management practices and performance of the firms.

The following research questions will be examined in the study:

1. What is the extent level of TQM tools and techniques?
2. What is the relationship between TQM practices with tools and techniques?

The hypothesis is set to test the relationship and the extent level between TQM practices with TQM tools and techniques. Quality tools act as an important role for product development process from commence phase until the very hindmost phase of production process [11]. The tools divided into basic tools and new basic tools. The basic tools which most implement by previous organization are flow chart, check sheet, histogram, Ishikawa diagram, scatter diagram and control chart. TQM practices as basis for implementing tools and techniques. By implementing quality tools, it can enhance business performance [11]. Based on the theoretical framework, the research hypotheses are formulated by:

***H1**: There is significant relationship between TQM practices and TQM tools and techniques*

2 Methodology

The population is observed from small and medium enterprise (SMEs). The unit of analysis is organisation. There are total of 400 samples have been selected for this study. Random sampling is used for this research. Survey method was used in this research. Finally, 80 questionnaires were received for final study. It represented 20.0% of response rate. Statistical Package for Science Social (SPSS) was used to analyse the data being collected. Descriptive and correlation test have been carried out to answer the research questions. Pearson and Spearman test have been used for correlation test.

3 Analysis Results and Discussion

3.1 Conbrach's alpha

Conbrach's alpha [12] was used to determine the reliability of the questionnaire as shown in Table 1. Cronbach's alpha is a measure of internal consistency, which able to determine how closely related a set of items are as a group.

Table 1. Cronbach's apha result

TQM Practices	Cronbach's alpha
Training	0.886
Top Management Commitment	0.898
Customer Focus	0.885
Continuous Improvement	0.872
Employee Involvement	0.864
Process Management	0.882

3.2 Extent level of TQM tools and techniques

Table 2 shows Ishikawa diagram is the most welcomed by SMES and score the highest mean which is 5.3125 among the rest of TQM tools and techniques. However, bottom five TQM tools and techniques which less practiced by SMES are design for manufacturability (DFM), matrix data analysis, mass customization, relation diagram and matrix diagram. Design for manufacturablity is the least conducted by Small and Medium Enterprise (SME) and score the lowest mean which is 3.1750 in between scale 1 to scale 7 among the rest of TQM tools and techniques.

Table 2. Extent level of TQM tools and techniques

	N	Mean	Interpretation	Standard Deviation	Rank
Flow Chart	**80**	**4.8375**	**High**	**1.67969**	**4**
Check Sheet	**80**	**4.9125**	**High**	**1.59266**	**2**
Histogram	**80**	**4.8875**	**High**	**1.66873**	**3**
Pareto Diagram	80	4.1500	Moderate	1.60773	10
Ishikawa Diagram	**80**	**5.3125**	**High**	**1.66569**	**1**
Scatter Diagram	80	3.8750	Moderate	1.66403	12

 DOI: 10.1051/matecconf/201713500044

Control Chart	80	4.1625	Moderate	1.65693	11
Affinity Diagram	80	3.5625	Moderate	1.71990	15
Tree Diagram	**80**	**4.8250**	**High**	**1.68219**	**5**
Progress Decision Program Chart	80	3.7875	Moderate	1.74075	13
Matrix Diagram	80	3.4250	Moderate	1.71202	16
Relation Diagram	80	3.2875	Moderate	1.85686	17
Matrix Data Analysis	80	3.2125	Moderate	1.87383	19
Arrow Diagram	80	3.5949	Moderate	1.75791	14
FMEA	80	4.5875	High	1.61240	7
QFD	80	4.2875	Moderate	1.67781	9
SPC	80	4.7625	High	1.56095	6
Mass Customization	80	3.2875	Moderate	1.63965	18
Design For Manufacturability	80	3.1750	Moderate	1.69717	20
DOE	80	4.3500	Moderate	1.75816	8
		Total:	82.2824		
		Mean:	4.1141	Moderate	

3.3 Correlation Test

From the Table 3, it shown the correlation coefficient, r=0.716, indicating a high positive correlation and significant where $p < 0.01$. Hence, H1 is accepted.

Table 3. Correlation between TQM practices and TQM tools and techniques

		TQM tools and techniques	Correlation
TQM Practices	Pearson Correlation Sig. (2-tailed) N	.716** .000 80	High

The results demonstrated on Table 4, all the results are significant. Statistical process control (SPC), flow chart, histogram, pareto diagram and check sheet have high correlation between TQM practices and TQM tools and techniques which shown r=0.623, r=0.618,r=0.605, r=0.601 and r=0.600. TQM practices are considered to be a significant predictor of TQM practices across TQM tools and techniques. The most effectiveness TQM practices toward TQM tools and techniques are statistical process control (SPC), flow chart, histogram, check sheet and pareto diagram. The highest correlated tools and techniques toward TQM practices is SPC. The reason from Paiva (2013) states that SPC has significant advantage over inspection as quality control techniques such as it emphasize on prevention of problem. Montgomery (2009) states that SPC is a powerful collection of problem-solving tools to maintain process stability and enhance capability through reduction of variability. Alternatively, the least effectiveness TQM practices toward TQM tools and techniques are matrix data analysis, relation diagram, affinity diagram, matrix diagram and process decision program chart.

 DOI: 10.1051/matecconf/201713500044

Table 4. Correlation between TQM practices and TQM tools and techniques

Hypothesis (H)	TQM tools and techniques		Mean TQM Practices (r)	Correlation	Result	Rank of Effectiv-eness
H1a	Flow Chart	Pearson Correlation Sig. (2-tailed) N	.618** .000 80	Moderate	Significant	2
H1b	Check Sheet	Pearson Correlation Sig. (2-tailed) N	.600** .000 80	Moderate	Significant	5
H1c	Histogram	Pearson Correlation Sig. (2-tailed) N	.605** .000 80	Moderate	Significant	3
H1d	Pareto Diagram	Pearson Correlati-on Sig. (2-tailed) N	.601** .000 80	Moderate	Significant	4
H1e	Ishikawa Diagram	Pearson Correlati-on Sig. (2-tailed) N	.569** .000 80	Moderate	Significant	6
H1f	Scatter Diagram	Pearson Correlati-on Sig. (2-tailed) N	.518** .000 80	Moderate	Significant	8
H1g	Control Chart	Pearson Correlati-on Sig. (2-tailed) N	.465** .000 80	Low	Significant	10
H1h	Affinity Diagram	Pearson Correlati-on Sig. (2-tailed) N	.403** .000 80	Low	Significant	18
Hypothesis (H)	TQM tools and techniques		Mean TQM Practices (r)	Correlation	Result	Rank of Effectiv-eness
H1i	Tree Diagram	Pearson Correlati-on	.463** .000 80	Low	Significant	11

DOI: 10.1051/matecconf/201713500044

Hypothesis (H)	TQM tools and techniques		Mean TQM Practices (r)	Correlation	Result	Rank of Effectiv-eness
		Sig. (2-tailed) N				
H1j	Progress Decision Program Chart	Pearson Correlati-on Sig. (2-tailed) N	.428** .000 80	Low	Significant	16
H1k	Matrix Diagram	Pearson Correlati-on Sig. (2-tailed) N	.411** .000 80	Low	Significant	17
H1l	Relation Diagram	Pearson Correlati-on Sig. (2-tailed) N	.400** .000 80	Low	Significant	19
H1m	Matrix Data Analysis	Pearson Correlati-on Sig. (2-tailed) N	.383** .000 80	Low	Significant	20
H1n	Arrow Diagram	Pearson Correlati-on Sig. (2-tailed) N	.441** .000 80	Low	Significant	14
H1o	FMEA	Pearson Correlati-on Sig. (2-tailed) N	.519** .000 80	Moderate	Significant	7
H1p	QFD	Pearson Correlati-on Sig. (2-tailed) N	.455** .000 80	Low	Significant	13
Hypothesis (H)	TQM tools and techniques		Mean TQM Practices (r)	Correlation	Result	Rank of Effectiv-eness
H1q	SPC	Pearson Correlati-on	.623** .000 80	Moderate	Significant	1

Hypothesis (H)	TQM tools and techniques		Mean TQM Practices (r)	Correlation	Result	Rank of Effectiv-eness
		Sig. (2-tailed) N				
H1r	Mass Customization	Pearson Correlati-on Sig. (2-tailed) N	.461 ** .000 80	Low	Significant	12
H1s	Design For Manufacturability	Pearson Correlati-on Sig. (2-tailed) N	.437** .000 80	Low	Significant	15
H1t	DOE	Pearson Correlati-on Sig. (2-tailed) N	.466** .000 80	Low	Significant	9

4 Conclusion

The study result shows that the extent level in TQM tools and techniques is in high and moderate level. Furthermore, there is a significant relationship between TQM practices and TQM tools and techniques in SMEs. The study is important for SMEs to gain insight on the effect of TQM practices towards TQM tools and techniques in organizational performance and identify the most important tools and techniques for their companies in future implementation.

5 Acknowledgements

Appreciation to MOHE and ORRIC, Universiti Tun Hussein Onn Malaysia for supporting this research (Vot: U429). Appreciation also to Manufacturing Technology Management (MTM) focus group, Faculty of Technology Management.

References

[1] C. Valmohammadi, "The impact of TQM implementation on the organizational performance of Iranian manufacturing SMEs," *TQM J.*, vol. 23, no. 5, pp. 496–509, 2011.

[2] M. A. Hameed, S. Counsell, and S. Swift, "Journal of Engineering and A conceptual model for the process of IT innovation adoption in organizations," *J. Eng. Technol. Manag.*, vol. 29, no. 3, pp. 358–390, 2012.

[3] M. F. Ahmad, N. Zakuan, A. Jusoh, S. M. Yusof, J. Takala, and M. S. M. Arif, "Comparative Study of TQM Practices between Japanese and Non-Japanese

Companies: Proposed Conceptual Framework," *Adv. Mater. Res.*, vol. 903, pp. 371–377, 2014.

[4] M. F. Ahmad, R. Z. R. Rasi, N. Zakuan, M. . Haji-Pakir, and J. Takala, "The Impact of ASEAN Free Trade Agreement as Moderator on TQM Performance Model in Malaysia: Survey Result," *Soc. Sci.*, vol. 11, no. 12, pp. 2932–2937, 2016.

[5] M. F. Ahmad, N. Zakuan, R. Z. R. M. Rasi, M. N. N. Hisyamudin, and J. Takala, "Mediator effect of total productive maintenance between total quality management and business performance: Survey result in Malaysia automotive industry," *Adv. Sci. Lett.*, vol. 21, no. 12, pp. 3723–3725, 2015.

[6] S. K. Breja, D. K. Banwet, and K. C. Iyer, "Quality strategy for transformation: a case study," *TQM J.*, vol. 23, no. 1, pp. 5–20, 2011.

[7] M. F. Ahmad, M. S. M. Ariff, N. Zakuan, M. Z. M. Saman, S. S. S. A. Rahman, T. A. R. Abdullah, and N. Fadzil, "The effect of demographics on quality management principles of ISO 9001:2008 amongst Malaysia Hajj Pilgrims," *Soc. Sci.*, vol. 11, no. 11, pp. 2748–2752, 2016.

[8] M. F. Bin Ahmad and S. M. Yusof, "Comparative study of TQM practices between Japanese and non-Japanese electrical and electronics companies in Malaysia: Survey results," *Total Qual. Manag. Bus. Excell.*, vol. 21, no. 1, pp. 11–20, 2010.

[9] M. F. Ahmad, M. S. M. Arif, N. Zakuan, S. S. S. A. Rahman, T. A. R. Abdullah, and N. Fadzil, "The Effect of Demographics on Customer Satisfaction amongst Malaysia Hajj Pilgrims: Survey Result," *Appl. Mech. Mater.*, vol. 660, pp. 1000–1004, 2014.

[10] M. F. Ahmad, M. S. M. Arif, N. Zakuan, S. Rahman, M. Latif, and M. Khalid, "The Mediator Effect of Customer Satisfaction between Quality Management Practices and Communication Behavior amongst Malaysia Hajj Pilgrims: Survey Result," *Appl. Mech. Mater.*, vol. 660, no. 2015, pp. 1005–1009, 2015.

[11] M. Ahmad, N. Zakuan, J. Ahmad, and J. Takala, "Meta–analysis of the TQM impact on business performance amongst regions and countries," *Int. J. Ind. Syst. Eng.*, vol. 20, no. 2, pp. 155–164, 2015.

[12] J. C. Nunally, *Psychometric Theory*, Second ed. New York: McGRAW-HILL, 1978.

Continuous Improvement and its Barriers in Electrical and Electronic Industry

Md Fauzi Ahmad[1,*], *Toh Li Yan*[1], *Chan Shiau Wei*[1], *Ahmad Nur Aizat Ahmad*[1], *Raja Zuraidah Raja Mohd Rasi*[1], *Nor Aida Abdul Rahman*[2], *Nik Hisyamudin Muhd Nor*[3], *Mohd Fahrul Hassan*[3], *Fatan Adibah Hashim*[4]

[1]Faculty of Technology Management, Universiti Tun Hussein Onn Malaysia, Malaysia
[2]Faculty of Aviation Management Department, Universiti Kuala Lumpur Malaysia Institute of Aviation technology (UniKL MIAT), Malaysia
[3]Faculty of Mechanical and Manufacturing Engineering, Universiti Tun Hussein Onn Malaysia, Malaysia
[4]Sharp Manufacturing Corporation Malaysia, Malaysia

Abstract. Continuous improvement is one of the core strategies for manufacturing excellent and it is considered vital in today's business environment. Continuous improvement is an important factor in TQM implementation. However, manufacturers in Electrical and Electronic Industry is facing variety of challenges such as, time constraint, quality issue, headcount issue, human issue and competition in domestic as well as the global market. This paper presents total quality management practices in Electrical and Electronic (EE) Industry. These manufacturers have to keep improving in key activities and processes to cope the challenges. Therefore, EE industry realize the importance of continuous improvement in helping the industries by setting clear goals and priorities for the area of improvement. The aims of this study are to determine the main factor in implementing continuous improvement practices, identify tools of continuous improvement that have been used and their obstacle in implementing continuous improvement practices. 200 questionnaires had been distributed to the employees in Electrical and Electronic Industry located at Bayan Lepas, Penang, Malaysia. A total of 41 questionnaires were answered which represented about 20.5% response rates. Survey result shows that teamwork and training and learning are respectively the importance factor and the most practices factor in Electrical and Electronic Industry. Most of the Electrical and Electronic Industry emphasis is on using Lean Manufacturing as the tool of continuous improvement practices. Besides that, Electrical and Electronic Industry faced the problems of lack of budget and lack of worker commitment.

* Corresponding author: mohdfauzi@uthm.edu.my

DOI: 10.1051/matecconf/201713500045

1 Introduction

Continuous improvement is planned, organized and systematic process of ongoing, incremental and company-wide change of existing practices to improve company performance. Continuous improvement is one of the core strategies for manufacturing excellent and is considered vital in today's business environment [1]. Continuous improvement has proved to be a useful way of business approach in centering low cost investment and high return improvements in both large, medium and small companies. Continuous Improvement method has become widely adopted as providing an important component for increasing company competitiveness. Continuous Improvement enables a company to be more successful in the pursuit of a specific strategy or set of objectives [2]. A well-known concept related to continuous improvement is Kaizen, which has its origin in Japan and means continuous change for the better by involving all employees [1]. Among the major potential benefits of continuous improvement are increased business performances and increased people performance [3][4].

Electrical and Electronic Industry is a leading sector in Malaysia's manufacturing sector, contributing significantly to the country's manufacturing output (29.3%), exports (55.9%) and employment (28.8%) (MIDA – Malaysian Industrial Development Authority)[5]. Nowadays, a lot of manufacturers in Electrical and Electronic Industry, especially in production department have faced heighted challenges such as time constraint, quality issue, headcount issue, human issue and competition in market. No doubt, these manufacturers are always embracing changes and improvements in their key activities and processes to cope the challenges. With that, many companies have realized the importance of continuous improvement, no matter is self-improvement or company improvement [6]. The improvement of employees themselves will be able to enhance their discipline and their knowledge [7]. Meanwhile, the improvement of company can help the company to increase the quality and productivity, eliminating the waste and reduce the headcount problems [8][9].

The following research questions will be examined in the study:

i. What is the main Critical Success Factors (CSFs) in implementing continuous improvement?
ii. What are the tools of continuous improvement being used?
iii. What is the main barrier factor in implementing continuous improvement?

2 Methodology

The population is observed from Electrical and Electronic Industry in Bayan Lepas, Penang. The unit of analysis is organisation. There are total of 200 samples have been selected for this study. Random sampling is used for this research. Survey method was used in this research. Survey forms have been sent to the respondents through postal mail. Finally, 41 questionnaires were received for final study. It represented 20.5% of response rate. Statistical Package for Science Social (SPSS) was used to analyse the data being collected. Descriptive has been carried out to answer the research questions.

 DOI: 10.1051/matecconf/201713500045

3 Analysis Results and Discussion

3.1 Cronbach's alpha

Cronbach's alpha [10] was used to determine the reliability of the questionnaire as shown in Table 1. Cronbach's alpha is a measure of internal consistency, which able to determine how closely related a set of items are as a group. Accepted value for Alpha value is 0.7 [10].

Table 1. Cronbach's alpha result

Variable	N	No. of items	Alpha value
Top Management Commitment	10	5	0.753
Training and Learning	10	5	0.878
Teamwork	10	5	0.923
Motivation	10	5	0.798
Resources Management	10	5	0.764

3.2 Normality Test

Based on the Table 2, the normality value is 0.710 at p <0.001, means that the distribution is normal distribution. Therefore, parametric test can be used.

Table 2. Normality Test

Variable	Shapiro-Wilk	Result
	Significant	
Practices of Factors	0.710, p<0.001	Normal Distribution

3.3 Extent level of Continuous Improvement

Five likert scale instrument has been used for final survey. Based on Table 3, training and learning had been rank as first. The mean score of training and learning is 3.9317. It is the highest factor. In reality, companies organized a lot of training program for their employees to enhance their knowledge. Without knowledge, employees will feel anxiety. Successful companies know that the more their employees learn about their jobs, the better they will become at every aspect of these jobs. Teamwork is ranked second important. Teamwork had contributed mean score about 3.7805. This followed by resources management, 3.7463 and top management commitment, 3.6341. Motivation is the least practiced factors, which only contributed about 3.3610. This may bring the meaning of employees were not motivated to improve their current job situation.

Table 3. Extent level of Continuous improvement

	Factors	Mean Score	Ranking
F1	Top Management Commitment	3.6341	4
F2	Training and Learning	3.9317	1
F3	Teamwork	3.7805	2
F4	Motivation	3.3610	5
F5	Resources Management	3.7463	3

3.4 Application of Tools

Based on the Table 4, lean manufacturing had high level of practices in companies among the 8 factors. The mean of degree of practices of lean manufacturing is 3.63. Lean Manufacturing is useful as well towards companies; mean score of degree of usefulness is 3.49. Instead of thinking ways to gain more profits, some manufacturers thinking to eliminate waste which also can save fund. For example, too much stock available in warehouse, it will increase the costs of storage. It should be ensure sufficient products available for customers. It is followed by ISO 9001, both the degree of practices and usefulness are rank as second. Its mean score for degree of practices is 3.41 and means for degree of usefulness is 3.32. Then, the third is FMEA. FMEA has 3.10 of the mean score for degree of practices and 3.07 of the mean score for degree of usefulness. Exceptional for the other tools, the seven-step problem solving is ranked as the seventh. The seven-step problem solving has the lower mean score for both degree of practices and degree of usefulness, which is 2.32 and 2.46. This maybe because of the processes are too tedious. Based on the means score in Table 4, Six-sigma had ranked as the fifth in both degree of practices and degree of usefulness. Means score for degree of practices is 2.63, which is lower than mean score of usefulness that is 2.78. The difference of mean is 0.14. This means that, employees found Six-sigma more useful than its being applied.

Table 4. Application of Tools

	Tools	Mean of Practices		Mean of Usefulness	
T1	The Seven-Step for Problem Solving	2.32	(7)	2.46	(7)
T2	Fish Bond Diagram	2.46	(6)	2.49	(6)
T3	Six-Sigma	2.63	(5)	2.78	(5)
T4	Lean Manufacturing	3.63	(1)	3.49	(1)
T5	Failure Mode and Effect Analysis (FMEA)	3.10	(3)	3.07	(3)
T6	Quality Control Circle (QCC)	2.98	(4)	2.88	(4)
T7	ISO 9001	3.41	(2)	3.32	(2)
T8	Others	0.24	(8)	0.32	(8)

Problems faced in Implementing Continuous Improvement have been evaluated based on 5 likert scale instrument. The result based on Table 5, there are two main causes of the problems in implementing continuous improvement are lack of budget and lack of worker commitment. It is normally difficult to get financial support and approval from top management. Top management only invests some programme that is profitable to the company. It is no doubt that, top management treat continuous improvement as an optional program. They may postpone continuous improvement program when there are a lot of workload. From the Table 5, it is obviously to know that, among forty-one respondents, there are only four continuous improvement members. Most of the respondents are engineer. This means that, engineers not only play their role as engineers, they also take part in continuous improvement activities. They need to finish their own job, at the same time; they need to help company to continuous improve to have better performance and more profitable. Most of the employees do not commit much in continuous improvement program due they do not have extra times and rewards. Employees' motivator is money or any other material wealth. It is important for top management to recognize their hard work.

Table 5. Problems faced in Implementing Continuous Improvement

	Barrier Factor	Mean Score	Ranking
B1	Failure to understand how change happen	3.32	7
B2	Lack of top management commitment / support	3.20	11
B3	Ego	3.21	10
B4	Employees resists changing	3.27	8
B5	Lack of time	3.46	3
B6	Lack of budget	3.51	1
B7	Lack of knowledge	3.37	5
B8	Lack of training	3.37	5
B9	Lack of worker commitment	3.51	1
B10	Lack of resources	3.20	11
B11	Lack of measurement	3.22	9
B12	Lack of focus action	3.44	4

4 Conclusion

Survey result shows that teamwork and training and learning are respectively the importance factor and the most practices factor in Electrical and Electronic Industry. Most of the Electrical and Electronic Industry emphasis is on using Lean Manufacturing as the tool of continuous improvement practices. Besides that, Electrical and Electronic Industry faced the problems of lack of budget and lack of worker commitment.

5 Acknowledgements

Appreciation to MOHE and ORRIC, Universiti Tun Hussein Onn Malaysia for supporting this research (Geran Kontrak, Vot: U429). Appreciation also to Manufacturing Technology Management (MTM) focus group, Faculty of Technology Management.

References

[1] M. F. Ahmad, N. Zakuan, a Jusoh, Z. Tasir, and J. Takala, "Meta-analysis of the relationship between TQM and Business Performance," *IOP Conf. Ser. Mater. Sci. Eng.*, vol. 46, no. 1, p. 12020, 2013.

[2] M. F. Ahmad, R. Z. R. Rasi, N. Zakuan, M. . Haji-Pakir, and J. Takala, "The Impact of ASEAN Free Trade Agreement as Moderator on TQM Performance Model in Malaysia: Survey Result," *Soc. Sci.*, vol. 11, no. 12, pp. 2932–2937, 2016.

[3] M. F. Bin Ahmad and S. M. Yusof, "Comparative study of TQM practices between Japanese and non-Japanese electrical and electronics companies in Malaysia:

Survey results," *Total Qual. Manag. Bus. Excell.*, vol. 21, no. 1, pp. 11–20, 2010.

[4] M. F. Ahmad, M. S. M. Arif, N. Zakuan, S. S. S. A. Rahman, T. A. R. Abdullah, and N. Fadzil, "The Effect of Demographics on Customer Satisfaction amongst Malaysia Hajj Pilgrims: Survey Result," *Appl. Mech. Mater.*, vol. 660, pp. 1000–1004, 2014.

[5] M. F. Ahmad, M. S. M. Arif, N. Zakuan, S. Rahman, M. Latif, and M. Khalid, "The Mediator Effect of Customer Satisfaction between Quality Management Practices and Communication Behavior amongst Malaysia Hajj Pilgrims: Survey Result," *Appl. Mech. Mater.*, vol. 660, no. 2015, pp. 1005–1009, 2015.

[6] M. F. Ahmad, N. Zakuan, R. Z. R. M. Rasi, M. N. N. Hisyamudin, and J. Takala, "Mediator effect of total productive maintenance between total quality management and business performance: Survey result in Malaysia automotive industry," *Adv. Sci. Lett.*, vol. 21, no. 12, pp. 3723–3725, 2015.

[7] M. F. Ahmad, M. S. M. Ariff, N. Zakuan, M. Z. M. Saman, S. S. S. A. Rahman, T. A. R. Abdullah, and N. Fadzil, "The effect of demographics on quality management principles of ISO 9001:2008 amongst Malaysia Hajj Pilgrims," *Soc. Sci.*, vol. 11, no. 11, pp. 2748–2752, 2016.

[8] M. F. Ahmad, N. Zakuan, A. Jusoh, S. M. Yusof, J. Takala, and M. S. M. Arif, "Comparative Study of TQM Practices between Japanese and Non-Japanese Companies: Proposed Conceptual Framework," *Adv. Mater. Res.*, vol. 903, pp. 371–377, 2014.

[9] M. Ahmad, N. Zakuan, J. Ahmad, and J. Takala, "Meta–analysis of the TQM impact on business performance amongst regions and countries," *Int. J. Ind. Syst. Eng.*, vol. 20, no. 2, pp. 155–164, 2015.

[10] J. C. Nunally, *Psychometric Theory*, Second ed. New York: McGRAW-HILL, 1978.

Muscular Activities Measurements of Forward Lean and Upright Sitting Motorcycling Postures via Surface Electromyography (sEMG)

Muhammad Izzat Nor Ma'arof[1,*], *Abdul Rahman* Omar[2], *Rohani* Haron[2], Hazran Husain[2] and *Muhammad Abdul Wahab*[1]

[1]Mechanical Engineering Department, Faculty of Engineering & Quantity Surveying, INTI International University, Persiaran Perdana BBN, Putra Nilai, 71800 Nilai, Negeri Sembilan, Malaysia
[2]Motorcycle Engineering Test Lab (METAL) of the Faculty of Mechanical Engineering, Universiti Teknologi MARA, 40450 Shah Alam, Selangor, Malaysia

Abstract. Motorcycling postures are generically speculated to be physical and physiologically demanding – which in-turn may lead to motorcycling fatigue, and then becoming a possible factor to road accident. The objective of this study was to measure the muscular activities of various motorcycling postures. High muscular activity reading will signifies that motorcycling is indeed physically and physiologically demanding to the motorcyclist. For this particular study, the following postures were tested: i) forward lean, ii) upright sitting, and iii) neutral sitting (as control). Surface electromyography (sEMG) measurement was conducted on the following muscles: i) extensor carpi radialis, ii) upper trapezius iii) latissimus dorsi, and iv) erector spinae. The results showed that for all test subjects, the muscular activities readings for the forward lean posture was actually close to neutral sitting's. Whilst, the upright sitting had showed much higher muscular activities measurement instead. Conclusively, this study had proven that any types of discomforts associated with the forward lean posture is not originated from muscular activities. Whereas, confirming that any discomforts in regards to the upright sitting is indeed related to muscular activities. Further studies are warranted to discover the actual risk factors that causes physical and physiological discomforts for the forward lean motorcycling posture.

1 Introduction

An ergonomics study by Ma'arof et al., (2012) had discussed in great detail that motorcycling posture i.e. the sitting posture practiced by motorcyclist whilst riding; is ergonomically unfitting. The study emphasized that the working posture is constituted to be "constrained, cramped, static and poor". Even so, as unfitting it may seem, global statistics has indicated that there is an increment in the number of registered motorcycles especially in the Asian region (Shell.com, 2013, Honda Annual Report, 2012). In the wake of global

* Corresponding author: muhammadIzzat.maarof@newinti.edu.my

petroleum crisis, the demands for economical mode of transportations are on the rise (Young, 2013). Most members of the society are seeking optional vehicles that have high fuel efficiency while simultaneously easy to be maintained. Indeed, the market for green vehicles does exist. Full-electric and hybrid automobiles are already mass produced in most countries. Yet, one could not deny the flexibility provided by motorcycles to tackle-down traffics on the overcrowded Asian cities and rural areas. However, with the rise of motorcycles usage, this number also shows a corresponding increase in global motorcycle roads accidents.

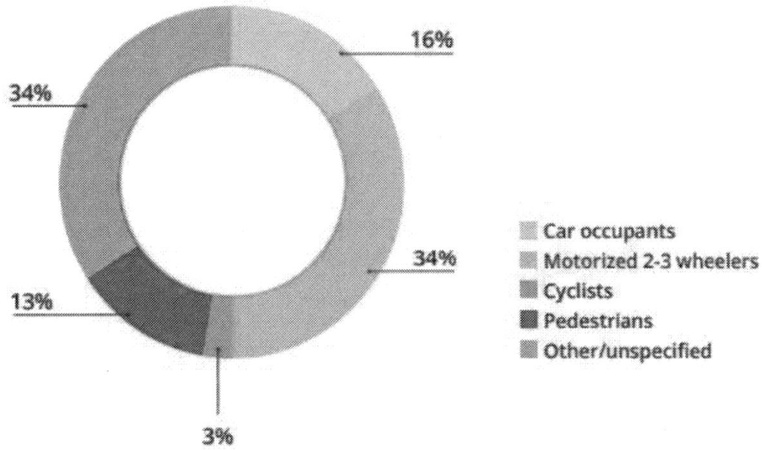

Figure 1.0(a): Deaths by road user type, South-east Asia region by "Road Safety in the South-East Asia Region" by World Health Organization (WHO) [35].

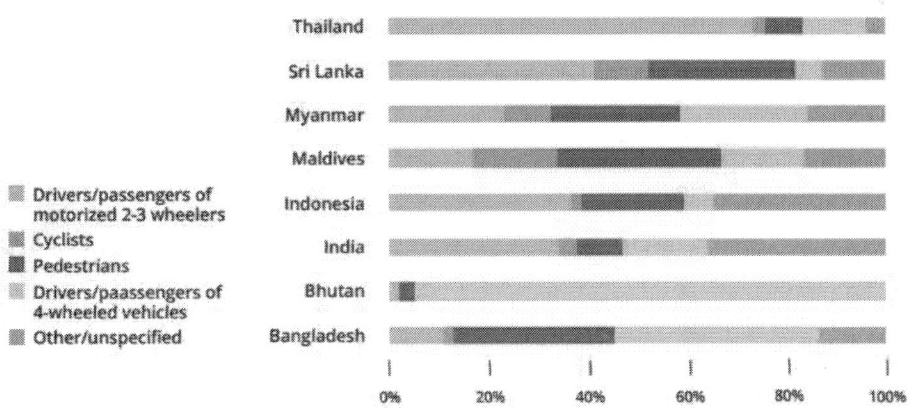

Figure 1.0(b): Distribution of road traffic deaths by type of road user by "Road Safety in the South-East Asia Region" by World Health Organization (WHO) [35].

The International Traffic Safety Data and Analysis (2013) had reported that in Malaysia; fatalities involving motorized 2-wheeled vehicles increased from 56% to 61% from the year 2003 to 2011. In the United States of America, the percentage of motorcycle road accidents has doubled from 7% to 14% in the interval of 21 years. While, United Kingdom, though, showed decrement in 2011 at 19% with respect to the year 2010 (22%), the percentage is still high in retrospect from the year 2000 (at 17%). In addition, the World

Health Organization had documented high percentage of fatal road accidents involving 2- and 3-wheeled vehicles in the Asian region. 74% was recorded in Thailand, 67% in Cambodia, 46% in Singapore, 36% in Indonesia, 35% in China and 32% in India. From the global accident reports, it could be concluded that most countries showed annual increment of road accidents involving motorized 2- and 3-wheeled vehicles.

Figures 1.0 (a) and (b) show the fatalities that occur in South-east Asia region that was recorded by World Health Organisation (WHO). Figure 1.0(a) shows that the two or three wheeled vehicles are major victims of such unfortunate incidents whereas, figure 1.0(b) shows that Thailand has the most number of accidents involving two or three wheelers, and then the number is decreasing progressively when compared with the other countries, as shown in the graph. Hence, proving that motorcyclist is indeed highly vulnerable towards road accidents.

This study upon reviewing several studies and articles on motorcycle road accidents, has reached to a conclusion that these accidents were commonly the cumulative or independent results of poor motorcycling behaviour, the non-usage of helmets, alcohol abuse, age factor, lack of motorcycling education, night-to-day riding and riding without license (Schneider et al., 2012; Keall and Newstead, 2012; Manan and Várhelyi, 2012; and Jou et al., 2012). Nonetheless, as cemented by Mr. Keith Code: globally recognized as a former motorcycle racer, writer, and founder of the California Superbike School while collective known as the ace on-track motorcycle instructor in the world; in his article Code (2010): "…horsepower wasn't the cause before, and was unlikely to be the cause now, of crashing motorcycles. Rather it is rider errors in applying core technical skills that cause them to go down." From this quote, it is clear that human error (in performing the motorcycling duties) is the main reason for motorcycle road accidents. Hence, the preceding question will be: how and in what working conditions are the motorcycling duties are performed? In following the rule of thumb of ergonomics, this study chose to start with the human operator i.e. the motorcyclist.

In examining the motorcyclist, several routes could be taken into account, for examples, the static and dynamic loadings whilst motorcycling, the psychological demands and even nutrition. Even so, this study will access the very essence of motorcycling which is the working posture. As noted earlier, Ma'arof et al., (2012) had clearly documented that the motorcycling riding posture is 'unfitting' for human operator. Yet, the same study still advocated that it is still the necessary working posture for motorcycling due to its advantages. Motorcycling postures are generically speculated to be physical and physiologically demanding – which in-turn may lead to motorcycling fatigue, and then later to be an eventual possible factor to road accident. Consequently, this study suggests for the improvement of the motorcycling riding posture – for this could improve riding safety and comfort.

The objective of this study was to measure the muscular activities (RMS values) of various motorcycling postures. High muscular activity reading will signifies that motorcycling is indeed physically and physiologically demanding to the motorcyclist. Henceforth, providing the necessary quantitative evidence that any sort of physical and physiological discomforts with respect to the particular motorcycling posture is indeed muscular related and not mere speculation.

1.1 Motorcycle Riding Postures

Motorcycles are manufactured with various designs and performance variations (Teoh and Campbell, 2010). For research purposes, motorcycles are commonly segregated and selected in various classifications systems (Ma'arof and Ahmad, 2012). Ma'arof and

Ahmad (2012) introduced the "Riding Posture Classification (RIPOC)" system – a motorcycle classification system specifically for ergonomics study. The system segregates motorcycle with respect to the human operator's trunk (upright or spinal flexion) and knee (flexed or extended) positioning. There are four riding postures introduced via the system. Table 1.1(a) provides the summary and general descriptions of each motorcycling posture.

Table 1.1(a): The General Description of the Riding Postures Designated by Ma'arof and Ahmad, 2012

RIPOC Types	Riding Posture Name	Details		
		Spinal flexion with respect to pelvis	**Arms & hands positioning**	**Knees positioning**
Type 1	Spinal Flexion Riding Posture	>20 degree	Below shoulder height	Flexed
Type 2	Upright Riding Posture	≤ 20 degree		Flexed, or ankles are perpendicular to the knees
Type 3	Upright-and-knees-extended riding posture	≤ 20 degree	Level or above shoulder height	Extended, ahead of the knees
Type 4	Spinal flexion-and-knees-extended riding posture	>20 degree		Extended, ahead of the knees

As a matter of consistency, this study will refer to the riding postures listed in Table 1.0 as motorcycling posture. In regards to the designated names, the Type 1 will be referred to as the forward lean, whilst, the Type 2 will be referred as the upright sitting. In examining the muscular activities via surface electromyography, this study had only focused on the first two motorcycling postures which were the Type 1 and Type 2. This was because these two working postures only has one common variable at the upper body region which is the spinal flexion.

Table 1.1(b): Body Parts Positioning Conditions According to the Riding Posture Establishment Conditions (RIPEC)

Human Operator Body Parts	Motorcycle Transmission Type		
	Manual	**Semi-Automatic**	**Automatic**
Head	Facing the direction where the human operator wanted to travel		
Right hand and fingers	Right hand-grip/throttle and front brake lever		
Left hand and fingers	Left hand-grip and clutch lever	Left hand-grip and clutch lever (or none)	Left hand-grip and rear-brake lever
Right foot	Foot-peg/board and rear-brake lever		Foot-peg/board or platform
Left foot	Foot-peg/board and gear lever		Foot-peg/board or platform
Buttocks	Seat		

In addition to the RIPOC system, Ma'arof and Ahmad (2012) also established the "riding posture establishment conditions" called the Riding Posture Establishment Conditions (RIPEC) (see Table 1.1(b)). Only by fulfilling all of the outlined conditions, the study noted that a riding posture is established. It was explained in the particular study that prior to the institution of the establishment conditions, it was observed that motorcyclists tend to be practicing various forms of sitting postures while motorcycling. Hence, for research consistency purposes, a standardized sitting posture has to be utilized. Therefore, the establishment conditions were formed. Thus, in measuring the muscular activities for this study, the "riding posture establishment conditions" were fulfilled.

From the RIPEC, Ma'arof et al., (2012) had discussed that motorcycling could be divided into two stages of human-motorcycle interfaces which are the "Human-Machine Interface (HMI) Stage" (first stage) and "Human-Machine-Environment Interface (HMEI) Stage" (second stage). The first stage is the initial stage where the motorcycle is static (no motorcycling activity) and the human operator assumes the riding posture. Only in the second stage the motorcycling activity takes place and the surrounding environment is now an integral element of the motorcycling ergonomics system. Since this study was performed in a laboratory, this study followed the outline listed for the HMI stage. The characteristics of the HMI stage as outlined by Ma'arof et al., (2012) are as follow:

I. The human operator has to successfully perform at least three (3) out of the four (4) operator's body parts positioning requirements of the RIPEC. At static, the human operator may need to independently support and balance the motorcycle.
II. Crucially, the motorcycle (workstation) must be at static.
III. The utilization of any device to keep the motorcycle at upright is negligible.

2 Research Methodology

For this study, a quantitative research approach was performed through experimental design using surface electromyography (EMG) measurement technique. The sEMG measurement was conducted to analyze the muscular activity of four selected muscle groups. The selected muscles were as follow: i) extensor carpi radialis, ii) upper trapezius iii) latissimus dorsi, and iv) erector spinae. The muscles were selected based on Velagapudi et al., (2010) and by consulting the present senior physiotherapist from the Faculty of Health Science of the Universiti Teknologi MARA (UiTM) during the sEMG pilot test (prior to this study).

2.1 sEMG Recording

2.1.1 Subjects

4 subjects (all male) had volunteery participated in this study. The subjects were informed that they could withdraw at any time and all results will be kept anonymous and confidential. Each subject was given a consent form prior to the experiment. The experimental procedure was approved by the Ethics Committee of the Research Management Institute (RMI) of the Universiti Teknologi MARA, Shah Alam, Selangor, Malaysia. Selection criteria of the subject for the True Experiment study are as follow:

i. Subjects were male with more than 1 year motorcycling experience (motorcycle of any engine capacity)
ii. Age between 18 – 25 years old
iii. Normal Body Mass Index (BMI) between 18.5 and 24.5.
iv. Subjects were healthy prior and during the experiment

v. Subjects have no history of severe physical or physiological trauma (Snijders et al., 1995)
vi. Subjects have no previous surgery through the abdominal wall (Snijders et al., 1995)
vii. Subjects have not participated in any heavy training or physical activity 24 hours before the experiments

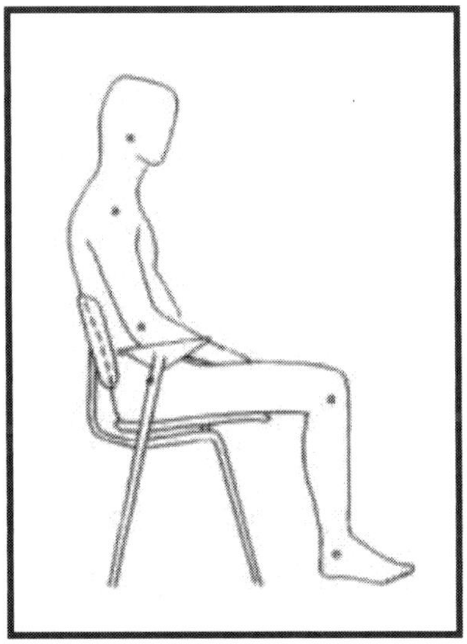

Figure 2.1.1(a): (i) Passive Sitting (Control Working Posture) Of The Subject During SEMG Recording (Source: Adapted From Snijders et al., (1995),

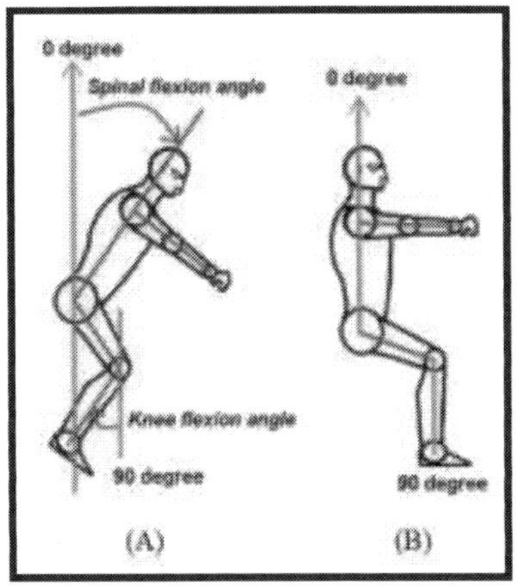

Figure 2.1.1(b): (A) Forward Lean Posture and (B) Upright Sitting Posture,

For the experiment, each subject performed three (3) working postures (passive sitting and the two motorcycling postures) as shown in Figure 2.1.1(a). The passive sitting (see Figure 2.1.1(a) (i)) was the control working posture. The control working posture provided the resting baseline for the muscle activity for the four selected muscles.

2.1.2 Experimental Procedure

The detail of the experimental procedure was as follow:

1) Each subject had performed 3 body postures throughout the experiment (1 Control posture and 3 motorcycling postures). For control, the subjects were seated on a chair (with full backrest) and positioned according to Snijders et al., (1995) (see Figure 1.(i)). The subjects were assisted with instruction to practice the passive sitting posture. Upon completing the EMG recording for the control sitting posture (passive sitting), the subjects were instructed to assume the motorcycling riding posture according to "riding posture establishment conditions" (Ma'arof and Ahmad, 2012) on 2 different motorcycles (2009 SYM 250cc Scooter and 2004 Kawasaki ZX2R). The motorcycles were positioned on a centre stand or dual paddock stand (front and rear). The experimental testing only replicates the "HMI" stage (Ma'arof et al., 2012) of the human-motorcycle interface stages.
2) All subjects were subjected to wear normal attire (shirt, pants and shoes) similar to the attire worn during the experiment conducted by Karmegan et al., (2012).
3) The lab was treated with controlled room attributes of having air-conditioning at 26°C and has sufficient lighting condition.
4) Surface electromyography (sEMG) reading of muscle activities (bilateral measurement) were taken via the Mega Biomonitor ME6000 sEMG device (with specification specifically designed for dermatological applications) for 5 minutes (with readings were taken every 0.5 seconds at 1000Hz). Therefore, each subject would generate 600 readings for the 5 minutes duration. The EMG signals were recorded with an 8-channel telemetric EMG system and a preamplifier (gain 1,000×), common mode rejection > 85 dB. The experimental time duration was very short since the objective was to measure the earliest muscular activity on the selected muscles (Kyoung and Nusbaum, 2008).
5) The four selected muscle were: i) extensor carpi radialis longus (forearm muscle), ii) upper trapezius (shoulder/neck), iii) latissimus dorsi (middle back), and iv) erector spinae (lower back).

2.1.3 Surface Electromyography (sEMG) Sensor Placement Procedure

The study procedure followed the six steps recommended by SENIAM (Hermens et al., 2000). Steps 1 and 2 were general in the sense that they were always valid, independent of the muscle on which the sensors are placed. Steps 3 to 6 were based on general starting points, which had been specified in detail for the individual muscles. The sequential procedures for sensors placement of sEMG in the study were as follow:

1) Selection of the sEMG sensors
   The size of the electrodes used is 10 mm. The sensors selection for the study had adhered to the recommendations by SENIAM. An electrode material forms the contact layer with the skin and gives efficient electrode–skin contact, low electrode–skin impedance and does not cause chemical reactions at the skin interface during muscle activation by EMG. For this purpose, the pre-gelled Ag/AgCl electrodes were used.

2) Test subject's skin preparation
   The skin of the subjects were prepared for good electrode–skin contact. This is important for obtaining accurate sEMG recordings in terms of amplitude

characteristics, fewer and smaller artifacts or electrical interference, less risk of imbalance between electrode and less noise. The shaving of the area of the skin surface at the sensor location was done to remove the hair. Then, the skin area was clean with alcohol and left dry by vaporization of alcohol before the sensors were positioned on the skin.

3) Positioning the subject in a starting posture
The starting postures before the placement of electrodes allow the determination for proper location of the sensors on the muscles. In the starting postures, the proper location of placement of electrodes were clearly determined by and the anatomical landmarks and palpation of the muscles. The descriptions of the starting postures were described in the experiment procedures.

4) Determination of the sensor location
The electrodes were placed with respect to the longitudinal location of the sensors on the muscles. Motor points and/or muscle tendons and the presence of other active muscles near the sEMG sensor (crosstalk) were avoided because these factors strongly influenced the stability of a sEMG recording. The centre-to centre distance between the conductive areas of two bipolar electrodes is also known as inter-electrode distance. The bipolar sEMG electrodes were applied at the sensor locations with an inter-electrode distance of 20 mm.

5) Placement and fixation of the sensors
The electrodes placements were as follow:

i. At upper trapezius slightly lateral to and halfway between the cervical spine at C-7 and the acromion (Cram et al., 1998)
ii. At latissimus dorsi three-finger width distal to and along the posterior axillary fold, parallel to the lateral border of the scapula (Lehman et al., 2006)
iii. At lumbar erector spinae at a two-finger width laterally away from the L1 spine (Guo et al. 2012)
iv. Over the belly of extensor carpi radialis (Laura et al., 1998)
v. The placement of ground electrodes were at the skin or bony prominence (Blanc and Dimanico, 2010)

The sensors placement procedures were assisted by a senior Physiotherapy lecturer from the Faculty of Health Sciences, UiTM, who had the experience regarding EMG procedures. The bipolar sEMG electrodes are placed at the sensor location of each muscle that has been identified with their orientation parallel to the muscle fibres. For each muscle, a ground/reference electrode was placed at a location in which the risk for a large common mode disturbance signal is minimal, so preferably an electrically inactive tissue was chosen.

6) Connection testing
After placement of the sensors and the reference electrodes, a test was performed by the senior Physiotherapy lecturer to determine whether the electrodes have been placed properly on the muscle and connected to the equipment so that a reliable sEMG signal can be recorded. This has been specified for each individual muscle. The clinical tests by isometric contraction of each muscle were also done to evaluate the activity of the tested muscles.

3 Results

Table 3.1 shows the demographic details for the test subjects in the experiment. All subjects satisfied the inclusion criteria.

Table 3.1: Demographic details for the test subjects in the experiment

	Mean	Standard Deviation
Age	24	1
Height	171	7
Weight	63.25	13
Motorcycling Experience	9	2

Table 3.2: Total average RMS (uV) (for the selected four muscles) and the percentage (%) of increment of total average RMS (uv) (in comparison to control) with respect to the tested postures for Test Subject 1

Test Subject 1			
Posture	**Total RMS (uV)**	**Control (uV)**	**Percentage (%) of Increment of Total Average RMS (uV) with Respect to Control**
Forward Lean	78.6	80.1	-1.8
Upright Sitting	209.1	80.1	161.3

Table 3.3: Total average RMS (uV) (for the selected four muscles) and the percentage (%) of increment of total average RMS (uv) (in comparison to control) with respect to the tested postures for Test Subject 2

Test Subject 2			
Posture	**Total RMS (uV)**	**Control (uV)**	**Percentage (%) of Increment of Total Average RMS (uV) with Respect to Control**
Forward Lean	127.0	51.6	146.1
Upright Sitting	210.8	51.6	308.5

DOI: 10.1051/matecconf/201713500046

Table 3.4. Total average RMS (uV) (for the selected four muscles) and the percentage (%) of increment of total average RMS (uv) (in comparison to control) with respect to the tested postures for Test Subject 3

Test Subject 3			
Posture	**Total RMS (uV)**	**Control (uV)**	**Percentage (%) of Increment of Total Average RMS (uV) with Respect to Control**
Forward Lean	124.4	96.5	28.87
Upright Sitting	148.6	96.5	53.94

Table 3.5: Total average RMS (uV) (for the selected four muscles) and the percentage (%) of increment of total average RMS (uv) (in comparison to control) with respect to the tested postures for Test Subject 4

Test Subject 4			
Posture	**Total RMS (uV)**	**Control (uV)**	**Percentage (%) of Increment of Total Average RMS (uV) with Respect to Control**
Forward Lean	97.2	87.3	11.38
Upright Sitting	162.8	87.3	86.57

The results for the sEMG measured were tabulated and displayed in Table 2, Table 3, Table 4 and Table 5. Each table shows the total average RMS (uV) (the total of the average reading measured for the selected four muscles) and the percentage (%) of increment of total average RMS (uv) (in comparison to control) with respect to the tested postures for the particular test subject. In analysing the result, it was evidence that the sEMG measurement for all subjects showed 2 similar trends in the total average RMS (uV). Firstly, all test subjects generally showed the increase in the total average RMS (uV) for all motorcycling postures in comparison to the average RMS (uV) reading given by the control posture. However, it is interesting to note that only test subject 1 showed the decrement of total average RMS (uV) for the forward lean motorcycling posture. Test subject 1 recorded 1.8% decrement in the total average RMS (uV, for the selected four muscles) with respect to the control. Secondly, the lowest increment of the total average RMS (uV) for test subject 2, 3 and 4 were given by the forward lean motorcycling posture; which were 146.1%, 28.87%, and 11.35% for test subject 2, 3 and 4 are respectively. Henceforth, it was proven that this riding posture showed the least amount of muscular activity in contrast to the upright sitting motorcycling posture. The muscular activity given by practicing, maintaining and holding the forward lean motorcycling posture during the first stage of the human-motorcycle interface was almost similar to be seated on a chair with a full backrest with the arms are rested. Yet, the reading was the opposite for the upright sitting posture instead which were

161.3%, 308.5%, 53.94% and 86.57% from the control for subject 1, 2, 3 and 4 respectively. Therefore, providing the quantitative evidence that the upright sitting motorcycling posture showed the highest muscular activity in comparison to the forward lean and the control sitting postures.

4 Discussion

The objective of this study was to measure the muscular activities of various motorcycling postures. High muscular activity reading will signifies that motorcycling is indeed physically and physiologically demanding to the motorcyclist. Henceforth, providing the necessary quantitative evidence that any sort of physical and physiological discomforts with respect to the particular motorcycling posture is indeed muscular related and not mere speculation.

The results had shown that the forward lean motorcycling posture showed the lowest RMS reading for all selected muscles during the first stage of human-motorcycle interface i.e. from the very intial stage of motorcycling. Hence, with the notion that this would be further carried into the second stage of motorcycling, the forward lean motorcycling posture – in contrasting to the upright sitting posture, is the least affected by muscular activity. Therefore, has proven that any types of discomforts associated with the forward leaning posture is not originated from muscular activities. Whereas, the results confirmed that any discomforts in regards to the upright sitting is indeed related to muscular activities. In addition to the data obtained via the experiment, it was also noted that the forward lean motorcycling posture has several additional work-related advantages.

Firstly, the forward lean motorcycling posture provide the necessary physical range of motions and leverage of the entire body for motorcycling duties. By practicing the forward lean motorcycling posture, the motorcyclist could easily change his body positionings in order to change the body weight distribution (static loading) on the motorcycle while performing the motorcycling duties. For instances, by varying the elbow flexion and extension, the motorcyclist is able to change the magnitude (force) of static loading at the wrists, thus, at the handlebar. This would aids the motorcyclist to have better control and manoeuvrability over the motorcycle. By having the knees flexed and tucked in, the motorcyclist could make full use of the physical leverage provided to apply the necessary amount of force on the foot-pegs with less efforts to manoeuvre the motorcycle. The physical range of motion and the use of leverage are simultaneously performed by the motorcycle and both work hand in hand. In addition to the motorcycling duties, as noted by notable research such as Peeters et al., (2013), O'Sullivan et al., (2012), Rohlmann et al., (2011), Seidel et al., (2011), Chen et al., (2010), Eger et al., (2008) – among others, having a capacity to vary the working posture from time to time is indeed ergonomically advantageous since no posture should be hold for prolonged session – regardless of its comfort level.

Secondly, the forward lean posture also facilitate for the motorcyclist to improve the aerodynamics characteristics of the motorcycling experience by being streamlined with the motorcycle. This is commonly achieved by the motocycliste by practicing very high degree of spinal flexion. For instances, the motorcyclist's upper body will be completely in contact with the motorcycle body works (e.g. fuel tank). By improving the aerodynamics characteristics, the motorcyclist is even least affected by the windblast during motorcycling. Thus, further minimizing the muscular activity needed to hold the motorcycling posture in-place.

Although it was apparent from this study that the forward lean motorcycling posture results in the least amount of muscular activities in order to be practiced, both physical and physiological discomforts were still reported by the subjects. Based on the result, it was speculated that the occurrence of discomforts might as well be the result of other physical and physiological risk factors such as postural positioning hazard, static loading and pressure distributions or even blood circulation issues. Further studies are warranted to discover the actual risk factor that causes physical and physiological discomforts for the forward lean riding posture. Therefore, ergonomics intervention could be formulated in order to improve the working condition for the motocyclist. Hence, improving comfort and overall motorcycling safety.

5 Conclusion

Conclusively, from the surface electromyography (sEMG) measurement in comparison with neutral sitting (control posture), the forward lean motorcycling posture showed the lowest RMS reading for all selected muscles during the first stage of human-motorcycle interface. Whilst, the upright motorcycling posture showed much higher muscular activities measurement instead. This quantitative evidence has confirmed that any types of physical and physiological discomforts associated with the forward leaning posture is not originated from muscular activities in practicing and holding the posture. Whereas, any discomforts in regards to the upright sitting is indeed related to muscular activities. This study also emphasizes that motorcycling provides a vast field for exploration and engineering development. Therefore, it is advisable to discover the actual risk factor that causes physical and physiological discomforts for the forward lean riding posture. Hence, ergonomics intervention could be formulated to minimize the discomforts. By minimizing the level of physical and physiological discomforts, the motorcyclist could ride in the most ergonomically safe manner, thus, minimizing the possibility of road accident which originated from human error.

References

1. Bertolasi, L., Priori, A., Tinazzi, M., Bertasi V., Rothwell, J. C. (1998), "*Inhibitory action of forearm flexor muscle afferents on corticospinal outputs to antagonist muscles in humans*", doi: 10.1111/j.1469-7793.1998.947bg.x September 15, The Journal of Physiology, 511, pp. 947-956. http://jp.physoc.org/content/511/3/947.full.pdf+html.
2. Blanc Y. and Dimanico U. (2010), "*Electrode Placement in Surface Electromyography (sEMG) "Minimal Crosstalk Area" (MCA)*". The Open Rehabilitation Journal, 3, pp. 110-126
3. Chen, S-H., Lee, Y-H., Chiou, W-K., and Chen, Y-L. (2010), "*A pilot study examining seat heights and subjective ratings*", International Journal of Industrial Ergonomics 40, pp. 41-46.
4. Code, K., (2010), "*Code Break – Electronic Rider Aids. How A Chip Can Save Your Bacon*". Motorcyclist July 2010, Retrieved August 22, from Sport Rider official website: http://goo.gl/hviXs8.
5. Cram J. R, Kasman G. S, Holtz J. (2008), "*Introduction to Surface Electromyography*". Gaithersburg: Aspen Publishers.
6. Eger, T., Stevenson, J., Callaghan, J. P., Grenier S., and VibRG. (2008), "*Predictions of health risks associated with the operation of load-haul-dump mining vehicles: Part*

2-Evaluation of operator driving postures and associated postural loading". International of Industrial Ergonomics 38, pp. 801-815.

7. Guo, L.-Y., Wang, Y-L., Huang, Y-H., Yang, C-H., Hou, Y-Y., Harn, H. I-C., and You, Y-L. (2012), "*Comparison of the electromyographic activation level and unilateral selectivity of erector spinae during different selected movements*", International Journal of Rehabilitation Research, 35, pp. 345–351.
8. Hermens H. J., Freriks, B., Disselhorst-Klug. C., Rau. G., (2000), "*Development of Recommendations for SEMG Sensors and Sensor Placement Procedures*". Journal of Electromyography and Kinesiology, vol. 10, pp. 10 361–374.
9. Honda. (2012), "*Honda Annual Report 2012*". pp. 16-17.
10. International Traffic Safety Data and Analysis, (2013), "*Road Safety Annual Report 2013*", pp. 296, 429, 441.
11. Jou, R. C., Yeh, T. H., and Chen, R. S. (2012), "*Risk factors in motorcyclist fatalities in Taiwan*", Traffic Inj Prev, vol. 13, pp. 155-62.
12. Karmegam K., Ismail, M. Y., Sapuan, S. M., Ismail, N., Shamsul Bahri, M. T. (2012), "Evaluation of motorcyclist's discomfort during prolonged riding process with and without lumbar support", Annals of the Brazilian Academy of Sciences, pp. 1169-1188.
13. Keall M. D. and Newstead, S. (2012), "*Analysis of factors that increase motorcycle rider risk compared to car driver risk,*" Accident Analysis and Prevention, vol. 49, pp. 23-29.
14. Kyung, G., Nussbaum, M. A., and Babski-Reeves, K. (2008), "*Driver sitting comfort and discomfort (part I): use of subjective ratings in discriminating car seats and correspondence among ratings*", International Journal of Industrial Ergonomics 38, pp. 516–525.
15. Lehman, G.J. , MacMillan, B., MacIntyre I. (2006), "*Shoulder muscle EMG activity during push up variations on and off a Swiss ball*", Dyn Med, 5.
16. Ma'arof, M. I. N., Ahmad, I. N., Abdullah, N. R., Karim, S. A., (2012), "*Motorcycling Riding Issues: Understanding the Phenomenon and Development of Ergonomics Intervention in Improving Perceived Comfort for Prolonged Riding*", iDECON, 2012.
17. Ma'arof, M. I. N., and Ahmad I. N., (2012), "*Proposed Standardized Method for Motorcycle Nomenclature System*". Southeast Asian Network of Ergonomics Societies Conference (SEANES), Road User Ergonomics, Paper 60.
18. Manan M. M. A., and Várhelyi, A. (2012), "*Motorcycle fatalities in Malaysia*", IATSS Research 36, pp. 30–39.
19. Mehta, C. R. and Tewari, V. K. (2000), "*Seating discomfort for tractor operators – a critical review*", International Journal of Industrial Ergonomics *25*, , pp. 661-674.
20. O'Sullivan, K., O'Sullivan, P., O'Sullivan, L., and Dankaerts, W., (2012), "*What do physiotherapists consider to be the best sitting spinal posture?*" Manual Therapy, pp. 1-6.
21. Organization, World Health. (2013), "*Global Status Report on Road Safety 2013 - Supporting A Decade of Action: World Health Organization*".
22. Peeters, GMEE (G.), Burton, N. W., and Brown, W. J. (2013), "*Associations between sitting time and a range of symptoms in mid-age women*", Preventive Medicine 56, pp. 135-141.

23. Robertson S. A. and Minter, A. (1996), "*A study of some anthropometric characteristics of motorcycle riders*", Applied Ergonomics, Vol. 27, No. 4, pp. 223-229.
24. Rohlmann, A., Zander, T., Graichen, F., Dreisharf, M., Bergmann, G. (2011), "*Measure loads on a vertebral body replacement during sitting*", The Spine Journal 11, pp. 870-875.
25. Schneider, W. H., Savolainen, P. T., Boxela, D. V., Beverleyc, R. (2012), "*Examination of factors determining fault in two-vehicle motorcycle crashes*", Accident Analysis and Prevention 45, pp. 669– 676.
26. Seidel, D., Hjalmarson, J., Freitag, S., Larsson, T. J., Brayne, C., and Clarkson, J. (2011), "*Measurement of stressful postures during daily activities: an observational study with older people*", Gait and Posture 34, pp. 397-401.
27. Shell, "*Shell Responds To Boom In Asian Motorcycle Market*". Retrieved on 22nd August 22, 2013 from Shell Global official website: http://goo.gl/bXADNe
28. Smith, T., Zellner, J., and Rogers, N. M. (2006), "*A three dimensional analysis of riding posture on three different styles of motorcycles*", International Motorcycle Safety Conference.
29. Snijders, C. J., Bakker, M. P., Vleeming, A., Stoeckart, R., and Stam, H. J. (1995), "*Oblique abdominal muscles activity in standing and in sitting on hard and soft seats*", Clinical Biomechanics, Volume 10, No. 2, pp. 73-78.
30. Stedmon, A. (2010), "HMI: Human-motorcycle interaction", Centre for Motorcycle Ergonomics & Rider Human Factors, Department of Mechanical, Materials, Manufacturing Engineering (M3), University of Nottingham, pp. 1-18
31. Swat K. and Krzychowicz, G. (1996), "*ERGONOM: Computer-aided working posture analysis system for workplace designers*". International Journal of Industrial Ergonomics 18, pp. 15-26.
32. Teoh, E. R., and Campbell, M. (2010), "*Role of motorcycle type in fatal motorcycle crashes*", Journal of Safety Research 41, pp. 507-512.
33. Young, E. (2013), "*Petrol Price: Why are they rising?*" Retrieved on 22nd August 22, 2013 from BBC News Official Website: http://www.bbc.co.uk/news/business-21547684. Blanc, Y., and Dimanico, U. (2010), "Electrode Placement in Surface Electromyography (sEMG) "Minimal Crosstalk Area" (MCA)", The Open Rehabilitation Journal 3, pp. 110-126.
34. Honda Motor Co., Ltd. [online] Available at: https://business.ohio.edu/media/152387/honda_annual_report.pdf. pp 12-17.
35. "Road Safety in the South-East Asia Region" by World Health Organization (WHO) Regional Office South-East Asia. [online] Available at: http://www.who.int/violence_injury_prevention/road_safety_status/2015/Road_Safety_SEAR_3_for_web.pdf. pp 5-6.

Printing and Curing of Conductive Ink Track on Fabric using Syringe Deposition System with DLP Projector and Hot Plate

Rd. Khairilhijra Khirotdin[1], Mohamad Izwan Saifudin Mohamad Zainuri[2], Afiqah Izzati Ishak[3], Nurhafizzah Hassan[4], Khairu Kamarudin[5], Mohd. Rasidi Ibrahim[6], Reazul Haq Abdul Haq[7], Mustafa Ibrahim[8], Nasuha Sau'de[9], Omar Mohd. Faizan Marwah[10]

[1,2,5,6,7,8,9,10]Faculty of Mechanical and Manufacturing Engineering, Universiti Tun Hussein Onn Malaysia, 86400 Parit Raja, Batu Pahat, Johor, Malaysia

[3,4]Faculty of Engineering Technology, Universiti Tun Hussein Onn Malaysia, 86400 Parit Raja, Batu Pahat, Johor, Malaysia

Abstract. Printing is a technique to transfer ink onto substrates to create pattern and syringe deposition system has shown some great potential in printing due to its ability to produce filamentary bead tracks which is important concerning conductivity and easily adopted on conformal surfaces which could not be realized by conventional technique. Fabrics with integrated electrical features able to create intelligent articles and may potentially open up new perspective areas of application in textile printing. However, the applicability of this technique on fabrics remains unknown which the ink used has to meet certain requirements including high electrical conductivity, resistance to oxidation, dry out without clogging, good adhesion with suitable viscosity and surface tension. Thus, there is a need to do this study which is to determine the feasibility of syringe deposition system to print a conductive ink tracks using silver epoxy-based conductive ink on fabric substrate via lycra material. This study is also aim to investigate the feasibility of using DLP projector with hot plate as another source of heat to be used in curing the ink tracks on fabric. The effect of printing and curing parameters to the characteristics and conductivity of the ink track is investigated. Several mechanical and electrical tests were also administered to determine the cure, hardness, adhesion and resistance level of the ink tracks. The results obtained were as expected which higher printing speed and lower deposition height used, a narrower and thinner ink tracks were produced. Sample with 4 mm/s of printing speed and deposition height of 1 mm resulted in dimension closer to the targeted dimension. The longer curing time and higher temperature used, a lower resistance is produced. The lowest resistance achieved is 0.9 Ω cured at 150°C for 60 minutes. The conductivity of the ink track was affected by curing process and cross-sectional area of the ink track. It is proven that a syringe deposition system is capable of printing the ink track and DLP projector with hot plate is suitable to cure the ink track properly on fabric.

---

*Corresponding author: khairil@uthm.edu.my

1 Introduction

Printing is a method used to transfer metallic conductive ink into a substrate to create electronics structure. Among all printing methods, syringe deposition system is one of the method to apply liquid solution with precision and high speed and capable of coping with variety of materials such as conductive ink and it has shown some great potentials in printing conductive ink due to its ability to produce filamentary bead tracks which is important with respect to conductivity [1]. Another prospect of this technology is it could be adopted especially on non-planar and curvature surfaces where conventional printing technique could not be realized [2]. However, the applicability of this technique on fabrics remains unknown which the conductive ink used has to meet certain requirements including high electrical conductivity, resistance to oxidation, dry out without clogging, good adhesion with suitable viscosity and surface tension. Therefore, there is a need to do this study in which it will focus on finding the right materials and technique to be used for printing electronics structures on fabrics.

In the mean time, conductive ink has been widely utilized in various applications of electronic components [3]. Since silver has high stability against air and one of the best conductivity, silver conductive ink is widely used to manufacture conductive patterns in electronic devices [4]. Conductive ink is mundanely come in liquid form and requires curing to expose its metallic contents. Curing is a process to crosslink the polymer materials and it is usually done after the conductive ink has been printed on substrates. The curing process is essential to eliminate the unwanted solvents and evaporate the binder in the substrate. In this study, it is focused on how to cure the conductive ink on fabric material. There are several type of curing process available using several technologies including oven, laser, ultraviolet (UV) radiation and electron beam. Some of these technologies suffer from drawbacks such as oven curing process is done in a closed system and the heat generated is presumably double (flash cure) from its designated temperature set if it is not properly controlled while laser is considered expensive system and requires a proper control especially on its power which could damage the substrate. Thus, a more suitable and simpler curing method is needed whereby a concept of gradually curing process is possible to properly cure the ink track without damaging the substrate. A Direct Light Projector (DLP) was seen capable of polymerizing photopolymer materials [5]. In addition, hot plate as well is promising since it was used previously to cure several conductive inks on different substrates [4]. Thus, by combining them might properly cure both areas of the inks since DLP projector is focused on curing the top part of the inks while hot plate cured the bottom part of the ink. This helps the heat to evenly distribute and a higher conductivity of the ink tracks could be realized without damaging the substrate. In addition, it could also decrease the cost as the projector and hot plate is quite cheap, easy to set up and portable. An investigation is performed to determine the optimum curing time and temperature to cure the ink track from syringe deposition system on fabric. Fabrics materials with integrated electrical features able to create intelligent articles with wide ranging of applications such as sports, work wear, health care, safety and others. These advances in technology may open new perspective for such areas of application in textile printing as sampling, strike-offs and customization.

2 Objective

The objective of this study is to determine the feasibility of syringe deposition systems to print a conductive ink tracks using silver epoxy -based conductive ink on fabric substrate via lycra material. This study is also aim to investigate the feasibility of using DLP projector with hot plate as another source of heat to be used in curing the ink tracks on fabric. The results obtained are used to determine the relationship between printing and curing

 DOI: 10.1051/matecconf/201713500047

parameters to the morphological, mechanical and electrical properties of the electronics structure printed.

3 Methodology

3.1. Pattern design

The pattern design to be printed onto the substrate is a strain gauge pattern. Flexibility and various sizes are the advantages for strain gauge as a chosen electronic structure for the study. The total length of the strain gauge is 219 mm with a 2 mm of ink track width as depicted in Figure 1. Taguchi's method was used to identify the number of experiments needed in order to save time and cost.

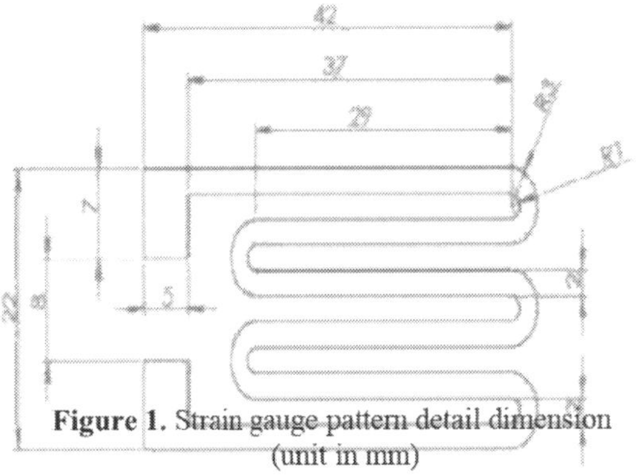

Figure 1. Strain gauge pattern detail dimension (unit in mm)

Figure 2. Syringe deposition system

3.2 Printing and curing process

Syringe deposition system (Model: FISNAR F4200N.1) was used as a deposition method in the study as shown in Figure 2. It is non -contact types of printing technique and using the syringe deposition system gave a slight advantage since it allows printing on conformal surfaces while most of printing technologies could not. Printing with syringe deposition system mostly resulted a semi- circular shape ink track in contrast with other printing technologies especially ink jet printing where a rectangular shape is always printed [6] as illustrated in Figure 3. Printing parameters including the nozzle diameter, pressure, printing speed and height deposition of the nozzle from the substrate is the influence factor to the physical characteristics of the ink droplet. Viscosity of the ink also played a major role and the printing parameters selected in this study are tabulated in Table 1 while the other parameters are set to be constant.

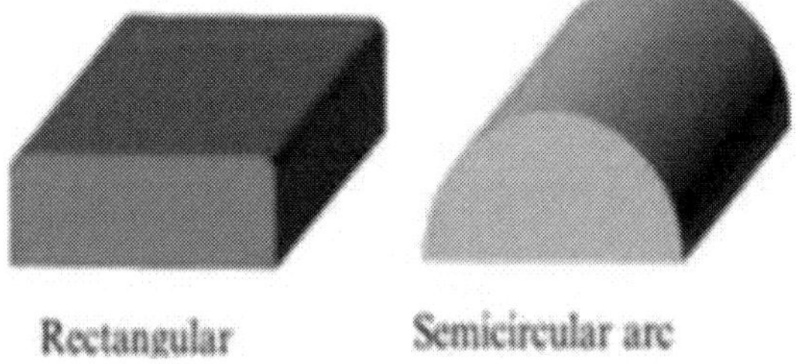

Figure 3. The ink droplet shape from inkjet and syringe deposition system [6].

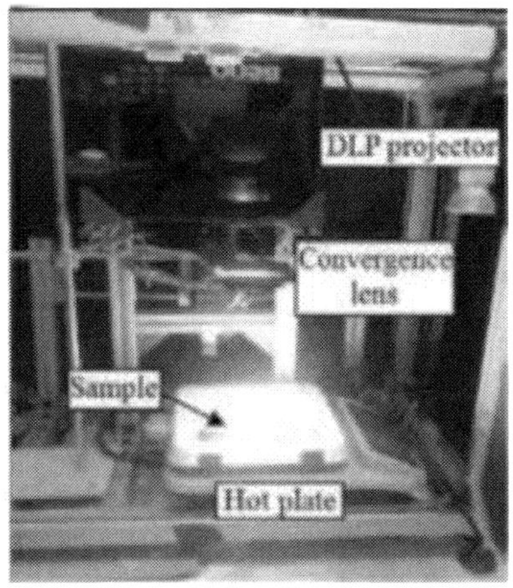

Figure 4. Experimental set up for curing process.

On the other hand, the curing process is a process which both chemical reaction and physical action takes place to make the ink become harder, tougher or more stable linkage. Since the conductive ink comes in liquid form and required curing to expose its metallic contents, the unwanted material is needed to be removed so that the conductive ink track could obtain better electrical properties. A combination of DLP and hot plate was used for curing process and the experimental set up is depicted in Figure 4 where DLP is set to focus on curing the top part while hot plate is set to cure the bottom part of the ink. Thus, by combining them might properly cure both areas of the inks (top and bottom) since the heat may evenly distributed and a higher conductivity of the ink tracks could be realized without damaging the substrate. The curing parameters selected including curing time and temperature are tabulated in Table 1 while the other parameters are set to be constant.

Table 1. Printing and curing parameters selected for the experiment.

No. sample	Printing Speed (mm/s)	Height deposition (mm)	Curing time (min)	Curing temperature (^oC)
1	3	0.5	15	110
2	4	1.0	30	120
3	5	1.5	45	130
4	6	2.0	60	140
5	7	2.5	75	150

3.3 Material

Viscosity is one of the most important properties when printing ink. Most conductive inks have surface tension and viscosity value which allow the droplet formation to be dispensed. This is due to the conductive ink mostly using water or polymer as the solvent in controlling viscosity of the ink. Viscosity of the ink is dependent on temperature and it is important to recognize that a large fluctuation in temperature can change viscosity and ultimately the ink performance. Silver epoxy-based ink (Model: AG 806) as shown in Figure 5 is the conductive ink used for the study and it is chosen since it has higher anti oxidation properties rather that copper based ink. A direct usage of the ink used is not possible due to high solid fraction thus an adjustment of viscosity via dilution with toluene solvent is necessary. A vibro-viscometer was used to measure the viscosity and the viscosity is required to be reduced to one sixth of the original viscosity to ensure the ink could be dispensed without clogging. The dilution process is performed by adding 0.5 ml of toluene for ten times as tabulated in Table 2 without altering too much its solid content.

Table 2. Result of the viscosity reductive ink with toluene.

Trial	Volume of toluene (ml)	Viscosity (mPa.s)
1	0	7880
2	0.5	6054
3	0.5	4527
4	0.5	3764
5	0.5	3000
6	0.5	2500
7	0.5	2365
8	0.5	2230
9	0.5	1730
10	0.5	1580
11	0.5	1230

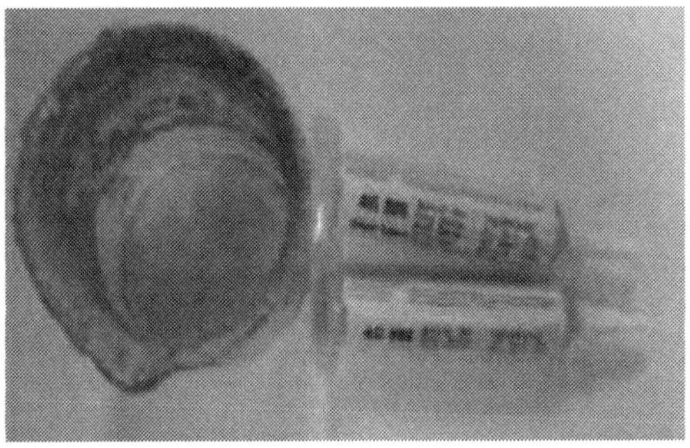

Figure 5. Silver epoxy-based ink (AG 806) Produced by Penchem Technologies.

3.3 Substrate

Lycra (spandex) as shown in Figure 6 (a) was used as a substrate material for the experiment and it is commonly used in sportswear due to it is light, smooth and soft. Lycra was chosen since the aim is to embed sensor on cloth and lycra materials is much suitable because it is a type of compression wear (highly stretching up to 500% of their original length) where under compression it clings strict to the skin and quite literally squeezing the muscles below without irritation hence helps to increase the sensor sensing capabilities. In addition, to have a good substrate in fabrics, all requirements must be fulfilled such as the smoothness of the

 DOI: 10.1051/matecconf/201713500047

substrate, the melting point of the substrate, the compressibility and the ink receptivity [7]. Rubber dye ink is coated on lycra surface (refer Figure 6 (b)) prior to printing process to prevent the fabric from absorbing the ink. The coating as well act as an insulator to cover any gaps presence in fabric structure (refer Figure 6 (a)) so that the ink would not dissipate much to the environment.

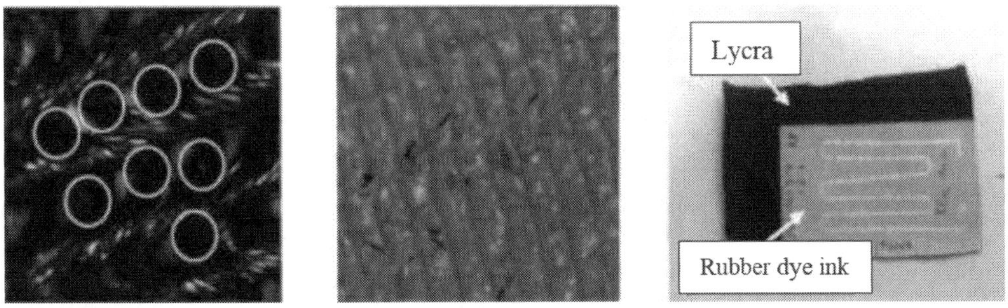

(a) (b) (c)

Figure 6. Micro-structure of lycra's surface before and after coating (a) gaps present (denoted as white circle) (b) gaps covered with rubber dye ink (c) lycra covered with rubber dye ink

3.4 Mechanical and electrical test

The electrical tests were conducted to obtain the resistance by measuring the resistance using digital millimeter. The mechanical tests were also performed including the manual scratch test, adhesion and hardness test. The manual scratch test is to determine the cure level of the ink track while the adhesion test via the tape test method is to assess the adhesion level of the ink track. HMV micro-hardness test is employed to determine the hardness level of the ink track.

4 Result and discussion

4.1 Printing process

All fabric substrates were first screen printed with rubber dye ink before the experiments begin. This is done to avoid the silver conductive ink from leaking into the fabric. The optimum parameters were selected from the preliminary experiments performed earlier and the other parameters were set to be constant including the viscosity of the conductive ink (1,230 mPa.s), the nozzle tip size (0.58 mm) and the pressure applied (0.8 bar). Table 3 and 4 below shows the result for experimental of printing using silver conductive ink.

Table 3. Result of the effect of variation of printing speed to morphology of the ink track at height deposition of 1 mm

Sample	Deposition Height (mm)	Printing Speed (mm/s)	Width (mm)	Designated width (mm)	Thickness (mm)	Length (mm)	Area (mm^2)	A/L (mm)
1	1	3	2.258	2	1.207	42	2.725	0.0648
2		4	1.965		0.231		0.454	0.0108
3		5	1.850		0.218		0.403	0.0096
4		6	1.425		1.125		1.603	0.0382
5		7	1.080		0.790		0.853	0.0203

Table 4. Result of the effect of variation of deposition height to morphology of ink track at printing speed of 4 mm/s

Sample	Printing Speed (mm/s)	Deposition Height (mm)	Width (mm)	Designated Width (mm)	Thickness (mm)	Length (mm)	Area (mm^2)	A/L (mm)
1	4	0.5	1.885	2	0.632	42	1.191	0.0284
2		1.0	1.965		0.711		1.397	0.0333
3		1.5	1.530		0.654		1.001	0.0238
4		2.0	1.618		0.944		1.527	0.0364
5		2.5	1.505		0.627		0.944	0.0225

All results were plotted as shown in Figure 7 (a) which it clearly seen that the line width and line thickness of ink track were affected by the variation of printing speed. When the printing speed is faster, the width of the ink tracks were smaller and far from the expected width size. This is due to the time for the ink to be dispensed is shorter and at the same time the thickness also thinner. The changes in thickness were not too obvious compared with the width of the ink track. In addition, the deposition height also gives an impact to the line width and line thickness. If the deposition height used is higher, the ink droplet is not stable result in a wrinkle condition of track and in some condition incontinuous ink tracks were produced. Thus, the distance between the nozzle tip and the substrate is needed to be kept close to ensure a continuous and a straight ink track is produced. Figure 7 (b) shows the changes in width and thickness of the ink track with variation of deposition height at constant printing speed of 4 mm/s. It was observed that the width of the track changed to a narrower ink track whereas not much changed occurred for the thickness of the ink track.

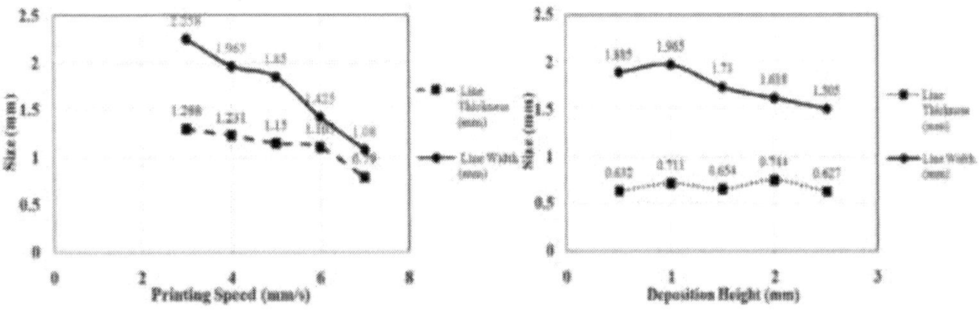

(a) (b)

Figure 7. Changes of width and thickness of the ink track with (a) variation of printing speed at deposition height of 1 mm (b) variation of deposition height at printing speed of 4 mm/s.

4.2 Curing Process

Figure 8 (a) shows the resultant changes in resistance obtained against variation of time and temperature selected. It was observed that the increase in temperature resulted in decrease in resistance. The trend also shows the resistance also decreases as the curing time increased. Furthermore, at temperature of 120°C shows a constant decreased in resistance for five different time intervals. However, there are some anomalies happened at temperature 110°C and 130°C where the resistance suddenly becomes higher at curing time of 45 minutes due to the temperatures were not properly controlled. At temperature of 140°C and 150°C, the trend shows decreasing in resistance for all curing time except at 75 times due to over cure phenomenon. The lowest resistance achieved is 0.9 Ω cured at 150°C for 60 minutes. Besides, Figure 8 (b) shows the changes in conductivity against variation of curing time and temperature selected where the conductivity was found by calculation using equation (1) where the conductivity, σ of the ink track is based on the size of cross-sectional area, *A* of the ink track and the resistivity, ρ as shown in the following equation (1) and (2).

$$A = Thickness\ x\ Width \tag{1}$$

$$Conductivity,\ \sigma = \frac{1}{\rho} \tag{2}$$

It shows that the conductivity increased as the curing time increased. At 60 minutes of curing time, the conductivity reached its peak for both curing temperatures at 140°C and 150°C. However, at curing time of 75 minutes, both lines had sudden dropped due to over cured phenomenon where some silver metals were burnt out and this decrease the conductivity.

 DOI: 10.1051/matecconf/201713500047

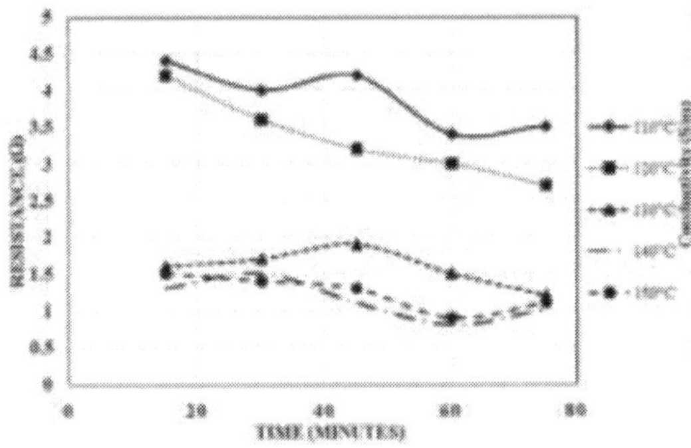

(a)

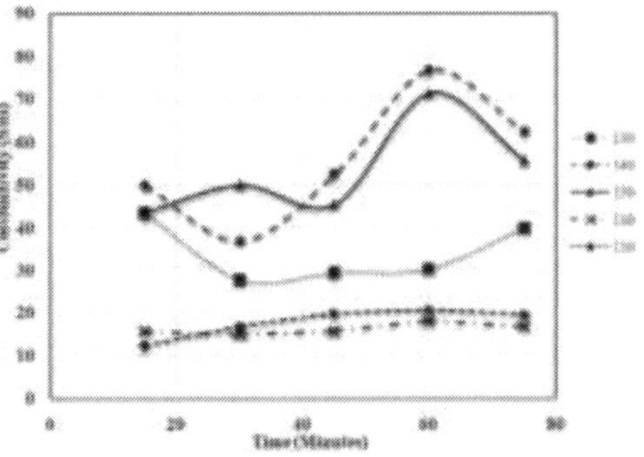

(b)

Figure 8. (a) Changes of resistance for variation of curing time at several curing temperatures (b) Changes of conductivity for variation of curing time at several curing temperatures.

4.3 Morphology analysis

The strain gauge ink track structure was analysed using optical microscope to determine its physical properties including its track width, thickness and cross-sectional shape after cured. The cross-sectional shape should form a semi-circular like shape and Figure 9 (a) shows a shape of semi-circular shape of the ink track structure as mentioned by [7]. The results of the physical properties of the ink track have been tabulated earlier in Table 3 and 4. The following equation (3) has been used to calculate the percentage of dimensional accuracy of the ink track dimension to its designated dimension. It is performed to determine the effectiveness of the syringe deposition system in printing conductive ink on lycra substrate. The results for the dimensional accuracy measurement are shown in Figure 10 (a) and (b). It was observed that the printing process by syringe deposition system produced a dimension closed enough to the designated design size due to the suitable parameter used. It was observed that the optimum parameter selected (deposition height of 1 mm and printing speed of 4 mm/s) is verified through the percentage of accuracy acquired since the selected parameters manage to achieve the physical characteristic of the designated ink track's size.

$$\text{Dimensional accuracy \%} = \frac{\text{sample result}}{\text{expected result}} \times 100\% \quad (3)$$

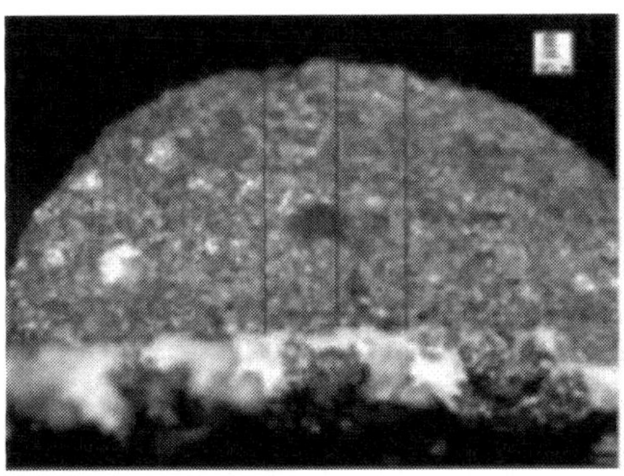

(a)

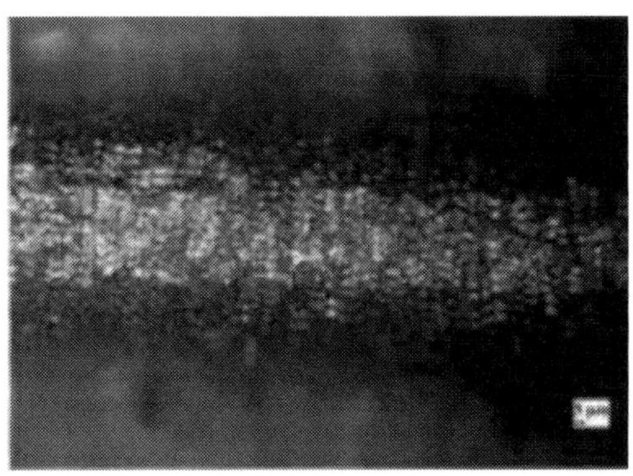

(b)

Figure 9. (a) Cross-section view of the ink track showing the thickness (b) A view of the width of the track showing a fine track edges.

DOI: 10.1051/matecconf/201713500047

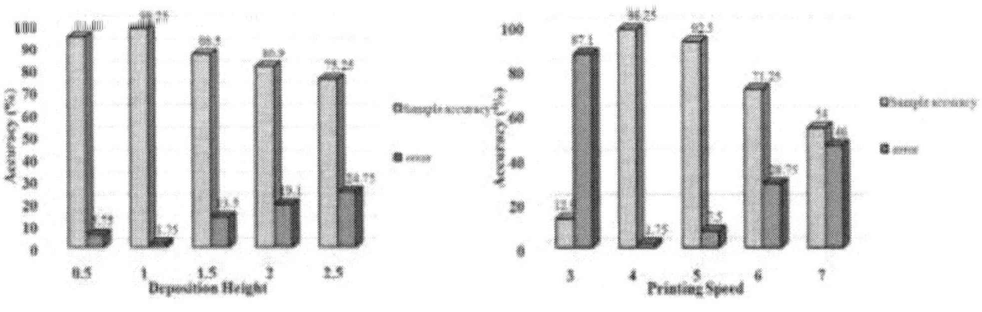

(a) (b)

Figure 10. Percentage of dimensional accuracy of the width with (a) variation of deposition height (b) variation of printing speed

4.4 Mechanical Test

Mechanical test is important to identify whether the ink track has been fully cured, its hardness and adhesion level after cured. The manual scratch test is performed to detect the smear from the ink track using a needle right after the sample was cured. The hardness test is conducted where the diamond indenter with certain load is applied onto the ink track resulted in a diamond shape. Lastly, the adhesion test is performed using the tape test method in which the 3M tape is applied on the ink track and smoothed by the eraser. The tape is pulled off rapidly at an angle of 180° and the adhesion level is assessed from the range 0B to 5B as stated by ASTM D3359. Based on the result from Figure 11, the graph shows a few fluctuated points at the beginning; however the overall trend shows a clear trend of increasing adhesion level as the curing time and temperature increased. Similar trend was observed on the results of hardness level as depicted in Figure 12 which the hardness level of the ink tracks showed inclination trend similar to the previous adhesion test. It was found that when the cured temperature and the time increased, the hardness level of the ink tracks was also increased accordingly.

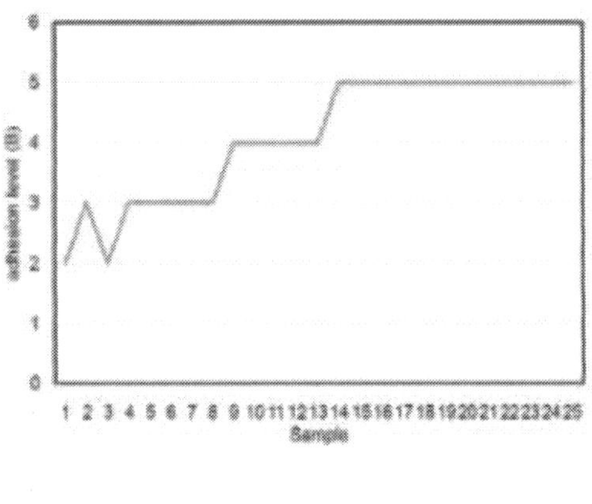

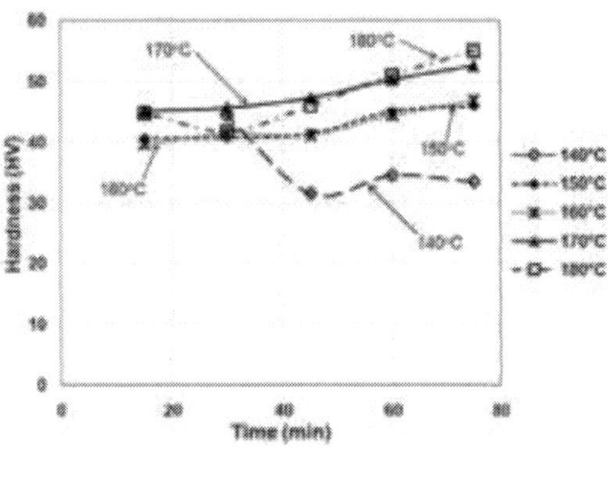

Figure 11. Changes of adhesion level for all samples. **Figure 12.** Changes of hardness level for all samples.

5 Conclusion

The objective of the study which is to investigate the feasibility of printing and curing the silver conductive ink tracks on fabric using syringe deposition technique and a combination of DLP projector and hot plate is achieved. The strain gauge pattern was successfully printed and properly cured on lycra substrate. The printing parameters involved which is the deposition height and the printing speeds were carefully controlled to achieve the desired ink track's size. The accuracy of the ink track dimension and cross-sectional shape of the ink track are depended on the printing parameters selected. The preliminary experiments are required to prevent any problem that may occur in the actual experiments. It also intended to save cost and time during the actual experiment. Besides, DLP projector with hot plate has proven capable of properly cure the silver conductive ink by means of evaporating the binders and unwanted solvent thus harden and cross linked the metallic particles with the substrate. The resistance is much depending on curing time and temperature in which the higher the temperature and the longer the cure time used, the lower the value of resistance. On the other hand, conductivity is inversely proportional to the resistance in which when the curing time and temperature increases, the conductivity obtained also increases.

Acknowledgement

The authors gratefully acknowledge the support to the Ministry of Higher Education of Malaysia and University of Tun Hussein Onn Malaysia under the Research Acculturation Grant Scheme awarded (RAGS-R052) and giving the opportunity to attend and present at this conference.

References

1. E. Fearon, T. Sato, D.Wellburn, Thermal Effects of Substrate Materials Used in the Laser Curing of Particulate Silver Inks, *Proc LANE Laser Assisted Net Shape Engineering,*c379-390 (2007).
2. R. Khirotdin, M. F. Ngadiron, M. A. Mahadzir, N. Hassan, Performance Evaluation of Strain Gauge Printed Using Automatic Fluid Dispensing System on Conformal Substrates, 3rd. International Research and Innovation Summit (2017), Malacca, Malaysia.
3. K. Khirotdin, M. A. Nasir, K. Kamarudin, M. Ibrahim, Investigation of DLP Projector as an Energy Light Source to Cure Silver Conductive Ink, *Applied Mechanics & Materials*, Vol. 773-774, p 13-19 (2015).
4. I. Kazani, C. Hertleer, G. De Mey, A. Schwarz, G. Guxho, L. V. Langenhove, Electrical conductive textiles obtained by screen printing. *Fibres & Textiles in Eastern Europe*, *20*(1), 57-63 (2012).
5. K. Khairu, I. Raman, M. A. S. Mohamed, M. Ibrahim, S. Wahab, Parameter Optimerization for Photo Polymerization of Microstereolithography, In *Advanced Materials Research* (Vol. 626, pp. 420-424). Trans Tech Publications (2013).
6. K. Suganuma, Introduction to Printed Electronics, 74, p.130 (2014).
7. K. Crippen, P. Tang, P. Mulready, DuPont Lycra shifts emphasis to global brand management, *Journal of Product & Brand Management*, Vol. 4 (1995).

Microstructure and wear behavior of friction stir processed cast hypereutectic aluminum silicon

Rosli Ahmad[1], *Samir* Sani Abdulmalik[1,*], *Ojonimi* Yusuf Usman[1], and *Sufyanu* Bashir Danjuma[1]

[1]Department of Manufacturing and Industrial Engineering, Faculty of Mechanical and Manufacturing Engineering, Universiti Tun Hussein Onn Malaysia, 86400 Parit Raja, Batu Pahat, Johor, Malaysia

Abstract. Hypereutectic as-cast Al-18Si-Cu-Ni alloy was subjected to friction stir processing (FSP). The resultant effect of FSP on the alloy was evaluated by microstructure analysis and wear tests (dry sliding). A significant microstructural modification and enhancement in wear behavior of Al-18Si-Cu-Ni alloy was recorded after friction stir processing. Wear resistance improvement was related to considerable modification in size, morphology and distribution of silicon particles, and hardness improvement. It was found that lower tool rotation speed was more effective to refine silicon particles and in turn increase wear resistance. Minimum Si particle mean area of about 47.8 μm^2, and wear rate of 0.0155 mg/m was achieved.

1 Introduction

Al-Si alloys are among the materials of tribological relevance, and continued to receive considerable attention due to their good castability and machining characteristics [1]. Past studies [2, 3] reported that Si particle morphology, size, and content, significantly affects the wear resistance of these alloys. The size and morphology of the Si particles are considered to be factors controlling the formation and propagation of cracks [4]. Alidokht et al. [2] reported that under dry sliding conditions the tendency toward cracking for larger Si particles is greater. It was noticed in [3] that the refinement of Si particles in Al-Si alloys increased the wear resistance. Silicon particles having even and rounded morphology can delay crack nucleation and growth processes in the subsurface region [2].

Hypereutectic Al-Si-Cu-Ni alloy is among the widely used commercial aluminium silicon alloys in automotive industries for piston application. However, coarse eutectic Si structure, segregation of the primary Si particles, and porosity affects the mechanical properties of Al-Si-Cu-Ni alloy particularly, fatigue, and wear resistance [5]. In order to solve these problems, a practical technique of microstructure modification which can effectively reduce the porosity and uniformly redistribute the fine Si particles in the aluminum matrix is imperative. Recently, friction stir processing (FSP), a variant of friction stir welding (FSW), in which the intense plastic deformation generated by traversing a welding tool [6] has been used as a technique to locally modify the microstructure in a

* Corresponding author: samiralhaji@yahoo.co.uk

DOI: 10.1051/matecconf/201713500048

number of Al-Si cast alloys [7-9]. In FSP a rotating tool with a designed pin and shoulder is plunged into a monolithic workpiece and produces a recrystallized fine particles microstructure within the stirred zone (SZ). Detail explanation of the process if given elsewhere [10]. Previous investigation [11] shows that FSP significantly refined the microstructure of cast Al-Si-Cu-Ni and created a uniform distribution of fine Si particles in the matrix, which is known to be very effective in enhancing wear resistance [2]. Thus, it is important to evaluate the wear behavior of FSPed Al-Si-Cu-Ni. In the present work, the evolution of microstructure during FSP at two different tool rotational speeds and its effects on wear resistance and hardness of Al-Si-Cu-Ni alloy was investigated.

2 Experimental procedures

In this study, hypereutectic Al-Si alloy with composition: 18Si, 0.9Cu, 1.15Ni, 0.2Mn, 0.16Mg, 0.06Zn, 0.19Fe, and 80.34Al (wt.%) was subjected to single pass FSP in longitudinal direction. The as received ingots are machined into plates having 30 mm × 150 mm × 7 mm dimensions. A non consumable cylindrical profile tool made from H13 steel was used to carry out the FSP. The tool has a shoulder diameter of 17 mm, pin diameter 6mm, and pin length 3.5 mm. The FSP experiment was performed using vertical milling machine at a fixed transverse speed of 42 mm/min, and tool rotating speeds: 1000 rpm, and 1400 rpm. For all the experiments, tool tilt angle was set to 3°, and tool down force was kept constant. The FSP setup is illustrated in Fig.

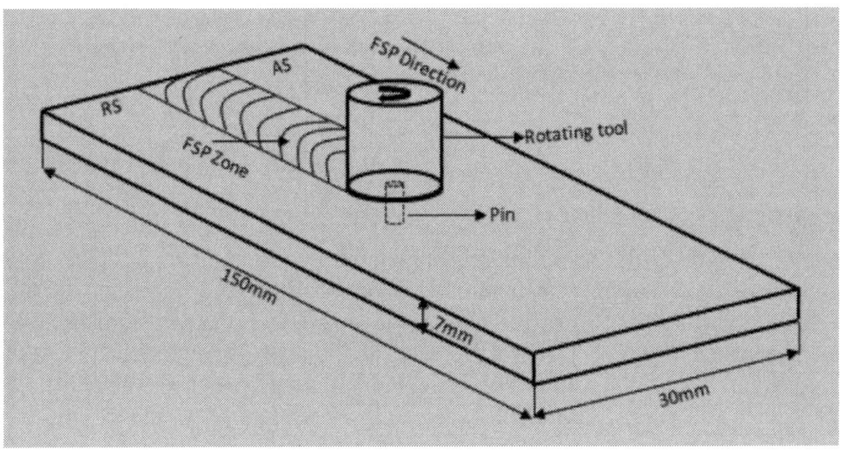

Fig. 1: Illustration of FSP setup

Microstructural evaluation was performed on the samples cross sections after FSP. Samples were ground using different grades of sand paper up to 1200 grit, polished with alumina paste, then etched in hydrofluoric acid 0.5 M solution to reveal the microstructures. The samples microstructure examination was accomplished using optical microscope. Image analysis technique was used for determining the sizes of the silicon particles, and there aspect ratio. Microhardness is measured across the FSP samples transverse cross-section, using a load of 500 g for 10 s.

Fig. 2 shows the schematic illustration of the wear test experiment, the tests were conducted on a pin-on-disk tribometer. Rectangular specimens (pin) of 4 mm × 3 mm dimension were carefully cut from the surface of FSPed samples, with their axis perpendicular to the tool travel direction. To have a uniform surface wear samples were ground using 1200 grit sand paper and polished with alumina solution, cleaned with acetone and dry before the test. The tests were performed against a counterpart stainless steel disc. Before each test the disc was polished on 1200 sand papers. The sliding speed and the normal load were fixed at 0.5 m/s and 20 N, and the sliding distances were varied up to 1000 m. At certain intervals wear tests were halted to determine the progress of wear. Each time, samples were cleaned in acetone and weighed to an accuracy of ± 0.1 mg. Rates of wear were measured as the ratio of mass loss to sliding distance, then examined the wear samples worn surfaces using scanning electron microscope (SEM).

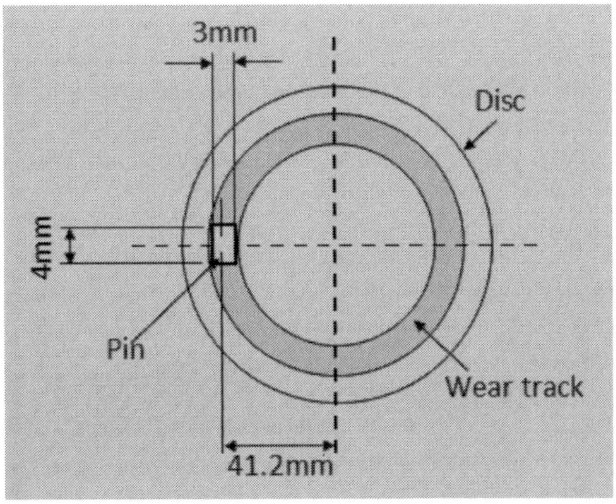

Fig. 2: Configuration of the pin-on-disc wear test; run for 1000 m sliding distance

3 Results and Discussion

Fig. 3 shows a micrograph of the as cast hypereutectic Al-18Si-Cu-Ni alloy. The base alloy consists of coarse primary silicon, and needle-like/platelet eutectic silicon morphology. The dispersal of the silicon particles was not uniform all through the aluminium matrix. Measured mean area of the eutectic and primary Si, was about 155.1 μm^2 with aspect ratio of 2.86.

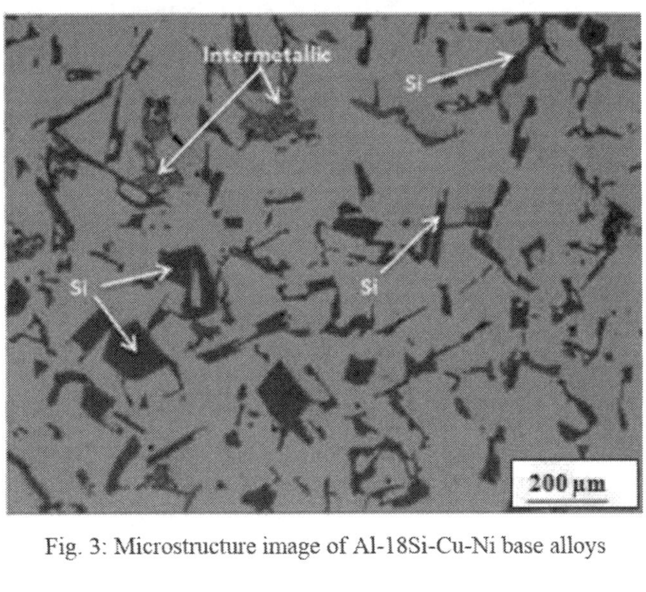

Fig. 3: Microstructure image of Al-18Si-Cu-Ni base alloys

Fig. 4(a-b) shows the optical micrograph at the center of the stir zone SZ of FSPed Al-18Si-Cu-Ni samples at various rotation speeds. It is clear that the stirring action of the tool during FSP brings about a significant breakdown of the coarser Si particles and creates a better dispersion of finer Si particles in the stir zone. This is related to thermal exposure and plastic deformation with a strain rate of 100 to 1000 s^{-1} and strains up to -40 during FSP [2]. The variation of silicon particles of FSPed, and as cast Al-18Si-Cu-Ni samples are presented in Fig. 5. In the whole stir zone SZ of FSPed Al-18Si-Cu-Ni samples, regardless of the tool rotation speed the silicon particles area size and aspect ratio decreases as compared to the as cast base alloy, as a result of the intense stirring effect. Such microstructure transformation is been reported in other Al-Si alloy [6]. However, raising the tool rotation speed above 1000 rpm does not cause further significant refinement of the silicon particle, which means, the mean area size of the silicon particles increases with increased heat input [9]. This is related with the enhanced flow-ability of the matrix as a result of high localized heat input conditions at higher tool rotational speeds, that soften the material around the tool pin and caused the Si particles flow with the plastic flow of the matrix [2, 8], and therefore the breakdown of the silicon particles would not intensified further [2]. In this study tool rotation speed of 1000 rpm gives the minimum mean area size and aspect ratio of 47.8 μm^2, 1.69 respectively.

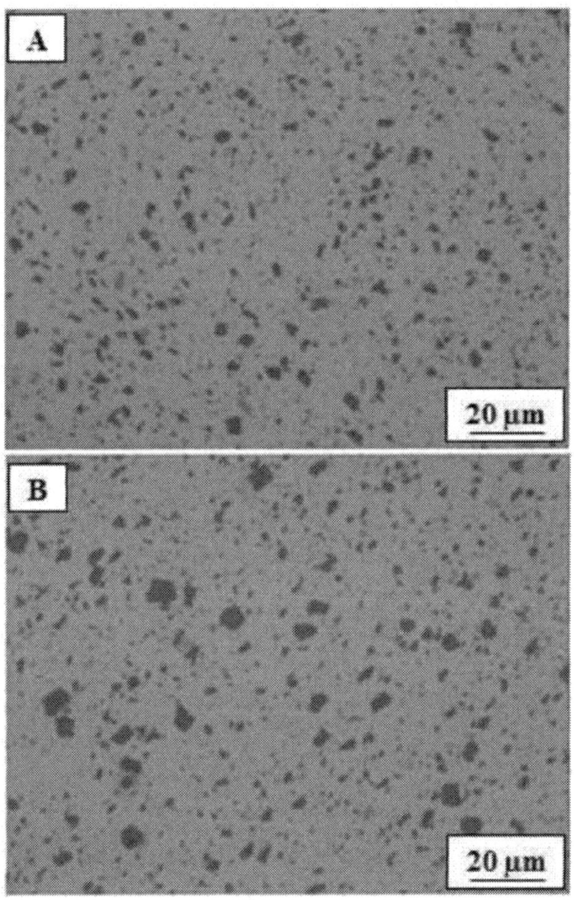

Fig. 4: Optical micrograph from the center of stirred zone of FSPed Al-18Si-Cu-Ni alloy at (a) 1000 rpm, and (b) 1400 rpm

 DOI: 10.1051/matecconf/201713500048

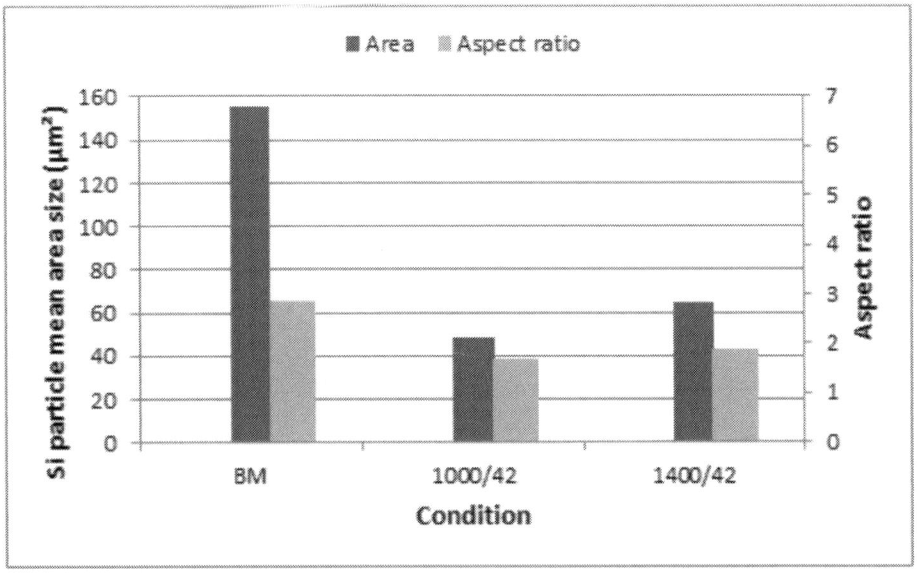

Fig. 5: Si particles mean area sizes, and aspect ratio of base alloy and FSPed Al-18Si-Cu-Ni samples, at various tool rotational speeds

Fig. 6 shows the microhardness values of the samples before and after FSP at various rotation speeds. The hardness enhancement within the SZ is considered due to fine dispersion of silicon particles in the aluminium matrix Fig. 3, and eliminating the as-cast defects [2, 12]. Alidokht et al., [2] reported the refinement in microstructure of friction stir processed cast Al-Si alloys. They observed that, hardness improved in the stir zone as a result of the modification, fragmenting and homogenization of the Si needles in the eutectic phase.

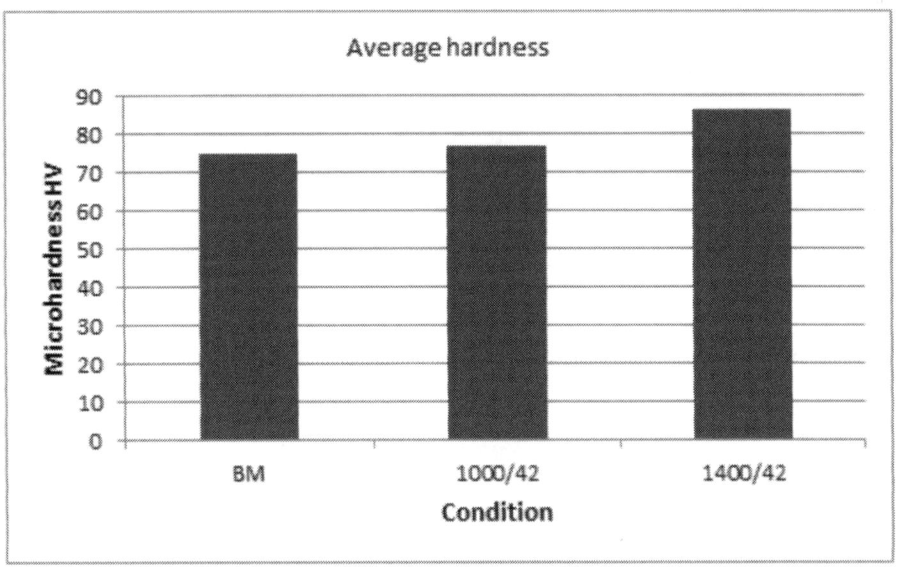

Fig. 6: variation of hardness

DOI: 10.1051/matecconf/201713500048

The relationship between sliding distance and the mass loss of as cast and FSPed Al-18Si-Cu-Ni samples is presented in Fig. 7. It is obvious the wear mass loss increases with sliding distance. Also from the figure, the extent of wear is considerably less in FSPed samples compared to the as-cast Al-18Si-Cu-Ni. Fig. 8 shows the graph of wear rate (obtained from; wear mass loss/sliding distance). At the initial stage of wear (sliding distance), the rate of wear for FSPed samples appear high, and decreases thereafter as the sliding distance increase. Whereas the mass loss curves of the FSPed samples is somewhat steady state at higher distances. For the as cast Al-18Si-Cu-Ni sample, it curve appeared increasing with a constant rate. It can be inferred from the results that after performing FSP the wear resistance of as-cast Al-18Si-Cu-Ni alloy significantly improved.

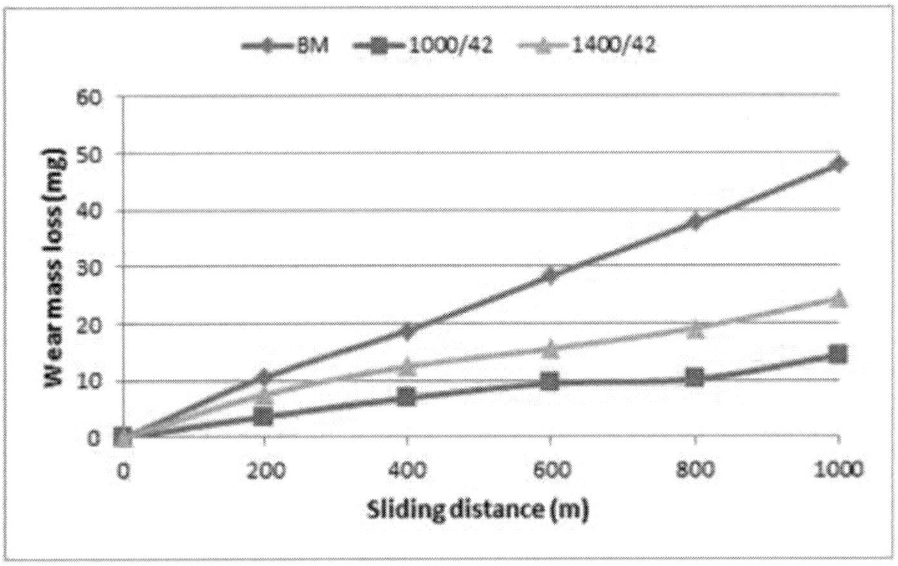

Fig. 7: Graph of wear mass loss

The sample processed at 1000 rpm tool rotation speed display slightly higher resistance to wear compared with the samples processed at 1400 rpm rotation speed. This result is similar to the one reported by Mahmoud and Mohamed [9], on FSP A143 Al-Si alloy.

SEM micrograph of Al-18Si-Cu-Ni cast alloy worn surface after 1000 m of dry sliding at linear velocity of 0.5 m/s, and force 20 N is shown in Fig 9a. Severe damage and long grooves can be seen on the worn surface. That deep craters and cracks on the worn surface show that delamination have taken place. This mean, on the side facing sliding direction, plastic deformation layer fractured to form small lumps and then delaminated [5]. Fig. 9 (b-c) presents SEM images of the FSPed worn surfaces. As can be seen the FSPed surface is less damaged compare with that of the as-cast sample. Wear mechanisms: both abrasion and delamination were observed in all the FSPed samples. It can be seen from Fig. 9 (b-c) that the deep grooves on the surface of both FSPed samples are less. This supports the lower wear rates of FSPed samples compared to that of the as-cast samples Fig. 8. The observed enhancement in wear results may be related to the enhancement of the alloy hardness as a result of microstructure refinement Fig. 4. Under abrasive wear condition, it is well known that the wear resistance of the material decreases with increasing hardness [2, 13]. Additionally the morphology, size and distribution of silicon particles in the FSPed Al-18Si-Cu-Ni samples contributed to the wear resistance.

 DOI: 10.1051/matecconf/201713500048

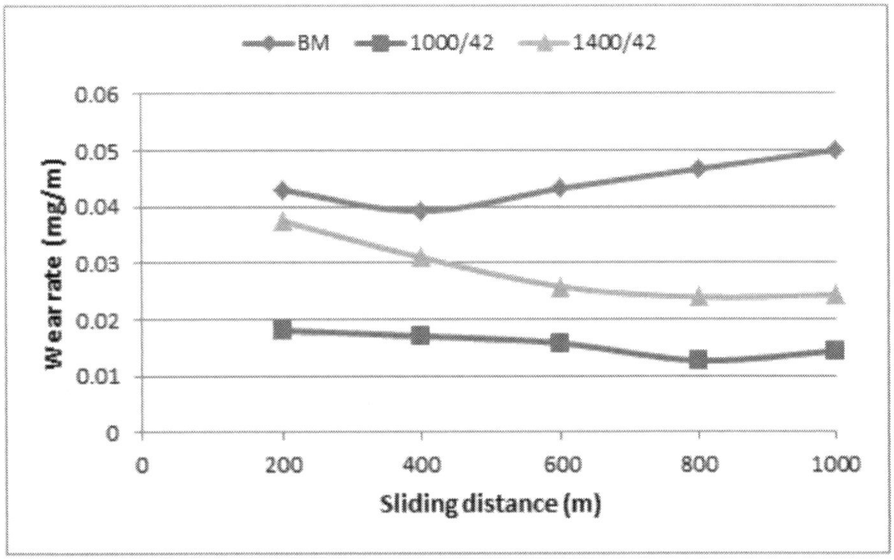

Fig. 8: Graph of wear rate

It is well known that larger, coarse irregularly shaped Si particles with large aspect ratio present in a material may lead to stress concentrations at the silicon matrix interface, and facilitates path for propagation of cracks [2].

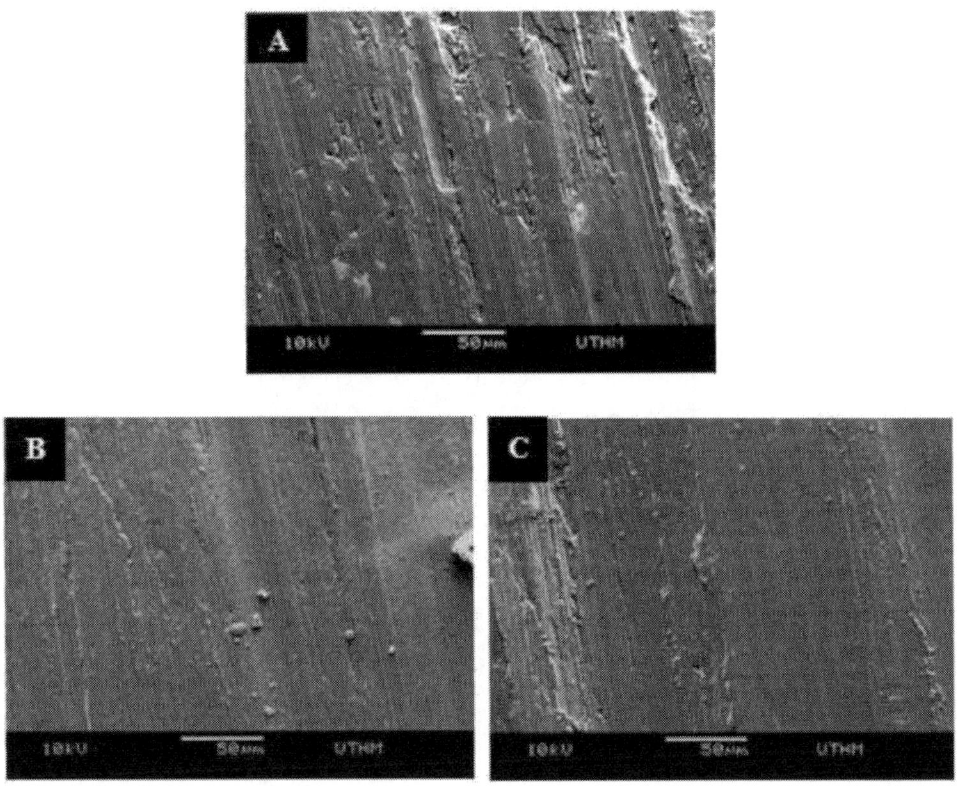

Fig. 9: SEM micrographs of the sample worn surfaces (a) as cast based metal, (b) as FSPed at 1000 rpm, (c) FSPed at 1400 rpm

4 Conclusions

In this study, cast hypereutectic Al-18Si-Cu-Ni alloy was subjected to FSP under different rotational speeds, and it microstructure, hardness and wear properties were investigated. FSP results in a significant breakup of large coarser eutectic and primary silicon particles, and subsequent uniform distribution of smaller Si particles throughout the Al-18Si-Cu-Ni matrix was achieved after FSP. Increasing tool rotation speed causes increase in Si mean area size, due to increased thermal exposure. At 1000 rpm tool rotation speed, mean area size and aspect ratio of the Si particles were measured as: 47.8 μm^2, and 1.69 respectively, compared to the 155 μm^2, 2.86 in as-cast Al-18Si-Cu-Ni. The FSPed samples exhibits higher hardness value compared to the as-cast base alloy. This is related to the considerable microstructure refinement, and homogenization. Resistance to wear in the FSPed samples was observed to be better than that of the cast Al-18Si-Cu-Ni. This improvement is related to increased value of hardness and a better dispersion of finer, near circular silicon particles. At lower tool rotational speed the difference in wear rate is more pronounced.

References

1. V.C. Srivastava, R.K. Mandal, S.N. Ojha, Materials Science and Engineering A, **304**, 555-558 (2001)
2. S.A. Alidokht, A. Abdollah-zadeh, S. Soleymani, T. Saeid, H. Assadi, Materials Characterization, **63**, 90-97 (2012)
3. M. Elmadagli, T. Perry, A. Alpas, Wear, **262**, 79-92 (2010)
4. H. Lashgari, S. Zangeneh, H, Shahmir, M. Saghafi, M. Emamy, Materials & Design, **31**, 4414-22 (2010)
5. T. Mahmoud, Surface and Coatings Technology, **228**, 209-20 (2013)
6. S.S. Abdulmalik, R. Ahmad, M.B. Asmael, In AIP Conference Proceedings, **1831**, 020037 (2017)
7. Z. Ma, A. Pilchak, M. Juhas, J. Williams, Scripta Materialia, **58**, 361-6 (2008)
8. Z. Ma, S. Sharma, R. Mishra, Metallurgical and Materials Transactions A, **37**, 11 (2006)
9. T. Mahmoud, S. Mohamed, Materials Science and Engineering A, **558**, 502-9 (2012)
10. R.S. Mishra, P.S. De, N. Kumar, Springer, 13-58 (2014)
11. F. Tsai, P. Kao, Materials Letters, **80**, 40-2 (2012)
12. N, Saini, D. Dwivedi, P. Jain, H. Singh, Procedia Engineering, **100**, 1522-31 (2015)
13. Y. Chen, K. Nakata, Materials characterization, **60**, 12 1471-5 (2009)

Understanding Factors of Distraction among Intrastate Bus Driver

Muhammad Nur Annuar Mohd Yunos[1,*], Jalil Azlis-Sani[1], Mohd Zamani Bin Ngali[1], Musli Mohammad[1], Mohd Saiful Izwan Aazizi[2] and S.M. Sabri S.M. Ismail[3]

[1]Faculty of Mechanical and Manufacturing Engineering, Universiti Tun Hussein Onn Malaysia, 86400 Batu Pahat, Johor, Malaysia

[2]Rapid Bus Sdn Bhd, Kompleks Rapid Bus Cheras Selatan, 71, Jalan Kb 2/15, Kampung Baru Balakong, 43300 Seri Kembangan, Selangor, Malaysia

[3]Prasarana Malaysia Berhad, Level 21, Tower B, Menara UOA Bangsar, No 5, Jalan Bangsar Utama 1, 59000 Kuala Lumpur, Malaysia

Abstract. The increase in number of the world population have affected the demand for public transport especially in larger cities. Intrastate buses are chosen as the main public transport by many people due to its affordable fares. Therefore, a better performance of the intrastate bus drivers would be significant to cater the high demand. Nevertheless, distraction among the intrastate bus drivers have been found as one of the major factors that could affect the performance of the drivers. Therefore, the objective of this paper is to provide a better understanding on the factors of distraction among intrastate bus drivers. This paper refers to findings from previous researches which are related to this field of study. In conclusion, this paper will provide a list of factors of distraction among the intrastate bus drivers which may degrade their performance.

1 Introduction

In this globalisation era, developing countries prioritised on improvement of public transport systems especially on bus services due to high demands. Department of Economic and Social Affairs (DESA) of United Nation projected that Asia population will increase approximately 1.4 million by 2050 and thus, around 86% of the world population will live in urban areas [1]. The rapid urbanisation will result in serious traffic congestion [2]. Throughout the world, public bus services are widely provided in many countries due to the popularity and affordable services for the citizens [3]. In addition, the usage of bus services could provide a significant reduction in traffic congestion [4]. With the increasing demand, public bus services are required to provide much more effective services to cater the growth of urban activities [5]. Therefore, the performance of the bus drivers may degrade due to greater workload [6].

* Corresponding author: muhdnurannuar@gmail.com

The performance of the bus drivers is crucial for the public transportation industry to ensure high quality of service as well as safety of all the passengers. 50 years back, there were a lot of researches on the performance of the bus drivers [7]. Bus drivers are often exposed to various factors which would affect their performance. Buses have been identified as one of the main public transport where large proportion are involved in Road Traffic Crashes (RTC) [8]. The drivers' behaviour contributed approximately 90 percent in road crashes as well as accidents [9]. Reports from World Health Organisation (WHO) shows that about 1.24 million fatalities and approximately 20 to 50 million injuries occur due to road crashes annually [10]. Based on the annual report from Social Security Organization Malaysia (PERKESO), the accidents reported which involved buses had been gradually increasing from 2009 to 2014. Therefore, it is significant for the bus operating companies to identify the factors of distraction that affect the performance of bus drivers in order to reduce the risk and fatality involving road crashes.

In Malaysia, bus services are being chosen and prioritized by citizens due to the safety attribute provided by the bus services [11]. Annual report of 2014 from Land Public Transport Commission (SPAD) found that the most often chosen public transport by Malaysian passengers are Rapid KL. Therefore, a better performance of bus drivers could increase customers' trust as well as improving the level of comfort of passengers while riding the buses.

2 Literature Review

Driver distraction is one of the key factors in traffic accidents. Driver distraction can be defined as any activities which would divert the attention and focus of the driver from driving safely [12]. There are many factors that could influence the drivers. In this research, all the factors that could distract the drivers are labelled as Factors of Distraction (FoD). Factors of Distraction is defined as any factors that is distracting the intrastate bus drivers and thus, affecting their performance [13]. Meanwhile, driver distraction is a situation where the attention of the driver is diverted to any other forms of activities which may affect the concentration of the driving activity as well as safety of the passengers [13].

2.1 Factors of Distraction (FoD)

There are many distractions that have been identified based on previous researches. Factors of distraction could arise either from inside or outside of the vehicle [14].

2.1.1 Technology Distraction

It is undeniable that technology has been developed to assist and improve human life in many ways. Nevertheless, technologies may often be a form of distraction to humans. In this context of intrastate bus drivers, technologies such as mobile phones, broadcast radio as well as ticketing machines are the common technology devices being used. There are many researches that are related to technology distraction especially mobile phones. The usage of mobile phone while driving has been proven to degrade the performance of the drivers [15].

2.1.2 Operational Distraction

Operational distraction is referred as any operation that is necessary to be conducted by the intrastate bus drivers while carrying out their job. In Malaysia, most of the prominent bus companies are giving extra tasks to their bus drivers which is not only limited to driving the buses. They are also required to run the ticketing machine when the passengers ride on the buses. Besides, there is also a form of network communication between the bus drivers and the bus traffic controller. Both these operations have been proven to be the factors of distraction to the intrastate bus drivers [14].

2.1.3 Passenger Distraction

Passenger distraction is defined as any form of distractions that arises from the passengers. For instance, situations where passengers are talking via mobile phone with a loud voice would cause distraction to the bus drivers. Some bus drivers may be distracted by the conversation of that particular passenger rather than focusing on driving the bus. In addition, the conversation between the passengers may distract the drivers as well. In another scenario, bus drivers could also be distracted due to passengers' behaviour such as fighting among passengers, babies crying loudly or rude and mischievous children[16].

2.1.4 Bus-Cabin Distraction

Bus cabin distraction could be distractions that occur from the cabin equipment such as seat belt, driver seat, bus window and others [17]. This kind of distraction is commonly related to ergonomics factors.

2.1.5 Personal-Belongings Distraction

Personal belongings of the bus drivers may be one of the factors that will cause distraction to the bus drivers. For instance, distraction could occur when the wallet of the bus driver has dropped under the seat when driving. Therefore, the response of the bus driver to reach the wallet would affect their focus and safety of the bus. Thus, personal belongings would affect the concentration of the bus drivers as well [17].

2.1.6 Environmental Distraction

Environment could also become one of the main distraction for the intrastate bus drivers when performing their duty. Environmental distraction can be referred as the change of weather including heavy rain or bright sunlight. The bus drivers would have to pay more attention ensure safety of the passengers on board. Therefore, the environment should be considered as one of the distraction that may occur to the bus drivers [14].

2.1.7 Infrastructure Distraction

Infrastructure distraction is a type of distraction that occur to the intrastate bus drivers due to the infrastructure outside of the vehicle. Several examples of infrastructure outside of the vehicle that may distract the bus drivers include roadside advertisement [18].

2.1.8 Road-Based Distraction

Road-based distraction is any form of distractions that may occur because of the roadway condition. This distraction is often related to the road conditions [19]. These distractions can occur when the drivers glance off-road while driving. The period of glance of the drivers is directly proportional to the risk of causing a road crash. The risk of exposure to road accident increases with the period of the driver's glance off the road.

2.1.9 Health Distraction

Health of the intrastate bus drivers could be a serious factor that require more attention as it can cause great distractions to the bus drivers. Bus driver has been identified as one of the professions with the worst health condition based on the health indicator including cholesterol level, blood pressure and body weight [7]. Several studies found that bus drivers are often exposed to cardiovascular disease [20].

Table 1. Comparison of previous researches

Author	Methodology	Findings
[21]	Survey (128 intra-city bus drivers)	Activities such as usage of mobile phone can distract the intra-city bus drivers and affect the safety of passengers.
[22]	Unobtrusive video (70 volunteer drivers)	The distraction of mobile phone on the performance of drivers by measuring the levels of no hand on the steering wheel, eyes directed inside rather than outside of the vehicle and lane wandering or encroachments.
[16]	Reviewing the previous research	The risk of having crash increase when there are activities related with mobile phones such as dialling and texting while driving.
[23]	Survey (Transit bus driver)	The rating of distraction for the driver via mobile phone is 91%.
[24]	Survey (295 Drivers)	Activities related with mobile phone which are dangerous and could distract the driver. - 73.9% sending SMS text - 61.3% that dialling mobile phone. - 46% talking on hand-held mobile phone.
[25]	Simulator (56 experienced driver)	Drivers glancing away from the roadway with less than 2 seconds. The distraction focused on In-Vehicle Information System (IVIS) such as smart phone where long glances of the drivers has become a major concern of the researcher.
[12]	Collective paper	Based on the NHTSA crash data, the police report about crashes which causes distraction are: - - Tuning the radio. - Speaking on the phone. - Looking at billboard
[19]	Video (28 Drivers)	The longer the period of the driver's eye glancing off the road, the higher the risk involved in a road crash. The road condition has a significant impact on the driver's distraction.
[26]	Simulator (31 Drivers)	Conducting an additional in-vehicle task will degrade the driving performance. The activities include: - - Tuning the radio. - Conversation with passenger. Driving performance is measured by: - - Maintaining speed - Reaction to hazardous situation Drivers over 60 years old are proven to be more cautious while driving as compared to the younger group of drivers.

DOI: 10.1051/matecconf/201713500049

[27]	Simulator (36 Drivers)	The findings of this research are: - - Younger drivers are more exposed to distraction as they tend to glance at IVIS more as compared to middle-age drivers. - Younger drivers tend to look off-road more than the experienced drivers.
[28]	Simulator (48 Drivers)	When there is the presence of billboards, the effect to the driver is: - - Change of visual attention of the driver. - Increase the time needed for the driver to respond to road signs. - Number of errors done have increased while driving.
[29]	Simulator (12 Drivers) Survey (160 drivers)	The findings of this research are: - - 22% of the drivers indicate that they are being distracted and in a dangerous situation due to the presence of advertising signs along the roadsides. - 50% of the respondents confirmed that they were distracted by the roadside advertisement sign at least once.
[17]	Reviewing previous research	The categories of distraction were identified in this research: - - New technology information devices such as electronic route guidance. - Entertainment devices such as Radio. - Mobile Phones. - Interaction with passengers. - Food consumption. - Other distraction such as sneezing, smoking, operating windows and reaching objects.
[30]	Survey (6578 Drivers)	The findings of this research are: - - 20% of drivers are engaged to secondary tasks while driving. - 11.3% is conversing with passengers. - 3.7% is smoking. - 1.3% are talking in a handheld mobile phone.

Based on previous researches, the most commonly used method are survey and simulator study. With reference to the findings from the previous researches, there are generally a lot of factors of distraction among the drivers while driving. The most common distraction being found from the previous research arises from the usage of mobile phones. In addition, other researchers also found that glancing outside of the vehicle especially when looking at the billboard could distract the drivers from concentrating on driving safely.

3 Conclusion

In conclusion, this paper provides an overview regarding the factors of distraction among intrastate bus drivers as well as the methodology used by previous researchers. Those studies have little methodologies as researches related to ergonomics would not have any specific methodologies. Nevertheless, this research would help future researchers to have a better understanding on the factors of distraction among the intrastate bus drivers.

Acknowledgment

The authors are supported by the Fundamental Research Grant Scheme (FRGS) Vot 1542, Office of Research, Innovation, Commercialization and Consultation (ORICC) Universiti Tun Hussein Onn Malaysia. The authors would also like to thank Rapid Bus Sdn Bhd and Prasarana Malaysia Berhad on the contribution and participation in this study.

References

[1] United Nations. (2014). *World Urbanization Prospects: The 2014 Revision, Highlights(ST/ESA/SER.A/352). New York, United.*

[2] Zhou, C., Dai, P., Wang, F., & Zhang, Z. (2016). Predicting the passenger demand on bus services for mobile users. *Pervasive and Mobile Computing, 25*(2013), 48–66.

[3] Ponrahono, Z., Bachok, S., Ibrahim, M., & Osman, M. M. (2016). Assessing Passengers' Satisfaction Level on Bus Services in Selected Urban and Rural Centres of Peninsular Malaysia. *Procedia - Social and Behavioral Sciences, 222*, 837–844.

[4] Le, T. P. L., & Trinh, T. A. (2016). Encouraging public transport use to reduce traffic congestion and air pollutant: A case study of Ho Chi Minh City, Vietnam. *Procedia Engineering, 142*, 235–242.

[5] Bachok, S., Osman, M. M., & Ponrahono, Z. (2014). Passenger's Aspiration Towards Sustainable Public Transportation System: Kerian District, Perak, Malaysia. *Procedia - Social and Behavioral Sciences, 153*, 553–565.

[6] Klein, M. C. A., van Lambalgen, R. M., & Treur, J. (2010). An Agent Model for Analysis of Human Performance Quality. *Web Intelligence and Intelligent Agent Technology (WI-IAT), 2010 IEEE/WIC/ACM International Conference on, 2*, 181–188.

[7] Tse, J. L. M., Flin, R., & Mearns, K. (2006). Bus driver well-being review: 50 years of research. *Transportation Research Part F: Traffic Psychology and Behaviour, 9*(2), 89–114.

[8] La, Q. N., Duong, D. Van, Lee, A. H., & Meuleners, L. B. (2015). Factors underlying bus-related crashes in Hanoi, Vietnam. *Transportation Research Part F: Traffic Psychology and Behaviour.*

[9] Ellison, A. B., Greaves, S. P., & Bliemer, M. C. J. (2015). Driver behaviour profiles for road safety analysis. *Accident Analysis and Prevention, 76*, 118–132.

[10]

[11] Feng, S., Li, Z., Ci, Y., & Zhang, G. (2016). Risk factors affecting fatal bus accident severity: Their impact on different types of bus drivers. *Accident Analysis and Prevention, 86*, 29–39.

[12] Ranney, T. a., Garrott, W. R., & Goodman, M. J. (2000). NHTSA Driver Distraction Research: Past, Present, and Future. *Driver Distraction Internet Forum*, (233), 9.

[13] Regan, M. A., Lee, J. D., & Young, K. L. (2009). Driver Distraction: Theory, Effects and Mitigation.

[14] Salmon, P. M., Young, K. L., & Regan, M. A. (2011). Distraction "on the buses": A novel framework of ergonomics methods for identifying sources and effects of bus driver distraction. *Applied Ergonomics, 42*(4), 602–610.

[15] Törnros, J. E. B., & Bolling, A. K. (2005). Mobile phone use - Effects of handheld and handsfree phones on driving performance. *Accident Analysis and Prevention, 37*(5), 902–909.

[16] Caird, J. K., Johnston, K. A., Willness, C. R., & Asbridge, M. (2014). The use of meta-analysis or research synthesis to combine driving simulation or naturalistic

study results on driver distraction. *Journal of Safety Research, 49*(February), 91–96.
[17] Stevens, A., & Minton, R. (2001). In-vehicle distraction and fatal accidents in England and Wales. *Accident Analysis and Prevention, 33*(4), 539–545.
[18] Crundall, D., Van Loon, E., & Underwood, G. (2006). Attraction and distraction of attention with roadside advertisements. *Accident Analysis and Prevention, 38*(4), 671–677.
[19] Peng, Y., & Boyle, L. N. (2015). Driver's adaptive glance behavior to in-vehicle information systems. *Accident Analysis and Prevention, 85*, 93–101.
[20] Wood, D. (2001). Established and emerging cardiovascular risk factors. *American Heart Journal, 141*(2), S49–S57.
[21] Olapoju, O. M. (2016). Culture of distracted driving among intra-city commercial bus drivers in Ile-Ife, South-western Nigeria. Transportation Research Part F: Traffic Psychology and Behaviour, 42, 425–432.
[22] Stutts, J., Feaganes, J., Reinfurt, D., Rodgman, E., Hamlett, C., Gish, K., & Staplin, L. (2005). Driver's exposure to distractions in their natural driving environment. Accident Analysis and Prevention, 37(6), 1093–1101.
[23] D 'souza, K. A., & Maheshwari, S. K. (n.d.). Research framework for studying public transit bus driver distraction, 130, 137–148.
[24] Young, K. L., & Lenné, M. G. (2010). Driver engagement in distracting activities and the strategies used to minimise risk. Safety Science, 48(3), 326–332.
[25] Borowsky, A., Horrey, W. J., Liang, Y., Garabet, A., Simmons, L., & Fisher, D. L. (2016). The effects of brief visual interruption tasks on drivers' ability to resume their visual search for a pre-cued hazard. Accident Analysis and Prevention, 93, 207–216.
[26] Horberry, T., Anderson, J., Regan, M. A., Triggs, T. J., & Brown, J. (2006). Driver distraction: The effects of concurrent in-vehicle tasks, road environment complexity and age on driving performance. Accident Analysis and Prevention, 38(1), 185–191.
[27] Donmez, B., Boyle, L. N., & Lee, J. D. (2007). Safety implications of providing real-time feedback to distracted drivers. Accident Analysis and Prevention, 39(3), 581–590.
[28] Edquist, J., Horberry, T., Hosking, S., & Johnston, I. (2011). Effects of advertising billboards during simulated driving. Applied Ergonomics, 42(4), 619–626.
[29] Bendak, S., & Al-Saleh, K. (2010). The role of roadside advertising signs in distracting drivers. International Journal of Industrial Ergonomics, 40(3), 233–236.
[30] Prat, F., Planes, M., Gras, M. E., & Sullman, M. J. M. (2015). An observational study of driving distractions on urban roads in Spain. Accident Analysis and Prevention, 74, 8–16.

The Hybrid Lean System to Improve Manufacturing Environment

Ahmad Nur Aizat *Ahmad*.[1], Te Chuan *Lee*.[1] Rohaizan *Ramlan*.[1], Md. Fauzi *Ahmad*. [1], Norafifah *Husin* [2] and Mustaqqim *Abdul Rahim*.[3]

[1]Department of Production and Operation Management, Faculty of Technology Management and Business, Universiti Tun Hussein Onn Malaysia (UTHM), Malaysia
[2]Department of Industrial Engineering, Faculty of Mechanical Engineering, Universiti Teknologi Malaysia (UTM), Malaysia
[3]Department of Civil Engineering, School of Environmental Engineering, Universiti Malaysia Perlis (UniMAP), Malaysia

Abstract. Lean manufacturing is a systematic approch could increase manufacturing competitiveness by identifying and elminating waste through lean tools. However, these tools are often categorized into different groups, continues to be difficult for companies to select the most appropriate lean tool to be adopted, especially in respect of the SMEs whose knowledge of lean tools and associated principles are limited. This study is to propose solution on how to implement appropriate lean tools in food manufacturing industry located at Rengit, Johor. Qualitative method is used as a methodology to see the changes after the implementation of lean tools. Interview session also a part of method to collect data from respondents. The finding from this paper shows that implementation of an appropriate lean tool will help to eliminate waste in industry sector, maintain and control the inventory, improve quality of the product and improve overall operational process.

1 Introduction

Lean manufacturing is one of the initiatives that many major manufacturing plants in Asia, especially in Malaysia have been trying to adopt in order to remain competitive in an increasingly competitive global market (Rahani AR, Muhammad al-Ashraf, 2012). The focus of the approach is on reduction of the cost and time through eliminating non value added activities by applying a management philosophy which focused on eliminating and identifying waste from each step in the production chain respective of energy, time, motion and resources alike throughout a product value stream, known as lean.. The term "lean" was first introduced as the "lean production system" by John Krafcik, after a review of the TPS. This was later on popularized as "lean manufacturing" by (Womack *et al.*, 1990) in the book

[1] Corresponding author: aizat@uthm.edu.my

The Machine that Changed the World. Lean was initially most widely used in manufacturing companies. In recent years, lean manufacturing have been developed and used in varied sectors such as public organizations, military and healthcare. (Liker, 2006). (Bhuiyan *et al.*, 2006) stated that lean production was one of the oldest improvement methodologies, providing high value to the customer by using of best practices such as Kanban, mistake proofing and 5S. Many authors have demonstrated on how and when used appropriately, lean philosophy can help to eliminate the waste in process industry, maintain better inventory control, increase and improve product quality and reduced operational costs and materials.(Gupta and Jain, 2013).

The achievement of a specific management strategy typically relies on organizational characteristics, which infers that all organizational should not or can't execute a comparative arrangement of strategy in their specific case (Shah and Ward, 2003).Wrong use of lean techniques brings about wasteful aspects of an association's assets and diminished employee trust in lean systems (Marvel and Standrigde, 2009). Therefore, applying the proper strategy at the suitable time for the right reasons is very important. The goals proposed in this work will be accomplish by identify lean factors and propose best solution to the company on how to implement appropriate lean tool.

Besides, do a continuous improvement which is combining of two lean elements its called hybrid of lean tools. The wide decision of lean tools accessible to minimize waste can be confused for those with restricted information of limited knowledge of lean tools and their proficiency, thus the introduction of lean tools as often as possible result in a negative effect on organization's effectiveness.

Moreover, Anand and Kodali (2009) indicated that there is a consolidated list of around 60 lean manufacturing tools, while Eswaramoorthi *et al.* (2011) highlighted that there are 36 different lean tools for machine tool industry. From the above, it can be noticed that there are various lean tools and a few researchers have classes them into various groups. Thus, a methodology to determine suitable lean tools will significantly help organizations in embracing the most appropriate lean tools for their specific aims and objective.

Besides that, there is a limited study about implement of hybrid lean tools. Hybrid of lean management system is a technique or tool of consolidation between two or more elements from lean management system. As significance role of continuous improvement today's organization and lack of adequate proof to demonstrate the positive effect of hybrid lean management system.

Thus, the aims propose in this project is achieved by propose solution and implementation of hybrid lean tools with respect to support manufacturing industry. In order to achieve the lean concept, a case study in manufacturing company to accomplish the objectives for this study: (1) To observe whether the organization applied lean management tool in their organization, (2) To combine the lean tools to become lean hybrid tools, and (3) To identify the benefits of lean hybrid system implementation with respect to the company's performance.

This study is important to give a better understanding for the manufacturing industry that faced of confusing to implement appropriate tools of lean management system. In addition, this research also contributed to the parties to use it as a source of references and guidance such as:

a) Academicians
Academicians could use this research as a source of reference for their further studies either related to this topic or other new upcoming topics. The information and knowledge provided are expected to be beneficial for them in order to improve their lean knowledge.

 DOI: 10.1051/matecconf/201713500050

b) Manufacturing Industry
This research is expected to propose comprehensive understanding on the challenge faced by the manufacturing company in getting to find an appropriate tool of lean management system to be implemented in the company. Suitable of lean tools can provide enhancement of company's performance and increase effectiveness of the workplace.

This study is focused at manufacturing industry located at Rengit, Johor in the food sector. The project approached an observation around the production line based on case study. The results shown the changes on the company's performance towards the implementation of lean hybrid tools. This study also examines the improvement of the changes towards a better workplace and environment.

2 Literature Review

Lean Manufacturing is a concept created to eliminate waste in manufacturing operations. There are various definitions of lean from different view of authors. Todd (2000), mentioned lean as initiative to reduce wastes in inventory, time, human effort and manufacturing space to become highly responsive to customer demand. As enhancement on mass production due to get the product right at time, continuous improvement effort, quality in product and process, flexible production and minimizing waste (Motwani 2003).

Pavnaskar *et al* (2003) defined lean as a practice to reduce the waste in human effort, inventory, time to market and manufacturing space to become highly responsive to customer demand. Taj & Berro (2006) defined lean as an approach to reduce waste (Muda) in value stream systematically. As the systematic removal of waste by all members of the organization from all areas of the value stream. (Worley 2004)

2.1 Waste

Figure 2.1: Waste of Lean

There are 8 wastes in manufacturing as stated by Taichi Ohno, the father of Toyota Production System. Figure 2.1 shown 8 wastes in lean.

2.2 Continuous Improvement

The Hybrid of lean management system is a combination of some elements from lean tools to be integrated into two or three elements. This system will help to organize a better workplace for efficiency and reduce waste. Some of the elements are working together to achieve the lean concept. The hybrid lean is the combining of lean elements as examples:

- Kanban, tagging and labeling
- 5S and visual control,
- poka yoke and shadow board.

There are similar major studies shown a benefit of implement of lean tools as done by Bevilacqua, Ciarapica, De Sanctis, Mazzuto, & Paciarotti, 2015) on their research A Changeover Time Reduction through an integration of lean practices: a case study from pharmaceutical sector highlighted that by using an integration of different lean practice can reduce the batch change and changeover time up to 50 per cent, increasing overall equipment effectiveness by 25 per cent.

3 Methodology

3.1 Primary Data

Primary data is the main data used to obtain information about the study. Primary data collected in this study observe the workplace environment by using pictures to get the real situation of the workplace. Besides, this research also involved interview session with 3 respondents to take their point of view about lean tools. Some data is analyzed using table and other methods for displaying the results in a format that is easy to understand.

3.2 Research Process Flow

Research methodology is the systematic way to solve the research problems. There are several steps that are generally adopted by a researcher to conduct a research. This study discussed on a few process part to implement lean tools including measurement, improvement, analysis and implementation as follow:

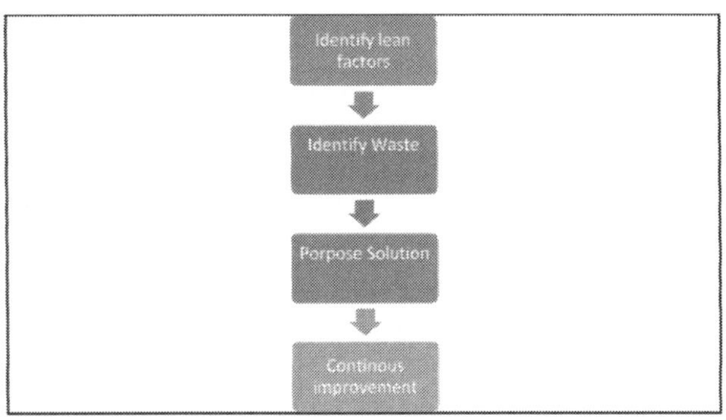

Figure 3.1: Process Flow Chart

From the process flow chart at studied company, there are 7 types of waste that are used to improve the process flow of the studied company. The waste are:

1) Defects
2) Waiting
3) Inventory
4) Overproduction
5) Transportation
6) Inappropriate processing
7) Unnecessary motion

3.2.3 Continuous Improvement

The transition of a lean environment cannot be done in a night. A continuous improvements are necessary to achieve company's goal. The hybrid lean management systems which are the combining of two or more lean method/tools are used to make the continuous improvement. This system helped and organized a better workplace for efficiency and reduced waste. Some of the elements are working together to achieve the lean concept. The hybrid that combining the lean elements as examples are: labeling, tagging, 5S, visual control, and poka yoke. The hybrid lean is the combination of leans element such as 5S system and tagging, visual control and 5S and etc. The details explanation of these elements as in table below:

Table 3.1 Definition of each element

Element	Definition
5S	Orderliness (seiri and seiton) – to maximise efficiency and effectiveness by reducing people's workload and human errors through simplifying processes. Cleanliness (seiso and seiketsu) – to maximise effectiveness by contributing to a healthier life, safety and wellbeing as well as enhancing transparency. Discipline (shitsuke) – through training and education to enhance the level of morale which leads to increased quality of work/life and work standards (Osada, 1991).
Visual control	Visual control is a business management technique employed in many places where information is communicated by using visual signals instead of texts or other written instructions. The design is deliberate in allowing quick recognition of the information being communicated. Santos et al. (2001) found that visual controls provided and enabled increased efficiency, and that conversely, a lack of visual controls inhibited increased efficiency.

Tagging	Tagging is a use visual communication to bring awareness of important workplace boundaries Tagging provide information to supplement hazard or handling labels. Carriers and consignees refer to markings when accepting hazardous materials to ensure proper handling of the material as well as regulatory compliance
Labeling	Labels may be used for any combination of identification, information, warning, and instructions for use, environmental advice or advertising. They may be stickers, permanent or temporary labels or printed packaging.
Poka yoke	Japanese term which means mistake proofing. A poka-yoke device is one that prevents incorrect parts from being made or assembled, or easily identifies a flaw or error (Bicheno, 2004).Or "mistake-proofing," – a means of providing a visual or other signal to indicate a characteristic state. Often referred to as "error-proofing," poka-yoke is actually the first step in truly error-proofing a system. Error-proofing is a manufacturing technique of preventing errors by designing the manufacturing process, equipment, and tools so that an operation literally cannot be performed incorrectly.

4 Results and Discussion

4.1 Introduction

All the proposed improvement activities are stated here. The proposed improvements are discussed based on the problems identified from the observation. This is the process flow of the process as shown in Figure 4.1:

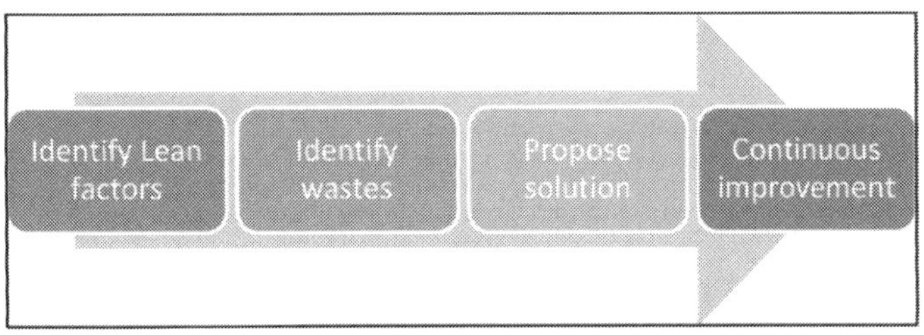

Figure 4.1: Process Flow

The research also covered of a production areas which from peeling, cutting, frying and packaging. Figure 4.2 shows the process flow of production process.

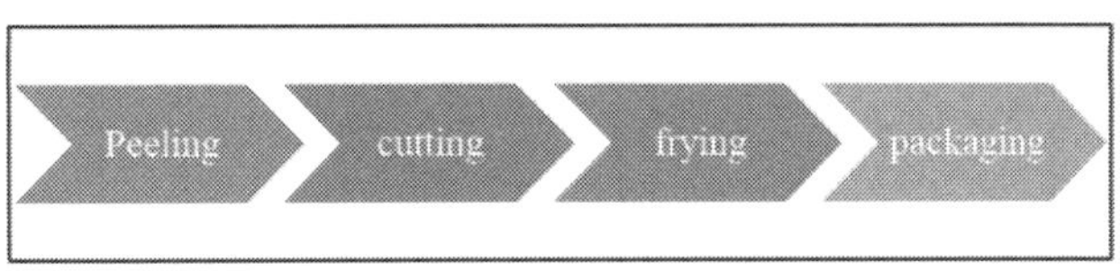

Figure 4.2: production flow of the studied company

4.2 Lean main factors Practices

Table 4.1: Lean Factors

Authors / Factors	Biggs (2009)	Puvanasaran et al	Nordin et al (2010)	Shah and Ward (2007)	Mackelprang and Nair	Wong et al (2009)	Singh et al.	Crabke (2007)	**Frequency**	**Rank**
5S	√	√	√			√		√	5	2
Kanban	√	√	√	√	√	√		√	7	1
Set up Reduction			√	√	√			√	4	3
Value stream mapping	√	√				√	√	√	5	2
Just-in-time	√	√	√	√	√	√		√	7	1
Total Productive Maintenance (TPM)	√	√	√	√	√	√		√	7	1

Based on the table 4.1, there are five main factors of lean practices based on different authors which are:

a. Just in time (JIT)
b. Kanban
c. Total productive maintenance (TPM)
d. Value stream mapping
e. 5S

4.3 Benefits of lean tools to the company

Interview session is done to prove that the objectives of this research are successfully achieved from the respondent's feedback. Table 4.2 shows the results of the respondents.

Table 4.2: interview respondent's opinion about benefits of implement lean tool

Respondents / Benefits	Opinion 1	Opinion 2
Respondent 1	Implementation of lean will encourage operators to produce only what, amount and when the product is needed by the next operation.	Identifying future improvement. The next improvement to be done is on plan to make sure the continuous improvement is adopted.
Respondent 2	Less space and manpower is required to manage when lean is implemented.	Ensure all the operators worked in proper and safely environment to avoid any risk.
Respondent 3	Shorter time is required in lean environment system to make sure less inventory in a value stream.	Employees are encourages to speak about the opinions, identify the problems and take corrective action to prevent problems repeated.

4.4 Wastes Identification

From the process flow, there are 7 waste are identified and occurred which are defects, waiting, inventory, over production, transportation, inappropriate processing, and unnecessary motion by interviewing of three respondents. From interview session found that the biggest problem that the company faced and the appropriate lean tool is suggested to the organization. The table 4.3 below shown how to implement appropriate lean tools based on problems identified. Table 4.3 shows the major wastes occurred in the process flow which are inventory, waiting, defect and overproduction based on the ranking system. Thus, the appropriate lean tools are proposed and suggested based on waste and problems occurred.

Table 4.3: Identify Wastes

Waste	Problem identify	Respondent	Respondent	Respondent	Frequency	Total frequency	Rank	Purpose solution
Defect	Over frying	√	√	√	3	5	3	Kanban 5S
	Machine breakdown				0			
	Not following SOP	√		√	2			
Waiting	Lead time	√	√	√	3	6	2	VSM
	Cycle time	√	√	√	3			
Inventory	Exceed raw material	√	√	√	3	9	1	JIT, VSM, layout improvemet
	Storing space	√	√	√	3			
	Product damage	√	√	√	3			
Over production	Produce more than customer demand	√	√	√	3	3	4	JIT
Transportation	Late supply					0		
	Length time from transportation to inventory							
Inappropriate processing	Failure to understand what customer want					0		
Unnecessary motion	Ergonomic					0		

4.5 Lean Implementation

The suggested solution for the company to implement lean tools has been identified as shown in the table 4.2 which are:

4.4.1 5S and Tagging System

The defects occurred because of over frying process. Not all raw materials have same qualities and specific standard. There are all different type of puberty, texture and size. In order to solve the following problem, 5S method and tagging is implemented.

4.4.2 Kanban

Kanban system is proposed to communicate production requirement between processes. Kanban is card system that contained just few basis information about the batch. The card

DOI: 10.1051/matecconf/201713500050

have the information about the process, number of product or parts required and also part numbers of each items. Hence, kanban is also used as signal movement of parts between processes. It is a signaling device that given a permission to withdraw the items in a pull system and gives authorization and instruction for the production to follow the guideline or quantities of items needed. Thus, employees be more alert and have a guideline on doing the process.

4.4.3 JIT

Just-in-time or JIT is the system of instantly replace or the production of only what is needed, the amount or quantity needed and when is needed. The wastes that have been found in are overproduction and inventory for JIT elements. This happened due to limitation of raw material, storing space. In this studied company, JIT problems occurred because of the unorganized raw material scheduling. So, employees produced the finish goods more that required or demanded by customer. Because of this JIT problems, there are more problems are occurred such as no space to put the access finished goods and huge bottlenecks created and product delivery aren't on schedule.

4.4.4 VSM

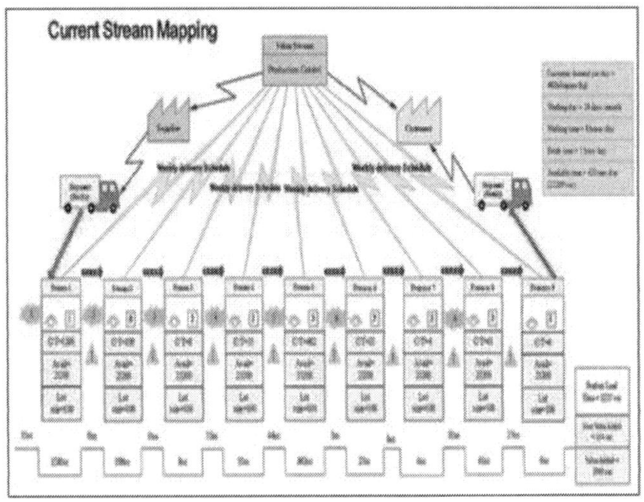

Figure 4.3: Current state stream mapping of studied company

During the interview session with the respective person in the studied company, VSM implementation is suggested. Table 4.3 shows VSM repeated twice as a problem solving for the reduction of lead time and cycle time in the production line. This is because the time taken from one process to the next process are exceed takt time. In one-piece flow production tasks are reduced to their simplest product, errors are reduced and continuous flow of goods enhanced. It is particularly suited for efficient repetitive production process; one-piece flow provides improved quality, bottom-line profits, and continuous output without the need for additional production capacity or staff and also can avoid bottlenecks in a process. Besides, one-piece flow production is able to reduce lead times and to adjust customer demand better than large batch production operations. Figure 4.3 and 4.4 shows the implementation of VSM

DOI: 10.1051/matecconf/201713500050

at the studied company. Table 4.4 show the comparison data and time between current state of VSM and future state of VSM at the studied company after VSM have been implemented.

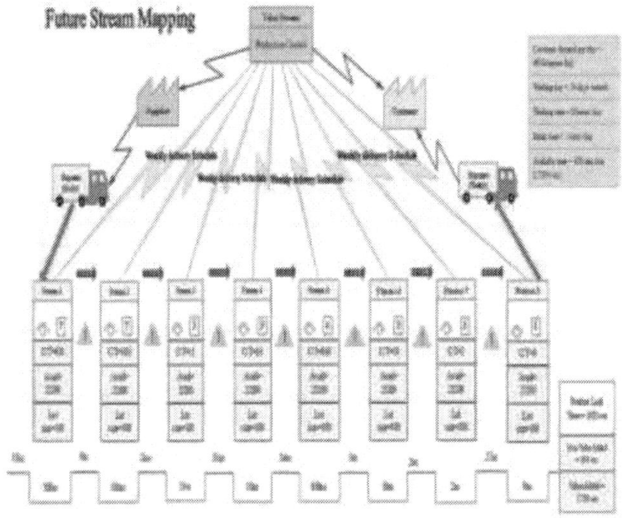

Figure 4.4: Future state value stream mapping of studied company

Table 4.4: Comparison between current and future value stream mapping

Process	Total cycle times			
	Current State (sec)	Future State (sec)	Total reduce	% reduced (increased)
No. of Operation	9 process	8 process	1	12.5
Process 1	1280	920	360	39.13
Process 2	109	101	8	7.92
Process 3	8	11	-3	-27.27
Process 4	55	10	45	450
Process 5	462	616	-154	-25
Process 6	25	50	16	32
Process 8	41			
Process 7	4	2	2	100
Process 9	9	9	0	0
Total	1993	1719	274	13.75

4.4.5 Layout Improvement

Layout improvement also been conducted at the studied company. The arrangement of new layout is proposed due to support the lean system by reducing unnecessary movement and to make sure product is delivered at a time through a various operations. Efficient layout is important and needed to make sure all the operation is in the proper arrangement and to eliminate not necessary movement of the items. Figure 4.4 and 4.5 shows the existing of the layout and new improvement of layout at studied company. This new layout applied the hybrid lean element such as tagging, lining, kanban and etc.

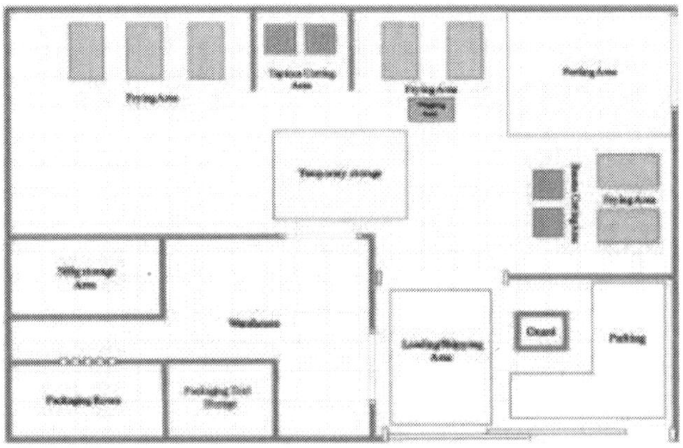

Figure 4.4: Existing Layout

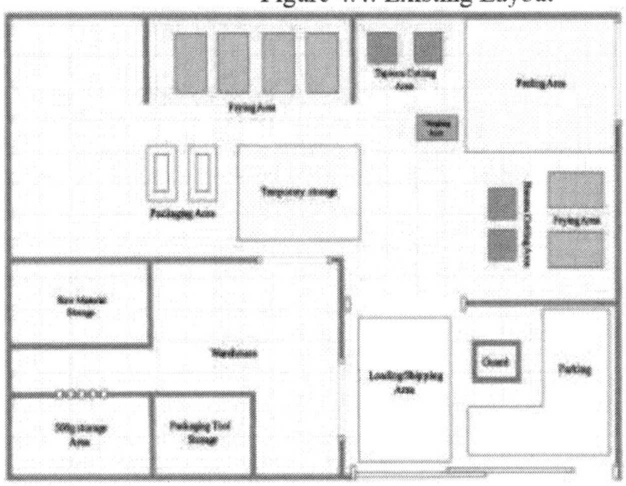

Figure 4.5: New improvement Layout

4.5 Continuous Improvement

The implementation to a lean environment cannot be done in a night. A continuous improvement method or procedures is needed to achieve lean environment system. The term of continuous improvement means improvement of service, products, processes, or method with the goal of increasing value and reducing non value added activities to improve product performance, customer service or workplace functionality. Continuous improvement principles, as practiced by the most dedicated manufactures give result and impressive improvement in performance. There is an interesting way to do continuous improvement by implementing lean using hybrid lean tools which is the combination of two or more lean tools into one process. In this research, it combined the element of lean which are 5S, poka yoke, tagging, labeling and visual control for continuous improvement.

Table 4.5: Pictures analysis of implementation of Hybrid lean tools

Before	After	Hybrid
No label and tools are placed not at proper place 	Labeling is done and tools are stored in their respective place identified. 	Hybrid of poka yoke and labeling.
Plastic bag and waste cotton unorganized accordingly 	Put in proper arrangement and kept systematically 	5S and label-ing
No safety awareness for employees. 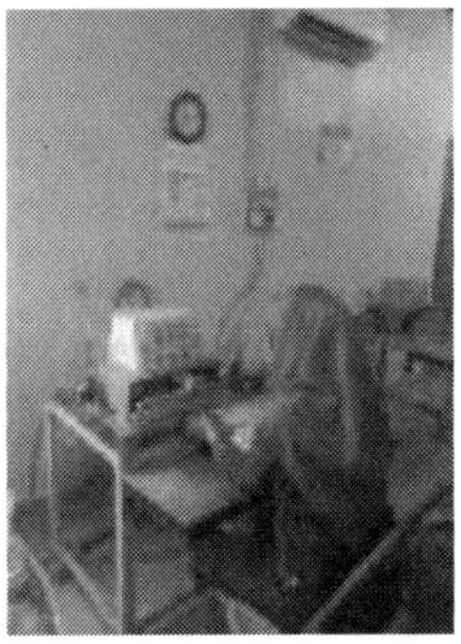	Caution sign and safety awareness for employee. 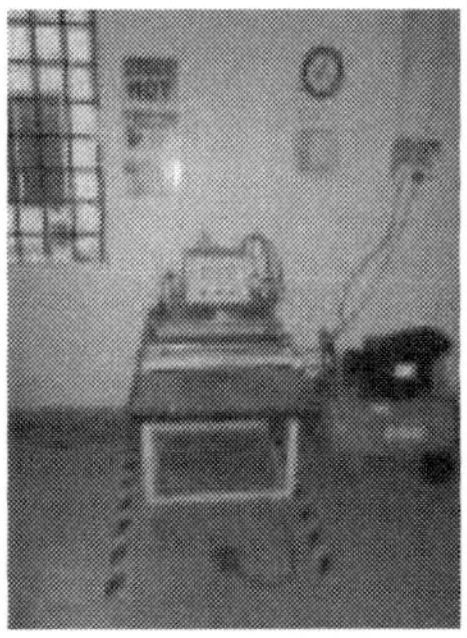	Tagging and visual control

No labeling and tag for forklift lane. 	Labeling and tag for forklift lane. 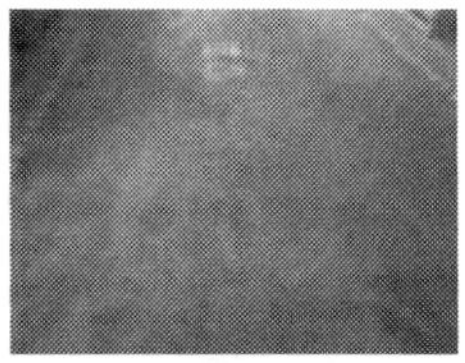 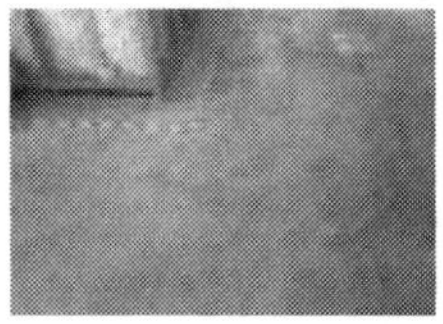	Visual control and tagging
Unsorted and dirty workplace.  	Comfortable and clean workplace. An awareness sign for slippery floor and using SOP. 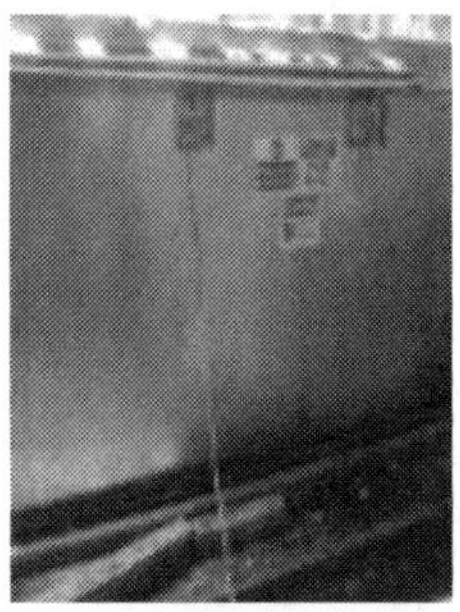 	Visual control and labelling

From Table 4.5, the company does not implement the lean tools appropriately. Table 4.4 shows the evident that the company has less awareness of the employee's safety and welfare. The company not provided safety precaution and full SOP for example, gloves during the frying process. Thus, with the hybrid of lean tools such as sign and a visual control, labeling and tagging improved the knowledge on taking a serious action to the safety precaution.

In addition, things that kept not in proper way affected and waste of the time because need to search for tools in unsorted places. This can delay the activity for the next process. Hence, unsorted things that randomly kept also caused injured to others. Besides, there is no labeling and the tools are place randomly and not in proper arrangement in the rack. The racks become proper arrangement and nice after labeling process is implemented. Employees can easily pick tool without searching and wasting their time. A tagging on the floor also can enhancing of awareness in the workplace. Since the company has no tagging or any visual control for forklift area, employees can easily passing around that area without noticed about the forklift was in a work progress.

5 Conclusion

This research is aim to observe whether the organization applied lean management tool in their organization is accomplished. The implementation of hybrid lean tools towards the improvement of the manufacturing company also has been done. Since this is a food company so the hygiene has to be taken care properly. As a result in the analysis, lean tools are not only improved a better workplace but it also reduced a cycle time in a process. In addition, the environment of the company become properly managed and avoided safety precaution. Major of the previous study mentioned that if the organization is providing a better workplace and care on employee's welfare and moral, employees will stay longer and loyal to the company.

Lean manufacturing is proven helps the company to reduce waste and increase the value of the operation. Applying lean philosophy also is needed to maximize the profit and revenue to the company.

References

1) Ahmad, A.N.A, Hassan S., Husin, N. (2015) Cyber Value Stream System Reduced Waste And Non-Value Added Of Manufacturing Operation, International Journal of Advancement in Mechanical And Aeronautical Engineering, Institute of Research Engineers and Doctors, USA , 4, 205, ISSN:2372-4153
2) Anand, G., and Rambabu Kodali. (2010), "Development of a Framework for Implementation of Lean Manufacturing Systems."*International Journal of Management Practice*, 4(1). 95-116.
3) Ballé, F., & Ballé, M. (2005). Lean development. *Business Strategy Review*.
4) Bicheno, J., & Bicheno, J. (2004). *The new lean toolbox : towards fast flexible flow* (3rd ed.). Buckingham: PICSIE Books.
5) Bhuiyan, Nadia, Amit Baghel, and Jim Wilson. 2006. "A Sustainable Continuous Improvement Methodology at an Aerospace Company." *International Journal of Productivity and Performance Management*, 55(8), 671–87.
6) Browning, T. R., & Heath, R. D. (2009). Reconceptualizing the effects of lean on production costs with evidence from the F-22 program. *Journal of Operations Management*, *27*(1), 23–44.
7) Gupta, S., & Jain, S. K. (2013). A literature review of lean manufacturing. *International Journal of Management Science and Engineering Management*, *8(*4) , 241–249.
8) Ho, S. K. M. (2010). Integrated lean TQM model for global sustainability and competitiveness. *The TQM Journal*, *22*(2), 143-158.

9) Liker, J. K. (2003). The 14 Principles of the Toyota Way : An Executive Summary of the. *The 14 Principles of the Toyota Way : An Executive Summary of the Culture Behind TPS*, 35–41.

10) Marvel, J. H., & Standridge, C. R. (2009). A simulation-enhanced lean design process. *J.H. Marvel; C.R. Standridge, 2(* 1), 90–113.

11) Motwani, J. (2003). A business process change framework for examining lean manufacturing: a case study. *Industrial Management & Data Systems, 103(*5), 339–346.

12) Rahani, A. R., & al-Ashraf, M. (2012). Production Flow Analysis through Value Stream Mapping: A Lean Manufacturing Process Case Study. *Procedia Engineering, 41*. 1727–1734.

13) Scherrer-Rathje, M., Boyle, T. A., & Deflorin, P. (2009). Lean, take two! Reflections from the second attempt at lean implementation. *Business Horizons, 52(* 1), 79–88.

14) Shah, R., & Ward, P. T. (2003). Lean manufacturing: context, practice bundles, and performance. *Journal of Operations Management, 21*. 129–149.

Investigating the Associations between Musculoskeletal Discomforts and Perceived Stress among Production Operators

Nor Hazana Abdullah*, *Nor Aziati* Abdul Hamid , *Eta* Wahab and *Alina Shamsuddin*

Department of Technology and Management, Faculty of Technology Management and Business, Universiti Tun Hussein Malaysia, Parit Raja Batu Pahat, Johor 86400 Malaysia

Abstract. . Both musculoskeletal pain and harmful stress have adverse health effects on production operators if they are not properly monitored or managed. However, the mechanism behind the association between these two variables is unclear especially when causal inference could not be deduced from majority of studies. Thus, there is a need for a better understanding of musculoskeletal pain and perceived stress among production workers in order to improve their health and safe working conditions. This is especially critical since majority of manufacturing companies in Malaysia are still labour intensive and any adverse effects of employees' health could impact the organizational performance. Therefore, the objective of this study is to determine the current prevalence of perceived stress and musculoskeletal pain among production workers and the possible linkage between perceived stress and musculoskeletal pain. A cross-sectional survey involving 400 shop-floor manufacturing workers was done using Cornell Musculoskeletal Discomfort Questionnaires (CMDQ) and Perceived Stress Questionnaires (PSQ). Data were analysed using hierarchal regressions. It is found that stress significantly affects MSDs albeit very weak (R^2= 0.048) after controlling for gender and age. Neck, lower back and lower legs (Right and Left) were reported to have high prevalence of MSDs compared to other parts of the body (M=2.96, SD= 1.342; M=2.72, SD=1.441; M(R)= 2.6, SD=1.418; M(L)=2.63, SD=1.415).

1 Introduction

Malaysia's Social Security Organization (SOCSO) statistics from year 1995 to 2009 has shown an exponential increase of cases in musculoskeletal disease from 5 cases (1995) to 161 cases in 2009 [1]. This trend continues as Borneo Post online reported that in 2010, there are 238 cases, followed by 268 cases in 2011 and 449 cases in 2012 [2]. Tan Sri Lee Lam claimed that such increase could be attributable to industrialization, aging workforce and lack of safety work practices at workplace [3]. Musculoskeletal diseases include injuries and illnesses that affect the human body's movement such as muscles, tendons, ligaments, nerves, discs, blood vessels, and many more. It is caused by ergonomics risk factors which include excessive force, repetition of movement or simply poor posture while working or risk factors which include poor work practices, poor fitness and poor health.

In similar vein, almost 70 percent of employees in Malaysia is claimed to experience escalated level of stress-related sickness [4] [5]. Failure to be adaptive to stress has ominous

* Corresponding author: hazana@uthm.edu.my

consequences for both individuals and organizations [6] [7]. Individual who are unable to cope with the stress are incline to suffer psychological related illnesses such as depression, schizophrenia and anxiety disorders. Excessive stress not only undermines individual performance but could spill onto the organizational productivity and competitiveness. Studies have shown that MSDs and stress not only has an impact on productivity, but also on job satisfaction [8], [9], and other organizational outcomes. In a nutshell, stressful and unhealthy employees tend to have low workplace presence which leads to low productivity and commitment to work.

Medical literatures considered stress and an important factor in the onset and maintenance of widespread musculoskeletal pain (see for example [10]). McFarland [10], in particular, claimed that stress-musculoskeletal pain linkage could be either explained through neurophysiological, psychophysiological, cognitive-behavioural or genetic-behavioural mechanisms. The neurophysiological mechanism explains how pain causes stress response in the interconnected nervous system, endocrine system and immune system (e.g. [11]) while psychophysiological highlights the impact of emotion and stress on pain (e.g. [12]. Cognitive-behavioural, on the other hand, emphasizes the roles of appraisals of danger, coping ability and choice of behaviour as causes of pain (e.g. [13]). Finally, genetic-behavioural regards each individual as having different threshold of pain and stress tolerance due to their genetics makeups (e.g. [14]).

As evidence from above-mentioned discussion, the stress-musculoskeletal pain linkage might have reciprocal causality rather than one-way cause and effect relationship. Inconsistent empirical evidence, which are mostly cross sectional in nature, limits our understanding and has led to much debates. . Moreover, since majority of these studies have been focusing on office employees [15], [16], farmers [17], laborer [18], medical staff [20]–[22], a study on manufacturing employees in developing country like Malaysia is warranted.. Das [17] further claimed that since both stress and MSDs are the two most common causes of work-related illnesses, there is a growing interest in their association, both in terms of common risk factors and the extent to which both are inter-related. Therefore, this study aimed to determine the current prevalence of perceived stress and musculoskeletal pain among production operators of manufacturing companies in Malaysia. The stress-musculoskeletal pain associations are also examined based on the psychophysiological mechanism. Specifically, we hypothesize the increase level of stress would increase musculoskeletal pain. Age and gender would be control variables in this study as the literatures have found that both variables could affect both musculoskeletal pain and stress [23]–[25].

Hans Seyle [30], the father of stress research, documented how stress affect the body in his 'General Adaptation Syndrome' (GAS) model. GAS consisted three phases called alarm reaction (Phase 1), resistance (Phase 2) and exhaustion (Phase 3). The alarm reactions activates physiological stress responses once the individual feel threatened. At this phase, individual's energy level and effectiveness is low due to the initial shocks. During the resistance phase, biochemical, psychological and behavioural mechanisms are activated to boost individual's ability to cope with the threat. If the cause of stress is not being eliminated or removed, the exhaustion phase commences where the immune system would deplete its ability to defend the system and individual is more likely to get sick. In the context of MCDs, muscles inevitably tense up in a reflect reaction against stressors and prolonged exposure to stress would cause the muscles to be in the prolonged state of guardedness. This will trigger illnesses such as headaches, migraine and muscle tensions in the area of shoulder, neck and head.

Schleifer [30] proposed hyperventilation theory of job stress where hyperventilation caused excessive loss of CO_2 which in return increases the blood acidity level. This imbalance acidity level triggers a chain of systemic physiological reactions such as muscle tension and muscle spasm.

Several studies have shown empirical supports on the psychophysiological mechanisms of stress-musculoskeletal pain associations succinctly described by Seyle and Scheleifer. For example, Gonzales [25] found that stress is significantly related to MSDs across four companies being studies. Lin [15] further claim that despite women having higher prevalence of MSDs, encountering difficult customers is regarded as very stressful event that lead to increased risks of MSDs. Zakerian [16] added that work factors such as job demand, negative social interaction and computer-related problems contributed to stress which in turn, increased MSDs among computer users at university's office. Cheng [31] did a large scale study involving 26-25 years old Taiwanese and found that higher level of stress increased multiple health problems such as MSDs even after controlling for age and education.

2 Methods

This cross-sectional survey research included production operators working in eleven medium-sized manufacturing companies in the state of Johor, Malaysia. The selection of companies and their respective employees were done based on their agreement to participate. No personal data was recorded to maintain confidentiality and privacy. The data were collected in October and November 2016.

2.1 Procedure

Four hundred and twenty production operators participated in this study. The survey was collected face-to-face since to ensure that data were recorded correctly. However, ninety-six responses were excluded due to incomplete data as these respondents were unwilling to answer some questions.

Male respondents were slightly higher at 56.2% compared to female (43.8%) with the majority of respondents aged between 23 to 27 years old (31.2%) followed by those aged 28-32 years old (21.9%). It could be concluded that more than 90% of the respondents aged below 42 years old. Majority of respondents (79.6%) had worked more than 2 years. In terms of types of production tasks, 45.4% worked in packaging, 28.4% worked in quality assurance while 13% handled machines as shown in Table 1

Table 3. Respondents' Profiles

Respondents' Profiles			
		Freq	%
Gender	Male	182	56.2
	Female	142	43.8
Age	18-22 years old	42	13.0
	23-27 years old	101	31.2
	28-32 years old	71	21.9
	33-37 years old	46	14.2
	38-42 years old	30	9.3
	Above 42 years old	34	10.5

Nationality	Malaysian	220	67.9
	Foreign	104	32.1
Years of Service	< 1 year	39	12.0
	1 year	27	8.3
	2 years	79	24.4
	3 years	57	17.6
	4 years	38	11.7
	> 4 years	84	25.9
Type of Tasks	Packaging	147	45.4
	Monitoring quality	92	28.4
	Handling of machine	42	13.0
	No response	37	11.4
Hours of Overtime	0-2 hours	89	27.5
	3-5 hours	78	24.1
	6-8 hours	65	20.1
	9-11 hours	37	11.4
	> 11 hours	54	16.7
	No response	1	0.3
Perceived Stress	Low	54	16.7
	Moderate	249	76.9
	High	21	6.5

2.2 Measures

Cornell Musculoskeletal Discomfort Questionnaires was used to measure the musculoskeletal pain [32]. The questionnaires include hand assessment and total body assessment. Perceived Stress Questionnaires (PSQ) by Cohen was used to measure stress [33]. There were 10 items in the questionnaires with the scale being modified from 1-5 rather 0-4. Both measurements had accepted reliabilities of more than 0.7 which are considered as acceptable [34].

2.3 Analysis

Both descriptive and hierarchical multiple regressions were used to analyse the data via the use of IBM SPSS Version 20.0. Descriptive analysis includes the tabulation of frequency, mean and SD. Prior to performing the hierarchical multiple regressions, assumptions such as linearity, multivariate normality, multi collinearity and homoscedasticity were examined. Upon satisfying all these assumptions, hierarchical multiple regressions were performed.

3 Results

Table 2 shows the frequency, mean, and standard deviation (SD) of musculoskeletal pain in terms of the frequency in the last work week. As evident from the Table 2, neck, lower back and lower legs (both left and right) have the highest frequency and mean of discomfort. Right hand is observed to have higher frequency of discomforts. This might imply that majority of respondent are right-handed. The prevalence areas of discomforts are consistent with previous studies such as [21][35] [36].

Table 3 shows the levels of discomfort. Highest frequency of 'Very Uncomfortable' was recorded for all these three areas. Similarly, Table 4 shows similar trend whereby all discomforts in all these three areas were reported to interfere with the respondents' abilities to perform their jobs.

Table 2. Frequency of Discomfort in Last Work Week

		During the last work week, how often did you experience ache, pain, discomfort in (%):						
		Never	1-2 times last week	3-4 times last week	Once every day	Several Times every day	Mean	SD
RIGHT HAND	Area A	30.6	35.2	13.6	8	12	2.38	1.35
	Area B	46.9	34.9	8.6	4.9	4.6	1.85	1.07
	Area C	38	31.8	13.9	6.5	9.9	2.19	1.28
	Area D	39.9	35	15.2	5.6	4.3	1.99	1.08
	Area E	32.4	39.5	15.1	6.5	6.5	2.15	1.11
	Area F	44.1	35.8	12.3	4.3	3.4	1.87	1.02
LEFT HAND	Area A	37.9	38.2	9.6	6.5	7.8	2.08	1.19
	Area B	48.1	37	6.8	5.9	2.2	1.77	0.965
	Area C	38.6	25.3	17.8	9.4	8.8	2.24	1.294
	Area D	40.7	36.3	12.7	6.8	3.4	1.96	1.057
	Area E	40.3	33.1	12.8	9.1	4.7	2.05	1.148
	Area F	46.3	28	14.3	4	7.5	1.98	1.201
	Neck	14.5	28.1	24.1	13.3	20.1	2.96	1.342
	Shoulder (Right)	17.8	42.4	13.1	11.8	15	2.64	1.314
	Shoulder (Left)	17	39.4	18.3	9.8	15.5	2.67	1.3
	Upper Back	25.3	29	16.7	15.4	13.6	2.63	1.367
	Upper Arm(Right)	35.8	29.6	18.2	8.3	8	2.23	1.244
	Upper Arm (Left)	34.7	31.6	18.4	8.1	7.2	2.22	1.209
	Lower Back	24.1	29.4	16.1	10.8	19.5	2.72	1.441
	Forearm (Right)	35.8	29.6	18.2	8.3	8	2.27	1.23
	Forearm (Left)	34.7	31.6	18.4	8.1	7.2	2.27	1.248
	Wrist (Right)	31.3	34.1	13.6	7.7	13.3	2.38	1.349
	Wrist (Left)	31.9	33.1	13.9	8.4	12.7	2.37	1.343
	Hip	39.4	25.9	14.7	7.8	12.2	2.38	1.371
	Thigh (Right)	35.8	25	18.2	11.4	9.6	2.34	1.322
	Thigh (Left)	35.9	24.4	20.6	10	9.1	2.32	1.64
	Knee (Right)	26.5	37.7	16	9.9	9.9	2.39	1.25
	Knee (Left)	26.7	37	17.4	9.9	9	2.38	1.23
	Lower Leg (Right	25.9	33	13.3	10.5	17.3	2.6	1.418
	Lower Leg (Left)	25.2	32.4	13.7	11.5	17.1	2.63	1.415

Table 3. Levels of Discomfort

		If you experience ache, pain, discomfort, how uncomfortable was this? (%)				
		Slightly Uncomfortable	Moderately Uncomfortable	Very Uncomfortable	Mean	SD
RIGHT HAND	Area A	62.3	25	12.3	1.51	0.719
	Area B	72.1	21.7	6.2	1.34	0.591
	Area C	59.6	29.6	10.5	1.52	0.693
	Area D	58.5	36.5	5	1.46	0.591
	Area E	63.7	27.3	9	1.45	0.655
	Area F	67.5	25.7	6.8	1.39	0.613
LEFT HAND	Area A	65.8	28.3	5.9	1.4	0.599
	Area B	72.3	23.1	4.7	1.32	0.56
	Area C	63.2	26.9	9.6	1.47	0.679
	Area D	61.3	34.4	4.1	1.43	0.589
	Area E	65.2	28.6	6.2	1.41	0.606
	Area F	65.4	28.1	5.9	1.4	0.62
	Neck	36.2	42.4	21.1	1.85	0.752
	Shoulder (Right)	43	41.4	15.3	1.73	0.736
	Shoulder (Left)	42.3	40.4	16.9	1.76	0.75
	Upper Back	37	45.7	17	1.81	0.73
	Upper Arm(Right)	53.8	35.3	10.9	1.57	0.682
	Upper Arm (Left)	53	37	10	1.57	0.669
	Lower Back	38.1	36.5	24.8	1.88	0.8
	Forearm (Right)	53.8	35.3	10.9	1.55	0.661
	Forearm (Left)	53	37	10	1.56	0.649
	Wrist (Right)	48.3	36.7	15	1.67	0.724
	Wrist (Left)	50.5	35.6	13.9	1.63	0.715
	Hip	54.9	31.7	13	1.59	0.723
	Thigh (Right)	49.1	38.2	12.7	1.64	0.698
	Thigh (Left)	49.5	36.8	13.4	1.64	0.719
	Knee (Right)	44.3	39.9	15.7	1.71	0.721
	Knee (Left)	44.5	41	14.5	1.7	0.708
	Lower Leg (Right	43.9	32.4	23.4	1.8	0.815
	Lower Leg (Left)	42.6	34.2	22.9	1.81	0.806

Table 4. Level of Interference with Work

		If you experience ache, pain, discomfort, did this interfere with your ability to work? (%)				
		Not at all	Slightly Interfered	Substantially Interefered	Mean	SD
RIGHT HAND	Area A	45.1	39.8	14.5	1.7	0.722
	Area B	59	32	8.7	1.51	0.68
	Area C	47.1	39	13.3	1.68	0.736
	Area D	48.5	43.5	7.7	1.6	0.644
	Area E	45.4	42.9	11.4	1.67	0.686
	Area F	53.9	36.8	9.3	1.55	0.659
LEFT HAND	Area A	50.3	42.2	7.5	1.57	0.629
	Area B	63.6	31.5	5	1.41	0.586
	Area C	48.1	38.2	13.7	1.66	0.708
	Area D	51.9	42.2	5.9	1.54	0.607
	Area E	50.2	38.3	11.5	1.61	0.685
	Area F	52.8	35.4	7.8	1.51	0.638
	Neck	30.6	46.6	22.8	1.92	0.728
	Shoulder (Right)	33.7	48.4	17.9	1.84	0.702
	Shoulder (Left)	37.4	43.1	19.5	1.82	0.734
	Upper Back	35.2	5.06	14.2	1.79	0.672
	Upper Arm(Right)	50.2	41.2	8.7	1.59	0.646
	Upper Arm (Left)	50.3	41.3	8.4	1.58	0.643
	Lower Back	31	48.6	20.4	1.89	0.71
	Forearm (Right)	50.2	41.2	8.7	1.64	0.695
	Forearm (Left)	50.3	41.3	8.4	1.62	0.686
	Wrist (Right)	43.9	40.2	15.6	1.72	0.73
	Wrist (Left)	44.2	41.7	14.1	1.7	0.703
	Hip	46.9	39.3	13.8	1.67	0.707
	Thigh (Right)	44.8	40.1	15.1	1.7	0.716
	Thigh (Left)	44.7	41.3	13.7	1.7	0.711
	Knee (Right)	36.6	44.7	18.8	1.82	0.723
	Knee (Left)	34.4	49.7	15.9	1.82	0.686
	Lower Leg (Right	36.3	37.9	25.5	1.9	0.799
	Lower Leg (Left)	36.3	38.8	24.5	1.89	0.792

Table 5 shows the hierarchical regression analysis on the effect of stress on MSDs. In the first model, age and gender were regressed against MSDs as control variables. The model is significant albeit very weak at $R^2=0.02$. In the second model, stress was regressed together and it is found that stress had significant role in prediction of MSDs. However, in terms of R^2, the value is still small at 4.8%. Whether this value has practical significance certainly could be debated.

Table 5. Summary of Hierarchical Regression Analysis for Predicting MSDs (N-324)

	Model 1			Model 2		
Variable	B	SE B	β	B	SE B	β
Age	-0.027	0.17	-0.087	-0.027	0.017	-0.085
Gender	0.105	0.053	0.109	0.091	0.052	0.094
Stress				0.187	0.046	0.220
R^2		0.02			0.048	
F for change in R^2		3.265*			16.472**	

* $p<0.05$ ** $p<0.001$

4 Discussions

The results clearly indicate neck, lower back and lower legs scored highly in terms of frequency of discomforts, extent of discomforts and level of discomforts' interference with the abilities to perform the jobs. The prevalence of MSDs at these areas might indicate the poor ergonomics working conditions of manual labour in participating companies. Furthermore, all of these respondents are not stationary while working, and might be involved in frequent lifting task that significantly impacted low back pain [35]. Da Costa and Vieira [37] claimed that heavy physical work, awkward postures, lifting and whole body vibration Lower legs discomforts might imply excessive standing of prolonged hours. Interestingly, neck pain is also a common discomfort in this study as being reported in studies of other types of occupations [36]. According to Matsudaira [38], neck and shoulder discomfort or katakori in Japanese is common among Japanese adults and might lead to somatization.

The regression result indicates that stress is an important consideration to reduce MSDs but the practical significance in this study is very small despite majority of respondents reported moderate level of stress (76.9%). Contrary to previous studies that highlight the significant moderate effect of stress on MSDs, this study indicates the presence of other risk factors. Moreover, since this study control for age and gender while others usually investigate direct relationship, this result is expected. Therefore, further replication of this study with inclusion of other risk factors might be needed.

5 Conclusions

Production operators experienced musculoskeletal discomforts at neck, lower back and lower legs. This is consistent with the nature of the production activities which are still largely manual and repetitive. Majority of them experienced moderate level of stress with 6.5% experience high level of stress. There is a significant but very weak association between stress and MSDs when gender and age were controlled.

This study is not without limitation. As the nature of this study is cross-sectional, causality inference is cautioned. Furthermore, all measures are based on employees' perceptions However, it is important to note that self-rating predicts better compared to objective measurement. Furthermore, subjective measures are also more widely use and easy to assess.

References

1. DOSH, [Online]. Available: http://www.dosh.gov.my/index.php/en/list-of-documents/statistics-socso/330-occupational-musculoskeletal-diseases-statistics-1995-2009/file. [Accessed: 19-Jun-2017].
2. P. Boon, BorneoPost Online | Borneo , Malaysia, Sarawak Daily News | Largest English Daily In Borneo," Borneo Post Online, 2013. [Online]. Available: http://www.theborneopost.com/2013/01/28/musculoskeletal-disease-increase-cause-for-concern/. [Accessed: 19-Jun-2017].
3. Lee Lam Thye, NST Online, 2016. [Online]. Available: https://www.nst.com.my/news/2016/04/142074/work-stress-and-mental-health. [Accessed: 20-Jun-2017].
4. Ain Ruzanna, Business Insider, 2014. [Online]. Available: http://www.businessinsider.my/scary-statistics-stress-silent-killer/#LXsZSHQGImbEYdr1.97. [Accessed: 20-Jun-2017].
5. A. Dewan, | Human Resources Online," Human Resources, 2013. [Online]. Available: http://www.humanresourcesonline.net/majority-malaysians-suffer-work-related-stress/. [Accessed: 20-Jun-2017].
6. P. K. Wilke, W. H. Gmelch, and N. P. Lovrich, Public Product. Rev., **9**, 4, pp. 342–356, 1985.
7. Jex and S. M., Sage Publications Ltd, 1998.
8. K. Fairbrother and J. Warn, J. Manag. Psychol., **18**, 1, pp. 8–21, 2003.
9. I. Maakip, T. Keegel, and J. Oakman, J. Occup. Rehabil., **25**, 4, 2015.
10. A. C. McFarlane, Best Pract. Res. Clin. Rheumatol., **21**, 3, pp. 549–565, Jun. 2007.
11. C. L. Cooper, Career Dev. Int., **10**, 5, pp. 396–399, Aug. 2005.
12. M. J. Sanders and C. M. Turcotte, Work, **35**, 4, 2010.
13. H. Busch, Scand. J. Caring Sci., **19**, 4, 2005.
14. K. Ekberg, J. Eklund, M.-A. Tuvesson, R. Örtengren, P. Odenrick, and M. Ericson, Work Stress, **9**, 4, 1995.
15. Y.-H. Lin, C.-Y. Chen, and S.-Y. Lu, Appl. Ergon., **40**, 4, 2009.
16. S. Zakerian and I. D. Subramaniam, Int. J. Occup. Saf. Ergon., **15**, 4, 2009.
17. B. Das and S. Gangopadhyay, Asia-Pacific J. Public Heal., **27**, 2, 2015.
18. T. Ghosh, B. Das, and S. Gangopadhyay, Indian J. Community Med., **35**, 2, 2010.
19. B. Das, " Int. J. Occup. Med. Environ. Health, **27**, 3, 2014.
20. H. Abdul Rahman, K. Abdul-Mumin, and L. Naing, Asian Nurs. Res. (Korean. Soc. Nurs. Sci)., **11**, 1, 2017.
21. K. Azma, A. Hosseini, M. H. Safarian, and M. Abedi, N. Am. J. Med. Sci., **7**, 7, 2015.
22. Y.-K. Kong, D.-M. Kim, K.-S. Lee, and M.-C. Jung, Appl. Ergon., **43**, 2, 2012.
23. C. R. Palliser, H. M. Firth, A. M. Feyer, and S. M. Paulin, Work Stress, **19**,. 4, 2005.
24. L. E. Rocha and M. Debert-Ribeiro, Rev. Saude Publica, **35**, 6, 2001.
25. E. L. González-Muñoz and R. Á. ChaurandProcedia Manuf., **3**, 2015.
26. N. Fold and A. Wangel, "Geografisk tidskrift," in *Geografisk Tidsskrift*, **97**, Kongelige danske geografiske selskabet, 1997.

27. "Ergonomics Guidebook for Manual Production Systems."
28. H. S. Loo and S. Richardson, J. Soc. Sci., **8**, 1, pp. 61–65, 2012.
29. Z. M. Makhbul, Int. J. Acad. Res. Bus. Soc. Sci., **3**, 1, pp. 183–195, 2013.
30. H. Selye, Annu. Rev. Med., **2**, 1, pp. 327–342, Feb. 1951.
31. Y. Cheng, Y.-L. Guo, and W.-Y. Yeh, Int. Arch. Occup. Environ. Health, **74**, 7, 2001.
32. A. Hedge, S. Morimoto, and D. McCrobie, Ergonomics, **42**, 10, pp. 1333–1349, 1999.
33. S. Cohen, T. Karmack, and R. Mermelstein, J. Health Soc. Behav., **24**, pp. 386–396, 1983.
34. J. C. Nunnally, *Psychometric theory*, 2nd ed. New York: McGraw-Hill, 1978.
35. A. M. Basahel, Procedia Manuf., **3**, 2015.
36. E. Pék, B. Bánfai, K. Deutsch, S. Jeges, and J. Betlehem, Lege Artis Med., **24**, 10–11, 2014.
37. B. R. Da Costa and E. R. Vieira, Am. J. Ind. Med. J . Ind . Med, **53**, 53, pp. 285–323, 2010.
38. K. Matsudaira, H. Konishi, K. Miyoshi, T. Isomura, and K. Inuzuka, PLoS One, **9**, 4, p. e93924, 2014.

DOI: 10.1051/matecconf/201713500052

Optimizing Woven Curtain Fabric Defect Classification using Image Processing with Artificial Neural Network Method at PT Buana Intan Gemilang

Shadika[1,]* *Tatang* Mulyana[2], and *Meldi* Rendra[3]

[1,2,3]Telkom University, Industrial Engineering, Bandung , West Java, Indonesia

Abstract. The textile industry is one of the industries that provide high export value by occupying the third position in Indonesia. The process of inspection on traditional textile enterprises by relying on human vision that takes an average scanning time of 19.87 seconds. Each roll of cloth should be inspected twice to avoid missed defects. This inspection process causes the buildup at the inspection station. This study proposes the automation of inspection systems using the Artificial Neural Network (ANN). The input for ANN comes from GLCM extraction. The automation system on the defect inspection resulted in a detection time of 0.56 seconds. The degree of accuracy gained in classifying the three types of defects is 88.7%. Implementing an automated inspection system results in faster processing time.

1 Introduction

Textile industry is one of the industrial sectors that ranks third largest exporter of export value that can increase foreign exchange. In order to produce products that conform to standards and quality based on consumer demand, it takes inspection process on the fabric. Inspection process at the company at this time is still done manually using the human eye with the help of lighting in the form of lights. After calculating the processing time to the manual inspection process obtained the average scanning time of 19.7 seconds. In addition, the limitations of human vision and human error affect greatly in the process of manual inspection, such as the defects that are missed when the fabric is inspected. Based on the company's production data, the volume of production and inspection volume are not balanced, since the total fabric produced is not entirely inspected on time.

This has an impact on the unfulfilled consumer demand within the specified time. This research designs a proposal system in the form of an integrated automation system that aims to optimize the inspection process with the help of image processing. The purpose of this research is to be able to classify defects in fabric using artificial neural network method by substituting the function of human vision into digital image processing. The use of artificial

*Corresponding author: shadikawanef@gmail.com

neural network methods is done to reduce the inspection time and increase the accuracy level in classifying defects.

2 Basic Theory

Quality is an important factor affecting products and services. The higher the quality of a cloth the less the defects are in the fabric and vice versa. Improving quality is a major factor in achieving business success, growth and enhancing the competitiveness of enterprises [1-7]. To determine the type of defects in the fabric based on Indonesian national standard (SNI) 08-0277-1989 consisting of 17 types of defects as contained in Table 1. The type of fabric defects observed generally found in textile companies based on SNI 08-0277-1989 data can be seen in Table 2.

Table 1. Types of defects on fabrics based on SNI 08-0277-1989

	Defect Type	Defect Definition
1	Nep	Nep
2	Sub	Yarn twisted, slab broken yarn, dirt plaster, knot yarn
3	Uneven yarn	Large yarn, small yarn
4	Broken yarn	Broken warp, broken feed
5	Tight yarn/saggy	Wrinkled woven fabric, arches
6	Line fold	Folded fabric
7	Warp line	Different yarn structure, comb lines, double weld, usi rapat, dense meetings, rare warp, different types of fibers, large yarns, small yarns.
8	Feed line	Double feed, stop mark, feed meetings, rare feed, large yarns, small yarn.
9	Incorrect pattern	Pattern defects, incorrect stamped shapes, wrong webbing, stamped marks, wrong color coractenun
10	Bare	Different thread structures, different types of fibers, feed meetings, rare feed
11	Stripped	Unevenness of color, color difference
12	Tear	Hole, tear
13	Unfinished yarn	Skipped yarn
14	Stains	Rust stains, color stains
15	Wide defects	
16	Feed bias	Feed bias include the curved feed
17	Flaw defects	

Artificial Neural Network (ANN) is one of the branches of Artificial Intelligence. ANN is an information processing system that has characteristics resembling a biological neural network. Neural networks are inspired by human biology and consist of several processing units called nerves. ANN can be interpreted as information processing that implies the input vector data and the output vector data.

Table 2. Similarity of type of fabric defect based on SNI and observation result

No	Defect Similarity	
	SNI	Observation Result
1	Nep	Hairy Feed
2	Slab	Brittle Feed

3	Broken Yarn	Broken Warp
4	Tight Yarn	Tight Wrap
5	Warp Line	Comb Defects
6	Incorrect Pattern	Broken Card
7	Unfinished Yarn	Plotting

The most significant advantage of ANN is the mapping function of the results can be determined through the training vector, as ANN can perform the learning process from the provided training data. There are two types of Artificial Neural Network architecture commonly used, namely: Single-Layer Networks and Multilayer Networks. Single - layer networks are used to classify and work well when classes are linearly separated. During the training phase, the input and output pairs are used to train the network. Figure 1 shows the inputs *(x1, x2 ... xn)* and output *(y1, y2, y3 ... ym)*. Using the input vector, the actual output is compared to the desired output, and the error between the actual output and the desired output is used to update the load.

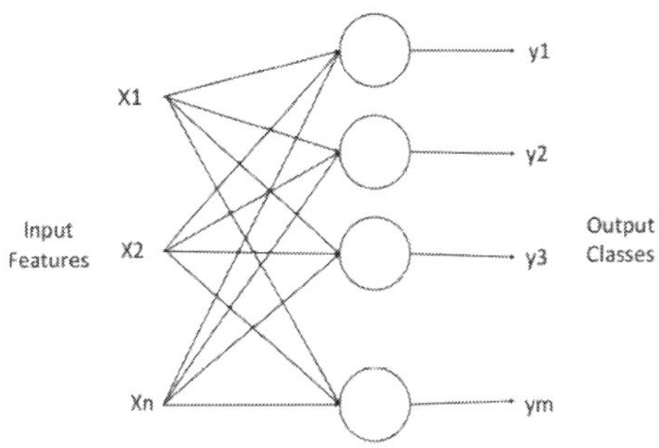

Fig 1. Single-Layer Networks

To make more complex decisions, a multilayer perceptron model is required. The architecture of the multilayer network in Figure 2 shows three types of layers: the input layer, the hidden layer, and the output layer. Multilayer network can be trained by using back propagation learning algorithm. The actual output vector is compared with the desired output vector, if there is no change then no load is changed, in contrast if any then the weight is updated [8-11].

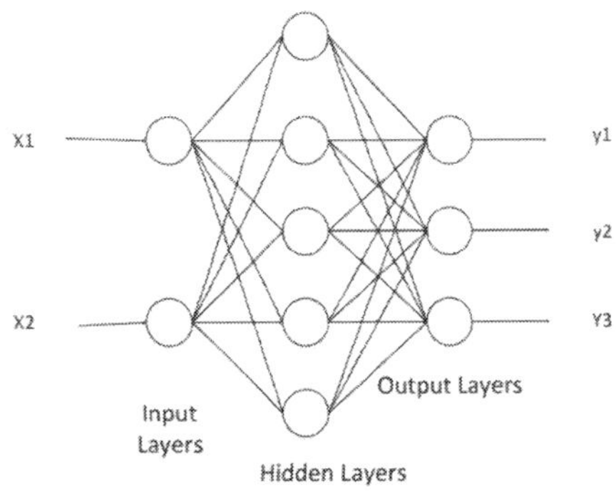

Fig. 2. Multilayer Networks

DOI: 10.1051/matecconf/201713500052

3 Methodology

To get the automated fabric defect classification using neural network some steps are needed. The step can be seen at Figure 3. First is capturing the defect image using camera, the camera that used in this research is webcam logitech C525. This research capture 30 data for each defect and 30 data for each normal fabric.After captured the fabric, then the captured image must be converted to grayscale image to make the defect classification process faster. Then the grayscale image is extracted using GLCM feature extraction as input for neural network. Next is train the network using the result of extraction feature. On this research the input is 12 factors of GLCM and the hidden layer that used is 20 hidden layer. The result then will be tested using the new data from defective fabric to make sure the system can detect the defect.

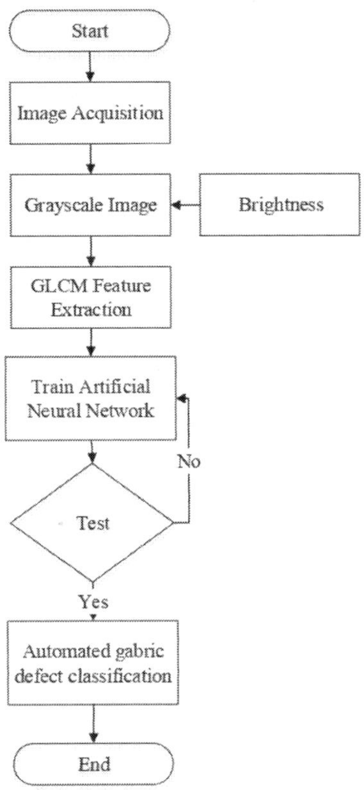

Fig. 3. Flowchart artificial neural network to obtain data

4 Results and Discussion

Extraction is done using Graphical User Interface (GUI) as Figure 5. The extraction results are used as ANN input. The amount of data used is as much as 120 data for training on ANN. The results of training data in the form of vectors, for defective spark plugs in the form of 1 0 0 0, defective feed 0 0 0 0, oil defects 0 0 0 1, and normal 0 0 1 0. The accuracy of ANN training can be seen in Figure 6. Accuracy ANN training is 95.8%, where in all confusion matrix there are five data that are in wrong class.

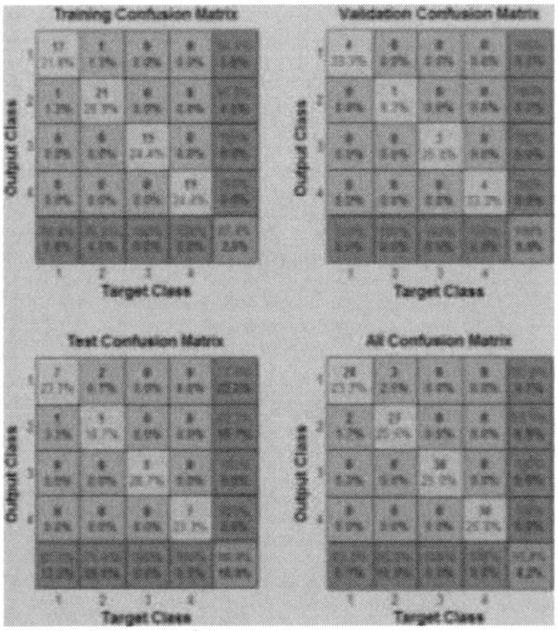

Fig.4. Accuracy of ANN training results

Furthermore, the results of the training will be the tested data as much as 80 data that is 20 defective warp, 20 defects of empty feed, 20 oil defects, and 20 normal. The result of classification test using ANN offline can be seen in Table 3. Accuracy is obtained by calculating using the following formula.

$$\text{Accuracy} = (\text{Amount of data true}) / (\text{Total data count}) \times 100\% \qquad (1)$$

Based on the offline classification results there are three incorrect data on warp defect and four incorrect data on empty feed defects. From the data, the accuracy of offline system to classified the defect on the fabric is 91%.

Table 3. Offline defect classification results

Input Type	Warp Defect	Empty Feed Defect	Normal	Oil Defect
Warp Defect	17	3	0	0
Empty Feed Defect	4	16	0	0
Normal	0	0	20	0
Oil Defect	0	0	0	20

After the inspection process, the processing time to classify three types of offline defects in the fabric can be seen in Table 4. The average processing time obtained to identify 80 types of defects is for 0.73 seconds.

DOI: 10.1051/matecconf/201713500052

Table 4. Offline classification time defects

Data Number	Process Time	Data Number	Process Time	Data Number	Process Time	Data Number	Process Time
1	0.770076	21	0.621023	41	0.535614	61	0.269596
2	0.592154	22	0.735082	42	0.524071	62	0.425395
3	0.571321	23	0.644239	43	0.528035	63	0.444889
4	0.646883	24	0.526157	44	0.572965	64	0.446793
5	0.576134	25	0.598328	45	0.641985	65	0.517335
6	0.541896	26	0.626834	46	0.620484	66	0.552104
7	0.626375	27	0.61664	47	0.642352	67	0.461299
8	0.570773	28	0.569393	48	0.630881	68	0.517712
9	0.286475	29	0.534994	49	0.600921	69	0.535475
10	0.334813	30	0.54379	50	0.637243	70	0.535781
11	0.530931	31	0.631401	51	0.581212	71	0.518743
12	0.299761	32	0.717383	52	0.653937	72	0.522677
13	0.575409	33	0.619605	53	0.587051	73	0.518001
14	0.340483	34	0.596116	54	0.638098	74	0.469349
15	0.516364	35	0.618333	55	0.582841	75	0.472082
16	0.651765	36	0.616771	56	0.566419	76	0.451548
17	0.609998	37	0.551801	57	0.631268	77	0.578512
18	0.743218	38	0.577327	58	0.652896	78	0.536244
19	0.653092	39	0.659496	59	0.546835	79	0.516837
20	13.4025	40	0.696695	60	0.553047	80	0.438055

Furthermore, the results of the real-time test will be tested data as much as 80 data that is 20 defective warp defect, 20 defects of empty feed, 20 oil defects, and 20 normal. The result of classification test using ANN offline can be seen in Table 5. Based on the result of real time classification there is one wrong data on the defect of breaks and eight wrong data on empty feed defect. Accuracy result obtained is equal to 88.75%.

Table 5. Results of real time defect classification

Input Type	Warp Defect	Empty Feed Defect	Normal	Oil Defect
Warp Defect	19	0	1	0
Empty Feed Defect	8	12	0	0
Normal	0	0	20	0
Oil Defect	0	0	0	20

After the inspection process, the process time to classify the three types of defects in real time on the fabric can be seen in Table 6. After testing of 80 data obtained average time classification process defects in real time for 0.56 seconds

Table 6. Timing of real time classification process defects

Data Number	Process Time	Data Number	Process Time	Data Number	Process Time	Data Number	Process Time
1	0.582776	21	7.4924	41	0.289456	61	0.304845
2	0.54779	22	0.736126	42	0.2976	62	0.295765
3	0.565936	23	0.632174	43	0.301907	63	0.279409
4	0.552999	24	0.633404	44	0.280222	64	0.293552
5	0.557814	25	0.54855	45	0.294041	65	0.270983
6	0.623224	26	0.635446	46	0.286486	66	0.289948
7	0.59373	27	0.694331	47	0.286644	67	0.282137
8	0.631674	28	0.647753	48	0.276272	68	0.276991
9	0.610841	29	2.96714	49	0.287542	69	0.309945
10	0.524721	30	0.620036	50	0.279349	70	0.270983
11	0.576364	31	0.657134	51	0.281802	71	0.281802
12	0.598072	32	0.697748	52	0.284347	72	0.284347
13	0.472274	33	0.7457	53	0.305232	73	0.305232
14	0.588562	34	0.559747	54	0.284995	74	0.291053
15	0.561275	35	0.517545	55	0.294033	75	0.301907
16	0.56842	36	0.521062	56	0.285294	76	0.280222
17	0.660343	37	0.623472	57	0.272616	77	0.294041
18	0.587955	38	0.68808	58	0.270013	78	0.286486
19	0.566684	39	0.519799	59	0.273119	79	0.293087
20	0.548564	40	0.51098	60	0.291053	80	0.275745

Results Recap data from MATLAB which is connected with excel software can be seen in Table 7. From the data recap, it can be known that the average data recap time for 80 data is 2.96. The recap time of the proposed data is faster than the time of recap of the existing data.

Table 7. Results of recapitulation of defect classification data

Data Number	Process Time	Data Number	Process Time	Data Number	Process Time	Data Number	Process Time
1	2.84671	21	2.78742	41	2.72373	61	3.08017
2	2.97526	22	2.82463	42	2.74481	62	3.16672
3	3.88056	23	3.07639	43	2.82579	63	3.1096
4	3.05133	24	2.89045	44	3.07321	64	2.7942
5	2.83511	25	2.93238	45	2.77318	65	3.05807
6	3.08693	26	2.80888	46	2.83453	66	3.05117
7	3.17767	27	2.99983	47	2.85975	67	3.20491
8	2.89033	28	2.81963	48	3.08696	68	3.0247
9	3.00807	29	2.96695	49	2.92583	69	2.93309
10	3.06618	30	2.97499	50	2.77245	70	2.74482
11	3.11074	31	2.84647	51	3.13556	71	3.06182

12	3.04461	32	2.90035	52	2.91712	72	2.96934
13	2.78062	33	3.21018	53	2.85838	73	3.0847

Table 7. Results of recapitulation of defect classification data (continued)

Data Number	Process Time	Data Number	Process Time	Data Number	Process Time	Data Number	Process Time
14	2.81724	34	2.80832	54	3.03216	74	3.06013
15	2.84362	35	2.72155	55	2.99249	75	3.13935
16	3.19836	36	3.08313	56	3.09187	76	3.02288
17	2.87577	37	2.79001	57	2.79276	77	2.75014
18	2.95698	38	3.05507	58	2.80411	78	2.89962
19	3.01999	39	2.83755	59	2.77501	79	2.74452
20	3.02631	40	2.89411	60	3.09436	80	2.92068

Based on the results of the research, the comparison of the existing manual process and the time of the proposed process can be seen in Table 8. In Table 8 it can be seen that scanning defects are faster than the existing process of 0.56 seconds, as well as the recap of the proposed defect faster than the existing process time is 2.96 seconds. The total process time of the proposed system is 3.5 seconds, 29.8 seconds faster than the existing system with a total processing time of 33.57 seconds.

Table 8. Comparison of total results of proposed and existing process time

No	Activity	Existing (second)	Automated (second)
1	Defect Scanning	19.87	0.56
2	Defect Recap	13.5	2.97
	Total (second)	33.37	3.52

5 Conclusion

The manual inspection process has several drawbacks so that an automated inspection process that minimizes manual inspection is needed. One way to use is to use image processing to improve process accuracy and reduce the average time of the inspection process. In this research used artificial neural network method with GLCM extraction feature and processed by using MATLAB software. In the extraction feature used 12 parameters to get the best results on the process of classification of fabric defects. The proposed system is capable of generating an overall accuracy of 88.75% and an average processing time of 0.56 seconds. The proposed system is capable of producing faster processing time than the existing time. The timing of the proposal is 29.8 seconds faster, so the use of defect classification with ANN can be applied to speed up the inspection process time.

References

[1] D.C.Montgomery. *Introduction to Statistical Quality Control, Sixth Edition* (John Wiley & Sons, Inc , United States of America, 2009)

[2] H Rachmat, T Mulyana, SH Hasan, MR bin Ibrahim, Design Selection of In-UVAT Using MATLAB Fuzzy Logic Toolbox, International Conference on Soft Computing and Data Mining, 538-545, (2017).

[3] E. Kurniawan, B. Rahmat, T. Mulyana and J. Alhilman, "Data analysis of Li-Ion and lead acid batteries discharge parameters with Simulink-MATLAB," 2016 4th International Conference on Information and Communication Technology (ICoICT). 1-5 (2017)

[4] H. Rachmat and T. Mulyana, "Website design of EMS-SCADA for AC usage on a building," 2015 3rd International Conference on Information and Communication Technology (ICoICT), 17-22 (2015)

[5] R. A. Anugraha and T. Mulyana, "Monitoring and controlling of EMS-SCADA via SMS gateway," 2015 3rd International Conference on Information and Communication Technology (ICoICT). 613-617 (2015)

[6] T. Mulyana, J. Alhilman and E. Kurniawan, "Data analysis using system identification toolbox of heat exchanger process control training system," 2016 4th International Conference on Information and Communication Technology (ICoICT). 1-6.(2016)

[7] M. Tatang. NNARX model structure for the purposes of controller design and optimization of heat exchanger process control training system operation. In: AIP Conference Proceedings. 020040. (2017)

[8] P. Banumathi, T. S. Sree, V.Priya. Artificial Intelligence Techniques in Textile Fabric.**4,** 6 (2015)

[9] M.Risaldi, Purwanto, H.Himawan. Classification of Coconut Wood Using Neural Network Backpropagation Algorithm.**10**, 10 (2014)

[10] A. Kadir, A. & A. Susanto, *Theory and Applications of Image Processing*. (Andi Publisher, Yogyakarta, 2013)

[11] A.D. Kulkarni, *Computer Vision and Fuzzy Neural Systems* (Prentice Hall PTR, New Jersey, 2001)

Design of Automation System for Ceramic Surface Quality Control Using Fuzzy Logic Method at Balai Besar Keramik (BBK)

Anggie Prahas Putri[1,*], *Haris* Rachmat[1] , and *Denny* Sukma Eka Atmaja[1]

[1] School of Industrial Engineering Telkom University, 40257 Bandung, Indonesia.

Abstract. Indonesia is the world's top 10 ceramic users. Currently, the process of ceramic quality inspection in Indonesia, particularly in Balai Besar Keramik (BBK) is still done manually by human vision. Therefore, it is necessary to design a visual inspection system for digital image processing of ceramic automation using a Fuzzy Logic method. Fuzzy model is one method that can be used to determine ceramic surface quality control. This study aims to apply the fuzzy model in the design of automation system for ceramic surface quality control and describes its accuracy rate and automatic database. To perform feature extraction using GLCM extraction method to obtain autocorrelation, the sum of square (variance), and a number of the object. The information is used as input for data processing using the fuzzy model in the identification of quality of ceramic surface defect. In this research used Fuzzy model with GLCM extraction feature and process by using MATLAB software. This research used 32 training data can produce the accuracy of 96.87% and based on the real time system of 92.31% used 13 real time test data.

1 Introduction

Ceramic is the most popular type of media used in Indonesia because there are various advantages possessed, and also available for high usage and can be applied to almost all parts of the house. In addition to strong, ceramic house also does not require polishing and easy to maintain. Based on Table 1 can be seen from the export side, the export volume growth has fluctuated. The export volume of ceramics in 2013 to 2014 has increased by 4.3%, and in 2014. On the import side, the growth of ceramic import volume always increases every year. The rate of import increase of 8.6% and in 2014 to increase by 1%. Factors that cause import volumes to be higher than export volumes are that ASAKI coach is the government's very easy to include imported products without any more in-depth examination of standards, and in terms of price the difference is also cheaper than the price of ceramics in Indonesia [1].

* Corresponding author: anggieprahas@outlook.com

Table 1. Volume Export and Import Ceramic [2]

Year	2013	2014	2015
Volume Export (Ton)	295,780	337,000	336,520
Volume Import (Ton)	733,510	966,000	991,910

To fulfil the demand for export and import of ceramic products required quality products that can compete with other companies. Quality is the overall nature of a product or service that has an effect on its ability to satisfy the stated or implied needs [3]. To obtain the best quality of a product needs to consider their quality control process. In SNI ISO 13006:2010 there are six variable characteristic requirements related to the quality of ceramic tiles, namely: length and width, thickness, straightness of side, rectangularity, surface flatness and surface quality [4].

Currently, the process of ceramic quality inspection in Indonesia, particularly in BBK is still done manually by human vision. The process of ceramic quality inspection manually will cause fatigue and saturation because process performed repeatedly. There are five groups causes of fatigue, namely monotonous work, load and duration work both physically and mentally, work environment, psychological, and disease [5].

The method applied in this research is a quality control process called the Fuzzy Logic method. The method is a methodology of problem-solving control systems, suitable for implementation on systems, from simple systems, small systems, embedded systems, PC networks, multichannel or workstations based data acquisition and control systems [6-8].

There are several reasons why people use fuzzy logic, among others (1) The concept of fuzzy logic is easy to understand. The mathematical concept underlying fuzzy reasoning is very simple and easy to understand. (2) Fuzzy logic very flexible (3) Fuzzy logic has the tolerance of improper data. (4) Fuzzy logic capable of modelling very complex nonlinear functions. (5) Fuzzy logic can build and apply the experiences of experts directly without having to go through the training process. (6) Fuzzy logic can cooperate with conventional control techniques. (7) Fuzzy logic based on natural language [6-8].

In the fuzzy set theory, the role of membership degree as the determinant of the existence of elements in a set is very important. The membership value or membership degree or membership function is the main characteristic of reasoning with the fuzzy logic [9-12].

Before the advent of fuzzy logic theory, it is known that a crisp logic has a true or an unambiguous value. Conversely fuzzy logic is a logic that has a value of fuzziness between right and wrong. In fuzzy logic theory, a value can be true or false at the same time but how much truth and error a value depends on the weight of its membership [9-12].

Based on research Atmaja error rate of measurement of tile surface area obtained at light intensity 300 lx with a distance of 50 cm by 0.0675%. Influencing factors for the percentage of ceramic tile measure wide error rate are camera distance, light intensity, and interaction of both factors (camera distance and light intensity). This research to measure the accuracy rate in process quality control of ceramic surface used Fuzzy Logic method. Fuzzy Logic method used Mamdani.

2 Gray Level Co-Occurrence Matrix (GLCM)

Gray Level Co-occurrence Matrix (GLCM) is a technique for obtaining image textures using second order calculations. First-order texture measurements using statistical calculations are based on pure image values only, such as variance, and do not pay attention to the neighboring pixels [7]. In this research used for extraction features are autocorrelation and sum of square (variance).

1. Autocorrelation:

$$\text{Autocorrelation}=\frac{\sum_{u=0}^{N}\sum_{v=0}^{N} I\,(u,v)\; I\,(u+x,v+y)}{\sum_{u=0}^{N}\sum_{v=0}^{N} I^2(u,v)} \quad (1)$$

The autocorrelation is combination pairs as one point, and define the correlation of this point between other combination pair series [8].
Where;
I is the mean value of I(u,v)
M and N are the image width and height

2. Sum of Square (Variance):

$$\text{Sum of Squares} = \sum_{i=0}^{G-1}\sum_{j=0}^{G-1} (i-\mu)^2 \,.P(i,j) \quad (2)$$

This features puts relatively high weights on the elements that differ from the average value of P(i,j). GLCM variance uses the GLCM, therefore it deals specifically with the dispersion around the mean of combinations of reference and neighbor pixels, i.e., encoding contextual (Order) information [8].

3 Methodology

3.1 Step Approach For Defect Detection

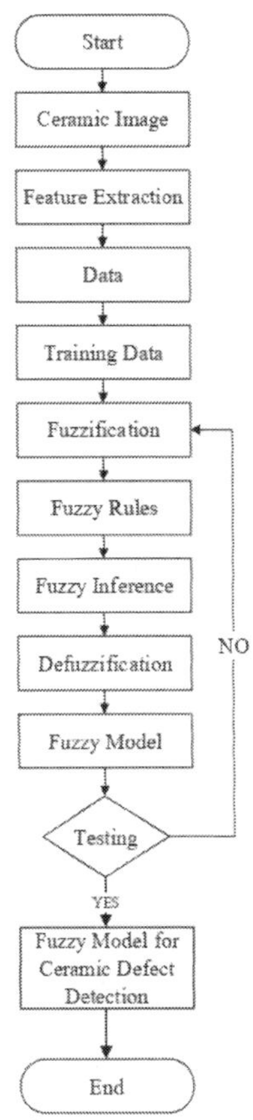

Fig. 1. Flowchart of methodology research

The complete flow chart of our step defect detection technique has been rendered in the following Figure 1. Based on Figure 1 had eight steps, consist of capture a ceramic image, undertake feature extraction,determine the universal set in input and output variable, define fuzzy set in input and output, create fuzzy rule, fuzzy inference, defuzzification, testing of Fuzzy Model.

The first step before data processing is taking a picture for training data and testing data (real time). Training data using 32 sample consist of 16 sample normal ceramics and 16 sample defect surface ceramics. Testing data (real time) using 16 sample consist of 8 sample normal ceramics and 5 sample defect surface ceramics. Capturing picture using Webcam Logitech C525 and Logitech Webcam Software with 1.3 MP camera resolution, 300 lx light intensity, and 50 cm camera distance.

The second step uses GLCM feature extraction and Prewitt Edge Detection to count the point object in an image. It will be obtained the value that will be used to Fuzzy Model, these are autocorrelation, a sum of square (variance), and a number of the object. The next step is converting the RGB image into grayscale image. Then images that have been converted into grayscale is the feature extraction process. The GLCM method is used to know value used in the fuzzy input, these are autocorrelation, and a sum of square (variance).

The third step is the classification of fuzzy logic as the basic fuzzy model. The universal set is value that allowed to be operated in a fuzzy variable. Universal set input is a range of minimum to a maximum value of all sample data. Universal set output is divided into 2 categories consist of a defect and normal. The output of this research is linguistic values that need to convert into the numerical value, choose number 1 as a defect, and 2 as normal.

The fourth step, fuzzy set for each input variable of feature extraction divide into 9 sets and output of fuzzy sets in this research was divided into two categories, defect and normal. The allocation of the fuzzy set is done by trial and error to get a good model.

The fifth step, extraction results obtained from 32 training data are used to create the fuzzy rules of the membership function that have been searched. The greatest value of a membership function is used to create a fuzzy rule based on the relationship between fuzzy sets. This research uses 3 input characteristics and 9 degree membership. The possible rules used are 9^3, total of training data used is 32 data using 729 rules. The largest membership value is used as a representation of rules in the fuzzy set. The process for obtaining fuzzy rules through calculations using MATLAB®.

The sixth step, fuzzy inference is the process s to get the output of each rule. Each rule gives a value appropriate with the input. This research is using fuzzy inference with Mamdani Model. Mamdani Model implement the implication functions and rule composition. Implication function performed to obtain an output modification of fuzzy area from each the rule. Implication function of Mamdani Model using implication function MIN and composition rule MAX, so that the Mamdani Model often known as MAX-MIN Model.

The seventh step, Defuzzification is the last step on a fuzzy model that is to convert the result of fuzzy set value from fuzzy inference into a crisp. Therefore, the fuzzy set obtained from fuzzy inference which is the extraction result of the sample image is recovered using de fuzzifier into crisp. Defuzzifier used in this research that is centroid method, because defuzzification value obtained is able to move smoothly, the conversion from fuzzy set also run smoothly. Defuzzification process will be calculated using MATLAB® software.

4 Results and Discussion

4.1 Feature Extraction

DOI: 10.1051/matecconf/201713500053

Feature extraction process is conducted with MATLAB® software. To perform the feature extraction of the image required a media interface that is contained in MATLAB® Library.

1. Convert RGB Image to Grayscale Image

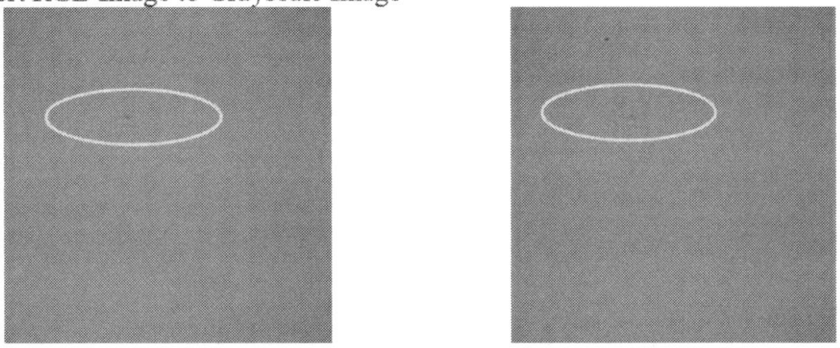

(a) RGB Image (b) Grayscale Image

Fig. 2. Example RGB Image to Grayscale Image

In Figure 2 image illustration convert RGB image to grayscale image.

2. Convert Grayscale Image to Binary Image

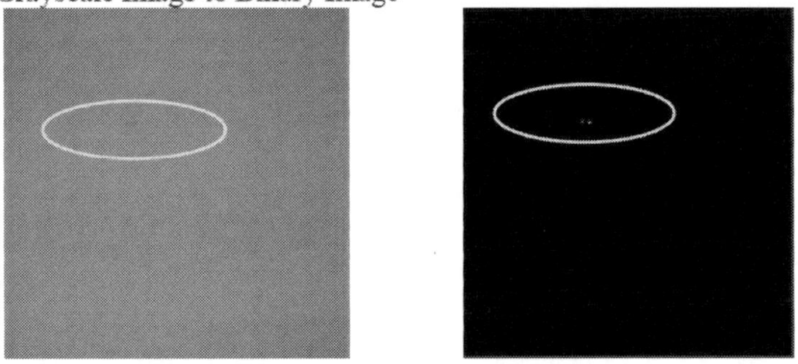

(a) Grayscale Image (b) Binary Image

Fig. 3. Example Grayscale Image to Binary Image

In Figure 3 is the result of the grayscale image into the binary image, and identify defect point using white color.

4.2 Identify Universal Set (U) for Input

Based on data obtained from the feature extraction then the universal sets of each image as follows:

1. Autocorrelation
   The universal set for autocorrelation from the training data of feature extraction is minimum 24.1814 and maximum 34.2239. Then the universal set for number of object is U_A = [24.1814 34.2239].
2. Sum of Square (Variance)
   The universal set for sum of square (variance) from the training data of feature extraction is minimum 24.0395 and maximum 34.0558. Then the universal set for number of object is U_B = [24.0395 34.0558].
3. Number of Object

The universal set for a number of object from the training data of feature extraction is minimum 0 and maximum 4. Then the universal set for a number of object is $U_C = [0\ 4]$.

4.3 Identify Universal Set (V) for Output

The minimum output value is 1 and the maximum value is 2, therefore the universal set for output V = [1 2].

4.4 Define Fuzzy Set for Input and Output

Here is the description of a fuzzy set of training data input from feature extraction and output.

1. Autocorrelation
   The membership to the variable autocorrelation function as follows. The membership function is illustrated in Figure 4.

$$\mu_{A1}[x] = \begin{cases} \dfrac{25.4367 - x}{1.2553}; & 24.1814 \le x \le 25.4367 \\ 0; & 25.4367 \le x \end{cases}$$

$$\mu_{A2}[x] = \begin{cases} 0; & x \le 24.1814 \\ \dfrac{x - 24.1814}{1.2553}; & 24.1814 \le x \le 25.4367 \\ \dfrac{26.6920 - x}{1.2553}; & 25.4367 \le x \le 26.6920 \end{cases}$$

$$\mu_{A3}[x] = \begin{cases} 0, & x \le 25.4367 \\ \dfrac{x - 25.4367}{1.2553}; & 25.4367 \le x \le 26.6920 \\ \dfrac{27.9473 - x}{1.2553}; & 26.6920 \le x \le 27.9473 \end{cases}$$

$$\mu_{A4}[x] = \begin{cases} 0, & x \le 26.6920 \\ \dfrac{x - 26.6920}{1.2553}; & 26.6920 \le x \le 27.9473 \\ \dfrac{29.2027 - x}{1.2553}; & 27.9473 \le x \le 29.2027 \end{cases}$$

$$\mu_{A5}[x] = \begin{cases} 0, & x \le 27.9473 \\ \dfrac{x - 27.9473}{1.2553}; & 27.9473 \le x \le 29.2027 \\ \dfrac{30.4580 - x}{1.2553}; & 29.2027 \le x \le 30.4580 \end{cases}$$

$$\mu_{A6}[x] = \begin{cases} 0, & x \le 29.2027 \\ \dfrac{x - 29.2027}{1.2553}; & 29.2027 \le x \le 30.4580 \\ \dfrac{31.7133 - x}{1.2553}; & 30.4580 \le x \le 31.7133 \end{cases}$$

$$\mu_{A7}[x] = \begin{cases} 0, & x \le 30.4580 \\ \dfrac{x - 30.4580}{1.2553}; & 30.4580 \le x \le 31.7133 \\ \dfrac{32.9686 - x}{1.2553}; & 31.7133 \le x \le 32.9686 \end{cases}$$

DOI: 10.1051/matecconf/201713500053

$$\mu_{A8}[x] = \begin{cases} 0, x \leq 31.7133 \\ \dfrac{x - 31.7133}{1.2553}; 31.7133 \leq x \leq 32.9686 \\ \dfrac{34.2239 - x}{1.2553}; 32.9686 \leq x \leq 34.2239 \end{cases}$$

$$\mu_{A9}[x] = \begin{cases} 0 ; x \leq 32.9686 \\ \dfrac{x - 32.9686}{1.2553}; 32.9686 \leq x \leq 34.2239 \\ 1 ; x \geq 34.2239 \end{cases}$$

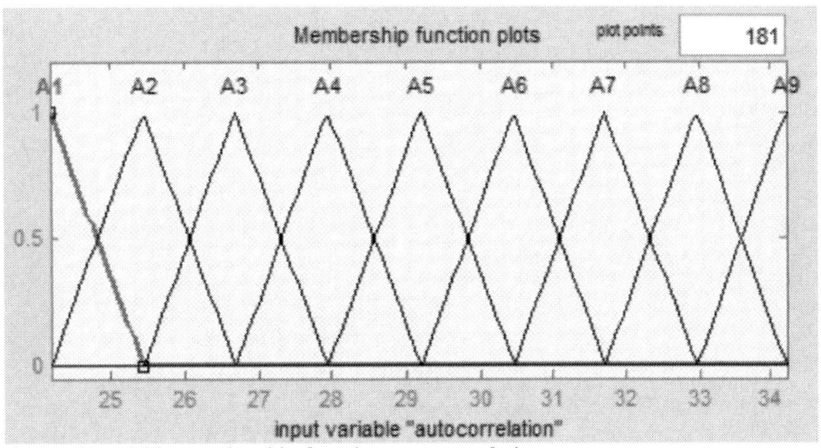

Fig. 4. Graph variable membership functions autocorrelation

2. Sum of Square (Variance)
   The membership to the variable sum of square (variance) function as follows, and the membership function is illustrated in Figure 5.

$$\mu_{B1}[x] = \begin{cases} \dfrac{25.2915 - x}{1.2520}; 24.0395 \leq x \leq 25.2915 \\ 0; x \geq 25.2915 \end{cases}$$

$$\mu_{B2}[x] = \begin{cases} 0; x \leq 24.0395 \\ \dfrac{x - 24.0395}{1.2520}; 24.0395 \leq x \leq 25.2915 \\ \dfrac{26.5436 - x}{1.2520}; 25.2915 \leq x \leq 26.5436 \end{cases}$$

$$\mu_{B3}[x] = \begin{cases} 0, x \leq 25.2915 \\ \dfrac{x - 25.2915}{1.2520}; 25.2915 \leq x \leq 26.5436 \\ \dfrac{27.7956 - x}{1.2520}; 26.5436 \leq x \leq 27.7956 \end{cases}$$

$$\mu_{B4}[x] = \begin{cases} 0, x \leq 26.5436 \\ \dfrac{x - 26.5436}{1.2520}; 26.5436 \leq x \leq 27.7956 \\ \dfrac{29.0477 - x}{1.2520}; 27.7956 \leq x \leq 29.0477 \end{cases}$$

$$\mu_{B5}[x] = \begin{cases} 0, x \le 27.7956 \\ \dfrac{x - 27.7956}{1.2520}; 27.7956 \le x \le 29.0477 \\ \dfrac{30.2997 - x}{1.2520}; 29.0477 \le x \le 30.2997 \end{cases}$$

$$\mu_{B6}[x] = \begin{cases} 0, x \le 29.0477 \\ \dfrac{x - 29.0477}{1.2520}; 29.0477 \le x \le 30.2997 \\ \dfrac{31.5517 - x}{1.2520}; 30.2997 \le x \le 31.5517 \end{cases}$$

$$\mu_{B7}[x] = \begin{cases} 0, x \le 30.2997 \\ \dfrac{x - 30.2997}{1.2520}; 30.2997 \le x \le 31.5517 \\ \dfrac{32.8038 - x}{1.2520}; 31.5517 \le x \le 32.8038 \end{cases}$$

$$\mu_{B8}[x] = \begin{cases} 0, x \le 31.5517 \\ \dfrac{x - 31.5517}{1.2520}; 31.5517 \le x \le 32.8038 \\ \dfrac{34.0558 - x}{1.2520}; 32.8038 \le x \le 34.0558 \end{cases}$$

$$\mu_{B9}[x] = \begin{cases} 0 ; x \le 32.8038 \\ \dfrac{x - 32.8038}{1.2520}; 32.8038 \le x \le 34.0558 \\ 1 ; x \ge 34.0558 \end{cases}$$

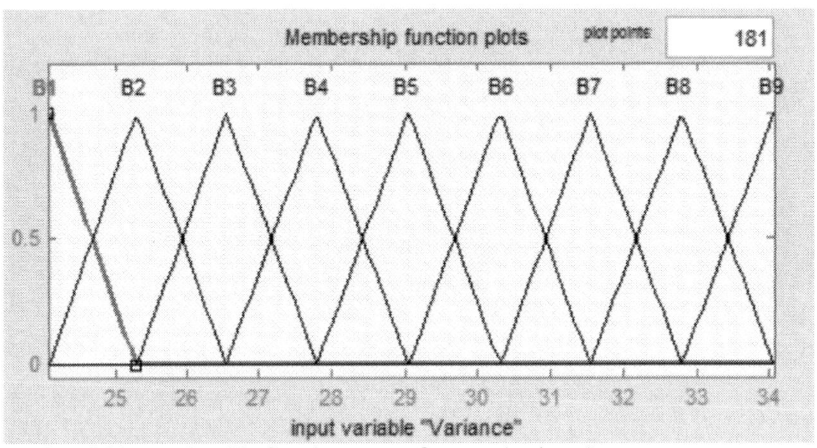

Fig. 5. Graph variable membership functions sum of square (variance)

3. Number of Object
   The membership to the variable number of object function as follows. The membership function is illustrated in Figure 6.

$$\mu_{C1}[x] = \begin{cases} \dfrac{0.5 - x}{0.5}; 0 \le x \le 0.5 \\ 0; x \ge 0.5 \end{cases}$$

$$\mu_{C2}[x] = \begin{cases} 0;\ x \le 0 \\ \dfrac{x - 0}{0.5};\ 0 \le x \le 0.5 \\ \dfrac{1 - x}{0.5};\ 0.5 \le x \le 1 \end{cases}$$

$$\mu_{C3}[x] = \begin{cases} 0,\ x \le 0.5 \\ \dfrac{x - 0.5}{0.5};\ 0.5 \le x \le 1 \\ \dfrac{1.5 - x}{0.5};\ 1 \le x \le 1.5 \end{cases}$$

$$\mu_{C4}[x] = \begin{cases} 0,\ x \le 1 \\ \dfrac{x - 1}{0.5};\ 1 \le x \le 1.5 \\ \dfrac{2 - x}{0.5};\ 1.5 \le x \le 2 \end{cases}$$

$$\mu_{C5}[x] = \begin{cases} 0,\ x \le 1.5 \\ \dfrac{x - 1.5}{0.5};\ 1.5 \le x \le 2 \\ \dfrac{2.5 - x}{0.5};\ 2 \le x \le 2.5 \end{cases}$$

$$\mu_{C6}[x] = \begin{cases} 0,\ x \le 2 \\ \dfrac{x - 2}{0.5};\ 2 \le x \le 2.5 \\ \dfrac{3 - x}{0.5};\ 2.5 \le x \le 3 \end{cases}$$

$$\mu_{C7}[x] = \begin{cases} 0,\ x \le 2.5 \\ \dfrac{x - 2.5}{0.5};\ 2.5 \le x \le 3 \\ \dfrac{3.5 - x}{0.5};\ 3 \le x \le 3.5 \end{cases}$$

$$\mu_{C8}[x] = \begin{cases} 0,\ x \le 3 \\ \dfrac{x - 3}{0.5};\ 3 \le x \le 3.5 \\ \dfrac{4 - x}{0.5};\ 3.5 \le x \le 4 \end{cases}$$

$$\mu_{C9}[x] = \begin{cases} 0\ ;\ x \le 3.5 \\ \dfrac{x - 3.5}{0.5};\ 3.5 \le x \le 4 \\ 1\ ;\ x \ge 4 \end{cases}$$

DOI: 10.1051/matecconf/201713500053

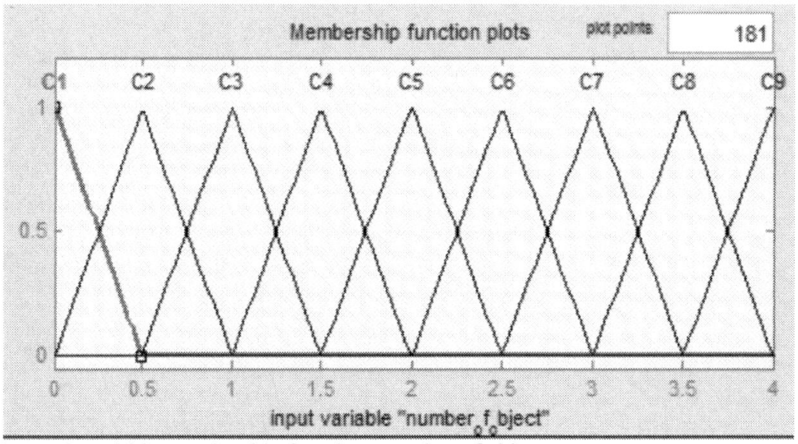

Fig. 6. Graph variable membership functions number of object

4. Output
   Output of fuzzy sets in this research were divided into two categories, defect and normal. It was presented using triangular membership function on the interval [1 2]: The fuzzy set is illustrated in Figure 7.

$$\mu_{DEFECT}[x]=\begin{cases}2 - x; & 1 \le x \le 2\\ 0; & x \ge 2\end{cases}$$

$$\mu_{NORMAL}[x]=\begin{cases}1 - x; & 1 \le x \le 2\\ 1; & x \ge 2\end{cases}$$

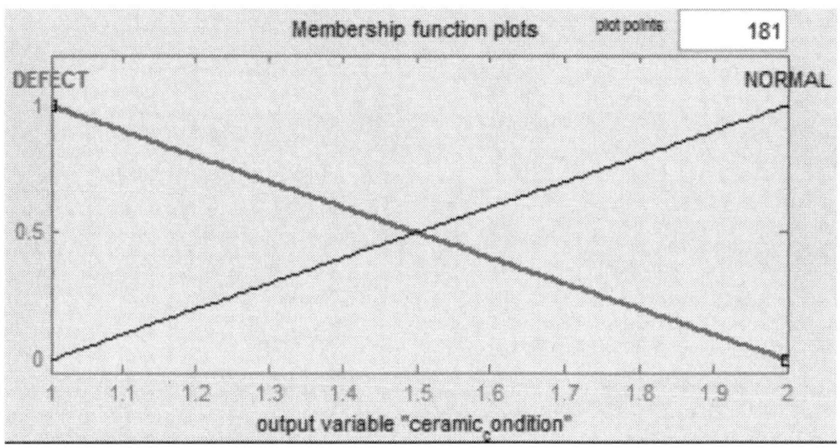

Fig. 7. Graph variable membership functions output

4.6 Create Fuzzy Rule (If – Then)

Here is an example of the fuzzy rule of figure nine. This rule is formed after the feature extraction, so it will produce 3 as input. Figure 8 is a sample training data image and the result of feature extraction shown in Table 2.

 DOI: 10.1051/matecconf/201713500053

Fig. 8. Sample Training Data Image

The following are the results of the sample training data image extraction.

Table 2 Results of The Sample Training Data Image Extraction

Data	Result of Extraction
Autocorrelation	24.9187
Sum of Square (Variance)	24.7652
Number of Object	2
Detection	Defect

Extracted from the images grouped in a fuzzy set of input

1. Autocorrelation

   Autocorrelation of the sample training data image extraction is 24.9187. Fuzzy set on the input autocorrelation divided into 9 fuzzy set A_1, A_2, A_3, A_4, A_5, A_6, A_7, A_8, A_9. X = 24.9187 can be included in the A1 and A2.

   $\mu_{A1}(24.9187) = \frac{25.4367\text{-}24.9187}{1.2553} = 0.4127$, so membership function $A_1 = 0.4127$

   $\mu_{A2}(24.9187) = \frac{24.9187 - 24.1814}{1.2553} = 0.5873$, so membership function $A_2 = 0.5873$

   $\mu_{A3}(24.9187) = 0$, membership function $A_3=0$

   $\mu_{A4}(24.9187) = 0$, membership function $A_4=0$

   $\mu_{A5}(24.9187) = 0$, membership function $A_5=0$

   $\mu_{A6}(24.9187) = 0$, membership function $A_6=0$

   $\mu_{A7}(24.9187) = 0$, membership function $A_7=0$

   $\mu_{A8}(24.9187) = 0$, membership function $A_8=0$

   $\mu_{A9}(24.9187) = 0$, membership function $A_9=0$

   Then,

   Max = (μ_{A1}, μ_{A2}, μ_{A3}, μ_{A4}, μ_{A5}, μ_{A6}, μ_{A7}, μ_{A8}, μ_{A9})

   Max = (0.4127, 0.5873, 0, 0, 0, 0, 0, 0, 0) = 0.5873

   Therefore, autocorrelation of the extracted sample training data image into the fuzzy set A_2.

2. Sum of Square (Variance)
   Sum of Square (Variance) of extracting the sample training data image is 24.7652. The set of fuzzy input sum of square (variance) is divided into 9 fuzzy set that ire B_1, B_2, B_3, B_4, B_5, B_6, B_7, B_8, B_9. X = 24.7652 can be included in the set B_1 and B_2.
   $\mu_{B1}(24.7652) = \frac{25.2915 - 24.7652}{1.2520} = 0.4204$, so membership function B_1=0.4204
   $\mu_{B2}(24.7652) = \frac{24.7652 - 24.0395}{1.2520} = 0.5796$ so membership function B_2=0.5796
   $\mu_{B3}(24.7652) = 0$, membership function B_3=0
   $\mu_{B4}(24.7652) = 0$, membership function B_4=0
   $\mu_{B5}(24.7652) = 0$, membership function B_5=0
   $\mu_{B6}(24.7652) = 0$, membership function B_6=0
   $\mu_{B7}(24.7652) = 0$, membership function B_7=0
   $\mu_{B8}(24.7652) = 0$, membership function B_8=0
   $\mu_{B9}(24.7652) = 0$, membership function B_9=0
   Then,
   Max = (μ_{B1}, μ_{B2}, μ_{B3}, μ_{B4}, μ_{B5}, μ_{B6}, μ_{B7}, μ_{B8}, μ_{B9})
   Max = (0.4204, 0.5796, 0, 0, 0, 0, 0, 0, 0) = 0.5796
   Therefore, sum of square (variance) of the extracted sample training data into the fuzzy set B_2.
3. Number of Object
   Number of Object of the training data image extraction is 2. The set of fuzzy on the number of objects is divided into 9 fuzzy set that are C_1, C_2, C_3, C_4, C_5, C_6, C_7, C_8, C_9..
   X = 2 can be included in the set C_4 and C_5.
   $\mu_{C1}(2) = 0$, membership function $C_1 = 0$
   $\mu_{C2}(2) = 0$, membership function $C_2 = 0$
   $\mu_{C3}(2) = 0$, membership function $C_3 = 0$
   $\mu_{C4}(2) = \frac{2 - 2}{0.5} = 0$, so membership function C_4= 0
   $\mu_{C5}(2) = \frac{2 - 1.5}{0.5} = 1$, so membership function C_5= 1
   $\mu_{C6}(2) = 0$, membership function $C_6 = 0$
   $\mu_{C7}(2) = 0$, membership function $C_7 = 0$
   $\mu_{C8}(2) = 0$, membership function $C_8 = 0$
   $\mu_{C9}(2) = 0$, membership function $C_9 = 0$
   Then,
   Max = (μ_{C1}, μ_{C2}, μ_{C3}, μ_{C4}, μ_{C5}, μ_{C6}, μ_{C7}, μ_{C8}, μ_{C9})
   Max = (0, 0, 0, 0, 1, 0, 0, 0, 0) = 1
   Therefore, number of object of the extracted training data image into the fuzzy set C_5.

Table 3 is the fuzzy set grouping of the training data image.

Table 3 Results of the extraction and classification of fuzzy sets of the training data image

Data	Result of Extraction	Membership Function	Fuzzy Set
Autocorrelation	24.9187	0.5873	A_2
Sum of Square (Variance)	24.7652	0.5796	B_2
Number of Object	2	1	C_5
Ceramic Condition	Defect	Defect	V_{Defect}

Based on Table 3 the rules from the feature extraction result of training data image as follows:
"IF autocorrelation is A_2 AND variance is B_2 AND number of object is C_5 THEN Ceramic Condition is V_{Defect}"

4.7 Fuzzy Inference

Here is an example of inference calculation manually. Example use the ninth image shown in Table 4.

Table 4 Result of Extraction Sample Image Data

Data	Result Extraction
Autocorrelation	24.9187
Variance	24.7652
Number of Object	2
Ceramic Condition	Defect

Based on the results of the ninth image obtained extract the greatest value of the implications is the value MIN 0.5796. Then the composition is modified into a fuzzy area. The area from the output is illustrated in Figure 18.

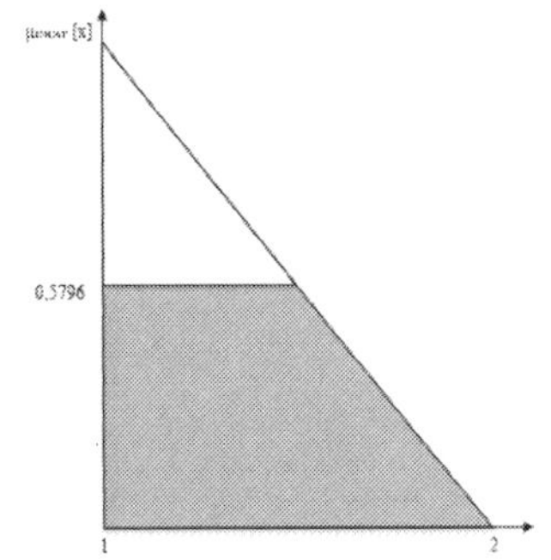

Fig. 9. MIN Results implications rules Sample Image Data

From the area results ninth data obtained would then be determined the implications of the membership function in the following ways:

1. The range is divided into two parts, namely D1 and D2 with a1 which is the boundary from the two areas as shown in Figure 10.

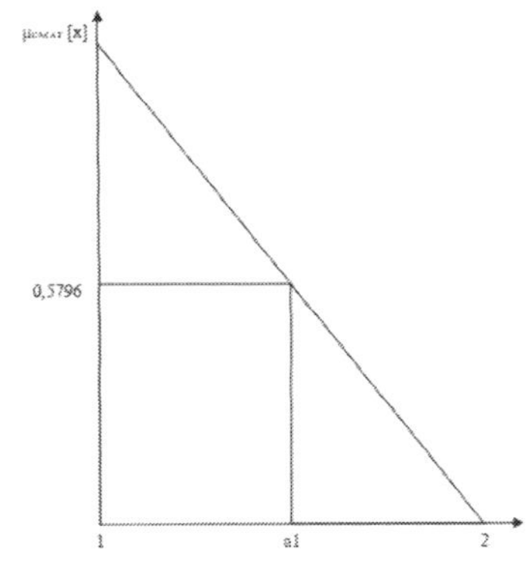

Fig. 10. Illustration Picture Implications Rules Division MIN

2. Then it would be determined value of a_1
   The point a_1 is between the intersection of the lines $y = 2 - x$ and $y = 0.5796$.

$$y_1 = y_2$$
$$2 - x = 0.5796$$
$$x = 1.4204$$

Therefore, the composition of the membership function results are as follows:

$$\mu[x] = \begin{cases} 0.5796 \; ; 1 \leq x \leq 1.4204 \\ 2\text{-}x \; ; 1.4204 \leq x \leq 2 \end{cases}$$

The result of fuzzy inference using Mamdani Method, the output still as a fuzzy set, therefore need to convert of fuzzy set into crisp using defuzzification.

4.8 Defuzzification

The centroid is calculating crips set by taking the center point of the fuzzy area. This is the formula:

$$x^* = \frac{\int_a^b z\mu(z)dz}{\int_a^b \mu(x)dz}$$

The result of fuzzy inference:

$$\mu[x] = \begin{cases} 0.5796 \; ; 1 \leq x \leq 1.4204 \\ 2\text{-}x \; ; 1.4204 \leq x \leq 2 \end{cases}$$

Using Centroid Method get formula:

$$x^* = \frac{\int_1^{1.4204} (0.5796)xdx + \int_{1.4204}^{2} (2\text{-}x)xdx}{\int_1^{1.4204} (0.5796)dx + \int_{1.4204}^{2} (2\text{-}x)dx}$$

$$x^* = \frac{0.2948 + 0.2710}{0.2436 + 0.1679}$$

$$x^* = \frac{0.5658}{0.4115}$$

$$x^* = 1.3749$$

The defuzzification result of the sample image is 1.3749, which means that the ninth image is DEFECT. Defuzzification process is calculated using MATLAB.

4.9 Defect Detection for Training Data

The following Table 5 shows the test results of 32 training data. The table data training consist of data training, real output, output model, and information. The result of identification of training data as much as 32 data resulted 1 wrong data, where 1 data that should be normally identified defect. This result is determined from Deffuzification values for defects ≤ 1.5 and Deffuzification values for normal > 1.5.

Table 5 Results of Training Data

Data	***Real Output***	***Output Model***	***Information***
Data Training 1	NORMAL	NORMAL	TRUE
Data Training 2	NORMAL	NORMAL	TRUE
Data Training 3	NORMAL	NORMAL	TRUE
Data Training 4	NORMAL	NORMAL	TRUE
Data Training 5	NORMAL	NORMAL	TRUE
Data Training 6	NORMAL	NORMAL	TRUE
Data Training 7	NORMAL	NORMAL	TRUE
Data Training 8	NORMAL	DEFECT	FALSE
Data Training 9	DEFECT	DEFECT	TRUE
Data Training 10	DEFECT	DEFECT	TRUE
Data Training 11	DEFECT	DEFECT	TRUE
Data Training 12	DEFECT	DEFECT	TRUE
Data Training 13	NORMAL	NORMAL	TRUE
Data Training 14	NORMAL	NORMAL	TRUE
Data Training 15	NORMAL	NORMAL	TRUE
Data Training 16	NORMAL	NORMAL	TRUE
Data Training 17	NORMAL	NORMAL	TRUE
Data Training 18	NORMAL	NORMAL	TRUE
Data Training 19	NORMAL	NORMAL	TRUE
Data Training 20	NORMAL	NORMAL	TRUE
Data Training 21	DEFECT	DEFECT	TRUE
Data Training 22	DEFECT	DEFECT	TRUE
Data Training 23	DEFECT	DEFECT	TRUE
Data Training 24	DEFECT	DEFECT	TRUE
Data Training 25	DEFECT	DEFECT	TRUE
Data Training 26	DEFECT	DEFECT	TRUE
Data Training 27	DEFECT	DEFECT	TRUE
Data Training 28	DEFECT	DEFECT	TRUE
Data Training 29	DEFECT	DEFECT	TRUE
Data Training 30	DEFECT	DEFECT	TRUE
Data Training 31	DEFECT	DEFECT	TRUE
Data Training 32	DEFECT	DEFECT	TRUE

To determine the accuracy of this training data, use the following formula:

$$\text{Accuracy Rate} = \frac{\text{Number of True}}{\text{Total Data Training}} \text{x} 100\%$$

$$\text{Accuracy Rate} = \frac{31}{32} \text{x} 100\%$$

$$\text{Accuracy Rate} = 96.87\%$$

Therefore, the accuracy rate used fuzzy model with triangle representation of membership function, Mamdani method of fuzzy inference, centroid method of defuzzification, and 729 rules the obtained accuracy rate 96.87%.

4.10 Defect Detection for Testing Data (Real Time)

Automatic ceramic inspection design based on ceramic defect detection, the results of the system are shown in Table 6.

Table 6 Result of System in Real Time Process

Data	*Real Ouput*	*Output Model*	*Deffuzification*	*Information*
Data Test 1	NORMAL	NORMAL	1.63277	TRUE
Data Test 2	NORMAL	NORMAL	1.62542	TRUE
Data Test 3	NORMAL	NORMAL	1.6361	TRUE
Data Test 4	NORMAL	NORMAL	1.63807	TRUE
Data Test 5	NORMAL	NORMAL	1.63451	TRUE
Data Test 6	NORMAL	NORMAL	1.63918	TRUE
Data Test 7	NORMAL	NORMAL	1.63201	TRUE
Data Test 8	NORMAL	NORMAL	1.63903	TRUE
Data Test 9	DEFECT	DEFECT	1.36873	TRUE
Data Test 10	DEFECT	DEFECT	1.3673	TRUE
Data Test 11	DEFECT	DEFECT	1.36595	TRUE
Data Test 12	DEFECT	DEFECT	1.36672	TRUE
Data Test 13	DEFECT	NORMAL	1.62537	FALSE

This research tested fuzzy logic model using Logitech C525 webcam, camera resolution of 1.3 MP or image size of 640 x 480 pixel, camera distance of 50 cm, and light intensity of 300 lx. The result of identification of training data as much as 13 data resulted 1 wrong data, where 1 data that should be normally identified defect. This result is determined from deffuzification values for defects ≤ 1.5 and deffuzification values for normal > 1.5. To determine the accuracy of this training data, use the following formula:

$$\text{Accuracy Rate} = \frac{\text{Number of True}}{\text{Total Data Training}} \text{x} 100\%$$

$$\text{Accuracy Rate} = \frac{12}{13} \text{x} 100\%$$

$$\text{Accuracy Rate} = 92.31\%$$

Therefore, the accuracy rate used fuzzy model with triangle representation of membership function, mamdani method of fuzzy inference, centroid method of defuzzification, and 729 rules the obtained accuracy rate 92.31%.

5 Conclusion

In this research used Fuzzy model with GLCM extraction feature and process by using MATLAB software. In the extraction feature used three variabel inputs consist of autocorrelation, sum of square (variance), and number of object. Accuracy rate of ceramic inspection process using design of automatic ceramic defect detection with light intensity 300 lx, 50 cm camera distance and 1.3 MP camera resolution or 640 x 480 pixel image size using fuzzy logic on ceramic defect with training data of 96.87% and accuracy based on the real time system of 92.31%.

References

1. Kanal Satu, Gas Masih Mahal, Industri Keramik dan Granit Indonesia Lesu, [Online] Available at: http://kanalsatu.com/id/post/49944, [Accessed 14 November 2016]
2. Bank Sentral Republik Indonesia, Volume Ekspor dan Impor Keramik, [Online] Available at: http://www.bi.go.id/id/statistik/seki/terkini/eksternal/Contents/Default.aspx, [Accessed 14 November 2016], (2016).
3. D. S. E. Atmaja, Optimasi Pengukuran Dimensi & Cacat Permukaan Ubin Keramik Menggunakan Pengolahan Citra Digital dan Full Factorial Design, Gadjah Mada University, **4**, 101-198 (2015).
4. Kotler, *Manajamen Pemasaran* (Erlangga, Jakarta, 2009).
5. A Muizzudin, Hubungan Antara Kelelahan Kerja Dengan Produktivitas Kerja Pada Tenaga Kerja Bagian Tenun Di PT. Alkatex Tegal, Semarang State University, 24-25, (2013).
6. S Wibowo, Penerapan Logika Fuzzy Dalam Penjadwalan Waktu Kuliah, Jurnal Informatika, 62, (2015).
7. I Ferdyansyah, Optimizing Woven Fabric Defect Detection For Inspection Using Image Processing And Fuzzy Logic At CV. Maemunah Majalaya, Telkom University, 13, (2016).
8. INF 4300. 2011. Digital Image Analysis. INF4300-2011-f02-texture.pdf, (2011).
9. H Rachmat, T Mulyana, SH Hasan, MR bin Ibrahim, Design Selection of In-UVAT Using MATLAB Fuzzy Logic Toolbox, International Conference on Soft Computing and Data Mining, 538-545, (2017).
10. H. Rachmat and T. Mulyana, "Website design of EMS-SCADA for AC usage on a building," 2015 3rd International Conference on Information and Communication Technology (ICoICT), Nusa Dua, 2015, pp. 17-22.
11. R. A. Anugraha and T. Mulyana, "Monitoring and controlling of EMS-SCADA via SMS gateway," 2015 3rd International Conference on Information and Communication Technology (ICoICT), Nusa Dua, 2015, pp. 613-617.
12. T. Mulyana, J. Alhilman and E. Kurniawan, "Data analysis using system identification toolbox of heat exchanger process control training system," 2016 4th International Conference on Information and Communication Technology (ICoICT), Bandung, 2016, pp. 1-6.

Cost Optimization on Energy Consumption of Punching Machine Based on Green Manufacturing Method at PT Buana Intan Gemilang

Ayudia Prillia[1], Haris Rachmat[1], and Tatang Mulyana[1]

[1] School of Industrial Engineering Telkom University, Bandung, Indonesia.

Abstract. PT Buana Intan Gemilang is a company engaged in textile industry. The curtain textile production need punching machine to control the fabric process. The operator still works manually so it takes high cost of electrical energy consumption. So to solve the problem can implement green manufacturing on punching machine. The method include firstly to identify the color by classifying the company into the black, brown, gray or green color categories using questionnaire. Secondly is improvement area to be optimized and analyzed. Improvement plan at this stage that is focusing on energy area and technology. Thirdly is process applies by modifying the technology through implementing automation system on the punching machine so that there is an increase of green level on the process machine. The result obtained after implement the method can save cost on electrical energy consumption in the amount of Rp 1.068.159/day.

1 Introduction

In this era with the development of technology and increasing in population, energy as one of resources that will always be needed for the whole societies especially electrical energy. With the rapid development of technology, informatiom, and industrial, there are problems for energy limitations. Fossil energy resources continuously vanished as it is a producer of electrical energy and fuel oil, it is necessary to find ways in order to save electric energy use and fuel oil so it would not wasting energy. From the data of national electricity power sales,the industrial sector is the largest user of electric energy. The use of technology in the industry will be continusly developed because the industry has to fullfill the costumer's needs, so the industry will need more energy for supply process. Studies of energy efficiency in machining process have been carried out from many aspects.

According to the research levels and techniques used in energy-reducing strategies, these efforts can be classified into three groups: approaches by improving functions of machine tools or selecting alternative machine tools for specific tasks; approaches by optimizing machining conditions like cutting parameters, cutter's material, etc and approaches by

*Corresponding author: [1]ayudiaprilliaa@gmail.com

reconfiguring machining systems. Improvement plan in terms will focus about consumption electricity energy in punching machine [1]. Implementation along with technology modifications automation so will generate efficiency cost savings. Green manufacturing is a sustainable approach to the design and engineering activities involved in product development and/or system operation to minimize environtmental impact [2]. The method of Green Manufacturing can efficiently be used by technology and save energy consumption on the industry. Implementation along with technology modifications automation so will generate efficiency cost savings. Automation technology is widely used for controlling the production process or the process of working punching machine and monitoring consumption energy. The use of designing automation technologies applied by the company to facilitate, user friendly, and also important because the system automation can reduce the impact of human error.

2 Basic theory

2.1Punching Machine Process

A punching mechanism is designed to cut a hole in some material such as paper, metal, or as in IBM machines, card stock. The original loom was first called a "treadle loom", later as "pattern loom" and finally a "draw loom". Now, Jacquard machine has developed into an electronic form with computer controlling, and Net communication is widely used in jacquard weaving sheds. Meanwhile, the number of hooks used in the electronic Jacquard machine has increased to 20,000 [3]. Punching process is designed to cut a hole in some material such as paper, metal, or card stock. There are three basic elements of punching device, a punch, a die, and a stripper. The punch is the piece which is driven through the card and cuts the hole. The die serves as a base, supports the card while it is being cut. The stripper serves as a guide for the punch and as a means of stripping the card from the punch after the latter has cut the card. Fig 1 show punching process the components consist of workpiece, die, scrap, clearance and punch.

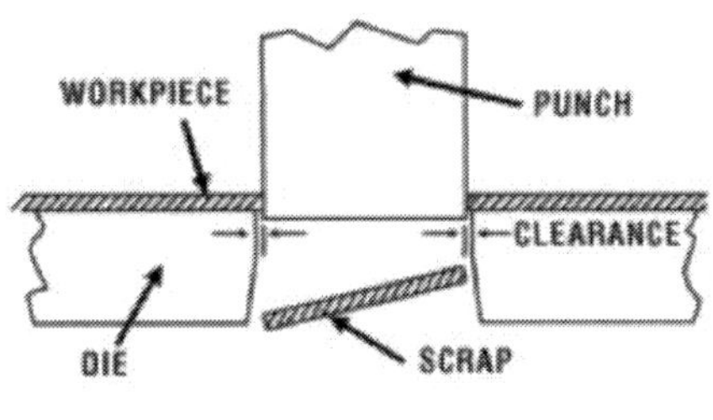

Fig.1. Punching Machine Process

2.2 Green Manufacturing

Green manufacturing is a method for manufacturing that minimizes waste and pollution. These goals are often achieved through product and process design [4]. Green manufacturing should be viewed as an opportunity to expand the local and global market share in this dynamic environment. A deeper understanding of green manufacturing strategies and techniques will enable manufacturers to realize that unlike other competing manufacturing strategies (like cost and time), being green positively impact all other manufacturing competitive edges. For example reducing material wastes and energy consumption will

DOI: 10.1051/matecconf/201713500054

reduce production cost and improve production time. Going green in manufacturing will also improve the quality of the production process which will in turn impact product quality and also will be more appealing to the growing number of customers looking for green manufacturers and products, Fig 2 show aspect manufacturing strategy.

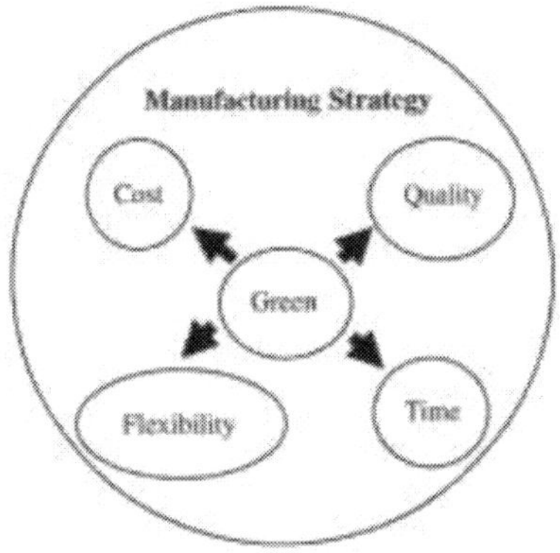

Fig.2. Manufacturing strategies [2]

Energy matter in green manufacturing not only because of economic costs added to the products but also because of environmental impacts associated with the energy production and supply [5]. Green manufacturing is a sustainable approach to the design and engineering activities involved in product development and/or system operation to minimize environmental impact. Fig 3 explain system model architecture for the design and control of the green manufacturing systems. The architecture is composed of two modules; the first module describes the design and planning processes of the green manufacturing systems and the second module describes the control process that controls the design and planning process at each level.

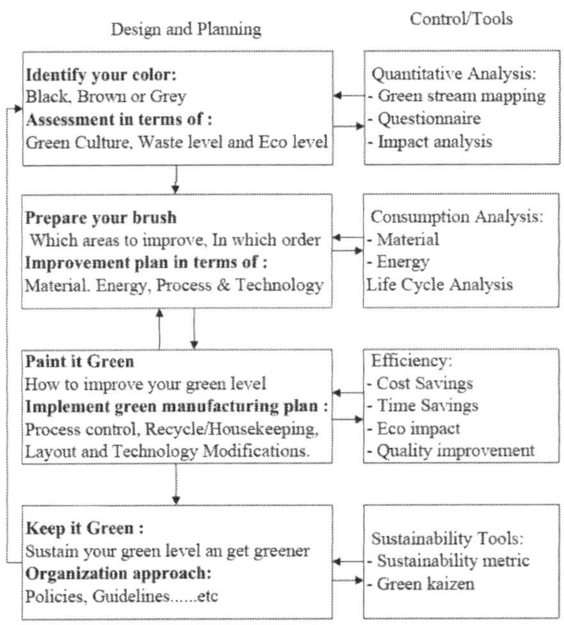

Fig.3. System model for green manufacturing [2]

2.3 Automation Components

Automation consists of three basic elements, actuator, controller, and sensor. The actuator which does the work, and controlled by the controller. The actuator in an automated process may in fact be several actuators, each of which provides an output that drives another in the series of actuator. The controller which tells the actuator to do work may be a simple digital system or an analog system. The sensor is provides feedback to the controller so that it knows the actuator is doing work which. These three elements have been integrated into a mechanical system either electronic, and computer-based systems to operate and control the production [6]. Fig 4 show elements of automated system consist of power, program of instructions, control system, process, and ouput.

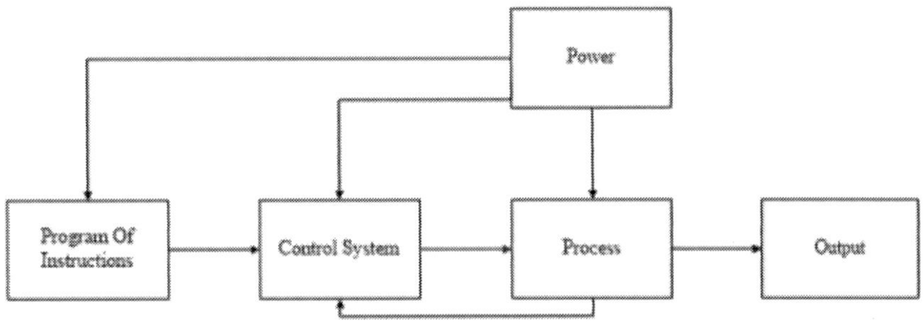

Fig.4. Elements of Automated System [6]

For the example control system is SCADA, that computer based that are used to controlling a process such as industrial processes on manufacturing, production and power generator [7]. SCADA integrated with PLC, based on [8] PLC use software program of Totally Integrated Automation. System monitoring would be done by utilizing the Human Machine Interface (HMI) that will be created by using Wonderware In touch 10.0. In the design of this system required a computer with Windows XP SP3. It also required software for the design of the Wonderware In touch 10.0, Active Factory 9.2, Active X Wonderware Generic Data Grid, Wonderware, SQL Server 2005 I/O Server DASSI direct, and Totally Integrated Automation (TIA). All hardware that is used as a conduit transmission needs between devices. UTP cable required with RJ 45 for liaison between switches with the server computer, the client computer and PLC. Sensors and actuators connected to a PLC by using cables.

3 Results and discussion

3.1 Description

The products of curtain and curtain fabric have various patterns that will make the motif more beautiful. To create some patterns that varies on the weaving machine required jacquard card so it can form the desired pattern. One type of fabric pattern is required to produce as many as 2500 pattern cards and for the pattern of curtains required the production of pattern cards as much as 600. The specification of jacquard cards have a width of 6,5 cm and length of 62,5 cm. The jacquard product consists of three parts that have its own hole patterns. Each part has a header with a big hole and 2 other holes on the next row. Below the header, there are 12 columns and 33 rows. Each hole of all parts has diameter of 0,45 while the big hole size on the header is 0,7. Jacquard Punching machine to produce jacquard card has several main components show Fig 5.

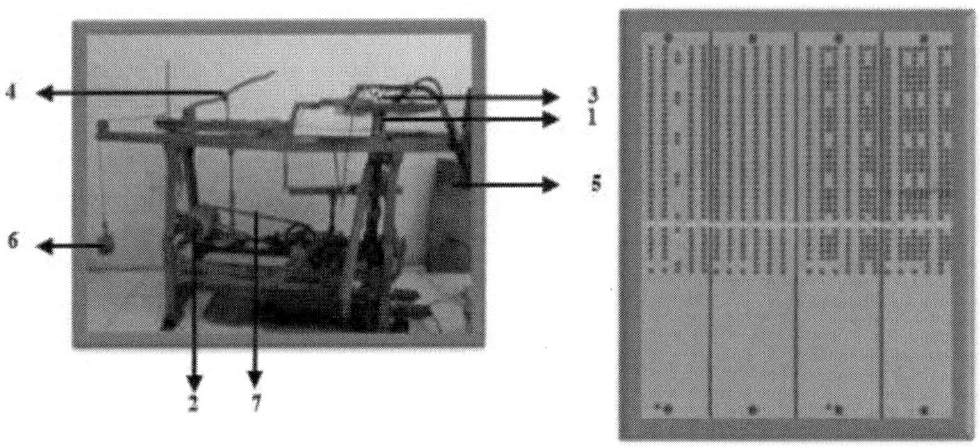

Fig.5. Jacquard Card & Punching Machine

Component number on punching machine

1. Drill
2. AC Motor
3. 12 Solenoid
4. Upper Handle
5. Lower Handle
6. Pendulumdur
7. Belt

3.2 Green manufacturing

1. Identify Color

To apply the method of green manufacturing in a company, the first step is to identify the color or culture on the company and the results is still not optimized. Quantitative analysis control tools is performed to measure the color of the company into the category of black, brown, gray, or green. If the results obtained by the company are green and still included in the black ,brown and gray then the green manufacturing method needs to be applied thus the improvement in getting the company reduces waste in the form of energy, material, process and technology. Thus it can be proceeded into the next stage. Fig 6 show model for the assessment layer to identify color.

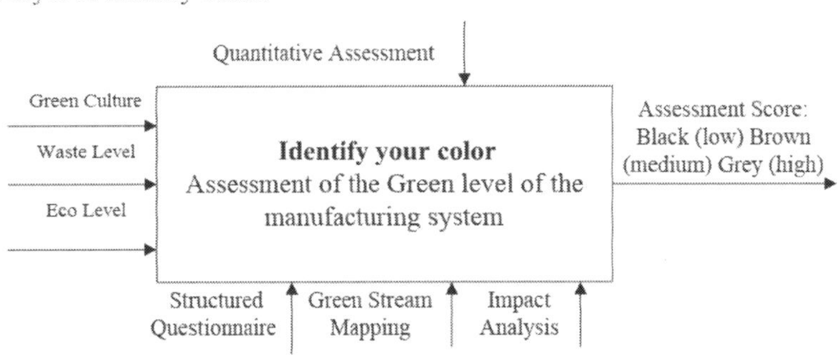

Fig.6. Model for the Assessment Layer [2]

Some tools that can be used in identifying companies with quantitative analysis are green stream mapping, questionnaire, and impact analysis tools. The control tool used to identify

 DOI: 10.1051/matecconf/201713500054

corporate colors in this study is uses questionnaires distributed to operators who operate jacquard punching machines, in order to classify companies into certain category. These are examples of forms of questions classified as focus into the use of energy and materials :

1. Can the rest of production process result be used again to become raw materials?
2. Is there any technology that can generate its own electricity? Such as solar panels, water turbines or windmills?

Table 1. Results of Weighted Questionnaires

Rank	Object	Weight	Persentage
1	Material	2,13333	41%
2	Energy	3,125	59%
Total		5,25833	100%

Questionnaires were given to 5 operators of PT Buana Intan Gemilang. After calculating recapitulation process based on questionnaires, the results obtained are percentage of materials as much as 41% and Energy of 59% in accordance to Table IV Results of Weighted Questionnaires. Based on the materials and energy percentage results, the company's category included in the category of score brown (medium).

2. Prepare your brush

After identifying the color of the company with the brown result (medium) it is necessary to impelement green manufacturing method so that the increasing color becomes better and can reduce the waste in energy, material, process and technology. This stage of the plan is to determine the objects will be analyzed to eliminate waste and optimization occurs in terms of costs incurred by the company. From the results of questionnaires distribution of energy needs in the company process is very influential. Thus, in this process of the object plan will study about energy, in the form of electrical energy used by companies, especially in the punching machine with the change of energy consumption and technological improvement, it is expected that green manufacturing can be implemented. Fig 7 show stage for model for green improvement/implementation planning layer.

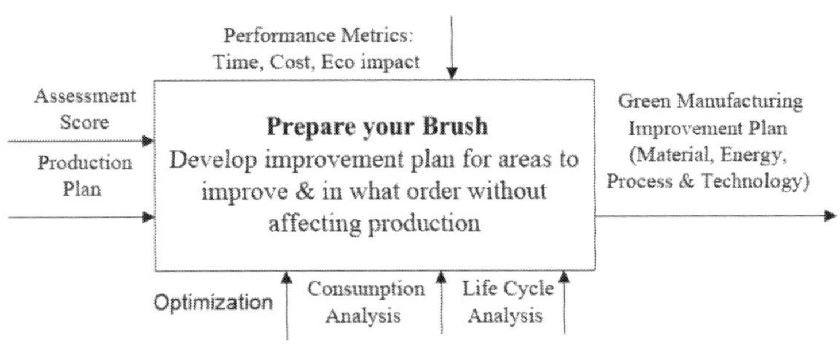

Fig.7. Model for Green Improvement/Implementation Planning layer[2]

3. Paint it Green

In the previous process (prepared your brush) has been determined that the use of energy and technological improvement is an object plan that will be analyzed according to the punching machine process.

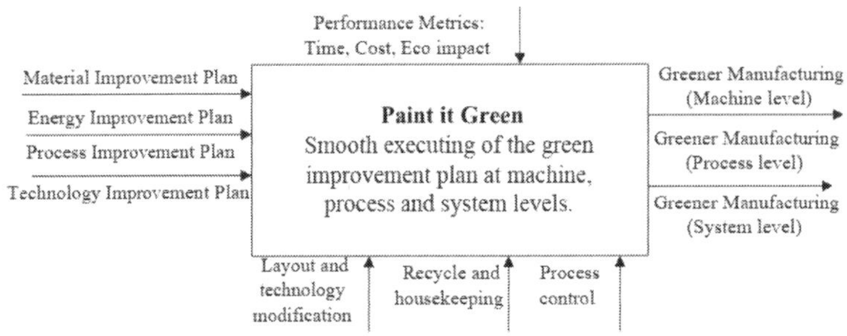

Fig.8. Model for Green Manufacturing Plan Implementation Layer [2]

The process that is done this time is paint it green if improvement plan on machine process punching machine has been done. With the modification of technology on the process machine by implementing the automation system is expected to reduce waste energy use and can perform optimization in terms of costs on the use of electrical energy. The following framework on the automation system punching machine can be seen in Fig 9.

Fig.9. Framework Automation System

3.3 Automation System

1. Image to Excel
   This software used to convert pattern image .bmp became binary data. After the pattern image became binary data, a row of binary data on microsoft excel automated calculated to Hexadecimal, then Hexadecimal data will be upload at Wonderware InTouch so PLC script can running it. Fig 10 show display microsoft excel 2007 image converter.

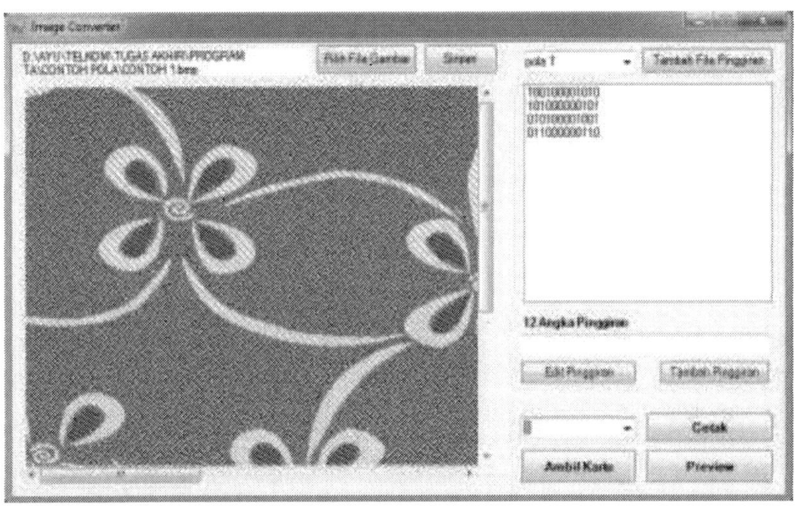

Fig.10. Microsoft Excel 2007 Image converter

2. Wonderware Intouch
   a. Login window
      The login page is the initial view when opening the HMI system. This login system is a security system provided only to workers who do have responsibility for operating a punching machine. On this page is required to fill the User ID and Password to be able to access other pages.

Fig.11. Log In window HMI

   b. Main window
      This window is a window that displays the result of hexadecimal number obtained from convert pattern in Microsoft Excel to be given to CX-Programmer. To create a whole pattern card there are 3 parts on the jacquard card. 1 part consists of 33 rows.

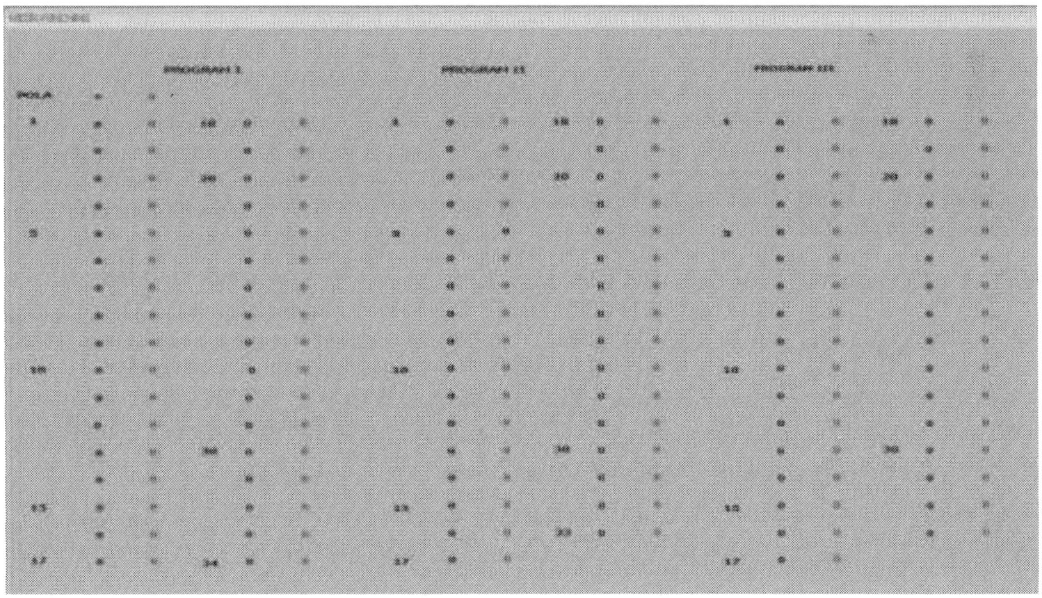

Fig.12. Main Window HMI

3. PLC Programming

1. Section General Process

a. Strating process

This process to start the program as a whole. To enable the Start input and will be active if the 15s TM seconds timer, timer function to delay the process and provide a time delay when Wonderware InTouch uploads the hexa code input for CX-Programmer. Once the TM timer is active and the start input is also active, then Run ouput will also be active.

b. Create header process

After the input Run active, then the process starts with the creation of the header on the jacquard card. The header on the first row that is marked with 2 holes 3 TICK and TICK 10 made by solenoid solenoid 3 and 10. To activate the output solenoid 3 and 10, the input of must be enabled. After the output is active then the next timer TM 001 will be active for 2 seconds to give time lag enable MAIN VALVE on the creation of a hole in the middle is a large header card.

c. Run Bagian 1

The next process this time is a ladder that will give an instruction to jaquard making process. A shift register is a function that activates internal relay. Internal relay is an input for each row which will activate the solenoid valve according to the hex rate of HMI. Run program part one is a program to create pattern on 33 first row on pattern card.

d. Spasi Bagian 1

After the program instructs to activate the solenoid, this script is to instruct the spacebar so that the next solenoid will create hole on the next row.

2. Bagian 1

This section is part of the MOV data transfer input process of the Wonderware InTouch hexadecimal, there are 35 steps or 35 rows in the jacquard card hole making section 1. There are input until STEP 35, Output MOV for STEP 1. This process of recording data from Hexadecimal code input from Wonderware InTouch.

3. Output Bagian 1

After the Hexadecimal code record process of the InTouch wonderware is stored according to the above process, then this process executes the solenoid which will create a hole in the jacquard card on the 1st part. There are 12 solenoid that will make a hole in the jacquarq card.

4. Output PLC

This section is a solenoid command that moves according to the section. Existing internal relays that have been previously described. This script program will run solenoid valve 1 until selenoid valve 12. These solenoid valves are the output to move solenoid one to solenoid valve 12. Solenoid will move simultaneously while each executes The step or row based on input hexa number.

5. End

On any program CX-Programmer should end with an end script. The function of the end is to close the program that has been created in one program.

3.4 Calculating Energy Consumption

Energy consumption on AC Motor punching machine, to calculate energy consumption need data AC motor is shown in Tabel 2.

Table 2. Data AC Motor

Name Motor	P(kW)	V(v)	I(A)		
	55	380	134,2		
	Fasa	V(i-i)	I(A)	Cos ϕ	μ
	R	387	112		
	S	385	118	0,86	0,95
Motorcycle pump ZD1200627	T	385	113		

Based on [9] by calculating the load as the ratio between input power (measured by power analysis tool) and power value at 100% loading. For a three phase motor, the step is to determine the input power is given in equation (1)

$$\mathrm{Pi} = \frac{\mathrm{V} \times \mathrm{I} \times \mathrm{Cos}\,\phi \times \sqrt{3}}{1000}\, kwh \tag{1}$$

Pi = three phase power (kW)
V = current (V)
I = current (A)

$$\mathrm{Pi} = \frac{385{,}67 \times 114 \times \mathrm{Cos}\ 0{,}86 \times \sqrt{3}}{1000}\, kwh = 49{,}7993$$

Determine the value of incoming power at full load, is given in equation (2)

$$\mathrm{Pr} = \frac{\mathrm{P}}{\eta_r}\, kw \tag{2}$$

Pr = incoming power at full load (kW)
P = power (kW)
η_r= efficiency is full load

$$\mathrm{Pr} = \frac{75}{0{,}95}\, kwh = 78{,}9474\, kwh$$

Then calculate the load in% is given in equation (3)

$$\mathrm{Load} = \frac{Pi}{P_r} \times 100\% \tag{3}$$

Where:
Load = Output expressed in% nominal power value
Load = 49,7993/78,9474 ×100% = 63%

Based on [10] After measuring the current and voltage, and calculating the power consumption of motor and then it can be calculated estimation of electrical energy consumption for each motor. It got enter power (Pi) and motor operation period every day is for 7 hours, then estimation of daily electrical energy consumptio is given in equation (4).

DOI: 10.1051/matecconf/201713500054

$$W = P \times t \quad (4)$$

Where:
W = Changes in electrical energy (kWh)
P = power used (kW)
T = time interval (hr)

Motor Load = 75 kW x 63% = 47.3093
$W = P \times t$
$= 49.7993 \times 7$ hours
= 348,5950096 kWh Energy consumption for 1 business day

Reduction of processing time would increase production capacity so that the reduce consumption energy. The following is a calculation of consumption energy in one year :
Known :
- Working time per shift = 7 hours x 3600 = 25200 seconds
- Cost for energy electricity = Rp 1.035,78 Rp/kWh
- Energy (W) = 348,6 kWh
- Cost Energy per day = Rp 361.068

To get the costs incurred by the use of the machine punching can be calculated with the equation below, based on the energy use and energy costs listrik/kWh.

$$\text{Cost Saving} = \text{kWh/tahun} \times \text{Rp/kWh} \quad (5)$$

Tabel 3. Data Result Energy Consumption

	Production Machine Punching Before In Automation	Production Machine Punching After In automation
Total Card Pattern Required	7200	7200
Production time to create	323	84
Time per day / second	25200	25200
Production Results/day	76	300
Total Days needed	95	24
Total Energy required kWh	33116,52591	8366,28023
Total costs incurred	Rp34.301.435	Rp8.665.626

4 Conclusion

Based on theresearch that had been conducted by implementing design green manufacturing method and automation system design on the punching machine, it can conclude as follows:

1. Has done identification the color at the company, and determine improvment area energy and automation system on the punching machine using Image Converter Microsoft Office 2007, CX Programmer v9.1, PLC Omron CP1E, and Wonderware Intouch was implemented at punching machine.
2. Automated system increased production jacquard card, it can produce 300 jacquard card in one day, previously it only produced 77 jacquard card/day. Then to produce 12 patterns, energy consumption of automated punching machines is 8.366 kwh, previously spent 33.116 kwh. So the cost to use of electrical energy consumption to

produce 12 patterns on an automated punching machine can save the cost Rp 1.068.159 / day

References

[1] Yingjie, Z. (2014). Energy efficiency techniques in machining process: a review.

[2] Deif, A. M. (2011). A system model for green manufacturing. *Journal of Cleaner Production* .

[3] Frankie M.C, NG (2006). *Digital Jacquard Textile Design In A Colorless Mode Hongkong*: Universitas Hongkong

[4] Foster, S. T. (2003). Managing quality : an integrative approach . London: Upper Saddle River.

[5] Dornfeld, D. A. (2013). *Green Manufacturing Fundamentals and Applications.* California : Springer.

[6] Groover, M. P. (2001). *Otomasi, Sistem Produksi, dan Computer-Integrated Manufacturing.* New Jersey: Pearson.

[7] Deshpande, V. S. 2014, Home Automation Using PLC and SCADA, Multidisciplinary Journal of Research in Engineering and Technology, Volume 1, Issue 1 (April 2014), 111-118

[8] Rachmat H. Andias Anugrah, Rino. Mulyana Tatang (2015). *EMS-SCADA Design Of Ac Usage On A Building*: Proceeding 8th International Seminar on Industrial Engineering and Management. ISSN : 1978-774X

[9] Agung, Raharjo, B (2014) *Studi Analisis Konsumsi dan Penghematan Energi di PT. P.G. Krebet Baru I.* Malang: Universitas Brawijaya

[10] Shantia, K, (2014) Analisis Pemanfaatan Energi Listrik pada Mesin-mesin Produksi Divisi Pabrikasi Di PT INKA Madiun. Malang: Universitas Brawijaya

Flexibility Improvement on Energy Consumption of Punching Machine Based on Green Manufacturing Method at PT Buana Intan Gemilang

Firda Mawaddah*, *Haris* Rachmat, and *Tatang* Mulyana

School of Industrial Engineering Telkom University, Bandung, Indonesia.

Abstract. These days, Green Manufacturing has established become a new technology for professional engineers and manufacturing workers to be an important part in the business world. Green Manufacturing was created to reflect the new manufacturing paradigm techniques to make something become more eco-efficient by reducing the energy consumption. PT Buana Intran Gemilang is one of Textile Company that produce curtain and prayer rug. In the production this company use a Jacquard punching machine designed to cut a hole in material such as card stock. The machine works manually with the flexibility less than 90 percent that led to the longest production time in meeting the demand, thus causing the electrical energy released for this machine is also very high. On the basis of these problems required a system to improve the flexibility of the machine in order to reduce energy consumption and make the company become greener environmentally friendly. From this research generated an automation system based on green manufacturing method which will be applied on a jacquard punching machine that can increase flexibility machine to be more than 90 percent. This system has saving the electricity consumption of 2092 KWh every making a complete fabric pattern.

1 Introduction

The continued development of the industrial revolution causes of result of the greater energy being used, but instead the availability of energy in the world is actually depleting. Statistics value of energy use in Indonesia 2014 shows that industry sector become the biggest user of electricity energy, followed by household in the second place, business in the third place and another general sector in the last place. Besides that, the use of electrical energy in the industrial sector continued to increase from year to year. PT. Buana Intan Gemilang is one of the textile company in Banjaran, Bandung that produce curtain and prayer rug as its product. PT. Buana Intan Gemilang have several production department, and one of the main part of this company is weaving and preparation department. Punching machine is one of machine that used in preparation department, the function of the machine is to make the jacquard card of the curtain and prayer rug to use in the weaving machine.

Reducing the energy consumption of electricity will impact to all other manufacturing competitive edges, such flexibility. The concept flexibility in manufacturing has become a

*Corresponding author : firda.maw@gmail.com

DOI: 10.1051/matecconf/201713500055

key consideration in the design, operation and management of manufacturing system [1]. By increasing the flexibility the same as make a better manufacturing system. Automation technique is use as one way to control and improve the flexibility of jacquard punching machine. Automation is the technology which can perform a series of process or procedure automatically without human assistance. It implemented with execute a list program of instruction which combine by a control system in perform the instruction [2].

The purpose of this research is improving flexibility of jacquard punching machine using automation system to be able to reduce energy consumption and turn the company to be a green manufacturing company.

2 Jacquard Punching Machine

Jacquard Punching Machine is machine that designed to cut a hole in some material such as paper, metal, or card stock. There are three basic elements of punching device, a punch, a die, and a stripper. The punch is the piece which is driven through the card and cuts the hole. The die serves as a base, supports the card while it is being cut. The stripper serves as a guide for the punch and as a means of stripping the card from the punch after the latter has cut the card.

In PT. Buana Intan Gemilang the purpose of jacquard punching machine is to make the pattern card (jacquard card) that will be represent the pattern of the fabric (curtain or prayer mat). The things that should be noted in this process is the hole making process and the precision location of the hole itself. The good quality of jacquard card is a card that does not have any defect. As increasing demand of textile sector and also supports from the government to develop local textile industries, the company must be able to compete with other local competitor by producing high number of product with high quality, and on the right time. The company usually makes the jacquard card with this punching machine in manual way, so its takes an expensive cost for the energy because the machine running in longer time and also low flexibility because the ability to change jacquard is very slow.

3 Research Methodology

Green manufacturing is a complicated system engineering problem that needs to be surveyed from the view of system engineering, its kind of sustainable development in the field of modern manufacturing. Making the same product using fewer resources and/or energy is a good strategy to make money. In other words, being efficient through preventing waste is both eco as well as money efficiency.

In designing the concept of green manufacturing, implement a system model as a tool to simplify its application [3]. This model system will be applied in case study at PT. Buana Intan Gemilang to be able to assist in applying the concept of green manufacturing so as to improve the flexibility of the punching machine.

Some of benefits of green manufacturing are (1) Controls and reduces material waste in manufacturing cycles. (2) Preserves capital and saves money. (3) Improve productivity and increases cost savings. (4) Helps drive and influence corporate behavior both internal and external to ensure sustainability. (5) Easy adaptability to changing rules in environmental regulations and legislation [3]. Figure 1 shows a step by step flow diagram of the process.

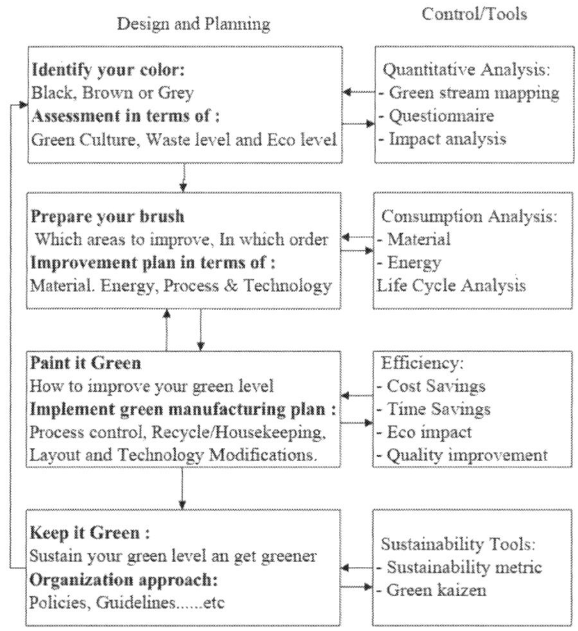

Fig. 1. System model for green manufacturing [4]

- Stage 1: Color Identification

The process of color identification in PT. Buana Intan Gemilang was doing by questionnaire method that distributed to some operator of jacquard punching machine. This process is done to find out the current state of the company, so it can easily determine what actions can be taking. Questionnaires distributed to 5 operators of punching machine that really know the condition of machine and company.

Base on the result of questionnaire it can be concluded that PT. Buana Intan Gemilang in the brown (medium) position, with the energy of the system have bigger percentage better than the material which means it needs an improvement in part of energy usage. The company has not enough energy-saving in the use of electricity but in terms of waste not too polluting because of all good material of raw materials as well as the rest of the result of production does not damage the environment.

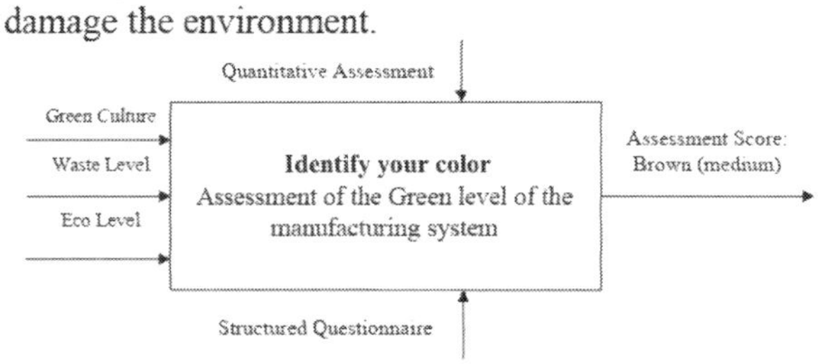

Fig. 2. Model for identification color [4]

- Stage 2: Improvement Object

In this part the process of planning on the improvement is about what will the company done on the way to make the company greener without affecting the results and production planning. The development process of this improvement will be focused on the energy consumption section of the punching machine up to the implementation of a new technology on the machine that can support the energy saving process itself. Jacquard

DOI: 10.1051/matecconf/201713500055

punching machine is a machine that is designed to cut a hole in some materials such as paper, metal or card. Aspects to be studied further from the results of this improvement application are the flexibility aspect of the punching machine.

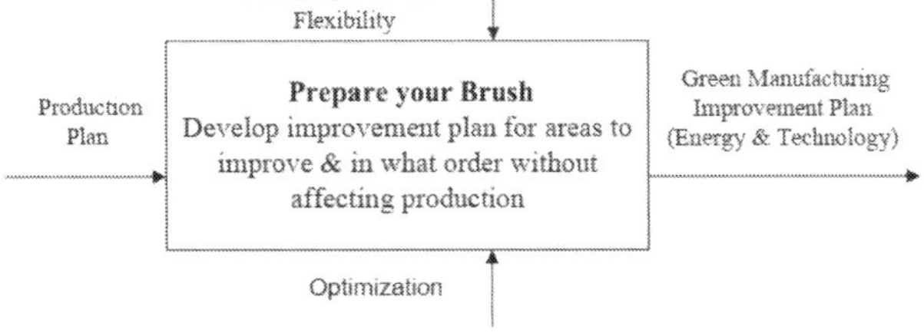

Fig. 3. Model for green improvement [4]

• Stage 3: Green Implementation

The next process is implementation of all the strategies that have been designed on the punching machine. Automation technology that applied to punching machines will be measured based on changes in production time and energy consumption of punching machines. If the automated punching machine has faster production time and less energy consumption with the amount of production that can meet the targets set by PT.Buana Intan Gemilang, the company has succeeded in increasing the color category from brown to green category.

Fig. 4. Frame work automation system [4]

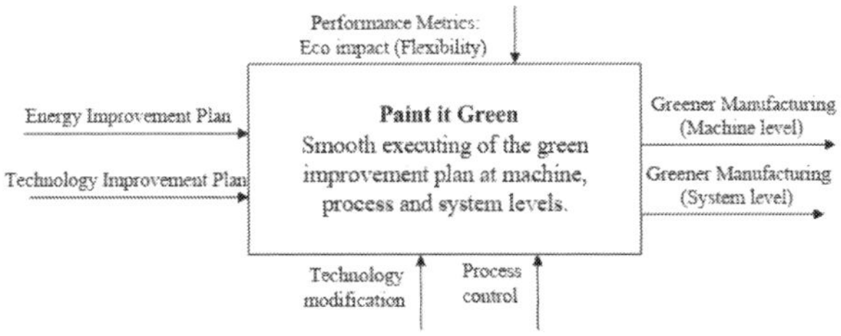

Fig. 5. Model for green implementation [4]

• Stage 4: Keep it Green

The last step is keep it green where green manufacturing realization should have sustainability as in an inherent component in any green planning activity. The expected output at this stage is some green manufacturing policies and guidelines on the operation of the machine. Some trainings conducted to educate operators on how to maintain the achieve improvements and also creation of standard work to sustain improve paint performance.

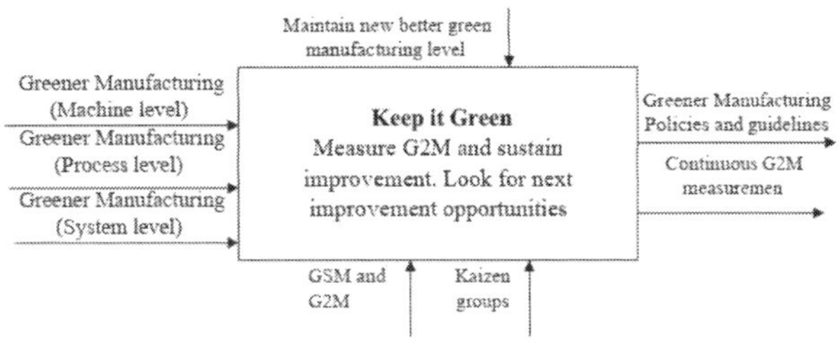

Fig. 6. Model for keep it green [4]

DOI: 10.1051/matecconf/201713500055

3 Results

The Energy consumption on AC Motor punching machine:

Table 1. Specification of AC Motor [14]

Name of Motor Pump ZD1200627	P(kW)	V(v)	I(A)		
	55	380	134,2		
	Phase	V(i-i)	if(A)	Cos θ	μ
	R	387	112		
	S	385	118	0,86	0,95
	T	385	113		

The load is calculated use by power analysis tool and power value at 100% loading. For three phase motor, the step is to determine the input power with the following equation:

$$\mathbf{Pi} = \frac{V \times I \times Cos\,\phi \times \sqrt{3}}{1000}\ kwh$$

$$Pi = \frac{385{,}67 \times 114 \times Cos\ 0{,}86 \times \sqrt{3}}{1000}\ kwh = 49{,}7993$$

Determine the value of incoming power at full load.

$$\mathbf{Pr} = \frac{P}{\eta_r}\ kwh$$

$$Pr = \frac{75}{0{,}95}\ kwh = 78{,}9474\ kwh$$

Then calculate the load in %

$$\mathbf{Load} = \frac{Pi}{P_r} \times 100\%$$

Load = Output power expressed in% nominal power value

$$\mathbf{Load} = \frac{49{,}7993}{78{,}9474} \times 100\% = 63\%$$

To get the cost incurred by the use of a punching machine can be calculated by the equation below, based on energy data used and the cost of electrical energy / kWh.

Cost Saving = kWh/year × Rp/kWh

$$\mathbf{Load} = \frac{49{,}7993}{78{,}9474} \times 100\% = 63\%$$

$$= 49{,}7993\ \text{kwh}$$

$$\mathbf{Pr} = \frac{75}{0{,}95}\ kwh = 78{,}9474$$

$$\text{Load} = \frac{49{,}7993}{78{,}9474} \times 100\%$$

Load = 63%

Motor Load = 75 kW x 63%

= 47,3093

W = P × t

= 49,7993 × 8 hour

= 398,3942967 kWh

Energy consumption for 1 day

Automation systems have been successfully applied to punching machines and have increased the flexibility of the punching machines to match the expected expectations. With automation systems, the production time and setup time of the punching machine can be reduced and the flexibility increases. Reduction of production time and setup time would increase the flexibility so that will reduce the consumption energy of the company.

From the observation it is known:

- Working time per day = 7 hours x 60= 420 seconds
- Cost for energy electricity = Rp 1467,8 Rp/kWh
- Energy (W) = 348,6 kWh
- Cost Energy per day = Rp 511.668

This formula is used to calculate the flexibility of the machine [6]:

$$\textbf{Flexibility of punching machine} = \left(1 - \frac{\text{switching time}}{\text{total work hour}}\right) \times 100\%$$

Switching time = setup time/day (minutes)

Total work hour = work hour/day (minutes)

a. Existing system

- Production time per product = 5.386 minutes
- Production per day = working time / production time

Production per day = 420 / 5.386 = 77 unit

$$\textit{Flexibility of punching machine} = \left(1 - \frac{62}{420}\right) \times 100\% = 85.2\%$$

b. Proposed System

- Production time per product = 1.4 minutes
- Production per day = working time / production time

Production per day = 420 / 1.4 = 300 unit

DOI: 10.1051/matecconf/201713500055

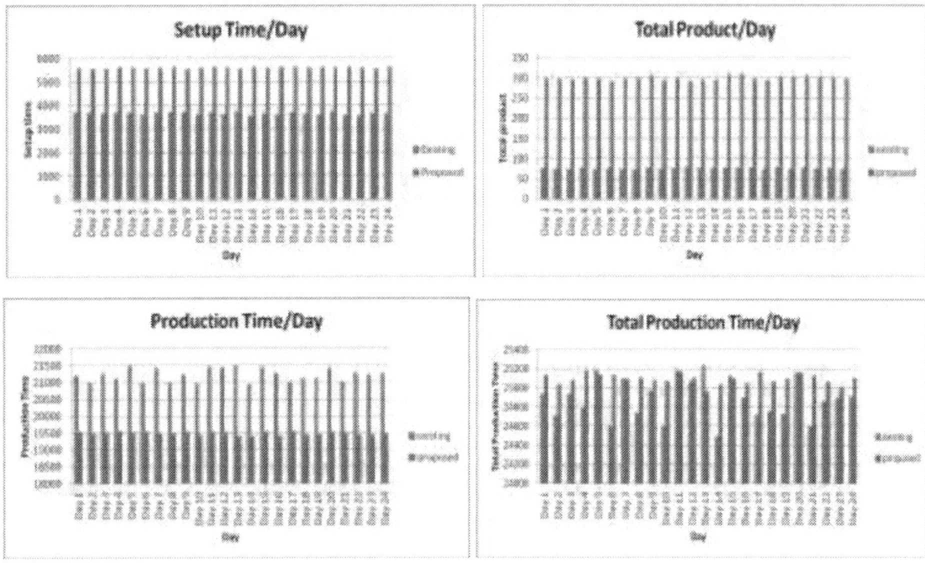

Fig. 7. Comparison of existing and proposed condition

Figure 7 shows the comparison of existing and proposed condition. The comparison is done to see the whether the proposed condition is better or not. Table 4 shows the comparison summary of the production condition.

Table 2. Comparison production condition

	Proposed System (Automation)	**Existing System (Manual)**
Working time/day	420	420
Production time (minutes/unit)	1.08	4.6
Production per day (unit)	300	77
Production per month (unit)	7195	1845
Flexibility of machine	94.2 %	85.2 %

Base on that calculation it can be conclude that automation system can increase the flexibility of machine as well the green manufacturing. The impact has successfully implemented and automation system has running as expected for increasing the flexibility of the jacquard punching machine. This also results in companies to be toward greener manufacturing, where companies can save more energy. The comparison for energy use can see below.

Table 3. Comparison of the use of electric energy

	Proposed System (Automation)	Existing system (Manual)
Total production (unit)	600	600
Total day	2	8
Electricity/day (KWh)	348.6	348.6
Total Electricity (KWh)	697.2	2788.8
Cost / KWh	Rp. 1,467.80	Rp. 1,467.80
Cost	Rp. 1,023,350.16	Rp. 4,093,400.64

Base on the calculation costs incurred by companies to produce a number of jacquard cards with punching machines that have been applied green manufacturing methods require a cheaper cost than before the automation. It can be concluded that with applied green manufacturing method using automation system can save cost of production and also save the usage of energy that make the company to be green manufacturing company.

4 Conclusion

The automation system was adopt to the punching machine to make the production time faster so it can significantly increase the flexibility and decrease the energy usage of the punching machine. Based on the analysis result of the system that has been designed and implemented, was conclude as follow:

1. The flexibility of punching machines increased 9 percent from before it was automated because the setup time was reduced.
2. The processing time of jacquard card making is 240 seconds faster than before it was automated due to decrease in setup time.
3. The machine setup time is 38 minutes faster than before it was automated because of the automation technology that has been applied to the machine.
4. The energy that can be saved to create a whole pattern of curtain fabric (600 jacquard cards) is 2092 Kwh.

So the company's color level increased to green after the adoption of the automation system and the company can reduce the energy consumption significantly.

References

1. A. K. Sethi, & S. P. Sethi, Flexibility in Manufacturing: A Survey. Kluwer Academic Publishers. (1990).
2. A. Gilchrist, Industry 4.0 the Industrial Internet of Things. Thailand: Apess. (2016).
3. P. S. Acharya, D. J. Vadher , & D. G. Acharya, A Review on Evaluating Green Manufacturing for Sustainable Development in Foundry Industries. International Journal of Emerging Technology and Advanced Engineering. (2014).
4. A. M. Deif, A system model for green manufacturing. Journal of Cleaner Production. (2011).
5. C. J. Bartodziej. The Concept Industry 4.0: An Empirical Analysis of Technologies and Applications in Production Logistics. Springer. (2016).
6. P. Chandra, & M. M. Tombak, Models for the evaluation of routing and machine flexibility. Decision Craft Analytics. (1991).
7. D. A. Dornfeld, Green Manufacturing Fundamentals and Applications. California: Springer. (2013).
8. M. P. Groover, Automation, Production System and Computer-Integrated Manufacturing. New Jersey: Pearson Education. (2001).
9. I.D. Paul. A review on Green Manufacturing: It's important, Methodology and its Application. (2014).
10. Y. Liu, An investigation into minimizing total energy consumption and total weighted tardiness in job shops. (2013).
11. W. MacDougall, Industrie 4.0. Germany Trade & Invest, 6. (2014).

12. K. E. Mineral, Statistik Ketenagalistrikan. Jakarta: Direktorat Jendral Ketenagalistrikan. (2014).
13. I. Ohno, Automatic Card Punching Machine for Jacquard Machine. (1967).
14. B. A. Raharjo, U. Wibawa, & H. Suyono, Studi Analisis Konsumsi dan Penghematan Energi di PT. P.G. Krebet Baru I. (2013).
15. D. R. Schlaepfer, & M. Koch, Industry 4.0 Challenges and solutions for the digital transformation and use of exponential technologies. Deloitte, 4. (2015).
16. J. W. Webb, *Programmable Logic Controllers Principles and Applications*. America: Prentice Hall Inc. (1999).
17. Z. Yingjie,. Energy efficiency techniques in machining process: a review. (2014).

MATEC Web of Conferences **135**, 00056 (2017) DOI: 10.1051/matecconf/201713500056
ICME'17

Inventory Policy Determination for Raw Materials in ILY Pharmaceutical using Periodic Review (R, s, S) and Periodic Review (R, S) Method to Minimize Total Inventory Cost

Puteri Mely Permatasari[1,*], *Ari* Yanuar Ridwan[1], and *Budi* Santosa[1]

[1]Industrial Engineering, Telkom University, Bandung, Indonesia

Abstract. ILY Pharmaceutical is one of the pharmaceutical industry in Bandung that produces various type of medicines. In managing the raw materials for production, ILY Pharmaceutical is not using any standardize calculation for the inventory policy of the raw materials. That condition caused the raw materials to be overstocked and made the total inventory cost to be higher. The objective of this research is to minimize the total inventory cost in ILY Pharmaceutical by giving an inventory policy proposal. Based on the classification of the raw materials using ABC analysis, two inventory policy will be used. Periodic review (R, s, S) for A category and periodic review (R, S) for B and C category. The proposed inventory policy able to reduce the total inventory cost by 42% from $12.194,98 to $7.029,29. The reduced cost is mainly caused by the quantity of the raw materials that is stored in the warehouse is smaller than the existing condition. Even though the frequency of the order is increased thus increasing the order cost but the quantity ordered is smaller because in the proposed condition, there are a maximum inventory level so, the quantity ordered cannot exceed that level.

1 Introduction

ILY Pharmaceutical is an industry in Bandung that produces various type of medicines such as caplet, tablet, capsule, dry syrup, syrup and external medicine fluids. The medicines then will be distributed to fulfil the demands of all health units associated with Indonesian Army all over Indonesia. One of the problems faced by ILY Pharmaceutical is regarding the inventory of raw materials. The stock of raw materials is always bigger than the demand thus causing overstock of the materials. ILY Pharmaceutical is procuring the raw materials quarterly and always order the raw materials in bulk to the supplier to cope with the fluctuating demand of the raw materials. Up until now, ILY Pharmaceutical does not have an exact calculation for the reorder point and maximum inventory level that is needed to help managing the inventory of raw materials in the warehouse. So, the raw materials ordered is done merely on the hunch of the personnel or when there is a stock out. This

* Corresponding author: puterimely@gmail.com

conditions then further caused the inventory of raw materials to be out of control. Fig. 1 shows that the gap of stock and demand of the raw materials is very wide.

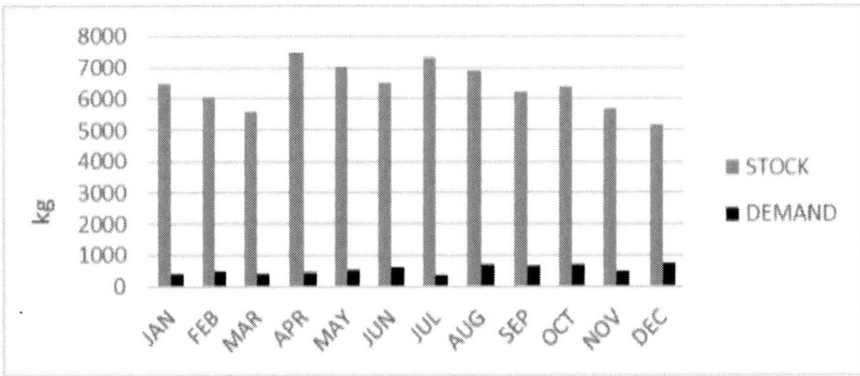

Fig. 1. Comparison of Raw Materials Stock and Demand in 2016

The overstock condition can further cause of loss to the company because it can make the holding cost higher. Furthermore, the probability of the raw materials that will be expired before used will be higher also. The expired materials will bring negative to the company because the materials cannot be stock anymore and must be disposed. In 2016, the total inventory cost has reached \$12.194. The overstock conditions then must be solved so the raw materials inventory can be well controlled and the total inventory cost can be minimized as well. Therefore, further research is needed to solve the problem by using inventory policy that will help determining the policy for the raw materials. So, the objective of this research is to minimize the total inventory cost in ILY Pharmaceutical by giving an inventory policy proposal.

2 Inventory Policy

Inventory is defined as an idle resource which waiting for further process such as production in manufacture system, marketing activity in distribution system or consumption activity in household, office etc. [1]. Inventory is helpful to anticipate the supply and demand uncertainty [2]. Based on the value addition or stage of completion, inventory can be classified into three types, which are raw materials, work-in-progress, and finished good [1]. In determining the inventory policy, there are some relevant cost that can be classified into four categories which are procurement costs, inventory holding costs, shortage costs, and obsolescence costs [3].

The items that are managed in the inventory can be classified based on the investment level in one period [4,5]. The classification is done using ABC Analysis. Using ABC Analysis, the items will be classified into category A, B, and C. Based on the classification results then the inventory policy that can be used are as follows [5]:

Table 1. Inventory Policy

Category	Continuous Review	Periodic Review
A	(s, S)	(R, s, S)
B	(s, Q)	(R, S)

For the C category, the usual EOQ or the same inventory as category B can be used [5]. In this research, periodic review inventory policy will be used.

The previous research conducted by Dayanti [6] shows that by classifying the raw materials using ABC Analysis and then use different inventory policy based on the category resulting in a lower total inventory cost. A category is using continuous review (s, S) while the B and C category is using continuous review (s, Q). Other research conducted by Arismawati [7] regarding the inventory policy of accessories part by using periodic review (R, s, S) shows that it can minimized the total inventory cost. Also, referencing from the research done by Merdalina [8], by using two periodic review policy of (R, s, S) and (R, S) can minimize the total inventory cost of the lubricant oil in PT. NYZ. From the previous research, it shown that both the continuous and periodic review policy can minimize the total inventory cost.

To compare between the continuous and periodic review policy, Setyaningsih [9] conducted a research in Public Hospital Bandung. The results shown that both the continuous and periodic review has their own advantages and disadvantages. Due to the characteristic of the items which has a low demand and low price the periodic review policy gave a lower total inventory cost than the continuous review.

3 Research Methodology

The data needed for this research were collected by doing observation and interview with the personnel in ILY Pharmaceutical. The data collected in this research are lead time, holding cost, order cost, shortage cost and raw materials demand. The data then used for calculating the parameter of the inventory policy. The total inventory cost of the proposed inventory policy then will be compared with the existing total inventory cost to see whether the proposed condition is better or not. The steps to solved the problem is described in conceptual model. Research methodology explains about the relationship between variables and the method that support the research and can be seen in Fig 2.

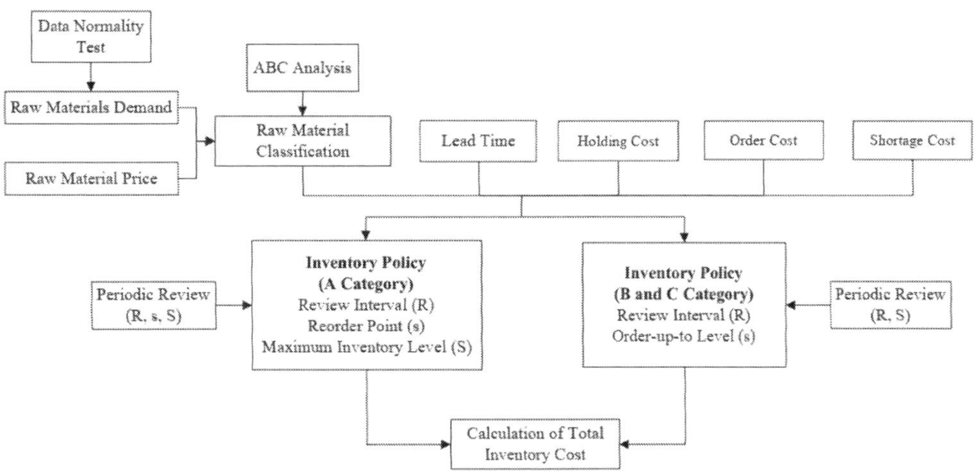

Fig. 2. Research Methodology

Looking at the previous research and the existing condition of ILY Pharmaceutical that is reviewing and procuring the raw materials quarterly so, in this research, periodic review (R, s, S) system and periodic review (R, S) system is suitable to be used for the proposed inventory policy. Periodic review (R, s, S) is when the inventory level at or below s then the

replenishment is done until it reaches the maximum inventory level (S) for every R period. If the inventory level is above s, then there is no ordering even though it already reaches R period. The algorithm formulation for calculating the reorder point (s) and maximum level (S) are as follows [5]:
The first step is calculating Q_p and S_p value.

$$Q_p = 1,3(B_3^{0,494})(A/vr)^{0,506}[1+(\sigma^2_{R+L}/x_{bar}^2)]^{0,016} \quad (1)$$

$$S_p=[0,973(R+L)D]+\{\sigma_{R+L}[(0,183/z)+1,603-(2,192z)\} \quad (2)$$

Where,

$$z = \sqrt{(Q_p\, r)/(B_3)} \quad (3)$$

Step 2. If $Q_p/x_{barR} > 1,5$ then s and S can be obtained by:

$$s = S_p \quad (4)$$

$$S = S_p + Q_p \quad (5)$$

If $Q_p/x_{barR} < 1,5$ then continue to the next step.

$$P_\mu(k) = r/(B_3+r) \quad (6)$$

Where the value of k is obtained from the table.

$$S_0 = x_{bar(R+L)} + k\sigma_{R+L} \quad (7)$$

Calculate the reorder point (s) and maximum stock (S)

$$s = minimum\ \{S_p, S_0\} \quad (8)$$

$$S = minimum\ \{S_p, Q_p + S_0\} \quad (9)$$

Notations,

R	: time interval between order
D	: raw materials demand
L	: lead time
r	: holding cost
B_3	: shortage cost
A	: ordering cost
v	: raw material price
σ_{R+L}	: standard deviation of demand during the review and lead time
$\hat{X}_{R+L}$	: average demand during the review and lead time
$\hat{x}_R$	: average demand during review time
Q_p	: order quantity perspective
$s_p + Q_p$	: maximum limit
S_0	: minimum unit
$p_{\mu\geq}(k)$	: probability of inventory shortage

In periodic review (R, S), the order is done in every R period to increase the inventory level up to S. The calculation for (R, S) is almost the same with (s, Q) but with some transformation [5]. The transformation are as follows:

(s, Q) (R, S)
s S
Q DR
L R+L

Calculate,

$$Gu(k) = (x_{bar(R+L)/} \sigma_{R+L})(r/rB_3) \quad (10)$$

Where,

$$S = x_{bar(R+L} + k\sigma_{R+L} \quad (11)$$

4 Results and Discussion

4.1 ABC Analysis

The ABC analysis classify the raw materials into three categories which are A, B, and C. From the analysis, from a total of 54 items, there were 12 items in A category, 11 items in B category and 31 items in C category. The inventory policy of every items then will be calculated based on the category. The review period is predefined which is 3 months.

4.2 Periodic Review (R, s, S)

After the classification, the parameter of periodic review (R, s, S) and periodic review (R, S) are calculated. In Periodic Review (R, s, S), the parameter that will be calculated is the reorder point (s) which is the where the order will be done if the inventory level is below that point. The second parameter is the maximum inventory level. The raw materials that will be calculated using this system is the A item. The calculation of Periodic Review (R, s, S) for raw materials MAT25 will be used as a sample calculation. It is known that:

R = 3 months,
L = 1 month,
D = 57,543 kg/year,
σ = 2,19 kg,
v = \$188,03/kg,
A = \$1,96/order,
R = \$2,05/unit/year,
B3 = \$11,36/kg short.

From the calculation using the equation (1) to (8), it is obtained that the reorder point (s) of MAT25 is 19,831 kg and the maximum stock (S) is 19,926 kg.

4.3 Periodic Review (R, S)

In Periodic Review (R, S), the order will be done every R period to make the inventory level reach the up-to-level S again. The raw materials that will be calculated using this system is the B and C item. The calculation of Periodic Review (R, S) for raw materials MAT01 will be used as a sample calculation. It is known that:

R = 3 months,
L = 1 month,
D = 98,245 kg/year,
σ = 6,958 kg,

r = \$2,05/unit/year,
B3 = \$2,86/kg short.

From the calculation using the equation (9) and (10), it is obtained that the order-up-to-level of MAT01 is 65,497 kg.

4.4 Total Inventory Cost

From the determination of periodic review (R, s, S) and periodic review (R, S) inventory policy, the existing and proposed cost then will be calculated and compared.

Table 3. Total Inventory Cost

Category	Existing	Proposal
Order Cost	\$ 12.087,10	\$ 6.605,64
Holding Cost	\$ 107,88	\$ 423,65
Shortage Cost	\$ -	\$ -
Total Inventory Cost	\$ 12.194,98	\$ 7.029,29

The proposed inventory policy has given a lower holding cost than the existing condition. The proposed inventory policy reduces the cost by 45%. The reduced cost is caused by the quantity of the raw materials kept in the warehouse is smaller than the existing condition. Furthermore, the quantity of the raw material that is purchased is smaller because in the proposed inventory policy, the maximum inventory level and the reorder point is determined so, ILY Pharmaceutical only done the replenishment if the inventory level has reached or below the reorder point and only replenished it up to the maximum inventory level.

The proposed inventory policy increase the order cost by 293% that was caused by the total order frequency of the raw materials was more than the existing condition. In the existing condition the replenishment for the raw materials is not done frequently but with a big quantity while in the proposed inventory policy, the order is done almost in every period but with smaller quantity.

Both the existing and proposed condition shows that there is no raw material shortage. The proposed inventory policy shown that with determining the parameter of periodic review (R, s, S) and periodic review (R, S), it can still fulfil the demand of the raw materials.

From the calculation results, it is obtained that by using periodic review (R, s, S) and periodic review (R, S), the inventory of the raw materials can be reduced because the item ordered is smaller in quantity.

5 Conclusion

The problem faced by ILY Pharmaceutical is the overstock of the raw materials that make a high total inventory cost. From the problem faced by ILY Pharmaceutical, this research then focused to minimize the total inventory cost in ILY Pharmaceutical by giving an inventory policy proposal. From the proposed inventory policy, the review period, maximum inventory level and reorder point were obtained. The proposed inventory policy able to reduce the total inventory cost by 42% from \$12.194,98 to \$7.029,29. The reduced cost is mainly caused by the quantity of the raw materials that is stored in the warehouse is smaller than the existing condition. Even though the frequency of the order is increased thus increasing the order cost but the quantity ordered is smaller because in the proposed condition, there are a maximum inventory level so, the quantity ordered cannot exceed that level.

References

1. S.N. Bahagia, Sistem Inventori, Bandung: Penerbit ITB (2006)
2. C. Bozarth, and R. Handfield, *Introduction to Operations and Supply Chain Management* (Pearson, 2008)
3. G. Ghiani, G. Laporte, and R. Musmanno, *Introduction to Logistics Systems Planning and Control* (John Wiley & Sons Ltd, West Sussex, 2004)
4. R. Nadkarni, and A. Ghewari, International Journal of Engineering Business and Enterprise Application, **1**, 24 (2016)
5. E.A. Silver, D.F. Pyke, and R. Peterson, *Inventory Management and Production Planning and Scheduling* (Wiley, New York, 1998)
6. S. Dayanti, A.Y. Ridwan, M.D. Astuti, *Telkom University e-Proceeding of Engineering*, **3**, 2982 (2016)
7. P. Arismawati, A.Y. Ridwan, and B. Santosa, *Telkom University e-Proceeding of Engineering*, **2**, 4877 (2015)
8. Merdalina. A.Y. Ridwan, and B. Santosa, *Telkom University e-Proceeding of Engineering* (2016)
9. S. Setyaningsih, and M. H. Basri, International Journal of Innovation Management and Technology, **4**, 2 (2013)
10. M. Gebicki, E. Mooney, S. Chen, and L. M. Mazur. Health Care Manag Sci, **17**, 215-229 (2013)

Issue in Information Sharing of Halal Food Supply Chain

Nor Ratna Masrom[1], *Raja Zuraidah* Rasi[2], *Badru At Tamam* Daut[3]

[1]Faculty of Technology Management and Technopreneurship, Universiti Teknikal Malaysia Melaka, Melaka, Malaysia.
[2, 3]Faculty of Technology Management and Business, Universiti Tun Hussein Onn Malaysia, Johor, Malaysia.

Abstract. Information sharing serves as an essential approach for the companies and antecedents of supply chain integration. Nowadays, with the highly competitive in halal food market, information sharing has become more conceivable. Furthermore, information sharing in supply chains has become more efficient by having meaningful relationship with the members of halal supply chain. There is a lack information regarding information shared within members in halal food supply chain literature. The information needed in order to shed the light of how companies nowadays can be more competitive. The purpose of this study is to investigate to what extent the information can be shared among the members of the halal food supply chain. This study elaborates the barriers of information sharing leading to enhanced supply chain integration among enterprises, as a result. The interview with four managers from halal certified food manufactures has been deployed to get rich data about the information sharing. The finding shows that most of cases halal certified food manufacturers has low communication with their suppliers. Trust is the key enablers within the members of halal food supply chain.

1 Introduction

Today's world offers opportunity to the organizations. Currently, the rising of Muslim population in the world has made the halal sector as one of promising profit. Malaysia is one of the halal food produces in the world. In order to survive and compete in halal sector, halal certified food manufacturing sector strongly needs to create, share and disseminate up-to-date and appropriate knowledge and information. A halal supply chain should connected by flows of information, finance and material by the suppliers, manufacturers, retailers, distributors and customers. Information sharing among supply chain partners creates information flow within supply chain management and this enables supply chain partners to make decision effectively. Information sharing is an important element that reflects cooperation in supply chain management, according to Li, Yan, Wang and Xia (2005). According to Simatupang and Sridharan (2002b), information sharing is "the ability to see private data in a partner's systems and monitor the progress of products as they pass

*Corresponding author: norratna@utem.edu.my

through each process in the supply chain. This activity includes monitoring (data capturing), processing, and dissemination of customer data, end-to-end inventory status and locations, order status, costs-related data, and performance status". These authors have argued that information sharing ensures that partners in the supply chain are able to fulfill demand within shorter order cycle times based on the shared information. Information systems improve the quality of forecasting, decision, planning and control, mitigate the risk caused by the external market environment and internal factors of the supply chain, and consequently improve the efficiency and effectiveness of supply chain operations.

However, there are always doubt about sharing information with trading partners due to the perceived complexities, risks, and costs (Kembro and Näslund, 2014). Willingness to share commercially sensitive information with trading partners is crucial if end-customers are to be completely satisfied and the total costs to the supply chain minimized (Ellinger et al., 2012; Yu et al., 2013).

Hence, an understanding of managers' perceptions of the information shared, the issues and to what extent information sharing is key to gaining deeper insights into the information- sharing process.

Tan et al 2015, said that information sharing will increase profit from reducing uncertainty and improve efficiency and effectiveness. Halal food sector is very competitive and it is asking halal certified food manufacturers to collaborate with other supply chain members in order to sustain as a world leader of halal food producer. The specific research questions are

RQ1. What information is shared across supply chain interfaces?

RQ2. What are the issues related to information sharing in halal food supply chain?

RQ3. To what extent that information can be shared in halal food supply chain?

2 Literature Review

Information sharing within business units and across supply chain partners such as suppliers and other strategic alliances is essential. Information sharing develops three major linkages known as the supplier linkage, internal linkage, and customer linkage. Integration through an effective and efficient information flow will eventually lead to a better performance of the firm and total supply chain to (Narasimhan and Kim, 2002; Palsson and Johansson, 2009; Lee et al., 2007). Past studies (Du, 2007; Gunasekaran and Ngai, 2004; Kim and Narasimhan, 2002) reported positive relationships between the level of information flow integration and performance. As such increasing the level of integration and information sharing among the supply chain members is necessary to improve the effectiveness of the supply chain. (Sezen, 2008; Trkman et al., 2007).

Halal food supply chain is a process of managing the procurement, movement, storage and handling of material parts, livestock, semi-finished or finished inventory both food and non-food, and related information and documentation flows through the organization and the supply chain in compliance with the general principles of Sharia law.

Muslims represent an estimated 23% of the global population or about 1.8 billion consumers with an average growth rate of 3% per annum. If the trend continues, Muslims will be 26% of the world's total projected population of 2.2 billion in 2030.

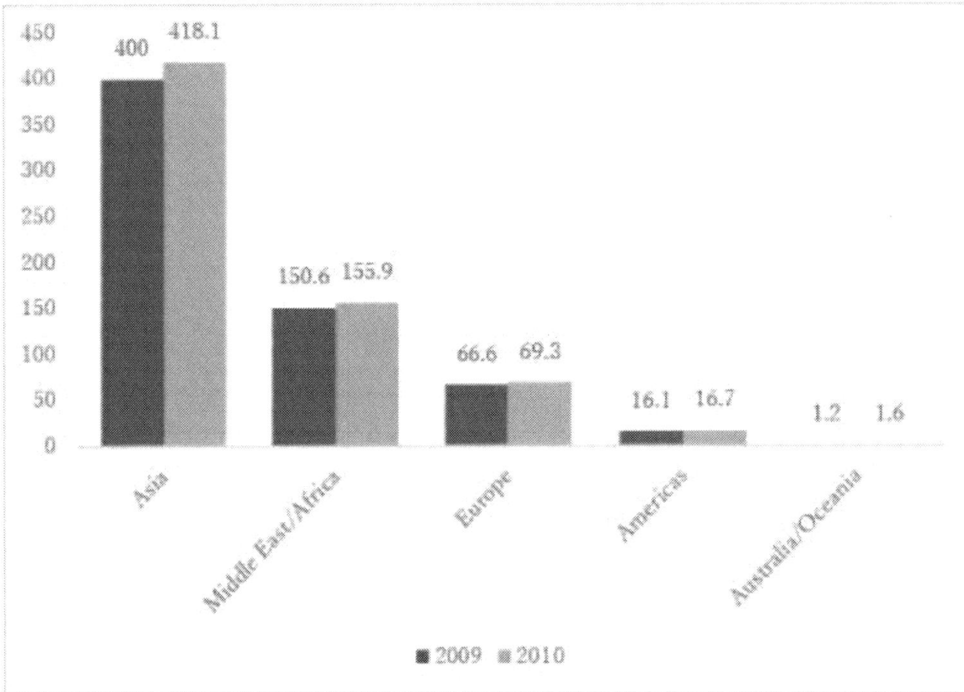

Figure 1: Global halal food market size by region

Source: World Halal Forum 2009 Post Event Report and the 6th World Halal Forum Presentation

The number in Figure 1 shows that halal industry is one of the fastest growing markets in the world. Halal market is not limit to Muslim only but it has been accepting by the non-Muslim. The demand for the halal certified products has gain popularity among them due to high quality, safe and ethical products. Halal food market has been estimated around USD667 million and represents 20 % of the entire global food industry. Many economists believe that based on four prevalent trends. Firstly, Islam is now the fastest growing religion in the world thus it will make big global demand for halal products. The annual growth in consumption of halal food is estimated at 16%. Secondly, the growing trend of consuming. Next the halal consumer power has been increasing because of the increasing number of Muslim and rising of their disposable income. Lastly, the increasing of awareness of consuming halal food among Muslims

Halal food is becoming an increasing part of the Western diet and has become a multi-billion-dollar global industry involving multinationals like Tesco, Unilever and Nestle who have aggressively expanded their halal-certified product lines. Halal is originated from Arabic word which means lawful and permissible according to Islamic Sharia law which refers to the teachings of Al-Quran and exemplary conduct of Prophet Muhammad (Peace and blessings of Allah be upon Him) as quoted in Hadith. The opposite of Halal is Haram which means forbidden, Roland (1996).

3 Methodology

This research was using a multiple case study approach (Yin 2003) to help explain manager perceptions of the challenges and security risks involved when exchanging information with other supply chain actors, tested in the first phase. Because the goal of the second, qualitative, phase was to explore and elaborate on the results from the first, quantitative, phase of the study (Creswell et al. 2003) information sharing were significant predictors of operational performance among halal certified food manufacturers in Malaysia. This qualitative phase is using multiple case study. Where case study is an exploration of a bounded system or a case over time through detailed, in-depth data collection involving multiple sources of information and rich in context (Merriam 1998). This method is an ideal to find deeper knowledge experience of practitioners in these emergent fields (Walsham, 1995). This phase has more than once case and the analysis involved two levels: within each case and across the cases (Stake 1995;Yin 2003).

The sample is purposive sampling of firms. Purposive sampling able to meet the objective of the research due to relevant characteristics for the research questions (Bryman and Bell, 2011). For the data collection, this research has collected the data from four participants. They are managers in halal certified food manufacturers in Malaysia that has involved in survey before. They have been selected because

1. They are responsible in exchanging information with supply chain members
2. They understand the process of halal food supply chain management in Malaysia

Furthermore, the purpose of selecting these respondents because to capture their experiences and thereby increase the validity and reliability of findings. And it is most likely to be the most informative sources.

It is focused on the data exchange between a case study and its main suppliers and retailers/distributors. Besides that, this phase also tend to understand the complex phenomena and contextually rich settings, as proven the current state of knowledge of each discipline being characterized as *"at an early formative development stage and subject to rapid change"* (Zsidisin, 2003). Table 1 provides a case company overview

3.1 Data Collection

This research used sources for collecting the data that will provide the richness and the depth of the case description (Stake 1995) including qualitative interviews triangulated with secondary sources. The participants has vast experience and knowledge of supply chain-related operations in halal food in Malaysia. Each interview usually has three stage of questioning as suggested from Seidman 2006. In this research the three stages are; description of information shared among the members in halal food supply chain, description of issues in information sharing and description of to what extent that information should be shared in halal food supply chain. Fieldwork has been done before the interviewing process. This is to understand the situation of the halal food activities and increase the validity of the research. The pilot test has been deployed to help to test the aspects of a final research. One manager from halal certified food manufacturer was voluntarily to see the clarity of the questions.

After the pilot, the appointment has been made before hand to set time and location of the interview process. All of the respondents preferred their office. The choice of location is really important because it is helping to get more information from the interview process. The time length of interviewing was between 35-75 minutes. The use of semi-structured interview and follow up questions has encouraged the flexibility in interview procedure. This procedure helps the research to get more information that relates to the objective of the central construct (Bryman and Bell, 2011). At the same time, the secondary sources has been referred. The secondary source including the including standard operating procedures and mission statements of each the case company.

Table 1: A case company overview

Case (company)	Function of respondent	Interview duration (min)	Date of interview	location	products
Manufacturer 1	Manager	75	20 March 2017	their office	noodles
Manufacturer 2	Managing director	40	23 March 2017	their office	Coco products
Manufacturer 3	Managing director	35	25 March 2017	their office	Virgin coconut oil
Manufacturer 4	Sales manager	40	27 March 2017	their office	Seafood products

3.2 Data analysis

The data has been recorded, transcribed and coded as proposed by Yin 2009. Yin also emphasizes that the analysis of the qualitative data utilized an explanation building process proposed. Then it will develop the finding through a series of iterations. Every participants were assigned to a generic name as in Table 1 and number to preserve anonymity.

Data analysis started with within-case analysis. This analysis to figure out description of information shared among the members in halal food supply chain, description of issues in information sharing and description of to what extent that information should be shared in halal food supply chain. Then followed by cross-case analysis to identify similar/contrasting themes and patterns. The data also has been verified through triangulating different sources of information and member checking (Creswell 2002).

4 Data Findings

RQ1. What information is shared across supply chain interfaces?

Table 2: Summary of information shared

Case company	Types of information shared
Manufacturer 1	*Place and received orders from supplier, product related information, some forecasting, delivery schedule*
Manufacturer 2	*Requirement for materials, forecasting, product related information, sales history*
Manufacturer 3	*Feedback on products quality and price, requirement for materials, Place and received orders, some forecasting*
Manufacturer 4	*Marketing-related information, some forecasting, Place and received orders*

Table 2 above shows the range of information that is being shared with the members in halal food supply chain in Malaysia. Most cases they shared is operational data such as delivery schedule, requirement for materials, some forecasting and feedback on products.

Most of the halal certified food manufacturers in Malaysia shared the information of their orders with their main suppliers. As a manufacturer 1 said

"We just place and received orders from supplier but not further communication. We specify our order to them specifically, the expected time arrival and quantity of the materials. When they send the wrong one, we will reject the orders"

Forecasting is very important in food supply chain because food is perishable product. Halal certified food manufacturers get the forecast from their customers (retailers/distributors) and shared to their supplier so that the supplier will prepare and deliver the raw material on right time and right quantity. Manufacturer 2 mentioned

"Our distributors always tell us the quantity that we are supposed to produce next week. Because our distributors are the closest to customers and they have been in this market for long period of time"

Due to stringent policy in halal food production, relevant information in the order is the policy that control the process and the quality of the process. As manufacturer 3 articulated

"In the order, we told them that we are producing halal food. So they must also comply with MS1500 procedure and their certificate is not expired"

To summarize, within case analysis emphasizes the types of information shared to every case company has been figured out and supported by Huang et al (2003). Huang et al (2003) specified that there are six categories of production information are often shared: product, process, resource, inventory, order and planning. Specifically, most of cases in halal certified food manufacturers shared process information with their main supplier. The process information includes lead time, the policy that control the process and quality of the process. This can be shown from the process of ordering and receiving materials. The ordering starts with the placement of the order and supplier agree to supply at certain amount on specific date. Besides the quantity and time, the relevant information are lead time in ordering and delivering the raw material and their expected quality.

Meanwhile, most of the halal certified food manufacturers in Malaysia shared order information with their retailers and distributors. They have better relationship with their retailer/distributor compare with supplier. The demand information flow from end customers to the supplier that creates forecasting and sales history. Lee et al (1997) showed that demand has an impact on the bullwhip effect. So, if they get the right information sharing the bullwhip effect will be decreased.

RQ2. What are the issues related to information sharing?

Table 3: Case by case summary of the issue information sharing

Case company	Issues
Manufacturer 1	Supplier does not care about us. Supplier just send order
Manufacturer 2	Supplier sometimes send the wrong products and it delayed my production process
Manufacturer 3	Supplier did not send the product on time especially in peak season due to high demand from competitors
Manufacturer 4	Suppliers do not share information about the stock level

The Table 3 above summarizes the interview data describing the issues related to information sharing. Most of respondents concern about their relationship with their suppliers. They do not have meaningful relationship with their suppliers. Manufacturers 1 mentioned as below

"We called our supplier many times to place in order but they hardly pick up my calls, I just place order and they send my order"

Furthermore, without meaningful relationship the uncertainty from the trading partner will result to inconsistent supplies in customer side as stated by manufacturers 2

"The distributor change their preference supplier without telling us…"

In order to secure their position, they have to deliver the right product on the right time. They have to guarantee their quality and freshness of the food. As manufacturer 1 said

"By hook or by crook, I have to fulfill the demand from my customers, if not, my business will be doomed"

In summarization, the main issue is to have meaningful relationship among the members in halal certified food manufacturers in Malaysia. With meaningful relationship, the uncertainty will be reduced. In addition, the information flow within halal food supply chain members also reduce the impact of bullwhip effect.

RQ3. To what extent that information can be shared?
The willingness of information sharing is depends on trust. As articulated by manufacturer 1,

"Trust is very important. If the distributor does not trust me, they will not share the information and business will face difficulty"

"Do you want to share your information without trusting them? I rather not"

The above statement has been mentioned by manufacturer 4, again shows that trust is very important in information sharing. If the trading partner agree to share their information, they have to restructure their information to be available in making good decision. Even though, they are willing to share their information but there are few things they still keep confidential if they have unique special process flow to produce their products or secret ingredients that can differentiate their product from others.

Mentioned by manufacturer 4

"I told my customer my ingredients and where I get my materials. However, I kept the process of how to produce my product because that is my marketing value"

Stated by manufacturer 3

"My kept my secret ingredients but do not worry, everything is halal compliance"

In summary, trust plays an important roles to what extent that the information can be shared among the members in halal food supply chain. The trust can be built through personal relationship and long-term collaboration that involved honesty. However, even though there is trust but if they have special process flow or special ingredients they will not share it to the trading partners.

5 Discussion

The members of supply chain place order decision based on the order received from downstream and member's own policy and forecast as in Huong et al (2003). This phenomena also happened in halal certified food supply chain and extend the literature by showing the information that shared in supply chain.

In line with finding from I˙pek Kocog˘lu et al.(2011),this research result shows retailer/distributor shared the information regarding demand forecast data with halal certified food manufacturers in Malaysia. The accurate demand forecast helps them to plan their inventory and distribution model.

The sharing information with supplier is also vital as mentioned in Zhang 2000 for the manufacturers to be highly competitive. However, halal certified food manufacturers has showed low information sharing with supplier. The uncertain relationship will hurt their business and hard for them to fulfil the demand from customers

They aware that they should have trust-based, transparent and strong relationships with their trading partners in order to share the information. They know that with useful information, they will be more competitive. However, there are few exceptional in the information sharing especially the information that gives advantage over competitors. Thus this research has added the literature by looking how trust should be encourage in halal food supply chain but there is still few information has to be kept.

6 Conclusion

Generally, the strategic relationships among members in supply chain, ought to be considered as the linkages constituting and sustaining a long-term common unity. Trust is the central goal of willingness to share the information. The finding shows that many cases in halal certified food manufacturers has very low communication with their supplier and low intensity of certainty with customer. They do not have meaningful relationship with the members in halal supply chain. Hence for improvement of information sharing, this research suggests the encouragement of using IT infrastructure by implementing real-time data transaction throughout the entire supply chain. Besides that, they should allow long-term, integrity and trust relationship with trading partners in halal food supply chain. This also show how halal certified food manufacturers try to have close relationship that are can be said as trustworthy partner. Setting internal and external meeting to review the relationship with trading partners in halal food supply chain.

This research is an explanatory of empirical data which required rich description of case information. The limitation of this research is that the sample is not entirely represent of all stages and business operation conducted in halal food supply chain. The data was gathered from four managers of halal certified food manufacturers in Malaysia that have limited range of products, the conclusion cannot be generalized to other countries or other sectors in Malaysia. Therefore, this opens to further research in this area. The recommendation of further research should involve suppliers and customers of halal food supply chain. Further research should look into what are the perception of information sharing from supplier, logistic provider and customer of halal food supply chain.

7 Acknowledgement

The author is grateful to Malaysian Ministry of Higher Education (MoHE) under Fundamental Research Grant Scheme (FRGS) to the Office of Research, Innovation,

Commercialization and consultancy (ORICC) Universiti Tun Hussein Onn Malaysia (UTHM) under Vot 1430.

References

Bryman, A., & Bell, E. (2015). *Business research methods*. Oxford University Press, USA.

Creswell, J. W. (2002). *Educational research: Planning, conducting, and evaluating quantitative* (pp. 146-166). Upper Saddle River, NJ: Prentice Hall.

Huang, G. Q., Lau, J. S., & Mak, K. L. (2003). The impacts of sharing production information on supply chain dynamics: a review of the literature. *International Journal of Production Research*, *41*(7), 1483-1517.

Koçoğlu, İ., İmamoğlu, S. Z., İnce, H., & Keskin, H. (2011). The effect of supply chain integration on information sharing: Enhancing the supply chain performance. *Procedia-Social and Behavioral Sciences*, *24*, 1630-1649.

Lee, H. L., Padmanabhan, V., & Whang, S. (1997). The bullwhip effect in supply chains. *Sloan management review*, *38*(3), 93.

Li, L., & Zhang, H. (2008). Confidentiality and information sharing in supply chain coordination. *Management science*, *54*(8), 1467-1481.

Seidman, I. Interviewing as qualitative research: A guide for researchers in education and the social sciences. 2006. *New York and London: Teachers College, Columbia University Google Scholar*.

Stake, R. E. (1995). *The art of case study research*. Sage.

Tan, K.H., Wong, W.P. and Chung, L. (2015), "Information and knowledge leakage in supply chain", Information Systems Frontiers, Vol. 18 No. 3, pp. 621-638.

Zsidisin, G. A., & Ellram, L. M. (2003). An agency theory investigation of supply risk management. *Journal of supply chain management*, *39*(2), 15-27.

DOI: 10.1051/matecconf/201713500058

Artefacts Removal of EEG Signals with Wavelet Denoising

Arjon Turnip[1,*], and *Jasman* Pardede[2]

[1]Technical Implementation Unit for Instrumentation Development Indonesian Institute of Sciences, Bandung, Indonesia
[2]Department of Informatics Engineering, Faculty of Industrial Technology, National Institute of Technology, Bandung, Indonesia

Abstract. The recording of EEG signals often still contains many contaminants electrical signals that originate from non-cerebral origin such as ocular muscle activity called artefacts. The amplitude of artefacts can be quite large relative to the size of amplitude of the cortical signals of interest. In this paper, an application of wavelet denoising method for artefacts removal of EEG signals is proposed. The experiment result shows that contaminant artifact of EEG signals can significantly removed.

1 Introduction

Biomedical signal in various form is a source of information that comes from human body and useful in medical interpretation. The Brain is an organ of the human body is composed of neuron that can generate the electrical potential energy known as *neuroelectric potentials* [1]. Electroencephalography (EEG) is a non-invasive measurement of brain electrical activity obtained by placing electrodes on the scalp in areas of the brain [2]. EEG provides brain signal information based on recording data result in non-invasive method to analyze the brain activity that is important for medical (i.e. diagnosis, monitoring, and managing diseases or disorders of the nerves) and research (i.e. neuroscience, cognitive science, cognitive psychology, neurolinguistics and psychophysiological research) uses. The EEG signals is often contaminated by artefacts that often comes from muscle activity, which may reduce its usefulness for clinical or research by disturbing interpretation of the signal [2 - 5].

The main sources of artefacts are the EOG artefacts, ocular artefacts (eye movement and eye blink), noise muscles (EMG), heart signal, and various kinds of noise which is mixed with brain signals and often be the artefacts in EEG recordings [6-10]. Removal of artefacts is the real solution for quantitative analysis in the EEG recordings. The researchers eliminate artefacts in the EEG signals called the raw EEG data [11] to obtain a signal that is "clean EEG signal" that can be further analyzed. Quantitative methods for the analysis of EEG has been developed by many researchers.

The method is widely used to treat and reduce the noise of the brain signals including bandpass filters, autoregressive models [arjon], Finite Impulse Response (FIR). In this paper, a method of using wavelet denoising with daubechies wavelet (db1) and 3rd decomposition level is proposed. Wavelet is a mathematical model that fits in detecting and analyzing the events that occur in different scales, providing information on the time and frequency domains [13]. The wavelet method has the

* Corresponding author: arjon.turnip@lipi.go.id

 DOI: 10.1051/matecconf/201713500058

ability to transform a time domain signal into time and frequency that helps to better understand the characteristics of the signal. Denoising has an important role in signal analysis by removing the signal noise and retain important information. In a state of statistical usually often associated with discrete and rarely sampled on continuous functions. Nowadays, the stationary wavelet transform (SWT) from the EEG signals mixed artefacts has been widely used in the denoising signal.

2 Methodology

Sample data of EEG signal obtained by experiments. Experiments in this study conducted by eight untrained subjects, who are males aged 20-22 years with good health, thin hair and without any abnormalities. The tools used in this experiment is the Emotiv EPOC wireless EEG Neuroheadset which has 14 electrodes and 2 reference and the package comes with an application programming interface (API). This will record EEG signals from the experimental activities in which there are 128 data per second of one epoch. Before the electrodes set on the tool, preferably sprinkled with a liquid electrolyte is inserted to improve conductivity of the electrode. The coated electrode attached to the subjects scalp based on channels F7, T7, O1, O2, T8 and F8, respectively (See Figure 1).

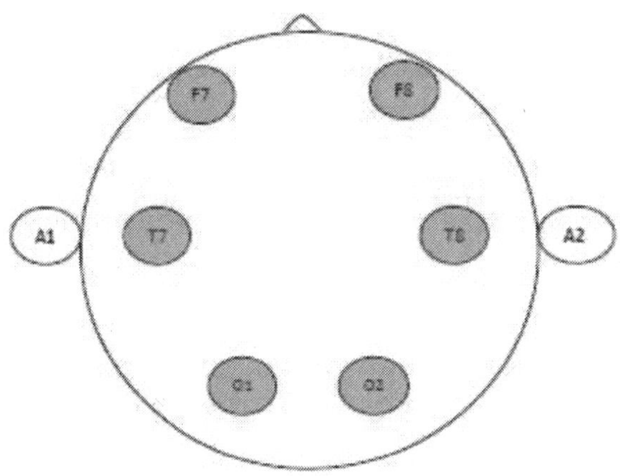

Fig. 1. A top view of the brain that shows the locations for EEG recording according to the 10-20 system.

Recording of sample data and design of the experiment with their stimulus were conducted with software embended with OpenViBE system. Stimulus scenario used in the experiment are given in Table 1 and 2.

Table 1. Stimulus (1st and 2nd) scenario of the Experiment with OpenViBE system.

Time (seconds)	Activity	Information
0-30	Normal	1st Stimulus
31-32	+ (preparation)	
33-63	Closed of eyes (trial 1)	2nd Stimulus
64-65	+ (preparation)	
66-96	Normal	
97-127	Closed of eyes (trial 2)	
128-130	Selesai	

Tabel 2. Stimulus (3rd) scenario of the Experiment with OpenViBE system.

Time (seconds)	Activity	Time (seconds)	Activity
0-18	Silent	**50-55**	Eye blinking
18-20	+ (preparation)	**55-60**	Eye blinking
20-25	Eye blinking	**60-65**	Eye blinking
25-30	Eye blinking	**65-70**	Eye blinking
30-35	Eye blinking	**70-75**	Eye blinking
35-40	Eye blinking	**75-80**	End
45-50	Eye blinking		

In the first scenario with the first stimulus, the data is recorded at normal conditions for 30 seconds. In the 31-32 seconds period, the sign '+' will appear in the screen to indicate a preparation for the second stimulus. When the arrow stimulus appear (second stimulus), subject must close their eyes for 30 seconds. For the third stimulus, start with the normal conditions for about 18 seconds. Then in 18-20 seconds periods, the sign '+' will appear again which is indicates the preparation for next stimulus. When the arrow stimulus appear, the the subject start to blinking. The detail of all scenario are given in Table 1 and 2.

2.1 EEG signal processing

Wavelet theory is a relatively new concept developed. The properties of the wavelet are given as follow: The time complexity is linear. Wavelet transform can be done perfectly with the time that is linear; Wavelet coefficients that are rarely elected. Practically, most of the wavelet coefficients of low value or zero. This condition is very beneficial, especially in areas of compression or data compression. Wavelet can be adapted to different types of functions, such as functions that are not continuous, and the function defined on bounded domains [15]. In general, a common function of wavelet is defined as the following equation [18 - 21].

$$\psi_{s,t}(t) = |s|^{\frac{1}{2}} \psi\left[\left(\frac{t-\tau}{s}\right)\right] \tag{1}$$

where s and τ , $s \neq 0$, denotes the scale and translation parameters, and t indicates time. In the continuous wavelet transform, the signal is analyzed using a set of basic functions that are related to scaling and simple transition. The development of the CWT is presented in the following equation.

$$X_{WT}(\tau, s) = \frac{1}{\sqrt{|s|}} \int x(t) \psi\left(\frac{t-\tau}{s}\right) dt \tag{2}$$

Wavelet decomposition is embodied in the input signal and the filtered, lowpass filter generates a waveforms called approximation and highpass filter generates a random waves called detail. The correlation of both filter with wavelet functions arranged in a hierarchy scheme called multiresolution decomposition, where decomposition separates the signal into "details" at different scales and "approximation". Decomposition and reconstruction schemes, and its procedure are given in Figure 2 and 3.

The wavelet denoising aims to remove noise in the form of artefacts in EEG signals recorded on while preserving the signal characteristics, regardless of the frequency content.

The process of denoising (noise reduction) based on the elimination or reduction of the data signal is considered noise. In the use of wavelet denoising, there are many methods used to reduce noise. Denoising is applied by downloading the form of signal threshold wavelet. Input coefficients used discrete wavelet transform. With thresholding, wavelet transform will be able to remove noise or other undesirable signals in the wavelet domain. Then, the desired signal will be obtained after performing the inverse of the wavelet. With this method, we need to understand the concept of wavelet coefficients that represent a measurement at a frequency between the signal and wavelet functions are selected. Wavelet coefficients calculated as a convolution of the signal and the wavelet function which is associated with a bandpass filter. When we analyze the signal at a high scale, we obtain the global information of a signal called approximation. And on a smaller scale, we obtain information from a signal called details [26-28].

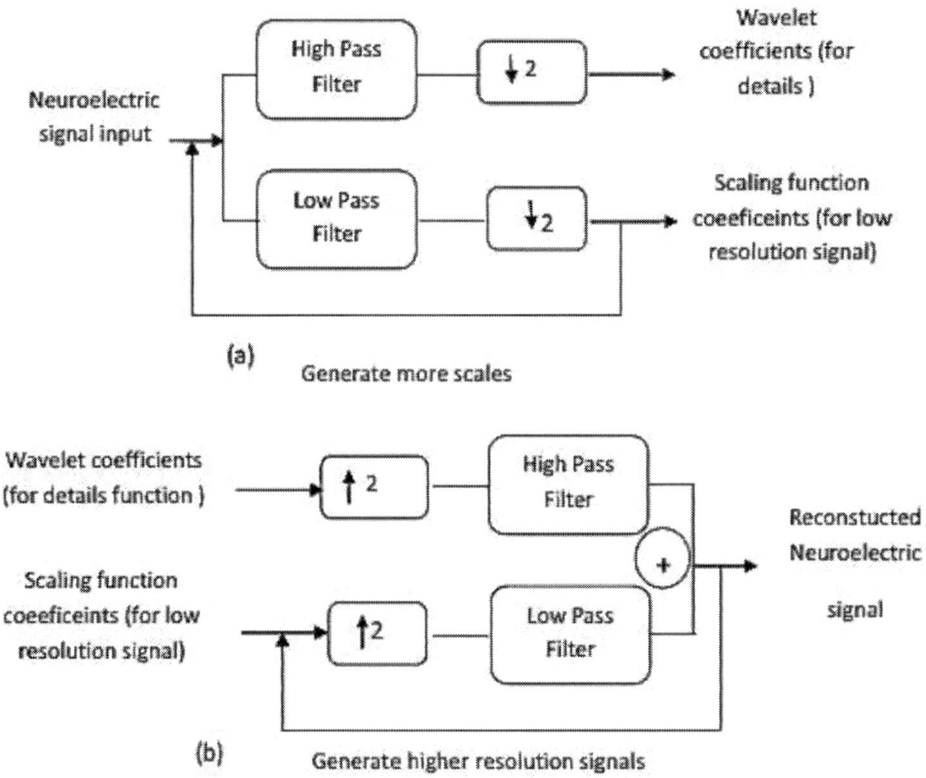

Fig. 2. (a) Schematic of discrete wavelet decomposition, (b) reconstruction of wavelet multiresolution.

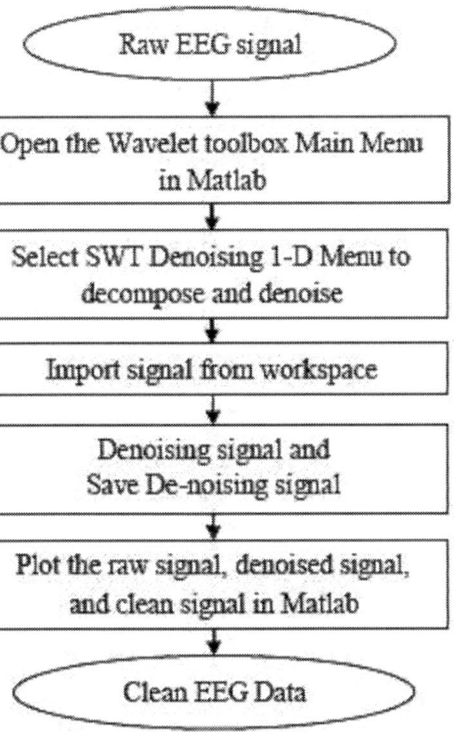

Fig. 3. Signal processing scheme using wavelet denoising method.

3 Results and discussion

In removing the artefact with wavelet denoising method, a description of the recording EEG signal denoising is described in the EEG model form with embedded noise as follows.

$$X[n] = S[n] + G[n] \tag{3}$$

with the **X** signals indicates the corrupted recorded signals with noise, **S** indicates the clean EEG signals, and **G** indicates the added noise and n is a point shots. The denoising procedure consists of three stages, namely: The decomposition, thresholding detail coefficients, and the reconstruction. The extracted signals of three subjects (subject 1, 4, and 6) are given below (Figures 4 – 12) to indicates the robustness of developed method.

DOI: 10.1051/matecconf/201713500058

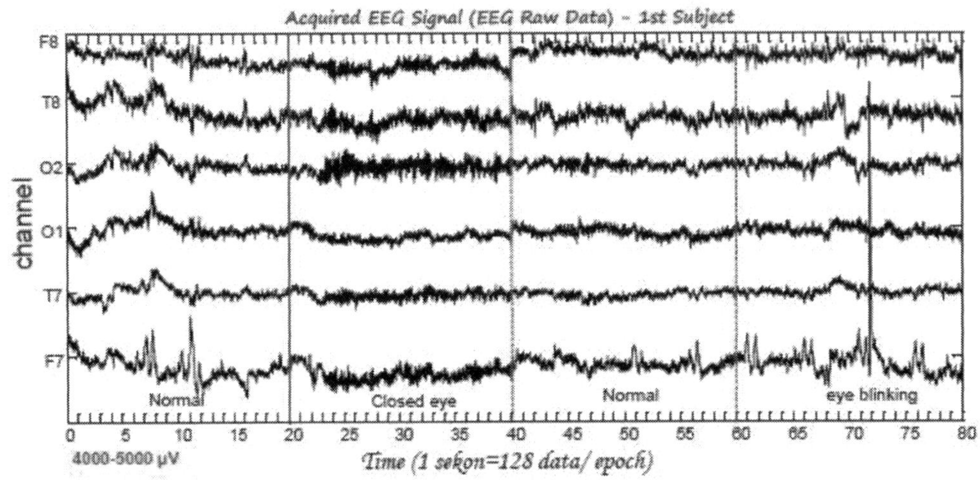

Fig. 4. Raw data EEG signals of the subject-1

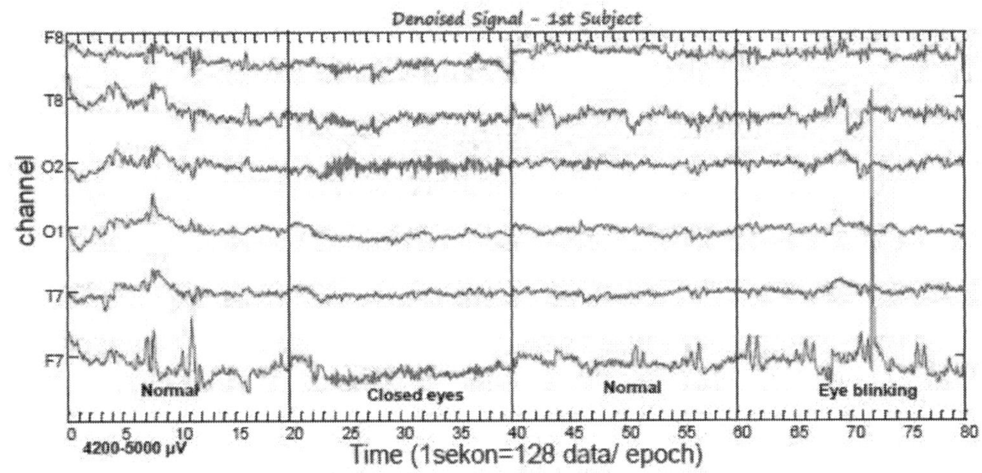

Fig. 5. The denoised EEG signals of the subject-1

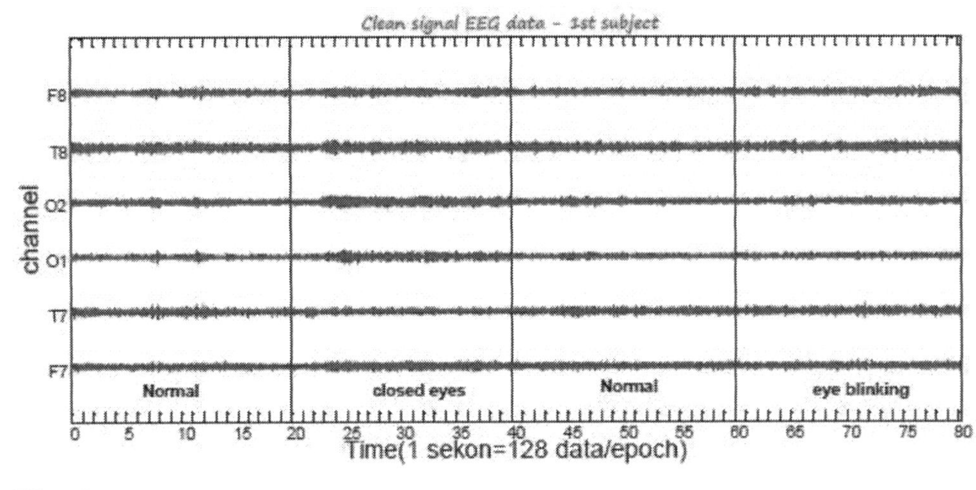

Fig. 6. Clean EEG Signals of the subject-1

DOI: 10.1051/matecconf/201713500058

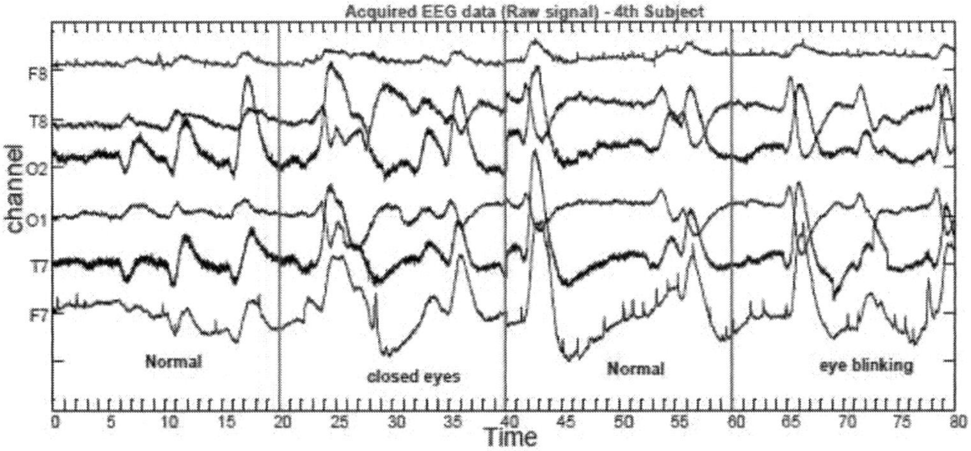

Fig. 7. Raw data EEG signals of the subject-4

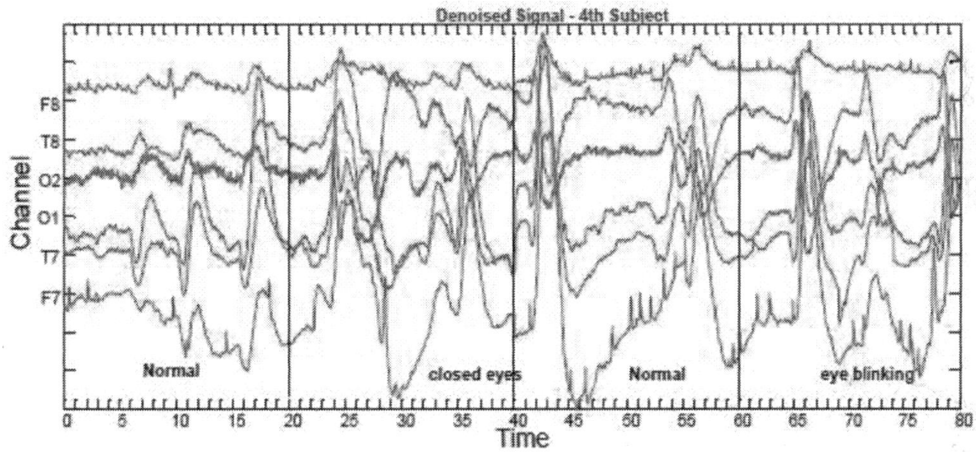

Fig. 8. The denoised EEG signals of the subject-4

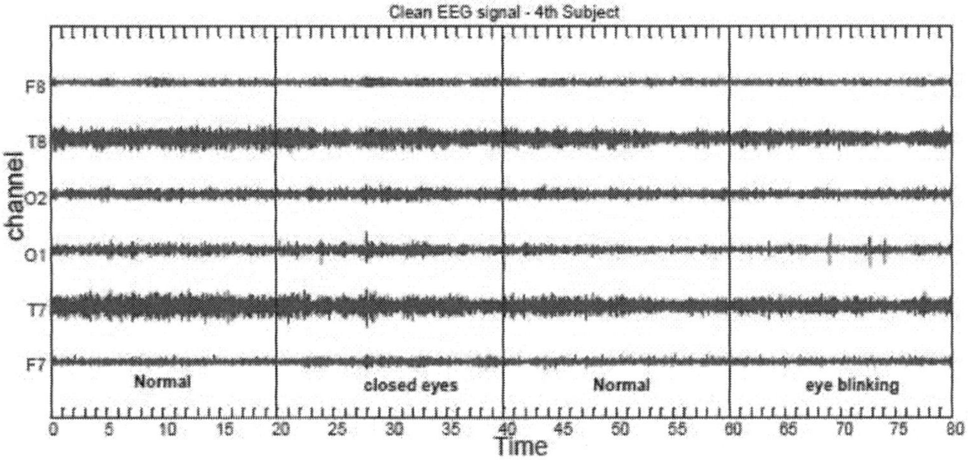

Fig. 9. Clean EEG signals of the subject-4

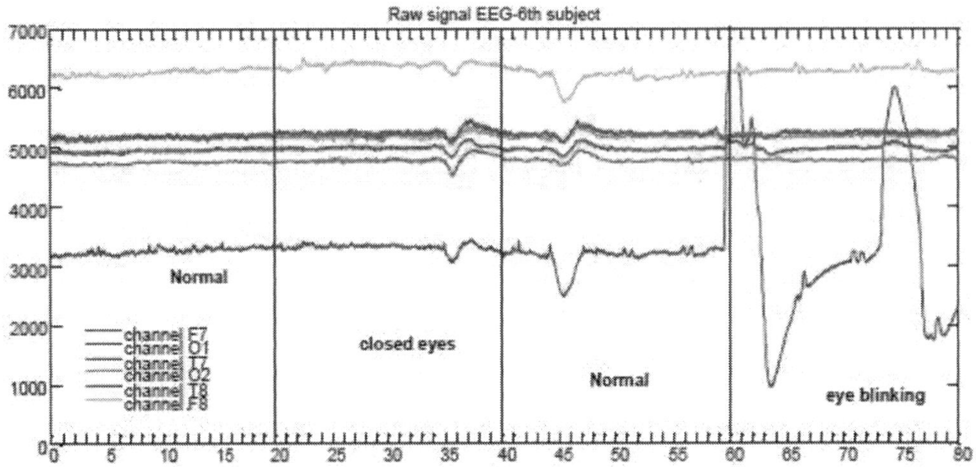

Fig. 10. Raw data EEG signals of the subject-6

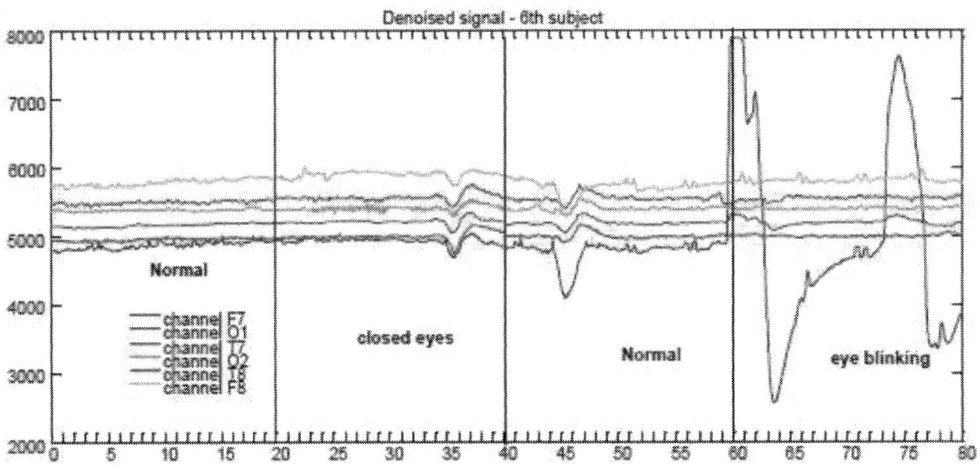

Fig. 11. The denoised EEG signal of the subject-6

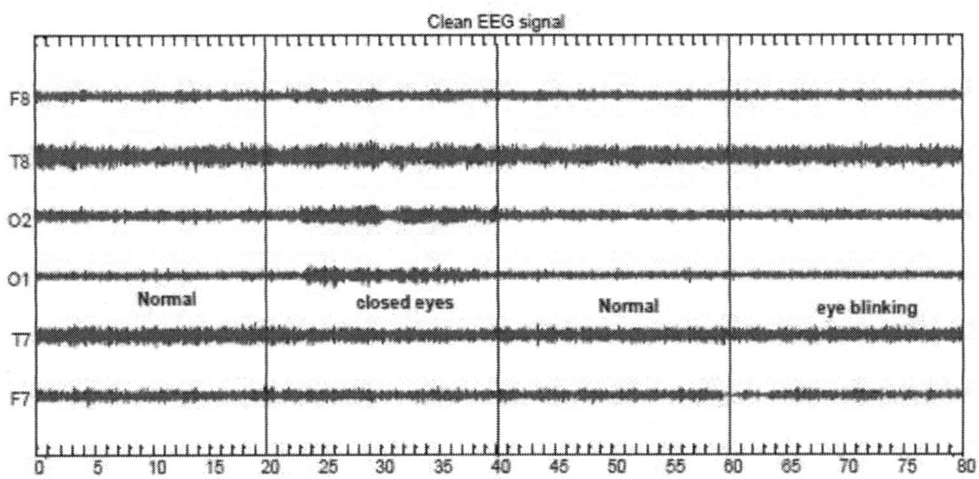

Fig. 12. Clean EEG signals of the subject-6

The input signal is the raw EEG data then decomposed by the decomposition of the signal with noise on wavelet bases where there is information on signal wavelet coefficient. Denoising can be obtained by thresholding wavelet coefficients. Denoising is done by separating the wavelet coefficients with the download threshold. Generally, wavelet coefficients associated with low frequency (hereinafter referred to as the coefficient of approximation). While the high frequency (hereinafter referred to as the coefficient of detail) is to be at denoise.

The wavelet denoising method using one of the families that Daubechies wavelet (dbN). Daubechies db1 used is the decomposition level 3. Wavelet group is used to pass the signal through wavelet transform blocks for the decomposition process, then the signal will be decomposed into wavelet coefficients which will be in thresholding and denoising where the process occurs. Thresholding function to eliminate noise and preserve the information that is important to the maximum signal.

Raw EEG data was still contains some high frequency where the artefacts are at this frequency. Wavelet method is performed to obtain signal finer and enhance noise removal of artefacts by eliminating high frequency in order to obtain the final result of clean EEG signals with amplitude 2-6 μV range of values where the value is smaller than the signal Raw EEG signal. So the value of the amplitude of the wavelet denoising result is closer to the EEG signal amplitude range.

4 Conclusion

The data sample has been obtained (hereinafter referred to as the raw EEG signal) and then processed the signal artefacts that tend to be the dominant high frequency, signal denoising with wavelet denoising. In order to obtain clean EEG signals with amplitudes of 2-6 μV and the range of values of the dominant frequency is in the frequency of 8-13 Hz which includes a group of alpha wave rhythm at the time of trial that showed subjects fully awake and relaxed. So the proposed research method produces a signal with amplitude and frequency ranges that approach the EEG signal. Wavelet denoising in this study using Daubechies db1 signal denoising level 3 of SWT denoising 1-D, the necessary analysis using different types of Daubechies to compare the results of denoising signal.

Acknowledgement

This research was supported by the thematic program through the Bandung Technical Management Unit for Instrumentation Development (Deputy for Scientific Services) and the flagship program through the Research Center for Physics (Deputy for Engineering Sciences) funded by Indonesian Institute of Sciences, Indonesia.

Reference

[1] S. Sanei, J.A. Chambers, "EEG Signal Processing", hal.12, UK: John Wiley&Sons, Ltd (2007)

[2] A. Jafarifarmand, M.A. Badamchizadeh, "Artefacts removal in EEG signal using a new neural network enchanced adaptive filter", Neurocomputing **103** (2013) 222-231

[3] S.O. Regan, S. Faul, W. Marnane, "Automatic detection of EEG artefacts arising from head movements using EEG and gyroscope signals", Medical Engineering&Physics **35** (2013) 867-874

[4] D. Looney, L. Li, T.M. Rutkowski, D.P. Mandic, A. Cichocki, "Ocular Artefacts removal from EEG using EMD"

[5] M. Fatourechi, Ali Bashashati, Rabab K.Ward, Gary E.Birch, "EMG and EOG artefacts in brain computer interface system: A survey", Clinical neurophysiology 118 (2007) 480-494

[6] T. Gasser, J. C. Schuller, U. S. Gasser, "Correction of muscle artefacts in the EEG power spectrum", Clinical Neurophysiology **116** (2005) 2044-2050

[7] M. T. Akhtar, W. Mitsuhashi, C. J. James, "Employing spatially constrained ICA and wavelet denoising, for automatic removal or artefacts from multichannel EEG data", Signal processing **92** (2012) 401-416

[8] H.-A. T. Nguyen, J. Musson, F. Li, W. Wang, "EOG artifact removal using a wavelet neural network", Neurocomputing **97** (2012) 374-389

[9] M. V. D. Velde, G. V. Erp, P. J. M. Cluitmans, "Detection of muscle artefact in the normal human awake EEG", Electroencepalography and clinical neurophysiology **107** (1998) 149-158

[10] P. K. Sadasivan, D. N. Dutt, "Minimization of EOG artefacts from corrupted EEG signals using neural network approach", Compt. Biol. Med.Vol. **24** No.6 pp 441-449 (1994)

[11] S. Olbrich, J. Jodicke, C. Sander, H. Himmerich, U. Hegeri, "ICA-based muscle artefact correction of EEG data:What is muscle and what is brain? Comment on McMenamin et al, Neuroimage **54** (2011) 1-3

[12] A. Turnip, K. S. Hong, Classifying Mental Activities From EEG-P300 Signals using Adaptive Neural Network, International Journal of Innovative Computing, Information and Control.,vol **8**, No.9 (2012)

[13] P. S. Kumar, R. Arumuganathan, K. Sivakumar, C. Vimal, "An adaptive method to remove ocular artefacts from EEG signals using wavelet transform", journal of applied science research, **5**(7): 741-745 (2009)

[14] C. E. Bire, B. Cahyono, "Denoising pada citra menggunakan transformasi wavelet" (2012)

[15] M. Z. B. Mohamad, "Denoising of an experimental acoustic emission (AE) data using one dimensional (1-D) Wavelet Packet Analysis (2009)

[16] J.-A. Jiang, C.-F. Chao, M.-J. Chiu, R.-G. Lee, "An automatic analysis method for detecting and eliminating ECG artefacts in EEG", computers in biology and medicine **37** (2007) 1600-1671

[17] W.-Y. Hsu, C.-H. Lin, H.-J. Hsu, P.-H. Chen, I-R. Chen, "Wavelet-based envelope features with automatic EOG artifact removal: application to single-trial EEG data", expert system with application **39** (2012) 2743-2749

[18] O. A. Rosso, M. T. Martin, A. Figliola, K. Keller, A. Plastino, "EEG analysis using wavelet-based information tools", journal of neuroscience methods **153** (2006) 163-182

[19] H. Leung, K. Schindler, A. Y. Y. Chan, A. Y. L. Lau, K. L. Leng, "Wavelet-denoising of electroencephalogram and the absolut slope method: A new tool to improve electroencephalographic localization and lateralization", clinical neurophysiology **120** (2009) 1273-1281

[20] R. Quian Quiroga, H. Garcia "Single-trial event-related potentials with wavelet denoising", clinical neurophysiology **114** (2003) 376-390

[21] L. Pasti, B. Walczak, D. L. Massart, P. Reschiglian, "Optimization of signal denoising in discrete wavelet transform", chemometrics and intelligent laboratory systems **48** (1999) 21-34

MATEC Web of Conferences **135**, 00059 (2017) DOI: 10.1051/matecconf/201713500059
ICME'17

Performance of Dual Depth Camera Motion Capture System for Athletes' Biomechanics Analysis

Wee Chang An[1], *Mohd Zamani* Ngali[1, *], *Zulhilmi* Kaharuddin[1], and *Siti Badriah Khairu* Razak[1]

[1]Faculty of Mechanical and Manufacturing Engineering Universiti Tun Hussein Onn Malaysia, Batu Pahat, Malaysia

Abstract. Motion capture system has recently being brought to light and drawn much attention in many fields of research, especially in biomechanics. Marker-based motion capture systems have been used as the main tool in capturing motion for years. Marker-based motion capture systems are very pricey, lab-based and beyond reach of many researchers, hence it cannot be applied to ubiquitous applications. The game however has changed with the introduction of depth camera technology, a marker-less yet affordable motion capture system. By means of this system, motion capture has been promoted as more portable application and does not require substantial time in setting up the system. Limitation in terms of nodal coverage of single depth camera has widely accepted but the performance of dual depth camera system is still doubtful since it is expected to improve the coverage issue but at the same time has bigger issues on data merging and accuracy. This work appraises the accuracy performance of dual depth camera motion capture system specifically for athletes' running biomechanics analysis. Kinect sensors were selected to capture motions of an athlete simultaneously in three-dimension, and fused the recorded data into an analysable data. Running was chosen as the biomechanics motion and interpreted in the form of angle-time, angle-angle and continuous relative phase plot. The linear and angular kinematics were analysed and represented graphically. Quantitative interpretations of the result allowed the deep insight of the movement and joint coordination of the athlete. The result showed that the root-mean-square error of the Kinect sensor measurement to exact measurement data and rigid transformation were 0.0045 and 0.0077291 respectively. The velocity and acceleration of the subject were determined to be 3.3479 ms^{-1} and −4.1444 ms^{-2}. The result showed that the dual Kinect camera motion capture system was feasible to perform athletes' biomechanics analysis.

*Corresponding author: zamani@uthm.edu.my

1 Introduction

Since the capturing of the motion of an athlete requires highly accurate system, both camera based (e.g. Vicon, UK) or inertia sensor based (e.g. XSens, Holland) were dubbed as the gold standard equipment for the porpose. However these types of systems are expensive; make it outside the reach of most users. Microsoft Kinect (Microsoft Corp., Redmond, WA, USA) sensor was then considered as the low-cost alternatives to the pricey systems. Previous study concluded that the Kinect sensor "showed good performance versus the clinical gold standard (Vicon and AMS/Visual-3D)" [1]. Another significant feature of the Kinect sensor is that it does not required any marker to be attached onto the subject – a marker-less motion capturing system. The Vicon system on the other hand requires reflective markers to be attached on the subject's body; the markers are definitely a huge disadvantage. Thus with the application of Kinect sensors as the dual depth camera motion capture system, it could facilitate the set-up process and phase out complex equipment set-up. The fast, low cost, portable system for capturing these biomechanical measures would facilitate the biomechanists or researcher to bring their work task to the athlete instead of bringing the subject into the lab. Biomechanics analysis is used to assess the performance of athletes in order to gain a greater understanding of human movement. Thus, appropriate quantitative solutions could be developed to improve athlete's techniques and thereby increase the performance, minimizing injuries and in turn, promoting career longevity. A biomechanics researcher uses quantitative biomechanical analysis methods to discover new techniques, which communicated to the teachers and coaches who will implement them [2].

2 Related work

There were numerous works have been done on motion analysis via depth cameras. Each of them yields outstanding results in term of performance of the Kinect sensor.

In the study conducted by Dutta [3], Kinect was used to capture three-dimensional image of a scene with cubes positioned at varying distances. The result of the study was compared to the gold standard motion capture system (Vicon), and the root-mean-square error (RMSE) in the investigation were reported to be 0.0169m, 0.0348m and 0.0141m in the x, y and z directions respectively, which relatively small and accurate.

As claimed by Seung-kook Jun [4], without any post-processing to the raw data captured by the Kinect, the application of such system tend to be limited. However with suitable data post-processing, such limitation was unbarred. The post-processing as suggested was to perform kinematic calibration and apply Kalman filter to the raw data captured by Kinect. As a result, it showed a good performance when compared to the gold standard equipment (Vicon and AMS/Visual-3D).

In Caon's [5] research paper, the writer employed multiple Kinect sensors to detect subject's posture and gesture estimation, in three dimensional smart environments. The system developed was able to track multiple users and reconstructs its situational information such as posture and coordinates. The system was used to smartly interact with objects around it, such as turning a lamp on and off, turning a TV on and off and so on.

Based on the literature review, most of the research works were only focusing on the subject's posture and the limitation and conflict of the usage of Kinect system. Quantitative analysis on how good dual Kinect sensor system can be applied as motion capture system for sports and biomechanics analysis was still questionable. Therefore, this work was initiated to obtain numeric measurement of dual depth camera system feasibility in these fields.

3 Methodology

This research was subjected to a systematic sequence. The process started with the data acquisition from each Kinect sensor, so that the skeletons data from the sensors could be retrieved. Two Kinect sensors were projected onto the subject. Each sensor streams its own skeletal data in its field of view (FOV). The Microsoft Kinect Software Development Kit (SDK) used in this study allows the skeletal tracking and data acquisition into the system. With the three coordinate spaces: the colour spaces; the depth spaces and; the skeletal spaces, the three-dimensional (3D) position data of the 20 predefined skeletal joints of human body were retrieved. Each sensor streamed the 3-dimensional space coordinates of the tracked skeleton into a custom formatted readable text file which later used for skeletons merging. The system however required a few calibration steps before fused into a single skeleton model. The calibration method is described in Figure-1 below.

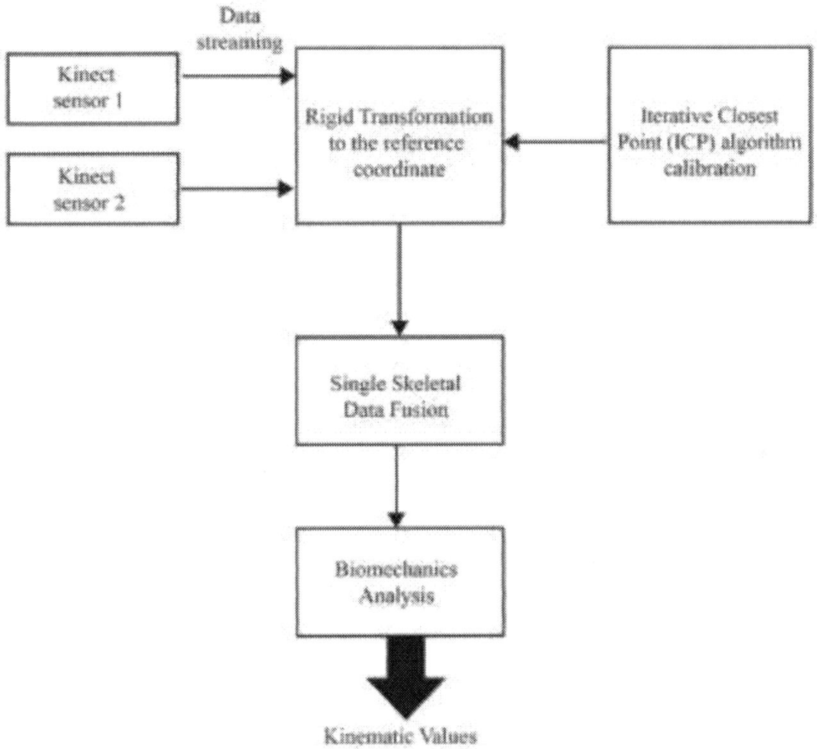

Fig. 1. Calibration method.

The overall research flow was described in Figure-2, from the beginning of data acquisition to data analysis. As illustrated in this figure, the skeletal data from each sensor has to be transformed to a common reference coordinate system before data processing were carried out. From the system, one sensor is selected as a reference sensor (principal camera, denoted by K_B) so as to provide a reference frame for other sensor (secondary camera, denoted by K_A). Next, the usage of the Iterative Closet Point (ICP) [3] algorithm was introduced to the K_A. It was fed into the ICP to estimate the rigid transformation (or the Rotation-Translation) that minimized the distance between the transition positions in the reference frame.

DOI: 10.1051/matecconf/201713500059

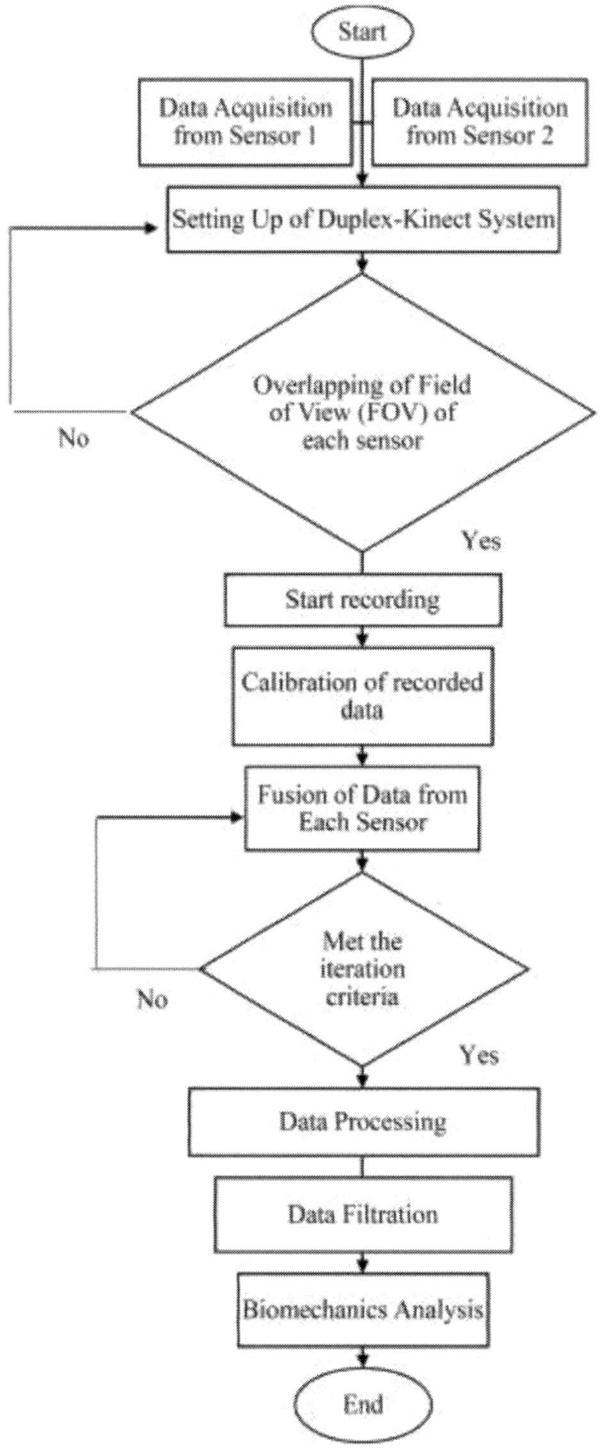

Fig. 2. Flowchart of research.

The skeleton was sliced into three sections, namely upper section, middle section and lower section. From each section, a joint was selected. The three selected joint was then used to compute the rigid transformation. The transformation obtained in term of rotation (R) and translation (t) was then used to register the skeletons acquired from each sensor in the reference coordinate system. The formula used to calculate the **R** and **t** based on the Eqn. (1).

$$B = \mathbf{R} * A + \mathbf{t} \quad (1)$$

Where **R**, **t** are the transforms applied to dataset A to align it with dataset B, as best as possible. To solve the equation, it involved three steps: computation the centroids of both dataset; moving of both dataset to the origin and then evaluate the optimal rotation, (matrix R) and; calculation of the translation, **t**. Figure-3 described the process graphically.

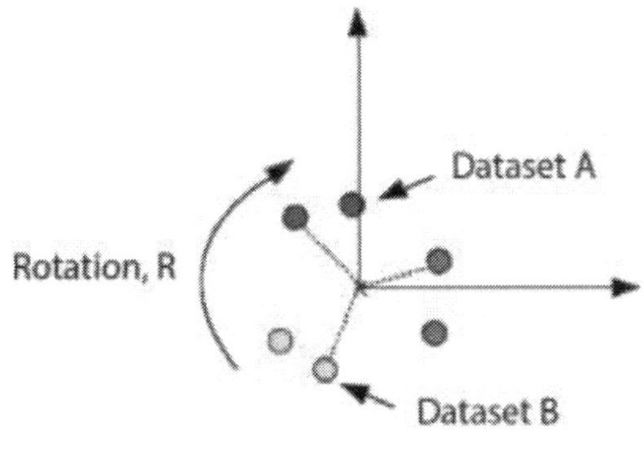

Fig. 3. Moving data set A and B to origin and rotate the data.

Once all the prerequisite conditions were met, whinc was the root-mean-square error (RMSE) of the rigid transformation was lower than the previous iteration in a number of given iterations; the merging of both skeletons was started. The merging of skeleton was illustrated in Figure-4.

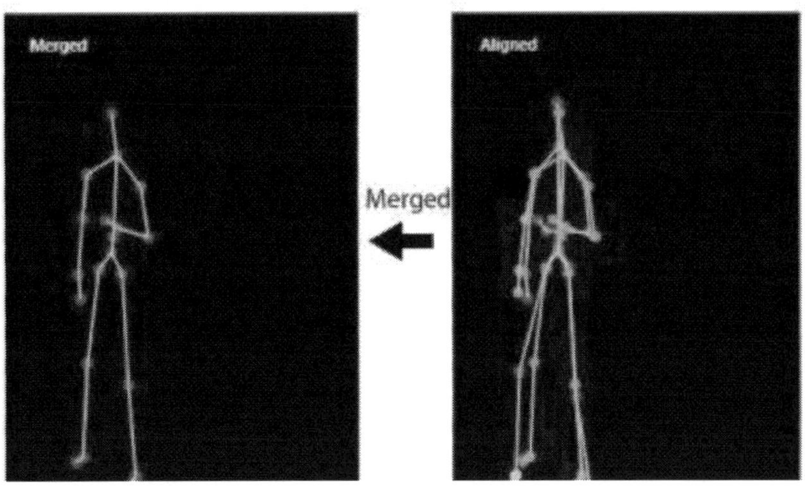

Fig. 4. Merged skeleton (left) and transformed skeleton (right).

DOI: 10.1051/matecconf/201713500059

Next, the data processing was done on the fused skeleton by extracting the joints coordinates in x, y, and y positions. Once the coordinates were successfully retrieved, biomechanics analysis was proceeded. However prior to analysis, Butterworth filter was applied to the data. Most system that measure physical quantity using a sensor return errors and noise, this includes the Kinect sensor used in this research. By using such filter, it had minimized the jittering and hence stabilized the joint positions over time. After several tuning were done on the recorded data, a 3Hz cut-off frequency of 5th order low pass Butterworth filter was chosen. Since a filter will delay the signal, thus, the data was filtered twice, in forward and backward direction to produce a zero lag filter [4]. After the filtration, linear and angular kinematics motion of an athlete were analysed using the fundamental equations, Eqn. (2), Eqn. (3), Eqn. (4) and Eqn. (5).

$$displacement, d_i = \sqrt{d_{x,i}^2 + d_{y,i}^2 + d_{z,i}^2} \quad (2)$$

$$velocity, v = \frac{\Delta d}{\Delta t} \quad (3)$$

$$acceleration, a = \frac{\Delta v}{\Delta t} \quad (4)$$

$$\mathbf{A} \bullet \mathbf{M} = |A||M|\cos\theta \quad (5)$$

4 Results and discussion

Analysis on the athlete was done using Mathworks' Matlab. The subject chosen was a long distance running athlete. The raw data showed that although the subject was standing statically in front of the Kinect sensor, it was observed that the joints of the skeleton jittered. This gave an unstable movement of the joints to be analysed. Figure-5 however showed both the data before and after the application of Butterworth filter.

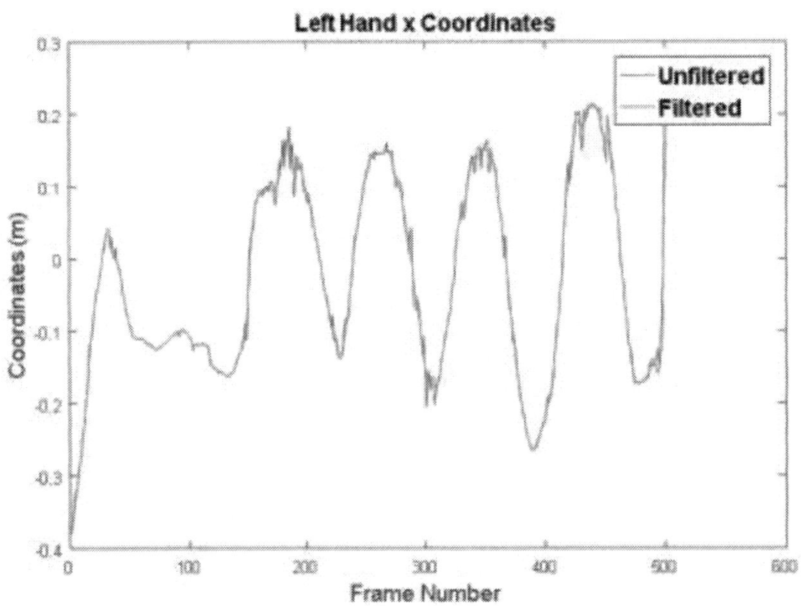

Fig. 5. Noise filtration.

All joints data were filtered as mentioned previously in methodology section with 5th order low pass Butterworth filter at cut-off frequency of 3Hz. The dual Kinect system used in this research was compared with the use of single Kinect sensor. Three types of basic human movements were tested in both systems, the flexion in sagittal plane, lateral in frontal plane and twist in transverse plane. The flexion result was illustrated in Figure-6.

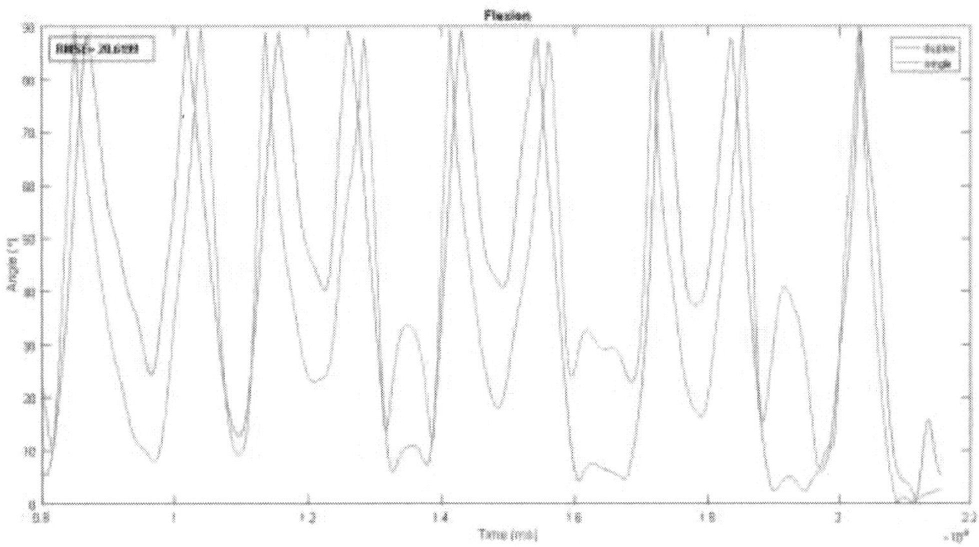

Fig. 6. Flexion angle of dual and single Kinect system.

After the comparison, it returned root-mean-square error (RMSE) of 20.8199, which is numerically large. The flexion angle for single Kinect system show there was a large angle jumping between one instance of time to another instance time, however for dual Kinect system, the differences was smaller. For lateral, the comparison is shown in Figure-7.

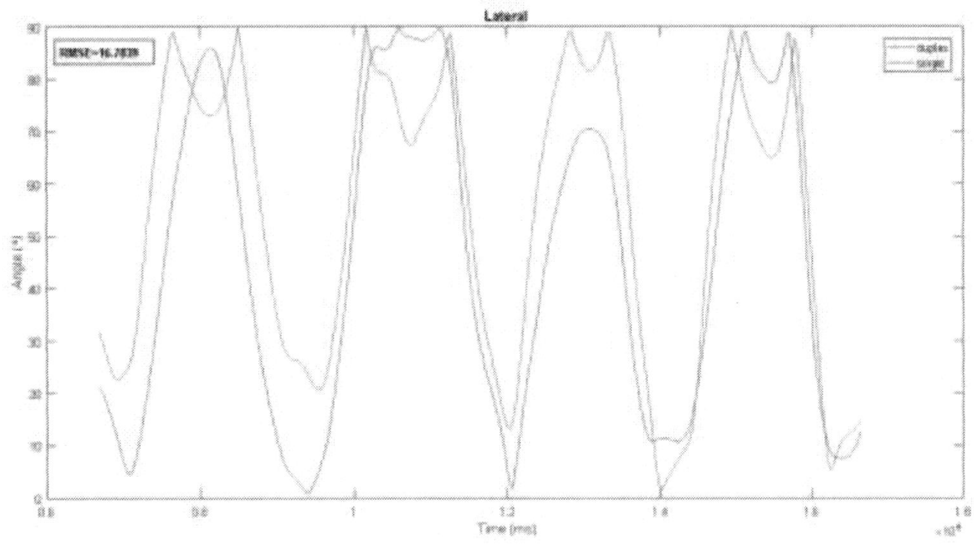

Fig. 7. Lateral angle of dual and single Kinect system.

The error for lateral movement for single Kinect system had a smaller RMSE compared to the flexion in sagittal plane. The RMSE was only 16.7839. This set of data had proved that the single Kinect system was taking the 'inferred' or 'not tracked' data as the actual data. The most significant feature of dual Kinect was that it could reduce occlusion problem. The twisting motion of subject is shown in Figure-8.

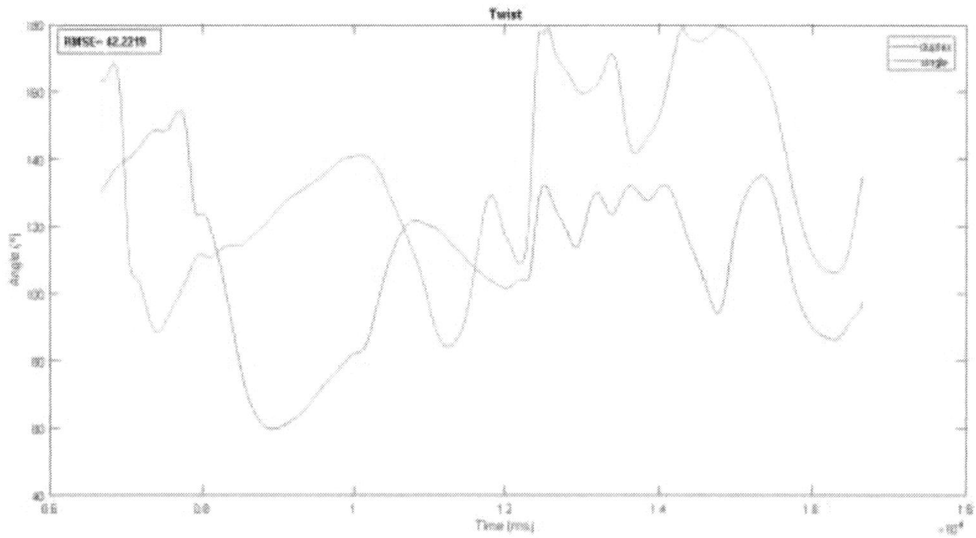

Fig. 8. Twist angle of dual and single Kinect system.

The RMSE error for single Kinect system in measuring twist angle was greatly deviated from the dual system with a high of 42.2219. As the subject twisting, the single Kinect was unable to track the joints that were occluded. However, dual system made it possible by considering the data from the other sensor. Next, the rigid transformation of the athlete running, return a RMSE of 0.0077291. Such transformation considered to be comparably accurate since the value is approximately close to 0. However, the transformation sometimes could be erroneous and a resolution was taken by increasing the number of iterations and set the maximum acceptable error to accommodate various set-up and recording.

By using the transformed data, linear and angular kinematics were analysed next. The centre of mass (CM) of a vertically standing athlete was assumed to be approximately at navel, which was estimated to be at the centre hip joint [5]. The linear kinematics motion of the athlete, the displacement, the velocity and the acceleration was described in Figure-9, Figure-10 and Figure-11 respectively.

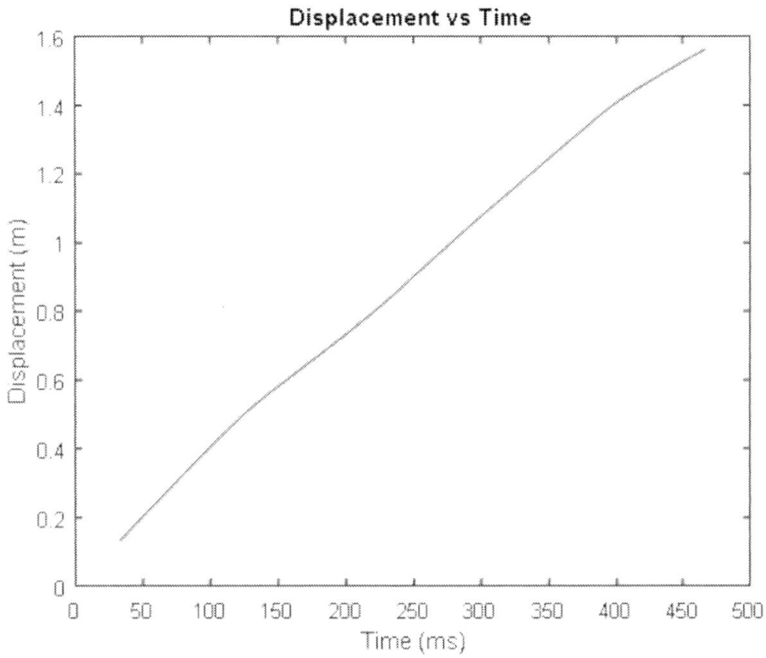

Fig. 9. Displacement of an athlete versus time.

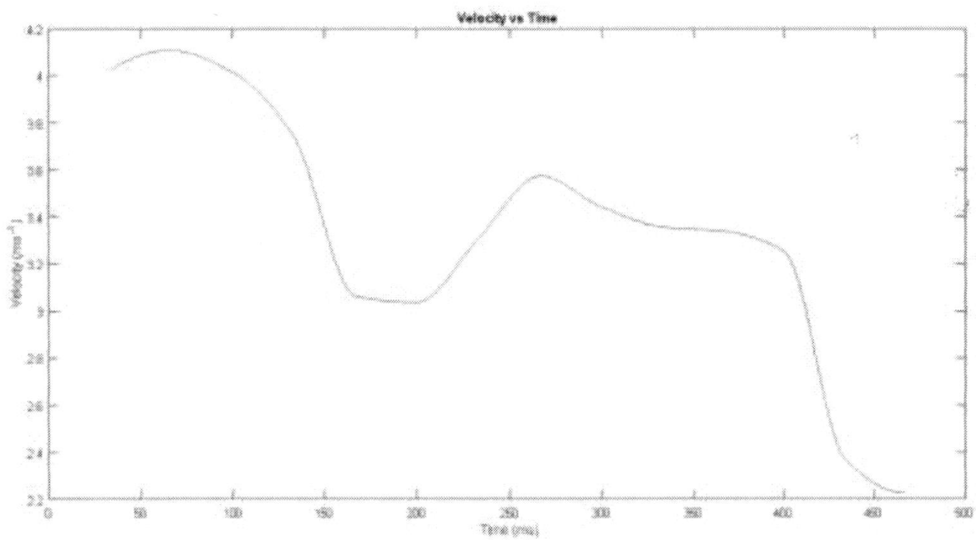

Fig. 10. Velocity of the athlete.

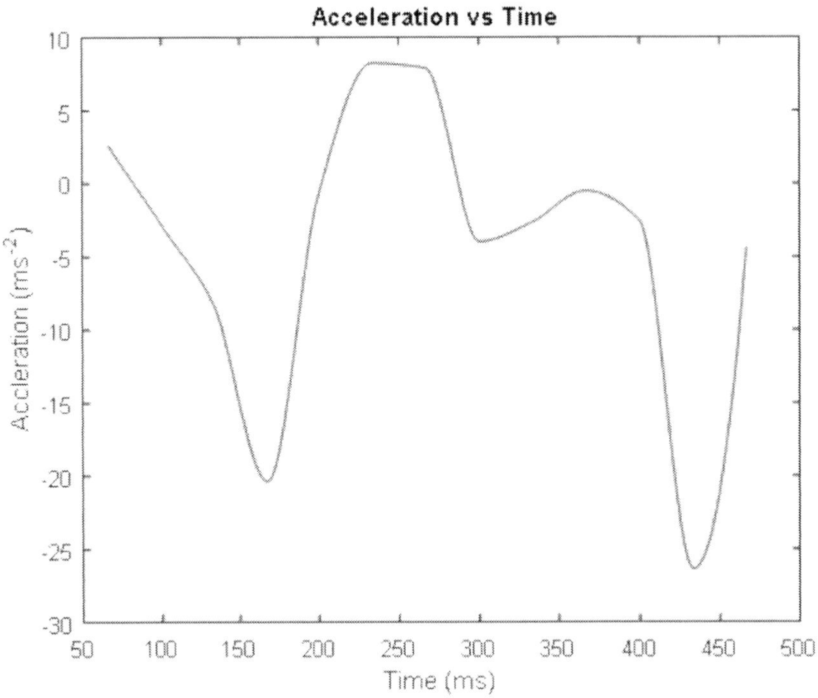

Fig. 11. Acceleration of the athlete.

The average velocity and acceleration of the athlete was calculated to be 3.3479ms^{-1} and -4.1444ms^{-2} respectively. Figure-12 below depicted the left knee angle over time when the athlete was running.

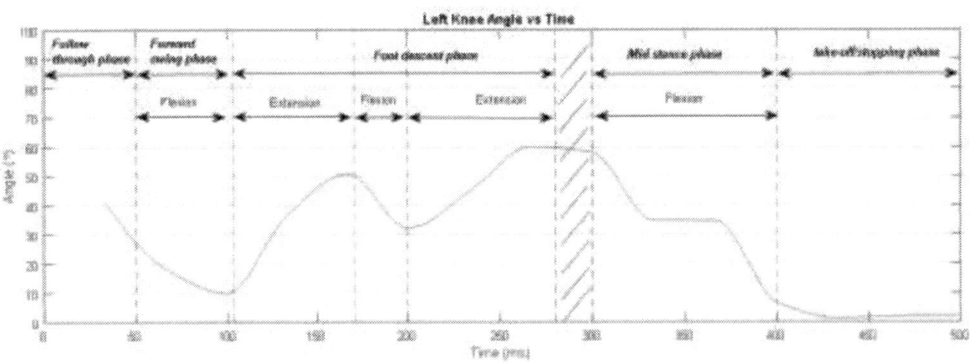

Fig. 12. Flexion and extension of left knee with respective phase.

The key events that mainly were analysed in this angular motion were the foot strike, mid-support and take-off, follow through, forward swing and foot descent, as described in Figure-13.

DOI: 10.1051/matecconf/201713500059

Fig. 13. The analysed stages of a running.

Foot strike, also known as the initial contact absorbs the impact, mid-support or mid-stance maintains the forward momentum and support the body weight, take-off critically responsible for the acceleration of the body, follow through or initial swing decelerating the running process for the recovery leg, forward swing also known as mid-swing coordinating the recovery leg movement, and foot descent or terminal swing prepare the foot for active landing. All the key events were identified in the Figure-12 for knee and Figure-14 below for hip.

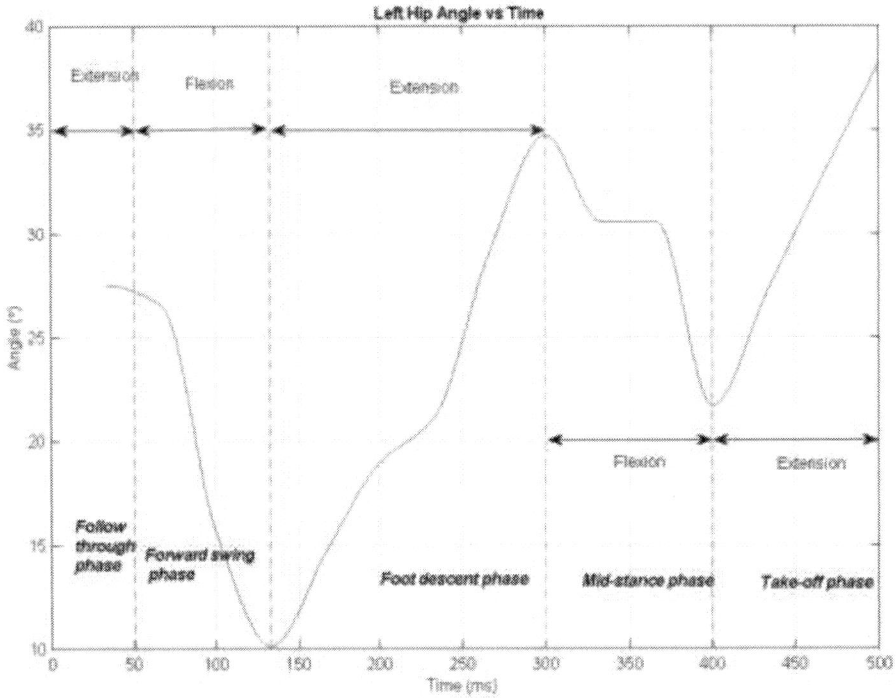

Fig. 14. Left hip angle during the running.

However to further provide insight into the interrelationship between joints when running, an angle-angle diagram (also known as cyclogram) easily interpreted the relation. The coordinations of the joints were plotted in Figure-15 for the hip and knee.

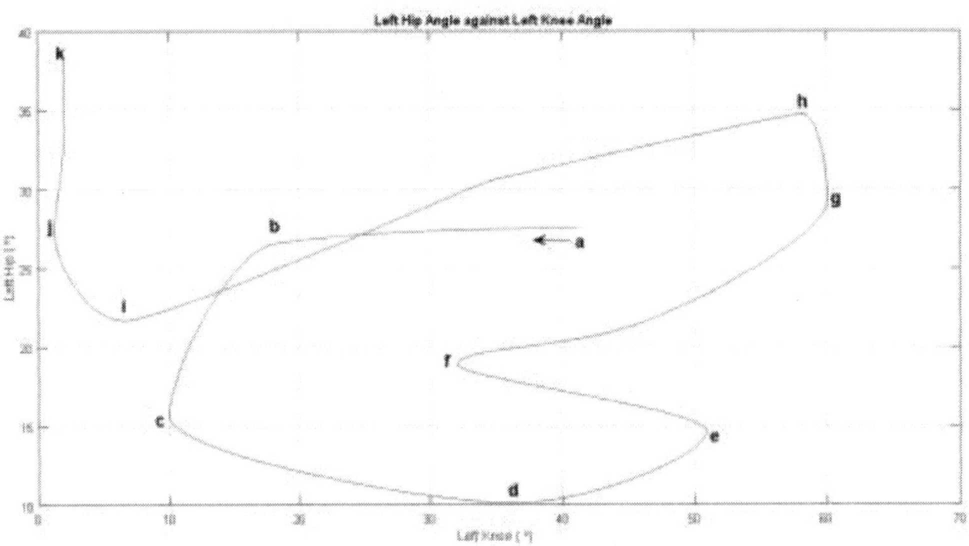

Fig. 15. Running angle-angle diagram performed by the athlete.

From Figure-15, the arrow mark indicated the movement direction of the hip and knee, with 'a' as starting point of the running. Next a continuous relative phase (CRP) was plotted for the differences of phase for both joints, as shown in Figure 16 below.

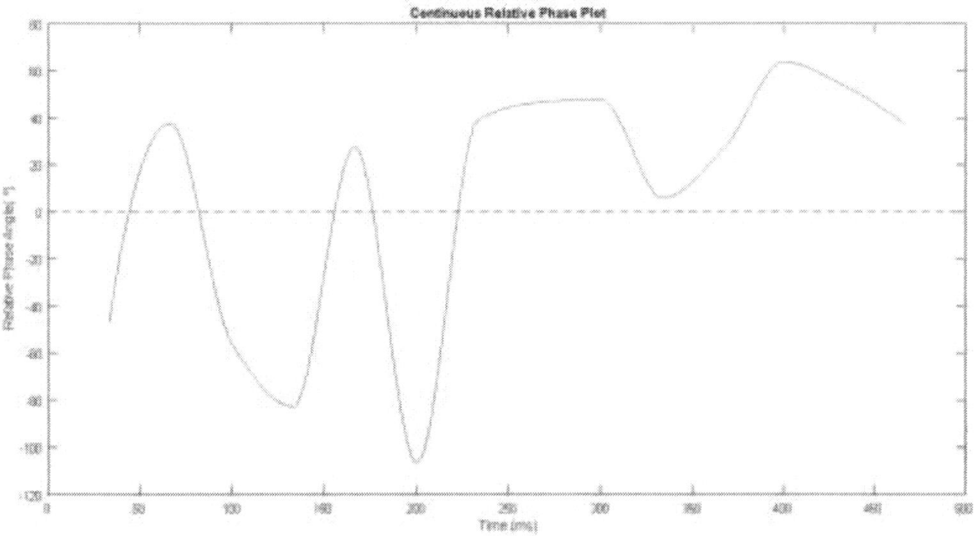

Fig. 16. Continuous relative phase of the runner between the hip and knee angles.

The two joints being in-phase when the CRP is equal to 0°, which was the instance at 44th, 82nd 155th, 176th and 222nd milliseconds. At these moments, the hip and knee undergoes a same action (either flexion or extension) at the same time. Referring to Figure-12 and Figure-14, both joints flexed at 44th and 82nd, extended at 155th and slightly flexed at 176th and back to extension at 222nd milliseconds. The anti-phase of joints occurred at 180°. However, from the graph, there were zero anti-phase of the hip and knee, but the phase is approaching the anti-phase and back once again (sinusoidal shape).

5 Conclusion

The depth camera motion capture system (Kinect) allows retrieval of skeleton joints coordinates of athletes. With the obtained coordinates, biomechanics analysis on the athlete can be done quantitatively to improve their performance while reducing the risk of injury. The biomechanics analysis includes linear and angular kinematics that consists of displacement, velocity, acceleration, angular velocity and angular motion (flexion, extension). This work verifies that the dual Kinect system is capable of performing the biomechanics analysis on athletes. The dual Kinect system is able to track the athlete's motion in a wider coverage as compared to single Kinect system. The system is portable, uncomplicated set-up, economical, and does not require substantial amount of time and effort for data acquiring and processing. In this work, linear kinematics analysis of running biomechanics which involves the centre of mass (CM) of the athlete were successfully computed. Displacement, velocity and acceleration of the athlete were calculated using the joint coordinates retrieved from the Kinect system. The three parameters were presented graphically. Angular kinematics analyses, which were much more complicated as compared to linear motion, were successfully calculated too using the joint coordinates and vectors. Angular velocity and joint coordination quantitatively defined the athlete's movement in

DOI: 10.1051/matecconf/201713500059

terms of angle-angle relation and in time series. The flexion and extension of joints were interpreted in term of time and their interrelationships were represented in a comprehensive continuous phase planes plot. Foremost, the system omits the need of markers attached to the subject's body, which allows more comfort and a lot of time saving in setting up the system, compared to conventional marker-based motion capture system which is expensive and lab-based. Referring to the performance of the dual depth camera system, higher number of depth cameras are expected to give better features to the motion caption system.

Acknowledgement

This research is supported by the Fundamental Research Grant Scheme (FRGS), Grant No: FRGS/1/2015 Vot 1543 by Ministry of Education, Malaysia.

References

1. J. Seung-kook, X. Zhou, D. K. Ramsey, V. N. Krovi, "A Comparative Study of Human Motion Capture and Analysis Tools A Comparative Study of Human Motion Capture and Computational Analysis Tools," in *The 2nd International Digital Human Modeling Symposium*, Michigan, USA (2003)
2. P. McGinnis, Biomechanics of Sport and Exercise With Web Resource and MaxTRAQ 2D Educational Software Access (3rd ed.), Illinois: Human Kinetics (2013)
3. D. T., "Evaluation of the Kinect Sensor for 3D Kinematic Measurement in the Workspace," *Applied Ergonomics* **43**(4), pp. 645-49 (2011)
4. J. Seung-kook, X. Zhou, D. K. Ramsey, V. N. Krovi, "A Comparative Study of Human Motion Capture and Analysis Tools A Comparative Study of Human Motion Capture and Computational Analysis Tools," in *The 2nd International Digital Human Modeling Symposium*, Michigan, USA (2003)
5. M. Caon, J. Tscherrig, E. Mugellini, O. A. Khaled, "Context-Aware 3D Gesture Interaction Based on Multiple Kinects," *First International Conference on Ambient Computing, Applications, Services and Technologies (AMBIENT),* pp. 7-12 (2011)
6. P. J. Besl, N. McKay, "A Method for Registration of 3-D Shapes.," *IEEE Trans. on Pattern Analysis and Machine Intelligence,* vol. **14**(2), pp. 239-256 (1992)
7. D. B. J., . D. J. Robertson, "Response Characteristics Of Different Butterworth Low-Pass Digital Filters.," in *Proceedings of the 2nd North American Congress on Biomechanics, (pp. 105–106)* (1992)
8. G. L. Moir, Strength and Conditioning, Burlington, Massachusetts, United States: Jones & Bartlett Learning (2016)

Correlation of Loaded and Unloaded Foot Area With Arch Index in Younger Flatfoot

Dwi Basuki Wibowo[1, *], *Gunawan Dwi* Haryadi[1], *Achmad* Widodo[1], and *Sri Puji* Rahayu[2]

[1]Mechanical Engineering Dept., Diponegoro University, Semarang, Indonesia
[2]Faculty of Public Health, HAKLI Institute of Health Science Semarang, Indonesia

Abstract. Harris & Beath claimed 23% of human kind population is indicated flatfoot. Identifying flatfoot is by using wet foot test. This footprint is not accurate because of the difficulty to make sure the patient stand upright. Another way is using x-ray to determine height of arch which is a distance from medial longitudinal foot arch to the ground. If the distance is less than certain level, so the foot type is included as flatfoot. Other method proposed by Kulkarni et al. using the footprint index (FPI) which is the ratio of B intercept to A intercept, where the footprint was obtained from pedobarography image. If FPI is lower than 0.63, it is categorized as flatfoot. Another method to determine arch type which is widely used is Cavanagh's Arch Index (AI) from division of mid foot area to entire footprint area (excluding the toes). If AI>0.26, then the foot type is flatfoot. This study is to learn the correlation between entire loaded and unloaded foot area with Cavanagh's AI. The entire loaded foot and footprint area for evaluating AI derived from a digital footprint modified from document scanner, while the entire unloaded foot area derived from a 3D scanner for foot orthotic. One hundred and two healthy asked voluntarily for doing footprint. From 102 subjects found 63 participants identified as flatfoot, 31 subjects are normal feet and 8 subjects identified as high arch. A series of 3 x 3 repeated measures ANOVAs were used to determine statistical differences ($\alpha<0.05$). A significant interaction existed between ratio of entire loaded and unloaded foot area (RFA) subject to all categories of AI ($p<0.05$) also a correlation coefficient of $r=0.67$ has found between RFA and AI on foot type of flatfoot which means that flatfoot can be indicated by RFA.

1 Introduction

Foot is one of the most important parts of the body which interacts with the ground when standing in an upright posture and walk. Human's foot and ankle are the mechanical structure which is complex and strong consist of 26 bones, 33 joints (20 joints are actively used), and more than hundreds of muscles, tendons and ligaments. These

*Corresponding author: ir.dwibasuki.ms@gmail.com

complex structures keep human body balance during standing and functioning as swing bar when walk or run [1].

Commonly human's foot divided by 3 parts that are rear foot, mid foot, and fore foot or more specifically as heel, arch, and metatarsal region (including the toes) [2]. Arch of the foot (which can elastically deform) has the function to absorb the impact loads during walking or sports activities to minimize injure at the foot. It means that there is a specific correlation between the type of arch and injuries caused by sports activities. Measuring changes in the deformation of foot arch during sports activities would be helpful to know more about the possible mechanisms of injury to the foot [3].

Flatfoot is a condition which arch of the foot inside is flat (called flat arch), so that most of the foot touching the ground [4, 5]. People with flatfoot in the world are quite a lot. Harris & Beat claimed 23% of the normal human population indicated flatfoot [6]. However, the numbers of flatfoot sufferers in many hospitals are not much. It because normally they did not feel the pain/hurt so much when doing activity, so they are rarely checks their foot to the specialist doctor of orthopedic [7]. People with flatfoot commonly sighing to easily tired and in some cases there were found patients that feel pain in ankle and knee caused by un-straight foot [8, 9]. The risks of pain at knee, foot, problems on the ankle (ankle weakness), or damage on the cartilage are increasing when influenced by obesity, injury at foot or ankle,.joint inflammation, aging, and diabetes [9]. Commonly foot type flatfoot can be determined by wet foot test which is printing the painted foot to scaled paper, as seen on Figure-1. This footprint is not accurate because of the difficulty to make sure the patient stand upright also the possibility of unprinted parts of foot because the paint applied on foot are not spread evenly [10]. Another way is using x-ray to determine the height of arch by using a distance of medial longitudinal foot arch (largest and most important parts of foot arch) to the ground. More specifically, the distance perpendicular from the lower rim of the navicular tuberosity to the ground which can be represented as an index: arch foot called high and low if the distance is too big or small. If the distance is smaller than a certain level, the feet is flatfoot categories [11].

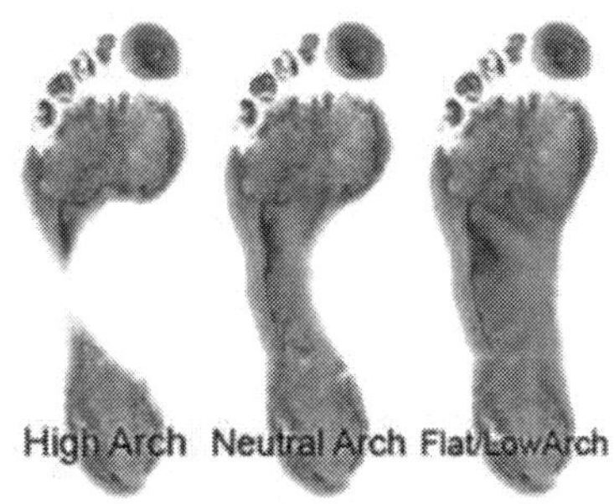

Fig. 1. Footprints of high arch, normal, and flatfoot from wet foot test.

A study by Kulkarni et al. has found another method to identify flatfoot by using footprint index from ratio of B intercept to A intercept where the footprint was found from pedobarography image. A intercept is a horizontal line from a to b through the center of mid foot, as shown in Figure-2. B intercept is the extension of line A intercept until cut the tangent line which is formed by connection of lateral minimum distance point and lower minimum distance point from Y axis, as shown in Figure- 2 by red dotted line. When this footprint index value is below 0.63, it is classifies as flatfootprint [12].

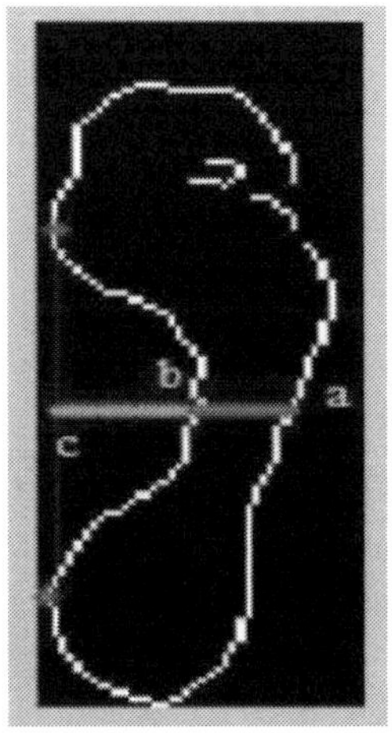

Fig. 2. MATLAB footprint image for calculating footprint index.

Another method to determine arch type which is widely used is Cavanagh's Arch Index (AI). This method is a division of mid foot area by the entire footprint area (excluding the toes). If AI less than 0.21, arch type of foot is called high arch. If AI between 0.21 and 0.26, arch type of foot is called normal arch. If AI greater than 0.26, arch type of foot is called flatfoot [13].

This study is to investigate the correlation between the ratio of loaded and unloaded entire foot area to the Cavanagh's AI which is very important for knowing the changing of shapes and deformation of foot when the sufferer is standing in an upright posture. The entire loaded foot area and footprint for evaluating AI derived from a digital footprint from the modified document scanner, while the entire unloaded foot area derived from a 3D scanner for foot orthotic [14].

2 Materials and methods

2.1 Subjects

One hundred and two healthy students of Mechanical Engineering Dept. Diponegoro University asked voluntarily for doing footprint. From 102 students found 63 participants identified as flatfoot (52 men and 11 women, aged 17-24 years, weight 45-129 kg, height 148- 179 cm) which each of them was no history of musculoskeletal injuries and has no physical defect on bones based on the check up at Diponegoro National Hospital Semarang.

The measurements were carried out at the Laboratory for Biomechanic of the Integrated Laboratory of Technical Service Unit (UPT Lab. Terpadu) at Diponegoro University. Table 1 shows the anthropometric data of the study group.

Table 1. Characteristics of the 102 participants.

	Range of Value	Mean and SD
Age (year)	17 – 26	20.0 ± 1.9
Gender (male : female)		87 : 15
Weight (kg)	39 – 129	63.3 ± 13.8
Height (cm)	147 – 185	167.3 ± 6.8
BMI (kg/m^2)	15.21 – 41.60	22.51 ± 4.16

2.2 Measures and procedures

In this study, measuring the entire unloaded foot area is derived from a 3D scanner which is a device that analyses a real-world object or environment to collect data on its shape and possibly its appearance (e.g. color). The collected data can then be used to construct digital three- dimensional models which are useful for a wide variety of applications.

3D Scanner Mini and Scansoft for Foot Orthotic made by Vismach Technology Ltd. China was used in this study. This scanner are active stereo vision 3D tech with white light pattern projection or no laser, so it is safe for eyes without protective glasses, quick capture (0.1 – 0.2 sec), clean 3D mesh, and ± 1.0 mm linier accuracy. The output 3D model of foot is in the standard language (dxf/stl/wrl/obj/ply/asc) format that can be exported into AutoCAD and fabricated by the (.stl) images to the rapid prototyping machine [15]. To obtain a 3D foot image is scanning soles of the feet as seen in Figure-3a. The output can be in the form of 3D plantar of foot, as shown in Figure-3b.

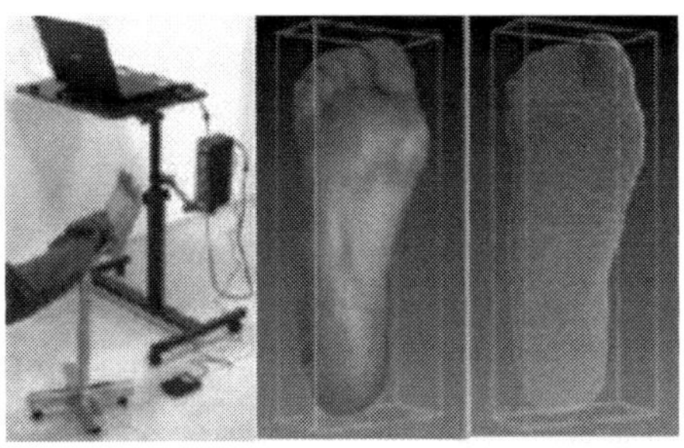

(a) (b)

Fig. 3. 3D scanning of the foot.

To obtain the entire loaded foot and footprint area for evaluating AI was done by scanning the foot of research subjects while standing in an upright posture above the glass window of digital footprint platform, modified from document scanner. Flatbed document scanner sized A4 with color depth 48 bits and optical resolution scanner 2400 ppi was used in this study. Modifications made to the frame and the glass scanner that can withstand loads up to 100 kg and glass used as the glass window is 10 mm thick clear glass, has a low reflection coefficients, and scratch-resistant. The scanner is designed for standing footprint for one foot at a time, as seen as Figure-4a. The scanning process occurs when the research subject is standing in an upright posture above the glass window and the foot is clean. Cleaning process of research subject's feet are by dipping into warm water (± 30o C) for about 2-3 minutes and then dried with tissue paper. Figure-4b is the scanning result of research subject's left foot.

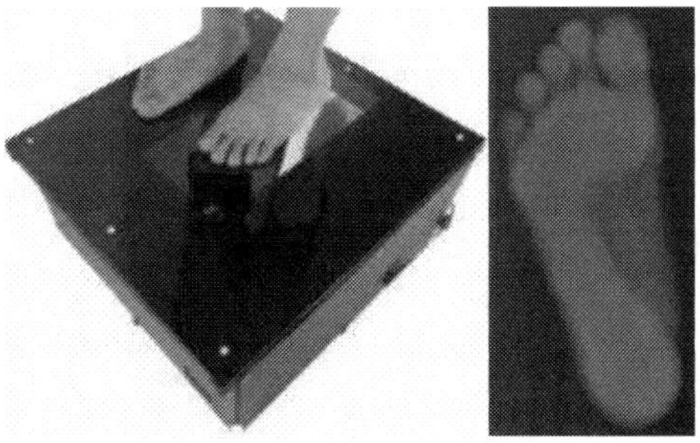

(a) (b)

Fig. 4. Footprint tool modified from document scanner.

To calculate contact area with and without the toes (to determine Cavanagh's AI), un-contacted foot (result of footprint image on Figure-4b) must be removed using MATLAB Software. The procedure is as follows: read the footprint image (Figure-4b) using *imread* function, (2) change the RGB image to gray scale image using *rgb2gray* function, (3) input the index level of image for filtering (index level value is gained from the point between contact and un-contact foot, normally between 88 to 96), (4) change the grayscale image to black and white image with *im2bw* function. The results obtained just images foot in contact as apparent in Figure-5b [15].

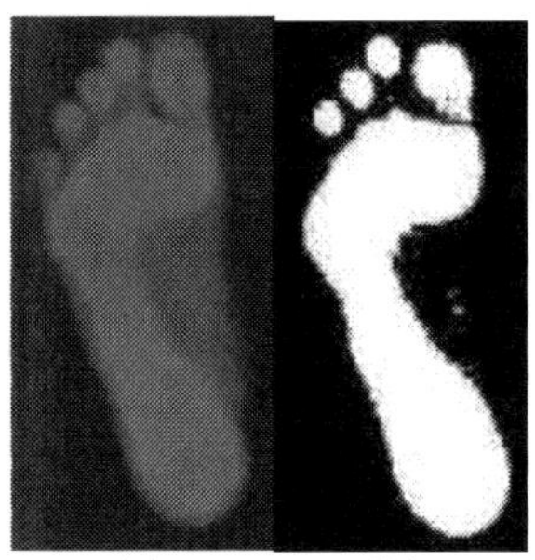

(a) (b)

Fig. 5. MATLAB evaluation removing un-contacted foot (a) result of RBG image to grayscale image, (b) black and white image after filtering with input of image index level.

2.3 Calculation and procedures

The entire unloaded foot area can be found by Rhinoceros 5.0 software with opening the .dxf file format 3D plantar of foot (Figure-3b) then choose command area. The result is shown at the upper left corner of the software, as shown in Figure-6.

DOI: 10.1051/matecconf/201713500060

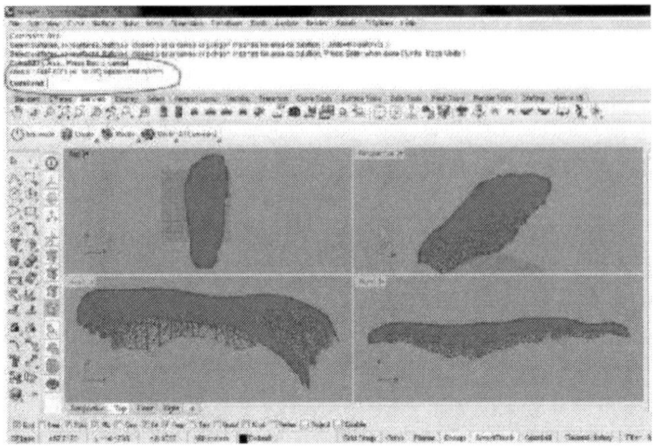

Fig. 6. Evaluating entire unloaded foot area using Rhinoceros 5.0 software.

Calculating the entire loaded foot area is done by using *bwarea* function in MATLAB, as shown in Figure-7. The result is shown in command window (Figure-8), but it is still in pixel unit [16].

```
LEFT.m
x = imread('EFLITA_LEFT.jpg');
y = rgb2gray(x);
figure(1)
imshow(y);
%%
[X1,map1]=gray2ind(y);
BW = im2bw(X1,map1,(88)/256);
figure(3)
imshow(BW);
bwarea(BW)
```

Fig. 7. MATLAB window for calculating foot area.

```
>> KIRI
Warning: Image is too big to fit on screen; displaying at 17%
> In imuitools\private\initSize at 71
  In imshow at 282
  In KIRI at 4
Warning: Image is too big to fit on screen; displaying at 17%
> In imuitools\private\initSize at 71
  In imshow at 282
  In KIRI at 8

ans =

      641506

>>
```

Fig. 8. MATLAB output window for pixels count of foot area.

To change the foot area in pixels to mm^2, scanner resolution data is needed. The scanner is set to scan in 200 ppi (default setting of scanner). The entire loaded foot area can be measured by using equation (1) as shown:

$$\left(\frac{\text{area in pixels}}{\text{scanner resolution}}\right) \times 645.16\ \text{mm}^2 \tag{1}$$

The Cavanagh's arch index is calculated through the following procedures. Draw the line from the center of the heel (at a distance of 15% of foot length, point I in Figure-9a) [17] to the tip of the second toe and extent this line until the heel cup. Mark the point of intersection between heel center lines with heel cup (point K in Figure- 9a). A second line perpendicular to the heel center line is then drawn tangentially to the most anterior part of the outline of the main body of the footprint in front of the metatarsal heads. Mark the point of intersection between these two lines (point j in Figure-9a). The line j-k then divided into equal thirds, as shown in Figure-9b, which divides the foot without the toes into three areas these are rear foot (A), mid foot (B), and fore foot (excluding the toes) (C) area. AI can be determined by equation (2) as follow:

$$\text{AI} = \frac{\text{B}}{(\text{A} = \text{B} + \text{C})} \tag{2}$$

where each area is calculated using MATLAB

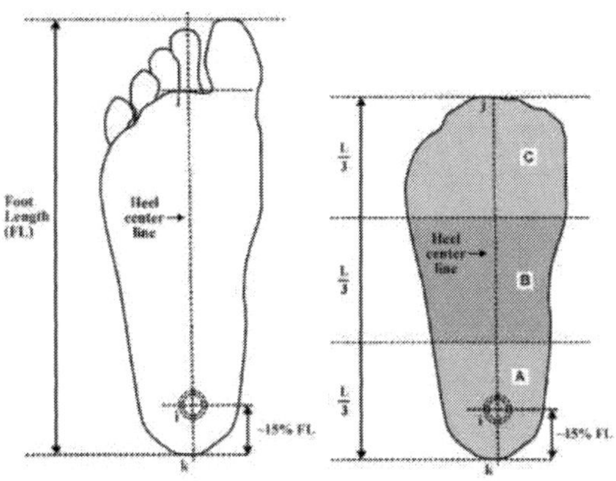

Fig. 9. Cavanagh's arch index.

3 Results

The three foot types (flat, normal, and high arch foot) provided by Cavanagh's AI as dependent variables and the three group of characteristics of the subjects (age, BMI, and ratio of foot area) as independent variables were compared using a 3 x 3 repeated measures ANOVA with Tukey's post hoc testing ($\alpha < 0.05$) using SPSS version 12. Ratio of foot area (RFA) is ratio between entire unloaded foot area and loaded foot area.

Thirty-one feet (30%) were classified as having normal foot, while 63 feet (62%) were classified as having flatfoot, and the remaining 8 feet (8%) were classified as a high arch. Table 2 shows the demographic information for all subjects.

DOI: 10.1051/matecconf/201713500060

Table 2. Demographic data.

	Age (year)	BMI (kg/m2)	RFA
Normal feet	20.3±2.2	21.78±3.56	0.521±0.050
Flat feet	19.8±1.8	23.12±4.34	0.609±0.091
High Arch	20.6±1.8	20.48±4.21	0.458±0.043
p-Value	0.409	0.008	0.001

In this study, the characteristic of age, BMI, and RFA are grouped into 3 groups, they are: (1) ages categorized by health department of Republic Indonesia: first adolescent (12 – 16 years old); last adolescent (17 – 25 years old); and adults (> 25 years old), (2) BMI: according to the standard rules, an individual is underweight if BMI < 18; normal if 25 ≥ BMI ≥18; and overweight if BMI > 25 [12], and (3) RFA: normal feet (RFA = 0.521 ± 0.050); flatfoot (RFA = 0.609 ± 0.091); and high arch (RFA = 0.458 ± 0.043).

There were significant interaction existed between BMI and AI (p = 0.008) and RFA and AI (p = 0.001). No significant difference was found in age (p = 0.409). From the results of different test between BMI and AI found the significant difference between underweight BMI group and normal BMI group with the result of AI (p< 0.001), underweight BMI group and overweight BMI group with the result of AI (p < 0.001), between normal BMI group and underweight BMI group with the result of AI (p < 0.001), and between overweight BMI group and underweight BMI group with the result of AI (p < 0.001).

From the results of different test between RFA and AI obtained significant difference between normal foot RFA group with flatfoot RFA group with the result of AI (p < 0.001), between flatfoot RFA group with normal foot RFA group with the result of AI (p < 0.001), between flatfoot RFA group with high arch RFA group with the result of AI (p < 0.001), and between high arch RFA group with flatfoot RFA group with the result of AI (p < 0.001).

Especially for flatfoot type of foot which is main focus in this study, there was a correlation between RFA and AI with coefficient of correlation r = 0.67 as shown in Figure-10. No correlation between BMI and age and AI because the coefficient of correlation are r = 0.33 and r =0.17 respectively.

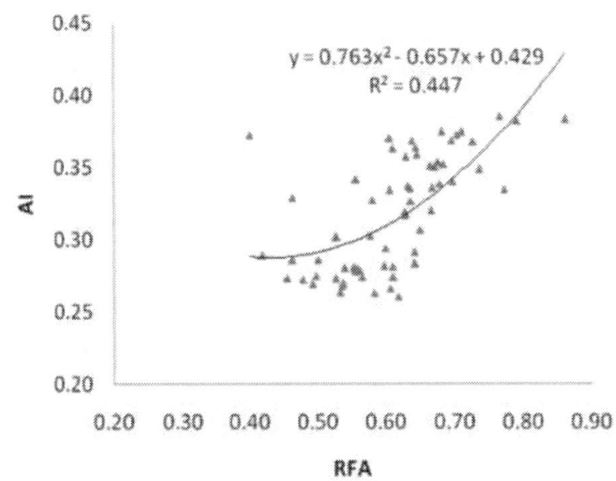

Fig. 10. Relationship between RFA to AI for flatfoot type.

4 Discussion

The result of this study indicate that no significant difference was found in age ($p = 0.409$), which means there is no difference between the age of subject with all the categories of AI. These results are consistent with research Chuckpaiwong et al. which resulted in no significant difference in age ($p = 0.9$) [18].

The correlation of BMI and AI is contrary. There are significant interaction existed between BMI and AI ($p = 0.008$), meanwhile in the flatfoot type of foot there is no significant correlation between BMI and AI with the coefficient of correlation $r = 0.33$. This is corresponding with statistical result which show that there is no significant differences were found between BMI normal and overweight with the result of AI ($p = 0.98$) and BMI over weight and normal with the result of AI ($p = 1.00$). Research conducted by Chuckpaiwong et al. which correlate 2 types of foot normal foot and flatfoot to height and weight also result no significant differences in weight ($p = 0.32$) or height ($p = 0.73$) between the normal and flatfeet [18].

Research in correlation between ratio of entire loaded and unloaded foot area to AI is new research and never done before. The statistical analyses show that there is significant difference between normal foot RFA and AI flatfoot ($p < 0.001$). Meanwhile, between normal foot RFA and AI high arch $p = 0.019$ ($p < 0.05$). It means there is significant differentiation between RFA groups with every categories of AI. From the result of polynomial regression for flatfoot type is shown that there is significant correlation between RFA to AI with the correlation coefficient $r = 0.67$ (see Figure-10).

In this study, digital footprint from the modified flatbed document scanner shows that there are difference between entire load foot area and area for evaluating AI with the result of wet foot test for about 7%. This result is corresponding to the research by Urry et al. which compared between AI from wet foot test with pressure platforms [10].
Determination of flatfoot's degree based on AI value also has not done before. Based on the Cavanagh and Rodgers research, people indicated as flatfoot if AI > 0.26 [13], in this study AI from flatfoot is grouped into 4 categories those are minor flatfoot (AI = 0.260 – 0.290); moderate flatfoot (AI = 0.291 – 0.334); severe flatfoot (AI = 0.335 – 0.364); and worst flatfoot (AI = 0.365 – 0.385). Table 3 shows the degrees of flatfoot with associate RFA values.

Table 3. Degrees of Flatfoot and associate RFA

Degrees of Flatfoot	Amount of Subjects	RFA	
		Range of Value	Mean and SD
Minor flatfoot	24 (38%)	0.417–0.642	0.543±0.057
Moderate flatfoot	13 (21%)	0.462–0.770	0.612±0.073
Severe flatfoot	15 (24%)	0.556–0.736	0.654±0.042
Worst flatfoot	11 (17%)	0.598-0.860	0.708±0.078

5 Conclusion

Wet foot test which is used widely in entire world (as illustrated in Figure-1) is proved to be un-accurate to identify flatfoot and high arch. The accuracy of footprint is very important due to handling of the flatfoot sufferers when designing orthotic shoes, especially for determining the dimension and contour of insertion of shoe insole in arch area [4].

Foot type identification using digital footprint can be done automatically (with helps of MATLAB software) after research subjects did scanning to determine foot area contact, type of foot (normal, flatfoot, and high arch), degrees of flatfoot, foot length, foot width, and shoes sizing. This digital footprint from modified flatbed document scanner also proved that it can produce an accurate footprint if the subject standing in an upright posture which can be assisted by the operator before scanning process is executed.

The statistical analyses shows there is a significant difference between RFA group and all categories of AI ($p < 0.05$), also coefficient of correlation $r = 0.067$ is found between RFA and AI. It means that flatfoot indication can be determined by RFA. More specific results can be seen from a study flatfoot categorization into 4 categories where the value RFA getting up from degree of flatfoot "minor flatfoot" (RFA = 0.543 ± 0.057) to "worst flatfoot" (RFA = 0.708 ± 0.078), as shown in Table 3. In relation to the deformation of the feet, worst flatfoot (AI = 0.365 – 0.385) means that the deformation of foot is bigger when the subject is standing in an upright posture above the platform. It is also contrary for degree of flatfoot "minor flatfoot" (AI = 0.260 – 0.290) the deformation of foot is small, which is shown by lower RFA value.

This study is advantageous to identify flatfoot and degrees of flatfoot through RFA with lower cost and easier compared to the method of determining height of the arch which research was done by Mall et al. [11], FPI method which was done by Kulkarni et al. [12], and Cavanagh's AI [13].

Acknowledgement

Our gratitude goes to the UNDIP Rector on PNBP Diponegoro University research fund Number: SP DIPA - 042.01.2.400898/2016 given. Acknowledgements also go to Orthopedhic Section of Tugurejo Hospital Semarang about technical identification of foot type by visual inspection and using wet foot test.

References

[1] C. G. Fontanella, Biomechanical Analysis of Heel Pad Tissues. Doctor Desertation. CMBM – Centre of Mechanics of Biological Materials, University of Padova, Italy (2009)

[2] R. S. Snell, Clinical Anatomy. (7th ed.). Philadelphia: Lippincott Williams & Wilkins (2004)

[3] Y.-W. Chang, H. W. Yi-Wen, H.-W. Wu, Y.-C. Chiu, H.-C.Hsu, Measurements of foot arch in standing, level walking, vertical jump and sprint start. International Journal of Sport and Exercise Sciences. **2**(2): 31-38 (2000)

[4] M. Root, et al., Normal and Abnormal Function of The Foot. Vol. II, Clinical Biomechanics Corp., Los Angeles (1977)

[5] S. Cubucku, M. K. Alimoglu, N. Balci, M. Beyazova, Plantar arch type and strength profile of the majory ankle muscle groups: a morphometric-isokinetic study. Isokinetics and Exercise Science, **13**, 217-222 (2005)

[6] Harris, J. Edwin, et al., Diagnosis and treatment of pediatric flatfoot. The Journal of Foot & Ankle Surgery. Clinical Practice Guideline. Volume **43**, Number 6 (2004)

[7] Wibowo, D. Basuki , Rudiansyah, Penanganan Nyeri Pada Telapak Kaki Secara Medik dan Bio-Mekanik. Seminar Teknologi Kesehatan dan Sosialisasi CBIOM3S, LPPM UNDIP Semarang (2015)

[8] Murley, S. George, Menz, B. Hylton, Landorf, B. Karl, A protocol for classifying normal-and flat-arched foot posture for research studies using clinical and radiographic measurements. Journal of Foot and Ankle Research, 2:22 (2009); doi:10.1186/1757- 1146-2-22

[9] Gross, K. Douglas, et al., Flat feet are associated with knee pain and cartilage damage in older adults. NIH Public Access. Arthritis Care Res. 2011 July; **63**(7) (2011); doi: 10.1002/acr.20431.

[10] Urry, R. Stephen, Wearing, Scott C, Arch indexes from ink footprints and pressure platforms are different. The Foot **15** (2005) 68-73. Elsevier (2005)

[11] N. A. Mall, W. M. Hardaker, J. A. Nunley, R. M. Queen, The reliability and reproducibility of foot type measurements using a mirrowed foot photo box and digital photography compared to caliper measurements. Journal of Biomechanics, **40**, 1171-1176 (2007)

[12] Kulkarni, S. Prachi, Kulkarni, B. Vinayak, Human footprint properties based on pedobarographic image analysis. International Journal of Advanced Research in Electronics and Communication Engineering (IJARECE). Volume **3**, Issue 11 (2014)

[13] R. Cavanagh, Peter, M. Rodgers, Mary, The Arch index: a useful measure from footprints. J. Biomechanics Vol. **20**, No. 5, pp. 547-551 (1987)

[14] Wibowo, D. Basuki, Haryadi, G. Dwi, A. Priambodo, Estimation of Foot Pressure from Human Footprint Depths Using 3D Scanner. AIP Conference Proceedings 1717, 040008 (2016); doi: 10.1063/1.4943451.

[15] ScanPod3D, 3D Scanner Mini and Scansoft for Foot Orthotic. Vismach Technology Ltd. http://www.scanpod3d.com (2013)

[16] C. R. Gonzalez, R. E. Woods, S. L. Eddins, Digital Image Procesing Using MATLAB. Second Edition. Gatesmark Publishing. A Division of Gatesmark, LLC (2009)

[17] A. S. Rodrigo, R. S. Goonetilleke, S. Xiong, Load distribution to minimise pressure-related pain on foot: a model. Ergonomics, Vol. **56**, No. 7, 1180-1193 (2014)

[18] B. Chuckpaiwong, J.A. Nunley, N. A. Mall, R. M. Queen, The effect of foot type on in-shoe plantar pressure during walking and running. Gait & Posture **28** (2008) 405 – 411, Elsevier

Pneumatic Muscles Actuated Lower-Limb Orthosis Model Verification with Actual Human Muscle Activation Patterns

M.A.M Dzahir[1, *], *N.S.M* Nor[1], *A.* Azaman[2], *M.* Hussein[1], and *M.A. Mat* Dzahir[1]

[1] Department of Applied Mechanics & Design, Faculty of Mechanical Engineering, UTM, Malaysia
[2] Department of Biotechnology & Medical Engineering, Faculty of Biomedical Engineering and Health Science, UTM, Malaysia

Abstract. A review study was conducted on existing lower-limb orthosis systems for rehabilitation which implemented pneumatic muscle type of actuators with the aim to clarify the current and on-going research in this field. The implementation of pneumatic artificial muscle will play an important role for the development of the advanced robotic system. In this research a derivation model for the antagonistic mono- and bi-articular muscles using pneumatic artificial muscles of a lower limb orthosis will be verified with actual human's muscle activities models. A healthy and young male 29 years old subject with height 174cm and weight 68kg was used as a test subject. Two mono-articular muscles Vastus Medialis (VM) and Vastus Lateralis (VL) were selected to verify the mono-articular muscle models and muscle synergy between anterior muscles. Two bi-articular muscles Rectus Femoris (RF) and Bicep Femoris (BF) were selected to verify the bi-articular muscle models and muscle co-contraction between anterior-posterior muscles. The test was carried out on a treadmill with a speed of 4.0 km/h, which approximately around 1.25 m/s for completing one cycle of walking motion. The data was collected for about one minute on a treadmill and 20 complete cycles of walking motion were successfully recorded. For the evaluations, the mathematical model obtained from the derivation and the actual human muscle activation patterns obtained using the surface electromyography (sEMG) system were compared and analysed. The results shown that, high correlation values ranging from 0.83 up to 0.93 were obtained in between the derivation model and the actual human muscle's model for both mono- and bi-articular muscles. As a conclusion, based on the verification with the sEMG muscle activities data and its correlation values, the proposed derivation models of the antagonistic mono- and bi-articular muscles were suitable to simulate and controls the pneumatic muscles actuated lower limb orthosis.

* Corresponding author: azuwan@mail.fkm.utm.my / azuwan@fkm.utm.my

1 Introduction

It is a general assumption that pneumatic muscles will play an important role in the development of assistive rehabilitation robotics system. In the last decade, the development of the pneumatic muscle actuated lower limb orthosis system was relatively slow compared to the motorized orthosis system. However, in recent years, the interest in this field had grown exponentially mainly due to the demand on a much compliant human-robotics system and advantageous attributes of the pneumatic muscles. Based on the review study, it could be understood that the suitable control schemes and strategies have yet to be found. Therefore, in this study, a novel control scheme and strategy for manipulating the anterior-posterior mono- and bi-articular muscle actuators was proposed using simplified pneumatic muscles actuated lower limb orthosis derivation model and sEMG muscle activities of the human lower limb muscles. From the literature review analysis on the pneumatic muscle actuated lower limb orthosis, it shows that around 16 prototyped had been evaluated based on the development of different orthosis systems in last 10 years [1 – 11].

Moreover, most of the existing pneumatic muscle actuated lower limb systems antagonistically controlled the leg orthosis with anterior-posterior mono-articular actuators at the hip, knee, or ankle joints [12 – 25]. However, compared to most of the published research articles, only few articles had introduced anterior-posterior bi-articular muscle actuators to simultaneously control the lower limb orthosis along with anterior-posterior mono-articular muscle actuators [29, 30]. The design of the proposed lower limb orthosis model was based on the human musculoskeletal system. This human musculoskeletal system consisted of different mono- and bi-articular muscles to support the movement of the joints. The muscles could be group into two different types of muscles which are mono-articular muscles (i.e., Gluteus Maximus, Gluteus Minimus, Gluteus Medius, Vastus Medialis and Vastus Lateralis) and bi-articular muscles (i.e., Gastrocnemius, Rectus Femoris, and Bicep Femoris). These antagonistic muscles (i.e., agonist and antagonist) exhibit co-contraction movements in order to initiate and precisely move the human exoskeleton system. The co-contraction of the human muscles also contributes in increasing the stiffness at the joints and improves the human balance. In addition, these set of movements also to prevent an injury at the joints. This concept will be utilised to design the control scheme and strategy of the lower limb leg orthosis.

Meanwhile, different suggestions and remarks also have been reached by the researchers to improve the design and control system of the pneumatic muscle actuated lower-limb orthosis, which why it might be concluded that the researchers' interest has been shifted to the implementation of the natural type of compliant actuators [2 - 11]. Although lots of researches have been investigated regarding the co-contraction movements of human antagonistic muscles. However, their model implementation in controlling the antagonistic muscle actuators of lower-limb orthosis has not been completely discovered [12 – 25]. In addition, research study which focuses on the implementation of mono- and bi-articular actuators using pneumatic muscles for the lower-limb rehabilitation orthosis has yet to be extensively investigated [1, 29]; thus, simply actuating the actuators might not give a good result on the joint's stiffness and stability of the lower-limb leg orthosis and its joint trajectories. Therefore, based on the related research findings, the simultaneous co-contractively like movements between the anterior and posterior actuators could be considered within the control system strategy [26 - 30].

Even though a considerable amount of work has now been done, this field is still rapidly evolving. The issue of which are the most effective control algorithms is still wide open. However, the randomized controlled trials are still necessary for identifying the suitable control algorithms even though it is expensive and time-consuming. Therefore, the purpose of this study is to validate the derivation model of the pneumatic muscle actuated lower limb orthosis with the actual human's muscle activities model during walking motion. Where, it is to identify the new method for controlling the orthosis system using the co-contractively control of the antagonistic mono- and bi- articular pneumatic muscle actuators.

2 Model derivation of the mono- and bi-articular muscle actuators

The contraction patterns of the antagonistic mono- and bi-articular artificial muscles of the pneumatic muscle actuated lower limb orthosis were found to be opposite to each other in between its anterior and posterior muscles [1]. Therefore, co-contraction models that generate input patterns for the antagonistic mono- and bi-articular actuators could be theoretically derived. To complete this theory, the model of the leg orthosis should be simplified before all the information could be group together, and then correlate all the important parameters into a mathematical equation as shown in Figure 1. The obtained model equations correlate the contraction of each of the antagonistic actuators with the positional data of hip and knee joints. The model derivation of the anterior-posterior contraction patterns could be referred from the previous journal publication [29]. Based on the derived mathematical model, the contraction of antagonistic mono-articular actuators for hip joint has a linear relationship with the hip joint angle. The contraction was also similar when compared at the antagonistic mono-articular actuators for the knee joint. Interestingly, the antagonistic bi-articular actuators have a linear relationship with a total of hip and knee joint angle as it manages the two joints simultaneously. The general idea for this model derivation was obtained based on the information that could be interpreted from the reference input data of the hip and knee joints.

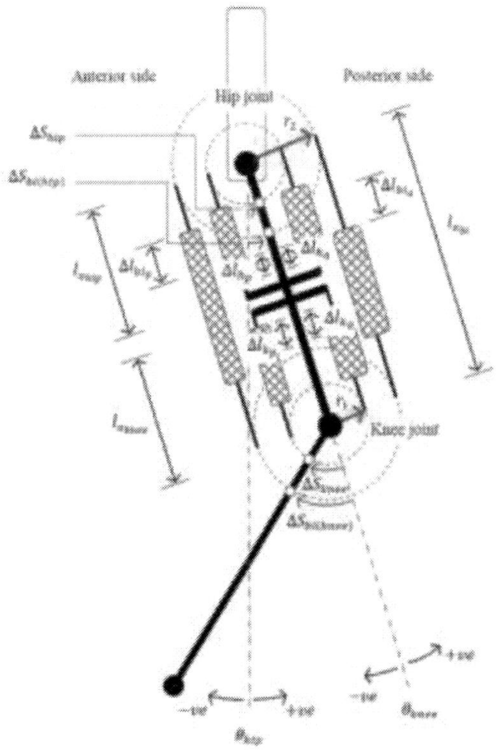

$$\varepsilon_{HP}(t) = \left(\frac{r}{L_o}\right).\beta_h.\theta_{hp}(t) \le 0.3 \quad ...(1)$$

$$\varepsilon_{HA}(t) = \left(\frac{r}{L_o}\right).\alpha_h.\theta_{ha}(t) \le 0.3 \quad ...(2)$$

$$\varepsilon_{KP}(t) = \left(\frac{r}{L_o}\right).\beta_k.\theta_{kp}(t) \le 0.3 \quad ...(3)$$

$$\varepsilon_{KA}(t) = \left(\frac{r}{L_o}\right).\alpha_k.\theta_{ka}(t) \le 0.3 \quad ...(4)$$

$$\varepsilon_{BP}(t) = \left(\frac{r}{L_o}\right).\beta_{bi}.(\theta_h(t) + \theta_k(t))_p \le 0.3 \quad ...(5)$$

$$\varepsilon_{BA}(t) = \left(\frac{r}{L_o}\right).\alpha_{bi}.(\theta_h(t) + \theta_k(t))_a \le 0.3 \quad ...(6)$$

Fig. 1. Pneumatic muscle actuated lower limb orthosis simplified model for mono- and bi-articular muscles between hip and knee joints; where, ε_{HA} and ε_{HP} are the anterior-posterior muscle contraction of the hip joint; ε_{KA} and ε_{KP} are the anterior-posterior muscle contraction of the knee joint; ε_{BA} and ε_{BP} are the anterior-posterior muscle contraction of the bi-articular muscle (hip and knee joint); α and $\beta$are muscle activation level for the anterior-posterior muscles; θ_H and θ_K are hip and knee joints.

3 Methodology

The first part of this research is to obtain a derivation model equations of the pneumatic muscle actuated lower limb orthosis as shown in Figure 1. These model equations represent the contraction value of each of the antagonistic mono- and bi-articular pneumatic muscles of the orthosis system. By introducing an input pattern of hip and knee joints during one complete cycle of walking motion into the mathematical model equations, the contraction-expansion value of each of anterior-posterior mono- and bi-articular pneumatic muscles can be obtained, and the verified with the actual human's muscle activities model. The second part of the research is to obtain sEMG muscle activities data model for both mono- and bi-articular muscles of human's lower limb using fifth order of polynomial equation. A healthy and young male 29 years old subject with height 174cm and weight 68kg was used as a test subject. The subject is an active sport player and has a strong lower limb muscles. Two mono-articular muscles which are Vastus Medialis (VM) and Vastus Lateralis (VL) were used to verify the mono-articular muscle models and muscle synergy between anterior muscles. Two bi-articular muscles Rectus Femoris (RF) and Bicep Femoris (BF) were used to verify the bi-articular muscle models and muscle co-contraction between anterior-posterior muscles. The sEMG system (TMSi system, Netherland) was used to obtain sEMG muscle activation data of four muscles during walking motion. The test was carried out on a treadmill with a speed of 4.0 km/h, which approximately around 1.25 m/s for completing one cycle of walking motion. Four sEMG electrodes were placed at the VM, VL, RF and BF muscles. The data was collected for about one minute on a treadmill and 20 complete cycles of walking motion were successfully recorded. The data were processed, filtered, and rectified to remove all the unrequired noise. A fifth order polynomial equations were used correlate the data into a model equations. The final part of the research is to verify and correlate the obtained derivation model with the actual human lower limb muscle's model for both mono- and bi-articular muscles.

4 EMG muscle activities

It is strongly believed that the human antagonistic muscles activate simultaneously, when one muscle (i.e., agonist or antagonist) is in contraction the other muscle (i.e., antagonist or agonist) will be in expansion and vice versa. We define this as muscle co-contraction or muscle synergy. Muscle co-contraction refers to when any movement occurs which involved two sets of muscles working around joint. Normally, the muscles on one side of the joint must relax so that the muscles on the other side can contract. However, in co-contraction is defined as both sets of muscles contract. Moreover, the definition of the co-contraction also could be explained as both muscles were simultaneously made contraction-expansion movements. When the muscles on one side of the joint are in contraction, the opposite muscles will not be in completely relaxed state. However, they are still in contraction but less than the opposite muscles. The muscle synergy refers to the interaction of two or more sets of muscles to produce a combined effect greater than the sum of their separate effects. Therefore, the result of this research study would verify the muscle activities of the human lower limb muscles both on the antagonistic mono- and bi-articular muscles. However, not all lower limb muscle could be easily obtained using sEMG muscle activity sensors due to small muscle's site and weak muscle activation. Thus only VM, VL, RF, and BF muscles were selected to represent the anterior-posterior mono- and bi-articular muscles.

5 Results and discussion

For the result evaluations, VM and VL muscles were used to verify the anterior mono-articular muscle model while, RF and BF muscles were used to verify anterior-posterior bi-articular muscle models. In this research, it was crucial to obtain good correlation value for the sEMG muscle activities of the antagonistic bi-articular muscles because these muscles activities were more complex when compared to the mono-articular muscles. Figure 2 shows the evaluation of the anterior mono-articular muscle model obtained from the mathematical derivation with VM (alone) and VM-VL (combined) muscle activities models. While, Figure 3 shows the evaluation of the anterior-posterior bi-articular muscle models obtained from the mathematical derivation with RF and BF muscle activities models.

Based on the results, it shows that the derivation models for both anterior-posterior mono- and bi-articular muscles were comparable to the actual muscles activities for the VM, VL, RF, and BF lower-limb muscles. In addition, the patterns of the sEMG muscle activities data were highly correspond with the derived antagonistic mono- and bi-articular artificial muscle models. These results might indicate that the muscle activities of the antagonistic mono- and bi-articular muscles followed a certain patterns which can be defined by suitable model derivation. The evaluation on the VM and VM-VL muscle activity patterns which were both anterior mono-articular muscle proved that muscle synergy was occurred between same anterior or posterior muscles. Figure 2 shows that there were an improvement in between the muscle activities data patterns when we compared the VM muscle activities alone and a combination of VM-VL muscle activities with the derivation models. Furthermore, the evaluation on the RF and BF muscle activities patterns proved that the human lower limb muscles exhibit co-contraction movements between anterior and posterior muscles. However, an existing delay less than 0.40 seconds occur in-between the activation of the anterior and posterior sEMG muscle activities. Albeit that, the patterns of the sEMG muscle activities for VM, VL, RF, and BF muscles were still agreeing with the derived antagonistic mono- and bi-articular muscle models.

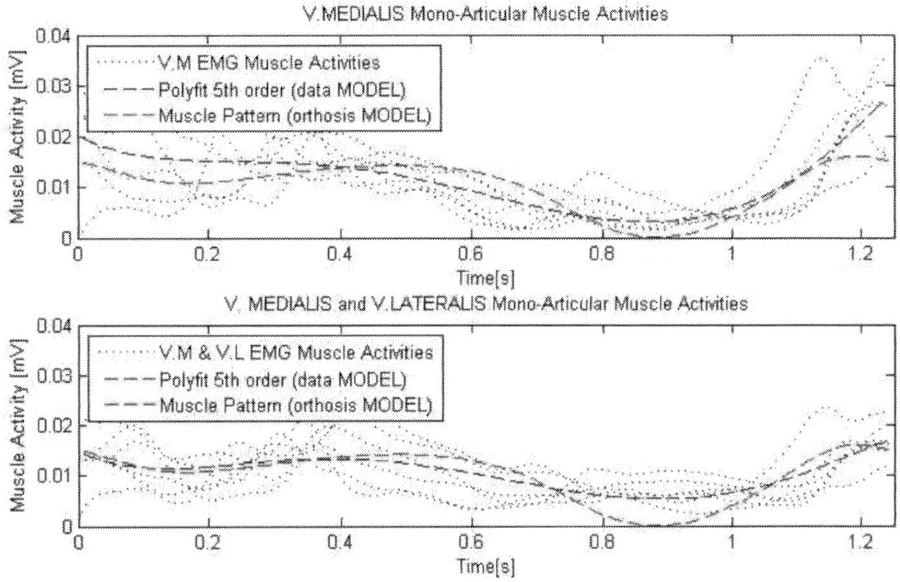

Fig. 2. Evaluation of the mono-articular muscle's mathematical model with the Vastus Medialis (VM) and Vastus Lateralis (VL) muscle activities.

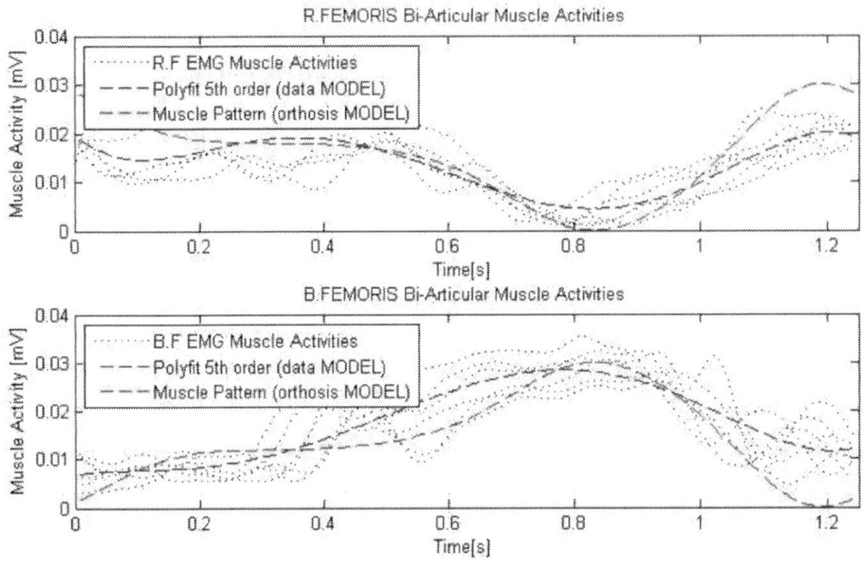

Fig. 3. Evaluation of the bi-articular muscle's mathematical model with the Rectus Femoris (RF) and Bicep Femoris (BF) muscle activities.

The correlation tests between the sEMG muscle activities for VM, VL, RF, and BF muscles models (data MODEL) with derivation models (orthosis MODEL) after rendering the existing delay produced high correlation coefficient values. The correlation value of the antagonistic mono-articular muscles varies between $r = 0.83$ for the VM muscle activities and $r = 0.93$ for the VM-VL muscle activities. On the other hand, the correlation value of the antagonistic bi-articular muscles varies between $r = 0.91$ for the RF muscle activities and $r = 0.86$ for the BF muscle activities. Therefore, it could be conclude that the proposed derivation models of the antagonistic mono- and bi-articular muscles were suitable and reliable to be implemented into the control scheme and strategy of the pneumatic muscles actuated lower limb orthosis.

6 Conclusions

In conclusions, the verification between sEMG muscle activities of VM, VL, RF, and BF muscles with derivation models shows that the antagonistic mono- and bi-articular muscle activities exhibit co-contraction movements in between anterior and posterior muscles, and muscle synergy in between multiple anterior or posterior muscles with respect to specific joints (i.e., hip and knee joints). Therefore, the hypothesis that the antagonistic muscles were activated and contract simultaneously could be implemented. However, there might be appropriate to include some delay as well in-between the activation of the anterior-posterior mono- and bi-articular muscle models. Based on the verification of the sEMG muscle activities data and its correlation values, it could be concluded that the proposed derivation models of the antagonistic mono- and bi-articular muscles were suitable to simulate and controls the pneumatic muscles actuated lower limb orthosis. Finally, it is crucial to obtain a good derivation model for the antagonistic mono- and bi-articular muscle models to successfully simulate and control the pneumatic muscle actuated lower limb orthosis comparable to human musculoskeletal system. In the future, an actual model of the pneumatic muscle actuated lower limb orthosis would be implemented.

Acknowledgements

This research study was supported by Potential Academic Staff (PAS) research grants Q.J130000.2724.02K16, Universiti Teknologi Malaysia (UTM).

References

1. M.A.M. Dzahir, S.I. Yamamoto, Recent trends in lower-limb robotic rehabilitation orthosis: control scheme and strategy for pneumatic muscle actuated gait trainers, Robotics, **3**(2), 120–148 (2014)
2. B.G. Nascimento, C.B. Vimieiro, D.A. Nagem, M. Pinotti, Hip orthosis powered by pneumatic artificial muscle: Voluntary activation in absence of myoelectrical signal, Artif. Organs, **32**, 317–322 (2008)
3. C.B.S. Vimieiro, B.G. Nascimento, D.A.P Nagem, M. Pinotti, Development of a hip orthosis using pneumatic artificial muscles, In Proceeding of TMSi, (2005) July18–1; São Paulo, Spain
4. K. Bharadwaj, T.G. Sugar, Kinematics of a robotic gait trainer for stroke rehabilitation, In Proceedings of the IEEE International Conference on Robotics and Automation, (2006) May 15–19;Orlando, FL, USA
5. D.P. Ferris, K.E. Gordon, G.S. Sawicki, A. Peethambaran, An improved powered ankle-foot orthosis using proportional myoelectric control, Gait Posture **23**, 425–428 (2006)
6. D.P. Ferris, J.M. Czerniecki, B. Hannaford, An ankle-foot orthosis powered by artificial pneumatic muscles, J. Appl. Biomechanics, **21**, 189–197 (2005)
7. K.E. Gordon, G.S. Sawicki, D.P. Fessis, Mechanical performance of artificial pneumatic muscles to power an ankle-foot orthosis, J. Biomech., **39**, 1832–1841 (2006)
8. N. Costa, M. Bezdicek, M. Brown, J.O. Gray, D.G. Caldwell, Joint motion control of a powered lower limb orthosis for rehabilitation, Int. J. Autom. Computation, **3**, 271–281 (2006)
9. T. Miyoshi, K. Hiramatsu, S.I. Yamamoto, K. Nakazawa, M. Akai, Robotic gait trainer in water: Development of an underwater gait-training orthosis, Disabil. Rehabilitation, **30**, 81–87 (2008)
10. P. Malcom, P. Fiers, V. Segers, I. Caekenberghe, M. Lenoir, D. Clercq, Experimental study on the role of the ankle push off in the walk-to-run transition by means of a powered
ankle-foot-exoskeleton, Gait Posture, **30**, 322–327 (2009)
11. P. Malcom, V. Segers, I. Caekenberghe, D. Clercq, Experimental study of the influence of the m. tibialis anterior on the walk-to-run transition by means of a powered ankle-foot-exoskeleton, Gait Posture, **29**, 6–10 (2009)
12. S. Galle, P. Malcom, W. Derave, D. Clercq, Adaptation to walking with an exoskeleton that assists ankle extension, Gait Posture, **38**, 495–499 (2013)
13. P. Malcom, W. Derave, S. Galle, D. Clercq, A simple exoskeleton that assist plantarflexion can reduce the metabolic cost of human walking, PLoS One, **8**, 0056137 (2013)
14. G.S. Sawicki, D.P. Fessis, A pneumatically powered knee-ankle-foot orthosis (KAFO) with myoelectric activation and inhibition, J. Neuro-Eng. Rehabil., **6**, 23:1–23:16 (2009)
15. T.T. Deaconescu, A.I. Deaconescu, Pneumatic muscle actuated equipment for continuous passive motion, IAENG Trans. Eng. Technol., doi:10.1063/1.3256258 (2009)

16. B. Pieter , V.D. Michael, V.H. Ronald, V. Bram, L. Dirk, Design and control of lower limb exoskeleton for robot-assisted gait training, Applied Bionics and Biomechanics, **6**(2), 229–243 (2009)
17. B. Pieter , V.D. Michael, V.H. Ronald, V. Bram, L. Dirk, Pleated pneumatic artificial muscle based actuator system as a torque source for compliant lower limb exoskeletons, IEEE/ASME Transactions on Mechatronics, **19**(3), 1046–1056 (2014)
18. T.J. Yeh, M.J. Wu, T.J. Lu, F.K. Wu, C.R. Huang, Control of McKibben pneumatic muscles for a power-assist, lower-limb orthosis, Mechatronics, **20**, 686–697 (2010)
19. J. Carberry, G. Hinchly, J. Buckerfield, E. Taylor, T. Burton, S. Madgwick, R. Vaidyanathan, Parametric design of an active ankle foot orthosis with passive compliance, In Proceedings of the Computer-Based Medical System (CBMS), (2011) June 27–30; Bristol, UK
20. Y. Park, B. Chen, D. Young, L. Stirling, R. Wood, E. Goldfield, R. Nagpal, Bio-inspired Active Soft Orthotic Device for Ankle Foot Pathologies, In Proceedings of the International Conference on Robots and Systems (IROS), (2011) September 25–30; San Francisco, CA, USA
21. Y. Park, B. Chen, C. Majidi, R. Wood, R. Nagpal, E. Goldfield, Active Modular Elastomer Sleeve for Soft Wearable Assistance Robots, In Proceedings of the IEEE International Conference on Robots and Systems (IROS), (2012) October 7–12; Vilamoura, Portugal
22. C.M. Teng, Z.Y. Wong, W.Y. The, Y.Z. Chong, Design and development of inexpensive pneumatically-powered assisted knee-ankle-foot orthosis for gait rehabilitation-preliminary finding, In Proceedings of the International Conference on Biomedical Engineering (ICoBE), (2012) February 27–28; Penang, Malaysia
23. T. Kawamura, K. Takanaka, Development of an orthosis for walking assistance using pneumatic artificial muscle-a quantitative assessment of the effect of assistance, In Proceedings of the International Conference on Rehabilitation Robotics, (2013) June 24–26; Seattle, WA, USA
24. S. Hussain, S.Q. Xie, P.K. Jamwal, Adaptive impedance control of a robotic orthosis for gait rehabilitation, IEEE Trans. Cybern., **43**, 1025–1034 (2013)
25. S. Hussain, S.Q. Xie, P.K. Jamwal, Robust nonlinear control of an intrinsically compliant robotic gait training orthosis, IEEE Trans. Syst. Man Cybern.: System, 43, 655–665 (2013)
26. K.D. William, Task-based methods for evaluating electrically stimulated antagonist muscle controllers, IEEE Trans. on Biomedical Engineering, **36** (1989)
27. M. Samer, F. Philippe, G. David, P. Philippe, E.M. Hassan, Towards a co-contraction muscle control strategy for paraplegics. IEEE Conf. on Decision and Control. (2005)
28. H. Stewart, F. Norm, B. Michael, Muscle co-contraction modulates damping and joint stability in a three-link biomechanical limb, Frontiers in Neurorobotics, **5** (2012)
29. M.A.M. Dzahir, S.I. Yamamoto, Design and Evaluation of the AIRGAIT Exoskeleton: Leg Orthosis Control for Assistive Gait Rehabilitation, Journal of Robotics, vol. **2013**, Article ID 535106, 20 pages, doi:10.1155/2013/535106 (2013)
30. P. Flavio, M.A.M. Dzahir, S.I. Yamamoto, Computed-torque method for the control of a 2 DOF orthosis actuated through pneumatic artificial muscles: a specific case for the rehabilitation of the lower limb, Medical Physics, arXiv: 1404.6968 [physics.med-ph] (2014)

Fabrication of Porous Stainless Steel 316L for Biomedical Applications

F. Mat Noor[1,2,*], *K. R.* Jamaludin[2] , and *S.* Ahmad[1]

[1]Faculty of Mechanical & Manufacturing Engineering, Universiti Tun Hussein Onn Malaysia (UTHM), 86400 Parit Raja, Batu Pahat, Johor, Malaysia
[2]UTM Razak School of Engineering & Advanced Technology, Universiti Teknologi Malaysia Kuala Lumpur, 54100 Jalan Semarak, Kuala Lumpur, Malaysia.

Abstract. Porous metals are very attractive materials for biomedical application as the physical and mechanical properties of these materials can be tailored similarly to the natural bone. In this work, porous stainless steel 316L has been fabricated by foam replication method. This method offers a lot of advantages including of easy processing technique, very economic, does not involve the use of toxic chemical and capable of producing porous structure that almost similar to natural bone. The porous stainless steel 316L (SS316L) samples were prepared by varying the SS316L composition from 40 wt% to 60 wt%. Sintering process was carried out at 1250°C in a vacuum furnace. The microstructure and pore size were observed and determined through Scanning Electron Microscope (SEM). Archimedes method was used to measure the samples density, while compression test was carried out to determine the compressive strength and elastic modulus. The average pore size for samples with 50 wt% and 60 wt% SS316L are 268μm and 299μm respectively. Samples with 40 wt% SS316L experienced the largest shrinkage which is 33% while the sample with 60 wt% SS316L experienced the smallest shrinkage which is 21%. The density and porosity of the porous SS316L with 50 wt% SS316L are 0.43g/cm3 and 93.6% respectively, and for porous SS316L with 60 wt% SS316L are 0.69 g/cm3 and 89.2% respectively. The modulus of elasticity and compressive strength for porous SS316L with 60 wt% SS316L are 0.46 GPa and 56 MPa respectively. All these properties are in the range of the natural bone properties. Besides, the cytotoxicity test showed that this porous SS316L does not have a cytotoxic potential for biomedical implant.

1 Introduction

Porous metals that are commonly used for biomedical applications are titanium and titanium alloys, Cobalt Chrome and nitinol shape memory alloys and stainless steel 316L [1]. Amongst all, stainless steel 316L is the longest being applied in the biomedical implant which is since 1920's [1, 2]. Stainless steel 316L are low cost, easy availability and processing, good workability, high fatigue life and fracture toughness [1, 3, 4]. However,

* Corresponding author: fazimah@uthm.edu.my

the modulus of elasticity of stainless steel 316L is much higher compared to natural bone which is 193 GPa for stainless steel 316L and 0.1-20 GPa for natural bone. One of the most effective ways to reduce the stiffness mismatches or stress shielding effect between the natural bone and medical implant is through the using of porous metals or metal foams [5].

Foam replication method is basically a simple, easy and cheap method. This method capable of producing metal foams with a porous structure that most resembles cancellous bone [6]. However, every single processing step needs for proper study and exploration in order to obtain defect-free structures with optimum properties [7]. Figure 1 shows that the microstructure of the natural bone is almost similar to the microstructure of polyurethane (PU) foam used as a sacrificial template for this work. The open and interconnected pores will provide a stable fixation through the bone ingrowth into the pores between the metal implant and natural bone [8].

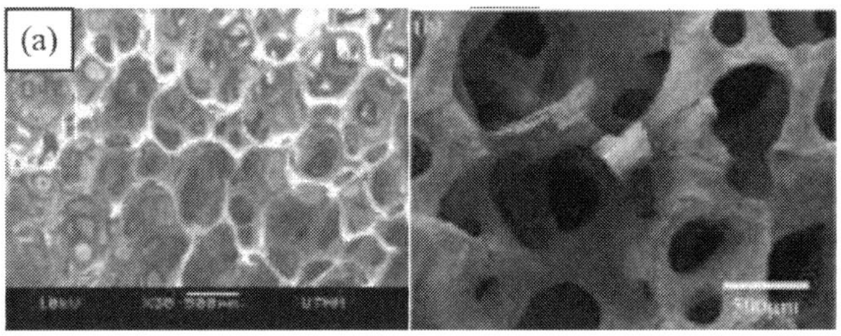

Fig. 1. Microstructure of the (a) polyurethane foam (PU) template and (b) natural bone [17] .

One of the major factors that determine the successful of a biomaterial or an implant is the properties of the biomaterials. The metal implants produced should have the properties similar to the natural bone. Table 1 listed some of the characteristics and properties of the natural bone.

Table 1. Properties of natural bone[8, 9]

Compressive strength (MPa)	Young's Modulus (GPa)	Pore size (μm)	Porosity (%)	Density (g/cm^3)
2-200	0.1-20	50-500	50-90	0.05-1.0

2 Methodology

The SS316L slurry with the composition of 40 wt.%, 50 wt.% and 60 wt.% SS316L was prepared by mixing with 2.5 wt.% polyethylene glycol (PEG), 2.5 wt.% methyl cellulose (CMC) and distilled water using a mechanical stirrer for 2 hours. PU foam templates were then impregnated in the slurry and the excessive slurry was manually squeezed out. After that, the coated PU foam templates were dried in the oven at 30°C for 24 hours. The dried templates were then sintered at a temperature of 1250°C for 90 minutes.

The microstructure of the porous SS316L was examined by using Scanning Electron Microscope. The density and porosity were measured by using Archimedes principle.

Compression test was performed according to ISO 13314: Mechanical testing of metals-Ductility testing – Compress a speed rate of 0.5 mm/min by using the universal testing machine (UTM), Shimadzu AG-1 10kN. Cytotoxicity test was evaluated according to ISO 10993-5:2009 "Biological evaluation of medical devices. Part 5-Tests for in vitro cytotoxicity" for the sample with the highest SS316L composition (60 wt.%). The test item was tested in triplicates at six concentrations of 0.2, 0.1, 0.05, 0.025, 0.0125, and 0.00625 g/mL. The treatment was carried out at 37±1 °C in a carbon dioxide incubator and assessment carried out after 24 hours incubation.

3 Results and Discussion

Figure 2 shows the digital images of the SS316L foam samples before and after sintering. The sample before sintering is shown in Figure 1 (a) with a cubic shape of 15mm. The sample with 40 wt% SS316L experienced the most significant shrinkage and shape deformation. It is followed by the sample with 50 wt.% SS316L. Only the sample with 60 wt.% SS316L compositions retains its original shape and slightly shrinks. Therefore only this sample has been forwarded for the compression test. Based on the samples dimension taken before and after sintering, it was found that the sample with 60 wt.% SS316L shrink about 20.62%, sample with 50 wt.% SS316L shrink about 29.76% while the sample with 40 wt.% SS316L shrink about 32.56%. The high shrinkage percentage for samples with 40 wt.% and 50 wt.% SS316L is due to the serious shape deformation. The shrinkage percentage increases with the decreasing of the SS316L composition. At lower SS316L composition, there were not enough SS316L particles to fully coat the PU foam struts. As the PU foam template burn out during sintering, the struts were seriously collapsed and large shrinkage occurs inside the SS316L foam [10]. This shrinkage was also due to the elimination of pores during the sintering process [11]. Most of the porous materials shrink and experienced net-shape deformation or distortion during sintering [12].

Fig. 2. Digital images of (a) PU foam template coated with SS316L slurry before sintering, (b) sintered of 40 wt% SS316L, (c) sintered of 50 wt% SS316L, and (d) sintered of 60 wt% SS316L

The effect of increasing the SS316L composition on the samples' density and porosity can be seen in Figure 3. From the figure, the density increased while porosity decreased as the SS316L composition has been reduced from 50 wt.% to 40 wt.%. It is related to the higher amount of closed pores found in the samples especially for sample with 40 wt.% SS316L composition. The samples were totally collapsed after sintering, no open pores produced and the samples become densified [13]. Nevertheless, the samples density increased while porosity decreased when the SS316L composition was increased from 50 wt.% to 60 wt.%, which was agreed with the result found by the previous researcher [13]. The density of the samples with 50 wt% and 60 wt.% SS316L are 0.43 g/cm^3 and 0.60 g/cm^3 respectively. While the porosity percentage for the sample with 50 wt.% and 60 wt.%

DOI: 10.1051/matecconf/201713500062

SS316L are 93.6% and 89.2% respectively. The struts size of sample with 60 wt.% SS316L in figure 4 (c) is thicker compared to sample with 50 wt.% SS316L as shown in figure 4 (b). The density and porosity of these samples are in the range of the natural bone's density and porosity [8, 9] .

Figure 4 shows the microstructure of the porous SS316L at different SS316L composition. There are more open pores found in the sample with higher SS316L composition. As can be seen in Figure 3 (a), the sample with 40 wt.% SS316L composition consisted of mostly closed pores due to the formation of liquid films bridging the pores. M.J. Matos et al [14] proved that the presence of closed cell windows contributes to the overall shrinkage. However, as the SS316L composition was increased to 50 wt.% and 60 wt.% as shown in Figure 4 (b) and 4 (c), almost all pores are open and interconnected throughout the samples with an average pore size of 268 µm and 299 µm respectively. The average pore size for samples with 60 wt.% SS316L is slightly higher than the samples with 50 wt.% SS316L since the shrinkage percentage for this sample is lower than the sample with 50 wt.% SS316L.

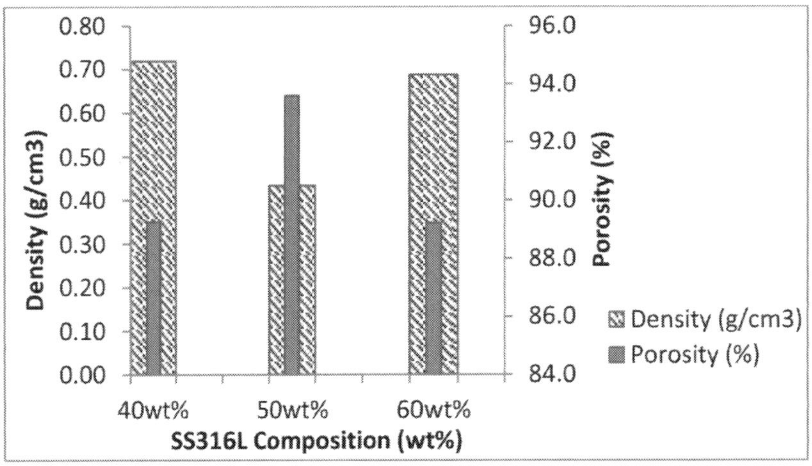

Fig. 3. Density and porosity percentage of porous stainless steel 316L at different SS316L composition

In general, the slurry should have a suitable viscosity to flow into the PU foam network and uniformly covering the PU foam and stay adhered on the PU foam struts until the sintering process. If the SS316L composition is too high, the slurry will become too thick. Thus, it will be more difficult for the slurry to enter the PU foam and the excessive slurry after the impregnation also could not be squeezed out easily. Nevertheless, if the SS316L composition is too low, the slurry will become too thin. Thus, there will be a limited amount of SS316L successfully coated and adhered on the PU foam struts and therefore it is difficult to maintain the shape of porous SS316L after the PU foam is burned out during the sintering [15].

According to J. Banhart et al [16], in order to evaluate the potential application of porous materials, it is very important to determine their compression behaviours. As for examples, the compression strength and the stress-strain curve. In this work, from the compression test, the modulus of elasticity and compressive strength obtained for sample with 60 wt.% SS316L are 0.46 GPa and 56 MPa respectively. This results are very promising since this properties are in the range of the natural bone properties.

DOI: 10.1051/matecconf/201713500062

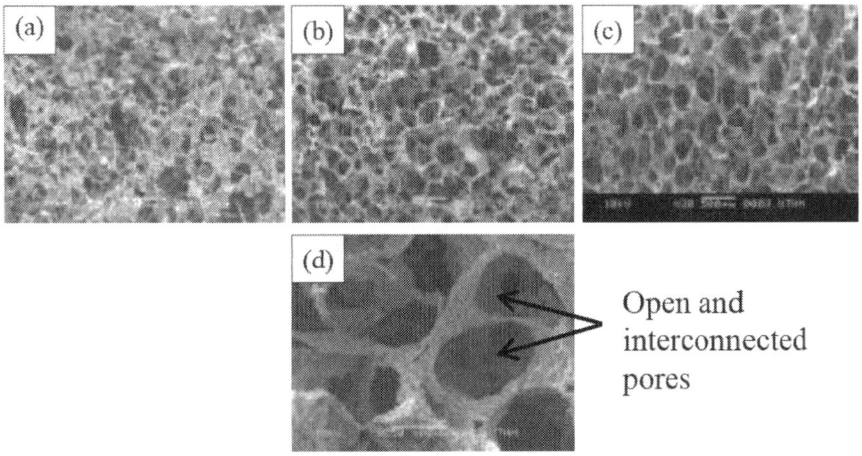

Fig. 4. Microstructure of the porous stainless steel 316L at (a) 40 wt% SS316L, (b) 50 wt% SS316L, (c) 60 wt% SS316L and (d) is the microstructure of 60 wt% SS316L at higher magnification.

Biocompatibility is associated with the corrosion behaviour of a material in the body and the tendency for this material to discharge possible toxic ions. Table 2 presents the results of the cytotoxicity tests of this porous SS316L. Positive and negative controls were included in this study to verify the proper functioning of the test system. The result showed that the toxicity levels were 0 after cultivation for 24 hours. This indicated that performance of SS316L foam is stable and without poisonous substance precipitation and other obvious side effects on cells. Cytotoxicity assay showed that this porous SS316L does not have a cytotoxic potential.

Table 2. Summary of cytotoxic grading on cultures

Test extracts and controls	Grade	Reactivity
Test extract, 0.2 g/mL	0-0-0	None
Test extract, 0.1 g/mL	0-0-0	None
Test extract, 0.05 g/mL	0-0-0	None
Test extract, 0.025 g/mL	0-0-0	None
Test extract, 0.0125 g/mL	0-0-0	None
Test extract, 0.00625 g/mL	0-0-0	None
Medium control	0-0-0	None
Negative control	0-0-0	None
Positive control	4-4-4	Severe

4 Conclusions

It is possible to produce porous SS316L by foam replication method for biomedical implants since the properties of the porous SS316L produced in this work similar to the natural bone properties and the cytotoxicity test also shows that this porous SS316L has good cytocompatibility. However, adding the appropriate amount of SS316L composition during the slurry preparation is very important. The slurry should not become neither too thick nor too thin. The density of the samples is in the range of 0.43 g/cm^3 to 0.60 g/cm^3. The porosity percentage of the samples was found the range of 89.2% to 93.6%. Whereas, the average pores size of porous SS316L is in the range of 268 μm to 299 μm. The modulus of elasticity and compressive strength for porous SS316L with 60 wt.% SS316L are 0.46 GPa and 56 MPa respectively. In the future work, the SS316L composition will be increased and the compression strength of the porous SS316L produced will be studied further.

Acknowledgement

The authors would like to acknowledge Malaysian Ministry of Education (KPM) for the financial support through Exploration Research Grant Scheme (ERGS-E028). The first author would also like to thank the University Tun Hussein Onn Malaysia (UTHM) for funding her PhD study.

References

[1] R. Singh, N. B. Dahotre, "Corrosion degradation and prevention by surface modification of biometallic materials.," *J. Mater. Sci. Mater. Med.*, vol. **18**, no. 5, pp. 725–51 (2007)

[2] M. Niinomi, M. Nakai, J. Hieda, "Development of new metallic alloys for biomedical applications.," *Acta Biomater.*, vol. **8**, no. 11, pp. 3888–903 (2012)

[3] K. Kato, A. Yamamoto, S. Ochiai, M. Wada, Y. Daigo, K. Kita, K. Omori, "Cytocompatibility and mechanical properties of novel porous 316L stainless steel," *Mater. Sci. Eng. C* (2013)

[4] M. Navarro, A. Michiardi, O. Castaño, J. A. Planell, "Biomaterials in orthopaedics.," *J. R. Soc. Interface*, vol. **5**, no. 27, pp. 1137–58 (2008)

[5] P. Quadbeck, K. Kümmel, R. Hauser, G. Standke, "Open Cell Metal Foams–Application-oriented Structure and Ma-terial Selection," *Ifam-Dd.Fraunhofer.De* (1966)

[6] Q. Chen, D. Mohn, and W. J. Stark, "Optimization of Bioglass® Scaffold Fabrication Process," *J. Am. Ceram. Soc.*, vol. **94**, no. 12, pp. 4184–4190 (2011)

[7] P. Quadbeck, K. Kümmel, R. Hauser, G. Standke, J. Adler, G. Stephani, B. Kieback, "Structural and Material Design of Open-Cell Powder Metallurgical Foams," *Adv. Eng. Mater.*, vol. **13**, no. 11, pp. 1024–1030 (2011)

[8] A. Nouri, P. Hodgson, C. Wen, "Biomimetic porous titanium scaffolds for orthopedic and dental applications," pp. 415–451 (2010)

[9] A. E. A. Maya, D. R. Grana, A. Hazarabedian, G. A. Kokubu, M. I. Luppo, G. Vigna, "Zr–Ti–Nb porous alloys for biomedical application," *Mater. Sci. Eng. C*, vol. **32**, no. 2, pp. 321–329 (2012)

[10] A. Alem, M. D. Pugh, and R. A. L. Drew, "Open-cell reaction bonded silicon nitride foams: Fabrication and characterization," *J. Eur. Ceram. Soc.*, vol. **34**, no. 3, pp. 599–609 (2014)

[11] S. Ahmad, N. Muhamad, and A. Muchtar, "Development and Characterization of Titanium Alloy Foams," *Int. J. Mech. Mater. Eng.*, vol. **5**, no. 2, pp. 244–250 (2010)

[12] S. Won, H. Jung, M. Kang, H. Kim, Y. Koh, Y. Estrin, "Fabrication of porous titanium scaffold with controlled porous structure and net-shape using magnesium as spacer," *Mater. Sci. Eng. C*, vol. **33**, no. 5, pp. 2808–2815 (2013)

[13] M. A. A. M. Nor, H. M. Akil, Z. A. Ahmad, "The effect of polymeric template density and solid loading on the properties of ceramic foam," *Sci. Sinter.*, vol. **41**, no. 3, pp. 319–327 (2009)

[14] M. J. Matos, S. Dias, F. A. C. Oliveira, "Macrostructural changes of polymer

 DOI: 10.1051/matecconf/201713500062

replicated open cell cordierite based foams upon sintering," vol. **106**, no. 5, pp. 209–215 (2007)
[15] J. P. Li, C. A. V. A. N. Blitterswijk, K. D. E. Groot, "Factors having in ¯ uence on the rheological properties of Ti6A14V slurry," vol. **5**, pp. 951–958 (2004)
[16] J. Banhart, J. Baumeister, "Deformation characteristicsof metal foams," *J. Mater. Sci.*, vol. **33**, pp. 1431–1440 (1998)
[17] Q. Fu, M. N. Rahaman, B. Sonny Bal, R. F. Brown, D. E. Day, "Mechanical and in vitro performance of 13-93 bioactive glass scaffolds prepared by a polymer foam replication technique," *Acta Biomater.*, vol. **4**, no. 6, pp. 1854–1864 (2008)

The Effects of Sintering Temperature on Densification and Mechanical Properties of Hydroxyapatite/Sodium Alginate Biocomposites

Nanthini Kanasan[1, a], *Sharifah* Adzila[1, b], *Nor Azimah* Mustaffa[1], *Sabriy Mad* Sidi[1], *Mohd Nasrull Abdol* Rahman[1], *N. F.* Nasir[1], and *P.* Gurubaran[2]

[1]Faculty of Mechanical and Manufacturing Engineering, Universiti Tun Hussein Onn Malaysia, Johor, Malaysia,
[2]School of Mechatronic Engineering, Universiti Malaysia Perlis, Pauh Putra Campus, Arau, Perlis, Malaysia,

Abstract. Hydroxyapatite (HA) has been studied for their excellent performances which are suitable for use in bone defects or regeneration of bone. Hydroxyapatite (HA) has poor mechanical properties which make it limited used in a load bearing area. In this case, sodium alginate (SA) was used to be mixed with hydroxyapatite (HA) by using precipitation method to increase the mechanical and physical properties. The mixing of hydroxyapatite/sodium alginate effectively increases the densification and hardness of hydroxyapatite. It could contribute to the development of the artificial implants of hydroxyapatite (HA) for bone implants applications in load bearing area. The behaviors during densification were studied together with the mechanical properties of samples. The influence of mixed powder and the sintering temperature on densification, microstructure and mechanical properties of hydroxyapatite (HA)/sodium alginate (SA) composite were studied. The maximum density were attained by 99.5/0.5% (HA/SA) at 1.47g/cm3 and 1.82g/cm3 for both temperature 1000°C and 1100°C respectively while, the maximum values of hardness were also attained by 99.5/0.5% for temperature 1000°C and 1100°C at 0.053GPa and 0.266GPa respectively.

1 Introduction

Hydroxyapatite (HA) $(Ca_{10}(PO_4)_6(OH)_2$ is the inorganic component of bones that have excellent biocompatibility with a lot of application in biomedical fields and act as filler due to its characteristics. These are similar to mineral component that present in bones and hard tissues [1]. The combination of ceramic based materials such as hydroxyapatite and natural polymer as sodium alginate will provide strength, stiffness and fracture toughness to the

Corresponding author: nanthini.kanasan@yahoo.com[a], adzila@uthm.edu.my[b]

DOI: 10.1051/matecconf/201713500063

bone [2,3].The composite materials have been used in load bearing application such as filler and bone defect.

Alginate (Alg) is a natural polymer which synthesized from brown algae. Alginate is more prefer to be used due to its biocompatibility, low cost and low toxicity. In this work, sodium alginates (SA) were used to improve the physical and mechanical properties of ceramic materials. Sodium alginate act as a hydrophilic and it can be dissolved in water but not entirely soluble [1].

The powder properties of hydroxyapatite are depends on the processing method and sintering temperature. This is important to improve the performance such as powder morphology and distribution of particles size. Several studies shows that, the porosity of composites are decrease by adding sodium alginate on hydroxyapatite [4,5].The densifications of hydroxyapatite are varies based on the powder characteristics where the carbonate content improved the densification of hydroxyapatite and sintering which are performed at a temperature lower than 1250°C to avoid the loss carbonate content [6]. Usually densifications are achieved by compaction and sintering method. The physical and mechanical properties can be improved by controlling the particle size and shape, the distribution of particles and agglomeration of powder [7]. In this work, HA and HA/SA were sintered at 1000°C and 1100°C for 2°C/min at heating and cooling rate with 2 hour soaking time. The structural and physical- mechanical properties of HA/SA are investigated.

The aim of this research is to identify the influence of mixed powder and sintering temperature on densification, microstructure and mechanical properties and to characterize the synthesized HA and HA/SA powder by using Field Emission Scanning Electron Microscope (FESEM). The material used is based on calcium phosphate group which is hydroxyapatite and also a natural polymer as sodium alginate. Since, it is the most common composite material that are suitable to be used in medical application due to it performances for regeneration of bone implants. The biomedical application of HA and sodium alginate were studied in term of behaviors, microstructure, densification and mechanical properties.

2 Experimental Method

The commercial HA nano powder $Ca_{10}(PO_4)_6(OH)_2$ and sodium alginate NaC_6H_7O were obtained from Q-Rec. The sample ratio is 100/0%, 99.5/0.5%, 97/3%, 95/5%, 92/8% and 90/10% (HA/SA).

Hydroxyapatite/sodium alginate (HA/SA) powder was prepared by using precipitation method. The mixed powders are sintered at two different temperatures from 1000°C to 1100°C. The mixing process was carried out by using hot plate and magnetic stirrer. The pure hydroxyapatites were mixed with distilled water and sodium alginate was put into the solution. The suspension powder was left overnight to get the precipitate solution. Next, the precipitation solutions were dried at 24 hours in 80°C.

Then, the dried sample was crushed and sieved to obtain the fine powders. 1 grams of mixed powder are compacted at 2.5MPa load by using stainless steel mould (13mm). The mould should be clean by using WD40 or silicon spray which act as a lubricant to remove dirt and to remove the sample easily after compaction. The compacted samples were sintered at 1000°C and 1100°C in soaking time 2 hours with 2°C/min of heating and cooling rates. Next, the bulk density of sintered samples were measured based on the Archimedes principle by using Mettler Toledo Densitometer. The sintered samples were grinded by using SiC paper (600-1000 grit) and polished by using diamond paste suspension.

For each composition, 10 indentations were made by using pyramidal diamond indenter (HMV,Shimadzu) for 10 seconds under 0.2HV. Thermal etched was done at 950°C and 1050°C in 30 minutes with 2°C/min of heating and cooling rates. The etched surface sample is coated with gold before characterized under Field Emission Scanning Electron Microscopy (FESEM).

3 Results and Discussion

3.1 Characterization of nanoparticles

The FESEM microstructure of HA and HA/SA sintered samples are analyzed at temperature about 1000°C and 1100°C as shown in Figure 1(a) and 1(b).The particle size growth slowly as the sintering temperature increases from 1000°C to 1100°C and the particles size was in irregular shapes. The particles size for 1000°C to 1100°C were increased due to the increase in percentage of sodium alginate (SA) while the porosity was decreased. The morphology of 1100°C shows the microstructures of grains are more compact, dense and has an irregular shape. The HA nanoparticle exhibit irregular shape with a grain size below 137nm. The different particles size and shape would affect the bonding between each other which it influence the efficiency of regeneration process. HA particle were fully dispersed through the SA. HA was not identify individually due to the fully covered of SA into HA which make a complete composite materials. Composite occur due to the interaction between hydroxyl and carboxyl of HA and SA. The reasons for this are related to strong ionic interactions between the alginate and Ca^{2+} content into HA structure. This study is similar to Choi and Kumta 2006, whereby the interaction between the alginate and carbonate content incorporated into hydroxyapatite structure [8].The rapid growth of HA and HA/SA particle size were observed in Figure 1(a) and 1 (b) for 1000°C and 1100°C. The microstructure of 1000°C shows that there are interconnected pores appear in the HA and HA/SA composite material while, for 1100°C the pores were decreased according to the increase in sintering temperature. The microstructure shows the better distribution of hydroxyapatite (HA) into the polymer reinforcement composite with good porosity that are required in growth of bone in implants process. Small spherical and irregular particles are observed in the agglomerates microstructure of hydroxyapatite (HA) and hydroxyapatite/sodium alginate (HA/SA).The percentages of porosity were affected when there is any changes in sintering temperature. The porosity will give the result to the decrease of strength in bone implants. The higher sintering temperature can control the microporosity of biocomposite materials. Nag et al., 2012 said that, the decrease of porosity resulting effect to the strength of implant materials drastically could bring challenge for load-bearing implants [9]. The comparison of HA and HA/SA sinter samples shows that the 1100°C have less porosity if compare with 1000°C. In comparison of particles size for sintering temperature of 1000°C were in the range of 0.26-0.72µm at ratio 100/0%, 99.5/0.5%, 97/3%, 95/5%, 92/8% and 90/10%.The temperature at 1100°C the size range is about 0.72-1.26µm at ratio 100%, 99.5/0.5%, 97/3%, 95/5%, 92/8% and 90/10%.It conclude that the ceramic-polymer experience problem due to aggregation of ceramic particles due to lack of bonding strength and faced difficulty in the intermixing of hydrophilic ceramic particle with hydrophobic polymer. Other than that, the mixing process and precipitation condition also influence the particle size where for 92% and 90% of HA faced some decrease in particle sizes. The results for 95/5% HA/SA for 1100°C, a lot of small size particles compared to other composition are due to high coating process.

DOI: 10.1051/matecconf/201713500063

Fig.1. (a) FESEM of HA and HA/SA at ratio composition about 100/0%, 99.5/0.5% and 97/3% shows the variety in crystal shape and size at about 1000°C and 1100°C of sintering temperature.

DOI: 10.1051/matecconf/201713500063

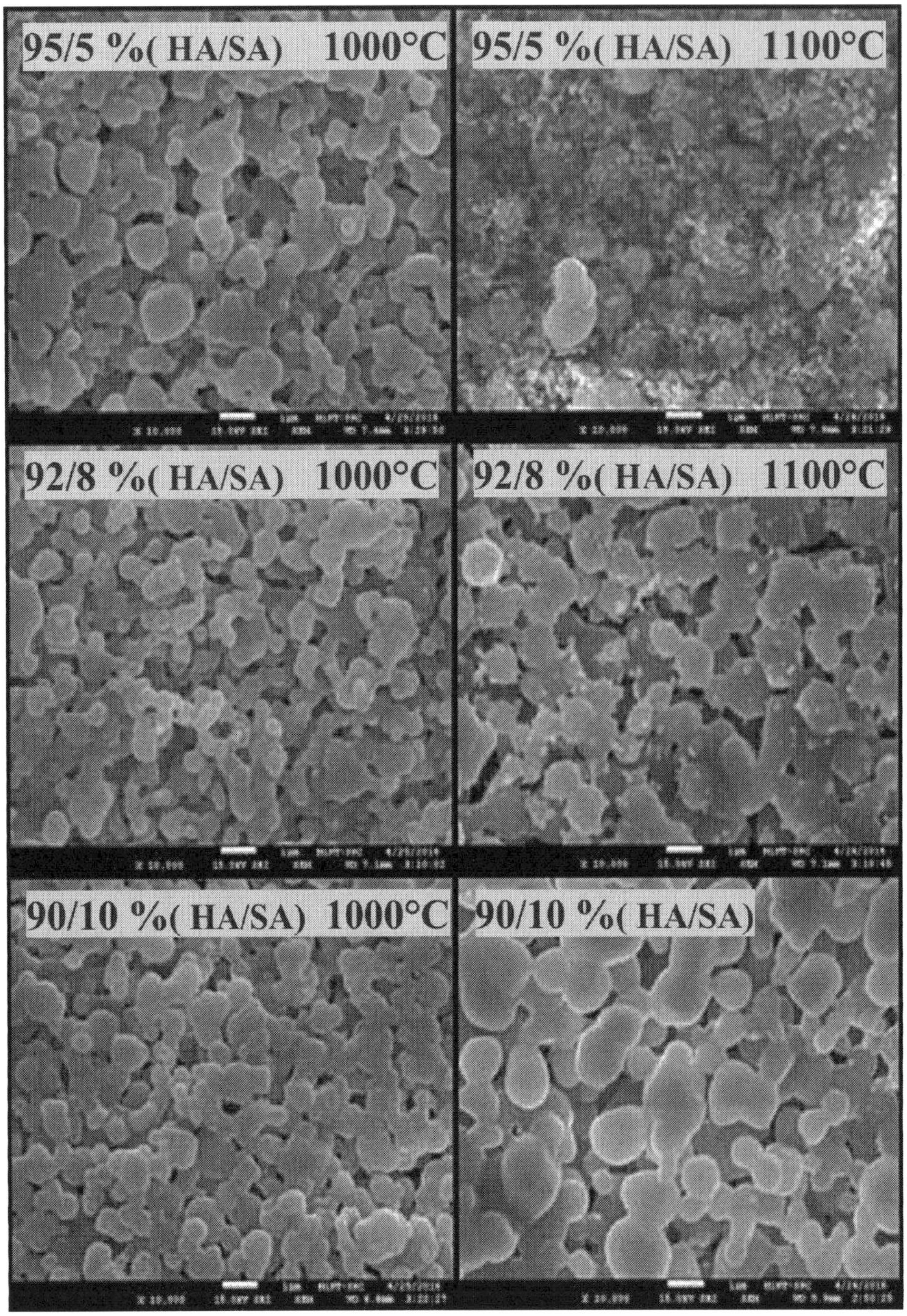

Fig. 1. (b) FESEM of HA and HA/SA at various ratio such as 95/5%,92/8% and 90/10% shows the variety in crystal shape and size at about 1000°C and 1100°C of sintering temperature.

3.2 Densification

The density tests are measure based on Archimedes principle. The densification for sintered HA and HA/SA sample are varies according to ratio composition are shown in Figure 2. The sintering temperatures are further increased by reduction of porosity and this resulting the rapid growth of HA particle size. The sintering temperature at 1000°C with ratio concentration of 100/0%, 99.5/0.5% (HA/SA) shows the densification was increased from 1.44 g/cm^3 to 1.47g/cm^3 and 1.8g/cm^3 to 1.82g/cm^3 for 1000°C and 1100°C each respectively. The maximum density are attained at 99.5/0.5% (HA/SA) about 1.47g/cm^3 and 1.82g/cm^3 for both 1000°C and 1100°C sintering temperature. It summaries that, HA are found to have a good compact structure at this composition. Further increase in the amount of sodium alginate (SA) decrease the density of samples. Then, the ratio concentration of sodium alginate (SA) at 1000°C for 5%, 8% and 10% were decreased at about 1.42g/cm^3, 1.28 g/cm^3 and 1.25 g/cm^3 respectively while, for 1100°C at 1.70g/cm^3, 1.55 g/cm^3 and 1.55g/cm^3. According to the previous results shows by Rajkumar, Meenakshisundaram, and Rajendran 2011, the increase in the amount of sodium alginate (SA) will decrease the density of samples [10]. The trends was unstable at 97/3% (HA/SA) where there is a slight decrement in value of 1.31g/cm^3 and 1.56g/cm^3 for 1000°C and 1100°C respectively.

3.3 Vickers Hardness

The mechanical properties of composites material are important due to which it act as a physical support for the regeneration of bone implants. Sodium alginate are prepared with hydroxyapatite to form a composites materials to achieve a certain mechanical properties. Figure 3 shows the Vickers hardness of sintered sample at various ratio concentrations. The hardness of hydroxyapatite (HA) and hydroxyapatite/sodium alginate (HA/SA) increased with an increase in sintering temperature. The increase of hardness with the increase of sodium alginate (SA) was observed up to 99.5/0.5%, whereas there were decrement of hardness values noticeable with further increase in SA. The increase in hardness happen due to the interaction between calcium ions and SA to form a strong matrix structures while, the decrease happen due to polymer that are broken when act by external forces. This finding are similar to Rajkumar, Meenakshisundaram, and Rajendran 2011 [10]. The maximum hardness is observed at 99.5/0.5% (HA/SA) as 0.053GPa and 0.266GPa for 1000°C and 1100°C respectively by which it indicates that the sintering temperature has an influence the hardness of sample. The particles size are closely packed to form dense materials when there is a higher sintering occurred and this is the reason for increase in hardness.

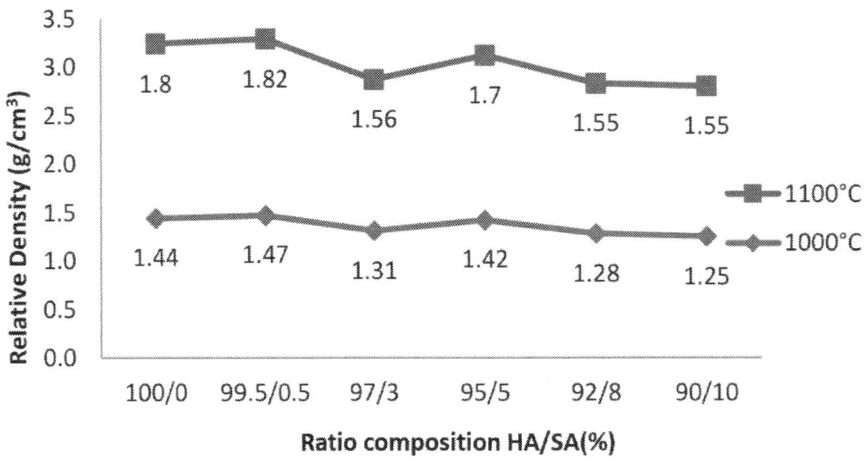

Fig. 2. Density (g/cm3) versus ratio composition of HA and HA/SA

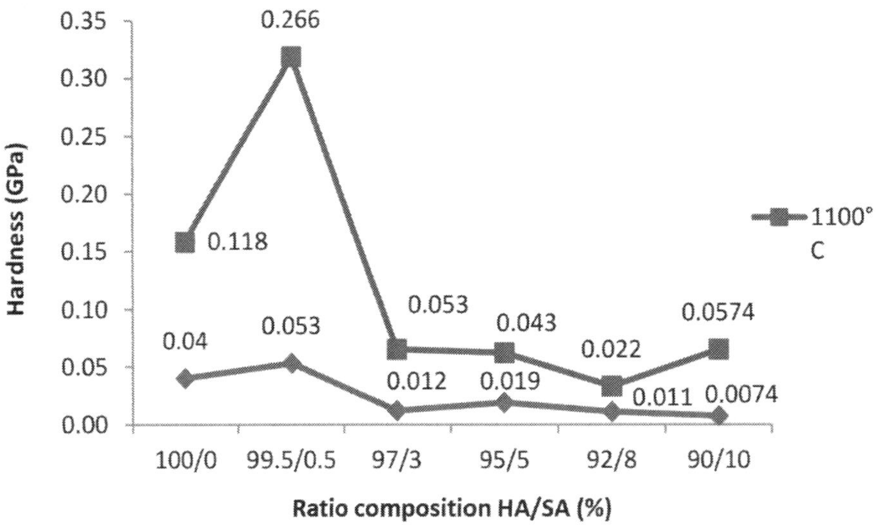

Fig. 3. Hardness (GPa) versus ratio composition of HA and HA/SA

4 Conclusions

In this work, the composite materials are investigated based on two precursors which are hydroxyapatite and sodium alginate. The parameter used are varies according to ratio composition between hydroxyapatite and sodium alginate. The HA and HA/SA composites sample shows that the sintering temperature have significant effect on the densification and mechanical properties. By characterization of HA and HA/SA composites, it indicates that the sintering temperatures has effect on the microstructure of composite materials. The FESEM analysis shows the distribution size of irregular and spherical particle was very close and no contamination happen. The maximum density were attained by 99.5/0.5%

(HA/SA) at 1.47g/cm^3 and 1.82g/cm^3 for both temperature 1000°C and 1100°C respectively while, the maximum values of hardness were also attained by 99.5/0.5% for temperature 1000°C and 1100°C at 0.053GPa and 0.266GPa respectively. The maximum hardness obtained were linearly similar with the density at 99.5/0.5% (HA/SA) at both sintering temperature. It indicates that hardness was influenced by densification. Higher density and hardness are usually controlled by bonding among the grains of sintered samples. The increase in densification and mechanical properties are promising and offer the potential biomaterials to be used in biomedical applications especially for bone tissues regeneration. Density and hardness play an important role where the behaviours of strength, ductility and toughness are investigated to identify the suitable materials to be used for bone implants.

Acknowledgement

This study was supported under the Short Termed Grant (STG), Vot number U124. The authors gratefully acknowledge the Office for Research, Innovation, Commercialization and Consultancy Management (ORICC), Universiti Tun Hussein Onn Malaysia and Centre Graduate of Study (UTHM).

References

1. A. Ilie, C. Ghiţulică, E. Andronescu, A. Cucuruz, and A. Ficai, New Composite Materials Based on Alginate and Hydroxyapatite As Potential Carriers for Ascorbic Acid., Int. J. Pharm. (2016)
2. S. Lala, M. Ghosh, P. K. Das, D. Das, T. Kar, S. K. Pradhan S., Magnesium Substitution in Carbonated Hydroxyapatite: Structural and Microstructural Characterization by Rietveld's Refinement., Materials Chemistry and Physics.**170**, 319–29(2016)
3. S. Adzila, I. Sopyan, R. Singh, E. Pusparini, M. Hamdi, Mechanochemical synthesis of sodium doped hydroxyapatite powder.,Indian Journal Of Chemistry.**52A** , 739–743(2013)
4. Thangamani, Nithyanantham, K. Chinnakali, and F. D. Gnanam, The Effect of Powder Processing on Densification, Microstructure and Mechanical Properties of Hydroxyapatite., Ceramics International. **28**(4), 355–62 (2002)
5. A. Dubnika, D. Loca, L. Berzina-Cimdina, Functionalized Hydroxyapatite Scaffolds Coated with Sodium Alginate and Chitosan for Controlled Drug Delivery., Proceedings of the Estonian Academy of Sciences. **61**(3), 193(2012)
6. E. Landi, A. Tampieri, G. Celotti, R. Langenati, M. Sandri, S. Sprio, Nucleation of Biomimetic Apatite in Synthetic Body Fluids: Dense and Porous Scaffold Development., Biomaterials. **26**(16), 2835–45(2005)
7. G. Muralithran, S. Ramesh, The Effects of Sintering Temperature on the Properties of Hydroxyapatite., Ceramics International. **26**, 221–230(2000)
8. Choi, Daiwon, P. N. Kumta, An Alternative Chemical Route for the Synthesis and Thermal Stability of Chemically Enriched Hydroxyapatite., Journal of the American Ceramic Society. **89**(2), 444–49(2006)
9. S. Nag, R. Banerjee, "Materials for medical devices, fundamentals of medical implant materials," ASM Handb., vol. **23**, pp. 6–16 (2012)
10. M. Rajkumar, N. Meenakshisundaram, V. Rajendran, Development of Nanocomposites Based on Hydroxyapatite/sodium Alginate: Synthesis and Characterisation., Materials Characterization. **62**(5), 469–79 (2011)

Powder Characterization of Calcium Phosphate/Collagen for Bone Implant Application

Nor Azimah Mustaffa[1, a], *Sharifah* Adzila[1, b], *Nanthini* Kanasan[1], and *Normayati* Nordin[1]

[1]Faculty of Mechanical Engineering, Material and Design Department, Universiti Tun Hussein Onn Malaysia (UTHM)

Abstract. Calcium phosphate (CaP) is a type of bioceramic material that is biocompatible and bioactive. It is usually used for bone implant application but it lacks mechanical strength. Therefore, sodium alginate (SA), a natural polymer, is combined with CaP to improve its properties of the CaP via precipitation method. The powder formed is then characterized by using FESEM, EDX, FTIR and DTA. Based on the FESEM result, it was confirmed that the SA particles were well embedded and homogeneously dispersed throughout the CaP matrix whereas the EDX result showed that the CaP and SA are pure and were not contaminated with other materials or substances.The FTIR result showed that the intensity of some peaks (3100-3600 cm-1 and 1585-1625cm-1) increased due to the addition of SA but for some peaks the addition of SA leads to a decrease in intensity (1650-1300cm-1 and 1190-976cm-1).This proves that the addition of SA in CaP influenced the intensity peaks and it was confirmed that chemical bonds were formed between these two substances. Meanwhile, the DTA result showed that CaP dehydroxylation occurred in the range of 650-13000C and the thermal degradation of SA happened at 245.80C but the SA thermal degradation temperature improved when it was added to the CaP matrix.

1 Introduction

Bones play an important part in the human body as they support body tissue and protect human organs.The human bone also acts as mineral storage (calcium and phosphorus) [1].Bone defectslike fractures, cancer, bone loss (osteoporotic) and other debilitating diseases can influence the function of bones in the human body. 60 to 70% of bone tissues consist of mineral calcium, phosphate and collagen, while the remaining percentage consists of inorganic elements and water [2]. Similarities between calcium phosphate (CaP) and the mineral content in bone tissue were the reason why calcium phosphate was chosen for this study [3].Calcium phosphate (CaP) has been used for repairing bone defects, dental and bone tissue applications in the human body since 1920 [4].Calcium phosphate (CaP) is

Corresponding author: [a])meemek90@gmail.com , [b])adzila@uthm.edu.my a type of ceramic material that is biocompatible, non-toxic and resorbable. Furthermore, calcium phosphate (CaP) is a bioactive and osteoconductive material that can promote new bone growth and promote cellular function and expression [3,5] . The lack of the calcium phosphate (CaP) makes it limited in the application of artificial bone implant [3,5,6]. Due to that, the combination of different materials known as composite material have been introduced to improve the propertiesof calcium phosphate (CaP) [7].

Biocomposites for bone implant application usually consist of a combination of calcium phosphates with polymer materials (collagen, HDPE, PLGA, alginate etc.) or metal (stainless steel, Ti, Mg and etc.) to enhance the properties of CaP. In this study, polymer material is used due to the mechanical properties of the polymer that can be tailored for specific applications and easily fabricated into any desired shape. This gives polymers an edge over ceramics materials (CaP) [8]. However, synthetic polymers like PLGA, PLA, PGA among others can inhibit the deformation of cell growth due to their hydrophobic surface whereas some synthetic polymers are harmful to our body system because they go through acidic degradation and invoke a chronic immune response [9].

Sodium alginate (SA) was chosen as it is anatural polymer that is extracted from brown algae and it is hydrophilic, biocompatible and biodegradable.It is usually used in the food and chemical industry but it has recently been developed for living tissue applications in wound healing, drug and protein delivery [10-12]. Sodium alginate (SA) is soluble in water and that possesses rheological properties. Sodium alginate has been used as a thickening or gelling agent, colloidal stabilizer and as a blood expander in the pharmaceutical industry for a long time [13]. A combination of CaP and SA was prepared using the precipitation method. Precipitation method is one of the simple methods that can easily be done and used to control the morphology and mean size of the powder [14].

The addition of sodium alginate to hydroxyapatite (HA) showed an increase in intensity at 962 and 471 cm^{-1}that leads to the growth rate of HA [10]. The alginate did not affect the subsequent modification of the characteristic band of calcium titanium phosphate (CTP) and HA.This showed that both ceramics were maintained in the microsphere [13].A study by Kui Zhou showed that porous HA/SA 3D material was formed afterweight loss occurredthrice at 100°C, 300 to 400°C and 600-1300°C respectively.This was confirmed by the DSC-TG graph [15].During the SEM analysis, it was found that the HA particles were homogenously dispersed and well embedded throughout the SA particles [10,13].

Based on the literature, it was showed that HA has been used inthe study of composite materials. However, in previous studies, calcium hydrogen phosphate wasnot widely used in combination with SA to produce composite material. The aim of this study is to explore the effect of calcium hydrogen phosphate in the SA filler as a CaP/SA composite with varying weight percentages of SA.

2 Sample Preparation

2.1 CaP/SA sample preparation

Calcium hydrogen phosphate ($CHPO_4$) from Fisher Scientific was used as the matrix material and combined with sodium alginate (SA, R&M Chemicals) with different weight percentages at 0.5, 1.0, 3.0, 5.0, 10 and 20 wt% by using the precipitation method. The CaP (99.5wt%) was stirred with distilled water by using a magnetic stirrer at 80°C for 2 hours and then the SA (0.5wt%) powder was added while stirring vigorously. The mixture is continuously stirred for another 2 hours until a paste is formed and the paste will be stored at an ambient temperature for 24 hours. Thereafter, the solution will be dried in the oven at80^{O}C for 24 hours and then grounded using a mortar and pestle to form a fine powder.

The same procedure was used to prepare other composite materials, namely 1.0, 3.0, 5.0, 10 and 20wt%.

2.2 Sample morphology and characterization

The microstructure of the powder was obtained by using field emission scanning electron microscope (FESEM) (JEOL JSM-7600F). The samples need to be prepared before running the FESEM analysis. The sample was coated with gold by using a sputter coater for 45seconds with 20-mA current and images of the microstructure were taken.

The Fourier transform infrared spectroscopy (FTIR) (Perkin Elmer Spectrum 100 Machine) was conducted to obtain the characteristic peaks of the chemical bond of the pure CaP and CaP/SA composites. The powder was analyzed by using the Perkin Elmer Spectrum 100 machineat a resolution of 4 cm^{-1}.

2.3 Thermal analysis

Differential thermal analysis (DTA) (DTA L62) was conducted to analyze the phase reaction and phase diagram at different temperatures. The sample powder was filled into 1/3 of a crucible and weighed.The crucible was put into the furnace where the temperature and heating rate wereset at 1300 ^{O}C and 10 min/^{O}C respectively.

3 Result and Discussion

3.1 Field emission scanning electron microscope

Figure 1 shows the FESEM image for the pure CaP, SA and the CaP/SA composite. The structure of the CaP/SA particles showed that when there is an increase in the SA content, more flakey structures were formed in the particles caused by the structure of SA particles (Figure 2). All the combinations showed that the SA was identified with CaP because the SA fully covered the CaP surface [10,13].The EDS results (Figure 2) showed that the sodium (Na) peak was identified in the composite and the extensive peak was formed when a higher wt% of SA added.

DOI: 10.1051/matecconf/201713500064

Fig. 1. FESEM image for all compositions of CaP/SA in X50,000

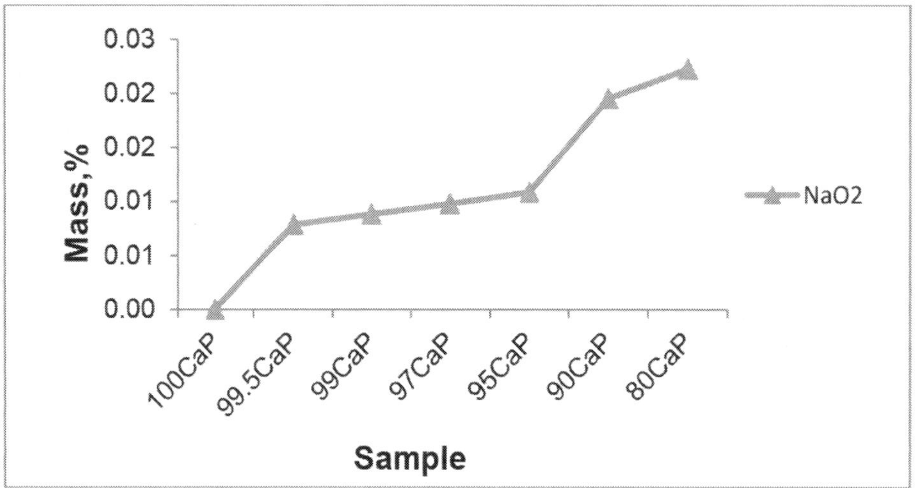

Fig. 2. Mass percentage of the Sodium (NaO_2) for CaP/SA composite

3.2 Fourier Transform Infra-Red (FTIR)

FTIR spectra for the composite (CaP/SA), pure SA and CaP are shown in Figure 3 and Figure 4.The broad peak was detected in pure SA at range 3100-3600 cm^{-1}due to water (H_2O) absorption and stretching mode of surface hydroxyl groups [10,16]. Carboxyl (C=O) stretching vibration was detected at 1590-1750 cm^{-1} [16] where the intensity of the peaks increased when there is an increase the SA content in the CaP/SA composite sample and the peak was more extensive for sample 80 CaP. The peak that was observed in the region between 1405 to 1456 cm^{-1} and 1025.4 cm^{-1} belongs to the carboxyl –COOH stretching band [16] and C-O stretching vibration [17] respectively where the peaks were detected in pure SA only. The peak V_3, hydroxyl stretch carbonate and phosphate were observed between 1300 and1650cm^{-1} [18] 976 and 1190cm^{-1} [18,19]respectively, where the highest peaks was observed for sample 99CaP which contains 1wt% of SA but the increasing wt% of SA led to a decreasein intensity. The peak for C1-H deformation manunuronic acid residue and hydroxyl stretch phosphate V_4 was observed between 878.9-882.4 cm^{-1} [17] and 520-600cm^{-1} [18, 20, 21] respectively where both of these peaks were observed in sample 97CaP. Based on the results,the addition of the SA influenced the intensity peaks of the composite where the formation of chemical bonds occurred between both materials.

DOI: 10.1051/matecconf/201713500064

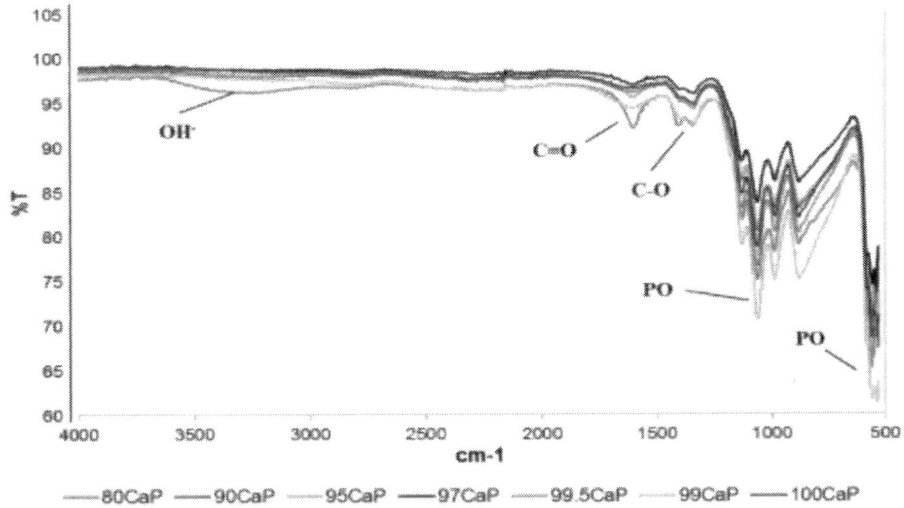

Fig. 3. FTIR result for pure CaP and CaP/SA for all compositions

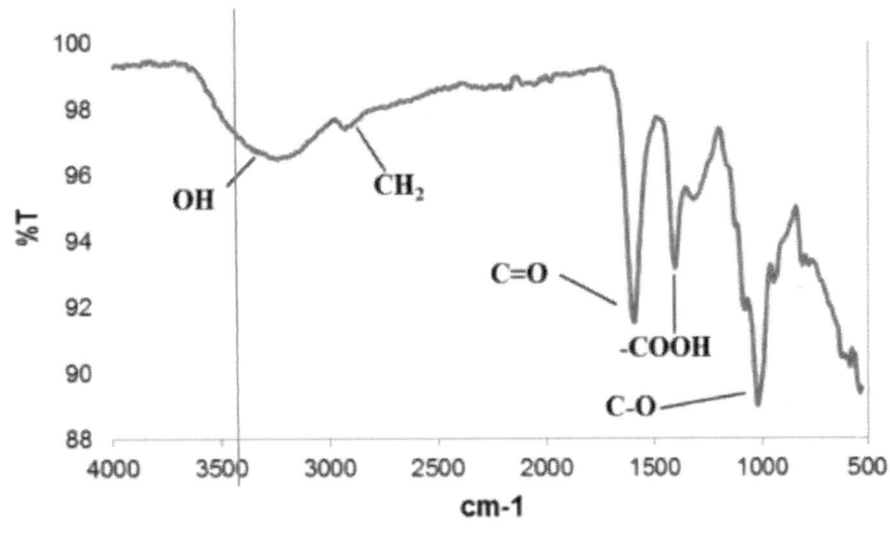

Fig.4. FTIR result for pure SA

3.3 Differential thermal analysis

DTA results for pure SA, CaP and CaP/SA composite are shown in Figure 5. Based on the figure,an exothermic peakwas detected between 100°C and 70°Cand the peak was also observed in pure SA at 150°C where oxidation occurred when the temperature started to increase and causeda change in the crystal structure of the SA as well as the loss of water in the samples [15, 21, 22]. The results showed the temperature approached150°C (SA sample) when the wt% of SA was increased in the CaP/SA composite. A range between 240 until 360°C was observed as endothermic peaks occurred for all samples. A pure SA peak was detected at 245.8°C where thermal degradation of intermolecular side chains took place [21]. This also affected the CaP/SA composite where 5wt% of SA achieved a higher temperature but the temperature decreased when a higher wt% of SA was added to the CaP. The crystal structure changed and chemical composition reduction was observed in pure CaP where the endothermic peak was detected at 357.8°C. The broad exothermic peak was induced at a range of 1100 to 1250°C where the highest peak was obtained at 100CaP and the peak gradually disappears when the wt% of SA increases. This peak occurred due to the dehydroxylation of CaP in the range of 650-1300°C and the break down reaction of SA [15].

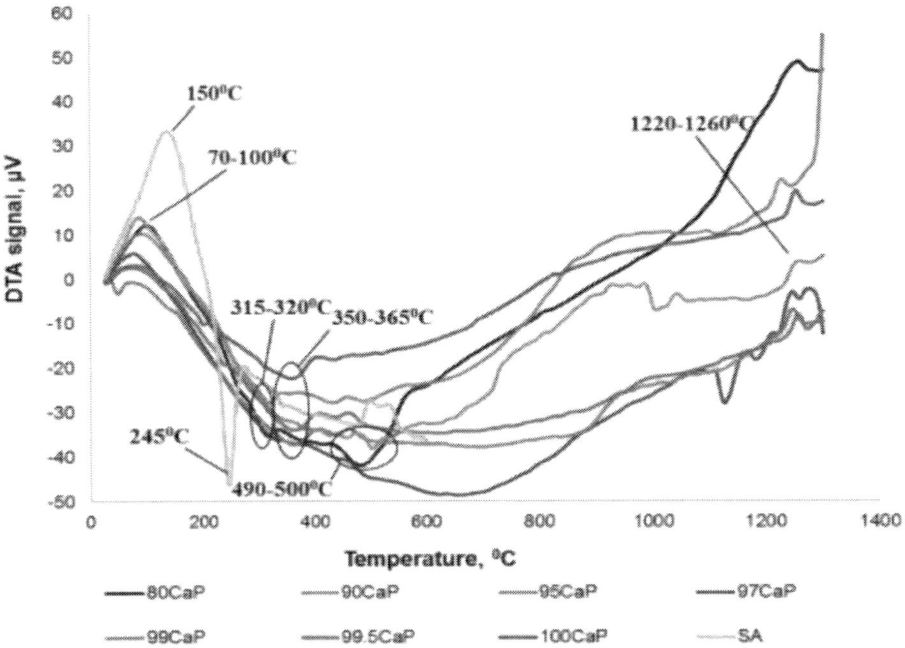

Fig. 5. DTA pattern for the pure CaP and SA and CaP/SA composite.

Conclusion

In conclusion, the addition of SA to CaP was successful as chemical bonds formed between the two substances as proven through the FTIR and FESEM results. Meanwhile, the thermal analysis results showed that CaP as the matrix can improve the properties of SA where the degradation temperature for SA increased when it was added to CaP.

Acknowledgement

This research was financially supported by a Short Termed Grant (STG), Vote number U124. The authors would also like to express their deepest appreciation to theOffice for Research, Innovation, Commercialization and Consultancy Management (ORICC), University Tun Hussein Onn Malaysia and to the Centre for Graduate Studies, UTHM for this research opportunity.

References

1. C. VanPutte, J. Regan, A. Russo, Seeley's Essentials of Anatomy and Physiology, Chapter 6. Skeletal System: Bone and Joints (2013)
2. T. M. Keaveny, E. F. M., O. C. Y., Chapter 8: Bone Mechanic. Standard Handbook of Biomedical Engineering and Design, 1-28 (2004)
3. S. Ghozal, Densification and Mechanical Characterization of Pure and Doped Hydroxyapatite, National Institurte of Technology, Rourkela, Degree Thesis (2012)
4. S. Sanchez-Salcedo, D. Arcos and M. Vallet-Regi, Upgrading Calcium Phosphate Scaffolds for Tissue Engineering Application, Key Engineering Materials. 377, 19-42 (2008)
5. C. Zhuo, X. Ye, Y. Fan, L. Ma, Y. Tan, F. Qing, X. Zhang, Biomimetic Fabrication of Three-Level Hierarchical Calcium Phosphate/Collagen/Hydrxyapatite Scaffold for Bone Tissue Engineering, Biofabrication. 6, 035013(2014)
6. L. Biolet, V. Lardot, A. Tricoteaux, A. Leriche, F. Cambier, M. Descamps. Processing and Properties of Calcium Phosphates Bioceramics by Hot Isostatic Pressing,. MATEC Web of Conference. 7, 04020 (2013)
7. J. Hashim, SainsBahan. 1stedidtion. Johor Bahru:UniversitiTeknikal Malaysia (UTM) Press, pp. 212-258
8. G. Turco, E. Marsich, F. Bellomo, S. Semeraro, I. DOnati, F. Brun, M. Grandolfo, A. Accardo, S. Paoletti, Alginate/Hydroxyapatite Biocomposite for Bone Ingrowth: ATrabecular Structure with High and Isotropic Connectivity. Biomacromolecules. 10, 1575-1583 (2009)
9. G. Chen, T. Ushida, T. Tateishi, A Biodegradable Hybrid Sponge Nested with Collagen Microsponges. John Wiley & Sons, Inc. (2000)
10. M. Rajkumar, N. Meenakshisundaram, V. Rajendran, Development of Nanocomposite based on Hydroxyapatite/Sodium Alginate: Synthesis and Characterization. Materials Characterization. 62, 469-479. (2011)
11. Y. L. Kuen, D. J. Mooney. Alginate: Properties and Biomedical Application. Progress in Polymer Science. 37, 106-126 (2012)
12. M. H. Sheridan L. D. Shea, M. C. Peters, D. J. Mooney, Bioabsorbable Polymer Scaffolds for Tissue Engineering Capable of Sustained Growth Factor Delivery. Journal of Controlled Release. 64, 91-102 (2000)
13. C.C. Ribeiro, C. C. Barrias, M. A. Barbosa, Calcium Phosphate-Alginate Microspheres as Enzyme Deliver Matrices. Biomaterial. 25, 4363-4373 (2004)
14. I. Mobasherpour, M. S. Heshajin, A. Kazemzadeh, M. Zakeri, Synthesis of Nanocrystalline Hydroxyapatite by using Precipitation Method. Journal of Alloys and Compounds. 430, 330-333 (2007)
15. K. Zhuo, X. Zhang, Z. Chen, L. Shi, W. Li, Preparation and Characterization of Hydroxyapatite-Sodium Alginate Scaffolds by Extrusion Freeforming., Ceramic Internationl. 41, 14029-14034 (2015)
16. Z. Li, H. R. Ramay, K. D. Hauch, D. Xiao, M. Zhang, Chitosan-Alginate Hybrid Scaffold for Bone Tissue Engineering, Biomaterials. 26, 3919-3928 (2005)
17. S. K. Papageorgiou, E. P. Kouvelos, E. P. Favvas, A. A. Sapalidis, G. E. Romanos, F. K. Katsaros, Metal-Carboxylate Interactions in Metal-Alginate Complexes Studied with FTIR Spectroscopy. Carbohydrate Research. 345, 469-473 (2010)
18. I. Rehman, W. Bonfield, Characterization of Hydroxyapatite and Carbonate Apatite by Photo Acoustic FTIR Spectroscopy. Journal of Materials Science: Materials in Medicine. 8, 1-4 (1997)

19. A. Ślósarczyk, Z. Paszkiewicsz, C. Paluszkiewicz, FTIR and XRD Evaluation of Carbonated Hydroxyapatite Powder Synthesized by Wet Methods. Journal of Molecular Structure. 744-747, 657-661 (2005)
20. L. Berzina-Cimdina, N. Borodajenko, Research of Calcium Phosphate using Fourier Transform Infrared Spectroscopy, Infrared Spectroscopy-Material Science, Engineering and Technology. Pp 123-148 (2012)
21. I. S. Neira, Y. V. Kolen'ko, O. I. Lebedev, G. V. Tendeloo, H. S. Gupta, F. Guitian, Y. Masahiro. An Effective Morphology Control of Hydroxyapatite Crystals via Hydrothermal Synthesis, Crystal Growth and Design. 9, 466-474 (2009)
22. X. Chaobo, L. Yongshang, L. Hongjuan, Z. Lina, Note: Preparation and Physical Properties of Blend Films from Sodium Alginate and Polyacrylamide Solutions. Journal of Macromolecular Science: Pure and Applied Chemistry. A37 (12), 1663-1675 (2000)
23. A. A. Said, Thermal Decomposition of Some Divalent Metal Alginate Gel Compound, Polymer Degradation and Stability. 39, 393-397 (1993)

MATEC Web of Conferences **135**, 00065 (2017) DOI: 10.1051/matecconf/201713500065
ICME'17

Antagonistic Mono- and Bi-Articular Lower-Limb Muscle Activities' Model Characterization at Different Speeds

M.A.M Dzahir[1,*], *N.S.M* Nor[1], *A.* Azaman[2], *M.* Hussein[1] , and *M.A. Mat* Dzahir[1]

[1]Department of Applied Mechanics & Design, Faculty of Mechanical Engineering, UTM, Malaysia
[2]Department of Biotechnology & Medical Engineering, Faculty of Biomedical Engineering and Health Science, UTM, Malaysia

Abstract. Nowadays, medical rehabilitation system has become a requirement due to increment in national rehabilitation centres and medical hospitals. An assistive rehabilitation orthosis becomes essential and was used for rehabilitation therapy, condition monitoring, and physical strengthening. This study focused on the lower limb assistive rehabilitation orthosis development using pneumatic artificial muscle. To successfully control this orthosis system which consists of antagonistic mono- and bi-articular muscle actuators, it is necessary to construct a reliable control algorithm. The suitable control scheme and strategy to manoeuvre this orthosis system similar to human musculoskeletal system have yet to be fully developed and established. Based on the review study, it is said that the co-contraction controls of anterior-posterior pneumatic muscles was able to improve the joint stiffness and stability of the orthosis as well as good manoeuvrability. Therefore, a characterization model of an antagonistic mono- and bi-articular muscles activities of human's lower-limb during walking motion will be necessary. A healthy young male subject was used as test subject to obtain the sEMG muscle activities for antagonistic mono- and bi-articular muscles (i.e., Vastus Medialis-VM, Vastus Lateralis-VL, Rectus Femoris-RF, and Bicep Femoris-BF). The tests were carried out at different speeds of 2km/h, 3km/h, and 4km/h for one minute walking motion on a treadmill. Then, the patterns of the sEMG muscle activities were modelled and characterised using fifth order polynomial equation. Based on the results, it is shown that the anterior and posterior muscles were exhibited a muscle synergy in-between multiple anterior or posterior muscles and muscle co-contraction between anterior-posterior muscles in order to control the movements at the joints during walking motion. As conclusion, it is proven that the sEMG muscle activities of the antagonistic mono- and bi-articular muscles were follow a certain contraction-expansion patterns during walking motion even when it were tested at different gait cycle speeds.

* Corresponding author: azuwan@mail.fkm.utm.my / azuwan@fkm.utm.m

1 Introduction

Based on the development of different orthosis systems in the last ten years, it can be concluded that the researchers' interest had been shifted into the implementation of the natural type of compliant actuators [1–10]. Although lots of researches have been investigated regarding the co-contraction movements of human antagonistic muscles. Their model implementations in controlling the antagonistic mono- and bi-articular muscle actuators of lower-limb orthosis have not been completely discovered [11–22]. In addition, research study which focuses on the implementations of mono- and bi-articular actuators using pneumatic muscles for the lower-limb rehabilitation orthosis also have yet to be extensively investigated [23-25]; thus, simply actuate the muscle actuators might not give a good result on the joint's stiffness and stability of the lower-limb leg orthosis and its joint trajectories. Therefore, based on the related research findings, the simultaneous co-contractively like movements between the anterior and posterior actuators could be considered within the control scheme and strategy [26–30]. Even though a considerable amount of works has been done, the field is still rapidly evolving. The issues on which are the most effective control algorithm is still wide open. However, the randomized controlled trials are still necessary for identifying the suitable control algorithms even though it is expensive and time-consuming.

Throughout the review articles, it can be hypothesis that the human antagonistic mono- and bi-articular muscles are simultaneously activated in-between anterior-posterior muscles. In other words, when the anterior muscle is contracts, the posterior muscle will be expand and vice versa. This is defined as a muscle co-contraction in-between antagonistic muscles (i.e., anterior and posterior) which is also could be explained as contraction-expansion movements between anterior and posterior muscles [23–25]. The muscle co-contraction refers to when any movement occurs which involved two sets of muscles working around a specific joint. Normally, it is required that the muscles on one side of the joint must be relaxed, so that the other muscles on the other side can be contract. The definition of the co-contraction also could be referred to both muscles were simultaneously contract [23, 25]. When the muscles on one side of the joint are in contraction, the opposite muscles will not be in completely relaxed state. There are still some contraction involves but it is less than the opposite muscles. In addition, there might be some existing delay occurs as well during the activation of the anterior and posterior muscles. Furthermore, the same side of anterior or posterior muscles should follow the same contraction-expansion patterns which is define as a muscle synergy. The muscle synergy of the same side either anterior or posterior muscles refers to the interaction of two or more sets of muscles to produce a combined effect greater than the summation of their separate effects.

The result of this research study will verify the isotonic muscle co-contraction patterns of the human lower limb muscles on both the antagonistic mono- and bi-articular muscles. Thus, in this paper a model characterization of the antagonistic mono- and bi-articular muscles of human's lower-limb during walking motion on treadmill system is proposed. The purpose of this study is to determine the characteristics and behaviours of the antagonistic mono- and bi-articular muscles contraction-expansion patterns when tested at different gait cycle speeds using a surface electromyography (sEMG). Through this, we are able to determine and verified the characteristics of the antagonistic mono- and bi-articular muscles' contraction-expansion patterns, whether it were followed a certain pattern or not when tested at different walking speeds. However, it is difficult to measure all of the lower-limb muscle activities using sEMG electrodes due to weak muscle activities, small muscle sites, and difficult location to place the electrodes. Therefore, only four muscles (i.e., VM, VL, RF, and BF muscles) were chosen to record the required muscle activities data.

DOI: 10.1051/matecconf/201713500065

2 Methodology

In this research, a healthy and young male subject aged 29 years old with weight 68kg and height 174cm is used as a test subject. The subject is an active tennis player and has a strong lower limb muscles due to flexible movements in multiple directions (i.e., forward, backward, side step, jumping etc.). For this preliminary study, four muscles are selected to represent the anterior-posterior mono- and bi-articular muscles. The mono-articular muscles are Vastus Medialis (VM) and Vastus Lateralis (VL). These muscles are used to verify the muscle synergy between multiple anterior muscles. Meanwhile, the bi-articular muscles are Rectus Femoris (RF) and Bicep Femoris (BF). These muscles are used to verify the muscle co-contraction between anterior and posterior muscles. Four sEMG electrodes (TMSi system, Netherland) are placed at VM, VL, RF and BF muscles location to obtain sEMG muscle activities of the selected muscles. The tests are carried out at some different speeds of 2km/h, 3km/h, and 4.0 km/h during walking motion on a treadmill. For each VM, VL, RF and BF muscles, the data was collected for duration of about one minute on a treadmill and minimum of fourteen complete cycles of walking motion were successfully recorded. The data were then processed, filtered, and rectified to remove all of the unnecessary noise using Microsoft Excel and MATLAB software. Then the scattered data of the muscle activities are obtained for one complete cycle of walking motion at different walking speeds. Finally, the fitting method using fifth order polynomial equation is used to characterise the patterns for the antagonistic mono- and bi-articular sEMG muscle activities. The characteristic of the antagonistic mono- and bi-articular muscles will be useful for the development of a control scheme and strategy of the pneumatic muscle actuated lower limb orthosis.

3 Results and discussion

The purpose of this research study is to determine the characteristics and behaviours of the antagonistic mono- and bi-articular muscles of lower limb during walking motion. It is strongly believed that each single mono- and bi-articular muscle have a specific isotonic contraction-expansion pattern which can be verified and modelled. The verified models of each anterior and posterior muscle will be useful for controlling the pneumatic muscle actuated lower limb orthosis. This is because, the pneumatic artificial muscle requires pressure as the input and a contraction value will be used to determine the amount of pressure input for manipulating the joint movements accurately. However, the hysteresis effects of the pneumatic artificial muscle during contraction-expansion should also be considered into the control scheme.

Figure 1 shows the VM muscle activities and combined VM-VL sEMG muscle activities at different gait cycle speeds of 2km/h, 3km/h and 4km/h. The VM and VL muscles were both an anterior mono-articular muscles which manipulate the knee joint movements. A fifth order polynomial fitting equation was used to represent a model for the contraction-expansion patterns of VM muscle and combination of VM-VL muscles at different walking speeds. Based on the interpretation of the results, it shows that the contraction-expansion of VM muscle gave an almost similar pattern at all gait cycle speeds. These prove that, each one single muscle will follow a certain contraction-expansion pattern and it was not affected by an implementation of different gait cycle speeds. This result can be verified with a derivation model of simplified pneumatic muscle actuated lower limb orthosis. In addition, the combination of VM-VL muscles was tested to prove that same anterior or posterior muscles could have a similar contraction-expansion pattern, and the combination of both muscles would produce a muscle synergy and improving the movement at the specific joint. To perform the analysis, both VM and VL muscles were combined and then its average values were determined. Based on the results at different gait cycle speeds, an improvement was spotted within the contraction-expansion patterns of combined VM-VL muscles. The obtained patterns show a much better result in terms of consistencies of the muscle activities. In addition, even though the fifth order polynomial fitting equation model of the VM muscle (black lines – first column) shows almost the same patterns, however its contraction-expansion patterns (blue lines) were not consistent enough at different gait cycle speeds. On the contrary, the fifth order polynomial fitting equation model of the combined VM-VL muscles (black lines – second column) were very consistent and it shows an exactly the same patterns even when tested at different gait cycle speeds.

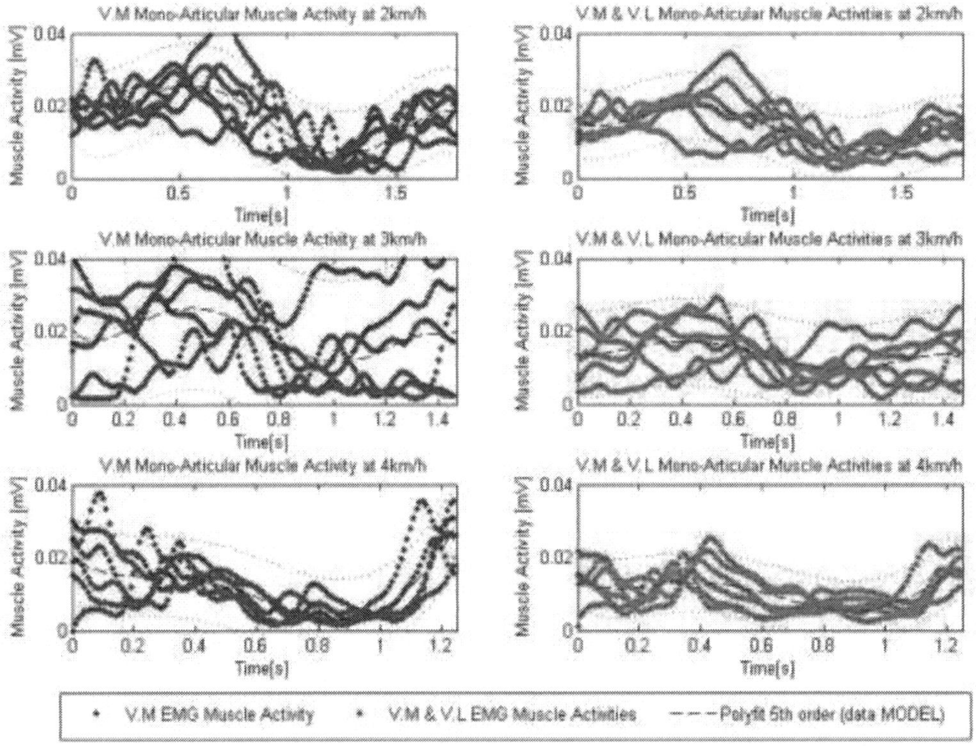

Fig. 1. Vastus Medialis (VM) and Vastus Lateralis (VL) sEMG muscle activities and its 5th order polynomial fitting equations at different walking speeds of 2, 3 and 4km/h.

Figure 2 shows the RF and BF sEMG muscle activities at different gait cycle speeds of 2km/h, 3km/h and 4km/h. The RF muscle is an anterior bi-articular muscle while the BF muscle is a posterior bi-articular muscle. These two muscles provide assistance to the mono-articular muscles in order to improvise and strengthen the movements at hip and knee joints. A fifth order polynomial fitting equation was used to model a contraction-expansion pattern of the RF and BF muscles at different walking speeds. Based on the evaluation of the results, it shows that the contraction-expansion of RF and BF muscle also gave an almost similar pattern when tested at different gait cycle speeds. These were very similar to the results shown by the VM and VM-VL sEMG muscle activities contraction-expansion patterns, where one single muscle will follow a certain contraction-expansion pattern during one complete cycle of walking motion. However, compared to the muscle synergy

DOI: 10.1051/matecconf/201713500065

that was produced by the VM-VL muscles (both were anterior muscles), the RF and BF muscles were not able to generate a muscle synergy between them because they were made of different anterior and posterior muscles. Instead of the muscle synergy, the muscle co-contraction was found in-between the anterior and posterior muscles. According to the result analysis, the RF and BF muscles exhibit co-contraction movements between them, where the contraction-expansion pattern of the anterior muscle was completely opposite to the posterior muscle. This was verified at different gait cycle speeds. However, there was an existing delay in between the activation of the anterior and posterior muscles which is approximately 35% of the walking motion depending on the gait cycle speeds. Furthermore, the contraction-expansion patterns of both RF and BF muscles become more consistent as the walking speed increased. This shows that, muscle reaction between anterior and posterior muscles becomes more active and reacts much faster to provide assistance to the joint's movements. As a final point, existing delay between the activation of the anterior and posterior muscles could also been reduced with the incremental of the walking speed.

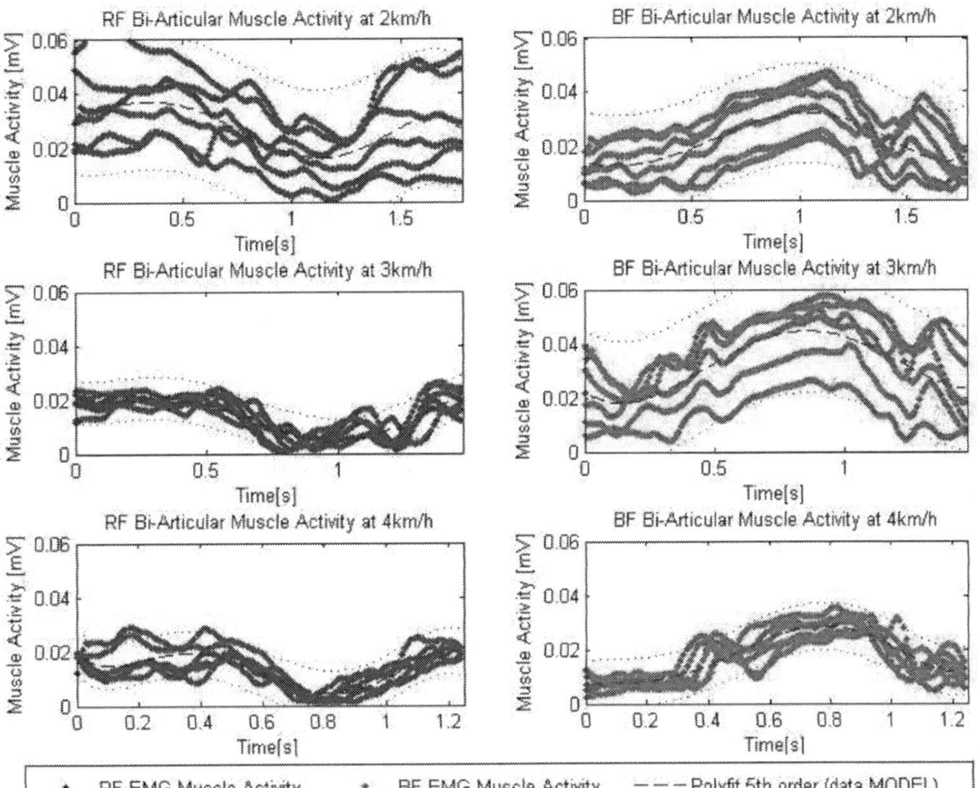

Fig. 2. Rectus Femoris (RF) and Bicep Femoris (BF) sEMG muscle activities and its 5th order polynomial fitting equations at different walking speeds of 2, 3 and 4km/h

4 Conclusions

As conclusions, the objectives of the research study have been fulfilled. According to the results, the sEMG muscle activities of the antagonistic mono- and bi-articular lower limb muscles follow a certain contraction-expansion patterns during walking motion even when it were tested at different gait cycle speeds. These patterns were defined as muscle synergy and muscle co-contraction. First is the muscle synergy, where multiple anterior or posterior muscles with same contraction-expansion pattern were combined together to produce more accurate movement at the joints. Based on the results, it shows that the fifth order contraction-expansion model of combined VM and VL sEMG muscle activities displayed an exactly similar pattern at different gait cycle speeds of 2km/h, 3km/h and 4km/h. These results conclude that each one of the anterior-posterior mono- and bi-articular muscles has a certain pattern to be followed during walking motion which was not affected by implementation of different gait cycle speeds. Secondly, muscle co-contraction where a set of anterior and posterior mono- or bi-articular muscles have a contraction-expansion patterns which were completely opposite to each other's. In addition, muscle co-contraction has an existing delay of approximately 35% of its corresponding gait cycle time. The contraction-expansion patterns of both RF and BF sEMG muscle activities become more consistent as the walking speed increased. This result demonstrates that the muscles reaction between anterior and posterior muscles becomes more active and able to react much faster in order to provide assistance to the joint's movements. With these results, it would be a huge benefit if the models could be verified with the simplified model of the pneumatic muscle actuated lower limb orthosis, and determine its significant value. Then, implementing the obtained models into the control scheme and strategy of the pneumatic muscle actuated lower limb orthosis.

Acknowledgements

This research study was supported by Potential Academic Staff (PAS) research grants Q.J130000.2724.02K16, Universiti Teknologi Malaysia (UTM).

References

1. B.G. Nascimento, C.B. Vimieiro, D.A. Nagem, M. Pinotti, "Hip orthosis powered by pneumatic artificial muscle: Voluntary activation in absence of myoelectrical signal", Artif. Organs, **32**, 317–322 (2008)
2. C.B.S. Vimieiro, B.G. Nascimento, D.A.P Nagem, M. Pinotti, "Development of a hip orthosis using pneumatic artificial muscles", In Proceeding of TMSi, (2005) July18–1; São Paulo, Spain
3. K. Bharadwaj, T.G. Sugar, "Kinematics of a robotic gait trainer for stroke rehabilitation", In Proceedings of the IEEE International Conference on Robotics and Automation, (2006) May 15–19;Orlando, FL, USA
4. D.P. Ferris, K.E. Gordon, G.S. Sawicki, A. Peethambaran, "An improved powered ankle-foot orthosis using proportional myoelectric control", Gait Posture **23**, 425–428 (2006)
5. D.P. Ferris, J.M. Czerniecki, B. Hannaford, "An ankle-foot orthosis powered by artificial pneumatic muscles", J. Appl. Biomechanics, **21**, 189–197 (2005)
6. K.E. Gordon, G.S. Sawicki, D.P. Fessis, "Mechanical performance of artificial pneumatic muscles to power an ankle-foot orthosis", J. Biomech., **39**, 1832–1841 (2006)
7. N. Costa, M. Bezdicek, M. Brown, J.O. Gray, D.G. Caldwell, "Joint motion control of a powered lower limb orthosis for rehabilitation", Int. J. Autom. Computation, **3**, 271–281 (2006)

8. T. Miyoshi, K. Hiramatsu, S.I. Yamamoto, K. Nakazawa, M. Akai, "Robotic gait trainer in water: Development of an underwater gait-training orthosis", Disabil. Rehabilitation, **30**, 81–87 (2008)
9. P. Malcom, P. Fiers, V. Segers, I. Caekenberghe, M. Lenoir, D. Clercq, "Experimental study on the role of the ankle push off in the walk-to-run transition by means of a powered ankle-foot-exoskeleton", Gait Posture, **30**, 322–327 (2009)
10. P. Malcom, V. Segers, I. Caekenberghe, D. Clercq, "Experimental study of the influence of the m. tibialis anterior on the walk-to-run transition by means of a powered ankle-foot-exoskeleton", Gait Posture, **29**, 6–10 (2009)
11. S. Galle, P. Malcom, W. Derave, D. Clercq, "Adaptation to walking with an exoskeleton that assists ankle extension", Gait Posture, **38**, 495–499 (2013)
12. P. Malcom, W. Derave, S. Galle, D. Clercq, "A simple exoskeleton that assist plantarflexion can reduce the metabolic cost of human walking", PLoS One, **8**, 0056137 (2013)
13. G.S. Sawicki, D.P. Fessis, "A pneumatically powered knee-ankle-foot orthosis (KAFO) with myoelectric activation and inhibition", J. Neuro-Eng. Rehabil., **6**, 23:1–23:16 (2009)
14. T.T. Deaconescu, A.I. Deaconescu, "Pneumatic muscle actuated equipment for continuous passive motion", IAENG Trans. Eng. Technol., doi:10.1063/1.3256258 (2009)
15. B. Pieter , V.D. Michael, V.H. Ronald, V. Bram, L. Dirk, "Design and control of lower limb exoskeleton for robot-assisted gait training", Applied Bionics and Biomechanics, **6**(2), 229–243 (2009)
16. B. Pieter , V.D. Michael, V.H. Ronald, V. Bram, L. Dirk, "Pleated pneumatic artificial muscle based actuator system as a torque source for compliant lower limb exoskeletons", IEEE/ASME Transactions on Mechatronics, **19**(3), 1046–1056 (2014)
17. T.J. Yeh, M.J. Wu, T.J. Lu, F.K. Wu, C.R. Huang, "Control of McKibben pneumatic muscles for a power-assist, lower-limb orthosis", Mechatronics, **20**, 686–697 (2010)
18. J. Carberry, G. Hinchly, J. Buckerfield, E. Taylor, T. Burton, S. Madgwick, R. Vaidyanathan, "Parametric design of an active ankle foot orthosis with passive compliance", In Proceedings of the Computer-Based Medical System (CBMS), (2011) June 27–30; Bristol, UK
19. Y. Park, B. Chen, D. Young, L. Stirling, R. Wood, E. Goldfield, R. Nagpal, "Bio-inspired Active Soft Orthotic Device for Ankle Foot Pathologies", In Proceedings of the International Conference on Robots and Systems (IROS), (2011) September 25–30; San Francisco, CA, USA
20. Y. Park, B. Chen, C. Majidi, R. Wood, R. Nagpal, E. Goldfield, "Active Modular Elastomer Sleeve for Soft Wearable Assistance Robots", In Proceedings of the IEEE International Conference on Robots and Systems (IROS), (2012) October 7–12; Vilamoura, Portugal
21. C.M. Teng, Z.Y. Wong, W.Y. The, Y.Z. Chong, "Design and development of inexpensive pneumatically-powered assisted knee-ankle-foot orthosis for gait rehabilitation-preliminary finding", In Proceedings of the International Conference on Biomedical Engineering (ICoBE), (2012) February 27–28; Penang, Malaysia
22. T. Kawamura, K. Takanaka, "Development of an orthosis for walking assistance using pneumatic artificial muscle-a quantitative assessment of the effect of assistance", In Proceedings of the International Conference on Rehabilitation Robotics, (2013) June 24–26; Seattle, WA, USA
23. M.A.M. Dzahir, S.I. Yamamoto, "Design and Evaluation of the AIRGAIT Exoskeleton: Leg Orthosis Control for Assistive Gait Rehabilitation", Journal of Robotics, vol. **2013**, Article ID 535106, 20 pages, doi:10.1155/2013/535106 (2013)

24. P. Flavio, M.A.M. Dzahir, S.I. Yamamoto, "Computed-torque method for the control of a 2 DOF orthosis actuated through pneumatic artificial muscles: a specific case for the rehabilitation of the lower limb", Medical Physics, arXiv: 1404.6968 [physics.med-ph] (2014)
25. M.A.M. Dzahir, S.I. Yamamoto, "Recent trends in lower-limb robotic rehabilitation orthosis: control scheme and strategy for pneumatic muscle actuated gait trainers", Robotics, **3**(2), 120–148 (2014)
26. S. Hussain, S.Q. Xie, P.K. Jamwal, "Adaptive impedance control of a robotic orthosis for gait rehabilitation", IEEE Trans. Cybern., **43**, 1025–1034 (2013)
27. S. Hussain, S.Q. Xic, P.K. Jamwal, "Robust nonlinear control of an intrinsically compliant robotic gait training orthosis", IEEE Trans. Syst. Man Cybern.: System, **43**, 655–665 (2013)
28. K.D. William, "Task-based methods for evaluating electrically stimulated antagonist muscle controllers", IEEE Trans. on Biomedical Engineering, **36** (1989)
29. M. Samer, F. Philippe, G. David, P. Philippe, E.M. Hassan, "Towards a co-contraction muscle control strategy for paraplegics". IEEE Conf. on Decision and Control. (2005).
30. H. Stewart, F. Norm, B. Michael, "Muscle co-contraction modulates damping and joint stability in a three-link biomechanical limb", Frontiers in Neurorobotics, **5** (2012)

DOI: 10.1051/matecconf/201713500066

Comparative Study of Machine Learning Approach on Malay Translated Hadith Text Classification based on Sanad

Syuhairah Rahifah Mohammad Najib[1], *Nurazzah* Abd Rahman[1], *Normaly* Kamal Ismail[1], *Nursyahidah* Alias[1,*], *Zulhilmi* Mohamed Nor[2], *Muhammad Nazir* Alias[3]

[1]Faculty of Computer and Mathematical Sciences,Universiti Teknologi MARA, Shah Alam, Selangor, Malaysia
[2]Fakulti Pengajian Quran dan Sunnah, Universiti Sains Islam Malaysia, Bandar Baru Nilai, Negeri Sembilan, Malaysia
[3]Fakulti Pengajian Islam, Universiti Kebangsaan Malaysia, Bangi, Selangor Malaysia

Abstract. Sanad is one of important part used to determine the authentication of hadith. However, very little research work has been found on classification of Malay translated Hadith based on sanad. There are some researches done using machine learning approach on hadith classification based on sanad but using different objective with different language. This research is to see how Machine Learning techniques are used to classify Malay translated Hadith document based on sanad. In this paper, SVM, NB and k-NN are used to identify and evaluate the performance of Malay translated hadith based on sanad. The performances are evaluated based on standard performance metrics used in text classification which is accuracy and response time. The results show that SVM has the highest accuracy and k-NN has the best response time (time taken in process for classification data) compare to other classifier. In future, we plan to extend this paper with the analysis on interclass similarity and also test on larger dataset.

1 Introduction

Sunnah (Hadith) is a second of fundamental sources in Islam after Qur'an [1][2] which is Muslims reference in any activities in their life [3]. Based on [2], the author said that hadith are related to actions and sayings of Prophet Muhammad by trustworthy narrators. It is essential in understanding Qur'an and Islamic jurisprudence [4]. However, [5] mentioned that hadith has been overlooked compared to Qur'an by most academicians in computer science. In the hadith, there are two main components which known as Isnador Sanad (the chain of narrators) and Matn (actual narrative or main text) [4][5][6]. The sanad contains of a chronological list of the narrators, each narrators stated the one from whom they heard the Hadith all the way to the main narrator of the Matn followed by the matn itself [4]. Sanad is essential in every hadith. It is used in the first step of checking the authentication of a

[1] Corresponding author: syahidah@pahang.uitm.edu.my

hadith [2][5][7]. To date, there is need to automatic classification Malay translated hadith [2] based on sanad.

Machine Learning is a wide area of Artificial Intelligence focused in design and development of algorithm that identify and learn patterns exist in data provided as input[8]. Text classification is an important issue which draws many researchers in machine learning and information retrieval techniques [6]. Referring to [8], the author also mentioned text classification is a key problem influenced by machine learning within information retrieval. However, very little research work has been found on classify Malay translated Hadith based on sanad. Classification hadith based on sanad has been done by [6][1][4] with different objective using different language. Review study on Malay translated hadith has been done by [9]to identify the authentication of narrator's name, improving Malay hadith retrieval system by [2] and retrieve Malay hadith text using mobile application by [3]. In this novel approach, Machine Learning techniques are used to classify Malay translated Hadith document based on sanad.

This paper is structured as follows. Section I cover on the introduction of Hadith and how the Machine Learning approach fill into the picture. Section II is focusing on some background information in Machine Learning and Malay Translation Hadith. Section III is focusing on the approach used in this paper. Section IV is a discussion on result and Section V which is contains the conclusion of the paper.

2.1 Research Background

2.1 Machine learning approach (Text classification)

In machine learning, there are three basically type algorithms used: (1) supervised; (2) semi-supervised; (3) unsupervised learning as shown in Figure 1. Supervised learning is required learning a function from training data provided as input. In the case of text classification, the training data are collected of document-class pair representing proper classes for given documents, according to human specialists (data are provided by human assistance as input data). Unsupervised learning is different from supervised learning which is no training data are provided. Semi-supervised learning combines large amount of unlabelled data with a small amount of labelled data [8].

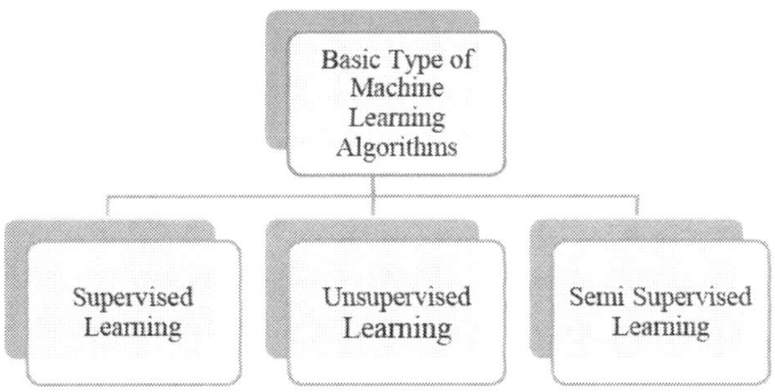

Fig. 1. Basic Type of Machine Learning Algorithm [8].

To classify document into a fixed number of predefined categories is the main reason of text classification. Each document can be categorised in multiple, exactly one, or no category at all. The classification of documents is recognised as a supervised learning task because the purpose is to use machine learning to automatically classify documents into categories based on previously labelled documents [10]. Based on [10], the author also mentioned that Naive Bayes, k-NN and Support Vector Machines (SVM) are popular techniques and common approach [11][12] in automatic text classification. This paper is focused on supervised classification using three most popular technique as stated by [10][11][12] as Figure 2.

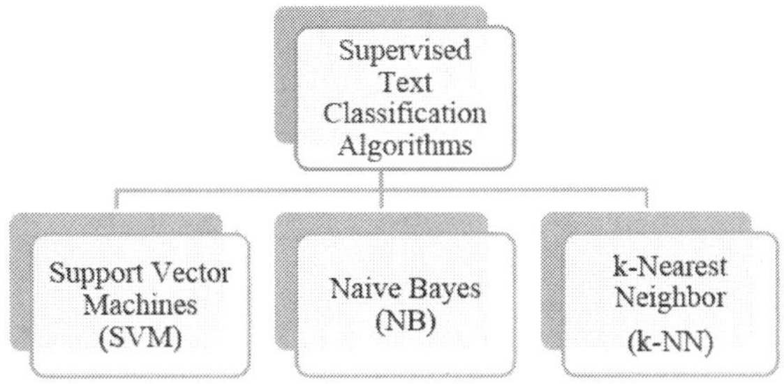

Fig. 1. Supervised Text Classification Algorithms [8].

2.1.1 Support Vector Machine

The Support Vector Machine (SVM) is a classification technique that was introduced by Vapnik and was first applied by Joachims for text classification [8][13][14]. It is a supervised learning algorithm that analyse the data and identify patterns used for classification [15]. The core fundamental of SVM is to determine the most appropriate border line in separating hyper plane [14]. In the set of training data, SVM creates a hyper plane for separating data in two categories (positive and negative) and classification in which data must be placed in this two categories [13][15]. Referring to [13], the author believed SVM is a powerful classifier based on the lowest structural risk principle. SVM also popular in text classification with better performance than other methods [14][16][17].

2.1.2 Naive Bayes

Naive Bayes algorithm is one of classification technique that makes exploit of statistical approach and based on the conditional probabilities for the problems of pattern recognition [16]. Naive Bayes uses Bayes Theorem concept [15][16][18] with strong independence assumptions. The classifier are named "naive" because the algorithm assumes that all terms occur independent from each other [19]. The independence assumptions of features do not depend and effect on each other in classification tasks. Although it is severely limited in its applicability, the computation of Bayesian classification approach is more efficient. It can be trained efficiently to estimate parameters for classification without requiring large amount of training data. The naive Bayes classifiers often work much better in many complex real-world situations than one might expect due to its apparently oversimplified assumptions. Under some specific conditions, naive Bayes classifier has been reported to perform surprisingly well for many real world classification applications [18].

2.1.3 K-Nearest Neighbor

K-Nearest Neighbor (k-NN) classifier is an on-demand (lazy) classifier. The classification is done only at the moment a new document is given to the classifier. The classification decision is computed based on the classes of the k "nearest" neighbour of the new document using a distance function in a predefined metric space. This is accomplished as follows: a) Determine the k nearest neighbours of the new document in a given document training set, b) Use the classes of the nearest neighbour to determine a class for the new document. This algorithm concentrate on specific features of the document to be classified [8]. It is effective and easy to implement [18] and also the most accepted algorithms for pattern recognition [16].

2.2 Malay translation hadith

The text of Malay hadith comes from translated version of Arabic hadith [9]. According to [7], there are six books contain reliable collection of Hadith text: Bukhari, Muslim, Abu Dawud, Termizi, Nasai and ibn Majah. At first, writing narration of Hadith was prohibited due to some religious reasons. However after the death of the prophet, Umar Ibn Abd al-Aziz initiated the writing project of Hadith to guarantee an integrity and uniformity of the text upon fearing that some of hadiths are being lost [6].

3 Methodology

Figure 3 shows the process of text classification in this research is referred to the framework based on [8]. Malay translated hadith dataset are used for this experiment. This dataset is focused only in sanad and the matn are removed.

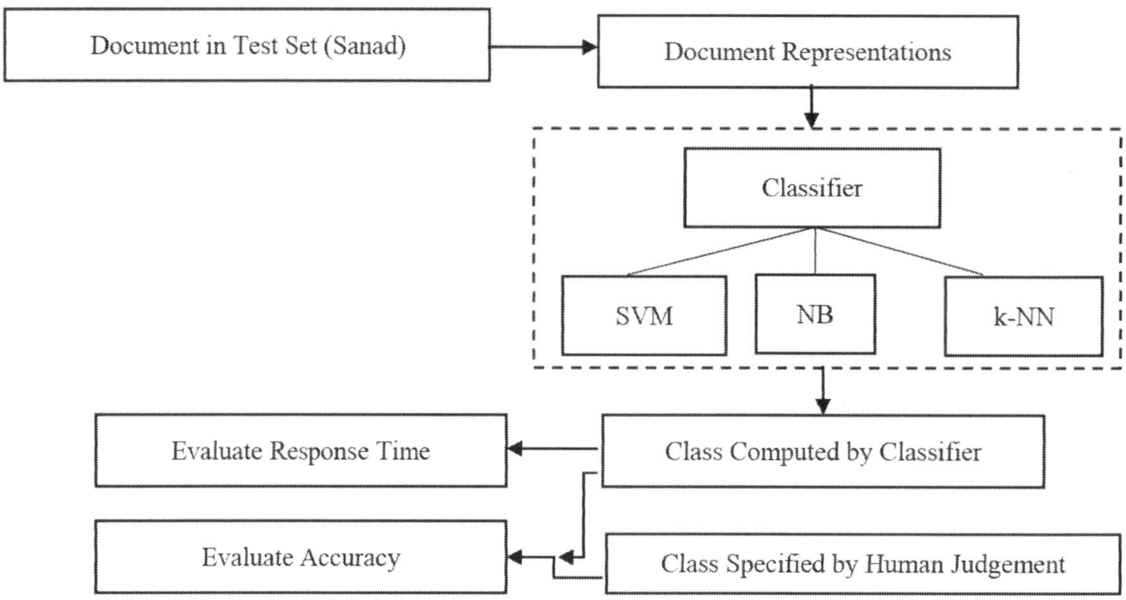

Fig. 2. Text Classification Process.

3.1 Document in test set (sanad)

The data used in experiment are data from Lidwa Pusaka [20] website for dataset hadith Sahih Bukhari and Mutiara Hadis [21] website for dataset hadith Sunan At-Termizi. 100 Hadiths are choose randomly for this experiment which divided by two categories: 50 hadiths from Sahih Bukhari and another 50 from Sunan At-Termizi. The data are labelled manually and the labels belong to two different category classes: 1. Sahih Bukhari (SB) and 2. Sunan At-Termizi (ST). Figure 4 shows a sample of sanad from Sahih Bukhari (SB). Figure 5 shows a sample of sanad from Sunan At-Termizi (SB).

> *"Telah menceritakan kepada kami Al Humaidi Abdullah bin AzZubair dia berkata, Telah menceritakan kepada kami Sufyan yang berkata, bahwa Telah menceritakan kepada kami Yahya bin Sa'id Al Anshari berkata, telah mengabarkan kepada kami Muhammad bin Ibrahim At Taimi, bahwa dia pernah mendengar Alqamah bin Waqash Al Laitsi berkata; saya pernah mendengar Umar bin Al Khaththab diatas mimbar berkata; saya mendengar Rasulullah shallallahu 'alaihi wasallam bersabda;"*

Fig. 3. Sample of sanad [20].

> *"Muhammad bin Basysyar menceritakan kepada kami, Ibrahim bin Abul Wazir memberitahukan kepada kami, Muhammad bin Musa memberitahukan kepada kami (yang berasal) dari Sa'd bin Ishaq bin Ka'b bin 'Ujurah dari ayahnya dari datuknya dimana ia berkata: "Nabi s.a.w. mengerjakan solat Maghrib di masjid Bani 'Abdil Asyhal kemudian orang-orang mengerjakan solat sunat, lantas Nabi s.a.w. bersabda;"*

Fig. 4. Sample of sanad [21].

3.2 Document representations

The dataset is represented as Figure 6. Row data referred to hadith document which contain 100 hadith from Sahih Bukhari and Sunan At-Termizi. The Column data referred to sanad and there are 272 data used for it. The documents are represented in two classes: Sahih Bukhari (SB) and Sunan At-Termizi (ST).

	2	3	...	271	272	class
	umarbinalkhat	alqamahbinw	...	islbinsufy	atha'	
1	1	1	...	0	0	SB
2	0	0	...	0	0	SB
3	0	0	...	0	0	SB
4	0	0	...	0	0	SB
5	0	0	...	0	0	SB
.	.	.	.	.	.	.
.	.	.	.	.	.	.
.	.	.	.	.	.	.
96	0	0	...	0	0	ST
97	0	0	...	0	0	ST
98	0	0	...	0	0	ST
99	0	0	...	0	0	ST
100	0	0	...	1	1	ST

Fig. 5. Document representation.

DOI: 10.1051/matecconf/201713500066

3.3 Classifier

The data were entered into classifier to predict the class of hadith. There are three most popular classifier are used in this experiment which are SVM, NB and k-NN.

3.4 Evaluation

Finally, the performances of classifiers are evaluated. Evaluation is a very important step to determine the performance of the classifiers used [8][22]. In this experiment, the evaluation are covered based on standard performance metrics: Accuracy [1][14][22]and response time [14]. Concisely, accuracy is related to the fraction of correct classifications [22]. Weka tool[23] is used to evaluate the response time (time taken in process for classification data) and for the accuracy, we compare the Class Computed by Classifier and Class Specified by Human Judgement.

$$Accuracy = \frac{TP+TN}{TP+TN+FP+FN} \quad (1)$$

where:

True Positive value (TP)
True Negative value (TN)
False Positive value (FP)
False Negative value (FN) [14]

4 Experiment and Result

Three classifiers are used to compare the performance of Malay translated hadith based on sanad. There are Support Vector Machines (SVM), Naive Bayes (NB) and k-Nearest Neighbor (k-NN). The results of experiment are shown in Table 1. Results are evaluated based on accuracy and response time.

Table 1. Result of experiment

Classifier	Accuracy (%)	Response Time (second)
NB	81	0.01
k-NN	62	0.00
SVM	82	0.16

Figure 7 shows the results of accuracy are 82% for SVM, 81% for NB and 62% for k-NN. We found that SVM shown the highest accuracy compare to other classifier.

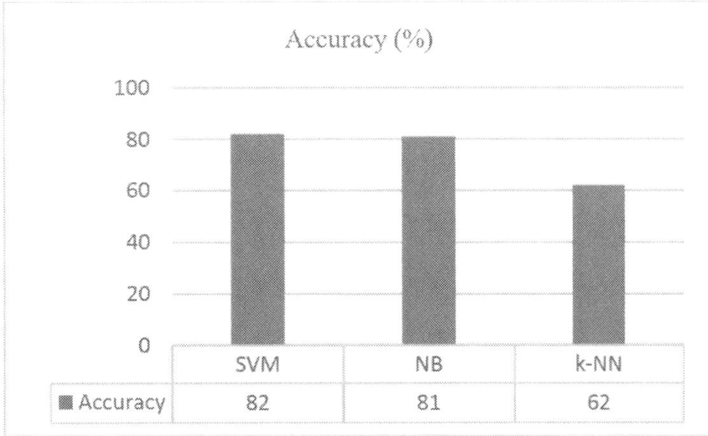

Fig. 6. Comparison the accuracy of classifiers.

Figure 8 shown the results of response time in second(s) are 0.16 s for SVM, 0.01 second for NB and 0.00 second for k-NN. We found that k-NN shown the best response time (time taken in process for classification data) compare to other classifier.

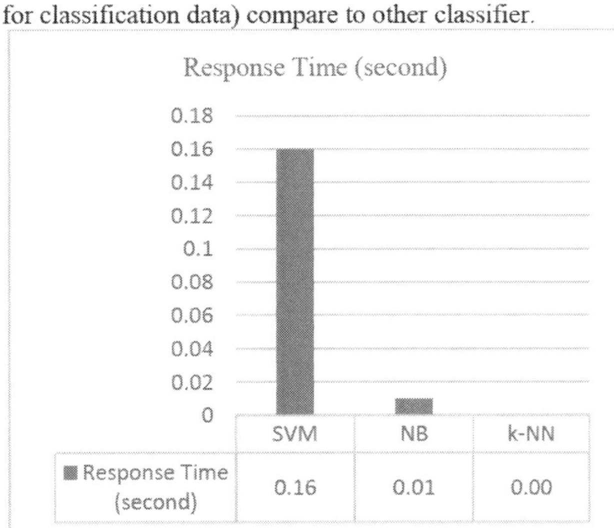

Fig. 7. Comparison the response time of classifiers.

5 Conclusions

The experimental results done on the Malay translated Hadith document based on sanad shown the best performance for accuracy is SVM classifier and response time is k-NN classifier. However, this paper only covers the accuracy and response time of the classification used without details explanation in analysis on interclass similarity. So in the future work, we plan to extend this paper with the analysis on interclass similarity and also test on larger dataset.

Acknowledgement

This research was funded by the Malaysian Government under Fundamental Research Grant Scheme (FRGS) (FRGS/1/2015/ICT01/UITM/03/1) in Universiti Teknologi MARA, Shah Alam.

References

1. K. A. Aldhlan, A. M. Zeki, H. A. Alreshidi, Novel mechanism to improve hadith classifier performance, *Proc. Int. Conf. Adv. Comput. Sci. Appl. Technol,* pp. 512–517 (2012)
2. N. A. Rahman, Z. A. Bakar, T. M. T. Sembok, Query expansion using thesaurus in improving Malay Hadith retrieval system, *Proc. nt. Symp. Inf. Technol. Syst. Dev. Appl. Knowl. Soc.,* **3,** pp. 1404–1409 (2010)
3. M. K. A. B. Zainudin, R. M. Rias, M-Hadith: Retrieving Malay Hadith text in a mobile application, *Proc 2012 IEEE Symp. Comput. Appl. Ind. Electron.,* pp. 60–63 (2012)
4. A. Azmi, N. Badia, iTree - Automating the construction of the narration tree of Hadiths (prophetic traditions), *Proc. 6th Int. Conf. Nat. Lang. Process. Knowl. Eng.* (2010)
5. M. A. Saloot, N. Idris, R. Mahmud R, S. Ja'afar, D. Thorleuchter, A. Gani, Hadith data mining and classification: A comparative analysis, *Artif. Intell. Rev.,* **46,** pp. 113–128 (2016)
6. M. Ghanem, A. Mouloudi, M. Mourchid, Classification of Hadiths using LVQ based on VSM Considering Words Order, *Int. J. Comput. Appl.* (0975 – 8887), **148,** pp. 25–28 (2016)
7. A. Bimba, M. A. Ismail, N. Idris, S. Jaafar, R. Mahmud, Towards Enhancing the Compilation of Al-Hadith Text in Malay, *Int. Proc. Econ. Dev. Res.,* **83,** pp. 76–81 (2015)
8. R. Baeza-Yates and B. Ribeiro-Neto, Modern Information Retrieval: The Concepts and Technology behind Search, *Inf. Retr. Boston.*, **82**, pp. 281–335 (2011)
9. N. A. Rahman, N. Alias, N. K. Ismail, Z. M. Nor, M. N. Alias, An identification of authentic narrator's name features in Malay hadith texts, *2015 IEEE Conf. Open Syst.,* pp. 79–84 (2015)
10. J. Lilleberg, Y. Zhu, Y. Zhang, Support Vector Machines and Word2vec for Text Classification with Semantic Features, pp. 136–140 (2015)
11. A. Harisinghaney, A. Dixit, S. Gupta, A. Arora, Text and Image Based Spam Email Classification using KNN, Naive Bayes and Reverse DBSCAN Algorithm, *Proc. 2014 Int. Conf. Reliab. Optim. Inf. Technol.,* pp 153–155 (2014)
12. Y. Lin, J. Wang, Research on text classification based on SVM-KNN, *Proc. IEEE Int. Conf. Softw. Eng. Serv. Sci.,* pp 842–844 (2014)
13. S. M. H. Dadgar, M. S. Araghi, M. F. Mastery, A Novel Text Mining Approach Based on TF-IDF and Support Vector Machine for News Classification*2nd IEEE Int. Conf. Eng. Technol.,* pp. 16–20 (2016)
14. J. Watthananon, The relationship of text categorization using Dewey Decimal Classification techniques, *12th Int. Conf. ICT Knowl. Eng.,* pp. 72–77 (2014)
15. B. Y. Pratama, R. Sarno, Personality classification based on Twitter text using Naive Bayes, KNN and SVM, *Proc. 2015 Int. Conf. Data Softw. Eng.,* pp. 170–174 (2016)
16. S. Brinda, K. Prabha, S. Sukumaran, A Survey on Classification Techniques for Text Mining, *2016 3rd Int. Conf. Adv. Comput. Commun. Syst.,* pp. 1–5 (2016)
17. C. Yin, J. Xiang, H. Zhang, J. Wang, Z. Yin, A New SVM Method for Short Text Classification Based on Semi-Supervised Learning, *Proc. - 2015 4th Int. Conf. Adv. Inf. Technol. Sens. Appl.,* pp. 100–103 (2015)
18. A. Khan, B. Baharudin, L. H. Lee, K. Khan, A Review of Machine Learning Algorithms for Text-Documents Classification, *J. Advance Inf. Technol.,* **1,** pp. 4–20 (2010)
19. V. Bijalwan, V. Kumar, P. Kumari, J. Pascual, KNN based Machine Learning Approach for Text and Document Mining, *Int. J. Database Theory Appl.,* **7,** pp. 61–70 (2014)

20. Perusahaan software Indonesia dan SIMRS Saltanera Software Kitab Hadits Online Terjemah Indonesia [Online] Accessed: 06-Oct-2016 http://app.lidwa.com/ (2016)
21. N. K. Ismail, N. A. Rahman, and Z. A. Bakar, Mutiara Hadis. [Online] Accessed: 06-Oct-2016, http://sigir.uitm.edu.my/webhadis/ (2007)
22. T. R. P. M. Rúbio, C. A. S. J. Gulo, Enhancing academic literature review through relevance recommendation: Using bibliometric and text-based features for classification, *2016 11th Iber. Conf. Inf. Syst. Technol.*, pp. 1–6 (2016)
23. M. Hall, E. Frank, G. Holmes, B. Pfahringer, P. Reutemann, I. H. Witten, The WEKA Data Mining Software: An Update *SIGKDD Explor.*, (2009)

 DOI: 10.1051/matecconf/201713500067

A Pattern for Concept Identification from English Translated Quran

Rohana Ismail[1,*], *Nurazzah* Abd Rahman[2], *Zainab* Abu Bakar[3]

[1]Faculty of Informatics and Computing, Universiti Sultan Zainal Abidin, Kuala Terengganu, Terengganu.
[2]Faculty of Computer and Mathematical Sciences, Universiti Teknologi MARA, Shah Alam, Selangor.
[3]Faculty of Computer and Information Technology, Al-Madinah International University, Shah Alam Selangor.

Abstract. Ontology development is time consuming and tedious task. The task can be minimized by automatic or semi-automatic ontology development. This minimizing task is a field of ontology learning. Ontology learning will be able to extract ontological elements to form ontology. Concept identifying is one of the important activities in ontology learning. Various methods can be used to find concepts. Thus, this experiment used the n-grams and JAPE pattern in identifying concepts. As the term Allah occurs very frequent in English translated Quran, this experiment considers the word surrounding the term Allah to find other related concepts. It is important because the occurrences produced the term Allah as a concept in ontology but at the same time ignore other related terms to term Allah. The strength connection between words surrounding term Allah has been analysed. Results show the significant terms related to term Allah can be extracted. Later, the term can be used as concepts in ontology.

1 Introduction

The availability of Semantic Web technologies makes the process of understanding Quran is much easier. It is because the Semantic Web technologies can help representing content from Quran as ontology. Ontology is a formal, explicit specification of shared conceptualization [1]. It can be used to form a knowledge based in order to increase the chances of knowledge sharing. Ontology is widely explored in Quran study [2][3][4][5][6]. For example, the ontology from Arabic Quran focusing on time purposely develop to represent semantic meaning of an Arabic word time [2]. It used semantic field theory and componential analysis. Another example is the form of Quran Ontology from Arabic Language Computing Research at the University of Leeds [5]. The comprehensive ontology is focusing on concepts, relations and verse that exist in Quran. However the ontology has insufficient concepts. For example, the Hajj is a concept and just be a part of a calendar event. It is actually many concepts exist related to Hajj.

[1] Corresponding author: rohana@unisza.edu.my

Currently, ontology from Quran is only cover for a few domains such as women, solat, time and animal [7][8][9]. Most of the ontology development is done manually. Manual extraction is a time consuming task, hence there is a need to used automatic or semi-automatic approaches to develop ontology [10][11]. This approach is called ontology learning. The ontology learning approach is able to produce concepts, relations and sometimes axioms from text [12][11]. Few researches attempts to develop ontology from Quran using ontology learning [7][9][13]. For instance, a research used semi-ontology building from Arabic corpus [13]. The research used the WordNet ontology as a model in their ontology development. Another example is using ontology learning for solat ontology development [9]. The ontology used English translated Quran as their corpus. It used statistical and linguistic methods in extracting concepts and relations.

Identifying concept is important in ontology learning. Concept can be a single term or multi term [14][15]. Linguistic and statistical methods are popular methods in ontology learning [12]. Such statistical method is Term Frequency (TF). The TF is simple but shows significant to find concepts [9]. But, sometimes it will ignore the multi term regarding the concepts. For example, the TF produced the term *Allah* as a concept but it will ignore multi terms regarding the term *Allah*. Hence, term surrounding *Allah* needs to be analyzed for getting more relevant terms to be concepts.

This paper proposed techniques to extract concepts related to term *Allah*. The concepts can be identified based on terms collocation with *Allah* and by using the JAPE pattern. The Log Likelihood measurement has been used in N-grams to analyzed terms collocations. Meanwhile, the JAPE pattern has been design to extract the concepts. Result shows the performance of the pattern extraction and the collocation term extraction.

2 Related Work

Extractions of ontological elements from text and building ontology from them is a field of ontology learning [12][16]. Ontology provides conceptual sharing of formal represented knowledge [1]. Ontology learning can be categorized into ontology learning layer cake that consists of layers of activities [17]. These layers consist of identifying terms, synonym, concepts, concept hierarchies, relations and rules. Various methods from established field such as Natural Language Processing and Information Retrieval could be used in ontology learning.

Concepts have no clear definition, but it should provide the intentional part of a domain for leading the way the structure of corresponding ontology[17]. In concepts identification from Quran study, it has varies either single terms or multi terms such as *Allah Bounties, mountain, Hajj, raging fire* [14]. Single terms can be identified using various statistical methods such as term frequency, tfidf, glossex, Average tf and C-Value [9]. Meanwhile, the multi terms can be identified using N-grams [18] or using linguistic method [19]. Linguistic method depends on Part Of Speech tagger [12]. A hybrid of both methods are also used to identify concept [9][18].

The use of linguistic technique in an ontology learning system such as CRCTOL, mainly depends on pattern and POS rule to detect multi terms from text to be concepts [15]. The performance of the extraction has been compared to other ontology learning system which is Text2Onto. The multi terms also can be identified by identifying terms collocation. The collocation will determine a sequence of terms that frequently appear more together than would be expected by chance collocation [20]. For example the noun *Fire* can be a collocation term when it occurs frequently with the term *raging*. The combinations of terms *raging Fire* that occurs frequently can be a collocation. However, not all the combination of terms can be a collocation. It can be analyzed using the statistical methods such as Log Likelihood.

Collocation can be identified using linguistic patterns such as verb-noun, noun-noun, adjective-noun and also using the N-grams method. For example, ontological elements from Curriculum Vitae (CV) has been extracted using patterns such as Noun-Noun, Adjective-Noun, Noun-Adjective, Noun-Preposition-Noun collocations [21]. Another example is the used of N-grams method. The method has been used in finding terms collocation in Arabic Quran corpus [22]. Based on the sequence of Part of Speech, the experiment has extracted collocation of N-gram words ranging from 2 to 6 grams Tagger. One more example in collocation extraction is the use of noun-adjective collocation. The collocation has been extracted from Arabic corpus [18]. The experiment used JAPE pattern in GATE application for extracting terms. These terms will be used for ontology building. Collocation has been widely explored to extract multi terms. It is considered as a phraseological perspective and from a frequency-based perspective [23]. Mostly, three criteria is used to identify collocations which is the distance, frequency and exclusivity [20]. The frequency identification is a simple method to determine the collocation of terms. In statistical, learning collocation has different methods [24]. The methods include i) Mutual Information, ii) Log-Likelihood Measures, and iii) T-test. The methods, used to measure the association between those extracted terms. It calculates the probability that a collocation would occur.

This paper proposed a hybrid of linguistic and statistical approaches. The proposed method has been tested on Hajj corpus. Apart from that, the proposed method tries to analyze specific terms regarding to *Allah*. It is proposed due to the highest occurrence of the term *Allah* in Quranic text translation. The highest term make the term *Allah* is a concept in ontology but at the same time can reduce the chances to get other relevant concept. Therefore, sequence of terms that co-occur with the term *Allah* has been analyzed to see the term collocation. Later, the collocation will be used in building ontology. In the initial extraction, the specific linguistic pattern has been produced for verb- *Allah* collocation. This is differ from extracting terms collocation in Arabic Quran corpus that used the N-grams method [22]. The pattern has been design in JAPE for extracting the terms and the Log Likelihood has been used as an association measure in learning collocation. The measurement used to filter the extracted terms.

3 Methodology

Experiment used English translated Quran. Among eight translations available for the English translation, Hillali Khan Translation has been selected because it contains more elaboration on verses. As the purpose is to construct the Hajj ontology, verses related to Hajj have been selected. The verses or data have gone through for a few processes to produce data such as preprocess, bi-gram application and GATE process.

In pre-processing the data, removing certain symbols from the data is required to get more accurate results when using NLP process. Apart from that, replacing the exist pronoun in the corpus in necessary. The detail process of extraction can be shown as in Fig. 1.

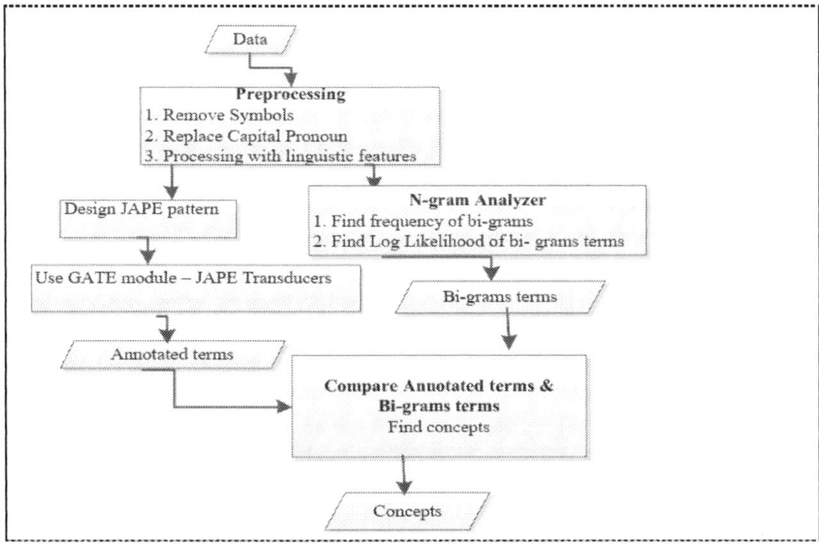

Fig. 1. The detail process of concept extractions.

The pre-processing includes the components as follows:

- Removing symbols, hyphen marks and unnecessary words.
  This component removes punctuation marks such as *Ka'bah, Fir'aun, 'Umrah to Kabah, Firaun, Umrah*. On top of that, it eliminates the hyphen marks such as *Al-Birr* to *AlBirr*. Lastly, it removes the unnecessary words such as *"etc."* and *"i.e"*.

- Replacing the capital pronouns exist in the middle of verses. Pronouns such as *Him, Me, My, Us, We, His Lord* were replaced by term *Allah*.

- Processing with linguistic features such as:
  Tokenizer: splits the text into tokens
  Sentence Splitter: splits the text into sentences
  POS Tagger: adds part-of-speech information to tokens based on Stanford tagger

After the pre-processing has been done, data needs to be run on two paths.

1. Run the data in GATE application.
   In GATE application, a module called JAPE transducers has been called to process the data. The JAPE transducer can annotates the data according to design patterns. Therefore, a pattern regarding the term *"Allah"* is designed using JAPE pattern, as shown in *Fig. 2.* The pattern matches a variant of verbs that have been identified via POS tagger from previous process, followed by term *"Allah"*.

```
Rule: AllahRule
(
({Token.category==VBZ}|{Token.category==VBP}
|{Token.category==VBG}
|{Token.category==VB, Token.kind==word})
({Token.string == "Allah", Token.kind==word})
)
```

Fig. 2. The JAPE pattern.

2. Run the data in N-gram application.
   The N-gram application will process the data into unigrams, bi-grams, tri-grams, 4grams and 5grams terms. The terms will be analysed based on their frequencies and log likelihood ratios. However, this experiment only cover bi-grams analysis since the intended output is bi-grams. Based on the bi-grams terms, frequencies and Log likelihood ratios are being employed in the terms to find collocation terms.

Next, the terms are being compared with the JAPE patterns to find concepts.

4 Results and Discussions

53 verses from Hajj domain has been used in this experiment. The verses contain different topics of Hajj. Among the topics, two topics have the highest number of verses. The topics are *"duty of Hajj and its moral"* which contains 42 verses and *"the honoured Makkah"* which contains 16 verses. Previous experiment produced 6,689 unique terms from 213,679 total terms of extracted from a whole English translated Quran. Meanwhile, 794 unique terms has been extracted from 3125 total terms from Hajj corpus. Based on that, the term Allah shows to be appeared as the highest term in whole English translated corpus (4761) and Hajj corpus (134). The highest term of Allah make the term Allah can be excluded as significant term. But this experiment shows that considering the term Allah can produce the significant terms related to term Allah.

Since the main purpose is to extract concepts for Hajj corpus, only terms related to *Allah* in Hajj corpus has been analyzed. The extractions produced 145 terms that relate to term *Allah*. Based on the total extracted terms, the data has pass through each steps shown as in fig 2. The first path is to use the n-grams analyser. It produces the extracted terms as shown in Table 1.

Table 1. Result of the extracted terms related to Allah.

Terms	No. of terms
Left term of Allah	76
Right term of Allah	69
verb -Allah	4
Allah-verb	38

Terms surrounding Allah which is the left and right terms of Allah has been analysed. Left terms of Allah produced highest number in the extraction which is 76(52.41%) and the rest are the right terms of Allah which is 69(47.58%). Apart from that, the extraction produced 4 terms from left term that is related to verb-Allah and 38 terms from right term that related to Allah-verb. Even though the Allah-verb is higher than verb-Allah, the used of the Allah-verb is common in text such as *Allah destroy, Allah made*. Thus, it cannot be used as a concept. The top 10 extracted terms is shown as in Table 2. The terms are sorted by Log Likelihood and frequency.

Table 2. Results of The top 10 terms related to Allah.

Terms	Log Likelihood (LL)	Frequency

fear Allah	50.4	9
Allah have	36.33	10
Allah has	21.6	6
Allah is	17.67	10
to Allah	16.24	11
of Allah	15.74	16
Allah Messenger	14.37	3
remember Allah	14.34	4
obey Allah	12.65	2
Allah wills	12.65	2

The above table shows that, the highest Log Likelihood value is the term *fear Allah*. Although the frequency of the term *fear Allah* is lower compared to the term *Allah have*, the Log Likelihood ratio will rate the term *fear Allah* as the highest. The Log Likelihood measure the collocation of the term or the strength between words to the term *Allah*. The higher the value, the more the term can be consider as significant terms to be concepts. Based on the extraction, only 3 terms are selected to be concepts which are *fear Allah*, *remember Allah* and *obey Allah*. The second path is using the GATE application. The extracted terms using the JAPE pattern is shown as in Table 3.

Table 3. Results of the pattern extraction.

Terms	Frequency	Terms	Frequency
fear Allah	8	ask Allah	1
remember Allah	4	glorifying Allah	1
have Allah	3	invoking Allah	1
obey Allah	2	magnify Allah	1
reaches Allah	2	purify Allah	1
unto Allah	2	sanctify Allah	1

Based on the table, it shows that the pattern is able to extract concepts that relate to term Allah. The extracted process has extracted 12 candidate concepts based on their frequency. However, only 6 terms are being selected as related terms and the rest are false extracted. The terms such as fear Allah, remember Allah, obey Allah, ask Allah is selected as concepts and invoking Allah, magnifying Allah as an instance of concept remember Allah. Other terms are false extracted because it relates to another word in the verse. For example, the glorifying Allah is actually related to glorifying Allah praises, purify Allah and sanctify Allah are related to purify Allah House and sanctify Allah House.

From both paths, it shows that the used of the statistical method such as Log Likelihood is able to extract collocation terms to be concepts but it cannot extract other relevant terms when having minimum frequency such as 1. Thus, the terms such as invoking Allah, magnify Allah cannot be extracted. Therefore, the Jape patterns can extract more the collocation terms to be concepts. The JAPE pattern can be used for even a small corpus.

5 Conclusions

The technique used to extract relevant terms related to Allah has been presented. The n-grams technique with Log Likelihood and the JAPE pattern are used to extract terms for ontology development. Both techniques are able to extract significant terms. Among the techniques, the JAPE pattern demonstrates that it is able to extract more collocation terms to be concepts. Future works includes extracting relations in the corpus to build hierarchy of the concepts.

References

1. T. R. Gruber, A translation approach to portable ontology specifications, *Knowl. Acquis.*, **5**, April, pp. 199–220 (1993)
2. M. Al-yahya and H. Al-khalifa, An Ontology Model for Representing Semantic Lexicons: An Application on Time Nouns in the Holy Quran, **35,** no. 2, pp. 21–35 (2010)
3. R. Iqbal and A. Mustapha, An experience of developing Quran ontology with contextual information support, **7,** no. 4, pp. 333–343 (2013)
4. H. S. Al-Khalifa, M. M. Al-Yahya, A. Bahanshal, and I. Al-Odah, SemQ: A proposed framework for representing semantic opposition in the Holy Quran using semantic web technologies, in *Proceedings of the 2009 International Conference on the Current Trends in Information Technology, CTIT 2009*, 2009, pp. 44–47 (2009)
5. K. Dukes, Ontology of Quran concepts. [Online]. Available: http://corpus.quran.com/ontology.jsp. [Accessed: 12-Jun-2014].
6. F. Harrag, A. Al-Nasser, A. Al-Musnad, R. Al-Shaya, A. S. Al-Salman, Using association rules for ontology extraction from a Quran corpus (2013)
7. S. Alrehaili and E. Atwell, Computational ontologies for semantic tagging of the Quran: A survey of past approaches, *Lr. 2014 Proc.* (2014)
8. A. J. Petiwala and S. S. Sathya, A Multi-Agent System to Learn Literature Ontology : An Experiment on English Quran Corpus, pp. 46–51 (2011)
9. S. Saad, N. Salim, and S. Tiun, Concept Extraction on Quranic Translation Text, *Int. J. Islam. Appl. Comput. Sci. Technol.*, **2,** no. 1, pp. 1–9 (2014)
10. M. F. Lopez, A. Gomez-Perez, J. P. Sierra, and A. P. Sierra, Building a chemical ontology using Methontology and the Ontology Design Environment, *IEEE Intell. Syst. their Appl.*, **14,** no. 1, pp. 37–46 (1999)
11. A. Maedche and S. Staab, The text-to-onto ontology learning environment, *Softw. Demonstr. ICCS-2000, Eight Int. Conf. Concept. Struct.* (2000)
12. W. Wong, W. Liu, and M. Bennamoun, Ontology learning from text, *ACM Comput. Surv.*, **44,** no. 4, pp. 1–36 (2012)
13. B. E. Benaissa, D. Bouchiha, A. Zouaoui, and N. Doumi, Building Ontology from Texts, *Procedia Comput. Sci.*, **73,** no. Awict, pp. 7–15 (2015)
14. S. Saad, N. Salim, and H. Zainal, Islamic knowledge ontology creation, *2009 Int. Conf. Internet Technol. Secur. Trans.*, no. NOVEMBER 2009, pp. 1–6 (2009)
15. X. Jiang and A. H. Tan, CRCTOL: A semantic-based domain ontology learning system, *J. Am. Soc. Inf. Sci. Technol.*, **61,** pp. 150–168 (2010)
16. M. Shamsfrad and A. A. Barforoush, The state of the art in ontology learning : a framework for comparison, *Knowl. Eng. Rev.*, **18,** pp. 293–316 (2003)
17. P. Buitelaar, P. Cimiano, and B. Magnini, Ontology Learning from Text : An Overview, pp. 1–10, 2004.
18. S. Zaidi, M. T. Laskri, and A. Abdelali, Arabic collocations extraction using gate, *2010 Int. Conf. Mach. Web Intell. ICMWI 2010 - Proc.*, pp. 473–475 (2010)

19. S. Ghadfi, N. Bechet, and G. Berio, Building ontologies from textual resources: A pattern based improvement using deep linguistic information, *CEUR Workshop Proc.*, **1302** (2014)
20. V. Brezina, T. Mcenery, and S. Wattam, Collocations in context A new perspective on collocation networks*, *Int. J. Corpus Linguist.*, **202,** pp. 139–173 (2015)
21. M. Roche and Y. Kodratoff, Pruning Terminology Extracted from a Specialized Corpus for CV Ontology Acquisition, *OTM Confed. Int. Conf. Move to Meaningful Internet Syst.*, pp. 1107–1116 (2006)
22. W. Alromima, I. F. Moawad, R. Elgohary, and M. Aref, Extracting N-gram terms collocation from tagged Arabic corpus, *2014 9th Int. Conf. Informatics Syst. INFOS 2014*, pp. NLP10–NLP15 (2015)
23. J. Parkinson, Noun-noun collocations in learner writing, *J. English Acad. Purp.*, **20,** pp. 103–113 (2015)
24. F. Xu and D. Kurz, Text Mining for the Extraction of Domain Relevant Terms and Term Collocations, *Proc. Int. Work. Computational Approaches to Collocations.* (2002)

MATEC Web of Conferences **135**, 00068 (2017) DOI: 10.1051/matecconf/201713500068
ICME'17

Natural Language Processing Approach for Searching the Quran: Quick and Intuitive

Abidah Zainal, *Noor Azah* Samsudin[1], *Mohammad Airin* Azman, *Aida* Mustapha

Faculty of Computer Science and Information Technology, Universiti Tun Hussein Onn Malaysia, Parit Raja, 86400 Batu Pahat, Johor, Malaysia

Abstract. The Quran is a scripture that acts as the main reference to people which their religion is Islam. It covers information from politics to science, with vast amount of information that requires effort to uncover the knowledge behind it. Today, the emergence of smartphones has led to the development of a wide-range application for enhancing knowledge-seeking activities. This project proposes a mobile application that is taking a natural language approach to searching topics in the Quran based on keyword searching. The benefit of the application is two-fold; it is intuitive and it saves time.

1 Introduction

The Quran is a book of guidance for all people all over the world. The Quran covers all aspects from economy to political, science, to astronomy. It has vast information from the beginning of time to the end of time. The Quran consists of 30 Juz, which was descended to Prophet Muhammad S.A.W in 23 years. The Quran serves as a record of time and act as it is most important source in tracing the development of Islam from Mecca to Madinah. Until today the Quran holds as the most trustable book in the world, therefore, the Muslim community regards the Quran as the highest literature for Arabic language. Many miracles in the Quran can be found such as the development of human embryo which recorded in Quran around 1400 years ago, which modern civilization found it in early 1970. Quran has also become as a guideline to time in history, which records actions that are precise and proven as a helping tool for anthropologist [1].

At the same time, Quran has proven to have no errors despite its diverse topics with great depth of knowledge. Based on this characteristic, a searching application for the Quran is proposed. The searching ability is based on natural language input and will be useful to shorten the time taken to search in the entire Quran. Although the digital Quran is convenient to perform searching because a normal search function is sufficient, the source of search is limited. This is because the knowledge is not comprehensive and only covers the searched keywords. The objective of the proposed application is three-fold; (1) to design a Quran search application based on keywords entered in natural language, (2) to develop the application based on mobile platform, and finally (3) to test the application

[1] Corresponding author: azah@uthm.edu.my

DOI: 10.1051/matecconf/201713500068

usability among the religious officer at Islamic Centre at Universiti Tun Hussein Onn Malaysia (UTHM).

The application uses tokenization to extract keyword when the user enters a sentence. The knowledge returned to users is based on the keywords and covers the tafsir, hadith and translation from a specific verse. The knowledge base is limited to the 30th surah in the Quran known as Juz Amma. The target users are common users with no age restriction. The application, however, will be evaluated by the Islamic knowledge expert at UTHM. This application is hoped to change the way users interact with the content of the Quran and attract young users to learn Quran in a more convenient manner. This application also provides the user specific information about Quran, complementary forms which are essential in finding information in urgent matter

The rest of the paper is organized as follows: Section II describes the related work on mobile-based Quranic applications, Section III presents methodology, Section IV presents the prototype implementation and testing, and finally, Section V concludes the work and highlights future research.

2 Related Work

This section reviews applications related to the Quran discovery using natural language processing. The review is important in order to understand how the system works and how it can be improved for future purposes.

2.1 iQuran Application by Isysway [2]

This application is invented to help user to read the Quran. It is very interactive which has bookmark for last recitation of the Quran. This Quran application by Isysway is available to download at Playstore for free. It also provides searching mechanism to search for verses in each surah that the users want to read. The searching function is using hardcode to display the result of the search and it is similar to the traditional method of reading the Quran but using mobile phone. Figure 1 shows the graphical user interface (GUI) of the iQuran application.

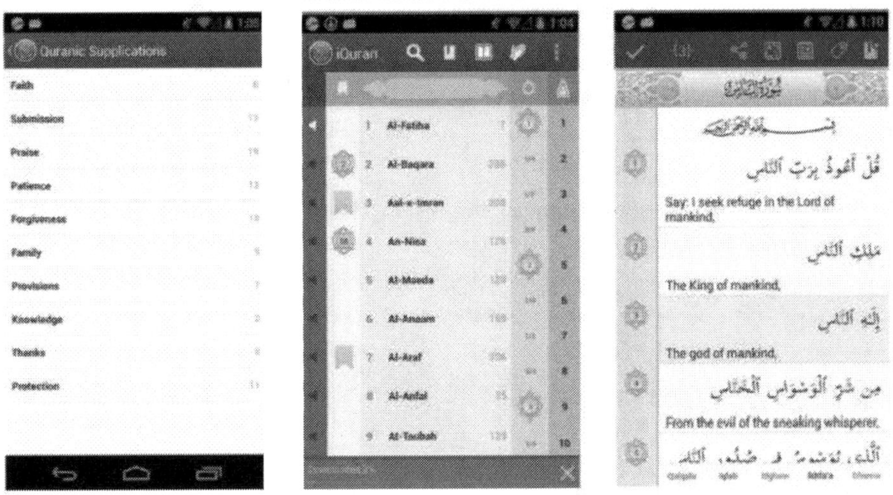

Fig. 1. Graphical user interface of iQuran.

2.2 Quran for Android by Quran.com [3]

Quran for Android is a free source application that available for android users to download using Google PlayStore. The main application functionality is to read the Quran in smartphone. The application requires Internet connection for user to download audio for the application. Due to this limitation, the application have problem in stopping the audio of the Quran when there are a call on the user phone. The application has a night mode that helps to lessen the stress of eyes when reading a text from a bright screen in dark area. It also uses bookmarks to mark the last verse of reading. This is to ensure users reads based on the correct continuity of the Quran. Figure 2 shows the interface for the Quran for Android application.

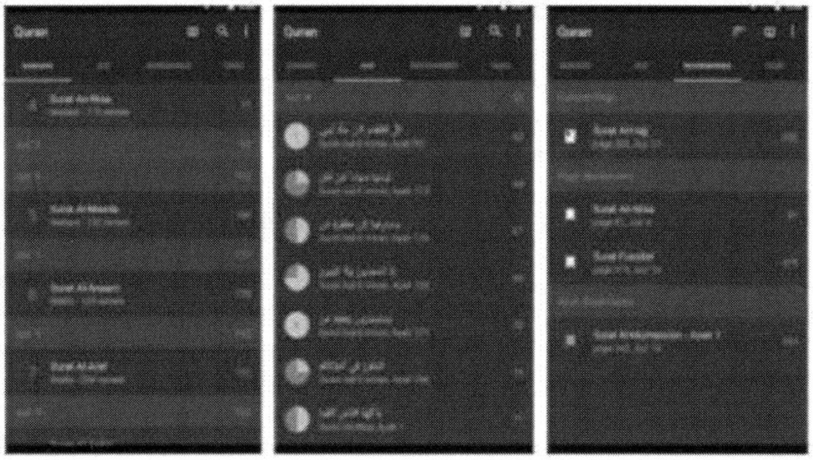

Fig. 2. Graphical user interface of Quran for Android.

2.3 Learn Quran by Badr Interactive [4]

Learn Quran is designed for users who want to start learning the way to read the Quran. Since learning Quran is very hard and some people have limited time on it, the application is developed to countermeasure this problem. This application aids the user to learn the Quran from the pronunciation to tajwid, with audio in helping the user to achieve the correct pronunciation for each verse. Although this application is free, it has premium feature which the user can get when paid for the application, such as no advertisement and more learning tutorial about the Quran. Figure 3 shows the user interface for the Learn Quran application.

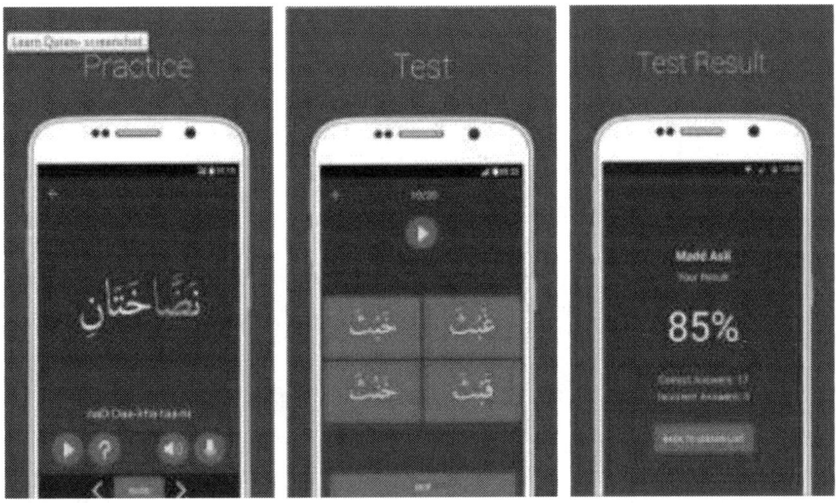

Fig. 3. Graphical user interface of PexPix.

The reviews concluded that all the proposed Quran Search Tool using Natural Language Processing offers a unique feature that can help user to enter input in a full sentence and the tool can still give answers to the search query.

3 Methodology

Methodology is an essential element for developing a work plan for the project in order to establish a mechanism to coordinate and control the flow of the project, and to ensure the project are finish within time. Research has shown that there are many techniques and models available for development of application such as the waterfall model, the prototype model, and the spiral development model. This project uses the prototype model to develop the proposed Quran Search Tool using Natural Language Processing. Figure 4 shows the steps in the prototyping model [5].

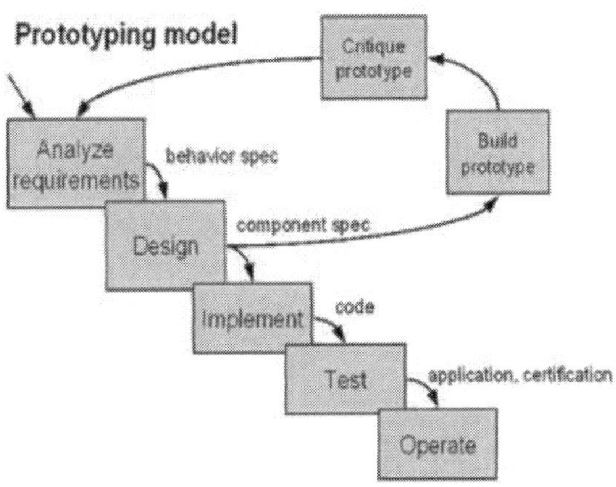

Fig. 4. Prototype development model.

The prototype model was chosen because the technique and flow of this model are repeatable, therefore allowing flexibility and efficiency. This is because the prototype will allow the users to have a hands-on experience of the product before it finishes. This will be a major boost in helping the users to understand what the application is capable of. At the same time, the users can request for changes since they already have a glance on the application before it is launched. The prototype model has six main phases needed to develop the Quran Search Tool using Natural Language Processing, which are planning, analysis, design, prototype development, and testing. The planning phase is a critical stage which will determine the outcome of this project, in this phase, the objective of this project will be determined, as well as model that suitable for the proposed project to be used.

Multiple method and research techniques have been used in ensure the development of this project is on the right track and while at the same time meets the requirements from the used. The activities that will be performed during this phase is to study Final Year Project Thesis from past year students, discuss with supervisor and review of the Internet. Gantt chart will be made to ensure the sustainability of this project running smoothly according to its schedule. At the same time, the project requirements were gathered to make the project a success. Analysis is a crucial which takes a lot of time such as identifying the hardware and software needed to make this application a success.

The goal of this phase is to provide the development of the application with much more understanding of the process to develop a prototype of the Quran Search Tool Using Natural Language Processing. The minimum requirements in this particular phase were collected from the interview with the religious officer in UTHM Islamic Centre. Furthermore the suggestion and recommendation from the center will be taken as guidance for the application. The traditional searching method in existing Quran application requires a longer period of time as compared to the natural language processing techniques that takes less time and makes the application more exciting.

In the analysis phases, the interface of the application will be designed, the navigation will be determined, the button on the application will be designed using Adobe Photoshop to make it more interactive for the users. At the same time, the flow and the way the application works will be specified using the Use Case Diagram, Flow Chart, and Activity Diagram. The design process of this application has been made using multiple hardware and software that are suitable for the development and success of this application. Implementation phase is where the prototype is developed and tested based on design, testing process need to be done before the launch of the proposed Quran Search Tool using Natural Language Processing application. One of the main reason to have a testing procedure is to ensure the application is able to perform according to what has been expected during the planning process, At the same time, in this phase testing is also carried out to remove any bugs or issues according to the scope and the application objective of the project.

4 Prototype Implementation

This section focuses on the analysis and design phases of Quran Search Tool using the Natural Language Processing. A research has been made to ensure the outcome of analysis and designed activities are based on the requirement of this project. At the same time discussion also act as important role to the overall project before the development. Requirement specifications of application need to be analyzed in order to make sure the details are correct, the output to meets the goal and objectives. Before the development process, the application flow needs to be designed based on information from the interview. The analyses also focused on the chosen keyword from the users, the most favorable

DOI: 10.1051/matecconf/201713500068

keyword in Quran, most common keyword are also identified, so that the project will be able to assist user in understanding the Quran.

4.1 Natural Language Processing

Natural Language Processing is an algorithm that manipulates string to maximize it use in retrieving data. Natural language processing has two primary roles to play in the storage and retrieval of large bodies of information: providing a friendly, easily-learned interface to information retrieval systems, and automatically structuring texts so that their information can be more easily processed and retrieved. Figure 5 shows the sample code excerpt from Android Studio IDE that describes Natural Language Processing.

```
input = (EditText) findViewById(R.id.editText);
String str = input.getText().toString();
bilangan = (TextView) findViewById(R.id.textView6);
no_surah = (TextView) findViewById(R.id.textView2);
content = (TextView) findViewById(R.id.textView4);
hadis = (TextView) findViewById(R.id.textView5);
String result = str.replaceAll("[\\-\\+\\.\\^:,?! ]","");


textout = (TextView) findViewById(R.id.textView3);
String trim=str.replaceAll("[\\-\\+\\.\\^:,?!><]","");
String[] s = trim.split(" ");
```

Fig. 5. Natural language processing.

Natural Language has been involved in numerous systems and games. One of Natural Language Processing's feature is removing unnecessary word, special character, and even symbols. This make user interaction much more interactive since when user enters symbols the application still recognize it require keyword, second is word separation the application can receive many input from user such as a sentence this way the application capable of chatting with the user. Natural Language Processing also involve artificial intelligent for retrieving data according to specific keyword. Quran Search Tool Using Natural Language Processing use trim function from Natural Language Processing, word separation and matches keyword to display the correct answer according to user input.

4.2 Use Case Diagram

Use case diagrams are useful in explaining the functionality of the system itself. In the development process, use case diagrams act as a reference for the activity in system. The use case diagram contains an actor, relationship, and use cases. Figure 6 shows the Use Case Diagram for the Quran Tool using Natural Language Processing.

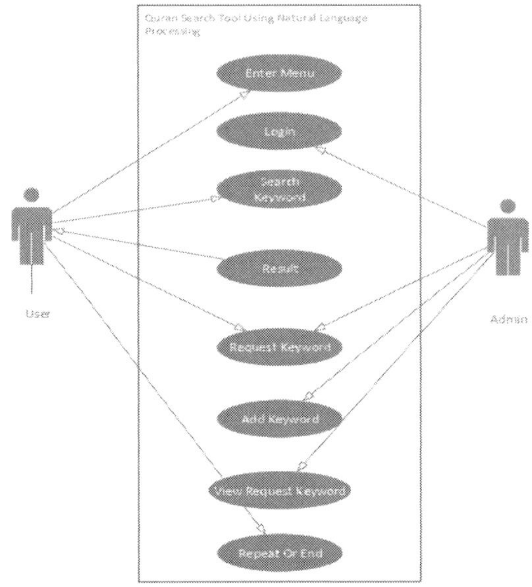

Fig. 6. Use case diagram.

4.3 Sequence Diagram

Sequence diagram for the proposed application shows the steps when the user enters the input sentence, which has a correct keyword. The system will return a result from the application and the user can continue session by entering new sentence or the user can end session. Figure 7 shows the sequence diagram for the entire system.

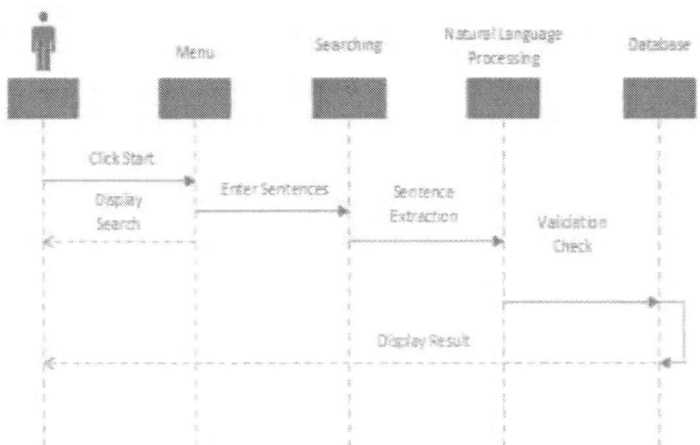

Fig. 7. Sequence diagram.

The application will send back a message shown that they keyword inserted exist. Meanwhile, the user can enter a new sentence to continue search or end the session.

4.4 Activity Diagram

Activity diagram acts as a guide to understand the flow and process of the application. The activity starts when the actor, which is the user, begins the application. After that, the user will enter keyword that they want and the natural language processing module will start. The natural language processing module is important to extract the correct keyword from the database. If the user enters a wrong keyword, no answers will be displayed. The user has a choice whether to start again the search or end the session. If succeed, the application will display the result of the keyword entered in another interface. The result will be generated on the page.

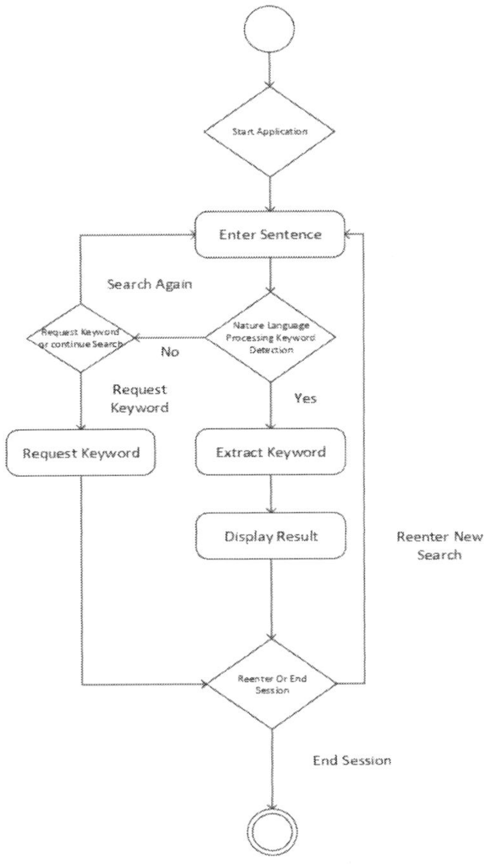

Fig. 8. Activity diagram.

Figure 9(a) illustrates the start interface of the application. The start interface consists of one button which consist a function to enter another activity which is the searching activity. Once the user click starts, the search interface will appear as shown in Figure 9(b). The search menu of this application consists of two buttons (1) input from user which is a sentence, and (2) text that will view the result of the searches. The button which has the word "CARIAN BARU" acts as a clear button if the user wants to clear the search sentence input by user. The green button function is to grab the sentence and initialize a process to generate the output for the two texts which will view.

Fig. 9(a). Welcome page.

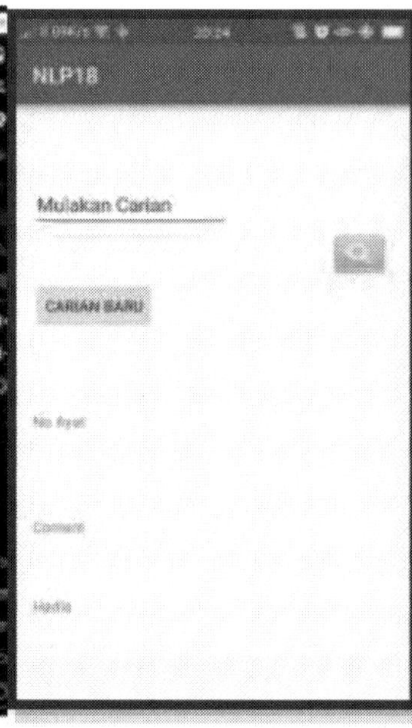

Fig. 9(b). Search page.

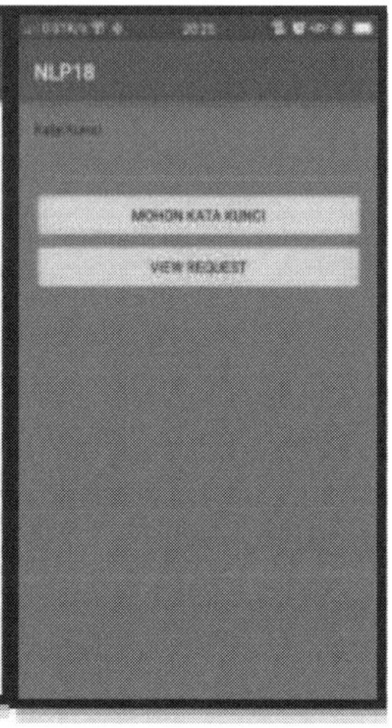

Fig. 9(c). Keyword request.

5 Conclusions and Future Works

The project started with the object-oriented analysis which defined all the requirements of the system. Then the object-oriented design phase which the application displays the main menu of the application, search interface, popup when a keyword input does not match, and a database which integrated from android to the website. Then enters the implementation and testing phase where the system is built and tested according to system requirements and objective, the overall of this chapter will focuses on limitation and future improvement of this application. One important limitation that need to be considered is that the nature language processing module in this application only can match with one keyword. If a keyword consists of space the application will not detected since the application remove the space character.

In order to improve its performance and functionalities, it is recommended to develop this application to other mobile platform such as Symbian for Nokia smartphone, IOS from Apple Inc., Windows phone for HTC and also Nokia from Microsoft, and Blackberry 11 from Blackberry. This will increase the flexibility and amount of users using this application since android only accommodate around 45 percent of all smartphone in this world. Moreover, the usability of this application can be improved by adding other functional ability such as to detect two keywords at the same time. Overall, it is hoped that this application is able to promote the culture of learning the Quran among the Muslim youngsters today by riding on the smartphone technologies.

Acknowledgements

This work was supported in part by a grant from the Ministry of Education under the Research Acculturation Grant Scheme (R045) and in part by a grant from Research Gates IT Solution Sdn. Bhd.

References

1. E. John, L. Esposito, The Oxford History of Islam (1999)
2. iQuran by Isyway [Online] Accessed on 21 October 2015 at https://play.google.com/store/apps/details?id=com.isysway.free.alquran&hl=en
3. Quran.com [Online] Accessed on 21 October 2015 at https://play.google.com/store/apps/details?id=com.quran.labs.androidquran&hl=en
4. Learn Quran [Online] Accessed on 21 October 2015 at https://play.google.com/store/apps/details?id= com.bi.learnquran
5. K. E. Lantz, The prototyping methodology. Prentice-Hall, Inc. (1986)

Analysis of translated query in Quranic Malay and English translation documents with stemmer

Mohd Amin Mohd Yunus[1], *Aida* Mustapha, and *Noor Azah* Samsudin

Faculty of Computer Science and Information Technology, Universiti Tun Hussein Onn Malaysia, Parit Raja, 86400 Batu Pahat, Johor, Malaysia

Abstract. Quranic documents result has a limited query due to focusing on exact words to retrieve those relevant documents. Therefore, there is variety of results to be useful for the target users to explore Quran documents in proper manner. Thus, this paper presents analysis according to conducted empirical experiments in 12 retrieval processes. Thus a system is needed to retrieve relevant documents across language boundaries as well as monolingual. Therefore, empirical experiments are conducted with the purposes to investigate English-Malay translation approach and vice versa against monolingual searching process. Furthermore, it is also conducted to investigate the performance between keywords and querywords based on total retrieve and relevant for each retrieval process. The retrieval however, included the unnecessary documents because of the translation polysemy. This research also is being applied in retrieving Quran English and Malay translated documents with queries compared to monolingual query searching retrieval. Furthermore, in order to produce more significant result, the comparison between stemmer and monolingual results are successfully analysed to evaluate precision and recall percentages. The most important findings are the use of stemmer more beneficial to the query and documents simultaneously regardless the experiments applied translation or not. It leads more and more relevant results displayed.

1 Introduction

It is observed that the dictionary-based method is a suitable method to disambiguate every translation of query rather than machine translation and corpus parallel document collection. Therefore, the goal of this research is to enable users to query in the Malay language against an English collection. Cross Language Information Retrieval (CLIR) systems can be evaluated automatically in much the same fashion as monolingual IR systems, using queries with known relevance judgments. However, in the multilingual context, there is a strong desire to compare the results against the optimal performance of

[1] Corresponding author: aminy@uthm.edu.my

the system if the query is perfectly translated. This gives rise to the following approach to evaluation in CLIR. Start with queries, documents, and relevance judgments in a single language. The queries have been translated into another language by human translators. These translated queries are helpful for the performance of the CLIR system. In order to conduct the experiments, data regarding the Malay and English Quran translation are gathered accordingly as two sets of collection respectively. The Malay and English dictionary files are useful for translating the query. A Malay and English stopword files are used for removing the unnecessary words in the query.

2 Related Work

Monolingual and multilingual are two different to be applied on query. Monolingual is easier to be understood as there is only single language performed in searching process. While multilingual can be applied to the query is used in translation for searching other languages in getting more multilingual documents with single query. This concept is called multilingual query expansion for searching more documents in various languages. There are several studies conducted to investigate the query to be useful for retrieval. Therefore Nie [9] stated the query expansion that needs to be improved in monolingual while Dang and Croft [10] present query formulation improving the search result. Another perspective, Strohmaier and partners [11] describe the suggested words for query so that users may choose the right query for expected results. On top of that, stemmer [2,6] is very important to apply to that translated query [12, 13,14,15,16].

3 Experimental Approach

Two al-Qur'an documents are used in these experiments, Malay Quranic documents collection [3] and English Quranic documents collection [4]. Each collection has 114 surahs and 6236 documents. Every document has numbers beginning with q which means query and followed by numbers with first three numbers denote chapter and last three number denote verse for every filename such as q034006 but in the filename the number of surah begins with "." or dot, followed by the number of surah and the number of ayat begin with , or comma, followed by the number of ayat as for example " .34,6". All documents are as flat files in ASCII or EBCDIC text and searching process is through pattern matching [1]. Short query is to type some queries for retrieving relevant documents. In this study, the words as input are called query for any matched words in the documents. In this research, the Malay query words are taken from Fatimah's collection [2] and the English query words are translated from the Malay query words. Fatimah has obtained them by considering several guidelines put forward by Popovic [5] and Salton [8]. Each query would be separated and broken into keywords and replaced by target language. For example, if query is Malay, so it is called as source language and the target language is English. Thus, English represents the translated word to retrieve English documents and if the query is English, the target language is Malay. The dictionary lists 1500 Malay words and 1325 English words including 36 Malay query words selected. The translation refers to the same index between Malay natural query languages [2] and the translation of the natural query language in English. When the keyword is Malay, then reference is to the English word at the same index or when the keyword is English, then reference is to the Malay word at the same index. It is considering word by word in the text files.

Stopwords are used for removing the meaningless words in the query during retrieval process. So, it is easier for the system for not processing meaningless words. By having Malay and English stopwords, it is easier to remove the common meaningless words and

English does the same. The Malay root words are used to make comparable to the query after stemming. Then, if stemmed words in the query are similar to root words, then retrieving process will be next to produce relevant documents and stemmed query [2]. Fatimah [2] calculated that the relevant document should be resulted from each query list with the Arab student assistance from the Universiti Kebangsaan Malaysia. The student searched the al-Quran, referring to well-known Islamic books regarding the same subject matter based on the natural language queries [2].

The flow of process is started by pre-processing, storing, retrieving, evaluation and presentation of the result. The query term can be Malay or English and also keywords or querywords according to words in the query. The query can be categorized into two which are keyword and queryword. If the query is keywords, the results retrieved according to word by word results and redundant document names existed if merged. But querywords, retrieved according to the whole words as one at all and only when no redundant or unique document names retrieved rankly. Query translation can replace the origin query in to another language of the query. This translation can be either of both languages.

When the query is conflated with stemming algorithm, this query can be translated in either Malay stemming Malay [2] or English [6] stemming algorithm according to the query keyed-in. Stemming algorithm removes the suffix, infix, and prefix of each word in the query to be root word in order to have more relevant documents in the results. Then, searching process is done according to the type of process of results required. All documents are in flat files and ASCII or EBCDIC text. For searching process, pattern matching is used in the process. Pattern matching refers to the words similarity between query and documents in retrieving process. A large amount of manual effort is needed to represent large document collection in the form of semantic relationship. File structures are stored in memory while collections or databases are usually stored on disk due to their large volumes. The query submitted to the system is also represented by a query signature that is used to search the signature file and unwanted documents are removed. Figure 1 explains the workflow of cross language information retrieval based on query translation in the retrieval process of Malay and English Quranic documents collection. Query is processed by removing the meaningless words or stopwords. Then, query can be translated into another language if needed by the dictionary. Next step is to stem the query by a stemmer if needed according to accurate language.

The query results consist of word by word result or keyword or the whole words or full phrase. This documents retrieval comes from al-Quran documents translation collection which are from Malay and English collection. For this research, when the query is Malay words, the words are translated into English, and vice versa. Then, there are two choices whether stemming words or not. For stemming process, it is going to remove suffix, prefix and infix of the query that contains word by word by using either Porter [6] for English and Fatimah [2] for Malay. Then, the stemming process is necessary for query and document in order to retrieve more relevant documents.

 DOI: 10.1051/matecconf/201713500069

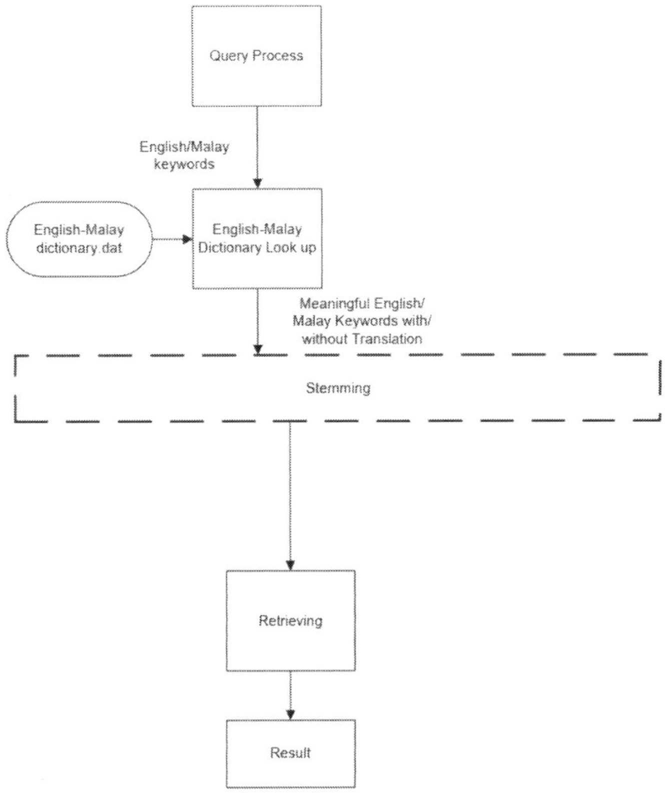

Fig. 1. Workflow of query translation information retrieval with stemmer.

4 Results and Discussion

The evaluation technique is used for precision and recall results [7]. Table 1 shows the formula to calculate the percentage of precision and recall. There are 12 processes of retrieving the results as shown in Figure 2. A few of processes consist of stemming approach and dictionary translation.

Table 1. Recall and precision formula.

$Recall\ (\%) = \frac{Number\ of\ relevant\ documents\ retrieved}{\text{Total number of relevant documents}} \times 100$
$Precision\ (\%) = \frac{Number\ of\ relevant\ documents\ retrieved}{\text{Total number of retrieved documents}} \times 100$

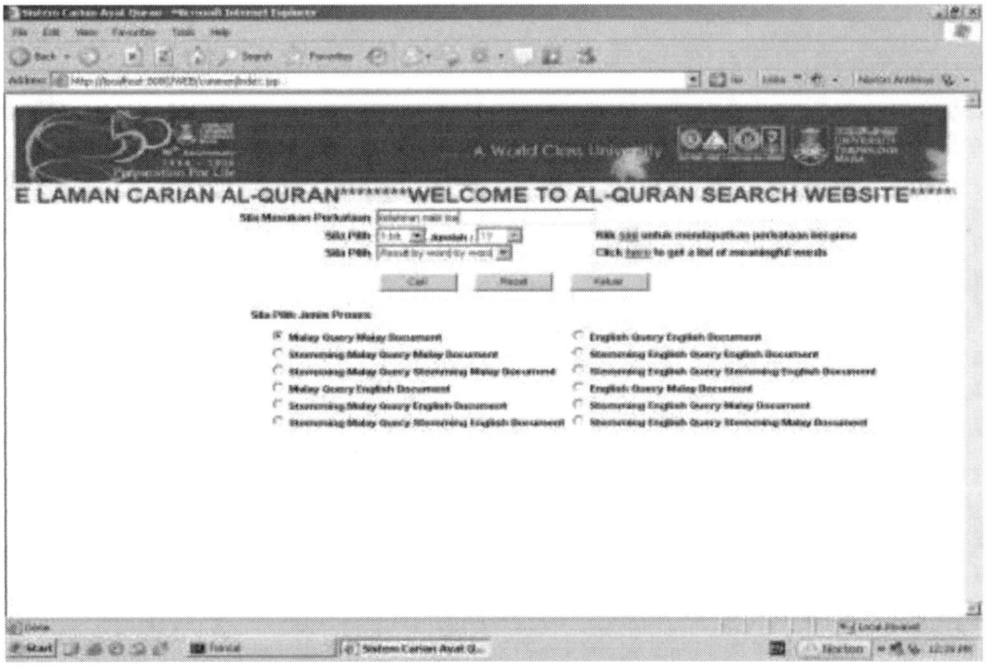

Fig. 2. Main menu of Al-Quran search website.

The results are shown for each of them according to non-stem both, stem query, stem both query and document. Their effectiveness and retrieval and relevant percentage are included as part of evaluation and analysis. Keywords (K) and querywords (Q) are different processes of retrieving results. K usually has redundant information and retrieving results word by word but not for Q. Total retrieve (TRT) is from the result and total relevant (TRE) is used from Fatimah's thesis [2]. Total retrieve and relevant is taken from comparison between TRT and TRE whereby totally matched documents. Table 2 shows that the experiments results in terms of average of recall and precision percentages. Translation of query is not affecting on recall and precision. Experiments on Malay documents show the better average Recall and Precision percentages in SMQSMD and SEQSMD. It means that the Fatimah [2] stemmer has more relevant documents at 44.62 for Queryword (Q). Its precision is 21.47 for and Q. Experiments on English documents show the lower performance than Malay documents because of its stemmer, Porter [6] has lower recall and precision, 16.60 and 14.69. Its performance with the stemmer shows the average recall and precision at 0.84 and 11.38 respectively. These values are lowest in the experiments. The high of recall and precision percentages depend on the efficient stemmer in order to help the queries to retrieve more relevant documents. Thus the queries can retrieve more relevant documents as the efficient stemmer taking part in the process in information retrieval. The queries explain keywords that repeat the same documents and lead to low effectiveness rather than querywords.

DOI: 10.1051/matecconf/201713500069

Table 2. Average percentages of recall and precision of experiments.

No	Experiments	Average of Recall		Average of Precision	
		K	Q	K	Q
1.	NSMQNSMD	34.30	30.12	23.47	23.90
2.	SMQNSMD	33.03	29.62	22.94	22.81
3.	SMQSMD	47.34	44.62	20.56	21.47
4.	NSEQNSMD	34.30	30.12	23.47	23.90
5.	SEQNSMD	33.03	29.62	22.94	22.81
6.	SEQSMD	47.34	44.62	20.56	21.47
7.	NSEQNSED	20.40	16.60	15.50	14.69
8.	SEQNSED	11.28	10.74	12.56	12.42
9.	SEQSED	0.84	0.84	11.38	11.38
10.	NSMQNSED	20.40	16.60	15.50	14.69
11.	SMQNSED	11.28	10.74	12.56	12.42
12.	SMQSED	0.84	0.84	11.38	11.38

5 Conclusions

The empirical result is presented to show the difference performance between monolingual and cross lingual. The results also show the benefits of applying keywords (K) and querywords (Q) in retrieval and relevant retrieval according to monolingual and cross lingua query. The results are concluded according to performance of query with or without translation, the benefits of K and Q, the stemming query itself or both.

The short query is quite suitable with the dictionary and retrieves better results and it helps to search more documents in other languages. These examples also include stemming words after translating the word and matching the words in every document in collection in order to retrieve relevant document required from the query given. It is called as CLIR compared to MLIR, only search specific language if given query with the same language. The relevant documents retrieval depends on combination of the efficient stemmers, dictionaries and stopwords to help the retrieval process.it has more documents retrieval when stemming both translated query and documents. While monolingual takes part of having high percentage of retrieving relevant and related documents. Then, querywords (Q) show unique documents rather instead of keywords (K) which show the redundant documents. Recall and precision percentages however, show better performance in K than Q. The queries need dictionaries to help and translate words in different languages for queries or documents. The difference languages but same meaning in queries can retrieve quite same documents as if it is monolingual search. The future study will be discussed on semantic approach related to performance of query translation. The study will reveal the effective way to use the approach to be applied on information retrieval application according to the empirical experiments.

Acknowledgement

This work was supported in part by a grant from the Ministry of Education under the Research Acculturation Grant Scheme (R045) and in part by a grant from Research Gates IT Solution Sdn. Bhd.

References

1. J. Elly, The Study Of Existing Malay Algorithm Performed On Words Beginning With 'D', B.Sc. Thesis, Universiti Teknologi MARA (2000)
2. A. Fatimah, *A Malay Language Document Retrieval System: An Experiment Approach And Analysis*. Tesis Ijazah Doktor Falsafah Universiti Kebangsaan Malaysia (1995)
3. H. Z. Hamidy, H. S. Fachruddin, Tafsir Quran. Translation. Klang: Klang Book Centre (1987)
4. T. A. Muhammad, M. K. Muhammad, Interpretation of the Meaning of the Noble Quran. Dar-us-Salam Publications. http.//www.amazon.com/Noble-Quran-Interpretation-Meanings-Language/dp/996074079X (1999)
5. M. Popovic, P. Willett, The Effectiveness Of Stemming For Natural-Language Access To Slovene Textual Data. *Journal Of The American Society For Information Science*, **43**(5), 384-390 (1992)
6. M. F. Porter, An Algorithm For Suffix Stripping, *Program,* **14**(3), 130-137 (1980)
7. G. Salton, M. J. Mcgill, Introduction To Modern Information Retrieval. New York: Mcgraw-Hill (1983)
8. G. Salton, Experiments In Automatic Thesaurus Construction For Information Retrieval, *Proceedings Ifip Congress 1971*, Ta-2, 43-49 (1971)
9. J. Y. Nie, Query expansion and query translation as logical inference. Journal of the American Society for Information Science, **54**(4), pp. 335-346 (2003)
10. V. Dang, W. B. Croft, Query Reformulation Using Anchor Text. *WSDM'10*, February 4.6, 2010, New York City, New York, USA (2010)
11. M. Strohmaier, M. Kröll, Christian, Intentional Query Suggestion:Making User Goals More Explicit During Search. *WSCD'09*, Feb 9, 2009 Barcelona, Spain (2009)
12. M. A. Yunus, R. Zainuddin, N. Abdullah, Semantic Method for Query Translation. The International Arab Journal of Information Technology (IAJIT), vol. **10,** no.3, pp. 253-259 (2013)
13. M. A. Yunus, R. Zainuddin, N. Abdullah, Semantic Query for Quran Documents Results. Proceeding of IEEE Conference On Open Systems (ICOS), pp. 1-5, IEEE (2010)
14. M. A. Yunus, R. Zainuddin, N. Abdullah, Semantic Speech Query via Stemmer for Quran Documents Results. Proceeding of International Conference On Electronic Devices, Systems and Applications (ICEDSA), pp. 17-21, IEEE (2011)
15. M. A. Yunus, R. Zainuddin, N. Abdullah, Visualizing Quran Documents Results by Stemming Semantic Speech Query. Proceeding of International Conference On User Science and Engineering (i-USEr), pp. 209-213, IEEE (2010)
16. M. A. Yunus, R. Zainuddin, N. Abdullah, Semantic Query with Stemmer for Quran Documents Results. Proceeding of IEEE Conference On Open Systems (ICOS), pp. 40-44, IEEE (2010)

Appendix: List of Abbreviations

NSMQNSMD	Non-Stem Malay Query Non-Stem Malay Document
SMQNSMD	Stem Malay Query Non-Stem Malay Document
SMQSMD	Stem Malay Query Stem Malay Document
NSEQNSMD	Non-Stem English Query Non-Stem Malay Document
SEQNSMD	Stem English Query Non-Stem Malay Document
SEQSMD	Stem English Query Stem Malay Document
NSMQNSED	Non-Stem Malay Query Non-Stem English Document
SMQNSED	Stem Malay Query Non-Stem English Document
SMQSED	Stem Malay Query Stem English Document
NSEQNSED	Non-Stem English Query Non-Stem English Document
SEQNSED	Stem English Query Non-Stem English Document
SEQSED	Stem English Query Stem English Document

An Ontological Approach towards Dialogue-based Information Visualization System: Quran Corpus for Juz' Amma

Mohd Amin Mohd Yunus[1,*] *Aida* Mustapha[1], *Rizwan* Iqbal[2] and *Noor Azah* Samsudin[1]

[1]Faculty of Computer Science and Information Technology, Universiti Tun Hussein Onn Malaysia, 86400 Parit Raja, Batu Pahat, Johor, Malaysia
[2]Faculty of Computer Science and Information Technology, Universiti Putra Malaysia, 43400 UPM Serdang, Selangor, Malaysia

Abstract. This paper presents a corpus that offers rich and authentic knowledge for Juz' Amma using the ontological engineering approach. The reason of the study conducted is because of lack of understanding of Juz' Amma in the form of listing result. At the same time, the listing result does not provide pictorial tidings for clustering the listing result. Therefore, the corpus covers knowledge from various authentic sources, including translation, annotation of keywords, tafsir, and related hadiths for every verse in English and Malay. The ontology is engineered and designed to effectively utilize the authentic knowledge extracted from the corpus intended to serve a dialogue-based information visualization system for Quranic text. A standard ontology engineering methodology was applied for developing the ontology. After creation, the ontology was evaluated and was found to fulfill the goals for which it was developed. The corpus and ontology are hoped to aid various semantic applications related to Quranic text in the future.

1 Introduction

With regards to studies in the Quranic text, most religious scholars agree that contextual information is considered mandatory for correct interpretation of the Quranic verses. In the contextual aspect, Quranic exegeses (tafsir) greatly and facilitate understanding by providing correct interpretation and meaning of the verses. Nonetheless, the tafsir is not verse-based of word-to-word translation. It is a lengthy and detailed explanation of the verses that may span from a few sentences to a full paragraph. In the same way, hadith may also become a potential source for providing keen understanding to the true meaning of some verses. Hadiths are the sayings or oral traditions related to the words and acts by Prophet Muhammad (pbuh).

[1] Corresponding author: aminy@uthm.edu.my

AQILAH approach is an ambitious effort to qualitatively enhance human experience in navigating the Quranic text using human-machine dialogues. The use of natural language dialogues via conversational agents offers higher understanding of the presented information [1]. The dialogue-based architecture of AQILAH facilitates human-computer interaction by using the ontology to rearrange the Quranic content based on need-to-know basis, as opposed to sequential order as in recitals. Hence, an ontological engineering approach to its core knowledge base is essential so the knowledge for Juz' Amma is accessible in a non-sequential form. Therefore, the objective of this study is to design and develop a Quran corpus and ontology that present rich, authentic knowledge for Juz' Amma of the Quran that encapsulates information revolving around the translations, tafsir, and hadiths. The ontology will be derived from the corpus using a standard ontology engineering methodology by Gruninger and Fox [4].

2 Related Work

Ontologies have gained considerable focus in the last two decades. At present day, they are extensively used in different areas like knowledge engineering, artificial intelligence, natural language processing, e-commerce, intelligent information integration, information retrieval, database design and integration, bioinformatics and expert systems. An ontology engineering methodology caters the methodological aspect of ontology development. It gives a set of guidelines and activities to develop a particular ontology. Among the major ontology engineering methodologies proposed to date include the methodology by Uschold and King [10] and Gruninger and Fox [4].

In the Quranic domain, researchers have worked on encapsulating the knowledge of Quran using ontologies [8,9, 11], which focus on ontology creation for Islamic concept such as prayers sourced from the Quranic text. Ontology developed by Dukes (2012) covers about 300 linked concepts with 350 relations, which primarily acts as a means of mapping different concepts to related verses in the entire Quran. However, all the ontologies discussed do not provide any contextual information related to the verses and do not report on the opted ontology engineering methodology.

3 Experimental Approach

This section discusses the details of the Quran ontology developed based on the corpus of Juz' Amma. The ontology design chose the methodology proposed by Gruninger and Fox (1995) because the motivation scenarios presented in the initial phase of this methodology facilitates the development team to identify intuitively possible solutions to the scenario problems [3,4].

3.1 Ontology design

The Gruninger and Fox's methodology is inspired by the development of knowledge-based systems using first-order logic [4]. The methodology proposes to develop the main scenarios, which describe the set of ontology's requirements that the ontology should satisfy after implementation. These scenarios provide a starting point to visualize the intended semantics of the objects and relations that will later become part of the

 DOI: 10.1051/matecconf/201713500070

computable ontology.

Subsequently, natural language questions or also known as the informal competency questions are derived from the motivation scenarios to help determine the scope of ontology. After formally implementing the ontology, the competency questions will be transformed into an equivalent formal form for evaluation purposes. This is to ensure that the ontology developed fulfills the requirements for which it is made. Figure 1 shows the processes involved in the methodology by Gruninger and Fox [4].

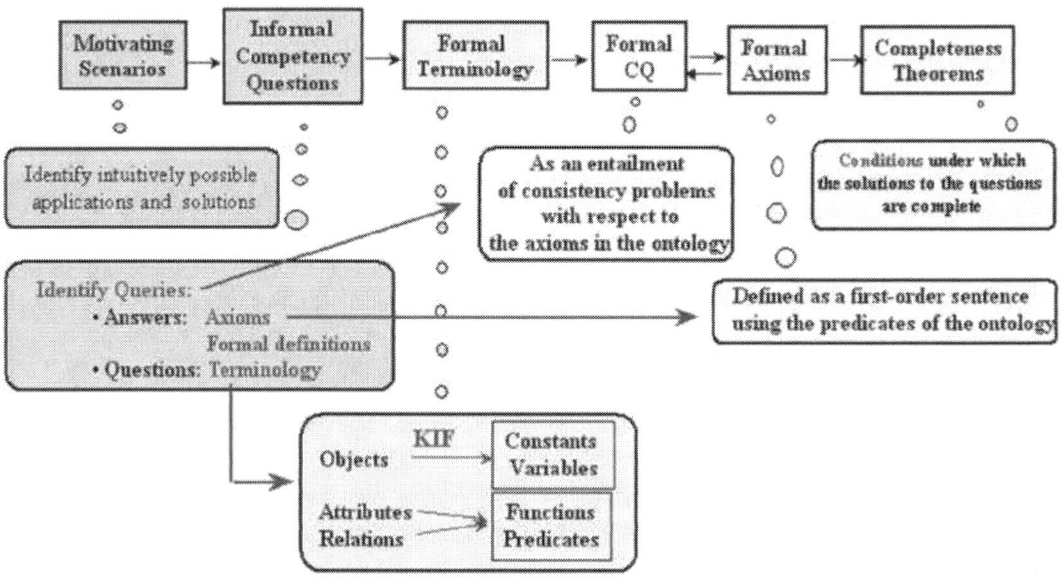

Fig. 1. Processes in the methodology (Source: Gruninger and Fox 1995 [4])

Based on Figure 1, the process begins by identifying the motivation scenarios. Such scenarios are related to the applications that will utilize the ontology. The scenarios define the set of requirements which the designed ontology should satisfy. They intuitively provide possible solutions to the scenario problems. The next process is defining the informal competency questions. They are derived from the motivation scenarios, specifying questions in natural language which the ontology should answer. The third process starts when the ontology engineer extracts the terminologies such as concepts, attributes, and relations in first order logic from the competency questions. Answers to these competency questions will guide the ontology engineer to extract the knowledge to be included in the formal definitions of the concepts, relations and formal axioms used.

Next, the competency questions are translated into their formal form followed by definitions of the ontology terms and constraints in their interpretation using axioms. The final process is to specify the completeness theorems, meaning defining the conditions under which the solution to the questions is complete. In this paper, evaluation of the Quran ontology will be carried out by running Description Logic (DL) queries and getting correct answers for it, which serve as the criteria for satisfying the completeness theorems.

3.2 Development of ontology

The Quran ontology for Juz' Amma has the following features which was distinguished from the existing Quran ontologies already mentioned in Section 2. The ontology is developed by extracting authentic knowledge from the Quran corpus. Gruninger and Fox methodology [4] is opted for developing the ontology, which is well-founded for building and evaluating ontologies [3]. The ontology covers the knowledge of Quran in the form of relationships between Juz', Surah and verse (Ayat). In order to provide contextual information, the verses of Quran have been linked with Quranic exegeses (tafsir), hadith and dual translation in English and Malay for correct interpretation of the verses. The ontology is engineered in a manner that it can be used as core knowledge base for a dialogue-based information visualization system and other related applications in the future. The design of

Quran ontology is flexible; it can be extended and modified as per need. The ontology development began by identifying the motivation scenarios. Motivation scenarios were created to discover the set of requirements for the Quran ontology that has to be formally implemented. It is often useful to use a template as a representational artefact to share the motivation scenarios. Use case templates in object-oriented methodology are useful for this purpose since it shows meaningful information in the most concise way [2]. Table 1 shows a sample of use case representing the motivation scenario developed for this ontology and it describes a user can input formal query in the query tab, provided in the Protege IDE.

Table 1. Use case for the Quran ontology.

Brief Description	**The use case describes that how a user can input formal query in the query tab, provided in the Protege IDE.**
Actors	User & Protege IDE
Preconditions	The Protege IDE is properly installed. The Qur'an ontology is loaded in the environment. The user is familiar with the formal query language, like SPARQL
Basic flow of events	The user initializes the Protege IDE. The user inputs a query in formal language (Like SPARQL) in the query tab. After inputting the query user presses the execute button. The system searches the Qur'an ontology for the appropriate answer, based on a triple (RDF) searching format. The answer/answers are populated on the screen If the user wants to run another query; he erases the previously written query and repeats the above procedure from step 2.
Alternative flow	Incorrect query syntax: If the user inputs the formal query in an incorrect syntax in step 2, then no results will be displayed. It is out of scope query, if the user inputs the query in correct syntax in step 2 but it is out of scope of the Qur'an ontology, then no results will be displayed.
Key scenarios	The system does not respond to the query and no results are displayed.
Post conditions	Successful query execution: The user successfully receives the correct answer for his query from the system. For failure condition, the user do not receive any response/answer for his query from the system.
Special Requirements	The user should be familiar with the formal language (SPARQL) and Protege IDE

After identifying the motivation scenarios, a set of informal competency questions are defined based on the motivation scenarios. These questions are prepared in natural language to identify the scope of ontology under development. The set of informal competency questions help the ontologist to extract content for the ontology, which will later become part of the computable ontology. Some informal competency questions in natural language defined for the Quran ontology may include which Surahs belong to Juz' 30, which verses (Ayat) belong to Surah 113, which Surahs were revealed in Makkah and have verses (Ayat) greater than 15, which Surahs belong to Juz' 30 and were revealed in Medina, what is the English Translation for verse 4, Surah 111, what is the tafsir (Quranic exegeses) of verse 2, Surah 114 and what is the English name for Surah 114.

Informal competency questions ease the work of the ontologist to identify concepts, attributes and relations to be included in the ontology. Furthermore, answers in natural language to the competency questions help the ontologist to extract knowledge to be included in the formal definitions for the ontology content [3]. Table 2 presents some relations and inverse relations for the Quran ontology. Some concepts, instances and attributes identified for the Quran ontology are presented in Table 3.

Table 2. Relations and inverse relations.

Relation name	Source concept	Source cardinality	Target concept	Inverse relatio
Belong to Juz	Surah	(1...*,1...*)	Juz	Has Surah
Belong to Surah	Verse	(1,1)	Surah	Has verse
Has Surah	Juz	(1...*,1...*)	Surah	Belong to Juz
Has verse	Verse	(1...*, 1)	Surah	Belong to Sural

Table 3. Concepts, instances and their attributes.

Concept name	Instances	Instance Attributes
Juz	78-Juz 99-Juz 80-Juz	Name of Juz' in Arabic
Surah	An-Naba At-Takwir	Surah name in Arabic Surah name in English No of verses
Verses	114-1 114-2 114-3	Belong to Surah Has translation Has Tafsir
Revelation Places	Makkah Medina	Not available
Translator	Hilali-Khan Abdel Haleem	Not available

Once all the ontological content is finalized, the ontology is transferred into a computable form using an ontology editor, which is the Protege IDE [7]. Protege is an open-source ontology editor that offers a wide range of useful plugins. Figure 2 shows a snap shot of the computable Quran ontology implemented in Protege. The wide range of plugins supported in Protege allows the ontologist and the entire development team to develop a clear understanding of the implemented ontology.

 DOI: 10.1051/matecconf/201713500070

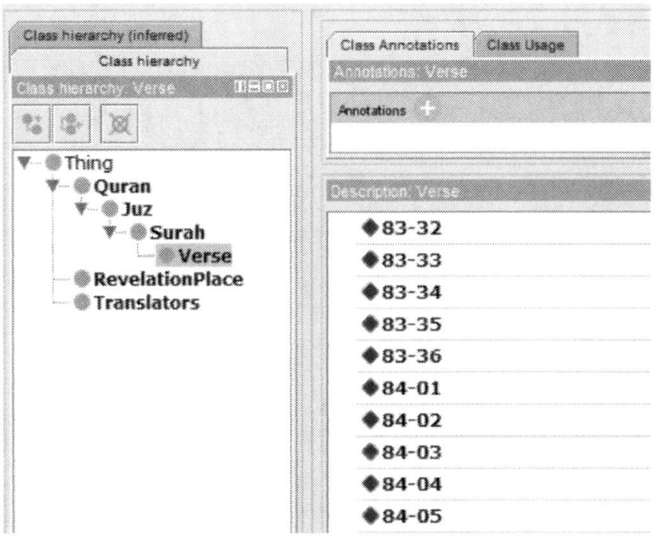

Fig. 2. View of class hierarchy of Quran ontology in Protege.

Besides many plugins available in Protege, OntoGraf was found to be very [6] effective in interactively navigating the Quran ontology. OntoGraf provides support for interactively navigating the relationships of the ontology using multiple layouts to automatically organize the structure of the ontology. It allows the user to create the desired view by filtering relationships [6]. Figure 3 illustrates a view of OntoGraf showing the relationships between classes and instances in the Quran ontology.

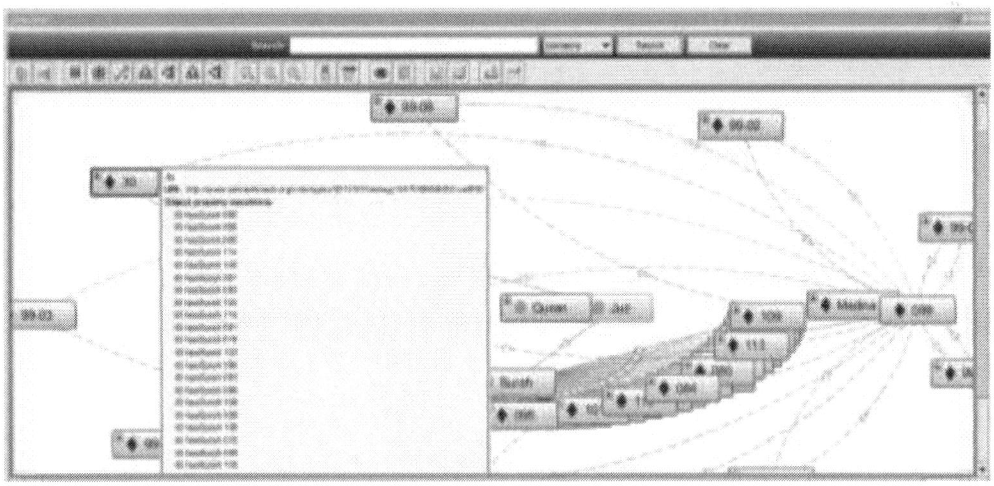

Fig. 3. OntoGraf showing relationships between classes and instances in the Quran ontology.

4 Result and Discussion

Table 1 shows the formula to calculate the percentage of precision and recall. After the ontology is implemented, it is important to ensure that the computable ontology fulfils the requirements and scope for which it is made. In order to perform evaluation, natural language competency questions are transformed into their machine level equivalent formal query language. For this research, competency questions for Quran ontology were converted into their equivalent queries in the form of Description Logic.

Description Logic (DL) is a family of formalisms that belongs to logic-based knowledge representation. DL empowers OWL ontology to provide precise meaning to the description and also helps to develop reasoning algorithms, which help to provide answers to queries about the domain [5]. Protege IDE provides a DL-Query Tab to run DL queries to retrieve knowledge from the loaded ontology. Table 4 represents some informal competency questions trans-formed in to their DL query equivalent, for evaluating the Quran ontology.

Table 4. Informal competency questions and their equivalent DL queries.

Informal competency questions	**Equivalent DL query**
Which Surahs were revealed in Makkah and have verses (Ayat) greater than 15?	Revealed in value Makkah and no of Verses some int [> 15]
Which Surahs belong to Juz 30?	Belong to Juz value 30
Which Surahs belong to Juz 30 and were revealed in Medina?	Belong to Juz value 30 and revealed in value Medina
Surahs which have verses greater than20 but less than 35 belong to Juz 30 and revealed in Makkah?	noOfVerses some int [> 15, <35] and belong to Juz value 30 and Revealed in value Makkah

Next, the Quran ontology was evaluated by running DL queries, using the DL-Query tab in Protege. It was found that the computable Quran ontology returned correct answers for the DL queries submitted, using the DL-Query tab. This manifests that the Quran ontology successfully fulfilled the requirements for which it was made and it can be effectively used for extracting the desired Quran related knowledge. Figure 4 shows a snapshot of a DL query running in Protege.

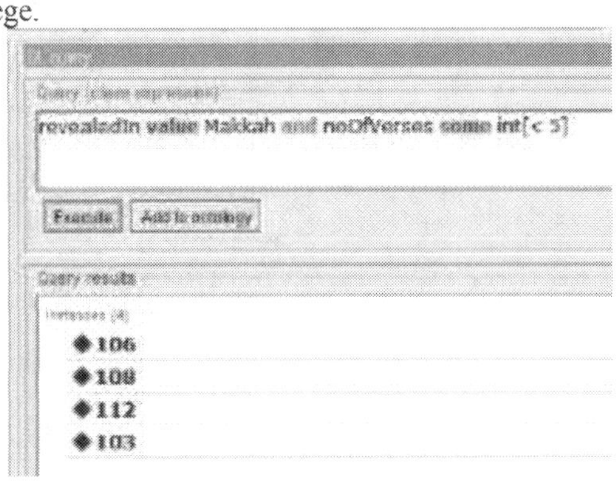

Fig. 4. A snapshot of Protege running a DL query on Quran ontology.

5 Conclusions

The corpus developed for Juz' Amma is based on individual verse or Ayat. Each verse in this corpus is annotated with literal translation, tafsir, and supporting hadith. All the knowledge in this corpus is based on authentic knowledge extracted from various authentic sources. The Quran ontology is developed by extracting knowledge from this authentic Quran corpus. The developed ontology encapsulates authentic knowledge of Quran in a machine understandable format, which makes it suitable to be used as the core knowledge base for a dialogue-based information visualization system such as AQILAH. The Quran ontology is developed using a standard ontology engineering methodology, which makes it authentic, robust and reliable. The ontology has been evaluated to ensure that it fulfils the requirements for which it was developed. Furthermore, the design of this ontology is flexible so it can be easily extended and modified in the future. It is hoped that this ontology will be able to serve AQILAH in aiding mental visualization when studying the Quran by enabling the users to communicate the content of Juz' Amma with clarity, precision, and efficiency. Both the corpus and ontology carry considerable potential to benefit upcoming Quran related semantic applications in the future.

Acknowledgement

This work was supported in part by a grant from the Ministry of Education under the Research Acculturation Grant Scheme (R045) and in part by a grant from Research Gates IT Solution Sdn. Bhd.

References

1. R. J. Beun, E. Vos, C. Witteman, Embodied conversational agents: Effects on memory performance and anthropomorphisation. *Intelligent Virtual Agents,* pp. 315-319 (2013)
2. G. Brusa, L. Caliusco, O. Chiotti, Towards ontological engineering: A process for building a domain ontology from scratch in public administration, *Expert Systems* **25**(5), 484-503 DOI 10.1111/j.1468-0394.2008.00471.x (2008)
3. A. Gomez-Perez, M. Fernandez-Lopez, O. Corcho, Ontological engineering: with examples from the areas of knowledge management, e-Commerce and the Semantic Web, Machine Learning and Data Mining for Computer Security, *Methods and Applications,* Springer-Verlag. http://fs1.bib.tiera.ru/content/dvd44/Maloof M.A (2005)
4. M. Gruninger, M. S. Fox, Methodology for the design and evaluation of ontologies, *Procs. Workshop on Basic Ontological Issues in Knowledge Sharing*, IJCAI, **95,** pp. 26-37 (1995)
5. I. Horrocks, Semantic web: the story so far, *Procs. 2007 Int. Cross-Disciplinary Conf. Web Accessibility* (W4A), ACM. pp. 120-125 (2007)
6. OntoGraf Plugin (2013). http://protegewiki.stanford.edu/wiki/OntoGraf (2016)
7. Protege IDE (2013). http://protege.stanford.edu/ (2016)
8. S. Saad, N. Salim, H. Zainal, Islamic knowledge ontology creation, *Procs. Internet Technology and Secured Transactions (*ICITST), pp. 1-6 (2009)
9. S. Saad, N. Salim, H. Zainal, S. A. M. Noah, A framework for Islamic knowledge *via* ontology representation, *Procs. Int. Conf. Information Retrieval and Knowledge*

Management (CAMP), IEEE. pp. 310-314 (2010)
10. M. Uschold, M. King, Towards a methodology for building ontologies, *Procs. Workshop on Basic Ontological Issues in Knowledge Sharing,* Montreal, Canada, **74,** pp. 23-42 (1995)
11. F. E. Zahari, Ontology application for the Quran, Master's thesis, Universiti Utara Malaysia (2009)

MATEC Web of Conferences **135**, 00071 (2017) DOI: 10.1051/matecconf/201713500071
ICME'17

A Comparative Study of Open-Domain and Specific-Domain Word Sense Disambiguation Based on Quranic Information Retrieval

Rehab Hasan Abood [1] and *Sabrina* Tiun[1,*]

[1]Faculty of Information Science and Technology, UKM, 43000 Bangi, Malaysia

Abstract. Information retrieval is the process of analysing typed query as well as to retrieve relevant document according to the user query. Several issues can significantly affect the effectiveness of information retrieval. One of the common issue is the ambiguity lies on the words where a single word could yield several meanings. The process of identifying the exact sense of word is called Word Sense Disambiguation (WSD). Quran is the holly book for nearly 1.5 billion Muslims around the world. In particularly, Quran contains numerous words that can undergone multiple meanings. Therefore, there is a vital demand to apply WSD approach on Quran, in order, to improve the information retrieval. Several WSD approaches have been proposed for Quranic retrieval. However, these approaches are divided into two main categories; open-domain WSD approach and specific-domain WSD approach. Open-domain WSD is an approach that utilizes an open-domain dictionary such as WordNet, that is exploited to provide the exact sense. Whereas, domain-specific WSD approach aims to utilize a restricted training data that contain specific senses related to the domain of Quran. Hence, this study aims to establish a comparative study to investigate the two WSD categories including domain-specific and open-domain. For the domain-specific approach, a predefined example data has been collected to train Yarwosky algorithm which is a semi-supervised machine learning technique. Then, based on the training, such algorithm can classify the exact sense for the words. In contrast, WordNet which is an open-domain dictionary has been used in this study with semantic distances, in order, to identify the similarity between the query word and the results of WordNet's concepts. That dataset that has been used in this study is a Quranic translation. The experimental results have shown the mixed superiority of Yarwosky algorithm and WordNet WSD approach.

1 Introduction

Word sense disambiguation (WSD) is the task of identifying the correct meaning of a word in context. In any language, there are many words that yield multiple meaning for example, the word 'address' has two meanings; first it refers to a geographic location, and second it

* Corresponding author: sabrinatiun@ukm.edu.my

refers to examining something. Hence, WSD has been introduced to resolve such problem by identifying the exact sense of word [1].

As a basic semantic understanding task at the lexical level, WSD is a fundamental problem in natural language processing. It can be potentially used as a component in many applications, such as machine translation (MT) and information retrieval (IR). Recently, the application of WSD to MT, research has shown that integrating WSD in appropriate ways significantly improves the performance of MT systems [2]. In the application to IR, WSD can bring two kinds of benefits. First, queries may contain ambiguous words (terms), which have multiple meanings [3]. The ambiguities of these query words can hurt retrieval precision. Identifying the correct meaning of the ambiguous words in both queries and documents can help in improving the retrieval precision. Second, query words may have tightly related meanings with other words not in the query [4]. Making use of these relations between words can improve retrieval recall. Overall, IR systems can potentially benefit from the correct meanings of words provided by WSD systems.

However, in the literature on the usage of WSD in IR, different researchers arrived at conflicting observations and conclusions. Some of the early research showed a dropped in retrieval performance by using word senses [5]. Some other experiments observed improvements by integrating word senses in IR systems [6]. This can be performed by identifying the exact sense for each word in the query which leads to increase the recall of retrieval [7].

There are less studies have been done on applying WSD on Quranic IR. This is probably due to the less of resources or the known mixed of results if WSD applied on IR. The recent studies of WSD on Quranic IR can be seen in Tiun et al [12] and Mohammad and Tiun [8]. An issue risen by Tiun et al [12], is that they concluded that the failed WSD on domain-specific is due to the use of external knowledge like WordNet or LDOC and supervised learning technique is used. Thus, this encouraged them to use minimal supervised WSD in applying WSD on Quranic IR [12][8], in which, the WSD is localised into the content of the information that need to be retrieved.

Nevertheless, it come to our conclusion that by applying WSD to IR, the IR accuracy should have some improvement since some believe that deep semantic analysis on the query process will give good impact on the IR performance, theoretically, and that could be the reason why WSD has been continuously studied. However, it is also wise to compare how much is the impact and how the results will behave when open-domain and closed domain of WSD studied on Quranic IR. Hence, this study will address this issue.

2 Related Works

Recently, researchers have focused on incorporating WSD with IR using various techniques. For example, Zhong and Ng [13] have proposed a novel language modeling approach to enhance information retrieval. Basically, the authors have used a unigram approach with negative Kullback-Leibler (KL) divergence as similarity distance function in order to link each document with its similar query. Then a supervised word sense disambiguation method has been used in order to expand the query words with their corresponding senses. In fact, the authors have used an open-domain source knowledge to train the WSD method on it using Chinese Treebank. The experiments have been performed on TREC dataset collection. As a result, the authors have reported a significant improvement when using the WSD with IR where the confidence level of retrieval was 95%.

Stevenson et al. [15] have proposed a domain-specific WSD approach for retrieving biomedical documents using the National Library of Medicine WSD dataset. Basically, the authors have used specific information related to the topic of each document where the

DOI: 10.1051/matecconf/201713500071

ambiguous words are frequently occurred. This has been performed by implementing multiple approaches to extract lists of key terms that related to the medical subject heading terms. Such key terms have been used to represent the document topic in a knowledge-based WSD system. The authors have reported a significant enhancement when using such WSD approach by achieving a 70% of f-measure.

Zouaghi et al. [7] have proposed an open-domain WSD approach where WordNet lexicon has been combined with Lesk algorithm for Arabic information retrieval. In fact, the authors have manually collected Arabic documents. In addition, they have used Part-Of-Speech Tagging in order to tag the targeted words. Then the nouns have been migrated to the WordNet lexicon and then the Lesk algorithm has been used to identify the exact sense of words. Similarly, the authors have reported an enhancement when using WSD for Arabic IR by achieving 73% of f-measure.

Finally, Mohamed and Tiun [8] have proposed a domain-specific WSD approach for Quran information retrieval using a semi-supervised Yarwosky algorithm as classification method. In fact, a domain-specific data has been collected for the purpose of training. This data consists of two groups of Quranic verses; first verses with sense A, and second group contains verses with sense B. Hence, Yarwosky algorithm has been trained on such data in order to classify a testing data. After that, an information retrieval system has been built based on Yarwosky algorithm to retrieved relevant verses. The proposed method has improved the information retrieval by obtaining 77% of F-measure.

3 Research Method

This phase aims to accomplish the research objectives by applying two WSD approaches including Yarowsky algorithm (i.e. specific domain) and Lesk algorithm with WordNet (i.e. open-domain) and then establish a comparative analysis. In order to achieve such objectives, the research design will be illustrated as shown in Figure 1. In fact, the research design consists of five main phases: The first phase is associated with the dataset that has been used in this study by identifying its source, details and containing instances. The second phase is associated with the normalization tasks which aims to turning the data into a suitable format by removing unwanted data such as stop-words. The third phase is associated with the query modification. Such process aims to expand the query with corresponding synonyms or matching in order to identify the word's sense. This process has been applied twice; first using an open-domain dictionary (i.e. WordNet), and second using Yarowsky algorithm which will be trained on a specific-domain example set. The fourth phase aims to apply information retrieval based on the resulted modified query from both of WordNet and Yarowsky algorithm. Finally, the fifth phase aims to evaluate the two approaches by establishing a comparison between the two approaches WordNet and Yarowsky algorithm.

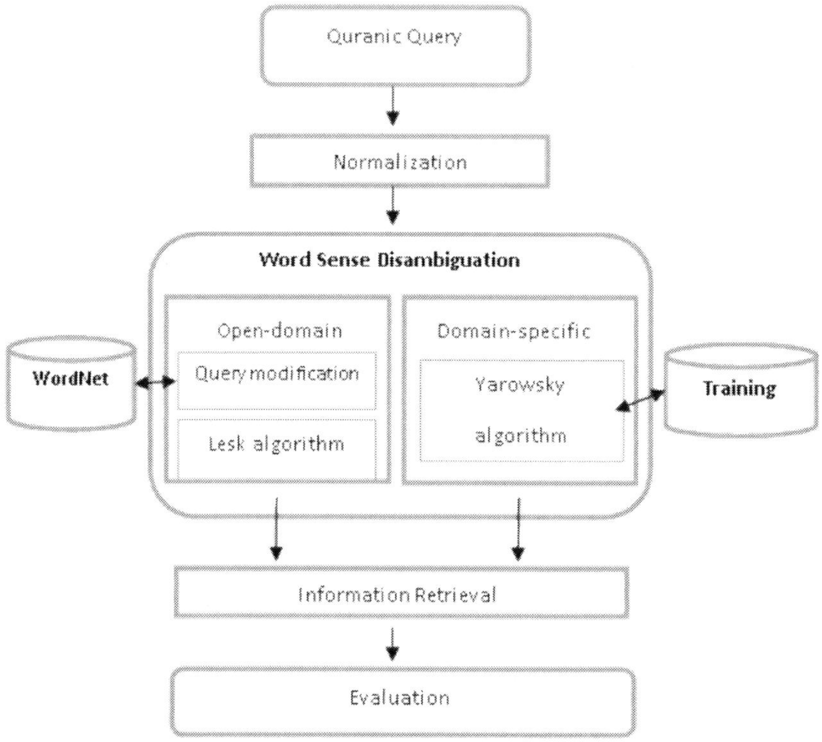

Fig. 1. Research Design Diagram

3.1 A Domain- Specific WSD (Yarowsky Algorithm)

Yarowsky algorithm is one of the machine learning technique that aims to classify the words' senses based on a relatively small set of training dataset [7]. Such small set makes Yarowsky algorithm to be categorized as a semi-supervised machine learning technique. In fact, this algorithm is mainly depended on created predefined examples, where the verses will be divided into two groups of senses. First group contains the verses that are related to sense 1, whereas second group consists of the verses that are related to sense 2. More explanation on how the Yarowsky algorithm is carried out in the phase can be obtained from Mohamed and Tiun [8].

3.2 Query Modification

Query modification or so-called query expansion is the process of identifying corresponding synonyms, hyponymy or any semantic relatedness matching instances of the query's words in order to solve the ambiguity of the query [9]. In fact, this phase is crucial because the process of word sense disambiguation is carrying out on it. For this purpose, WordNet dictionary has been used. Following sub-sections will illustrate the mechanism of modifying the query.

3.2.1 A WordNet Dictionary

This phase aims to identify the corresponding synonyms of the query/word. Such tasks would facilitate the process of identifying the exact sense for the query/word. For this manner, an open-domain dictionary which is WordNet has been used to provide the semantic correspondences. Table 1 shows the sample results of using WordNet.

Table 1. Synonym matching using WordNet.

Word	WordNet results
day	S: (n) day, twenty-four hours, twenty-four-hour period, 24-hour interval, solar day, mean solar day (time for Earth to make a complete rotation on its axis)
charity	S: (n) charity, brotherly love (a kindly and lenient attitude toward people)
pray	S: (v) pray (address a deity, a prophet, a saint or an object of worship; say a prayer) "pray to the Lord"

As shown in Table 1, the results of WordNet provide a definition for each concept in the query. Usually, WordNet bring various definitions that are related to multiple domains. For example, retrieving the definition of the word 'mouse' will be as 'quite person', 'animal' and 'electronic device'. In order to identify the exact definition, Lesk algorithm will be applied. Such algorithm can be described in the following section:

3.2.2 Lesk Algorithm

This algorithm is used to provide assumption that words in a given "neighborhood" tend to share a common topic [11]. Lesk algorithm is usually used for identifying the most appropriate sense of definitions that brought from dictionary [12]. For example, the word 'pray' has two senses located in the WordNet that can be shown in Table 2.

Table 2. Senses of the word 'pray'.

Word	WordNet senses
pray	Pray (address a deity, a prophet, a saint or an object of worship; say a prayer) "pray to the Lord".
pray	Beg, implore, pray (call upon in supplication; entreat) "I beg you to stop!".

Whereas, the word 'worship' has three senses that can be seen in Table 3.

Table 3. Senses of the word 'worship'

Word	WordNet senses
worship	Worship (show devotion to (a deity)) "Many Hindus worship Shiva"
worship	Worship, adoration (a feeling of profound love and admiration)
worship	Idolize, idolise, worship, hero-worship, revere (love unquestioningly and uncritically or to excess; venerate as an idol) "Many teenagers idolized the Beatles"

Each of the two senses of the word 'pray' is compared with each of the three senses of the word 'worship' and, it is found that the words 'deity' and 'worship' occurs in one sense each of the two words. These two senses are illustrated as follows:

(i) *Pray (address a deity, a prophet, a saint or an object of worship; say a prayer) "pray to the Lord".*

(ii) *Worship (show devotion to (a deity)) "Many Hindus worship Shiva".*

These senses will be declared to be the most appropriate senses when the words 'pray' and 'worship' are used together.

4 Result and Evaluation

Since the evaluation will be based on retrieval effectiveness thus, the metrics that have been used is precision, recall and f-measure. Precision refers to the correctness of retrieval answers, while recall refers to the correctness of not retrieved answers. Such metrics can be calculated as the following Equations (1 and 2):

$$Precision\ P = \frac{TP}{TP+FP} \quad (1)$$

$$Recall\ R = \frac{TP}{TP+FN} \quad (2)$$

where TP is the number of correct retrieved answers, FP is the number of incorrect retrieved answers, and FN is the number of correct answers but not retrieved. Hence, it is possible to calculate the overall accuracy which called f-measure based on the following Equation (3):

$$f - measure = 2 \times \frac{P*R}{P+R} \quad (3)$$

DOI: 10.1051/matecconf/201713500071

4.1 Results

4.1.1 Evaluation of the 'day' verses

This section will address the results of the two WSD approaches when using the 'day' verses. The evaluation has been performed based on the effectiveness of retrieval. Table 4 shows such results.

Table 4. Results of the two WSD approaches using the word 'day'.

Senses	WSD approach	TP	FP	TN	FN	F-measure
Sense A	Yarowsky	16	2	2	20	0.89
	WordNet	17	2	1	20	0.92
Sense B	Yarowsky	5	9	6	20	0.41
	WordNet	5	10	5	20	0.39
Average	Yarowsky	21	11	8	40	0.65
	WordNet	22	12	6	40	0.65

4.1.2 Evaluation of the 'charity' verses

This section will address the results of the two WSD approaches when using the 'charity' verses. The evaluation has been performed based on the effectiveness of retrieval. Table 5 shows such results.

Table 5. Results of the two WSD approaches using the word 'charity'.

Senses	WSD approach	TP	FP	TN	FN	F-measure
Sense A	Yarowsky	14	4	2	20	0.81
	WordNet	11	7	2	20	0.70
Sense B	Yarwosky	11	8	1	20	0.73
	WordNet	12	8	0	20	0.75
Average	Yarwosky	25	12	3	40	0.77
	WordNet	23	15	2	40	0.72

4.1.3 Evaluation of the "charity' verses

This section will address the results of the two WSD approaches when using the 'pray' verses. The evaluation has been performed based on the effectiveness of retrieval. Table 6 shows such results.

Table 6. Results of the two WSD approaches using the word 'pray'.

Senses	WSD approach	TP	FP	TN	FN	F-measure
Sense A	Yarowsky	5	10	5	20	0.40
	WordNet	3	6	11	20	0.29
Sense B	Yarwosky	4	11	5	20	0.38
	WordNet	2	13	5	20	0.18
Average	Yarwosky	9	21	10	40	0.39
	WordNet	5	19	16	40	0.23

4.1.4 Evaluation of the 'light' verses

This section will address the results of the two WSD approaches when using the 'light' verses. The evaluation has been performed based on the effectiveness of retrieval. Table 7 shows such results.

Table 7. Results of the two WSD approaches using the word 'light'.

Senses	WSD approach	TP	FP	TN	FN	F-measure
Sense A	Yarowsky	3	12	5	20	0.21
	WordNet	5	10	5	20	0.42
Sense B	Yarowsky	13	2	5	20	0.79
	WordNet	16	0	4	20	0.87
Average	Yarowsky	16	14	10	40	0.50
	WordNet	21	10	9	40	0.64

4.1.5 Comparison between the Four Words

In this subsection, a comparison has been taking a place in order to compare each mentioned word using the two WSD approaches Yarowsky and WordNet. The evaluation has been performed based on F-measure and the average for the two senses. Table 8 shows the results for each word.

Table 8. Comparison between all the words

Words	Yarowsky	WordNet
Day	0.65	0.65
Charity	0.77	0.72
Pray	0.39	0.23
Light	*0.50*	*0.64*

DOI: 10.1051/matecconf/201713500071

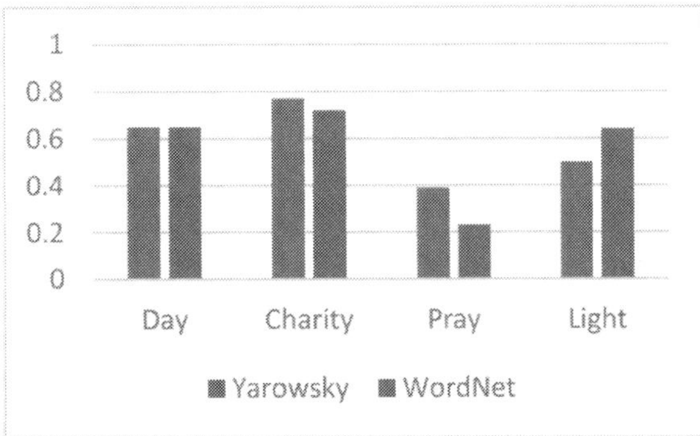

Fig. 2. Results of the comparison between all the words.

As shown in Figure 2, Yarowsky WSD approach has outperformed the WordNet approach for two words including 'charity' and 'pray' by obtaining 77% and 39% of F-measure respectively. This is because these words contain specific senses that are related to the domain of Quran. Since the Yarowsky has been trained on a domain-specific data thus, the superior of Yarowsky results is expected. In contrast, WordNet WSD approach has outperformed Yarowsky for the word 'light' by achieving 64% of F-measure. This is because such word contains senses that tend to be more domain-less. Finally, the classification of the sense word 'day' has shown a similar performance for both Yarowky and Lesk, with WordNet by achieving a 65% of F-measure. This is because such word has two senses; the first is domain-specific which is 'judgment day' (i.e. related to Quran), and the second is domain-less which is 'day and night'.

5 Discussion and Conclusion

The experimental results have shown a superiority for Yarowsky WSD approach in terms of retrieving words that have specific sense in the Quran, meanwhile do not have specific sense in the WordNet. For instance, the word 'day' has several senses in WordNet but only one sense could be occurred in Quran which is 'day and night', however in Quran the word day has been associated with 'judgment day' such sense is not located in the WordNet. In contrast, WordNet approach has shown a superiority in terms of retrieving words that have been used in Quran generally which indeed WordNet dictionary will identify its sense. For instance, the word 'light' has many senses in WordNet in compared to the Yarowsky which trained on two senses only. That is why WordNet is able to achieve higher results for the word 'light'. This emphasizes that the domain-specific WSD approach has a significant impact on improving the retrieval using restricted domains such as Quran. To sum up, the research finding of this study conclude the mixed of results of open-domain WSD when applied on Quranic IR. Therefore, probably a careful study of designing WSD technique should be applied on domain-specific IR system, in a way the mixed of both resources; external knowledge like WordNet and localised semi-supervised can be considered in order to have the best of both approaches.

Acknowledgement

This project is funded by Malaysian Government under code FRGS/1/2016/ICT02/UKM/01/14 and ERGS/1/2013/ICT07/UKM/03/1.

References

1. C. Laorden, I. Santos, B. Sanz, G. Alvarez, P. G. Bringas, Word Sense Disambiguation for Spam Filtering, *Electronic Commerce Research and Applications*, **11**(3), 290-298 (2012)
2. M. Carpuat D. Wu, Improving Statistical Machine Translation using Word Sense Disambiguation. *EMNLP-CoNLL*, 61-72 (2007)
3. R. Navigli, Word Sense Disambiguation: A Survey. *ACM Computer Survey*, **41** (2), 1-69 (2009)
4. E. Agirre, P. G. Edmonds, *Word Sense Disambiguation: Algorithms and Applications*. Springer Science & Business Media, (2007)
5. E. M. Voorhees, Query Expansion Using Lexical-Semantic Relations. *SIGIR'94*, 61-69 (1994)
6. V. Lopez, V. Uren, E. Motta, M. Pasin. Aqualog: An Ontology-Driven Question Answering System for Organizational Semantic Intranets. Web Semantics: Science, Services and Agents on the World Wide Web, **5**,72-105, (2007).
7. A. Zouaghi, L. Merhbene, M. Zrigui, Combination of Information Retrieval Methods with LESK Algorithm For Arabic Word Sense Disambiguation. *Artificial Intelligence Review*, **38**(4), 257-269, (2012).
8. O. J. Mohamed, S. Tiun, Word Sense Disambiguation based on Yarowsky Approach In English Quranic Information Retrieval System. *Journal of Theoretical and Applied Information Technology*, **82** (1), 163-171, (2015)
9. B. Hammo, A. Sleit, M. El-Haj, Effectiveness of Query Expansion in Searching the Holy Quran. *CITALA'07*, 1-10, (2007).
10. S. Banerjee, T. Pedersen, An Adapted Lesk Algorithm for Word Sense Disambiguation Using Wordnet. Computational *Linguistics and Intelligent Text Processing,* 136-145, (2002)
11. F. Vasilescu, P. Langlais, G. Lapalme, Evaluating Variants of the Lesk Approach for Disambiguating Words. *LREC'04*, (2004).
12. S. Tiun, H. Zakr, N. Z. Mohd, Z. Abidin, A. I. I. Hisham, Word Sense Disambiguation for English Quranic IR System. Taibah University International Conference on Advances in Information Technology for the Holy Quran and Its Sciences, 199-202 (2013)
13. Z. Zhong, N. T. Ng, Word Sense Disambiguation Improves Information Retrieval. *Proceedings of the 50th Annual Meeting of the Association for Computational Linguistics* (ACL 2012), 273 – 282 (2012)
14. M. Stevenson, E. Agirre, E., A. Soroa, Exploiting domain information for Word Sense Disambiguation of medical documents. *Journal of the American Medical Informatics Association*: JAMIA, **19** (2), 235–240 (2012)
15. F. Vasilescu, P. Langlais, G. Lapalme, Evaluating Variants of the Lesk Approach for Disambiguating Words. *LREC* (2004)

Query Translation and Quran Result in TreeMap

Mohd Amin Mohd Yunus[1], *Aida* Mustapha, and *Noor Azah* Samsudin

Faculty of Computer Science and Information Technology, Universiti Tun Hussein Onn Malaysia, Parit Raja, 86400 Batu Pahat, Johor, Malaysia

Abstract. This paper presents the view of TreeMap representation with retrieval results more attractive on eyes. It is also found that TreeMap more friendly to present Quran results for keyword (K) and queryword (Q). This research also is being applied in retrieving Malay verses Quran translation.

1 Introduction

Cross Language Information Retrieval (CLIR) systems results can be represented automatically with the use of TreeMap (University of Maryland, 2000). This gives rise to the following approach to present the result in CLIR. Start with queries, documents, and relevance judgments with result visualization representation in a single language. In order to come out with TreeMap, the queries have been translated into another language by translation dictionary. These translated queries are helpful for the performance of the CLIR system. In order to conduct the experiments, data regarding the Malay and English Quran translation are gathered accordingly as two sets of collection respectively. The Malay and English dictionary files are useful for translating the query. A Malay and English stopword files are used for removing the unnecessary words in the query.

2 Related Work

Information visualization is a branch of computer graphics and user interface which are concerned with the presentation of interactive or animated digital images to users to understand data. For example, scientists interpret potentially huge quantities of laboratory or simulation data or the results from sensors out in the field to aid reasoning, hypothesis building and cognition. The field of data mining offers many abstract visualizations related to these visualization types. They are active research areas, drawing on theory in information graphics, computer graphics, human-computer interaction and cognitive science. This is quite similar to this paper in which the use of TreeMap is presented and discussed. The flexibility of their system to visualize different terms will be discussed. The development of the system involves two stages. The first stage is the processing of 6236 documents of the Malay translated Quran to create a database of all terms. The second stage is the creation of the web-based system using the preprocessing data created in the first

[1] Corresponding author: aminy@uthm.edu.my

stage. Both stages are explored. The visualization will help to improve the understanding of relationship between the terms in this specific domain [3].

The tools of computer interface design are familiar to most computer users today such as windows, menus, icons, dialog boxes, and so on. These make use of bitmapped display and computer graphics to provide a more accessible interface than command-line-based displays. A less familiar but growing area is that of information visualization, which attempts to provide visual depictions of very large information spaces. TreeMap aids in this process by providing an abstract visual representation of the compressors of interest in a single screen display that can be grouped and arranged by geographic region, asset, field or platform. Each individual compressor is represented in the TreeMap by rectangular area whose characteristics are mapped to attributes of the compressors chosen by engineer [5].

3 Experimental Approach

This paper focus on the testing Malay query words are taken from Fatimah's collection as natural language queries [1,2]. Fatimah has obtained them by considering several guidelines. Each query would be separated and broken into keywords and replaced by target language. For example, if query is Malay, so it is called as source language and the target language is English or Arabic. The translation refers to the same index between Malay natural query language [1] and the translation of the natural query language in English. When the keyword is Malay, then reference is to the English word at the same index or when the keyword is English and then reference is to the Malay word at the same index. It is considering word by word in the text files. The query can be categorized into two which are keyword and queryword. If the query is keywords, the results retrieved according to word by word results and redundant document names existed if merged. But querywords, retrieved according to the whole words as one at all and only when no redundant or unique document names retrieved rankly. Query translation can replace the origin query in to another language of the query. This translation is most important for those languages to investigate those information retrieval results. Figure 1 explains the workflow of cross language information retrieval based on query translation in the retrieval process of Quranic documents collection. Query is processed by removing the meaningless words or stopwords. Then, query can be translated into another language if needed by the dictionary. Next step is to stem the query by a stemmer if needed according to accurate language.

DOI: 10.1051/matecconf/201713500072

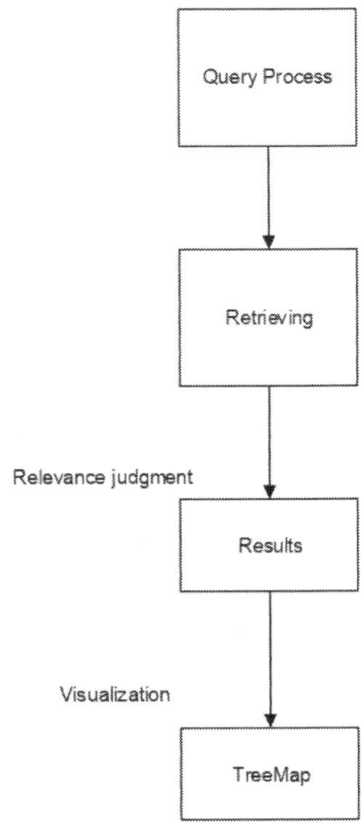

Fig. 1. Workflow of query translation retrieval and visualization results.

The query results consist of word by word result or keyword or the whole words or full phrase. The documents retrieval comes from al-Quran documents translation collection which is from Arabic, Malay and English collection. For this research, when the query is Malay words, the words are translated into English or Arabic vice versa. TreeMap display rows of data as groups of rectangles that can be arranged, sized and colored to graphically reveal underlying data patterns [4]. Thus total retrieve and relevant can be presented in terms of Surah names and document sentences or ayats or verses.

4 Result and Discussion

All results of the query(ies) translation are referred to the natural language queries of Malay [1] and then translated into Arabic and English queries in this study. In this paper actually concentrate on the comparison between translated query and normal query in retrieving documents by presenting TreeMap with K and Q accordingly as shown in Figure 3and 4 based on Table 1 query and Figure 2 Normal Query Result.

Table 1. Natural language queries.

No.	Malay	English	Arabic
1	Kelahiran nabi isa	Prophet jesus birth	مولد النبي عيسى

Fig. 2. Normal query result for "kelahiran Nabi Isa".

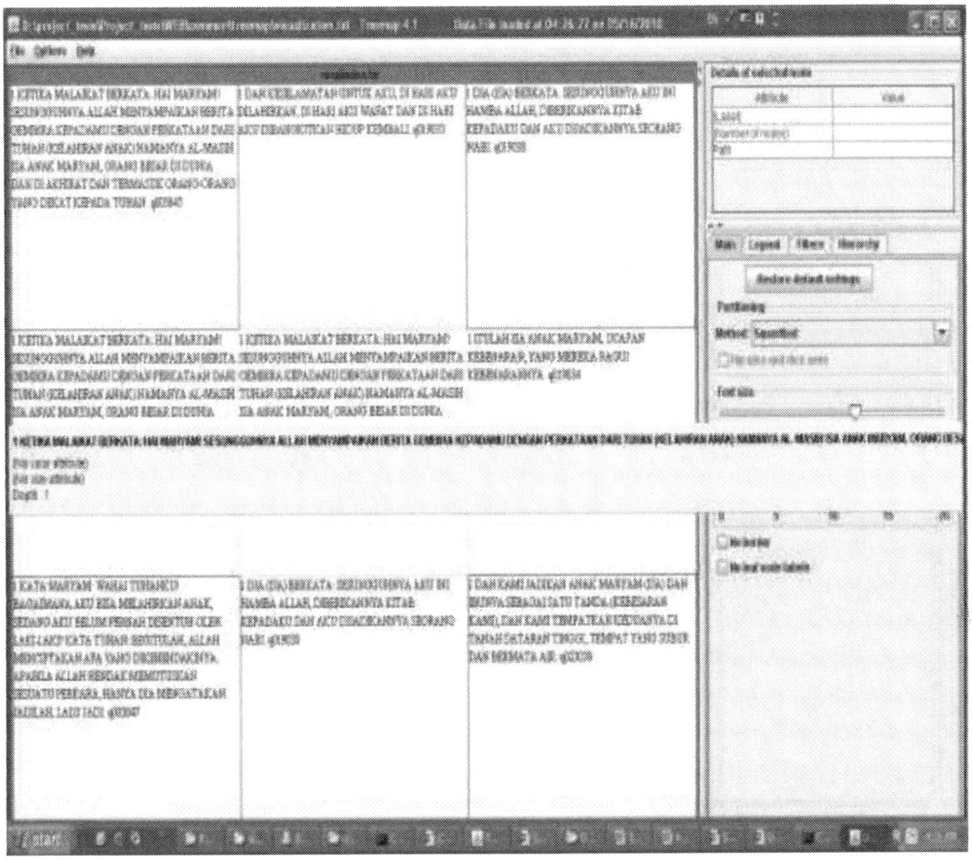

Fig. 3. Data visualization of K Result using TreeMap for "kelahiran Nabi Isa".

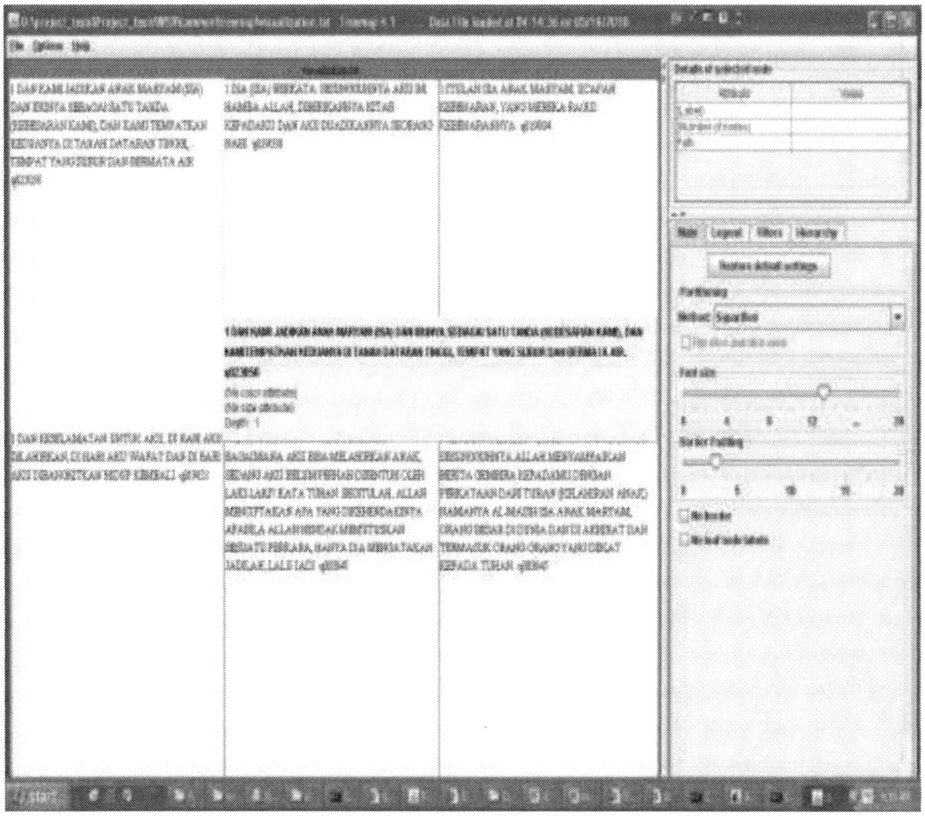

Fig. 4. Data visualization of Q Result using TreeMap for "kelahiran Nabi Isa".

5 Conclusion

It is found that TreeMap to be used to understand result given according surah name and ayat. The visual result is depending on given translated query with the same language. The relevant documents retrieval depends on combination of the efficient stemmers, dictionaries and stopwords to help the retrieval process. The results are better with stemming both query translation and documents collection. Then, querywords (Q) show unique documents rather instead of keywords (K) which show the redundant documents. Recall and precision percentages however, show better performance in K than Q. The queries need a dictionary to help and translate words in different languages for queries or documents. The difference languages but same meaning in queries can retrieve quite same documents as if it is monolingual search.

Acknowledgement

This work was supported in part by a grant from the Ministry of Education under the Research Acculturation Grant Scheme (R045) and in part by a grant from Research Gates IT Solution Sdn. Bhd.

References

1. A. Fatimah, A Malay Language Document Retrieval System: An Experiment Approach And Analysis, Tesis Ijazah Doktor Falsafah Universiti, Kebangsaan Malaysia (1995)
2. A. M. Y. Mohd, Short Query Translation: A Dictionary-Based Approach to Cross Language Information Retrieval. Tesis Master of Computer Science, Universiti Teknologi MARA. Malaysia (2008)
3. K. I. Normaly, A. R. Nurazzah, A. B. Zainab, Terms Visualization for Malay Translated Quran Documents, *Proceedings of the International Conference on Electrical Engineering and Informatics*, Institut Teknologi Bandung, Indonesia, 17-19 (2007)
4. University of Maryland, TreeMap. www.cs.umd.edu/hcil/treemap (2000)
5. C. Plaisant, G. Chintalapani, C. Lukehart, D. Schiro, J. Ryan, Using Visualization Tools to Gain Insight Into Your Data, The SPE Annual Technical Conference and Exhibition. Colorado: Denver (2003)

A Survey: Framework of an Information Retrieval for Malay Translated Hadith Document

Nurul Syeilla Syazhween Zulkefli[1,*], *Nurazzah* Abdul Rahman[1], and *Mazidah* Puteh[2]

[1]Faculty of Computer and Mathematical Sciences, Universiti Teknologi MARA, Shah Alam, Selangor, Malaysia
[2]Faculty of Computer and Mathematical Science, Universiti Teknologi MARA, Kampus Dungun Terengganu, Sura Hujung Dungun, Terengganu, Malaysia

Abstract. This paper reviews and analyses the limitation of the existing method used in the IR process in retrieving Malay Translated Hadith documents related to the search request. Traditional Malay Translated Hadith retrieval system has not focused on semantic extraction from text. The bag-of-words representation ignores the conceptual similarity of information in the query text and documents, which produce unsatisfactory retrieval results. Therefore, a more efficient IR framework is needed. This paper claims that the significant information extraction and subject-related information are actually important because the clues from this information can be used to search and find the relevance document to a query. Also, unimportant information can be discarded to represent the document content. So, semantic understanding of query and document is necessary to improve the effectiveness and accuracy of retrieval results for this domain study. Therefore, advance research is needed and it will be experimented in the future work. It is hoped that it will help users to search and find information regarding to the Malay Translated Hadith document.

1 Introduction

Hadith plays an important role as a significant source of reference throughout the world for the Muslims as well as the non-Muslims since it is a second fundamental source in Islam after the holy book, Al-Quran [3]. The actions, characteristics, stories and the sayings of the Prophet Muhammad are forms of Hadith [8,9]. Therefore, this source of reference is used by various parties for various reasons [1]. With the advancement of technology today which serves the platform in searching information more effectively, it is now easy for the users to search for Hadiths through online. Thus, there has been an increasing search activity for this information [12]. Most of the online searching activities on the web are based on the Information Retrieval (IR) framework. IR is the process of "representation, storage, organization of an access to information items" [2]. It aims to provide easy access to information based on the search request. Basically, user need to type in keyword (query text) for the information they want and the system will immediately respond by displaying the information sought in the user screen. In IR query, natural language is usually used and

*Corresponding author: syeilla.syazhween@gmail.com,

text from query and document collection is not always well structured. In retrieving Hadith information, users normally prefer to search for Hadith information in the language that they understand. Nevertheless, the collection of Hadiths that are published either in the form of books or digital resources are mostly in Arabic, English and Indonesian and for other languages are very limited. Therefore, the accurate retrieval of the Hadith information that they want would be difficult. To illustrate, the web offers a variety of application to retrieve relevant information with regards to Hadith in Arabic, English and Indonesian [1]. Even so, in retrieving Hadith information written in Malay language, the IR which was designed for Arabic, English and Indonesian may not work as effectively for the searching of Malay Translated Hadith document [12]. Despite this limitation, little amount of research has been made concerning Hadith searches in Malay language. Based on the past literature, Mutiara Hadis (*http://sigir.uitm.edu.my/webhadis/*) and JAKIM e-Hadith (*http://www, islam.gov.my/portal/e-hadith.php*) are two examples of on-line searches for Malay Translated Hadith documents based on Malay query. Both systems utilise keyword searches but results are significantly limited due to the word used in the query text. In addition, both retrieval systems do not support complex queries. Therefore, this limitation should be reviewed and extended in order to improve the IR in retrieving Hadith information in Malay languages (both query and document retrieved). The IR methods must not only be able to interpret the information, both in the query text and documents content, but also rank them accordingly to the degree of relevance based on user's information need (query). Therefore, this paper offers a review and proposed an improvement of IR framework for Malay Translated Hadith information retrieval to help people who are interested in searching and retrieving Malay Translated Hadith information for knowledge and references.

2 Proposed Framework

In this paper, searching information based on the text of Hadith written in Malay language is focussed. Therefore, this paper has been carried out to review and analyse the limitation of the simple and basic IR framework in retrieving Malay Translated Hadith documents. Based on the limitation, proposed framework is presented and described. Figure 1 below illustrates the basic IR framework and also the proposed framework for further investigation for Malay Translated Hadith document retrieval.

2.1 Dataset

In IR, it is very crucial to prepare a dataset. The dataset that is needed is a list of query text, document collection and a set of relevance judgement by the domain experts. The list of query text must be related to the information needed from the document collection. While, the document collection used must be the ones that have been evaluated by the domain experts (relevance judgement). A set of relevance judgement is referring to the number of relevant documents in a collection for each query. In this paper, a benchmark dataset related to the Shahih Bukhari collection written in Malay language is used for analysis and discussion purposes regarding to the effectiveness of IR in retrieving Malay Translated Hadith Document.

2.2 IR Process

Figure 1 is designed based on the basic IR framework, as presented in [2]. Basic IR framework consists of three steps; (1) Input query text by user; (2) query processor; (3)

retrieval and ranking process of relevant document related to a query. Meanwhile, document processing and indexing process has been done offline, before query is inputted. Due to the limitation of basic IR in retrieving Malay Translated Hadith information (discussion as in Section 3), the proposed framework is presented.In the proposed framework, sematic information extraction method from query and document will be explored. Also, conceptual similarity of information is focused. Therefore, semantic understanding of query, document and collections is necessary in order to improve the effectiveness of IR in retrieving accurate information from Malay Translated Hadith document.

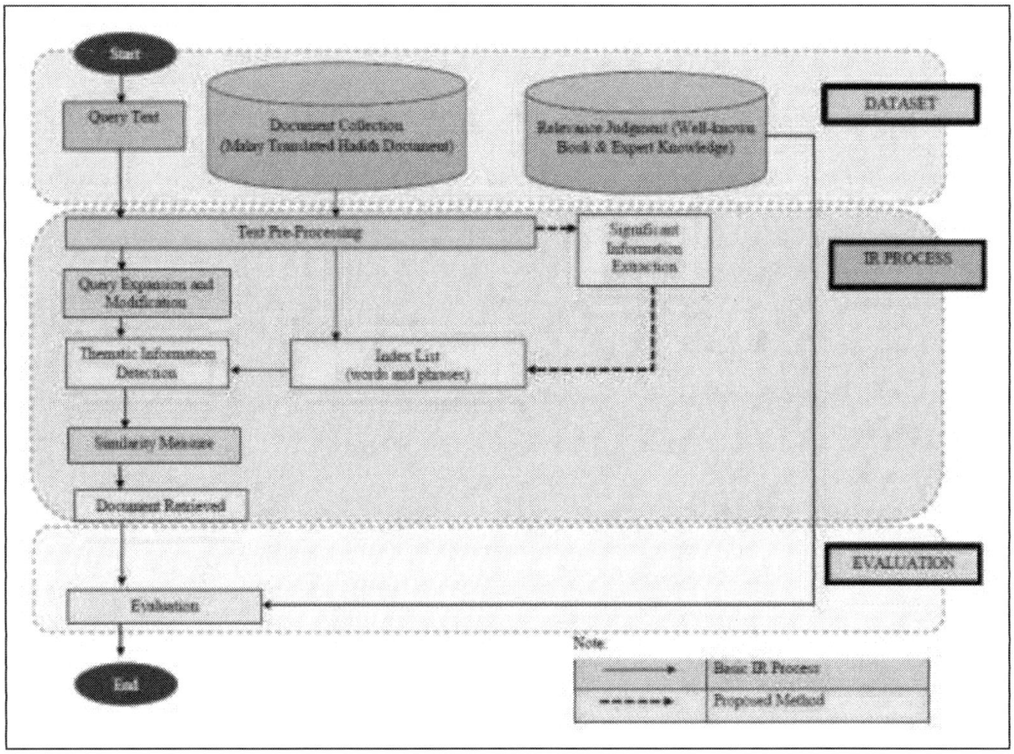

Fig. 1. Proposed framework for Malay Translated Hadith Information Retrieval

2.3 IR Evaluation

A set of documents are retrieved based on the input query text. The documents retrieved for each query is saving for the evaluation purpose. Most of the previous researches evaluate the effectiveness of IR methods using Precision and Recall scores based on the search retrieval results. Precision score measures the likeliness of a document to be relevant to the query while Recall score measures the retrieval of relevant documents. Others measures found in literature include E-Measure, F-Measure and MAP (Mean Average Precision scores for each query). However, this paper only looking for Precision and Recall scores for the review, analysis and discussion purposes.

3 Review and Discussion

This section presents the analysis and discussion of existing method applied for Malay Translated Hadith document retrieval. The analysis is based on the experimental results

 DOI: 10.1051/matecconf/201713500073

from previous studies, which are using similar dataset (list of query to be testing, document collections, and set of relevance judgment).

3.1 Analysis of Retrieval Results

In investigating the effectiveness of IR in retrieving Malay Translated Hadith documents based on query search, the benchmark dataset from Nurazzah [12], consists of 36 numbers of queries, 2028 digital Hadith documents from Sahih Bukhari collection in which written in Malay language and a set of relevance judgement for 36 queries is used. This paper chooses to make a direct comparisons from previous experimental results [10-12]. Therefore five (5) similar queries used from previous studies are selected. There are three (3) methods were used in the experiment: VSM, LSI and Clustering Method. Table 1 illustrates the Precision (P) and Recall (R) scores using these three methods in retrieving Malay Translated Hadith document. A set of document are retrieved based on the input query text. The document retrieved for each query is saved for the evaluation purpose. This paper however only measures the efficiency of the retrieval process using P and R scores.

Table 1. Previous experimental results: Precision (P) and Recall (R) scores for three different methods in retrieving Malay Translated Hadith Document

Query No	Query Text	VSM		LSI		Clustering	
		P	R.	P	R.	P	R.
Q1	*Adab-adab berkaitan makan dan minum*	0.04	0.75	0.07	0.20	0.10	0.39
Q2	*Tuntutlah ilmu hingga ke liang lahad*	**0.50**	0.65	0.07	0.20	**0.50**	0.50
Q3	*Apakah hokum berhias bagi kaum wanita*	**0.01**	0.18	**0.01**	0.27	0.08	0.32
Q4	*Bagaimana cara solat jenazah*	0.07	**0.79**	**0.01**	0.46	0.13	0.33
Q5	*Pembahagian harta mengikut faraid*	0.04	0.68	0.02	0.67	0.17	**0.16**

The results show the low scores for P and R. The highest P is 0.50 which is from Q2 based on VSM and Clustering method. For the R scores, 0.79 is the highest score, which is from Q4 based on VSM. In the other hand, the lowest P score is 0.01, captured from LSI method for Q3 and Q4. The R score using clustering method for Q5 is the lowest score. Meaning that: (1) in the most cases, all the relevant hadith are not retrieved; (2) the sequence of retrieved Hadith does not meet the requirement from the user; and (3) some irrelevant Hadith documents are also retrieved. Therefore, the advance research on IR for this domain study should be extended since searching information from Hadith text is important for Muslims especially to receive their inspirations from the reliable source of their religious beliefs and daily practice from the sayings of their Prophet Muhammad, called Hadith. Due to the importance of Hadith, it is evident that more intelligent searching-based information for Hadith document is needed.

3.2 Limitation of Existing Method

In retrieving Malay Translated Hadith documents, the application enables its users to search through their website based on IR process [12]. The theoretical foundation of Malay Translated Hadith IR is the Vector Space Model (VSM) [10-14]. It is a model to represent the documents content [2]. Baeza-Yates and Ribeiro-Neto (2011 and 1999), Frakes and Baeza-Yates (1992), Jurafsky and Martin (2000), Manning and Schutze (1999) and Van Rijsbergen (1979) are some of the authors that described the IR process and vector space model in their publications well. In the VSM, each query text and Hadith document in the corpus is represented by a vector. Each query and document gets weights based on what words appear in them. These weights model is corresponding to the content of the query and document. They are dependent on how often the word appears in the query, document and in the entire collection. Documents whose vectors are close to each other in this space are considered as similar document content to the query.

Some modification of the vector space model has been proposed in retrieving Malay Translated Hadith documents. Very common changes used are based on Malay stopword list and Malay stemming rules in creating the document representation during the text pre-processing stage. Stopword list is a list of words which does not contribute to the content of the documents such as "*dan*" (and), "*atau*" (or), "*adalah*" (is), etc. The words in the stopword list are simply removed. Malay stemmer in the other hand can describe the Malay morphological in the Malay Translated Hadith documents.

Both stopword list and stemming rules can usefully improve the search retrieval results. Most keywords put in the query text are simple and unstructured. Therefore, it is hard to retrieve a number of relevant documents, where the documents do not contain the few query words. Dealing with this limitation, authors from [14] have expanded user's query using Malay Thesaurus in the process of retrieving Malay Translated Hadith document. The authors come to a conclusion that thesaurus does improve the retrieval effectiveness in this domain although the improvement is very low in percentage.

Latent Semantic Indexing (LSI) approach has also shown to give improvements in the retrieval results for Malay Translated Hadith documents [10]. In LSI process, a more compact representation is presented. Also, the unimportant data can be eliminated. The reduction in the dimension takes related words and documents close to each other.

There has also been proposed research on Malay Translated Hadith documents retrieval using clustering method during indexing process [11,12]. Clustering is an unsupervised learning method, where it is automatically group the documents with similar subject [5,6]. The results are based on the document representation, the similarity measure and the clustering algorithm [2]. In Malay Translated Hadith document clustering, VSM is used to represent the information in the document. Then, cosine similarity measure is reported as an effective method to compute the similarity among Hadith documents [12]. Previous researcher used Agglomerative Hierarchical Clustering Algorithm in the experiments of grouping similar Hadith documents [12]. Among five linkage method, the complete linkage is outperformed [11].

Observed that the existing method used only depends on the words information extraction and it ignores word sequences (phrases) information contained in the query text and document collections. Depending only on the words extraction of information from the query and documents will not fully describe the Hadith contents and this will result in the low quality of retrieval results. Even though, the previous researchers did use various method in retrieving Hadith documents however, there are still issues to be solved. Previous research have not focussed on semantic extraction from query and document. Existing method was unable to handle the issue of conceptual similarity of information from the query, document and collections. For instance, there are two documents with similar subjects, but both documents contain two distinct contents of information. They are considered as two different documents. Based on the analysis, the reason for the limitation of IR in retrieving Malay Translated Hadith document is due to the complexity of information contained in Malay Translated Hadith document, which is use various of expression to deliver the similar content. Also, the increase of space dimensions while applying the word extraction method has made it hard for the data to be analysed. Moreover, significant information extraction from document is ignored since the method has not considered the word sequences information of documents. Therefore, it seems problematic for IR application to extract the significant information from the collection using the basic and simple method. In addition, the limitation is due to the limited digital resources that capture related information for Malay translated Hadith documents. Presently, Malay language resources are not provided with a large corpus. Furthermore, linguistic database for Malay language that is similar to WordNet is not yet available [16]. Also, the existing digital Malay Thesaurus used by [14] is not completed yet and can be extended.

In addition, it is reviewed that the re-evaluation for the document collections based on the query set is needed. This is because the existing relevance judgement is based on the well-known book [17] only. Existing relevance judgement for each query is based on the topic listed in the reference book [17]. For example, in query text of *"Bagaimana cara solat jenazah"*, the researcher will select Hadith document titled as *'solat jenazah'* found from the table of contents in the book and regarded them as relevance judgement for the query. Even though there were other relevant documents contained in the book regarding to this query but, they were categorised under other topics in the book. Thus, the researcher assumed that they were all irrelevant to the query. By merely referring to books of Hadith collection, researchers will unintentionally overlook the relevant documents which are placed under other topics which then will be discarded. Therefore, this paper suggests the re-evaluation process for the relevance judgement. Re-evaluation process involves well-known Hadith book and domain expert knowledge. Both relevance judgements from the two distinct methods are compared. The missing relevant document in benchmark dataset is updated and the redundant relevant document is eliminated.

3.3 Proposed Method

In fact, the retrieval application should be able to understand the contents of the documents and query information need. It is apparent now that the retrieval of Malay Translated Hadith documents is needed to be supported by a more efficient method. The method must not only be able to interpret the content of information from a collection but also be able to identify the related information from the collection based on the search request. The real content of a document is totally different from the words in the vector space model. Thus, this paper designs the proposed framework for Malay Translated Hadith information retrieval. The sections below describe a detailed explanation regarding the methods that are to be utilised in the proposed framework. The proposed methods involved are Significance Information Extraction and Thematic Information Detection.

3.3.1 Significant Information Extraction

Significant information in this paper is referring to the words and phrases, also called as keyphrase [15] which are important and meaningful information for a specific subject in the entire collections. A keyphrase is described as important and significant information as a greatly shortened summary for a document [4]. However, it has received less attention in the IR study, even though it is an important research topic. Keyphrase extraction is given a more focus from the previous research for journal or conference articles summarizations. Also, the researchers mostly applied for English language dataset [15] in NLP study. This paper however uses Keyphrase Extraction in IR methods. It is mainly used for Malay language dataset. Malay Translated Hadith Document is chooses since it is one of the most important sources of Islamic knowledge and it have no author-assigned keyphrases. Due to its advantage, this paper chooses to exploit Keyphrase Extraction method in the proposed IR framework for Malay Translated Hadith information retrieval. Therefore, query and document content analysis should be observed in future study, so that we can get a criteria and characteristics of significant information from Malay Translated Hadith document.

3.3.2 Thematic Information Detection

In the existing methods, the documents are independent of each other. The word extraction has no interactions for each document in the entire collections. However, some subject-related documents actually contain useful information that can be used as clues to extract significant information from each other. For example, two documents of about the same topic "*Zakat Fitrah*" would share a few common phrases, e.g. "*segantang gandum*", "*segantang kurma*", "*sedekah*", etc. They can provide additional knowledge for each other to better evaluate and extract relevant information from each other in the entire collection. Therefore, given a specific query or document, IR can retrieve a few similar documents in the entire collections through Thematic Information detection. In the literature study, thematic information can be captured based on external resources; corpus-based and knowledge-based resources [16]. However, there are no external resources available for Malay languages [16], and digital resources specifically for Hadith information in Malay language is also not yet available. Therefore, in the proposed framework, this paper chooses to focus on method that will automatically detect the Thematic Information based on Malay Translated Hadith document. Thematic information in this paper is referring to the group of the related information which is significant for a specific subject. The words and phrases are significant if they can deliver the content of query and document in collections [5,6].

4 Conclusion and Future Work

This paper has discussed the limitation of IR methods in retrieving the most relevant Malay Translated Hadith document regarding to the Malay query (information need). This paper then discusses method which aims to improve the effectiveness of IR in retrieving the most relevant information regarding to the Malay Translated Hadith document. The discussion shows that the advance research for IR methods is needed and can be extended. Existing method used in IR framework have not focussed on semantic extraction from query and document. Most of the methods are based on the words extraction from query text and documents. Depending on the words only, the significant information which is consisted of the sequence of words (phrases) is ignored. Also, it does not consider the semantic interpretation of the query text and documents and hence, the subject-related information is discarded. The bag-of-words representation is often unsatisfactory as it ignores the conceptual similarity of words with various expression of information. The limitation has produced low scores for Precision and Recall measures, as it is reported in the experimental results in previous studies which have used the similar dataset. The low scores mean that the current methods applied for IR framework in retrieving information from Malay Translated Hadith document is less effective and can should be reviewed, observed and improved. A more efficient IR framework is needed and the proposed method is presented and discussed to be experimented in future work. The methods involved are semantic information extraction and thematic information detection. It is hoped that the proposed method is able to improve the effectiveness of IR in retrieving Hadith information from Malay Translated Hadith document in order to help people in searching and retrieving accurate Hadith information written in Malay language.

References

1. S. Abdul Rauf, J. S. Waqar, Information Mining from Muslim Scriptures. *In: International Joint Conference on Natural Language Processing (WSSANLP)* pp. 66-71 (2013)
2. R. Baeza-Yates, B. Ribeiro-Neto, *Modern Information Retrieval: The concepts and technology behind search* (*Second Edition*), Pearson Education Limited (England: Pearson) Chapter 1-3 pp. 1-130 (2011)
3. J. Dakir, F. A. Shah, A contextual approach in understanding the prophet's hadith. *Applied Sciences*, **8**(7): 3176-3184 (2012)
4. Hammouda KM, Kamel MS (2004) *Efficient phrase-based document indexing for web document clustering. In: IEEE Transactions on knowledge and data engineering*, **16**(10), pp. 1279-1296 (2004)
5. M. Ilic, P. Spalevic, M. Veinovic, Suffix Tree Clustering–Data mining algorithm. *In: Twenty-Third International Electrotechnical and Computer Science Conference (ERK 2014)*, pp. 15-18 (2014)
6. A. K. Jain, S. Maheshwari, Phrase based Clustering Scheme of Suffix Tree Document Clustering Model. *Computer Applications*, **63**(10): 30-37 (2013)
7. K. Jbara, Knowledge discovery in Al-Hadith using text classification algorithm. *American Science*, **6**(11): 409-19 (2010)
8. N. Moath, A. Abdelkarim, A. Musab, O. Abdelrahman, A Lexicon for Hadith Science Based on a Corpus, *Computer Science and Information Technologies*, **6**(2): 1336-1340 (2015)
9. A. S. Mohammad, I. Norisma, M. Rohana, J. Salinah, D. Thorleuchter, G. Abdullah, Hadith data mining and classification: a comparative analysis. *Artificial Intelligence*, 46(1): 113-128 (2016)
10. T. M. R. Amirah, *Latent Semantic Indexing (LSI) Using Parallel Programming Technique For Malay Hadith Translated Document Retrieval.* Unpublished master disertation thesis, Universiti Teknologi Mara (UiTM), Malaysia (2014)
11. N. A. Rahman, Z. A. Bakar, N. S. S. Zulkefli, Malay document clustering using complete linkage clustering technique with Cosine Coefficient. *In: Open Systems (ICOS 2015)*, pp. 103-107: IEEE (2015)
12. N. A. Rahman, *Evaluating The Effectiveness of Clustering Techniques In Retrieving Malay Translated hadith Text.* Unpublished doctoral thesis, Universiti Teknologi Mara (UiTM), Malaysia (2011)
13. N. A. Rahman, Z. A. Bakar, T. M. T. Sembok, N. K. Ismail, Cluster-Based Hadith Retrieval System. *In: Proceedings of the International Conference on ICT for the Muslim World (ICT4M)*, Kuala Lumpur (2006)
14. N. A. Rahman, Z. A. Bakar, T. M. T. Sembok, Query Expansion using Thesaurus in Improving Malay Hadith Retrieval System, *In: Proceedings of the 21st annual international ACM SIGIR conference on Research and development in information retrieval* pp. 46-54: ACM (2010)
15. I. Sahmoudi, H. Froud, A. Lachkar, A new keyphrases extraction method based on suffix tree data structure for Arabic documents clustering. *Database Management Systems*, **5**(6):17-33 (2014)
16. S. A. Noh, N. Omar, A. Y. Amru, Evaluation of Lexical-Based Approaches to the

Semantic Similarity of Malay Sentences. *Qualitative Linguistics,* **22**(2):135-156 (2015)
17. H. Zainuddin, H. S. Fachruddin, T. Nasharuddin, A. Johar, M. A. A. R. Zainuddin, *Terjemahan Hadis Shahih Bukhari Jilid I,II,III,IV (Cetakan Keenam),* Kuala Lumpur Malaysia: Klang Book Centre, pp. 1-220 (2005)

DOI: 10.1051/matecconf/201713500074

Application of A Computer Animation Technique to Assist the Teaching of Pre-Handwriting Skills to Children with Coordination Difficulties/Dyspraxia

Muhammad Fakri Othman[1,2], *Wendy Keay* Bright[2], *Norhalina* Senan[1], *Suriawati* Suparjoh[1]

[1]Department of Multimedia, Faculty of Computer Science and Information Technology, Universiti Tun Hussein Onn Malaysia, Parit Raja, 86400 Batu Pahat, Johor, Malaysia
[2] Department of Creative Communications, Cardiff School of Art and Design, Cardiff Metropolitan University, Cardiff, United Kingdom

Abstract. We have proposed a method to assist children with coordination difficulties or dyspraxia to improve their handwriting skills. We have chosen an animation technique called 'Rotoscopy', a method that normally been used in animation and film production and adapted it to Rotoscopy Pre-handwriting Interface (RPI) prototypes using the interactive whiteboard (IWB) as interaction device. The motivation of this research is to discover how efficient if Rotoscopy is used beyond its normal purposes? Does it give benefits in terms of behavioural and motivational aspect rather than commercial and profit point of view? Implementation of RPI prototypes has taken place through series of workshops with a teacher and a group of children with handwriting difficulties at a special education school in Caerphilly, Cardiff, United Kingdom. In the workshops children were given pre-handwriting activities in two different environments. They have been trained to use RPI prototypes and IWB as well as using pen and paper. Their activities and action has been observed and recorded using video camera. Evaluation method is based-on video analysis of children's pre-handwriting result and their reaction and motivation during the workshop. It was learnt that majority of children who used RPI prototypes and IWB have produced better results in terms of accuracy of the drawing as compared to results of pen and paper activities. Furthermore the children are more motivated to use the prototypes and IWB rather than using pen and paper. The study's contribution includes offering a new way to improve children's pre-handwriting skills using computer animation technique and touch-based devices.

*Corresponding author: fakri@uthm.edu.my, faothman@cardiffmet.ac.uk

1. Introduction

Rotoscopy is a traditional animation technique that has been comprehensively adopted in computer animation. It captures human movement in such a way that it can be animated in a simplified form to convey naturalistic actions [1][2]. It is also a technique in which animators trace over live-action film movement for use in animated films [3]. The process involves frame-by-frame projection of moving images to allow the animator to copy every single frame to create a natural-looking animation.

A famous film that used rotoscopy is *A Scanner Darkly,* produced in 2006 by Richard Linklater using a software called *Rotoshop.* Originally, rotoscopy technique was invented in 1917 by Max Fleisher for his film, *Out of the Inkwell.* It is named after projection equipment called the *Rotoscope;* where pre-recorded live-action film images were projected onto a frosted glass panel and re-drawn by an animator. Recently the device has been replaced by computers, and the technique has been referred to as interpolated rotoscopy. In visual effects industry, rotoscopy refers to the technique of manually creating a matte for an element on a live-action plate to be composited over different background [4]. Figure 1 shows examples of rotoscopy whilst a short animation can been produced through the rotoscopy process using a live action video. In this example, graphics images as well as human motion and body shape from a video sequences were traced manually using rotoscopy system to get the image silhouettes, which are then been transformed to 2D graphics and animation. From the output, we can create a new 2D graphics or character that are different from original object but inherits its shapes and motion.

Figure 1. Example of rotoscopy process: from a video to 2D graphics and animation

2. Related Works

2.1. Rotoscopy and Tracking

Early studies in Rotoscopy were focused on techniques for 2D graphics and animation, such as *Contour Tracking* [2] and *SnakeToonz* [5]. *Keyframe-based Tracking* [3] was proposed as a technique for rotoscopy and animation, which has been improved by *Bi-directional*

Tracking [6]. However, both methods produced 2D results and faced similar difficulties. Rotoscopy typically produces non-photorealistic images and animation as it requires more human effort to trace object contours from captured video sequences, whilst tracking is a process in rotoscopy to capture video to be transformed to 2D or 3D graphics and animation. Though a few tracking algorithms have been introduced, such as *Keyframe-based Tracking* and *Bi-directional Tracking*, previous research in rotoscopy and animation focuses more on video and 2D image rather than producing 3D output.

2.2. Rotoscopy, 3D Graphics and Animation

For 3D graphics and animation, the studies involved production of 3D models from video, as well as to improve the image quality. A real-time 3D model acquisition system [7] was developed to assist user to rotate an object by hand and see a continuously-updated model as the object is scanned. However the system has limited functions and required much human efforts. Another research, 3D objects from photographs and video clip [8] has emerged, but it was only for non-animated 3D images.

More recently, a new method to capture 3D mesh from 2.5D video [9] has been used, but the result has several limitations such as it cannot distinguish between different objects in the scene, the output cannot be viewed from any direction and there is a flickering problem. A 3D sketching system to create 3D models [10] was developed in 2006 but it has problem to generate correct skeleton and accurate sampling of the silhouette. Generally, research of rotoscopy has been established in the area of computer graphics, in particular, tracking, video and animation, though it needs more improvement in terms of quality of output and production time.

2.3. Rotoscopy and Dyspraxia

Dyspraxia is a motor learning disability that affects movement and co-ordination due to brain messages not being properly transmitted to the body [11]. A motor learning disability is a condition where people have difficulty planning and organising smooth coordinated movements. It is fairly common in children and adults and is sometimes known as Clumsy Child Syndrome. Other terms, such as developmental co-ordination disorder (DCD), minimal brain dysfunction, and perceptuo-motor dysfunction are also used to describe dyspraxia [12]. According to The World Health Organisation, dyspraxia affects 6% of all children and the percentage is estimated to increase as high as 10% [13]. However only a few study related to computer animation technique such as rotoscopy have been undertaken to investigate the potential benefits for people with learning difficulties. This gap provides the main objective and motivation of our research. An application allowing children to be creative and expressive with movement could be beneficial since children with dyspraxia naturally experience movement disorder and may respond well activities that will encourage them to practice their movement and gesture ability. Research on the efficiency of e-learning and multimedia computer technology for children with learning difficulties strongly recommended such approach as it has emerged positive results to the children [14].

2.4. *Rotoscopy, Computer System and Handwriting*

Rotoscopy may also appropriate due to the highly visible nature of the graphical output, which is critical for dyspraxic children who have difficulties in hand movement as well as handwriting. A study of sign language co-articulation has been undertaken using the rotoscopy technique [15] as it promotes movement to perform signs using hands. Walden's research [16] found that technical advances of rotoscopy have had aesthetic consequences not just for the quality of image but also to the nature of the actor's performance within animation. In terms of handwriting, Boyle [17] has revealed children with *Moderate Learning Difficulties* (MLD) like dyspraxia can be helped to improve their handwriting using simple intervention programme like speed of handwriting exercise, with support of other general gross motor coordination test. Meanwhile an investigation by Snape and Nicol [18] using a pen-based computer writing interface, has given positive improvement to children, though it hasn't employed rotoscopy for their system. As shown in Table 1, previous research in rotoscopy has not focused to handwriting, whilst handwriting recognition system hasn't employed rotoscopy as its main solutions nor supporting children with handwriting difficulties. All are different to our approach that applies rotoscopy to assist the teaching of handwriting for children with dyspraxia.

Table 1. Present and proposed rotoscopy systems

Projects	**Characteristics**			
	Function	***Rotoscopy***	***Hand-writing***	***Dyspraxia***
3D avatars animation [15]	French sign language coarticulation	YES	NO	NO
Double Take [16]	Cinema's actor and performance relation	YES	NO	NO
Handwriting recognition systems [18]	Pen-based computer writing interface	NO	YES	NO
Speed of handwriting exercise [17]	Handwriting using simple intervention programme	NO	YES	YES
Proposed system	Assist handwriting for dyspraxic children	YES	YES	YES

2.5. *Our Research*

Underpinned by theoretic influences on neurological development, animation and movement, and contextualised through a user-centred design approach [19], the research explores the potential of rotoscopy as a motivational tool of learning for children with dyspraxia [20]. A significant aspect of the methodology is to develop a rotoscopy interface using iterative process of continuous prototyping [21]. For the proposed system, the interface is designed to assist dyspraxic children in practising their handwriting skills through movement and performance, rendered visible to them through the projection of actions onto a large surface. We use the combination of 2D and 3D animation tools that supports rotoscopy functions such as *Adobe Flash* and *Adobe AfterEffects*; as well as pen tablet technologies for handwriting tools. Evaluation methods measure the extent to which the animation of movement in a child-friendly graphic form can engage children's motivation [14] and lead to improved handwriting skills [18]. The target group for this study are children at reception class and children with specific learning difficulties, based on the Early Years Foundation System (EYFS) standard [22], as well as their teachers and parents. As a summary, there has been little research investigating the potential of Rotoscopy as a learning tool. However the significance of movement and motor stimulation in child development is well understood. The success of *ReacTickles®* [23] project as well has given us the idea on interface design for the perceptually mapping of input and output. *ReacTickles* software was designed as a relaxing and playful experience with technology for children with autism. Figure 2 shows the relation of our research with previous studies in rotoscopy, dyspraxia, and handwriting.

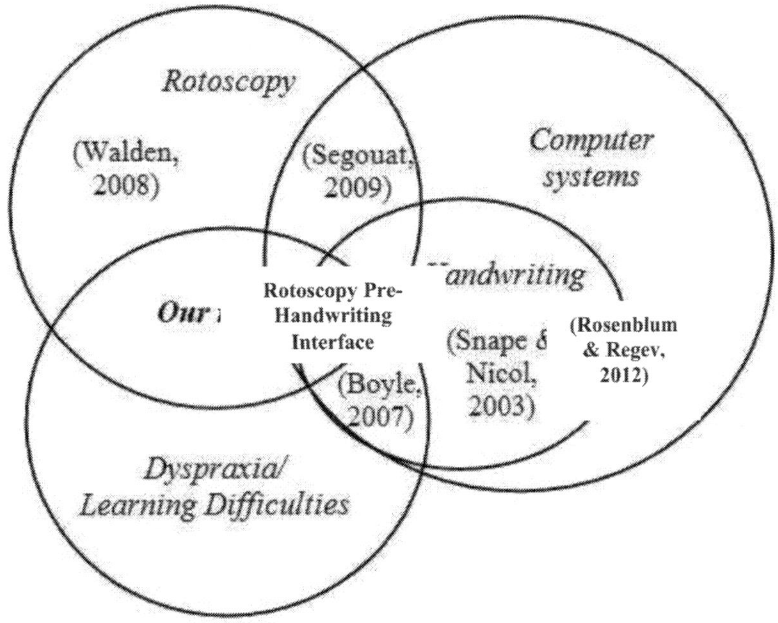

Figure 2. Previous studies in rotoscopy, dyspraxia and handwriting and its relation with our research

3. User-Centred Design

Our research has implemented a user centred design approach, which is a combination of ethnography, user observation and informant design [24]. Ethnography and user observation involve user participation in familiar activities in natural settings, which may include exploring prototypes. It can take place at the outset, during, and on completion of a project. Informant design allows children or teachers to be valued as experts or 'native informants', informing designers of key issues that relate to their experience as well as helping to develop early design ideas and testing prototypes in development [25].

The rationale for choosing this method relies on understanding the characteristics of children with learning difficulties, like dyspraxia, who are very subjective in terms of knowledge, skills and cognitive processing; and handwriting is a particular problem that they experienced. User centred design also can reduce the likelihood of system failure as the direct

DOI: 10.1051/matecconf/201713500074

relationship between the users and the suppliers means that outcomes are more likely to meet learners' and teachers' real needs [26]. User-centred design approach must be organised to meet with the needs of learners, and with the staff who are likely to include the resulting technology in their pedagogical practices, rather than to meet commercial development requirements [27]. The crucial component for successful practice is ensuring that the adult participants, as well as the children, perceive themselves as learners in the process of designing new technology [28]. By breaking down the distinction between adult 'expert' and child 'user', this method can lead to innovative solutions, and, in the process, engage all parties in the active construction of knowledge. Furthermore, benefits of involving users designing process may include:-

- First-hand experience of end user's needs, interest and requirement.
- Identify design challenges and difficulties at early stage, thus will save time and reduce the cost.
- Achieving educational aims as user's participation gives opportunity to create resources that are embedded in teaching strategies and educational contexts.

3.1. Contextual Analysis

We have conducted a contextual analysis by working with experts and offering appropriate activities with prototypes to selected dyspraxia children (end-users). The testing process is fully-guided by suitable experienced professionals and captured on video for further analysis using standardized observation scales. The implementation of prototypes and the training or testing process has taken place at a special education school in Caerphilly, Cardiff, United Kingdom. We have also established a connection with Dyscovery Centre, University of Wales, Newport (http://dyscovery.newport.ac.uk), a special research centre for children with dyspraxia and other learning difficulties. The metrics of testing are video and observation, involving series of different version of prototypes; since all these are techniques applied in ethnography and user observation and informant design approach. We also used *Wechsler Intelligence Scales for Children®* (WISC) [29] as the evaluation standard. WISC is a special intelligence test designed for children age 6 to 16 to generate their IQ score as well as to diagnose children's learning disability. The results are measured on children's performance, and how the application helps users improve their gross-motor skills, as well as their interest and motivation towards learning.

3.2. Interactive Whiteboard

The interactive whiteboard or IWB has been used in classroom to test the prototypes as well pre-handwriting activities. It has been chosen as it is easy to use as it has touch screen mode that allows children to make free hand movements on the screen. The concept is quite similar to pen and graphics tablet but in more spacious movement area. Previous research using pen-based computer writing interface [30] has considered it was as efficient as, and produced comparable writing to the pencil and paper. Figure 3 is an example of activities using interactive whiteboard and *ReacTickles* [31] software for children with autism at the Dyscovery Centre.

DOI: 10.1051/matecconf/201713500074

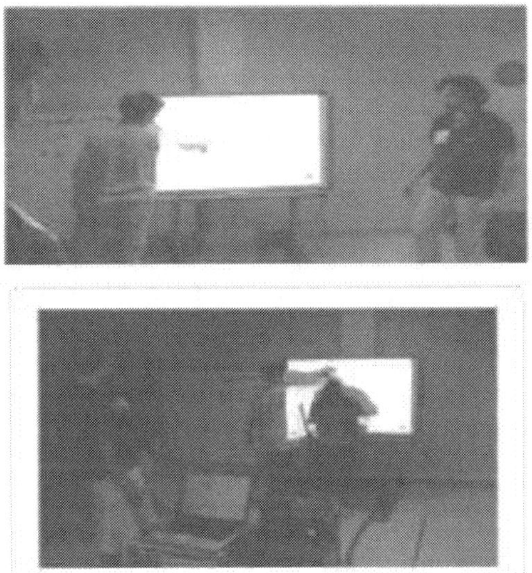

Figure 3. Example of activities using interactive whiteboard

4. Rotoscopy Pre-Handwriting Interface (RPI) Prototypes

As our research aims to give children with dyspraxia a playful and physical experience in developing their handwriting skills, we are investigating rotoscopy technique to identify whether motivation and engagement can be increased by using a performance-led embodied approach, which would allow children to practice handwriting using gross motor skills, for example bodily movement and gesture. We also use user-centred design approach to identify the requirements of the system and a context for its application as a learning tool. The development and implementation of the systems are focused on the benefits of using interactive whiteboards [32], as large-scale environments that support gross-motor skills and performance. Rotoscopy technique has been used to trace a projected image on interactive whiteboard screen. The system is developed using 2D and 3D animation software and are utilised been and adapted to respond to user activity in such a way that the mirrored or abstracted image enables the child to visualise the activity and to recognize the projected actions own. This mirroring capacity of the system is vital for the development of social communication and learning [19]. For this reason rotoscopy is chosen to be used as a technique to assist children with dyspraxia, who are known to experience problems with low self-esteem as a result of co-ordination difficulties [33].

The RPI prototypes are designed to assist children with dyspraxia in practising their handwriting skills through movement and performance, rendered visible to them via the projection of actions onto a large surface. Evaluation methods have measured the extent to which the animation of movement in a child-friendly graphic form can engage children's motivation and lead to improved handwriting skills [31]. In particular the RPI prototypes that have been designed and developed are based-on Ripley's Methods for teaching of handwriting to children with dyspraxia [12]. Ripley's method consists of different stages of skills: Prototype 1: Group 1 to Group 4 Symbols, Prototype 2: Numbers, Prototype 3: Circle Letters, Prototype 4: Wiggly Letters and Prototype 5: Child's Name. Meanwhile the RPI activities consist of squiggling, drawing basic shapes, drawing pictures and writing letters

DOI: 10.1051/matecconf/201713500074

(Figure 4). These are actions required in children, not only as a pre-handwriting activity before they learn how to write, but to improve their gross-motor skills by using a large screen, such as an interactive whiteboard (IWB), as the interface.

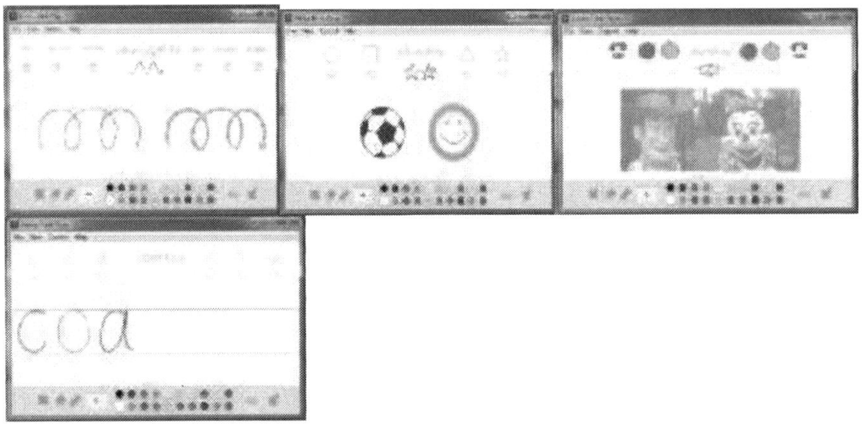

Figure 4. Examples of RPI prototypes: squiggling, shapes, drawing and writing letters

5. The trial of RPI prototypes

Figure 5. Trial of the first prototype, RPI (*Squiggles*) and pen and paper (PP) tasks

The trials of the first two of the RPI prototypes, RPI (*Squiggles*) and RPI (*Shapes*) has taken place to a group of children with handwriting difficulties at a special school in Caerphilly, Cardiff, United Kingdom. The participants consisted of seven boys and two girls, with writing levels P5 (intermediate) and P6 (advance). The study has obtained an ethical approval from Cardiff Metropolitan University, UK. The participant's consent has also been acquired for research publication. The objectives of the trials are to introduce the system to the target audience, to expose the children to the rotoscopy technique and to adapt it to the system environment as well as to test the functionality of the system prototypes to identify syntax, logical errors and any technical problems. The activities in three trial workshops are introducing the system to the children, trial of the first prototype, RPI (*Squiggles*) and pen

DOI: 10.1051/matecconf/201713500074

and paper (PP) tasks (Figure 5) and the trial of the second prototype, RPI (*Shapes*) and PP tasks (Figure 6)

Figure 6. Trial of the second prototype, RPI (*Shapes*) and pen and paper (PP) tasks

6. Results and Discussion

6.1. Results of squiggling using RPI (Squiggles) for all children

The first study examined how children squiggle using the first module of RPI prototypes called RPI (*Squiggles*). There were six activities in this prototype, but only three were tested due to the children's ability, after advice had been taken from the teacher. The first task was *Squiggle 1* (Task 1), where the children needed to draw curly lines or squiggles, followed by *Squiggle 5* (Task 2), which allowed the children to draw straight lines. Table 2 shows the thumbnail of the results for the squiggling activities (Task 1 and Task 2) drawn by every children using RPI (*Squiggles*).

Table 2. Results of squiggling using RPI (*Squiggles*)

Task 1: Reference Image 	Task 2: Reference Image 	Child 5: Task 1 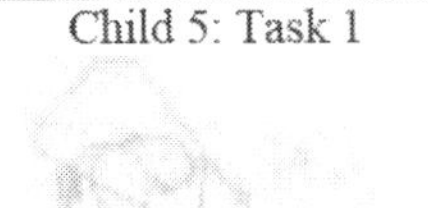	Child 5: Task 2
Child 1: Task 1 	Child 1: Task 2 	Child 6: Task 1 	Child 6: Task 2
Child 2: Task 1	Child 2: Task 2	Child 7: Task 1 	Child 7: Task 2 

Child 3: Task 1	Child 3: Task 2	Child 8: Task 1	Child 8: Task 2
	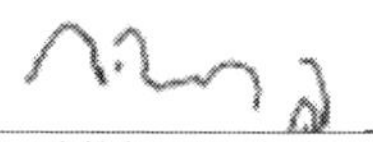		
Child 4: Task 1	Child 4: Task 2	Child 9: Task 1	Child 9: Task 2
		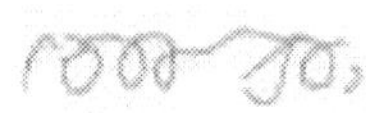	

6.2. Results of squiggling using Pen and Paper (PP) for all children

The second study was undertaken to assess children's response to squiggles and shapes in a different environment. The study focused on how the children perform basic squiggles using traditional pen and paper (PP) methods with the same squiggle lines they used in the first study, squiggling using RPI (*Squiggles*). Table 3 shows the thumbnails of the squiggling results for every child involved.

Table 3. Results of squiggling using PP (Children 1 to 9)

Child 1	Child 2	Child 3
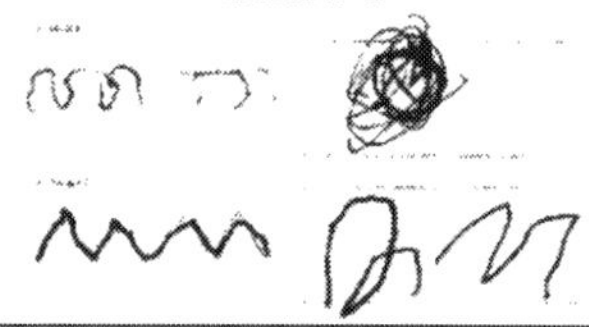		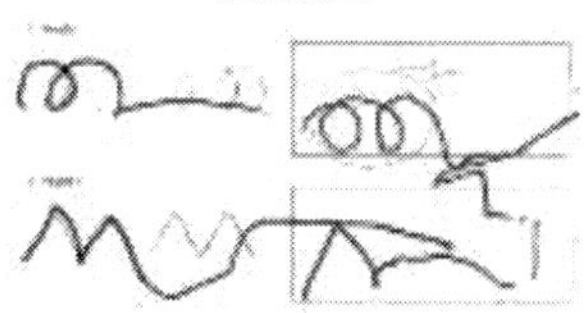
Child 4	Child 5	Child 6
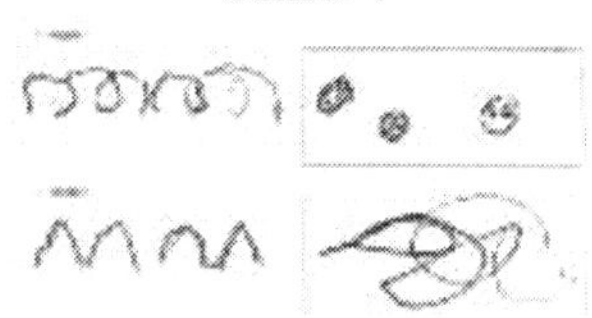	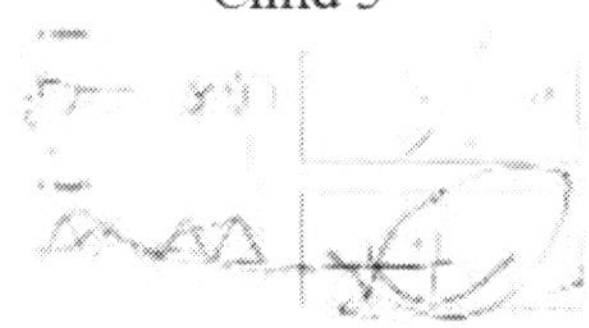	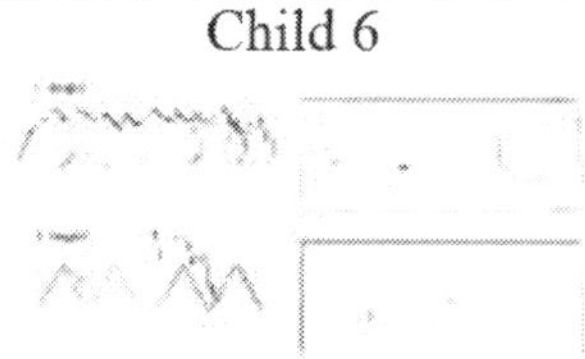
Child 7	Child 8	Child 9
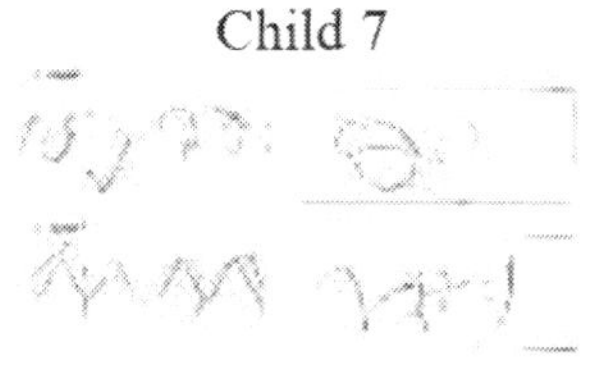	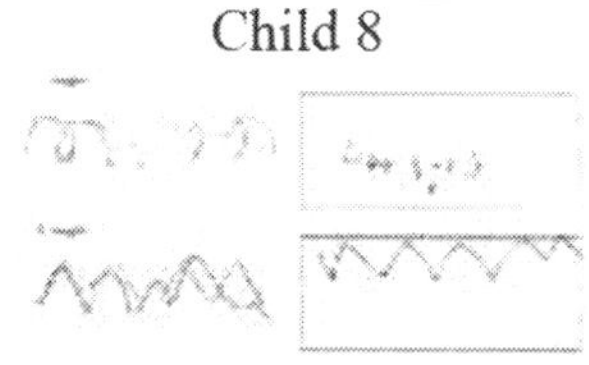	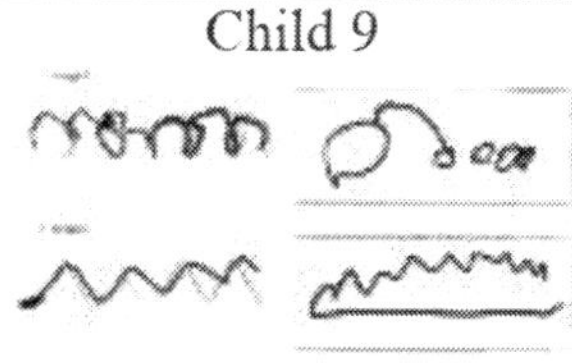

6.3. Statistical analysis of children's results using RPI (Squiggles) (Task 1 & 2)

The data obtained from children's squiggling using RPI are then analysed using the SPSS package to determine the results' mean score and to compare with the reference images for every task in order to measure the accuracy of children's squiggling or drawings. SPSS is used to identify whether there is a significant difference between the results from children's squiggling compared with the results from the pen and paper (PP) tasks. The spatial parameters of the results, which include the width and height of the squiggles and drawings, as well as the number of squiggles or lines for Task 1 and Task 2, are presented in Table 4.

Table 4. Spatial parameters of squiggling results for Task 1 and Task 2

Child	Width (pixel)	Height (pixel)	No. of Squiggles	Width (pixel)	Height (pixel)	No. of Lines
1	535	131	4	457	123	4
2	538	182	4	432	107	3
3	518	144	1	474	133	1
4	511	95	4	452	91	4
5	500	158	2	447	122	2
6	516	116	1	440	119	2
7	511	129	4	411	97	4
8	501	134	3	465	117	3
9	532	113	3	453	112	3

Table 5. Mean of spatial parameters for Task 1 and Task 2 results

		Width1	Height1	Squiggles	Width2	Height2	Lines
N	Valid	9	9	9	9	9	9
	Missing	0	0	0	0	0	0
Mean		518.00	133.56	2.72	447.89	113.44	2.83

The mean for the spatial parameters of both tasks and references are shown in Table 5 above. The comparison of the mean for Task 1 and Task 2 with the mean of the reference images for both tasks (Task 1R and Task 2R) is presented in Table 6 and Table 7 respectively. Both tables show that the mean of children's squiggling parameters for both tasks are close to the parameters of the reference images provided by RPI.

Table 6. Mean of spatial parameters for Task 1 and Task 1R

		Width1	Height1	Squiggles1	Width1R	Height1R	Squiggle1R
N	Valid	9	9	9	9	9	9
	Missing	0	0	0	0	0	0
Mean		518.00	133.56	2.72	525.00	100.00	4.00

Table 7. Mean of spatial parameters for Task 2 and Task 2R

		Width2	Height2	Lines2	Width2R	Height2R	Lines2R
N	Valid	9	9	9	9	9	9
	Missing	0	0	0	0	0	0
Mean		447.89	113.44	2.83	455.00	95.00	8.00

DOI: 10.1051/matecconf/201713500074

6.4. Analysis of Task 1 results

A Paired Samples Test, or ***t*-test,** has been implemented to compare the mean of the children's results spatial parameters with the mean of the spatial parameters of the reference images for Task 1 (Task1R), as depicted in Table 8.

Table 8. Paired Samples Test (*t*-test) for Task 1 and Task 1R

Paired Samples Test		Paired Differences					T	Df	Sig. (2-tailed)
		Mean	Std. Deviation	Std. Error Mean	95% Confidence Interval of the Difference				
					Lower	Upper			
Pair 1	Width - Width1R	-7.000	14.142	4.714	-17.871	3.871	-1.485	8	.176
Pair 2	Height - Height1R	33.556	25.735	8.578	13.774	53.337	3.912	8	.004
Pair 3	Squiggles – Squiggle1R	-1.278	1.176	.392	-2.181	-.374	-3.261	8	.012

The *t*-test results in Table 8 show that there was no significant difference for width of Task 1 ($M = 518$, $SD = 14.142$) and width of Task 1R ($M = 525$, $SD = 0$) condition $t(8) = 1.485$, $p = 0.176$. These results suggest that the width of the children's squiggling results is quite similar to width of the squiggling reference. Specifically, the study results suggest that the accuracy of the children's squiggling is close to the reference image, in terms of the width of the children's squiggling results. However, there was a significant difference in the height of Task 1 (M=133.56, SD=25.735) and the height of Task 1R (M= 100, SD=0) conditions: $t(8)=3.912$, $p = 0.004$. There was also a significant difference in the number of squiggles (M=2.72, SD=1.176) and the height of the reference (M=455, SD=0) conditions: $t(8)=3.912$, $p = 0.012$. Both results suggest that the height of the children's squiggling and the number of squiggles drawn were different when compared with the squiggling reference. These results suggest that the children's squiggling was less accurate in terms of height and number of squiggles. More important the findings show that the efficiency of the Rotoscopy function provided in RPI (*Squiggles*) has helped children to perform better than in pen and paper-based activity.

7. Conclusion and Future Works

Rotoscopy may appropriate for children with dyspraxia as it supports activities that require hand movement. In this case, the input and output device play important role as a medium of communication between children and the methods; as this tangible technologies have a close correspondence in behavioural meaning between input and output. The research provides a contribution to knowledge in terms of offering an interactive tool for developing pre-handwriting skills through a bespoke animation technique. For future works, the proposed method may also be easily adapted for teaching handwriting in any languages, for instance the Jawi handwriting, as it has the further benefit of being responsive to the widespread availability of touch-based devices, affording direct interaction by tracing lines on background images.

Acknowledgement

The research publication is seconded by Universiti Tun Hussein Onn Malaysia (UTHM) under Short Term Grant (STG), Vot.U367 and also supported by Gates IT Solution Sdn. Bhd. We would like to thanks Cardiff School of Art and Design, Cardiff Metropolitan University, Cardiff, United Kingdom, Mr Anthony Rhys and pupils of Trinity Fields School and Resource Centre, Caerphilly, Cardiff, UK and the Dyscovery Centre, University of Wales Newport, UK.

References

[1] M. F. Othman, and W. Keay-Bright, "Using Rotoscopy Technique to Assist the Teaching of Handwriting for Children with Dyspraxia," Proc. The 2010 Third international Conference on Advances in Computer-Human interactions (ACHI 2010). ACHI. IEEE Computer Society, Washington, DC, Feb. 2010, pp. 175-178, doi:http://dx.doi.org/10.1109/ACHI.2010.39

[2] V. Garcia, E., Debreuve, and M. Barlaud, "Contour Tracking Algorithm For Rotoscopy," Proc. ICASSP 2006: IEEE International Conference On Acoustic Speech.

[3] A. Agarwala, A. Hertzmann, D. Salesin, and S. Seitz, " Keyframe-Based Tracking for Rotoscoping and Animation," Proc. SIGGRAPH, Aug. 2004, pp. 584—591, doi: http://doi.acm.org/10.1145/1186562.1015764.

[4] J. Bouldin, "The Body, Animation and the Real: Race, Reality and the Rotoscope in Betty Boop," Proc. Affective encounters: rethinking embodiment in feminist media studies. Eds. A. Koivunen and S. Paasonen. Turku, FI: Media Studies, 2001, pp. 48-54.

[5] A. Agarwala, "SnakeToonz: a semi-automatic approach to creating cel animation from video," Proc. 2nd ACM International Symposium on Non-photorealistic Animation and Rendering 2001. pp 139-ff.

[6] J. Sun, et al, "Bi-directional tracking using trajectory segment analysis," Proc. SIGGRAPH 2005.

[7] K. Nickel, et al, "3D-tracking of head and hands for pointing gesture recognition in a human-robot interaction scenario,". Proc. Sixth IEEE International Conference on Automatic Face and Gesture Recognition, 2004, pp 565-570.

[8] J Ponce, et al, "3D photography from photographs and video clips," Proc. International Symposium on Core Research, 2003.

[9] N. Snavely, et al, "Stylizing 2.5-D Video," Proc. NPAR 2006, pp 63-69.

[10] F. Levet, et al, "3D Sketching with Profile Curves," Proc. International Symposium on Smart Graphics, 2006.

[11] M. Boon, Helping Children with Dyspraxia., Jessica Kingsley Publishers, 2001.

[12] K. Ripley, B. Daines, and J. Barrett, Dyspraxia: A Guide for Teachers and Parents., David Fulton Publishers, 1997.

[13] M.M. Portwood, Understanding developmental dyspraxia: a textbook for students and professionals., David Fulton Publishers, 2000.

[14] A. Savidis, and C. Stephanidis, "Developing inclusive e-learning and e-entertainment to effectively accommodate learning difficulties," SIGACCESS Access. Comput. , 83, Sep. 2005, 42-54.

[15] J. Segouat, "A study of sign language coarticulation," SIGACCESS Access. Comput. , 93, Jan. 2009, 31-38, doi:http://doi.acm.org/10.1145/1531930.1531935.

[16] K. L. Walden, "Double Take: Rotoscoping and the processing of performance," Refractory: a journal of entertainment media, Dec. 2008.

[17] C.M. Boyle, "An Analysis Of The Efficacy Of A Motor Skills Training Programme For Young People With Moderate Learning Difficulties," International Journal Of Special Education 2007, Vol 22, No. 1,pp 11-24.

[18] L. Snape, and T. Nicol, T., "Evaluating the effectiveness of a computer based letter formation system for children," Proc. The 2003 Conference on Interaction Design and Children, 2003.

[19] K. Facer, and B. Williamson. "Designing Educational Technologies with Users", Futurelab, 2004, Available at http://www.futurelab.org.uk/resources/documents/handbooks/designing_with_users.pdf

[20] W. Keay-Bright, "Tangible Technologies as Interactive Play Spaces for Children with Learning Difficulties", The International Journal of Technology, Knowledge and Society, 2008, Volume 4, 111-120.

[21] M. Buchenau, and J. F. Suri, "Experience prototyping," Proc. The 3rd Conference on Designing Interactive Systems: Processes, Practices, Methods, and Techniques; August 17 - 19, Aug. 2000, doi: http://doi.acm.org/10.1145/347642.347802.

[22] The Early Year Foundation Stage Standards. Available at: http://www.standards.dfes.gov.uk/eyfs/site/index.htm.

[23] W. Keay-Bright, "Can Computers Create Relaxation? Designing ReacTickles© Software with Children on the Autistic Spectrum," CoDesign, 3, (2), 2007, pp. 97—110.

[24] J. Gulliksen, B. Goransson, I. Boivie, S. Blomkvist, J. Persson, A. Cajander, "Key principles for user-centred systems design," Behaviour & Information Technology, Vol. 22, no. 6, Nov.-Dec. 2003, pp. 397-409.

[25] E. Mazzone, J. C. Read, and R. Beale, "Understanding Children's Contributions during Informant Design," The 22nd BCS British-HCI 2008, Liverpool, UK.

[26] A. Druin, "Cooperative Inquiry: Developing New Technologies for Children with Children," Proc. The SIGCHI Conference on Human Factors in Computing Systems: the CHI Is the Limit, ACM, New York, 1999, pp. 592—599, doi:http://doi.acm.org/10.1145/302979.303166.

[27] E. Branton, "Developing a Child-Centred Design Process," Futurelab, 2003, Available at: http://www.futurelab.org.uk/resources/publications_reports_articles/web_articles/Web_Article592.

[28] B. Williamson, "The Participation of Children in the Design of New Technology," Futurelab, 2003, Available at:

http://www.futurelab.org.uk/resources/publications_reports_articles/discussion_papers/Discussion_Paper396.

[29] E. Kaplan, D. Fein, and J. Kramer, "Wechsler Intelligence Scale for Children® - Fourth Edition Integrated," 2004, Available at: http://www.pearsonassessments.com/HAIWEB/Cultures/en-us/Productdetail.htm?Pid=015-8982-800

[30] J.C. Read, S. MacFarlane, and M. Horton, "The Usability of Handwriting Recognition for Writing in the Primary Classroom," People and Computer XVIII - Design for Life, Springer, London, 2004, doi:10.1007/b138141.

[31] W. Keay-Bright, "Designing Playful Sensory Experiences with Interactive Whiteboard Technology: The Implication for Children on the Autistic Spectrum," EAD 07, 2007.

[32] M. Horton, and J. C. Read, "Interactive Whiteboards in the Living Room? – Asking Children about their Technologies," The 22nd BCS British-HCI 2008, Liverpool, UK.

[33] S.E. Henderson, and L. Henderson, "Toward An Understanding of Developmental Coordination Disorder: Terminological and Diagnostic Issues," Neural Plasticity 2003, Vol.10, No. 1-2.

Axiomatic Ontology Learning Approaches for English Translation of the Meaning of Quranic Texts

Saidah Saad[1], *Bahari* Idrus[2]

[1,2]Centre for Artificial Intelligence, Faculty of Technology and Information Science, Universiti Kebangsaan Malaysia

Abstract. Ontology learning (OL) is the computational task of generating a knowledge base in the form of an ontology, given an unstructured corpus in natural language (NL). While most works in the field of ontology learning have been primarily based on a statistical approach to extract lightweight OL, very few attempts have been made to extract axiomatic OL (called heavyweight OL) from NL text documents. Axiomatic OL supports more precise formal logic-based reasoning when compared to lightweight OL. Lexico-syntactic pattern matching and statisticsal one cannot lead to very accurate learning, mostly because of several linguistic nuances in the NL. Axiomatic OL is an alternative methodology that has not been explored much, where a deep linguistics analysis in computational linguistics is used to generate formal axioms and definitions instead of simply inducing a taxonomy. The ontology that is created not only stores the information about the application domain in explicit knowledge, but also can deduce the implicit knowledge from this ontology. This research will explore the English translation of the meaning of Quranic texts.

1 Introduction

Ontology learning (OL), which is the process of creating an ontology and populating it, has been the subject of intensive studies for the past decade. Researchers in this field have been motivated by the possibility of automatically building a knowledge base on top of text documents so as to support the extraction of reasoning-based knowledge. Ontology learning (OL) is the computational task of generating a formal knowledge base in the form of an ontology, given an unstructured corpus whose content is in a natural language (NL). This approach is based on the OL layer cake [1], [2]. Several works can be found in this area, most of which are limited to statistical and lexico-syntactic pattern matching-based techniques (lightweight OL) [3]. These techniques do not lead to very accurate learning, mostly because of several linguistic nuances in the NL, which become more challenging due to the complex structure of texts in literary documents [4], [5]. Formal OL is a hard task as it involves an

[1]Corresponding author: saidah@ukm.edu.my

accurate NL, and the understanding and conversion of the domain context into an equivalent formal presentation. This knowledge base formally represents both assertive facts as well as general truth statements expressed in some NL in a textual document. Formal knowledge supports semantic IR [6], [7], Q&A [8], [9] and reasoning services.

Ontologies are knowledge-based and play an important role in merging the semantic web and axioms. An axiom is one of the important components in an ontology that can create and describe the relationship between concepts, either within or across ontologies. Besides the taxonomy and structure of concepts, axioms represent more information about categories and their relationship to each other, as well as constraints on the properties and roles of each category. It is important to note here that there is a similar way of representing a concept or a fine line between an internal concept structure and axioms. First, a category can be represented using a formal frame, with roles and properties represented by the slots of a frame. Second, the same facts can also be expressed using axioms. Axioms represent knowledge in terms of objects, concepts, and roles. Concepts formally describe notions in an application domain, e.g. we could define the concept of being a father as "a man having a child". A knowledge base can be used to store the information we have about the application domain. Besides this explicit knowledge, we can also deduce implicit knowledge from a knowledge base. For example, from the fact that *Fatimah* belongs to the concept of *Woman* and the axiom $Man \equiv \neg Woman$, it can be deduced that *Fatimah* does not belong to the concept of *Man*.

Ontological axioms are assertions in a logical form that put some constraints into an ontology or are used to deduce new information. An example of ontological axioms is the assertion of the concept of subsumption or equivalence.

Axiom construction constitutes one of the layers in the OL layer cake [1] that has become a main building block for fixing the semantic interpretation of the concepts and relations of an ontology. T.Berners Lee said, "For the semantic web to function, computers must have access to structured collections of information and sets of inference rules that they can use to conduct automated reasoning". Currently, there are not that many real-world ontologies that make substantial use of axioms.

Understanding the literary documents that focus on the Quran is a grand challenge in the computational world for knowledge representation and reasoning. It is a challenging task due to the nature of Quranic texts, which have a scattered organization of knowledge, unique patterns of structure and style, and concepts that are interrelated across the entire document. Most of the current research has focused only on the concepts or relations extraction level. However, for these documents, the axiom level is necessary to explicitly define the full meaning of the concept of the Quran. The axiomatic method, which comprises principles for the development of theories, is aimed at the foundation, systematization, and formalization of a field of knowledge about a domain of the world. If knowledge of a domain is assembled in a systematic way, it gives rise to a set of categories that can be stipulated as being primitive or basic. Primitive categories are not defined by explicit definitions but by axioms that define their meaning implicitly.

2 Related Work

Knowledge representation, which is a description of logic-based ontologies, is a key role of semantic technologies. It can be used to describe the intended meaning from document sources and to exploit the powerful description of logic reasoning tools to facilitate the machine understandability of document sources, especially literary documents.

Ontology ideally consists of a TBox and an ABox. A TBox contains intentional knowledge in the form of a terminology, while the ABox contains extensional knowledge that is specific to the individuals of the domain of discourse. A TBox is a conceptual definition of the upper layer components, while the ABox is more about the components of the lower layer. Meanwhile, the process of ontology creation can be categorized into seven main features (refer to Figure 1) forming increasingly complex subtasks of a "layered cake" [1], [2], which consists of:

i. Relevant terminology acquisition and extraction in the domain.
ii. The identification of synonymous terms or linguistic variants that can be across languages.
iii. The formation of a concept, which can be abstract or concrete, elementary or composite, real or fictitious.
iv. The concept of a hierarchical organization or concept hierarchy (taxonomy)
v. Learning about relations, properties, attributes and also the proper domain and range.
vi. Axiom schemata instantiation using a special logical axiom system such as disjoints, equivalences, symmetries, and others. It can also be the minimal or maximal cardinality of the relation.
vii. A general axiom, which depends strongly on the logical formalism being used in the background knowledge.

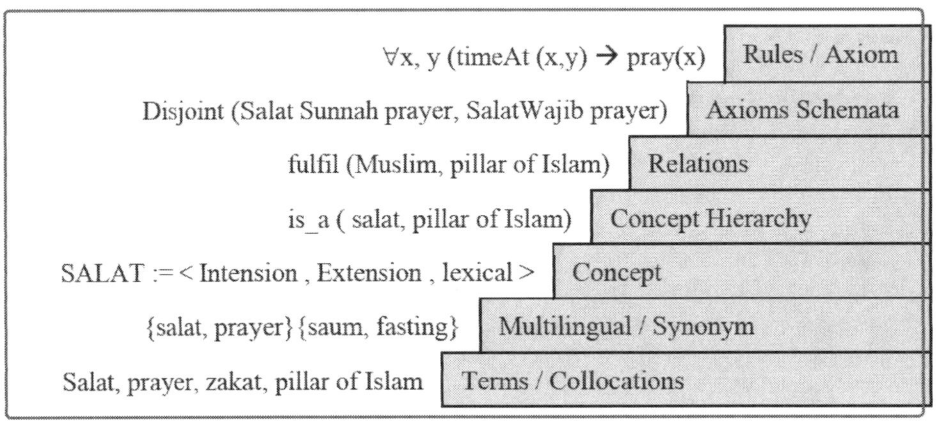

Fig. 1. Layered cake of ontology learning.

Among all those layers, ontology learning based on automatic rule acquisition is probably the least addressed. This rule-based systems are really important applications of semantic where the main reason to narrowing the semantic gap is to complement feature extraction which based on those layer cake with the formalized knowledge using description logic and rule based mechanism and implemented in ontology languages such as RDF, RDFS, Web Ontology Language (OWL) and rule language such as SWRL.At present, rule acquisition is still at a bottleneck,and is an obstacle to the wide propagation of rule-based systems.

State-of-the-art research on the automated learning of ontologies from texts is currently focusing on inexpressive ontologies. The acquisition of complex axioms involving logical connectives, role restrictions, and other expressive features of the OWL remains largely unexplored [10]. At present, the availability of computational techniques for the understanding of Quranic and Islamic knowledge is still limited [11], and most of the current research is just focusing on the level of the extraction of concepts/relations.

In order to provide a higher level of expressiveness to learned ontologies, several approaches have been proposed for extending ontology learning tools. The first approaches were manual, consisting of frameworks and tools such as Prot´eg´e, OntoEdit, NEON, and

KAON, which allowed axioms to be added by users or domain knowledge engineers. The more recent approaches include some kind of automation in order to add axioms under the evaluation and supervision of a knowledge expert. Approaches such as LExO [12], LEDA [13], and ReLExO [14] use a sequence of linguistic analysers. LExO starts by analysing the syntactic structure of an input sentence. The resulting dependency tree is transformed into a set of OWL axioms (concept inclusion, transitivity, role inclusion, role assertions, concept assertions and individual equalities) by means of manually-engineered transformation rules. ReLExO supports the acquisition and refinement of complex class descriptions in order to identify passages from the text that indicate the validity of certain knowledge. Given that the text can contain inconsistencies, LeDA allows the automatic generation of disjointed axioms based on machine learning classifications. The classifier, which determines the disjointedness of any given pair of classes, is trained, based on a gold standard baseline ofmanually-created disjointed axioms.

In [15], an automatic axiom-learning algorithm starts from a set of non-taxonomic relations. It uses the Web as a corpus, and linguistic techniques based on text patterns and a statistical analysis from the distribution of web information. But most of these are limited to axiom schemata provided by OWL, not on the general axiom, which is based on the description logic.

Baumann et al [10] presented an approach to the phenomenon of time. The basic relations are a temporal part of either among time regions or among time boundary regions, a temporal coincidence of time boundaries, and relations that link a time boundary (region) and a time region (including chronoids), declaring the former as a boundary of the latter, such as being the first or the last time boundary of a region. These are formalized by a set of axioms, specifying logical interrelationships between the categories and the relations. Their ontologies are axiomatized as formal theories in first-order logic and are analysed methodologically. They prove the consistency of both ontologies, and the completeness and decidability for one.

At present, the availability of computational techniques for the understanding of Quranic and Islamic knowledge is limited. Kais [16], for example, used Name Entity Extraction from the Quranic text to form the ontology for the Quran. Al-Yahya [17] presented the design and implementation of the ontological model, and the result of this application was the "time nouns" vocabulary of the Quran. Abbas [18] manually augmented a corpus of the Quran with an ontology or index of the key concepts taken from the 'Mushaf Al Tajweed' (Al Tajweed is another name for the Quran). It contains a comprehensive hierarchical index or ontology of nearly 1200 concepts in the Quran. Al-Kabi et al. [19] used an automatic classification technique to classify Quranic verses based on certain surah (chapters), according to the classification made by Islamic scholars. However, the basic ontology of Islamic knowledge in terms of the classification of the main topics in the Quran was already built decades before (based on the Quranic indices). The entire topic was classified manually by domain experts based on their understanding of the context of the Quran. Saidah [11, 21] describes the extraction of ontological component (concept, relation extraction) and the use of natural language pattern in extracting the knowledge of Quranic English translation texts (based on Salah subject) using natural language processing techniques. Research done by Saidah [11, 21] will be our main reference for continuing this research. This knowledge can be used as a general guide to the construction of a computer-automated ontology based on Islamic knowledge.

3 Methodology

The methodology for processing this kind of approach consisted of three phases as shown in Figure 2.

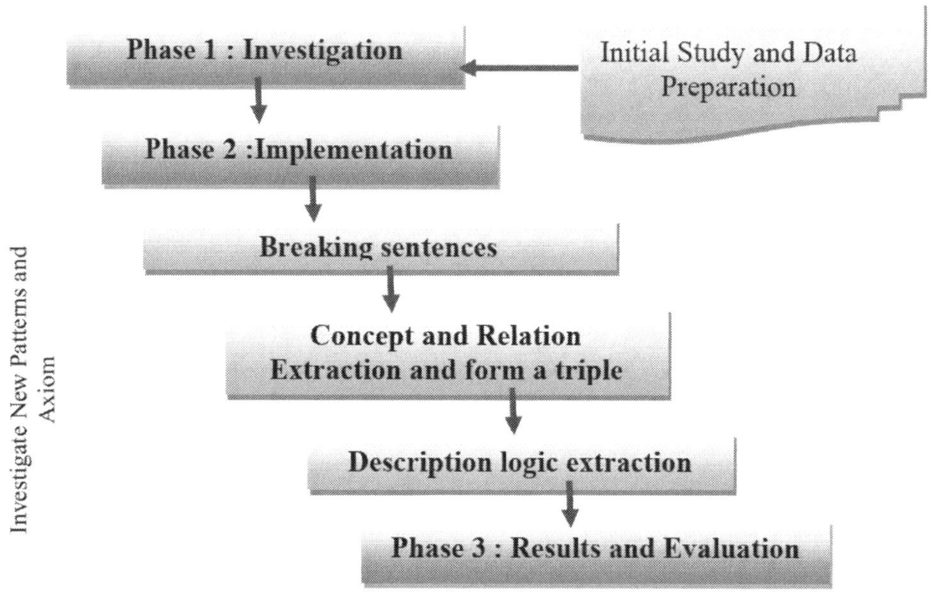

Fig. 2. Research method.

Phase 1 : Investigation

In this phase, the investigation was based on the current state of the:

i. Ontology learning and population
ii. Quranic understanding
iii. Analysis of Quranic content structure and populate Quranic gold standard

This research used the English Extended Quranic Translation Text by Dr. Muhammad Taqi-ud-Din Al-Hilali, and Dr. Muhammad Muhsin Khan [20]. It is quite a complete version that summarizes the information according to the tafsir of At-Tabari, Al-Qurtubi, and Ibn Kathir, with comments based on Sahih Al-Bukhari. Among the features available in this translation are that it defines all the unique words used in the Quran such as Muttaqun, Muzammil and Garden and so on, it gives extra information about certain verses, and it sometimes explains the mutashabihat verses and also the Asbabun Nuzul. This enriched the data and helped in the creation of the axiom rules.

Phase 2 : Implementation. This was based on the following steps (as shown in Figure 3):

Step 1. The document extraction to find the ontological component, as mentioned in Figure 1, began by splitting the document into sentences. The pre-processing stage of the document, such as tagging and identifying the syntax dependencies, was important

DOI: 10.1051/matecconf/201713500075

before proceeding to the next step. In this stage, a phrase that needed to be parsed had to undergo a temporary definition replacement process to simplify the on-going parsing process. For example: 'Those who believe' or 'You who believe' was replaced with Mukminun, and the words 'He', 'His', 'Our' and 'We' (where the first characters are in capital letters) were replaced by 'Allah'. This was obtained from a predetermined REPLACEMENT list. Next, the natural language pattern of the term had to be identified in order to proceed to the next stage.

In the extraction stage, all the identified patterns that matched the SPO (subject-predicate-object) were extracted. After identifying the ontological components such as concepts/instances, attribute/properties and relations using natural language patterns (SPO) – which based on stages in ontology layer cake from term extraction until relation extraction as mentioned in [11, 21],these were then used to analyse the description logic approach in the computational semantic domain such as lightweight ontology learning and formal ontology learning

Step 2. Triple extraction is a process to transform and convert complex and compound sentences to a set of simple sentences: "The believers, men and women, are Auliya", can be simplified to "The believers men are Auliya" and "The believers women are Auliya". This triple extraction, being a simplification process, helped in the easy translation of NL to DL at a later stage. This transformation can be done using a systematic parsing algorithm such as syntax dependency.

Step 3. All the lexical variations needed to be normalised into proper triple statements. A normalizer automatically identified all the lexical variations of the sentences, before translating them into their standard normal form. This process was necessary to identify the most generalized concept or relation for the mapping process, such as believer: NOUN + suffix (er) was shown as a person, man and woman was also a person; this extra knowledge could be coined from other structured knowledge such as a glossary or WordNet. However, WordNet is just for common terms and is not enough for Islamic terms [5],[11].

Step 4. DL Translator for TBox and ABox components. Those previously processed sentences had to have expressive equivalency in the DL language, as follows:

- $\mathcal{AL}$: Attributive Language supports the atomic concept definition, concept intersection, full value restriction, limited role restriction, and atomic concept negation.
- [$\mathcal{U}$]: Union supports the concept union
- [$\mathcal{E}$]: Existential supports the full role restriction
- [C]: Complement supports the concept negation
- [$\mathcal{H}$]: Role Hierarchy supports the inclusion axioms of roles
- [$\mathcal{O}$]: Nominal supports the concept creation of unrecognized Named Entity
- ($\mathcal{D}$): Data Type supports the range concepts to become the data type

Phase 3 : Every step in phase 2 was supported by a domain expert evaluation in order to get the 'truth' and the right concept according to the understanding of the Quranic domain and the proposed 'gold standard'.

For example, the concept of AlBirr has a different meaning or definition according to the Quran, where:

- Al-Birr is piety, righteousness… (AlBaqarah, 177)
- AlBirr is righteousness (AlMujadilah, 9)
- AlBirrmeans: a person who is the one who believes in Allah and the Last Day, the Angels, the Book, the Prophets, and gives his wealth, in spite of his love for it, to the kinsfolk, to the orphans, and to the Al-Masakin (the poor), and to the wayfarer, and to those who ask, and to set slaves free, performs as-Salat (Iqamat-as-Salat), and gives the Zakat, and who fulfils his covenant when he makes it (AlBaqarah, 177)

So, this information is stored in the form of rules (axioms) to facilitate the extraction of knowledge, and also the translation of the actual concept based on the context of the verse. Examples of knowledge representation in the form of rules (axioms) are the following (as mentioned in [21]):

AlBirr(?a) ⇒piety(?s)
AlBirr(?a) ⇒righteousness(?m)

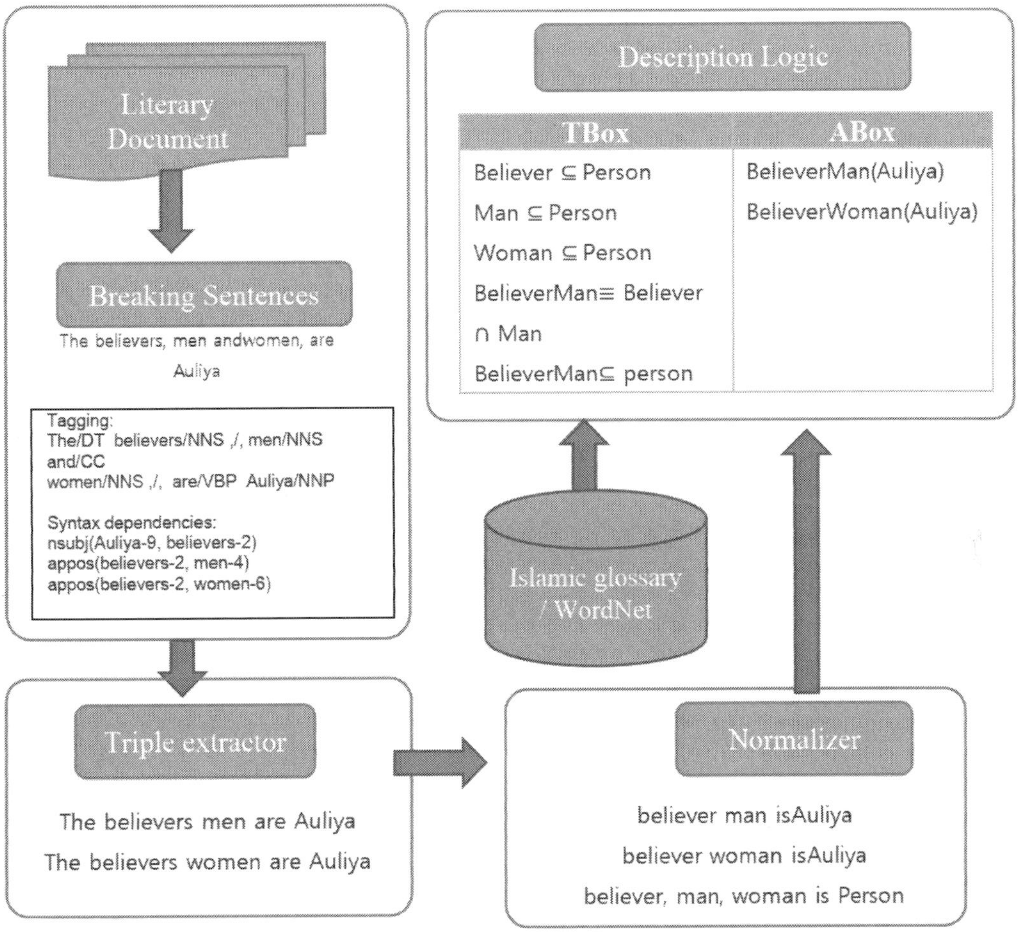

Fig. 3. Research activities.

A concept that has a definition with a combination of several other concepts and instances is:

AlBirr(?a) $\Longrightarrow$ person (?p) $\wedge$ [believe(?p,Allah) $\wedge$ believe(?p,LastDay) $\wedge$ believe(?p,angle) $\wedge$ believe(?p,Book) $\wedge$ believe(?p,Prophet)] $\wedge$ [givewealthto (?p,kinsfolk) $\vee$givewealthto (?p,orphan) $\vee$givewealthto(?p,masakin) $\vee$givewealthto(?p,wayfarer) $\vee$givewealthto(?p,thosewhoask) $\vee$givewealthto(?p,setslavefree)] $\wedge$ perform(?p,salat) $\wedge$ give(?p,zakat) $\wedge$ fulfil(?p, covenant)

From other knowledge constructions, part of the concept of AlBirr actually has a formation of other concepts such as iman, which consists of belief in Allah, Last Day, Angel, Book and Prophet, and also the concept of 'sedekah':

- Iman(?c) $\Longrightarrow$ believe(?p,Allah) $\wedge$ believe(?p,LastDay) $\wedge$ believe(?p,angle) $\wedge$ believe(?p,Book) $\wedge$ believe(?p,Prophet)

- Sedekah(?s) ⇒givewealthto (?p,kinsfolk) ∨givewealthto (?p,orphan) ∨givewealthto(?p,masakin) ∨givewealthto(?p,wayfarer) ∨givewealthto(?p,thosewhoask) ∨givewealthto(?p,setslavefree)

So, a new knowledge can be deduced from that: AlBirr is a person that has Iman, does sedekah, salat, zakat and fulfils the covenant, as follows:

AlBirr(?a) ⇒ person (?p) ∧ has(?p, ?c) ∧ give(?p, ?s) ∧ perform(?p,salat) ∧ give(?p,zakat) ∧fulfill(?p, covenant)

In order to do that, the semantic web rule language (SWRL)can be used. The SWRL was recently proposed to improve the power of ontology. The SWRL rules provide procedural knowledge power to lift the limitations of the ontological inference, particularly in discovering semantic relations among instances. The SWRL is based on a combination of OWL and rule mark-up language. It extends the set of OWL axioms to include Horn-like rules, thereby enabling the SWRL rules to be combined with an OWL knowledge base. The most significant benefits of rules are their ability to chain characteristics and infer the existence of new facts. The SWRL rules are then utilized to infer new facts in accordance with semantic relations between properties inside instances and known facts in ontology [22].

In this research, a sound ontology learning technique that maps English language sentences into their equivalent axiomatic Description Logic (DL) expressions was proposed in order to automatically generate a consistent pair of T-box and A-box, thereby forming both a regular (definitional form) and generalized (axiomatic form) DL ontology.

4 Discussion

In this paper, we described an approach for the automatic development of expressive ontologies from definitions provided in a text of English translation to a meaning of Quranic content, and to automatically generate an axiomatic ontology based on this content. This research is still in progress to identify the automatic construction and pattern of the axiomatic element. The axiomatic method is comprised of principles for the systematization and formalization of a field of knowledge about a domain of this content, and if the knowledge of this domain is assembled in a systematic way, a set of categories is stipulated as primitive or basic. Primitive categories are not defined by explicit definitions but by axioms that define their meaning implicitly. With this approach, all the important concepts defined in the Quran can also be understood and can be related to each other, where one concept not only can be defined by just one other concept but is comprised of several concepts.

Acknowledgement

This research is supported by the UKM University Research Grant GUP-2015-003.

References

1. P. Buitelaar, P. Cimiano, B. Magnini, Ontology learning from text: An overview. ontology learning from text: Methods, evaluation and applications. Frontiers in Artificial Intelligence and Applications Series 123 (2005)
2. P. Cimiano, Ontology Learning and Population from Text. Evaluation, doi:10.1007/978-0-387-39252-3 (2006)

3. S. Dasgupta, A. Padia, K. Shah, R. KaPatel, P. Majumder, DLOLIS-A: Description Logic based Text Ontology Learning. CoRR abs/1303.5929 (2013)
4. S. Saad, N. Salim, H. Zainal, Islamic knowledge ontology creation. Internet Technology and Secured Transactions, 2009. ICITST 2009. International Conference for, hlm.1–6. Inproceedings (2009)
5. A. B. Sharaf, E. Atwell, A Corpus-based computational model for knowledge representation of the Qur'an, Proceedings of the 5th Corpus Linguistics Conference, in proceedings (2009)
6. A. M. Pejtersen, Semantic information retrieval, Communications of the ACM 41 (4) 90–92 (1998)
7. S. Remi, S. C. Varghese, Domain Ontology Driven Fuzzy Semantic Information Retrieval, Procedia Computer Science 46 (2015) 676 – 681 (2015)
8. D. Moldovan, C. Clark, M. Bowden, Lymba's power answer 4 in trec 2007, in: TREC-2007, Association for Computational Linguistics (2007)
9. M. Sheker, S. Saad, R. Abood, M. Shakir, Domain-Specific Ontology-Based Approach For Arabic Question Answering, *Journal of Theoretical And Applied Information Technology*, E-ISSN 1817-3195 / ISSN 1992-8645). Vol **83** (2016)
10. L. Baumanna and Herreb, Axiomatic Theories of the Ontology of Time in GFO. Applied Ontology, IOS Press, **8** (2013).
11. S. Saad, Ontology Learning And Population Techniques For English Extended Quranic Translation Text. PhD Thesis. UTM Skudai. Johor (2014)
12. J. Volker, P. Hitzler, P. Cimiano, Acquisition of owl dl axioms from lexical resources, In Enrico Franconi, Michael Kifer, and Wolfgang May, Editors, The Semantic Web: Research and Applications, volume 4519 of Lecture Notes in Computer Science, pages 670–685. Springer Berlin / Heidelberg, 10.1007/978-3-540-72667-8 47 (2007)
13. J. Volker, D. Vrande˘ci´c, Y. Sure, A. Hotho, Learning disjointness, In Proceedings of the 4th European conference on The Semantic Web: Research and Applications, ESWC '07, pages 175–189, Berlin, Heidelberg, 2007. Springer-Verlag (2007)
14. J. Volker, S. Rudolph, Lexico-logical acquisition of owl - dl axioms. In Raoul Medina and Sergei Obiedkov, editors, Formal Concept Analysis, volume 4933 of Lecture Notes in Computer Science, pages 62–77. Springer Berlin/Heidelberg, 2008. 10.1007/978-3-540-78137-0 5 (2008)
15. L. D. V. Terrientes, A. Moreno, and D. S´anchez, Discovery of relation axioms from the web. In Proceedings of the 4th international conference on Knowledge science, engineering and management, KSEM'10, pages 222–233, Berlin, Heidelberg, Springer-Verlag (2010)
16. K. Dukes, Ontology of Quranic Concepts. Retrieved January 2012, from http://corpus.quran. com/ontology.jsp (2010)
17. M. Al-Yahya, H. Al-Khalifa, A. Bahanshal, I. Al-Odah I., N. Al-Helwah, An Ontological Model For Representing Semantic Lexicons : An Application On Time Nouns In The Quran. The Arabian Journal for Science and Engineering. Volume 35, Number 2C, 21–35 (2010)
18. N. Abbas, Quran 'Search for a Concept' Tool and Website. Master Thesis. School of Computing, University of Leeds (2009)
19. M. N. Al-Kabi, G. Kanaan, R. Al-Shalabi, Statistical Classifier of the Holy Quran Verses (Fatiha and Yaseen Chapters). Journal of Applied Sciences **5**(3): 580-583 (2005)

20. T. Al-Hilali, M. K. Khan, Interpretation of the meaning of the Qur'an in the English Language. King Fahd Quran Printing Complex, Madinah (1998)
21. S. Saad, N. Salim, H. Zainal, Ontology Learning and Population from Quranic Translation Texts. Taibah University International Conference on Advances in Information Technology for the Holy Quran and Its Sciences, pp. 407–415 (2013)
22. Chi & Chen, Ontology and semantic rules in document dispatching, *The Electronic Library*, **27**(4), 694-707 (2009)

MATEC Web of Conferences **135**, 00076 (2017) DOI: 10.1051/matecconf/201713500076
ICME'17

Embedding Repetition (*Takrir*) Technique in Developing Al-Quran Memorizing Mobile Application for Autism Children

Norhalina Senan[1], *Wan Azizah* Wan Ab Aziz, *Muhammad Fakri* Othman, *Suriawati* Suparjoh

Department of Multimedia, Faculty of Computer Science and Information Technology, Universiti Tun Hussein Onn Malaysia, Parit Raja, 86400, Batu Pahat, Johor, Malaysia

Abstract. Nowadays, there are various types of learning materials used in the process of teaching and learning of Al-Quran including the use of mobile application. However, the features of mobile application that are appropriate for the process of memorizing the Al-Quran, especially for the needs of children with autism is still limited. Thus, this paper proposes an interactive Al-Quran mobile application namely *iHafaz* to facilitate autism children recite and memorizing Al-Quran. A *takrir* (repetition) technique in Islamic learning approach is embedded in this mobile application in order to assist autism children memorizing the Al-Quran easily. This mobile application consists of two main modules which are *Hafaz* (Memorize) and Latihan (Exercise). Result from the user testing shows that 72.4% of respondents agree that the *takrir* technique embedded in the mobile application able to improve the usability of the mobile application in helping the autism children to recite and memorize the Al-Quran easily.

1 Introduction

Al-Quran is the religious text that became as the main guidance in Muslim's daily life. Various efforts have been made to uphold the Al-Quran including producing more young generation who can recite and memorize Al-Quran in proper way, including children with autism. However, most of the children with autism are associated with learning difficulties [1] which cause autism Muslim children face a problem in learning Al-Quran as well. Basically, the autism children can be divided into two categories which are LFA (*Low-Functioning Autism*) and HFA (*High-Functioning Autism*) [2]. Based on their learning ability, children with LFA present with limited language and severe cognitive impairment, while children with HFA have better language development and cognitive ability.

With the difference learning disabilities, autism children require special techniques and methods that correspond to their cognitive level especially in learning Al-Quran. In Islamic

[1]Corresponding author: halina@uthm.edu.my

pedagogical study, there are a few methods that can be used to facilitate Muslim to learn Al-Quran such as *al-Baghdadi* and *Iqra* which focusing on reciting Al-Quran, Jibril method to memorize Al-Quran and color technique in Mushaf Al-Quran to highlight *Tajweed* rules [3]. In addition, with the rapid development of IT, contemporary methods on learning Al-Quran were introduced. The invention of Al-Quran apps or online version encourages Muslim to learn Al-Quran by self-learning approach. However, these methods applicable to all people regardless of their special needs especially people with learning disabilities.

Therefore, in this paper, interactive Al-Quran mobile application known as *iHafaz* is proposed in order to fulfill the special needs of children with autism in reciting and memorizing the Al-Quran. The target user is Muslim autism children under HFA category who had basic knowledge of Al-Quran and also minimum ability to read, write, speak (pronounce) and high intelligence level to memorize Al-Quran. For that, an Islamic learning approach called *takrir* (repetition) technique is embedded in this mobile application as the main feature to assist autism children reciting and memorizing the Al-Quran easily. In order to increase the usability of this mobile application, visual and auditory elements such as presenting Al-Quran verses with Uthmani scripts, background color selection mechanism based on user preference, audio recitation for the autism children to imitate the right Al-Quran pronunciation, simple interface design and iconic button design are also provided.

The rest of this paper is organized as follows: Section 2 presents briefly on the related works of this study. Section 3 describes the details of the development process. A proposed mobile application *iHafaz* is presented in Section 4 followed by the discussion and conclusion in Section 5.

2 Related Work

In this section, the discussion on Al-Quran Memorization Technique, the utilization of Information Technology in learning Al-Quran for autism children and the characteristic of high-functioning autism (HFA) are presented.

2.1 Al-Quran Memorization Technique

Memorizing Al-Quran is the process of committing Al-Quran verses to memory. Al-Quran contains all knowledge in this world including sciences, history, medical facts, provisions, laws and many more [4]. Abduldaem [4] claims that people who memorizing Al-Quran is the memorizer of the largest encyclopedia. With that awareness, many Muslim parents nowadays send their children to Islamic school for learning and memorizing Al-Quran.

There are several techniques for memorizing Al-Quran. Based on the study in [5], there are four phases of memorizing Al-Quran implemented at *Madrasah Tahfiz Al-quran*, Terengganu, Malaysia which are *Sabak, Para Sabak, Ammokhtar, Halaqah Dauri.* In *Sabak* phase, they are required to memorize verse by verse based on their ability up to the point determined by the teacher. After the student able to memorize the specific verse of Al-Quran, they will go for second phase which is *Para Sabak*. *Para Sabak* is weekly memorization process in pair. The pair must be at similar level for compatibility purposes. After that, in phase *Ammokhtar*, the student have to memorize the Al-Quran more than one juzu' in the back of recent verses. In this phase, the order and actual arrangement of the Al-Quran in their reading is emphasized. Thus, canning stimulus is implemented if they forget or make mistake in their reading. Finally, after the students accomplished memorizing the whole Al-Quran, they will enter the last phase which is *Halaqah Dauri*. In this phase, they need to repeat the memorizing process in exact order and the performance is monitored by their teacher. The process will be repeated several times until they master their reading well.

Study in [6] reviews two main techniques in memorizing the Al-Quran applied at Islamic school in Indonesia which are *tahfidz* and *takrir* techniques. In *tahfidz* technique, face-to-face approach between student and teacher is applied during memorizing process. The teacher will check the correctness of the Al-Quran reading from the basic skill of *Qiro'ti*, memorizing *Juz 'Amma, Bin Nadzor Juz 1* till *musyafahah*. While, in *takrir* technique, the student need to repeat the memorization process whether by *musyafahah* (in pair) method, by group method, *orestafet* (spinning) method within a group and *seaman* method. *Takrir* technique is also conducted in front of the teacher.

Another research on the Al-Quran memorization techniques was conducted in [7] where the effectiveness of the techniques applied in several tahfiz schools in Malaysia is compared. The finding shows that three techniques which are *tasmiq* (reciting the new memorization of Al-Quran verse), normal memorization and verse writing book are the most effective memorization techniques used among the *tahfiz* students. However, the most preference technique used by the teacher is the *takrir* technique. Therefore, it is suggested that the most preferred techniques used by both teachers and students need to be combined in order to increase the effectiveness of memorizing the Al-Quran. This is due that *takrir* technique is still important in the process of memorizing the Al-Quran among *tahfiz* student to ensure they can improve their memorization consistently.

2.2 Utilization of Information Technology (IT) in Learning Al-Quran

With the advancement of Information Technology (IT), many online and mobile applications regarding Al-Quran have been developed to facilitate Muslims on learning Al-Quran [8][9][10]. In their study, the current technology in expert system, human-computer interface (HCI) and persuasive multimedia were applied in enhancing the efficiency of Al-Quran application in order to fulfill the needs of different target user especially disable people. Instead of the technology requirement, the right way of reciting the Al-Quran is very crucial in developing the Al-Quran application in order to preserve the originality and purity of the Al-Quran. For that, in [11], the important features to be applied in Al-Quran application are highlighted including word-by-word learning system to allow the user to learn the pronunciation of Al-Quran word from the basic level, contain transliteration to pronounce the words correctly, contain audio recitation and translation of the words.

Recently, several studies have been conducted in utilizing the Information Technology to help disability people especially autism children learning Al-Quran [12][13][14][15]. In [12], a research on the literature of the mobile application for disability student in Malaysia environment from Islamic perspective has been conducted. Based on their finding, it shows that the use of IT facilities in Malaysian education is still at early stage. Thus, they belief that all parties including government authority, school and developers must cooperated and committed in the implementation of IT facilities especially the needs of Islamic teaching and learning materials for disability children especially in Malaysia. For that some effort has been done in enhancing Al-Quran learning capabilities among autism children by using IT. Kamaruzaman and Jomhari [13] proposed digital game-based learning integrated with five LACIP skills (listening, arranging, constructing, imitation and pronunciation) to assist LFA autism children learning Al-Quran. Then, in [14][15] they expand their research in assisting autism children learning Al-Quran by developing serious games based on the LACIP skills and ABAHATA Al-Jabari Al-Quran. Their finding shows that the learning interest on learning Al-Quran is increased after using the serious game.

2.3 High-Functioning Autism (HFA)

An autism spectrum disorder is a brain developmental disability that affects the social skill, communication skill, social imagination skill, sensory issues and routines [16]. In addition, people with autism also have restricted, repetitive and stereotypes patterns of social behavior. With these limitations, most of the autism children experience with learning disabilities. The cognitive abilities of each autism children are different and basically can be categorized into two types which are Low Functioning Autism (LFA) and High Functioning Autism (HFA).

As compared to LFA, children with HFA have developed language and cognitive abilities. Most of HFA children have a high degree of intelligence and able to engage well in many aspects of their daily activities almost similar to normal children [17]. In learning perspective, children with HFA have basic skill to read, write and speak. However, with behavior difficulties, the way they think and learn are different from normal children. They often having problems in interacting and communicating with teachers and friends appropriately, understand the instruction and struggling in giving their attention to teaching and learning activities due to the difficulties in interpreting and understanding gestures, facial expressions or tone of voice. Therefore, they need a special treatment and approach in learning activities in order to improve their learning performance including the use of visual aids such as laminated photos or illustrations and computer software which able to show both text and image simultaneously.

3 User Requirement Analysis

In this study, the user requirements are analyzed based on the following activities:

a) *Interview with Subject Matter Expert.* An interview with Subject Matter Expert which are the teachers and Project Leader of Faqeh, Dr Nazean Jomhari was conducted at *Yayasan Pendidikan Al-Quran* specifically for children with special needs. The purpose of this interview session is to understand the education background of autism children including the existing teaching and learning methods used to learn and memorize Al-Quran, methods on giving appropriate instruction and guidelines on designing Al-Quran memorization mobile application that can fulfill the needs of autism children.

b) *Observation on target user.* The main objective of this activity is to understand the behavior and learning style of autism children especially on memorizing Al-Quran. This is important to ensure the mobile application developed for them able to help them in learning and memorizing Al-Quran.

c) *Review on the existing mobile applications.* A review on the existing mobile applications available at Google Play was performed in order to identify the appropriate features that can be applied in the proposed mobile application.

From the activities, the following user requirements are identified and used as the guideline in developing the proposed mobile application, *iHafaz*:

a) *Autism Children Learning Style.* Each of the autism children at *Yayasan Pendidikan Al-Quran* has their own learning style. Some of them are auditory learner and some are visual learner. In this institution, both visual and auditory methods were applied in learning and memorizing Al-Quran. Currently, cardboard was used as learning material to display the Al-Quran verses. Simultaneously, the teachers will recite the Al-Quran verses and followed by the students. These methods are used to support both visual and auditory learners.

b) *Al-Quran Memorization Technique*. *Takrir* (repetition) was applied in this institution for learning and memorizing Al-Quran. In average, they need to repeat 60 times of each verse in order for them to memorize it.
c) *Content of Mobile application*. 10 chapters from Juz Amma which are Al-Fatihah, Al-Nas, Al-Falaq, Al-Ikhlas, Al-Masad, Al-Nasr, Al-Kafirun, Al-Khautar, Al-Maun and An-Quraish are selected. These chapters are short chapter which not more than 7 verses. Simple activities should be provided in order to attract autism children engage with this mobile application. Malay language is utilized (except for the recitation of Al-Quran verses) for better understanding since this language is used in their daily communication.
d) *Multimedia elements*. Multimedia elements used in this mobile application are simple text from Comic Sans MS family for instruction and button, Al-Quran text with Uthmani scripts, audio with .wav file format for reciting the Al-Quran verses, and graphics with .png file format for button and background design.
e) *Interface Design*. The interface design for autism children should be simple, clear and easy to learn. It is also suggested to apply adaptable user interface where the autism children are allowed to change background colour or size of text based on their own preference.

4 *iHafaz* Prototype

The proposed application namely *iHafaz* is developed on mobile platform using Adobe Flash Professional CS6 with Actionscript 3.0. A simple interface design concept with less graphic and animation is applied in this prototype. The purpose is to attract the attention of autism children to focus on important information and not be distracted by non-essential content. In addition, to improve the usability of this mobile application, all the instructions and contents are delivered using audio and displayed in a large size and well-organized. In general, this prototype consists of 3 main modules which are *Hafaz* (Memorize), *Latihan* (Exercise) and Audit as shown in Figure 1.

Fig. 1. Main interface.

In *Hafaz* module, a list of 10 selected chapters in Juz Amma as shown in Figure 2 is provided. Each chapter button is assigned a number from 1 to 10 to help autism children

DOI: 10.1051/matecconf/201713500076

easily recognize the sequence of the chapter and select the desired chapter.Al-Quran text with Uthmani scripts is utilized since autism children were familiar with this text which used in printed version.

Fig. 2. A list of 10 chapters from Juz Amma.

The selected chapter is displayed as shown in Figure 3. The verses of the chapter are segmented and displayed line by line to make it easier for autism children to recite and memorize it.Then user can press on each verse and the selected verse will be displayed in bigger size (see Figure 4). In this module, a *takrir* technique is embedded to facilitate autism children to recite and memorize the chapter. It is done by providing the audio of verse recitation and repetition button at the bottom of each verse. The user needs to follow the recitation and memorize the verse. To help the autism children give their attention and focus on the recitation, the pointer on the highlighted verse is provided simultaneously with the audio. After that, the repetition button need to be pressed and the number of repetition will be displayed at the upper corner right of the interface as the indicator to the user. The maximum number of repetition in this mobile application is 60 as determined by the Subject Matter Expert.

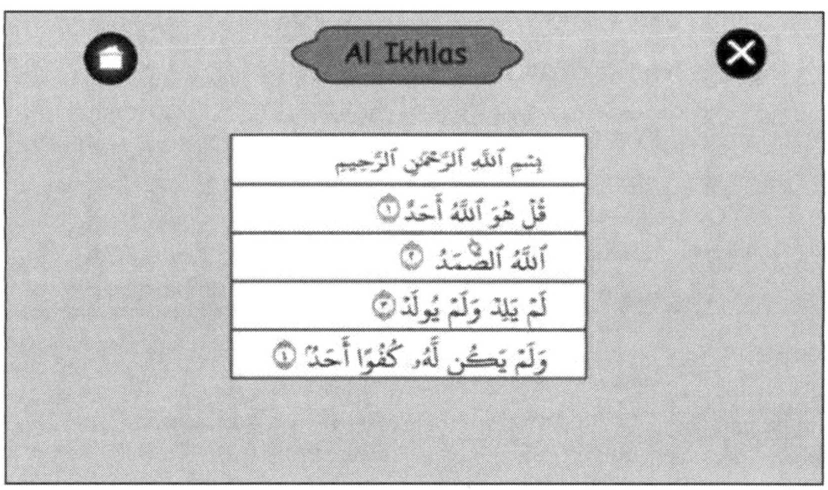

Fig. 3. Surah Al-Ikhlas.

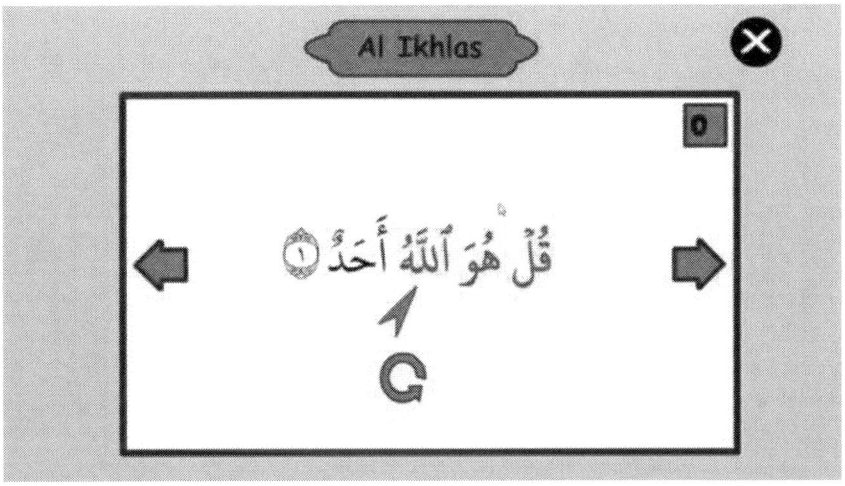

Fig. 4. Embedding *takrir* technique for assisting autism children memorizing Al-Quran.

In *Exercise* module, an exercise using serious game with 'drag and drop' approach is provided. The exercise comprises of 3 levels which are easy, medium and difficult. In easy and medium level presented in Figure 5 and Figure 6, the cue of the verses is provided to facilitate autism children do the exercise with fun approach. This is important to avoid them feel frustrated at early stage and eventually refuse to use this mobile application if they feel that the exercise are too difficult to be solved. The ability of the user to memorize the verses of each chapter can be tested in Difficult Level where the user needs to rearrange the verses by the correct sequence without any cue (See Figure 7). In each level, the correct answer will be marked and the praise audio such as "Congratulation" is produced as the rewards to motivate the autism children. The last main module is *Audit* module shown in Figure 8. In this module the user can tick on the chapter that has been memorized. At the same time, the teachers also can monitor the progress of each student through this module.

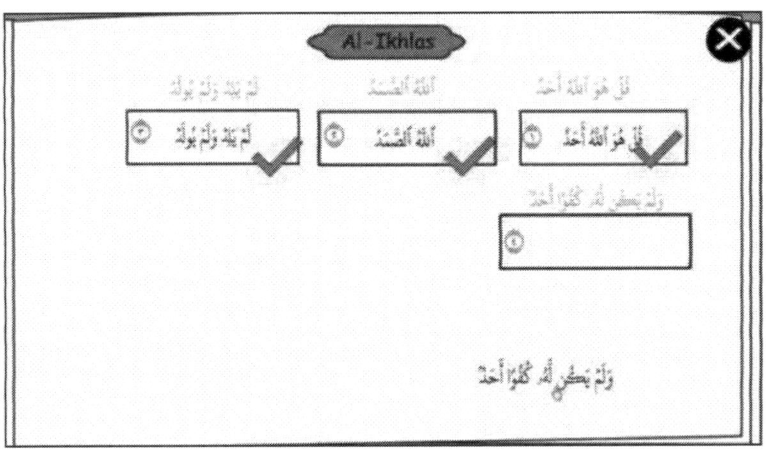

Fig. 5. Easy level of memorization exercise.

DOI: 10.1051/matecconf/201713500076

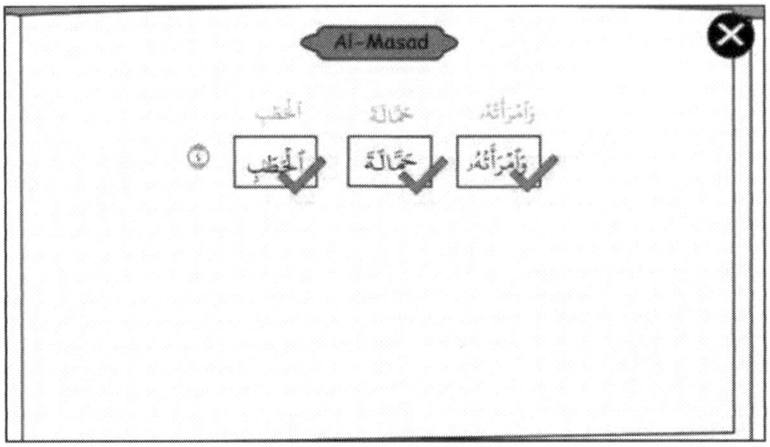

Fig. 6. Medium level of memorization exercise.

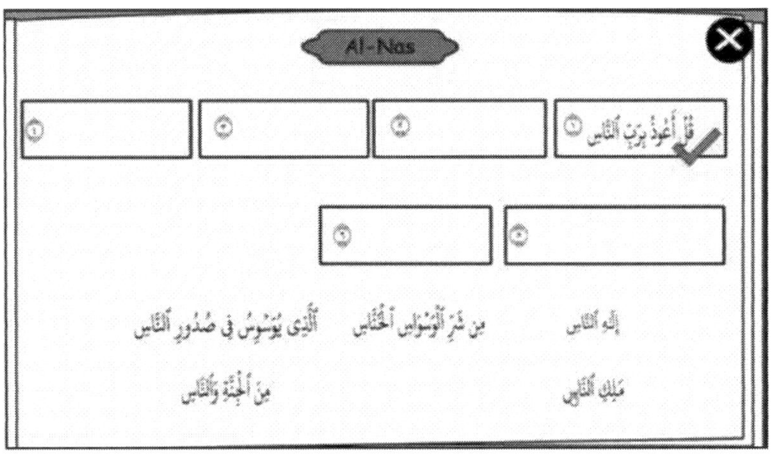

Fig. 7. Difficult level of memorization exercise.

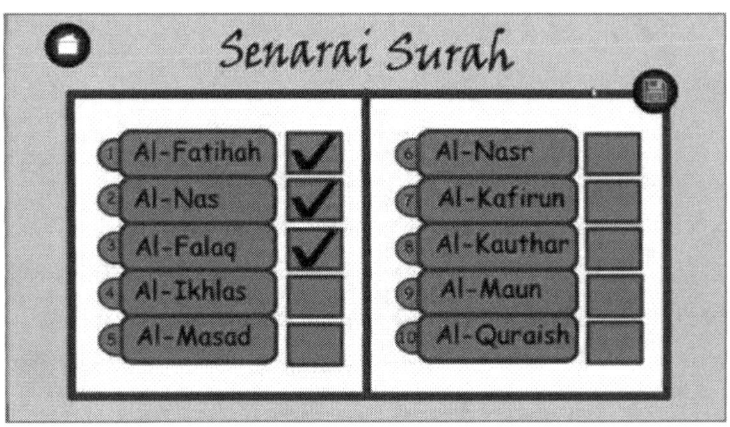

Fig. 8. Audit module.

Instead of these 3 main modules, other modules provided in this prototype are *Setting* and *Info* modules. In order to handle individual preference, user is allowed to change the colour of the background in this module. There are 5 different soft colours which are white, purple, green, blue and grey provided in this prototype. The user only need to click on the preferred colour and then the background colour of the whole prototype will be changed accordingly. The colour is selected based on the observation on the target user and the existing applications.

5 Result and Discussion

In order to identify the advantages and disadvantages of the prototype, user testing is conducted for two groups of target user which are autism children and teachers as Subject Matter Expert. The total number of respondent is 13 where the number of teachers and autism children are 6 and 7 respectively.

The teacher's acceptance rate is measured in term of their satisfaction level on the use of *takrir* technique, adaptable user interface approach, the selected colours, the size of text and the quality of the audio used. The result presented in Figure 9 shows that 67% of the teachers agree that use of takrir technique and the recitation of the Al-Quran in *iHafaz* able to facilitate autism children in learning and memorizing the Al-Quran. However, 37% of the teachers feel that the audio of the recitation is too fast but agree with the use of *takrir* technique in this prototype. While only 50% of users agree that the size of the text used in this prototype is suitable to be read by autism children and the rest feel that the size of the text used is quite small and difficult to read by autism children. It is suggested to use larger text size for better reading performance. Surprisingly, more than 80% of the teachers agree with the utilization of adaptable user interface approach, the selected colours and the quality of the audio used in this prototype.

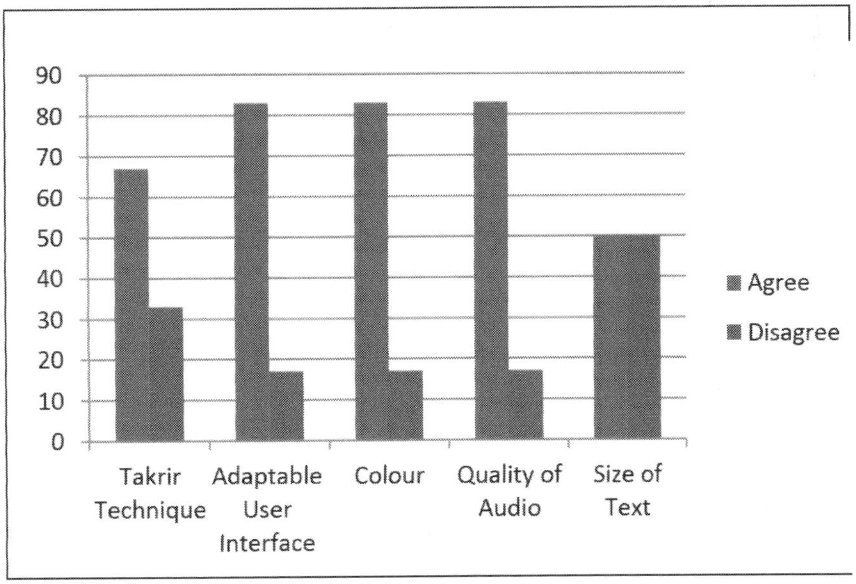

Fig. 9. Teacher's acceptance rate.

Figure 10 shows that more than 85% of the autism children agree on the use of the *takrir* technique and recitation audio in assisting them memorizing the Al-Quran. They are also agreeing that the prototype is attractive, the instruction is clear and the exercise provided is fun and easy to use. However, most of them feel that the text size used is too small and hard to be read. In average 84.5% of both teachers and autism children agree that the *takrir* technique embedded in *iHafaz* capable to assist them in learning and memorizing the Al-Quran. Whereas, overall 72.4% of both respondents agree that the interface design for *iHafaz* mobile application is easy to use, easy to learn and able to provide an enjoyable experience.

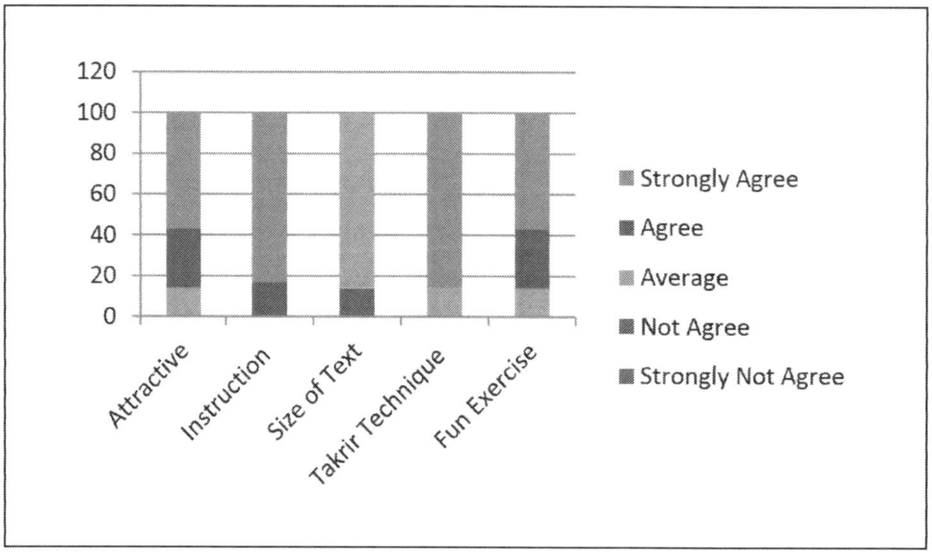

Fig. 10. Autism children's acceptance rate.

6 Conclusion and Future Work

In conclusion, most of the autism children involved with learning disability including in learning and memorizing Al-Quran. Thus, appropriate methods which able to facilitate them in learning and memorizing Al-Quran are required. In this study, the *takrir* technique embedded in *iHafaz* prototype able to assist autism children in reciting and memorizing the Al-Quran easily. The simple interface design with adaptable user interface approach; high quality audio for recitation and instruction; well-structured content and fun learning approach are important features to be considered in attracting autism children in reciting and memorizing Al-Quran using new technology. Results from user testing show that both teachers and autism children satisfied with the performance of the *iHafaz* prototype. However, some improvements on interface design such as identifying the appropriate size of text and enhancing the features of adaptable user interface where the autism children has ability to change the size of text based on their own preference are required. This is important in order to increase user's satisfaction level.

References

1. R. Jordan, Autism with Severe Learning Difficulties, Souvenir Press (2013)
2. K. A. Feather, Low Functioning to High Functioning Autism: A Prescriptive Model for Counselors Working with Children across the Spectrum, the *56th Annual European Branch of the American Counseling Association Conference*, Naples, Italy (2015)

3. S. P. B. Mohamad, Z. B. M. Yusoff, D. S. B. H. Adli, Pengajaran Dan Pembelajaran Al-Quran Bagi Golongan Kelainan Upaya Mental: Analisis Dari Sumber Tradisi Islam. *Jurnal Islam dan Masyarakat Kontemporari*, **8**, 153-168 (2014)
4. A. Al-Kaheel, Innovative Way to Memorize the Quran. Retrieved from www.hakeel7.com/eng (2017)
5. S. Ariffin, M. Abdullah, I. Suliaman, K. Ahmad, F. Deraman, F. A. Shah, M. Y. Z. M. Yusoff, M. M. A. Razak, M. M. M. Noor, J. T. Meftah, A. K. Kasar, S. Amir, M. R. M. Nor, Effective Techniques of Memorizing the Quran: A Study at Madrasah Tahfiz, Al-Quran, Terengganu, Malaysia, *Middle-East Journal of Scientific Research* **13** (1): 45-48 (2013)
6. Sukati, Studi Komparatif Tentang Metode Menghafal Al Qur'an. ITERASI **5**(2) (2015)
7. A. H. Abdullah, A. Safar, M. I. Mustari, A. Muhammad, I. Ismail, Keberkesanan Kaedah Hafazan Di Pusat Tahfiz. Pusat Pengajian Islam Dan Pembangunan Sosial, Universiti Teknologi Malaysia (2003)
8. A. Muhammad, W. Z. Qayyum, S. Tanveer, A. Martinez-Enriquez, A. Z. Syed, E-hafiz: Intelligent system to help Muslims in recitation and memorization of Quran, *Life Science Journal*, **9**(1), 534-541 (2012)
9. F. Razaly, N. H. Ajizan, S. Saad, M. S. Yaacob, M. M. Zain, M. Hussein, A. R. Musa, Dedicated graphical user interface system for the visually impaired users in learning Al-Quran, In *Proceedings of the 9th WSEAS international conference on Applications of computer engineering*, pp. 210-213, World Scientific and Engineering Academy and Society (2010)
10. A. F. Rosmani, N. A. Wahab, i-IQRA': Designing and constructing a persuasive multimedia application to learn Arabic characters. In Humanities, Science and Engineering (CHUSER), 2011 IEEE Colloquium on pp. 98-101, IEEE (2011)
11. Quran Reading, Islamic Knowledge: The Modern Ways of Leaning Quran. Retrieved March 1, 2017, from www.quranreading.com (2010)
12. N. Z. M. Zain, M. Mahmud, A. Hassan, Utilization of mobile apps among student with learning disability from Islamic perspective, In *Information and Communication Technology for the Muslim World* (ICT4M), 2013 5th International Conference on, pp. 1-4, IEEE (2013)
13. N. N. Kamaruzaman, N. Jomhari, N. Kamarulzaman, M. Z. M. Yusoff, Digital Game-Based Learning for Low Functioning Autism Children in Learning Al-Quran, In *Advances in Information Technology for the Holy Quran and Its Sciences* (32519), 2013 Taibah University International Conference on, pp. 184-189, IEEE (2013)
14. N. N. Kamaruzaman, N. Jomhari, N. Kamarulzaman, M. Z. M. Yusoff, Engaging Children with Severe Autism in Learning Al-Quran through the Serious Game, *Indian Journal of Science and Technology*, **9**(40) (2016)
15. N. N. Kamaruzaman, N. Jomhari, N. Kamarulzaman, M. Z. M. Yusoff, Digitizing Abahata Al-Jabari Method using LACIP Skill Theory to Support Learning Al-Quran for Children with Low-Functioning Autism: A Review towards Al-Quran Perspective, *QURANICA, International Journal of Quranic Research*, **7**(1) pp. 1-16 (2015)
16. The National Autistic Society, Autism Spectrum Disorders: A Resource Pack for School Staff. Retrieved March 9, 2017 from www.autism.org.uk (2011)
17. M. L. Secor, High Functioning Autistic, Retrieved March 9, 2017 from www.autismspeaks.org (2013)

DOI: 10.1051/matecconf/201713500077

The effect of white etching layer patches on a larger rail defect

Shahmir Hayyan Sanusi[1,*], and *David* Fletcher[2]

[1]Department of Mechanical Engineering, Universiti Tun Hussein Onn Malaysia, Parit Raja, 86400 Batu Pahat, Johor, Malaysia
[2]Department of Mechanical Engineering, University of Sheffield, S1 3JD Sheffield, UK

Abstract. Manufacturing defects such as porosity that formed in the rail could result in catastrophic failure if it remain in service. In this study, a crack that initiated from a large void was modelled using boundary element model to investigate the effect of white etching layer patches on larger defects in rail. The configuration studied represents a large defect in the rail that was observed from a rail removed from service. These investigation has provided a better understanding of how a crack initiated from a larger defect grows under the effect of white etching layer. The results highlight that a continuous white etching layer region will dominates the growth of the crack compared to a patches white etching layer region for a smaller crack size while patches of white etching layer will accelerates the crack growth for a larger crack size. The crack growth predicted in this study for a small crack size is at least 6 times higher than the wear rate mentioned in the literature and will increase further if the expansion of a transformed layer is considered making it impossible to be removed by natural wear.

1 Introduction

Porosity, voids, and inclusions are undesirable imperfections that developed in a metal casting process. The term porosity is widely used to describe any hole or void formed in a cast product [1]. In the rail manufacturing process, porosity could be formed in rails due to improper control during the vacuum degassing process. As the molten steel changes state from liquid to solid during the manufacturing process, trapped gas that is dissolved in the molten steel would remain in bubble form as the steel solidifies [2] which could result in sudden rail fracture. Due to a greater demand in the rail transport industry, the increase in train capacity, larger axle loads, and higher speed have resulted in a greater variety and frequency on the formation of rail defects across the rail network [3]. Smaller defects may remain in track until evidence of their growth led to rail removal, but for larger defects, immediate rail replacement is required [4].

---

* Corresponding author: shahmir@uthm.edu.my

Numerous studies has been done in the past to investigate the effect of inclusions in rails [5]–[10]. If inclusions remain in the rail, they may establish a favoured region of stress concentration. Potential damage process might develop (depending on the level of stress and wear rate) which could result in catastrophic failure. Observation from recent work [11] suggests that the metallurgical transformed regions sometimes form as patches (not continuous) with various length and thickness. Therefore it would be interesting to study the effect of this transformed region on cracks that initiated from larger defect in rail. To demonstrate these issues, a simple two dimensional (2D) boundary element (BE) model was developed within the Beasy software package [12] to investigate the influence of patches white etching layer region (Figure 1) on a larger defect in rail.

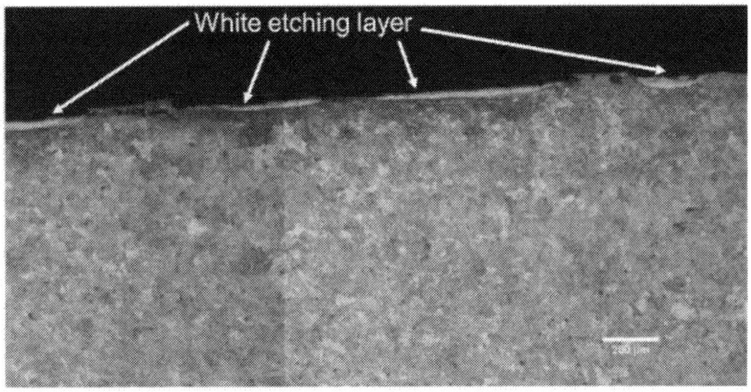

Fig. 1. Patches of white etching layer lying on the rail surface

2 Modelling method and conditions

A specific defect (Figure 2) described in the morphological study [11] is considered in the modelling. The model variables and conditions were decided to match the operating condition of a high speed traffic. An in-depth case study is made of a large void (maximum length of 2.9 mm and a maximum height of 0.9 mm) that was modelled to investigate the effect of white etching layer patches on a larger defect in rail. The defect was located approximately 2 mm below the rail surface with two inclined cracks (referred as left and right crack) initiated from the left and right side of the void. The left crack has an inclination angle of 4.65° while the right crack has an inclination angle of 15.68°. Different crack lengths (0.5 mm, 1 mm, 2 mm, and 4 mm) were considered to examine the crack growth rate. The real geometry of the defect was modelled by extracting the defect perimeter coordinates from the metallographic image using an open source data extractor package known as the G3data [13].

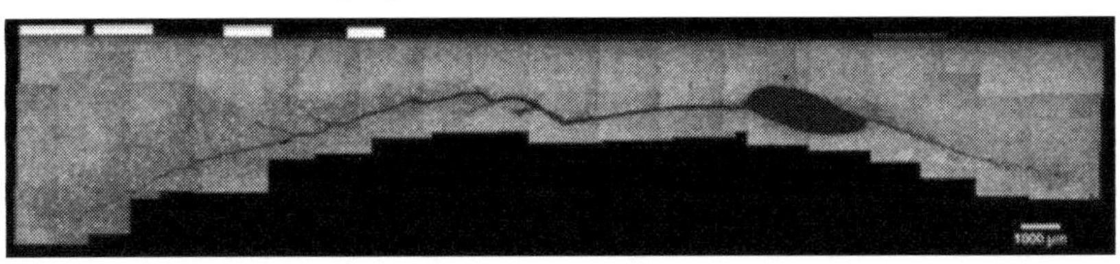

Fig. 2. Example of massive void morphologhy below the running band in a longitudinal cross section from a high speed mixed traffic line. The central portion of the defect lies approximately 2 mm below the rail surface. Incline crack initiated from the left and right side of the void. The white marker above the rail surface indicates the location of a patch of white etching layer with a thickness ranging between 20 to 250 μm.

Table 1. Conditions modelled.

Case	Consideration
1	No metallurgical transformed layer
2	Continuous thin (0.25 mm) metallurgical transformed layer (length: 4 mm)
3	Patches of thin (0.25 mm) metallurgical transformed layer (length: 6 mm)

Three cases have been considered throughout this study as indicated in Table 1. For case 1, no metallurgical expansion layer has been considered and case 1 has been chosen as the baseline case to identify the effect when a metallurgical transformed layer with a different configuration exist above the larger defect. The configuration of the model for case 1 is shown in Figure 3a. For case 2, a thin metallurgical transform layer with a thickness of 250 µm lying above the defect was considered where the total length of the transform region was set to be 4 mm covering the whole length of the void (Figure 3b). Case 3 represents patches of metallurgically transformed region where two additional transform regions with a length of 1 mm were placed on the left and right side of the main transform region with a spacing of 1 mm between the additional transform region and the main transform region (Figure 3c). The white etching layer was represented in the model by applying a thermal body load to the whole region of transformed layer which symbolizes an expansion of a surface layer due to volume change. The value of expansion due to transformation was taken as 0.8% which was calculated based on the changes in density from pearlite to martensite as described in [14].

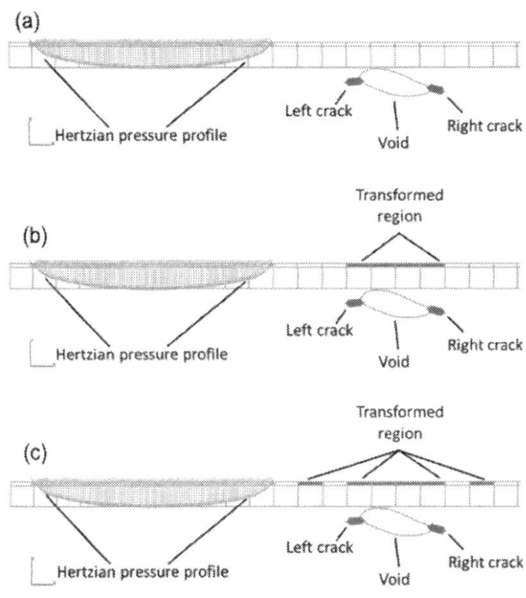

Fig. 3. Representation of BE model of rail–wheel contact for 2 incline cracks initiated from the left and right side of a larger defect (named as left and right crack). (a) case 1 - no metallurgical transform region considered. (b) case 2 – thin metallurgical transform region lying above the void with a continuous length of 4 mm and a thickness of 250 µm. (c) case 3 – similar with case 2 where thin metallurgical transform region lying above the void with a continuous length of 4 mm and a thickness of 250 µm. Two additional transform region with similar thickness and a length of 1 mm was placed on the left and right side of the main transformed region with a spacing of 1 mm to represent patches of transformed region.

3 Results and discussions

All the cases listed in Table 1 were investigated for a 0.5 mm crack length to examine a crack that has initiated from a void but not yet grown significantly. Modelling was also performed for a range of crack lengths of 1 to 4 mm to investigate the effect of a different configuration of the metallurgical transformed layer on propagated cracks. For each condition modelled, the range of stress intensity factor was assessed by considering the movement of a wheel incrementally across the defect (Figure 4), generating a series of SIFs. The combined action of traction and normal stress is indicated in the figure, and results in asymmetry of the results for left and right crack tips. Results are presented as stress intensity factors in Figure 5 to Figure 7, in which the origin is at the void centre. Stress intensity factors are converted to indicative crack growth rates as display in Figure 8.

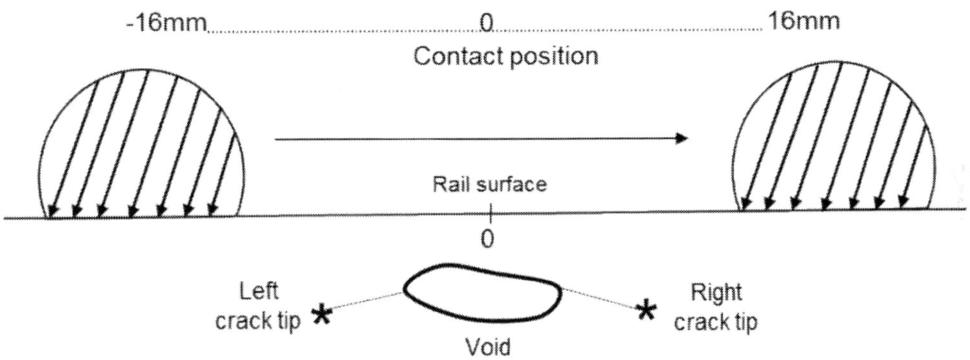

Fig. 4. A series of analyses are performed with incremental movement of the contact across the defect. The origin of the position is measured from the centre of the void.

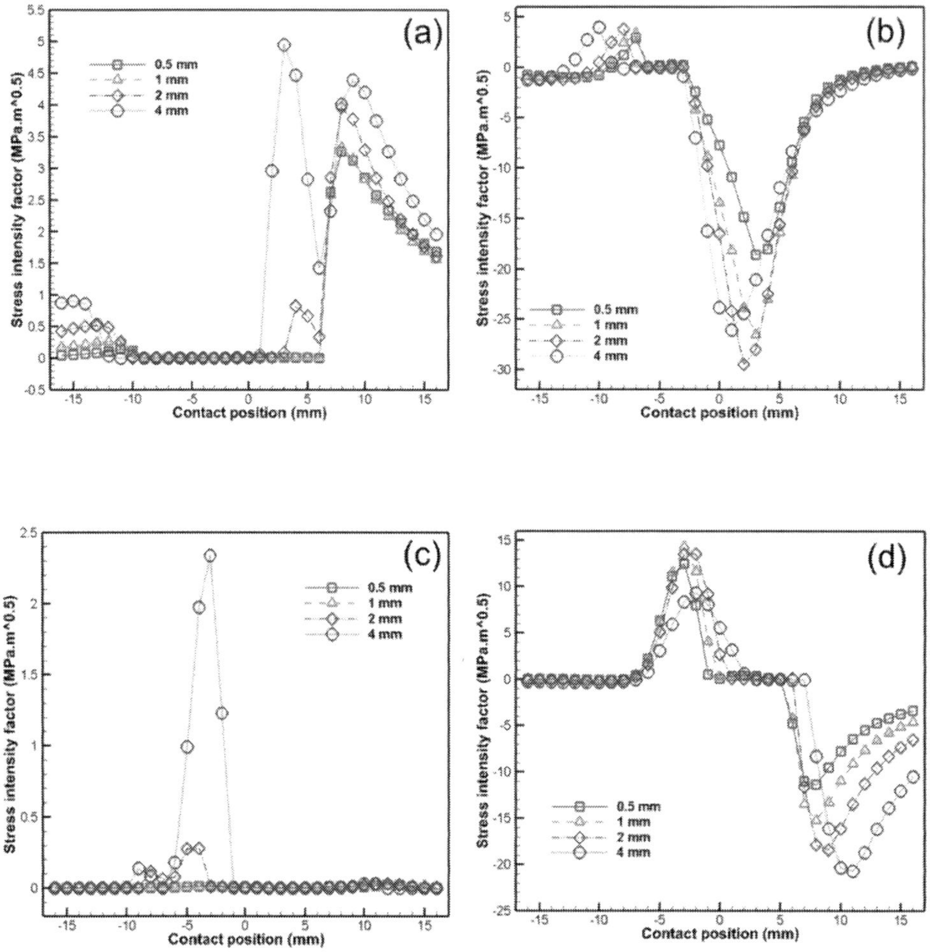

Fig. 5. Stress intensity factor dependence on contact position. (a) left tip mode I, (b) left tip mode II, (c) right tip mode I, and (d) right tip mode II

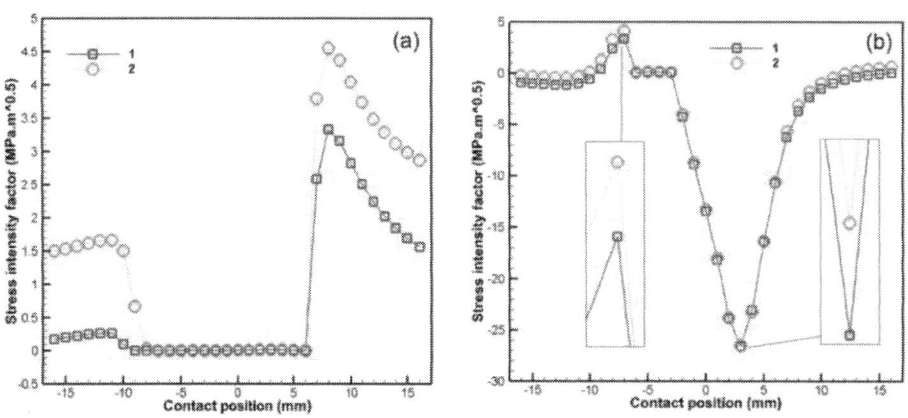

Fig. 6. Effect of a thin metallurgical expansion layer (represented as density change) on stress intensity factor for 1 mm crack, left crack tip. (a) Mode I, and (b) mode II. Legend entries refer to cases in Table 1.

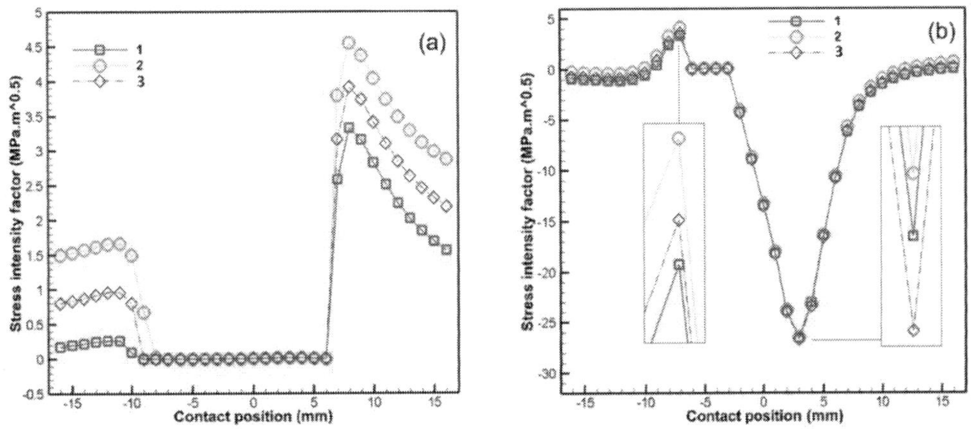

Fig. 7. Effect of thin metallurgical expansion layer patches (represented as density change) on stress intensity factor for 1 mm crack, left crack tip. (a) Mode I, and (b) mode II. Legend entries refer to cases in Table 1.

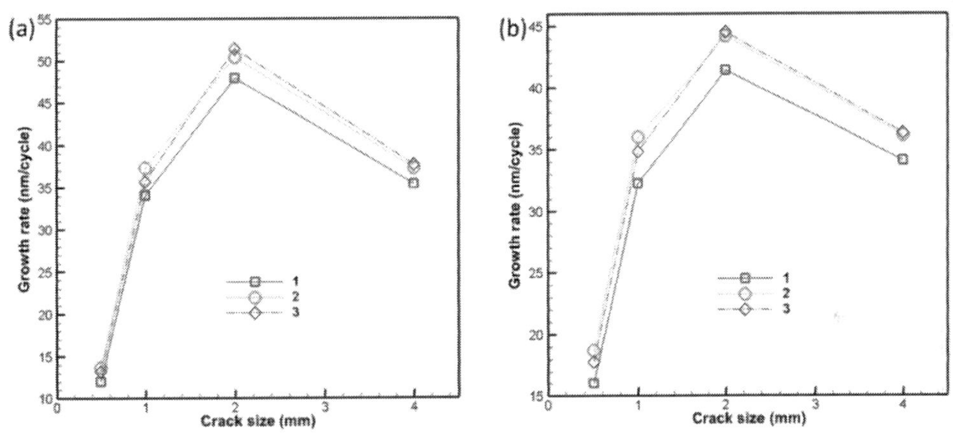

Fig. 8. Crack growth rate dependence on crack size for all cases considered. (a) Left crack tip, and (b) right crack tip. Legend entries refer to cases in Table 1.

3.1 Stress intensity factor dependence on crack size

The trends in mode I and mode II SIF for the left and right crack tips that initiated from a larger defect is shown in Figure 5 where consideration was made for case 1. This case only considered mechanical load (without considering any metallurgical transformed layer above the void). The mode I SIF shows a rise with increasing crack size for the left crack tip (Figure 5a), but the values are small even for larger cracks (4mm) where the peak KI is still below the threshold SIF (6 MPa.m1/2 for tensile growth of cracks in carbon steel [15]). Data for the mode II SIF (Figure 5b) shows higher stress intensity factor values even for smallest crack where the range exceeded the shear mode threshold of 1.5 MPa.m1/2 proposed by Otsuka et al. [15]. Even though the value of mode I SIF does not exceed the opening mode threshold, crack growth would still be expected since the range of mode II SIF exceeded the shear mode threshold value.

For the right crack tip, the range of KI and KII show rising trends with increasing crack size (Figure 5c and d). However, there was some reduction in the range of KII once the crack reaches a size of 4 mm for the left and right crack tip. This reflects the trend in the outcome of crack growth under mixed mode developed for inclined surface breaking cracks (Figure 8 and Table 2) where there is an increase in the growth with an increase of crack size, but decreased once both cracks exceeded a size of 2 mm.

Table 2. Crack growth rates (nm/cycle) for all the cases and sizes of crack modelled.

Case	0.5 mm		1 mm		2 mm		4 mm	
	L	R	L	R	L	R	L	R
1	12.0	16.0	34.1	32.3	48.0	41.5	35.4	34.1
2	13.7	18.7	37.3	36.0	50.4	44.2	37.3	36.1
3	13.2	17.8	35.7	34.9	51.4	44.5	37.7	36.3

As the crack size increase, the tip moves deeper into the material where the stress region become lesser with an increasing depth. A similar observation can be seen from the shear stress plot (Figure 9) where the magnitude of the shear stress starts to decrease with an increasing depth. As the crack reach some critical crack length and depth, this results in a reduction in stress intensity value hence resulting in a drop in the growth rate [16]. Therefore, the propagation of the crack may tend to slow even though the crack elongate. Place the figure as close as possible after the point where it is first referenced in the text. If there is a large number of figures and tables it might be necessary to place some before their text citation.

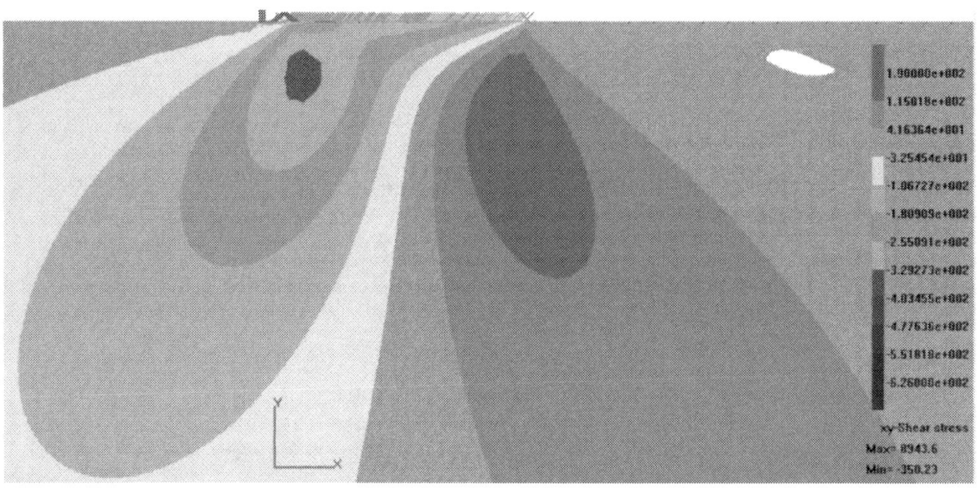

Fig. 9. Shear stress plot at -16mm contact (far away from the defect) for case 1. Units in MPa

3.2 Expansion effect of a thin surface layer on cracks initiated from void

The effect of the thin expansion layer on mode I and II SIF for the left crack tip is shown in Figure 6 and the baseline (case 1) is included for comparison. The trend (Figure 6a) shows that the presence of the metallurgical transformed layer has a significant effect on the mode I SIF where its value will always be above the baseline case if the contact is far from the crack tip. However, this effect is only valid for short crack size (0.5 – 2 mm) whereby for the largest crack (4 mm), the expansion of the transformed layer gives very minor effect on the KI value. This phenomenon is similar to the case for horizontal crack [14] where the effect will diminish as the crack lengthens. Even though the expansion of the transformed layer give a rise on the KI value, the range is still below the opening tensile threshold even for the largest crack size. For mode II data (Figure 6b), the trend shows that the expansion of the transformed layer increases the magnitude of the negative and positive peak but the effect is small. These differences can be seen numerically in Table 3. The changes are small and reduce as the crack extends. Results for the right crack tip exhibit similar behaviour where expansion of the transformed layer only affects the range of KI value for shorter crack and gives very little effect on the range of mode II SIF.

Table 3. Mode I and Mode II SIF range for all case modelled.

Case	**0.5 mm**				**1 mm**				**2 mm**				**4 mm**			
	L		**R**		**L**		**R**		**L**		**R**		**L**		**R**	
	ΔK_I	ΔK_{II}	ΔK_I	ΔK_{II}	ΔK_I	ΔK_{II}	ΔK_I	ΔK_{II}	ΔK_I	ΔK_{II}	ΔK_I	ΔK_{II}	ΔK_I	ΔK_{II}	ΔK_I	ΔK_{II}
1	3	21	0	23	3	29	0	29	4	33	0	32	4	30	2	30
2	4	22	1	24	4	30	1	30	4	33	1	32	4	30	2	30
3	4	22	1	24	3	30	1	30	4	33	1	32	4	30	2	30

The expansion of the metallurgical transform layer lying above the defect will introduce an opening stress region below its whole length until a certain depth as shown in Figure 10. It can be seen that the region of the maximum opening stress occured at the side of the defect where crack initiation point is located. For the shortest crack (0.5 mm), the whole length of the crack is situated at the region where maximum opening stress occured (e.g. y-direction direct stress). As the crack lengthens, the crack tip had exceeded these maximum stress regions where only some part of the crack length is affected by the opening stress and this will result in a very minor reduction in the KI value as the crack lengthens.

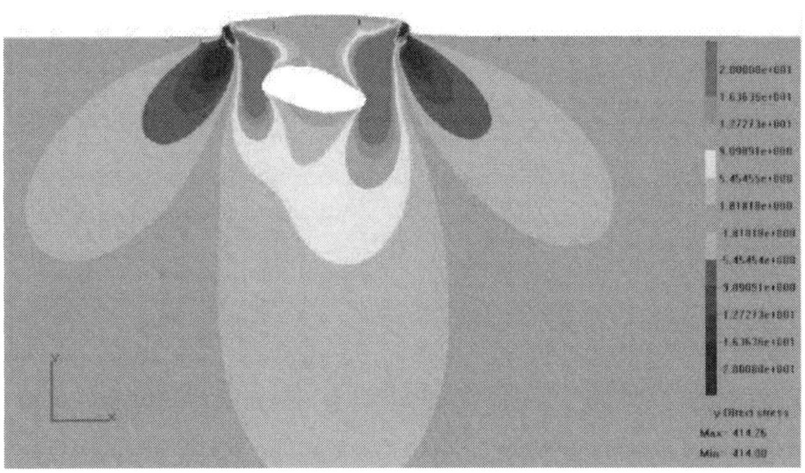

Fig. 10. Deform vertical stress plot for the case where thin expansion layer is present above the defect (case 2). Mechanical load (contact) and crack were absent. The deformation plot scale factor was set to 100 for clarity. Units in MPa.

DOI: 10.1051/matecconf/201713500077

3.3 Effect of patchy transform region

The effect of a patchy transformed region on mode I and mode II SIF for 1 mm crack focusing on the left crack tip can be seen from Figure 7. The trend in Figure 7a shows that the value of KI rises above baseline (case 1) but fall below case 2 throughout all positions of the contact. Looking at the numerical changes in Table 3, the trend change where the range of mode I SIF for case 3 raise above ΔKI for case 2 as the crack lengthens from 2 mm to 4 mm. Although the SIF changes are small, they do influence the trend of the crack growth (Figure 8) where patches of white etching layer will drive to a higher crack growth in case 3 compared to case 2 as the crack lengthens. For the mode II data (Figure 7b), the patches of the transform region give minimal but variable effect on the range of mode II SIF. For all sizes of crack modelled, it was found that patchy white etching layer has an insignificant effect on the ΔKII value compared to case 2.

Looking at the results overall, when patches of transform region formed on the surface, they will introduce an opening stress region below the transformed region (similar to case 2) due to volume expansion. However, the presence of gap/spacing between the transform region will introduce an additional compressive stress region below the gap (Figure 11). Since the transform region expands due to volume expansion particularly noticeably when the contact is not present, the non-transformed region of the gap will compress due to the reaction movement (Newton 3rd law) and this will result in the formation of the compressive stress region under the gap. Therefore any crack tip that lies in this region will experience compressive stress that will reduce its KI value. Focusing on the left crack tip, Figure 12 show the deformed plot for case 3 when the contact is not present. For the 1 mm crack size, some part of the crack is located at the tensile stress region while some part especially the crack tip is located at the compressive stress region (Figure 12a). This results in a small reduction in the ΔKI value compared to case 2 for similar crack size. However, as the crack lengthens to a size of 2 mm, some part of the crack especially near the crack tip is located at the opening stress region (Figure 12b) and this attribute to higher ΔKI value compared to case 2. Hence although minor, the variation in SIF observed can be seen to be caused by the physical region modelled and not by artifacts of the modelling process.

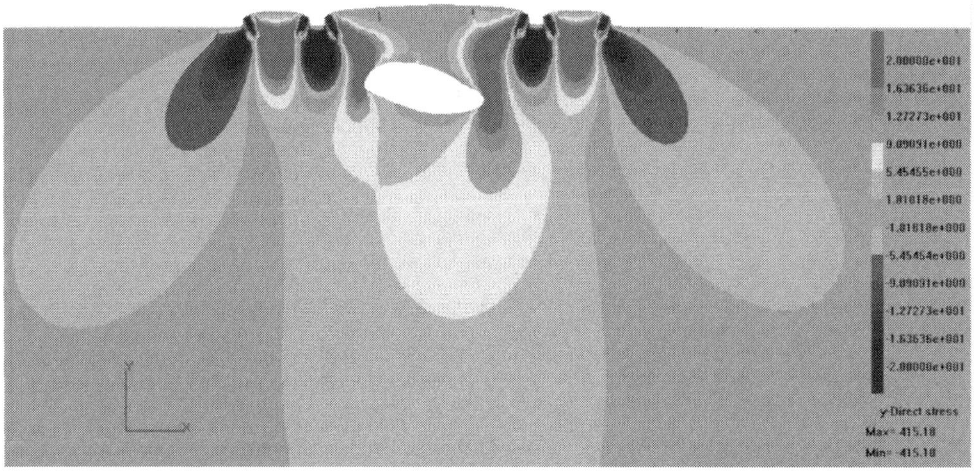

Fig. 11. Deform vertical stress plot for the case where patches of thin expansion layer are present above the defect (case 3). Mechanical load (contact) and crack were absent. The deformation plot scale factor was set to 100 for clarity. Units in MPa.

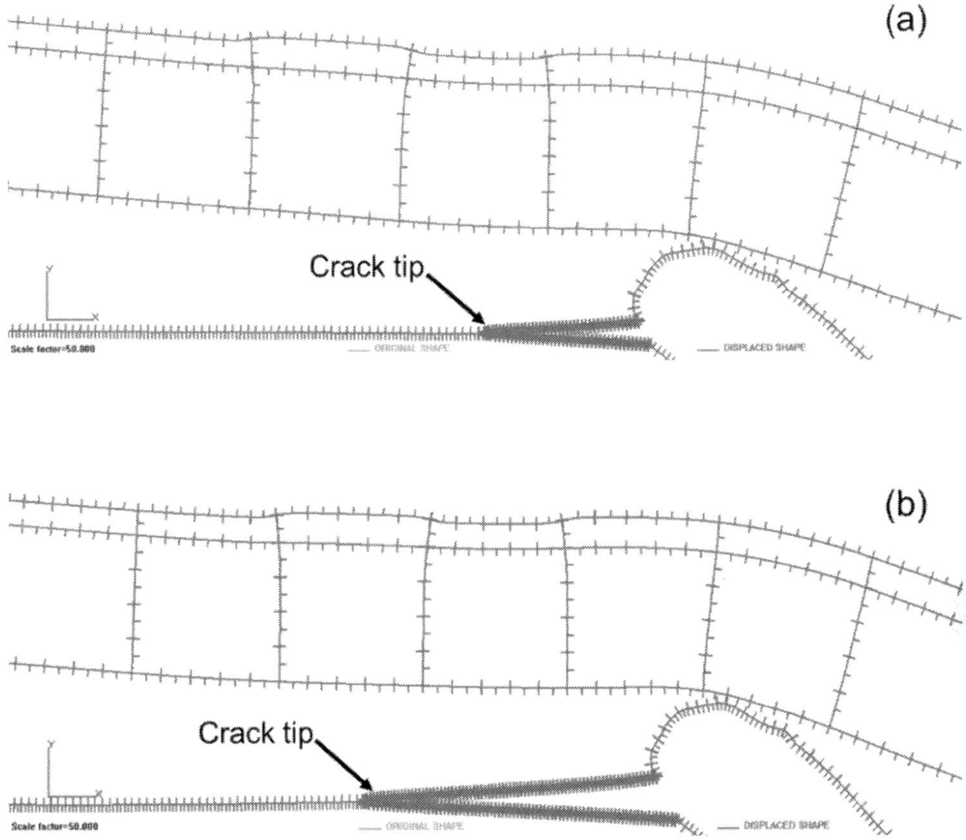

Fig. 12. Deformed plot for the case where patches of thin expansion layer are present above the defect (case 3) focusing on the left crack tip. Mechanical loads (contact) were absent. The deformation plot scale factor was set to 50 for clarity. (a) 1 mm crack size. (b) 2 mm crack size.

3.3 Real world consideration

It is known that there is an interaction between wear and fatigue crack whereby some rail steel grades wear more rapidly, and small cracks are always removed when wear process dominates. Typical rail wear rate measured from twin disc test under a dry condition with a maximum pressure of 1500 MPa is around 1.05 nm/cycle [17] where the value would rise up to 2.5 nm/cycle when decarburised layer are present on the rail surface [18]. However the crack growth predicted in this study for a small crack size is at least 6 times higher than the wear rate mentioned in the literature, and its growth rate will rise if the expansion of a transformed layer is considered.

4 Conclusions

Severe thermal loading for example caused by poor traction control has resulted in a formation of a thick metallurgical transformed layer on the rail surface which is known as the white etching layer. The presence of white etching layer on a rail remove from service (normally associated with squat type defect) comes in various type of form where sometimes the layer is not continuous and appear in the form of patches with various thickness and sizes. The effect of different WEL configurations on an inclined crack initiated from larger defect have been studied leading to the following conclusions:

a) The presence of thin (0.25 mm) metallurgical transformed layer / white etching layer lying above the defect will give an increase in the growth rate for an inclined crack. However, the effect reduces with an increasing crack size as the crack tip moves away from the region of maximum stress.

b) Patches of white etching layer will have lower crack growth compared to a continuously transformed layer while the crack tip is below region which is not transformed. However, the rate will increase once the crack tip enters the next region of a following metallurgical transform layer.

Since the crack considered in the model is subsurface, therefore there is no interaction of the crack and wear. But to give some context to its growth rate, it was found that the growth rate is at least 6 times higher than the wear rate mentioned in the literature, and its growth rate will rise if the expansion of a transformed layer is considered. Looking at the overall behaviour of the crack (Figure 2), there is a tendency for this crack to grow transversely potentially leading to rail fracture under bending stress. Therefore, immediate rail replacement is required once the crack grows to its critical crack length [17].

References

1. AFS, "Understanding Porosity," American Foundary Society, 2016. [Online]. Available: http://www.afsinc.org/content.cfm?ItemNumber=6933. [Accessed: 17-Oct-2016].
2. R. D. Pehlke, "Formation of Porosity During Solidification of Cast Metals," in Foundry Processes: Their Chemistry and Physics, S. Katz and C. F. Landefeld, Eds. Boston, MA: Springer US, 1988, pp. 427–445.
3. S. Kumar, S. Gupta, B. Ghodrati, and U. Kumar, "An approach for risk assesment of rail defects," Int. J. Reliab. Qual. Saf. Eng., vol. 17, no. 4, pp. 291–311, 2010.
4. D. F. Cannon, K. O. Edel, S. L. Grassie, and K. Sawley, "Rail defects: An overview," Fatigue Fract. Eng. Mater. Struct., vol. 26, no. 10, pp. 865–886, 2003.
5. H. K. Jun, J. W. Seo, I. S. Jeon, S. H. Lee, and Y. S. Chang, "Fracture and fatigue crack growth analyses on a weld-repaired railway rail," *Eng. Fail. Anal.*, vol. 59, pp. 478–492, 2016.
6. D. M. Fegredo, M. T. Shehata, A. Palmer, and J. Kalousek, "The effect of sulphide and oxide inclusions on the wear rates of a standard C-Mn and a Cr-Mo alloy rail steel," *Wear*, vol. 126, no. 3, pp. 285–306, 1988.
7. Z. Mouallif, B. Radi, and I. Mouallif, "The thermomechanical modeling of aluminothermic welds affected by different defects Summary :," *Int. J. Eng. Res. Develoment*, vol. 11, no. 10, pp. 44–48, 2015.
8. Z. Mouallif, B. Radi, and I. Mouallif, "Effect of the Inclusion Defect on the Mechanical Behavior of Thermite Welds," *Adv. Theor. Appl. Mech.*, vol. 9, no. 1, pp. 11–20, 2016.
9. I. Mouallif, Z. Mouallif, A. Benali, and F. Sidki, "Finite element modeling of the aluminothermic welding with internal defects and experimental analysis," in *MATEC Web of Conferences*, 2012, vol. 1, p. 12.
10. J. E. Garnham, R.-G. Ding, and C. L. Davis, "Ductile inclusions in rail, subject to compressive rolling–sliding contact," *Wear*, vol. 269, no. 11–12, pp. 733–746, Oct. 2010.
11. D. I. Fletcher and S. H. Sanusi, "SBB Rail Analysis," Sheffield, 2013.

12. "BEASY Boundary Element software." [Online]. Available: www.beasy.com. [Accessed: 04-Aug-2016].
13. G3data, "G3data," 2010. [Online]. Available: https://github.com/pn2200/g3data.
14. D. I. Fletcher and S. H. Sanusi, "The potential for suppressing rail defect growth through tailoring rail thermo-mechanical properties," *Wear*, 2015.
15. A. Otsuka, K. Mori, and T. Miyata, "The condition of fatigue crack growth in mixed mode condition," *Eng. Fract. Mech.*, vol. 7, no. 3, pp. 429–439, Sep. 1975.
16. A. Kapoor, D. I. Fletcher, and F. J. Franklin, "The role of wear in enhancing rail life," *Tribol. Res. Des. Eng. Syst. Proc. 29th Leeds-Lyon Symp. Tribol.*, vol. Volume 41, no. 0, pp. 331–340, 2003.
17. "Rolling contact fatigue in rails: a guide to current understanding and practice," in *RT/PWG/001 Issue 1*, Railtrack PLC, 2001, p. 46.

AUTHOR INDEX

AUTHOR INDEX

AUTHOR INDEX